THE TECHNOLOGY OF ARTIFICIAL LIFT METHODS

Volume 2b

THE TECHNOLOGY OF ARTIFICIAL LIFT METHODS

Volume 2b
Electric Submersible Centrifugal Pumps
Hydraulic Pumping—Piston Type
Jet Pumping
Plunger Lift
Other Methods of Artificial Lift
Planning and Comparing Artificial Lift
Systems

Kermit E. Brown

The University of Tulsa

Gene Riling
Clarence Dunbar
Don Rhoads
Satish Goel
Luiz Couto
Phil Wilson
Forrest E. Chancellor

Hal L. Petrie
Bolling Abercrombie
Phil Pattillo
Carlos R. Canalizo
Robert H. Gault
George L. Thompson
Bill Waters

Kenneth C. McBride
W. G. Skinner
L. A. Smith
Hank Arendt
Rusty Johnson
Tom Doll
John T. Dewan

PennWell Books
PennWell Publishing Company
Tulsa, Oklahoma

Library of Congress Cataloging in Publication Data   (Revised)

Brown, Kermit E.
     The technology of artificial lift methods.

     Includes bibliographical references.
     CONTENTS: v. 1.   Inflow performance, multiphase
flow in pipes, the flowing well.—v. 2.   Introduc-
tion of artificial lift systems beam pumping.
     1.   Oil wells—Artificial lift. 2. Pipe—
Fluid dynamics. I. Title.
TN871.B819   622'.3382   76-53201
ISBN 0-87814-031-X (v. 1)
Printed in the United States of America

     4 5 85

Contents

Chapter 5

Hydraulic pumping—piston type, by Phil Wilson 357

x   Contents

Chapter 6

Jet pumping, by Hal L. Petrie 453

Chapter 9

Planning for and comparison of artificial lift systems, by Tom Doll, Kermit E. Brown, and John T. Dewan   567

Chapter 4

Electric submersible centrifugal pumps

by   Gene Riling
     Clarence Dunbar
     Don Rhoads
     Satish Goel
     Luiz Couto
     Kermit E. Brown

4.1   INTRODUCTION

The first submersible pumping unit was installed in an oil well in 1928 and since that time the concept has proven itself throughout the oil-producing world. Presently, it is considered as an effective and economical means of lifting large volumes of fluids from great depths under a variety of well conditions. Submersible pumping equipment is used to produce as low as 200 b/d and as high as 60,000 b/d of fluid from depths up to 15,000 ft. The oil cut may also vary within very wide limits, from negligible amounts to 100%.

A typical submersible pumping unit (Fig. 4.11) consists of an electric motor, seal section, intake section, multistage centrifugal pump, electric cable, surface installed switchboard, junction box, and transformers. Additional miscellaneous components of installation will include means of securing the cable alongside the tubing and well head supports. Optional equipment may include a pressure sentry for sensing bottom-hole temperature and pressure, check and bleeder valves.

In its operating position, the standard down hole equipment is suspended from discharge tubing and submerged in well fluid. The setting depth, or bottom hole pressure, creates no problems as the seal section equalizes internal pressure in the motor with the submergence pressure in the well. Such installations in directional wells are common.

The submersible pump is also used in producing high viscosity fluids, gassy wells, and high temperature wells. With additional experience and improved technology, wells that were once considered non-feasible for submersibles are now being pumped economically.

The electric motor turns at a relatively constant speed and the pump and the motor are directly coupled with a protector or seal section in between. Power is transmitted to the subsurface equipment through a three-conductor electrical cable which is strapped to the tubing on which the unit is run into the well.

The surface equipment for a typical submersible installation consists of a bank of three single-phase transformers, a three-phase transformer or an auto-transformer, a motor controller (or switchboard), a junction box, and a submersible wellhead where the round or flat power cable can be packed off.

The pump performs at highest efficiency when pumping liquid only. It can and does handle free gas along with the liquid. The manner in which the pump handles gas is not completely understood, and high volumes of free gas cause a very inefficient operation.

The motor and the pump rotate at 3,475 to 3,500 rpm for 60 Hz power and 2,900 to 2,915 rpm for 50 Hz power. The unit is a precision-built piece of equipment and, under normal operating conditions, can be expected to give from 1 to 3 years of good operating life with some units operating over 10 years.

1

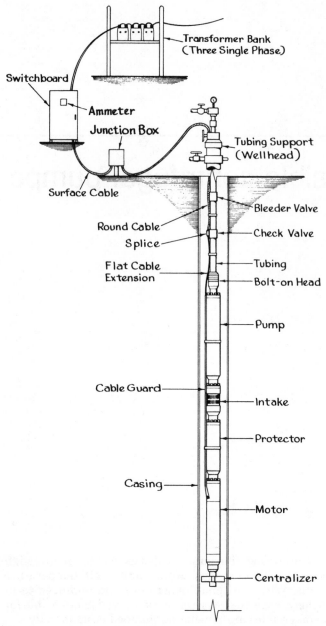

Fig. 4.11 Submersible centrifugal pumping unit (Courtesy TRW-Reda)

The electrical pump is becoming more popular and is presently being used in a greater percentage of the wells that are eligible artificial lift candidates. In particular it is well suited for off-shore applications and is used in many countries other than the U.S. where high volumes are produced. Continued improvements that lengthen the running time will make the electrical pump more and more attractive. Practically all of the present installations are tubing retrievable to replace the pump; however, the cable suspended unit (CSPS) is being manufactured and is presently installed in some areas. The cable suspended pump may be used to kick-off a well (initial unloading) and then be retrieved if the well comes in flowing.

This chapter covers the following topics: (1) basic fundamentals necessary to understand the pump, such as electric and hydraulic fundamentals, (2) description of the pumping system, (3) types of installations, (4) design of pumping systems, (5) operation of the pumping system, and (6) analysis and trouble shooting.

4.2 BASIC FUNDAMENTALS

4.21 Introduction

Certain basic fundamentals are very important in understanding the operation of an electrical submergible pump. For example a basic understanding of electrical fundamentals, hydraulic fundamentals, and certain certrifugal pump principles is needed. This section discusses those fundamentals deemed necessary for this understanding.

4.22 Electrical fundamentals

4.221 Introduction

This section will not present a course in electrical engineering but will review various basic terms and concepts required for better comprehension of the performance and operation of electrical machines. Basic formulas used for calculations of electrical parameters and for sizing the equipment are given without going into the theoretical details of electrical and mechanical design of electrical machines. The reader is advised to refer to specialized literature for a more detailed study of the subject.

4.222 Atomic structure of Matter

Electricity is a form of energy and can be produced from other types of energy such as heat and light. It is achieved through processes which convert one type of energy into electricity, e.g., pressure, friction, chemical reaction, electromagnetic reaction. The quantity of electricity in a body is known as the electric charge and is measured in coulomb in the practical system of units. Contemporary physics identifies electricity with matter and its structure.

Everything is made up of atoms and there are more than 100 types of atoms which are known as elements. Various combinations of these elements give rise to different materials, e.g., salt is obtained by combining sodium and chlorine atoms.

The atom is not a solid mass and consists of smaller particles which are known as electrons, protons and neutrons. These particles are the same for all elements, but their quantities, number of electrons, protons and neutrons, differ in the atoms of various elements. The protons and neutrons are grouped in the center of the atom known as the nucleus, and the electrons orbit around the nucleus like the planets revolving around the sun. Depending on the number of electrons and protons contained in an atom, it can be positively charged, negatively charged, or neutral.

The electron has a negative charge equal to 1.602×10^{-19} coulomb and is the most elementary charge, i.e., no charge of electricity can be produced which is not an integral multiple of the charge carried by an electron. Experiments have shown that an electron has a mass equal to 9.1066×10^{-31} kg. Every neutral atom of matter contains at least one electron which is held in position by forces analogous to elastic forces, i.e., an electron may oscillate within the atom or may be displaced by an electrostatic field.

An electron may be forced out of an atom by the mutual repulsion of another electron moving in its immediate vicinity at a high velocity. This process is usually known as bombardment or collision, although the "colliding" electron may not necessarily hit the atom. When a neutral atom loses an electron, it attains the properties of a positively charged body and is called a positive ion. On the other hand, if a neutral atom gains an electron, it behaves like a negatively charged body and is called a negative ion. In addition to the electrons within an atom, every substance has a certain number of "free" electrons which move freely in the interatomic spaces or pass from one substance to another.

The proton is a positively charged particle with the charge numerically equal to that of an electron, i.e., 1.602×10^{-19} coulomb and has a mass equal to 1.6734×10^{-27} kg. The neutron is electrically neutral and has almost the same mass as that of a proton. The protons and neutrons are concentrated in the nucleus of an atom and account for most of its mass.

4.223 Flow of electrons—electric current, electromotive force

The nucleus of an atom is quite stable and remains disintegrated except in atomic energy applications when the nucleus is "bombarded" with different particles, e.g., in nuclear reactors to produce nuclear energy. However, the "bombarded" electrons, especially those in the outer orbit of the atom, are not as stable and are free to move around.

If an atom loses one of its electrons, it gets a positive charge and attracts one of the electrons of a neighboring atom or one of the electrons floating around in the air. If a group of atoms placed side by side pass their electrons from one atom to another, a constant flow of electrons is established. Such a flow of electrons is known as electric current.

The flow of electrons from atom to atom is due to two fundamental characteristics: atoms and electrons are always in motion, and as soon as an atom loses an electron, it attracts other electrons to restore the balance.

The electric current (flow of electrons) through a material is measured in amperes, which is equal to the flow of 1 coulomb of charge per second. In terms of electrons, one ampere is equal to the flow of 6.242×10^{18} electrons per second.

As the flow of current is always between two points with unequal charges, the basic requirement of achieving or maintaining a flow of current is the creation of the difference in charge. It can be done by moving electrons to cause either a surplus or lack of electrons at one point.

To utilize current flow, the flow of electrons must be directed from the source (means causing the difference in charge) through a conductor (a material which allows passage of electrons from one atom to another) and back to the source. This complete cycle of the flow of electrons is called an electric circuit.

The agent which tends to produce or maintain an electric current in a circuit is called an electromotive force (EMF). The terms commonly used in practice are "difference of potential" and "voltage." Electromotive force is not a force in the mechanical sense but is the electrical energy developed per unit charge. The electromotive force is measured in volts, which is the energy of a charge equal to 1 coulomb. In other words, two points have a potential difference of one volt if the

work done in moving 1 coulomb charge from one point to the other is equal to one joule (1 joule = 1 watt–second = 1 newton–meter = 10^7 ergs).

In addition to primary cells and storage batteries, the primary sources of electromotive force are electromagnetic induction (electric generators) and contact of two dissimilar bodies (thermoelectric effect, pyroelectric effect, piezoelectric effect).

The direction of the emf in a circuit is taken as the direction in which a positive charge would be forced around a circuit containing only this one source of emf. If there are several sources of emf in a circuit, the resultant emf is taken as the algebraic sum of all these emf's. The emf's acting around the circuit in one direction are taken as positive and those acting in the opposite direction are negative. The emfs acting in a direction opposite to that of the resultant emf are called "back" or "counter" emf.

The terminal of a device which is at a higher potential is called the positive terminal and the other terminal is called the negative potential. The drop of potential is always from the positive to the negative terminal, irrespective of the direction of flow of electricity.

4.224 Conductors and insulators

As mentioned above, an electric current in an electric circuit is due to flow of electrons through a medium which allows passage of electrons from one atom to another. The materials which offer little resistance to the flow of electrons are known as conductors. Such materials include metals such as gold, silver, copper, and aluminum. A continuous flow of electric current is established by applying a continuous emf across a conductor. On the other hand many other materials, such as glass, paper, rubber, ceramics, and plastics, offer a high resistance to the flow or movement of electrons. Such materials are known as insulators or dielectrics. In these materials, application of a continuous emf does not result in a continuous current.

According to the electron theory of conduction, all substances contain a certain number of free electrons in addition to electrons and protons constituting atoms. Further, these free electrons are in a constant state of violent agitation. On applying an emf to the substance, these free electrons drift towards the point of highest potential and establish the flow of current.

According to this theory, good conductors have a large number of free electrons, and poor conductors and dielectrics have very few. On applying an emf to dielectrics, there is a transient flow of current which is believed to be due to displacement of bound electrons in the atoms without their actual removal from the atom. The displacement of these electrons is directly proportionate to the applied emf. Upon increasing the value of emf, the displacement increases until a value is reached, at which time the electron is knocked out of the atom. At this emf, the dielectric breaks down and becomes a partial conductor. The value at which a dielectric breaks down is known as the dielectric strength.

4.225 Ohm's law, resistance, and conductance

When a steady difference of potential is applied across a conductor which is held at a constant temperature and in which there is no internal source of emf, a steady current proportional to the potential difference flows through the conductor. This relationship between potential difference and current is known as Ohm's law and can be written as:

$$V = RI \qquad (4.21)$$

where:

V = the difference of potential in volts
I = the resultant current in amperes.

The proportionality factor R is called the resistance of the conductor and is measured in mhos. The reciprocal of resistance is known as the conductance of the conductor. The practical unit of conductance is ohms.

These definitions of resistance and conductance hold only if there is no emf in the circuit under consideration. It is possible only if the current remains constant and the conductor is of uniform material and at constant temperature. Also, the same current must flow through each cross-section of the conductor and the drop of potential must be the same between all the points in the two end surfaces, i.e., the end surfaces are equipotential surfaces. Under such conditions, the resistance of a homogeneous conductor of a small cross-section as compared to length is given by the relation

$$R = \rho \frac{L}{A} \qquad (4.22)$$

where:

L = the length of the conductor
A = the area of cross-section.

The factor ρ is called the specific resistance or resistivity of the material. Its units depend on the units of length and cross-section (ohm meter or ohm ft). The reciprocal of resistivity is called specific conductance or conductivity (mho/meter or mho/feet).

The resistance of practically all materials changes with temperature. The resistance of most of the metallic conductors used in electrical machines increases with an increase of temperature. On the other hand, most of the nonmetallic conductors and dielectrics show a decrease of resistance with temperature increase.

For most practical purposes, the resistance of metallic conductors at any temperature can be determined from the approximate formula

$$R_T = R_0(1 + \alpha_0 T) \qquad (4.23)$$

where:

R_T and R_0 = the resistances of the material at temperature T and zero degree centrigrade respectively
α_0 = the zero degree temperature coefficient of resistance.

For copper, $\alpha_0 \approx 0.00427$. The relation between resistance and temperature of dielectrics and nonmetallic conductors is usually nonlinear.

4.226   Electricity and magnetism—inductance

Electricity and magnetism are closely related to each other and the study of one must include a study of the other. All electric generators and motors are based on the linkage between these two types of energy.

Any body which possesses the property of attracting pieces of iron or steel and which, when suspended freely, takes up a definite orientation with respect to the geographical meridian is known as a magnet. The phenomenon by virtue of which a magnet attracts pieces of iron or steel and orients itself into a definite position when suspended freely is called magnetism. Many substances, when placed near a magnet or near a conductor carrying electric current, acquire the properties of a magnet. Such substances are known as magnetic substances and the body which acquires the properties of a magnet is said to be "magnetized."

A magnet or magnetized body, when suspended freely, orients itself along the north-south meridian. The end pointing towards north is called the north pole and the end pointing towards the south is called the south pole. The two poles cannot be separated from each other, i.e. each magnet has both poles. Magnetic properties are not uniform over the surface of a magnet; they are stronger at the poles. However, the two poles of a magnet have exactly the same strength. Also, unlike poles attract each other while like poles repel.

Any region in which a magnetic substance when placed therein becomes magnetized is a magnetic field. A magnetic field always exists in and around every magnetized substance. All magnetic substances and magnetic poles experience a force in this region; the magnitude and direction of the force varies, depending on the position of the substance in the magnetic field which shows that the magnetic field has a direction as well as intensity. This fact can be seen by drawing the imaginary lines of force between the poles of a magnet (Fig. 4.21). The lines of force are directed away from the

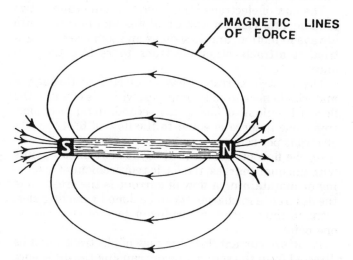

Fig. 4.21   Magnetic lines of force due to a permanent magnet. (Courtesy OiLine-Kobe)

north pole, showing the direction in which a north pole will be repelled if placed at that point. Also, the lines of force are more concentrated near the poles, indicating higher intensity of magnetic field at these points. Keep in mind that the lines of force never cross each other.

A magnetic field also exists around a current of electricity (Fig. 4.22). In this case, the magnetic lines of force are in a circular path around the current. The magnetic field associated with a current depends upon

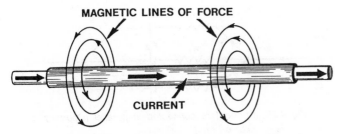

Fig. 4.22   *Magnetic lines of force due to flow of current through a conductor. (Courtesy OiLine-Kobe)*

the amount of current flowing: the higher the current the stronger the magnetic field.

Faraday discovered that an emf is produced in a coil if, for any reason, the magnetic field through a coil varies. Later, it was found that the emf can also be produced by (a) a conductor moving or cutting across a stationary magnetic field, (b) a moving magnetic field cutting across a stationary conductor or (c) a change in the number of magnetic lines enclosed by a loop or coil. These principles are used in the operation of generators and transformers. This phenomenon of producing emf by varying magnetic field is known as the electromagnetic induction and the emf is called induced emf.

When the current in an electric circuit varies with time, so does the magnetic field accompanying the current. This change in magnetic field associated with the circuit gives rise to an induced emf in such a direction so as to oppose the change in the current. It has been found that the induced emf is proportional to the rate of change or current, i.e.,

$$e = -L\frac{di}{dt} \qquad (4.24)$$

where:

e = the induced emf
di/dt = the rate of change of current.

The negative sign shows that the induced emf tends to counteract the change of current. The factor of proportionality L is the coefficient of self induction or, simply, inductance. On measuring emf in volts, current in amperes, and time in seconds, the inductance is obtained in henrys.

4.227   Capacitance

Any two conductors separated from each other by an insulation (usually known as dielectric) form an electric condenser or a capacitor. On connecting the two conductors to a source of emf and thereby creating a potential drop across the two conductors, each conductor stores a certain amount of charge. The amount of charge is directly proportional to the potential drop, i.e.,

$$Q = CV \qquad (4.25)$$

where:

Q = the charge
V = the potential drop between the conductor surfaces.

The coefficient of proportionality C is the capacitance of the condenser and is expressed in farads on measur-

ing potential drop in volts and charge in coulombs. The unit farad is a very large unit for most practical purposes; therefore, in practice, a unit equal to one-millionth of a farad (a microfarad) is used.

4.228   Direct current and alternating current

The current (flow of electrons) can be characterized by two parameters—magnitude and direction. If the magnitude and direction of flow of current does not change with time, it is called a continuous current. If a current flows in the same direction but varies in value, the current is known as the direct current. In practice, the term "direct current" designates either a continuous current or a current whose value varies only by an inappreciable amount, e.g., current obtained from a battery or direct-current generator. The direct current whose values fluctuate within appreciable limits, e.g., current from a rectifier, is the pulsating current. An alternating current reverses in direction, first positive and then negative, and alternates between constant positive and negative values. Thus, magnitude and direction vary in alternating current.

The variation of alternating and direct currents or voltages with time can be graphically represented as shown in Fig. 4.23. The shape of these curves is known as waveforms. Although theoretically alternating current can have any waveform as long as it changes direction and its magnitude varies between constant positive and negative values, in practice, especially in power systems, the alternating current undergoes gradual

ALTERNATING CURRENT

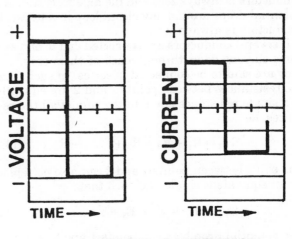

DIRECT CURRENT

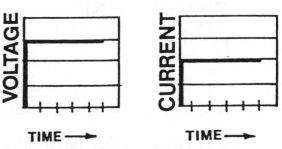

Fig. 4.23   *Wave shapes (Courtesy OiLine-Kobe)*

ALTERNATING CURRENT

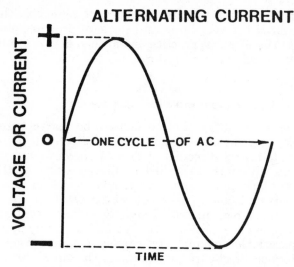

Fig. 4.24 *Sinusoidal alternating current or voltage (Courtesy OiLine-Kobe)*

changes in magnitude and direction and follows a sine curve (Fig. 4.24).

4.229 Direct current circuits

The direct current circuits consist of several conductors connected in an orderly manner. The analysis of these circuits involves determination of the currents flowing through each conductor and the potential drop across each conductor. It can be achieved by applying Ohm's law to each conductor and by applying two general laws, known as Kirchhoff's law, to the network. According to the Kirchhoff law, the algebraic sum of the currents coming to any junction in a network of conductors is always zero and the algebraic sum of the potential drops around any closed loop in a network of conductors is always zero.

If several conductors are connected end to end so the same current flows through each of them (Fig. 4.25), they are said to be connected in series. In such a case, the resistances between points 1 and 2 are equivalent to a single resistance equal to the sum of all the resistances, i.e.,

$$R_{1\text{-}2} = R_1 + R_2 + R_3 + \ldots + R_n. \qquad (4.26)$$

The emf's between points 1 and 2 can also be replaced by an equivalent single emf such that

$$E_{1\text{-}2} = E_1 + E_2 + E_3 + \ldots + E_n \qquad (4.27)$$

The potential drop between points 1 and 2 will be

$$
\begin{aligned}
V_{1-2} &= R_1 I_{1-2} - E_1 + R_2 I_{1-2} - E_2 \ldots + R_n I_{1-2} - E_n \\
&= I_{1-2}(R_1 + R_2 + R_3 \ldots + R_n) - \\
&\quad (E_1 + E_2 + E_3 \ldots + E_n) \\
&= R_{1-2} I_{1-2} - E_{1-2} \qquad (4.28)
\end{aligned}
$$

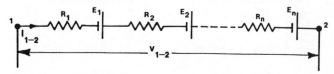

Fig. 4.25 *Conductors connected in series. (Courtesy OiLine-Kobe)*

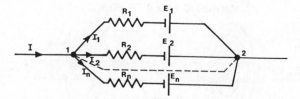

Fig. 4.26 *Conductors connected in parallel (Courtesy OiLine-Kobe)*

When several conductors are connected to two common junction points so that the same potential drop is established through each, they are connected in parallel (Fig. 4.26). Currents in various branches in such a circuit are calculated by applying Kirchhoff's law. If there are no emf's in various branches, the combined resistance of the several branches from points 1 to 2 is given by the relation

$$\frac{1}{R_{1-2}} = \frac{1}{R_1} + \frac{1}{R_2} + \frac{1}{R_3} + \ldots + \frac{1}{R_n} \qquad (4.29)$$

$$I = I_1 + I_2 + I_3 + \ldots + I_n \qquad (4.210)$$

$$V_{1-2} = I_1 R_1 - E_1 = I_2 R_2 - E_2 = \ldots = I_n R_n - E_n \qquad (4.211)$$

from which currents in various branches can be calculated.

In practice, many circuits consist of several conductors, some of which are in series and some in parallel. Such a circuit is called a series-parallel circuit. The total resistance of such a circuit can be calculated from the constants of several branches by applying successively the formulas for series and for parallel circuits.

It follows from the definition of emf and potential drop that the total work in moving charge Q from one point in a circuit to another point is W = VQ. Also, from the definition of current, the current is the amount of charge carried per unit time, i.e., Q = It. Thus, upon establishing a flow of current I through a device for a time t under the action of a voltage V, the energy input to the device is

$$W = VQ = VIt \qquad (4.212)$$

or the power input (energy per unit time) is

$$P = VI \qquad (4.213)$$

If V is measured in volts and I in amperes, the power input P is obtained in watts, and, on measuring t in seconds, the energy input W is in joules or watt-seconds. For practical purposes, the watt is a very small unit of power and consequently in electrical systems the power is measured in kilowatts, which is equal to 1000 watts. Also, the energy in power systems is expressed in kilowatt-hours.

4.2210 Alternating current circuits—general definitions

The following definitions are applicable to currents, electromotive forces, potential differences or any other functions of time.

An *oscillating quantity* is a quantity which, as a function of some independent variable (such as time), alternately increases and decreases in value, always remaining within finite limits.

A *periodic quantity* is an oscillating quantity, the values of which recur for equal increments of the independent variable.

The *period* of a periodic quantity is the smallest value of the increment of the independent variable which separates recurring values of the quantity.

A *cycle* is the complete series of values of a periodic quantity which occur during a period. A cycle per second is designated by the unit Hertz or Hz.

The *frequency* of a periodic quantity, in which time is the independent variable, is the reciprocal of the period. Frequency is expressed as a number of Hz.

The *angular velocity* of a periodic quantity is the frequency multiplied by 2π.

An *alternating quantity* is a periodic quantity which has alternatively positive and negative values.

The *instantaneous value* of an alternating current is the value of the current at any instant. Instantaneous values of current, potential difference, and emf will be designated by lower-case letters throughout this chapter, e.g., i, v, and e.

The *maximum value* of an alternating current is the numerical value of its maximum instantaneous value. The maximum value is often called the amplitude of the alternating current as well. It is designated by capital letters with the subscript $m(I_m, V_m, E_m)$.

The square root of the means of the squares of the instantaneous values of an alternating current over a complete period is called the *root mean square* (RMS) or the effective value of the alternating current. In specifying the value of an alternating current as so many amperes, this RMS value is always meant unless specifically stated otherwise. In the same manner, the square root of the mean of the squares of the instantaneous values of an alternating potential difference over a complete cycle is called the RMS value of the alternating potential difference. When the value of an alternating potential difference is specified as so many volts, this RMS value is always meant unless specifically stated otherwise.

The reason for selecting the above particular function of the instantaneous values of an alternating current or a potential difference as a measure of the current or the potential difference is that the average power dissipated as heat in a resistance R when an alternating current of RMS value I flows through it is RI^2. A similar argument exists for the use of RMS values of potential.

RMS values are designated by capital letters without subscripts. The general expression for the RMS value of an alternating current is

$$I = \sqrt{\frac{1}{T}\int_0^T i^2 dt} \qquad (4.214)$$

and similarly for an alternating potential difference.

Let v be the instantaneous value of the potential drop from any point 1 to any other point 2, and let i be the instantaneous value of the current from 1 to 2 at the same instant; then the power input at this instant is

$$p = vi.$$

When v and i are both positive or when they are both negative, the power input is positive but when the two parameters have opposite signs, the power input is negative, i.e., there is an actual power output.

The average value of the product vi over a complete period is the average power input or output, usually called simply power input or output. It is given by the relation

$$P = \frac{1}{T_0}\int^T p\,dt = \frac{1}{T_0}\int^T vi\,dt \qquad (4.215)$$

where T = the time for a complete period.

In certain special cases, the average power output P equals the product of the RMS value V of the potential difference by the RMS value I of the current; it can never be greater than this product and, as a rule, is less. The ratio of the average power P to the product of the RMS value V of potential difference by the RMS value I of the current is called the power factor of the circuit between the terminals considered, i.e.,

Power factor = P/VI.

When V is expressed in volts and I in amperes, power P is obtained in watts. On expressing V in kilowatts and I in amperes, the power P is in kilowatts.

The product of the RMS volts across the terminals of a circuit and the RMS amperes through it is called the volt-amperes, or apparent power taken by the circuit; this product divided by 1,000 is called the kilovolt-ampere (KVA) input.

Volt-amperes = VI
KVA = VI/1000

4.2211 Alternating current circuits: sinusoidal currents and voltages

A simple sinusoidal current (simple harmonic current) is an alternating current whose instantaneous values are equal to the product of a constant and the sine of an angle having values varying linearly with time. Thus

$$i = I_m \sin(\omega t + \Theta) \qquad (4.216)$$

where:

t = time in seconds, measured from any arbitrarily chosen instant
I_m = maximum value (amplitude) of the current
$\omega = 2\pi f = 2\pi/T$
f = frequency in cycles per second (Hertz)
T = period as a fraction of a second
Θ = constant which depends upon the instant chosen as the zero of time.

The RMS value of sinusoidal current is related to the amplitude as follows:

$$I = I_m/\sqrt{2} \qquad (4.217)$$

The phase of a periodic current for a particular value of the independent variable is the fractional part of a period through which the independent variable has advanced, measured from an arbitrary origin. For a simple sinusoidal current, the origin is usually taken as the last previous passage through zero from the negative to the positive direction. The phase angle is the angle obtained by multiplying the phase by 2π if the angle is to be expressed in radian, or by 360 if the angle is to be expressed in degrees.

When a sine-wave emf is impressed on a circuit having constant parameters, the resulting steady-state current is likewise a sine function of time, having the same frequency but whose emf and current do not reach their maximum values simultaneously. Let the current be represented by the relation

$$i = I_m \sin (\omega t + \Theta) \qquad (4.218)$$

and the voltage by

$$v = V_m \sin \omega t \qquad (4.219)$$

where t is the time measured from the instant when v = 0 and is increasing in the positive direction. The voltage reaches its maximum value when t = $\pi/2\omega$; the current reaches its maximum value when t = $(\pi/2\omega) - (\Theta/\omega)$. Hence, when Θ is positive, the voltage or potential drop reaches its maximum value Θ/ω seconds after the current reaches its maximum, or the current reaches its maximum value Θ/ω seconds before the potential drop reaches its maximum. When Θ is negative, the current reaches its maximum value Θ/ω seconds after the potential drop reaches its maximum. In the first case, the current "leads" the potential drop; in the second case the current "lags" the potential drop. The angle Θ is called the angular phase difference between the current and the potential drop.

When the phase difference is zero, the current and potential drop are in phase; when the phase difference is $\pi/2$ radians (90°), the current and potential drop are in quadrature; when the phase difference is π radians (180°) the current and potential drop are in opposition.

Let the voltage drop from terminal 1 to terminal 2 of a circuit be v = $V_m \sin (\omega t + \Theta_v)$ and the current from terminal 1 to terminal 2 be i = $I_m \sin (\omega t + \Theta_i)$. Then the instantaneous power input is

$$p = vi = \frac{V_m I_m}{2}[\cos (\Theta_v - \Theta_i) - \cos (2\omega t + \Theta_v + \Theta_i)]$$
$$= VI [\cos (\Theta_v - \Theta_i) - \cos (2 \omega_t + \Theta_v + \Theta_i)]$$
$$(4.220)$$

where V and I are the RMS values of voltage and current. The average power input is

$$P = VI \cos (\Theta_v - \Theta_i) \qquad (4.221)$$

where $(\Theta_v - \Theta_i)$ is the angular phase difference between the current and voltage. The instantaneous values of current, voltage, and power as well as the average power is graphically shown in Figure 4.27.

Putting the phase difference $\Theta_v - \Theta_i$ equal to Θ

$$P = VI \cos \Theta \qquad (4.222)$$

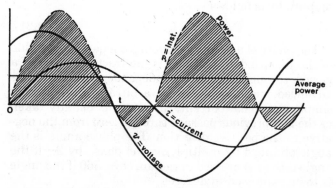

Fig. 4.27   Current, voltage and power in an alternating-current circuit. (Courtesy OilLine-Kobe)

Hence the power factor of the load is

$$\text{Power factor} = P/VI = \cos \Theta \qquad (4.223)$$

Since with sine-wave currents and voltages the power factor is equal to the cosine of the angle which expresses the difference in phase between them, this difference in phase is frequently called the power-factor angle. If the wave shape is not a pure sine curve, the power factor cannot be interpreted as the cosine of the phase difference since phase difference has no definite meaning except in reference to sine waves. A non-sinusoidal voltage and current may both reach their zero values at the same time and, in a sense, may be said to be in phase. But the power factor as defined above may be far from unity.

The reactive power in a circuit in which a sinusoidal current is flowing is equal to the effective emf times the effective current times the sine of the phase difference between them. When the emf and current are in volts and amperes respectively, the reactive power is obtained in vars.

The flow of a current i through a resistance R is always accompanied by a voltage drop in the direction of this current. At each instant, the voltage drop across the resistance is given by the relation

$$v = Ri$$

It is obvious from this relation that the current through a pure resistance and voltage drop across it are always in phase and, consequently, the power factor is equal to one. Also, the reactive power in a circuit consisting of resistances is zero.

In the case of an inductance L, the potential drop across it due to the flow of a current through it is obtained from the expression

$$v = L \frac{di}{dt} \qquad (4.224)$$

The instantaneous value of the current can be obtained by solving this differential equation. For the case of sinusoidal currents and voltages, the current lags the potential difference by $\pi/2$ (90°). The power factor in this case is equal to zero, i.e., no active power is consumed by a circuit consisting of only inductances and all of the power input is reactive.

The values of instantaneous currents and voltage drops in circuits consisting of only capacitance are related as follows:

$$\frac{dv}{dt} = iC \qquad (4.225)$$

By solving this differential equation, the current leads the voltage drop by $\pi/2$ (90°) and power factor equals zero. In this case as well, the power input is purely reactive.

The quantities' resistance, inductance, and capacitance are known as the electric circuit parameters. All of the practical alternating circuits consist of one or more parameters connected in various configurations (series and/or parallel). The combined effect of these parameters in an alternating-current circuit is known as impedance, which is expressed in ohms.

Depending on the impedance of a circuit and the parameters constituting the circuit, the value of power factor lies between zero and one. Also the current can lag or lead the voltage. However, in practice the current usually lags the voltage in most of the cases and the power factor associated with such circuits is called lagging power factor. Since a lagging power factor is more general, the word power factor alone generally means a lagging power factor. The power factor is of great significance in power consumption since the higher the power factor, the more efficiently the power is being consumed. Therefore, the load should have as high a power factor rating as possible.

The alternative-current circuits can be analyzed by applying Kirchhoff's law just like direct-current circuits. Only in this case, the term resistance is replaced by the impedance of the circuit. However, it should be pointed out that mathematical treatment of alternative current circuits is more complicated than that of direct-current circuits and requires knowledge of the theory of complex variables and vector analysis. A study of these topics is beyond the scope of this section and the interested reader should refer to specialized literature on this subject.

4.2212 Alternating current generator

All electrical machines are based on the principle of electromagnetic induction according to which an emf is induced in a coil if, for any reason, the magnetic field through the coil varies. Fig. 4.28 shows an elementary AC generator in the form of a loop of wire arranged to rotate in a magnetic field. The ends of this loop are connected to slip rings which rotate with the loop and are equipped with stationary brushes. Leads to the brushes enable electrical contact to be maintained between an exterior circuit and the rotating loop which acts as an elementary armature winding.

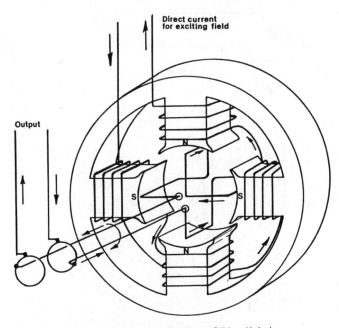

Fig. 4.29   Four-pole generator. (Courtesy OiLine-Kobe)

As the loop rotates, it cuts the magnetic field and an emf is induced in the loop. The instantaneous magnitude of the induced emf depends on the rate at which the loop cuts across the stationary magnetic field and, as such, depends on the position of the rotating loop. By considering the position of the loop at various instances of time, we can see the induced voltage follows a sine wave.

A complete sine wave of 360° represents 1 cycle. If the loop is rotating 60 times per second, or 3,600 rpm, it will generate a 60 cycle voltage. Thus, the frequency of the AC power supply depends on the speed of the generator at the central station.

Fig. 4.29 shows an elementary AC generator with four poles, alternately north and south. As the armature coils move past one north and one south pole, the voltage will pass through one complete cycle. Completing one revolution will generate two cycles. Therefore, to obtain a 60 cycle voltage, the coils must turn at only half of the former speed, or 1,800 rpm. In general,

$$\text{Frequency, cps (hz)} = \frac{\text{pole pairs} \times \text{rpm}}{60}$$

In a two-pole generator the voltage goes through 360° of the sine wave while rotating 360° in space. On the other hand, in a four-pole generator, the voltage goes through 360° of its cycle in rotating only 180°. Similar conclusions can be drawn for generators of various poles.

In Fig. 4.210, a second loop has been added with terminals connected to additional slip rings. These loops are 90° apart. When the voltage of coil 1 is maximum, that of coil 2 is zero and vice versa. The voltage waves are shown in Fig. 4.211. Such a generator is called a two-phase alternator. Two-phase AC power is not popular in power supply systems as it leads to odd and nonstandard voltages.

Fig. 4.212 shows an alternator in which three coils are rotated on a common shaft. They are located 120° apart and as these coils rotate each coil gives rise to a voltage wave. In all, three voltage waves are generated

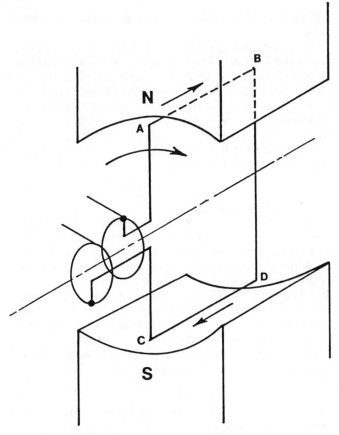

Fig. 4.28   Elementary AC circuit (Courtesy OiLine-Kobe)

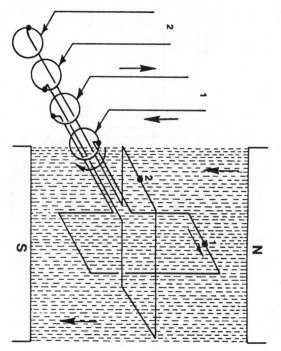

Fig. 4.210 Two-phase generator (Courtesy OiLine-Kobe)

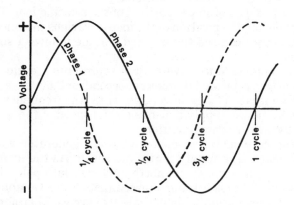

Fig. 4.211 Voltage produced by a two-phase generator. (Courtesy OiLine-Kobe)

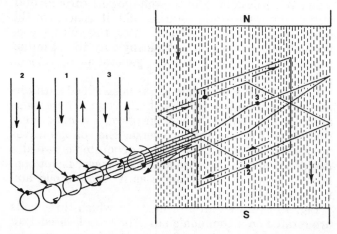

Fig. 4.212 Elementary three-phase generator. (Courtesy OiLine-Kobe)

which have a phase difference of 120°. Such a generator is known as a three-phase alternator.

Six slip rings are shown in Fig. 4.212. However, it is possible to connect one terminal of each of the three sets of coils to a common point (called the neutral) and bring out only the three remaining terminals to three slip rings. This is called a Y or star connection. The alternate connection is known as the Δ (delta) connection which appears to short-circuit all of the windings unless we consider the phase displacement between them.

Whether a three-phase alternator is designed to be connected internally as a Y or Δ machine is of importance to the designer but is rarely of any significance to the power user.

The load (a motor for instance) can also be internally connected as Y or Δ. Regardless of the way load or alternator is internally connected, the power in a three-phase circuit is given by the relation

$$\text{Power} = \sqrt{3} \times \text{voltage} \times \text{current} \times \text{power factor} \tag{4.226}$$

4.2213 Polyphase induction motor

The polyphase induction motor is the most common type of motor used. It consists of a wound stator which produces a rotating magnetic field in the air gap when a polyphase current flows through it. The stator has the same number of winding coils as the number of phases of the alternating current to be used (3 coils for a three-phase motor). The number of poles of the rotating magnetic field also depends on the way the stator is wound. The speed of the rotating field or the synchronous speed is

$$N = 120 \text{ f/p} \tag{4.227}$$

where f is the frequency and p is the number of poles.

There are two general types of rotors. The squirrel-cage type consists of heavy copper bars short-circuited by end rings. The wound rotor has a polyphase winding of the same number of poles as the stator and the terminals are brought out to slip rings so that external resistance may be introduced. The rotor conductors must be cut by the rotating field and hence the rotor cannot run at synchronous speed and must slip. The slip is given by the relation

$$s = (N - N_2)/N \tag{4.228}$$

where N_2 is the rotor rpm.

One disadvantage of induction motors is that they take lagging current and the power factor at half load and less is low. The speed-load curve of a typical submersible motor is shown in Fig. 4.213. The speed slightly decreases to full load and the slip is usually close to 3%. The power factor, depending on the size of motor, varies from 0.8 to 0.9 at full load. The direction of rotation of any three-phase motor may be reversed by interchanging any two stator wires.

The induction motor inherently is a constant speed motor, the rotor speed being

$$N_2 = 120f(1-s)/p \tag{4.229}$$

The speed can be changed only by changing the frequency, poles or slip.

In general, the number of poles can be changed by

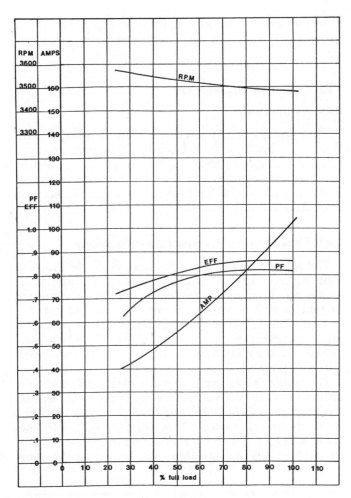

Fig. 4.213   Typical motor performance.

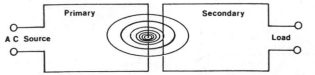

Fig. 4.214   Induction of current in secondary circuit. (Courtesy OiLine-Kobe)

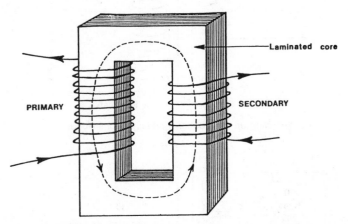

Fig. 4.215   Basic elements of a transformer (Courtesy OiLine-Kobe)

employing two distinct windings or by reconnecting a single winding. Design complications prevent more than two speeds being readily obtained in this manner. Another objection to changing the number of poles is because the design is a compromise and sacrifices of desirable characteristics usually are necessary at both speeds.

The slip of a motor can be changed by introducing resistance into the rotor circuit. This is possible only in motors with a wound rotor. This method of control is only practical in very large units and its applications are limited. Also, the wound rotor, controller, and external resistance make it more expensive than the squirrel-cage type.

4.2214 Transformers

A transformer changes the voltage in an alternating-current circuit. Like other electrical machines, transformer operation is also based on the principle of electromagnetic induction. Fig. 4.214 illustrates the principle of operation of a transformer. When a wire is carrying alternating current, the magnetic field moves by reason of the pulsating or alternating nature of the current. If another conductor is placed so the magnetic lines of force move back and forth across it, a current will be induced in the second conductor. The conductor carrying the generated current is called the primary, while the conductor carrying the induced current is called the secondary.

The expanding and contracting field of the primary circuit (Fig. 4.214) cuts across the secondary and induces in it an alternating voltage of the same frequency as the primary voltage. The magnetic coupling of the two conductors can be improved by providing a path for the magnetic lines of force in the form of an iron ring. The magnetic field can further be intensified by forming the primary and secondary conductors into coils. The basic elements of a transformer, thus, are the metal core serving the purpose of the iron ring described above and the primary and secondary coils (Fig. 4.215).

The ratio of the number of turns in the secondary coil to the number of turns in the primary is called the turn ratio of the transformer and determines the voltage output of the secondary in relation to the primary voltage supply. If the number of turns in the secondary coil is larger than the number of turns in the primary coil, the voltage output of the secondary will also be greater than the primary voltage. Such a transformer is called a step-up transformer. On the other hand, if the number of turns in the secondary coil is less than that in the primary, the secondary voltage will be less than the primary voltage and such a transformer is known as a step-down transformer.

As an example, Fig. 4.216A shows a step-up transformer. The primary coil has 100 turns while there are 1,000 turns in the secondary coil. Therefore, the voltage induced in the secondary coil will be 10 times greater (1,200 v) than the voltage applied to the primary (120 v). Similarly, a step-down transformer is shown in Fig. 4.216B with 100 turns in the primary coil and 10 turns in the secondary coil. In this case the voltage induced in the secondary coil will be one-tenth (12 v) of the voltage applied to the primary (120 v).

In practice, it is sometimes necessary to change the voltage ratio by changing the turn ratio in the trans-

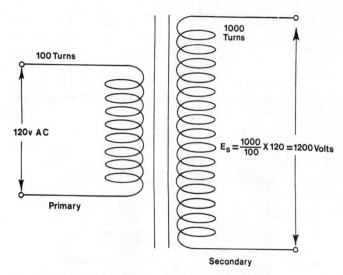

Fig. 4.216A   Step-up transformer (Courtesy OiLine-Kobe)

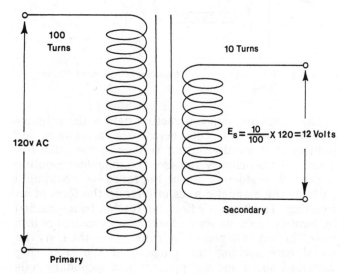

Fig. 4.216B   Step-down transformer (Courtesy OiLine-Kobe)

former. This is done by placing taps on various turns of one or both of the windings (Fig. 4.217). Then the voltage ratio is changed by using different taps.

Another type of transformer frequently used in oil fields is known as an autotransformer. It differs from other transformers in that it has only one winding rather than two or more (Fig. 4.218). Part of this winding is used for both the primary and secondary while the rest of the winding acts as either primary or secondary, depending on whether the transformer is used to step-up or step-down voltage.

Current flows in the secondary coil only when the secondary circuit is connected to a load. The current in the secondary of a transformer is stepped up or down inversely to the change in voltage. Although a transformer is not 100% efficient, it may be considered so for practical purposes. Therefore, the product of the voltage and current in the primary circuit is always equal to the product of current and voltage in the secondary circuit.

Transformers are available in a wide range of sizes to meet various needs. According to use, transformers range from giant power transformers, distribution

transformers in various sizes, and hand-held instrument transformers to very small specialty transformers.

4.23   Fluid flow fundamentals

4.231   Introduction

This section briefly reviews the principles of fluid flow as applied to pumps. No attempt is made to cover various concepts in detail; only the terms frequently used in submergible pump applications are defined. Refer to various standard textbooks on the subject for more detailed information.

4.232   Specific weight, density, specific gravity

The specific weight of a substance is the weight of a unit volume. In the English system of units, it is expressed as pounds per cubic foot, and in the metric system as grams per cubic centimeter (other derived units such as kilograms per cubic meter, tons per cubic meter, pounds per gallon or pounds per barrel are also used in industry.) It is commonly denoted by the Greek letter γ (gamma).

The density of a substance is the mass contained in a unit volume. It has the units of slugs per cubic foot in the English system and gram-mass per cubic centimeter in the metric system. It is usually denoted by the greek letter ρ (rho) and is related to the specific weight by dividing by g ($\rho = \gamma/g$) where g is the acceleration due

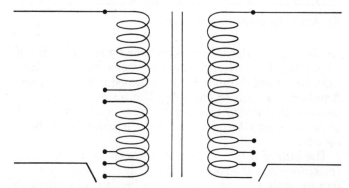

Fig. 4.217   Variable voltage ratio transformer (Courtesy OiLine-Kobe)

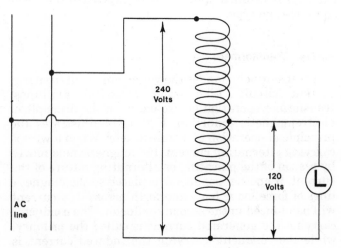

Fig. 4.218   Auto transformer. (Courtesy OiLine-Kobe)

to gravity (usually taken equal to 32.17 ft/sec² or 981 cm/sec²).

Although specific weight and density are two different concepts and have separate units, the specific weight is commonly referred to as density by engineers.

The ratio of the weight of a given volume of a substance to the weight of the same volume of another substance taken as standard is known as specific gravity. In the case of liquids and solids, fresh water at 60°F.(62.4 lbs/ft³ or 1 gm/cc) is taken as the standard substance, whereas air at the same temperature and pressure is regarded as the standard for gases. Thus, specific gravity of a liquid or solid is the ratio of the weight of a given volume of the liquid or solid to the weight of the same volume of water taken at 60°F. The specific gravity of a gas is the ratio of the weight of a given volume of gas to that of an equal volume of air at the same temperature and pressure (normally at 14.7 psia and 60°F). Specific gravity, being a ratio, has no units.

The specific gravity of liquids is frequently measured in different special scales in various trades and industries. The most common of these are the API and Baume scales. The API scale has been approved by the American Petroleum Institute, the American Society for Testing and Materials, the U.S. Bureau of Mines and the National Bureau of Standards. It is used exclusively by the Petroleum Industry in the United States. In this system, fresh water has been arbitrarily given an API gravity of 10.0 and higher numbers designate proportionately lighter liquids. The specific gravity of a liquid is related to the API gravity by the following formula

$$\text{Sp. gr.} = \frac{141.5}{131.5 + °\text{API}} \qquad (4.230)$$

Table 4-H (Appendix H) gives the values of specific gravity and specific weight (commonly referred as density by engineers) in various units for different values of API gravity.

Specific gravity of a mixture of different fluids can be determined from the formula

$$SG_m = \sum_{i=1}^{n} \frac{C_i \times SG_i}{100} \qquad (4.231)$$

where:

C_i = the concentration in percent of the ith component
SG_i = the specific gravity of ith component
n = total number of components
SG_m = resultant specific gravity of the mixture.

4.233 Pressure, head, and pressure gradient

Pressure exerted by a fluid on a surface is the force per unit area and is expressed in PSI in the English system and kg/cm² in the metric system. In a liquid at rest, the pressure at any point is equal to the pressure acting on the free surface plus the pressure from weight of the liquid above the point under reference. The pressure at a point due to a column of fluid is given by the relation γH where γ is the specific weight and H is height of fluid column above the point. The height of the liquid column is called the static head and is expressed in the units of length (ft, meter).

Pressure and head represent the same values in different units and are related to each other by the following relation:

$$\text{Head} = \frac{K \times \text{pressure}}{\text{sp. gr.}} \qquad (4.232)$$

where K is a constant of proportionality and its value depends on the units of head and pressure. If the pressure is expressed in terms of PSI and head in feet of the liquid, then K=2.31. In the metric system, on representing pressure in kg/cm² and head in meters, K=10.

In petroleum engineering, the term pressure gradient is frequently used in connection with artificial lift design. Pressure gradient is the pressure due to a column of liquid of unit height and is thus equal to sp. gr./K.

The pressure at a point is usually measured by pressure gauges installed at that point. These gauges indicate the pressure above atmospheric pressure which is known as the gauge pressure. Therefore, the absolute pressure is equal to the gauge pressure plus atmospheric pressure (usually 14.73 PSI or 1 kg/cm²). However, there are several types of pressure transducers available which are calibrated to indicate absolute pressure directly.

4.234 Velocity of a fluid, laminar and turbulent flow

The velocity of a fluid is defined with respect to some system of coordinate axes which is usually stationary with respect to the earth's surface. If the velocity of a fluid is low, the particles move in parallel layers and the velocity at any point is constant in magnitude and direction. Such type of flow is called laminar.

On the other hand, if the velocity is high, the motion is not steady and the velocity changes in both direction and magnitude. This is called turbulent flow. The velocity at which the flow changes from laminar to turbulent is known as critical velocity and the flow corresponding to that velocity is called critical flow or transition flow.

4.235 Viscosity

When the flow is laminar the applied shear stress, according to Newton's viscosity law, is proportional to the velocity gradient perpendicular to the velocity, i.e.,

$$\tau \alpha \frac{dv}{dy}$$

or

$$\tau = \mu \frac{dv}{dy} \qquad (4.233)$$

where:
τ = the applied stress
dv/dy = the velocity gradient.

The constant of proportionality μ is called absolute or dynamic viscosity. The relation between shear stress and velocity gradient is linear in the case of Newtonian fluids.

Dynamic viscosity in the metric system is measured in poise which is gm/sec.cm or dyne sec/cm². The unit most frequently used in industry is centipoise, which is one hundredth of a poise. In the English system of units, dynamic viscosity is measured in slug/ft-sec or lb-sec/ft².

Sometimes, it is more convenient to use kinematic viscosity which is equal to absolute viscosity divided by the mass density, i.e.,

$$\nu = \frac{\mu}{\rho} \qquad (4.234)$$

where:

ν = the kinematic viscosity
ρ = the mass density.

In the metric system, kinematic viscosity is measured in stokes having dimensions of cm²/sec. In industry, kinematic viscosity is usually expressed in centistokes which are equal to one hundredth of a stoke. In the English system, kinematic viscosity has the dimensions of ft²/sec.

The viscosity is usually measured by observing the time necessary for a given volume of fluid to flow through a hole of standard dimensions. Empirical correlations are used to convert this time into kinematic viscosity. The Saybolt Universal viscometer is commonly used in the United States for determining the kinematic viscosity of petroleum products. For heavy fluids, Saybolt Furol is used.

Refer to Appendix 4I for Tables 4I(1), 4I(2), 4I(3), and 4I(4) and Fig. 4I(5) for quick conversions of absolute and kinematic viscosities from one set of units to another.

4.236 Reynolds number

The resistance to flow of a fluid is related to a dimensionless number N_{Re} known as the Reynolds number:

$$N_{Re} = \frac{\rho \, v \, d}{\mu} \qquad (4.235)$$

where:

ρ = density of the fluid
v = velocity of the fluid
μ = absolute viscosity of the fluid
d = some characteristic dimension of the passage (e.g. diameter).

Since the Reynolds number is dimensionless, any set of units can be used as long as they are consistent.

It has been determined that the flow is laminar at Reynolds number up to about 2,000. On the other hand, the flow is turbulent at Reynolds numbers above 4,000.

N_{Re} can be used for comparing flow of fluids under different conditions of velocity, viscosity, density and fluid passage type for similarly shaped channels.

4.237 Streamlines and streamtubes

The concept of streamlines and streamtubes is often used for analyzing the flow conditions (Fig. 4.219). A streamline is an imaginary line drawn through a moving fluid so that it is always tangent to the velocity vector of the fluid at any given time. There is no velocity component normal to the streamline, and no two streamlines ever cross or intersect each other.

A streamtube is an imaginary closed area bounded by streamlines. The cross section of the tube may be any closed curve and is usually not constant in shape or area. As the tube consists of streamlines, no fluid

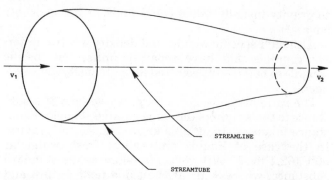

Fig. 4.219   Streamlines and streamtubes (Courtesy OiLine-Kobe)

crosses the walls of the streamtube, consequently the weight flow is the same at all sections of a tube. Also if the losses are neglected, the total energy at any point in the tube is constant.

If the properties of a fluid such as pressure, density, velocity, etc. do not change with time at a given section of a streamtube, the flow is steady flow. If the velocity of a fluid is constant over the cross-section but varies only in the direction of flow, the flow is one-dimensional. On the other hand, if the velocity varies over the cross-section as well as in the direction of flow, the flow is two or three-dimensional. One-dimensional flow is rarely encountered in practice, but the analytical solution of such a flow is very simple. Therefore, in actual practice, the two and three-dimensional flows are approximated to one-dimensional flow by considering average velocity, and the one-dimensional analytical solutions are adjusted to meet the requirements of two or three-dimensional flows by means of empirical or experimentally determined correction factors.

4.238 Analysis of steady flow—conservation of mass, momentum, and energy

Using average velocity of the fluid, steady flow can be analyzed by applying one or more of the principles of conservation of mass, momentum, and energy. For the cases of fluids, these can be written as:

a) Conserving mass (continuity equation) for a liquid or in incompressible flow
$$Q = AV = A_1 \, V_1 = A_2 \, V_2 \qquad (4.236)$$

b) Conserving momentum
$$\Sigma F_x = \rho Q(v_{x_2} - v_{x_1}) \qquad (4.237)$$
$$\Sigma F_y = \rho Q(v_{y_2} - v_{y_1}) \qquad (4.238)$$
$$\Sigma F_z = \rho Q(v_{z_2} - v_{z_1}) \qquad (4.239)$$

c) Conserving energy

$$J_{q_{12}} + W_{12} = J(U_2 - U_1) + \frac{P_2 - P_1}{\gamma} + \frac{v_2{}^2 - v_1{}^2}{2g} + Z_2 - Z_1 \qquad (4.240)$$

where:

Q = volumetric flow rate
A = area of cross section
v = average velocity of fluid
ρ = mass density of the fluid
$v_x, v_y, v_z,$ = components of average velocity in x, y and z directions.
F_x, F_y, F_z = components of momentum with respect to x, y and z coordinate axes

J = mechanical energy equivalent of heat
U = internal energy
P = pressure
γ = specific weight of the fluid
Z = elevation above some reference datum.

(The subscripts 1 and 2 denote the two sections of an element.)

In the case of an adiabatic flow $q_{12} = 0$ and for a special case in which no work is done on or by the fluid and in which there is no change in internal energy, the energy balance equation can be written as

$$\frac{P_2}{\gamma} + \frac{v_2^2}{2g} + Z_2 = \frac{P_1}{\gamma} + \frac{v_1^2}{2g} + Z_1 \qquad (4.241)$$

and is known as Bernoulli's equation. Here the term P/γ has the units of length and is called the static pressure head. The components $v^2/2g$ and Z, also having the dimensions of length, are known as velocity head and potential head.

The classical methods of hydrodynamics usually apply to only ideal fluids and are not of much use in solving flow problems. Energy losses cannot be calculated for turbulent flow without recourse to empirical relations. The energy equation is of great significance in solving flow problems. The law of conservation of momentum can be used only in a few cases in which the forces acting on the fluid are known or are to be calculated. According to the continuity equation for steady flow, the flow rate across every cross section is the same. In the case of liquids, considering the volumetric flow rate is sufficient since the density remains practically constant. However, in the case of compressible flow in which the density changes from one section to another, the continuity equation should be written in terms of mass or weight flow rate.

4.239 Flow of liquids in conduits

When a liquid flows through a conduit, it has no velocity at the walls of the conduit while it has some velocity within it. Consequently, the liquid is subjected to shear stresses at the wall and the energy losses due to shearing forces are known as frictional losses.

Several expressions have been developed for calculating the energy (pressure or head) losses due to flow of fluids in conduits. One of the more convenient, frequently used expressions is Darcy and Weisbach's

$$h = f \frac{L}{4m} \frac{v^2}{2g} \qquad (4.242)$$

where:

h = the energy or head loss
f = known as friction factor
L = the length of the conduit
v = the average velocity of fluid
g = the acceleration due to gravity

The component 4m in the equation is known as hydraulic diameter of the conduit and is equal to the ratio of four times the cross sectional area of flow divided by the wetted perimeter. In the case of a circular pipe 4m = d. Therefore,

$$h = f \frac{L}{d} \frac{v^2}{2g} \qquad (4.243)$$

The friction factor f depends on the nature of flow—laminar or turbulent—and is expressed as a function of the Reynold's number N_{Re} and relative roughness (ϵ/d) of the conduit. The roughness ϵ is a linear dimension characterizing the effective roughness of the pipe from a hydraulic point of view and is determined experimentally. Fig. 4J(1) of Appendix 4J shows the values of relative roughness for some of the commonly used pipe materials.

In laminar flow, the shear is entirely due to molecular forces. The velocity is zero at the walls and is maximum at the center. The velocity distribution follows a parabolic law, such that the maximum velocity is twice the average velocity. The shearing stress is proportional to the dynamic viscosity and transverse velocity gradient. The friction factor in laminar flow does not depend on surface roughness, as the losses due to surface roughness are insignificant compared to the energy required to overcome shear forces, and is equal to $64/N_{Re}$.

In turbulent flow there is violent mixing or eddying in the fluid, which is the principal cause of energy loss in addition to shear due to molecular forces. The velocity distribution is much more uniform than that in laminar flow. The friction factor in this case depends on Reynolds number and on the relative roughness of the pipe.

Several correlations have been developed for determining the value of friction factor in the turbulent region:

a) Drew, Koo and McAdams for smooth pipes and $3,000 < N_{Re} < \times 10^6$

$$f = 0.0056 + 0.5\ Re^{-0.32} \qquad (4.244)$$

b) Nikuradse's correlation

$$\frac{1}{\sqrt{f}} = 1.74 - 2\log_{10}\left(\frac{2\epsilon}{d}\right) \qquad (4.245)$$

c) Colebrook equation

$$\frac{1}{\sqrt{f}} = 1.74 - 2\log_{10}\left(\frac{2\epsilon}{d} + \frac{18.7}{N_{Re}\sqrt{f}}\right) \qquad (4.246)$$

The Colebrook equation is the basis for various friction factor charts. The equation is solved by trial and error method. The solution of the equation is graphically shown in Fig. 4J (2), Appendix 4J.

d) Jain correlation

$$\frac{1}{\sqrt{f}} = 1.14 - 2\log\left(\frac{\epsilon}{d} + \frac{21.25}{N_{Re}^{0.9}}\right) \qquad (4.247)$$

Being an explicit relation, the Jain equation does not require a trial and error solution as is the case with the Colebrook equation. Compared to the Colebrook equation, the Jain correlation gives an error of $\pm 1\%$ in the range $10^{-6} \le \epsilon/D \le 10^{-2}$ and $5 \times 10^3 \le N_{Re} \le 10^8$. The maximum error is 3% for Reynolds numbers as low as 2,000.

4.24 Centrifugal pump fundamentals

4.241 Introduction

The pump is probably the earliest invention of mankind for converting natural energy to useful work. Depending upon the culture recording the description, a pump has been known as a water wheel, Persian wheel, norias, etc. These were used mainly for drawing water from wells and have continued to exist in some countries in the East even up to these days.

Pumps are manufactured in a variety of sizes and types and are used in many applications. A voluminous amount of literature dealing with various aspects of pump design, operations and manufacturing has been published over the past several years.

This section does not detail centrifugal pump design or operation but familiarizes the reader with the basic terms and concepts frequently used in the pump industry. The reader is advised to refer to specialized literature on the subject matter for a more detailed treatment.

4.242 Classification of pumps

Pumps are classified in several ways on the applications they serve, the materials from which they are made, the liquids they handle, their orientation in space, or the type of driving system. All such classifications are limited in scope and, in many instances, overlap each other. Another way of classifying the pumps is based on the principle of transfering the energy to the fluid. According to this system, the pumps are classified into two basic groups—dynamic pumps and displacement pumps. Each of these groups can be further classified into several subgroups depending upon design features and characteristics. Fig. 4.220 shows such a classification of pumps into various subgroups.

In dynamic pumps, energy is continuously added to the fluid and is utilized to increase the velocity of the fluid. The velocity difference is subsequently converted into pressure energy. The centrifugal pump basically consists of a moving part, known as an impeller, which is mounted on a rotating shaft and a stationary part, called a diffuser, which is a series of stationary passages with gradually increasing cross-sectional areas. The rotation of the impeller with appropriately shaped blades sets the fluid particles in motion from the inlet towards the discharge. As the fluid flows through the impeller, the particles are accelerated; thus, their kinetic energy increases. This energy is partially converted into potential energy (pressure or head) in the impeller and in the diffuser. Fig. 4.221 shows the stage layout of a centrifugal pump.

In displacement pumps, energy is periodically added by application of force to one or more movable boundaries of any number of enclosed fluid-containing volumes. Under the action of the force, pressure of the volume increases sufficiently to force the fluid through valves and other resistances into discharge section.

The submersible pumps, used presently for production of petroleum crude, belong to the category of closed impeller, multistage, self-priming, single suction, radial and mixed-flow centrifugal pumps. Most of the discussion and principles outlined in this section are directed at these types of pumps although they apply to several other types as well.

4.243 Flow through an impeller, velocity triangles

The flow of a fluid through an impeller is extremely complex in nature. The velocity vectors are not parallel to the walls and there is considerable mixing and secondary flow near the impeller discharge. Most of the

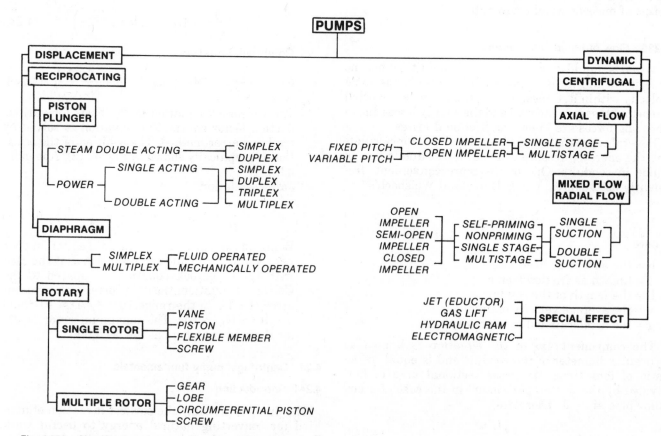

Fig. 4.220 Classification of pumps (Courtesy OiLine-Kobe)

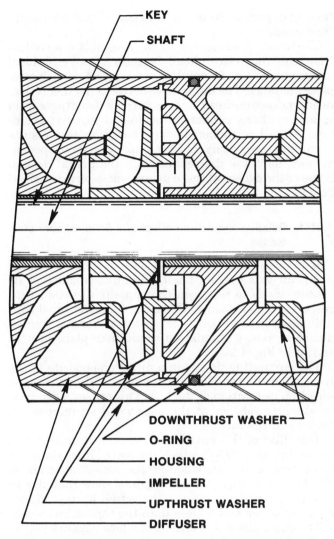

Fig. 4.221 Stage layout

KEY

SHAFT

DOWNTHRUST WASHER

O-RING

HOUSING

IMPELLER

UPTHRUST WASHER

DIFFUSER

4.244 Theoretical head developed by an impeller

The calculation of head developed by an impeller is based on three assumptions, none of which are satisfied in actual practice. These assumptions are (1) the fluid leaves the impeller tangentially to the blade surfaces, i.e., the fluid is completely guided by the blade at the outlet which is possible only in the case of an infinite number of blades, (2) the impeller passages are completely filled with the flowing fluid at all times which means there is no separation or void spaces, and (3) the velocities of the fluid at similar points are the same on all the flow lines. The head which is calculated based on these assumptions is known as the theoretical head.

To determine the pressure distribution, consider a small element of fluid (Fig. 4.224) of width b, thickness dR and circumferential length $R d\phi$ rotating with the impeller with an angular velocity ω.

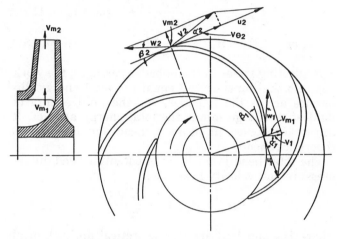

Fig. 4.222 Inlet and outlet velocity diagrams of an impeller having backward curved vanes. (Courtesy OiLine-Kobe)

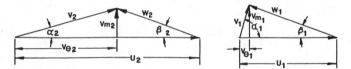

Fig. 4.223 Virtual inlet and outlet velocity diagrams of the impeller of Fig. 4.222. (Courtesy OiLine-Kobe)

pump design is based on the assumption of one-dimensional flow in which secondary motions are neglected. It is also assumed that the fluid exactly follows the impeller passages. The errors caused by such assumptions are corrected by the introduction of experimentally determined factors.

The absolute velocity V of a fluid particle flowing through a rotating impeller can be resolved into two components—peripheral velocity U of impeller and relative velocity W. The analysis is carried out graphically by the method of velocity triangles. The velocity triangles can be drawn for any point in the impeller passage but the inlet and discharge triangles are of maximum interest. Such triangles for an impeller are shown in Figs. 4.222 and 4.223. In these figures, subscript 1 refers to the inlet and subscript 2 to the discharge. The component of absolute velocity normal to the peripheral component (V_m) is usually referred to as the meridional component. The angle between absolute velocity V and tangential velocity U is usually designed by α and is called the absolute angle. The angle between W and U extended (or negative U) is designated by β and is called blade angle—it is the angle made by a tangent to the impeller and a line in the direction of motion of the blade.

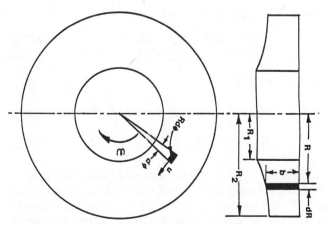

Fig. 4.224 Rotating container filled with liquid. (Courtesy OiLine-Kobe)

The elementary mass of the section is

$$dm = \frac{\gamma}{g} bRd\phi dR \qquad (4.248)$$

The centrifugal force acting on this elementary mass will be

$$dF = dm \, R \, \omega^2 = \frac{\gamma}{g} bR^2 \omega^2 d\phi dR \qquad (4.249)$$

and the pressure increase due to the action of the centrifugal force is given by the expression

$$dP = \frac{dF}{dA} = \frac{\gamma bR^2 \omega^2 d\phi dR}{g bRd\phi} = \frac{\gamma}{g} R\omega^2 \, dR \qquad (4.250)$$

Integrating this between R_1 and R_2, we get

$$P_2 - P_1 = \frac{\gamma}{g} \omega^2 \left(\frac{R_2^2 - R_1^2}{2} \right)$$
$$= \frac{\gamma}{g} \frac{U_2^2 - U_1^2}{2} \quad \text{since } R\omega = U$$
$$\text{or} \quad H_2 - H_1 = \frac{U_2^2 - U_1^2}{2g} \qquad (4.251)$$

Neglecting losses in the impeller passages, the work put into the impeller will equal the total energy difference across the impeller, i.e., the total energy of fluid at discharge minus the total energy at inlet. It can be expressed in terms of head and is equal to theoretical head.

The total head is made up of potential head due to pressure and velocity head, i.e.,

$$H_{th} = H_{pot} + \frac{V_2^2 - V_1^2}{2g} \qquad (4.252)$$

where H_{th} and H_{pot} are the theoretical and potential heads respectively.

The potential head increase across the impeller consists of the head due to action of centrifugal force and due to change in the relative velocity of the fluid. Thus

$$H_{pot} = \frac{U_2^2 - U_1^2}{2g} + \frac{W_1^2 - W_2^2}{2g} \qquad (4.253)$$

and consequently

$$H_{th} = \frac{U_2^2 - U_1^2 + W_1^2 - W_2^2 + V_2^2 - V_1^2}{2g} \qquad (4.254)$$

From Figure 4.223 we note:

$$W_1^2 = U_1^2 + V_1^2 - 2U_1V_1 \cos \alpha_1 \qquad (4.255)$$
$$W_2^2 = U_2^2 + V_2^2 + 2U_2v_2 \cos \alpha_2 \qquad (4.256)$$

and thus we get

$$H_{th} = \frac{1}{g} (U_2 V_2 \cos \alpha_2 - U_1 V_1 \cos \alpha_1) \qquad (4.257)$$

$$\text{or} \quad H_{th} = \frac{1}{g} (U_2 V_{\theta 2} - U_1 V_{\theta 1}) \qquad (4.258)$$

Equation 4.258 is the fundamental equation used in centrifugal pump design and is called the Euler equation. Note that the equation does not contain specific weight of the fluid. Consequently, an impeller operating at a given speed will generate the same amount of head irrespective of the fluid specific gravity since

head is expressed in terms of height of that particular fluid column.

The actual head developed by a pump is always less than the theoretical head because of several ideal assumptions, e.g., frictionless or lossless flow with complete guidance. The differences between the theoretical head and actual head due to various deviations from ideal conditions cannot be evaluated exactly; therefore, overall correction factors, based on test results of similar previously built pumps, are developed to account for these discrepancies in the actual design. Some of the causes of these discrepancies are described below.

4.2441 Relative circulation: effect of a finite number of blades

If a closed container filled with a fluid is rotated about an axis outside itself, the fluid, because of its inertia, tends to rotate in the opposite direction with respect to the container. Thus, if an impeller filled with a fluid is closed at both inlet and outlet and is rotated about its axis, a circulation would take place in it as shown in Fig. 4.225.

The circulating flow is in a direction opposite to the rotation of the impeller at discharge and in the same direction as rotation at the hub. This causes a decrease in absolute velocity at the outlet and an increase in velocity at the inlet.

The effect of this circulation on velocity triangles is shown in Fig. 4.226 in which the dotted lines show the theoretical diagram, whereas the actual diagram is shown in solid lines. These diagrams show that, due to relative circulation of fluid, the actual head produced by an impeller will be less than the theoretical head.

The exact amount of circulation flow depends on the shape of the impeller passage. The relative circulation is less in an impeller with a large number of blades as the passage becomes narrower and provides better guidance to the fluid. Also, the circulation is smaller in a narrow impeller than in a wide one. That is why for the same impeller diameter, the head produced by a high flow rate pump (at the best efficiency point) is less than that by a low flow-rate pump.

To estimate the impeller head slip, i.e., the difference between the theoretical head for an infinite number of blades as given by the Euler equation and the actual

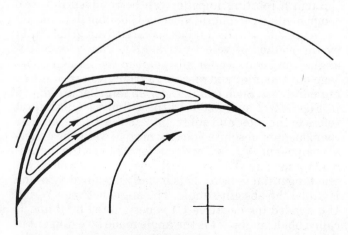

Fig. 4.225   Circulation of liquid in a closed rotating impeller passage. (Courtesy OiLine-Kobe)

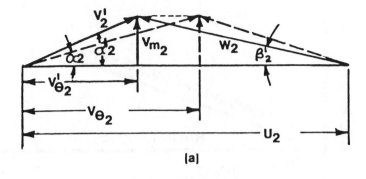

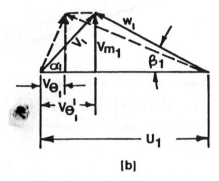

Fig. 4.226 Inlet and outlet velocity diagrams with circulatory flow correction. (Courtesy OiLine-Kobe)

stage pump, or through axial thrust balancing holes. The actual amount of leakage depends on the pressure difference across the clearance, amount of clearance, and seal design and on many other design parameters such as the distance between impeller and diffuser walls and pressure distribution in this gap.

Leakage lowers the head produced at any specific flow because the actual flow through the impeller is greater than the flow that comes out of the pump. This moves the head-capacity curve horizontally to the left. The leakage is characterized by the volumetric efficiency of the pump which is equal to the ratio of the measured capacity of pump to the capacity through the impeller.

4.2444 Mechanical losses

Mechanical losses external to an impeller include disk friction and frictional losses in bearings. Of these, the disk friction is by far the most important.

The power required to rotate a disk in a fluid is the disk friction. The impeller usually has closed shrouds which rotate in a fluid. Certain power is required to rotate this disk in the fluid and it must be supplied by the driver. Internally, this power is converted into heat and raises the temperature of the fluid.

The disk friction loss is due to two simultaneously occurring processes: friction of fluid on the disk which is comparatively small and a pumping action of the disk. The fluid bounded by the external surfaces of the rotating disk and stationary walls is thrown outward under the action of centrifugal forces. This is replaced by the fluid flowing in towards the center along the diffuser walls. Thus, circulatory flow is established on both sides of the rotating disk (Fig. 4.227). Power loss due to disk friction is given by the relation:

$$HP_{DF} = Kn^3D^5\gamma \qquad (4.259)$$

where:

n = rpm
D = the outer diameter
γ = the specific weight of the fluid
K = a constant of proportionality.

theoretical head produced by the impeller, is one of the most difficult problems in pump design. A number of methods based on experimental data or complicated theoretical considerations have been proposed to calculate the slip.

4.2442 Hydraulic losses

The flow of a fluid through an impeller is associated with various types of losses; as a result, the actual head obtained from an impeller is less than the head imparted to the fluid. These losses include friction loss in the impeller passage, diffusion loss due to divergence or convergence of the passage, fluid shock loss at the inlet, mixing and eddying loss at the impeller discharge, turning loss due to turning of the absolute velocity vector, and separation losses. All these losses are estimated on the basis of test data obtained on previously built pumps.

The ratio of the actual head obtained from an impeller to the head imparted to the fluid is known as the hydraulic efficiency of the impeller. The conversion of kinetic energy into potential energy in the diffuser is also associated with similar losses and, consequently, the total head obtained from a pump is less than the head produced by the impeller. The ratio of the head obtained from a pump to the total head imparted to the fluid in the impeller is called the hydraulic efficiency of the pump.

4.2443 Leakage losses

Leakage loss is a loss of capacity through the running clearances between the rotating element (impeller) and stationary element (diffuser). Depending upon pump design, the leakage can take place at several places, e.g., between the diffuser and impeller at the impeller eye, between two adjacent stages in a multi-

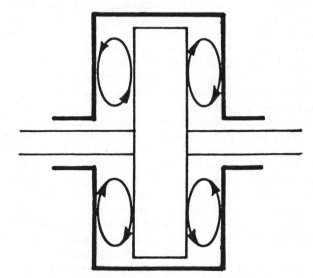

Fig. 4.227 Pumping action due to disk friction. (Courtesy OiLine-Kobe)

The value of K depends on the side clearance between the rotating disk and the stationary wall, surface roughness and disk diameter, and accounts for the differences between an impeller and an enclosed disk. The losses in bearings depend on the radial clearances, surface roughness, materials, lubrication conditions and the forces acting on the bearings.

Mechanical losses do not have any effect on head and capacity of a pump but increase the brake horsepower (load on the driver). The ratio of the power delivered to the fluid by the impeller to the brake horsepower supplied to pump shaft by the driver is known as the mechanical efficiency.

4.2445 Overall pump efficiency

All the losses taking place in a pump can be characterized in terms of hydraulic, volumetric and mechanical efficiencies, depending upon the nature of the losses. The product of these three components is called overall pump efficiency and is the ratio of useful power output delivered by the pump to the brake horsepower input of the driver.

If the maximum head possible for a given pump (virtual head) is plotted as a function of the quantity of fluid pumped for impellers of the same diameter operating at the same speed, straight lines are obtained as shown in Fig. 4.228.[1]

The slope of the line is a function of the outlet vane angle. A forward-curving vane, with a negative outlet vane angle, gives a positive slope on Fig. 4.228 or higher heads for higher throughputs. A backward-curving vane (positive outlet vane angle) gives a negative slope on Fig. 4.228 and a radial vane gives constant head. At zero flow, Q = 0, and

$$\text{Virtual head} = \frac{U_2^2}{g}$$

regardless of vane angle.

4.2446 Actual developed heads

An oil well pump impeller is normally designed for a set rate and the maximum head compatible with good efficiency. If the pump is operated at a lower or higher rate, the inlet and outlet vane angles are not the best for that flow and introduce turbulence which increases as the rate changes more and more from the design value. Figure 4.229 shows the effect of various factors on the head developed by a centrifugal pump.

Experimental values of the developed head at zero capacity agree well with the theoretical head resulting from centrifugal force alone ($U_2^2/2g_c$). This value of 1/2 can vary with pump design and one company reports values ranging from 0.47 U_2^2/g to 0.67 U_2^2/g. This means that the kinetic energy of the fluid as it leaves the impeller is expended in friction and turbulent losses. Brown discusses these principles in detail.[2]

Equation 4.258 shows that the head varies with the quantity of fluid pumped, but not with the density. The discharge pressure, however, does vary with the density. In other words, a given centrifugal pump develops the same head at a given capacity, irrespective of the fluid being pumped. If a pump normally handling water is filled with gas instead of water, the discharge

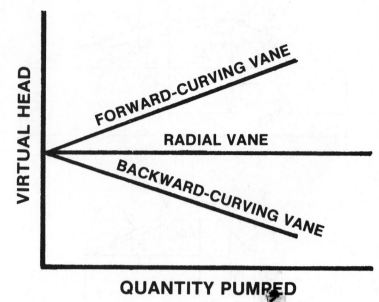

Fig. 4.228   Virtual head vs. capacity for different outlet vane angles (After Church)

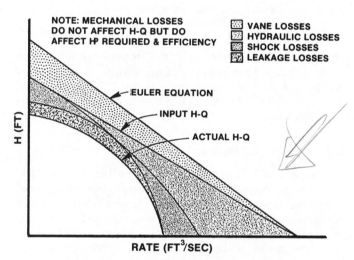

Fig. 4.229   Graphical representation of idealized performance vs. actual performance (After Church)

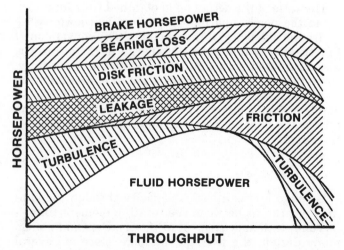

Fig. 4.230   Factors increasing fluid horsepower to brake horsepower. (After Church, Ref. 4.21)

pressure immediately drops so low that no gas can be delivered against the pressure of water filling the discharge piping. The gas remains in the impeller because of its low density, resulting in cessation of flow through the pump. This is known as gas locking. It can be alleviated in an ordinary pump by stopping the pump and allowing the gas to rise to the top.

4.2447 Brake horsepower

The power required must overcome all the losses and supply the energy needed to pump the fluid. These losses include the friction of flow through the impeller and turbulent losses, the disk friction or energy required just to rotate the impeller in the fluid, the leakage of fluid from the periphery back to the eye of the impeller, and the mechanical friction losses in the bearings, and stuffing boxes (Fig. 4.230).

The fluid horsepower is the energy absorbed in the fluid leaving the pump. The brake horsepower is the energy requirement of the pump per unit of time.

4.2448 Efficiency

The efficiency of a centrifugual machine is the ratio of fluid horsepower to brake horsepower.

4.2449 Pump performance curves

In practice, a pump is tested by running it at a constant speed and varying the flow by throttling the discharge. During the test, the flow rate, pressure increase across the pump, and brake horsepower are measured at several points. The pressure increase is subsequently converted into head, and the overall efficiency of the pump is calculated. Based on these data, curves showing head, brake horsepower and overall efficiency as a function of flow rate are drawn. These curves (Fig. 4.231) describe the performance of the pump under test conditions.

In summary the one very important concept is that the head in psi developed by a centrifugal submergible pump depends upon the peripheral velocity of the impeller, and depends upon the weight of the liquid pumped. The head developed converted to feet however would be the same whether the pump was handling water with a specific gravity of 1.0, oil with a specific gravity of 0.80, a brine of a specific gravity of 1.35, or any other fluids with various specific gravities.

The pressure reading on a pressure gauge would differ although the impeller diameter and speed would be identical in each case.

Figure 4.232 illustrates the relationship of identical pumps handling liquids of different specific gravities.

4.245 Specific speed

In the early stages of pump development, pumps were classified according to their hydraulic type ratios, such as the ratio of impeller width at discharge to the impeller outside diameter or the ratio of the impeller eye diameter to the impeller outside diameter. In 1915

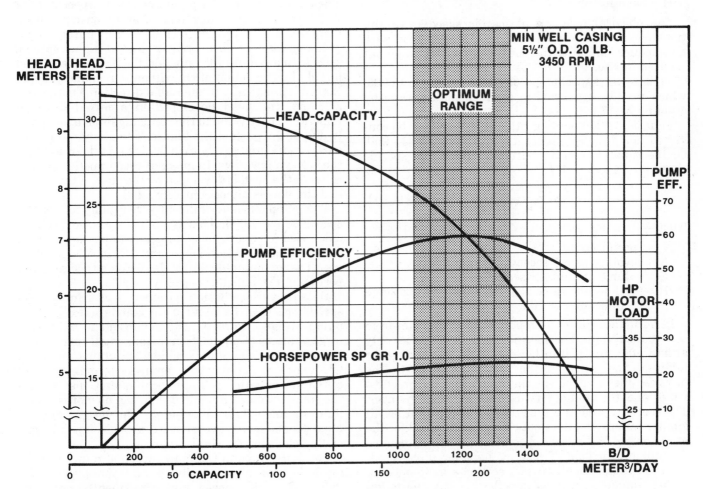

Fig. 4.231 Typical pump performance curve

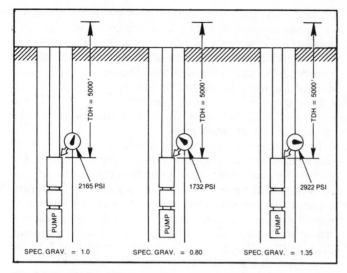

Fig. 4.232 Effect of fluid specific gravity on pressure head developed

a new characteristic was introduced to describe the performance and design of water turbines. Later on the same parameter was applied to centrifugal pumps. It is called specific speed and is defined in equation form:

$$N_s = \frac{n\sqrt{Q}}{H^{0.75}} \qquad (4.260)$$

where:

n = revolutions per minute
Q = capacity in gallons per minute
H = head in feet

The physical meaning of specific speed is revolutions per minute to produce 1 gpm at 1 ft head with an impeller similar to the one under consideration but reduced in size.

In these units specific speeds vary from 500 to 15,000, which may be converted to a dimensionless ratio by dividing by 17,200.

Equation 4.260 omits g_c; this equation may be converted to a dimensionless ratio by dividing by 17,200. Or the equation can be written as follows for specific speed:

$$N_s = \frac{N\sqrt{Q}}{g_c^{3/4} H^{3/4}} \qquad (4.261)$$

where:

N_s = specific speed
N = revolutions per second
Q = volume of fluid per second
H = total developed head or work done by one stage on unit mass of fluid

Performance curves for a pump show flow rates from zero to maximum. The specific speed term applies only at best efficiency point; therefore it doesn't change for different flow rates on the performance curve. The physical meaning of specific speed has no practical value and the number is used as a "type" number. The specific speed as a type number is constant for all similar pumps and does not change with speed for the same pump. With each specific speed are associated definite proportions of leading impeller dimensions, such as ratio of impeller width at discharge to impeller outside diameter, ratio of impeller eye opening to impeller outside diameter, and other hydraulic parameters.

A good way to relate specific speed to pump performance is to compare two pumps of different specific speeds but the same outside diameter and running at the same speed (rpm). The best efficiency point of the higher specific speed pump will occur at a higher flow and the head produced at best efficiency will be slightly lower. The flow passages of the higher specific speed pump will be larger, the impeller diameter smaller, the impeller vane angles larger, and for vertical pumps it will be more of a mixed flow type.

In the study of pump performance and classification of all important design constants, specific speed is a criterion of similarity for centrifugal pumps in the manner that Reynolds number is a criterion for pipe flow. When used as a type number, specific speed is calculated for the best efficiency point. For a multistage pump, specific speed is calculated on the basis of the head per stage. All important pump design and performance characteristics are so connected with the specific speed that it is impossible to discuss certain features without reference to it.

Fig. 4.233 shows performance of various specific speed pumps for different flow rates. The curves were developed experimentally by Worthington Pump Co. and are more or less taken as "state-of-the-art" by the pump industry. It is inadvisable to extrapolate the curves for other conditions and the results of any such extrapolation must be taken with caution. It is obvious from these curves that, for a given flow rate, maximum efficiency is attained by pumps of specific speeds in the 2,000-3,000 range. Also note that as specific speed increases, the pump design changes from purely radial at low specific speeds to strictly axial at higher specific speeds. In a radial-type impeller, the head is developed largely by the action of centrifugal force, and the discharge is practically in the radial direction. On the other hand, the head developed in a mixed-flow-type impeller is partly due to the centrifugal force and partly due to the push of the blades. The discharge is also partly radial and partly axial and this is the reason for the name mixed-flow. In an axial-type impeller, practically all the head is developed by the push of the blades and the flow is almost entirely axial.

The pump efficiency falls off very rapidly for specific speeds below 1,000. This is mainly because low specific speed impellers have long, narrow passages which result in large friction losses and greater disk friction loss. The amount of leakage also becomes a significant portion of the impeller capacity.

Data on various design parameters as a function of specific speed can be found in any standard text book on pump design. However, the requirements of lifting large volumes of liquids from rather small casing sizes impose extremely unfavorable restrictions on pump OD and force the hydraulic engineer to make several compromises with the design techniques. Consequently the standard data on pump design can no more be used for designing submersible pumps; the design engineer must depend on his experience and on design data specially developed for similar pumps.

For a given pump OD, flow rate, and rotative speed (as is the case in submersible pumps) there is only one value of specific speed and consequently only one

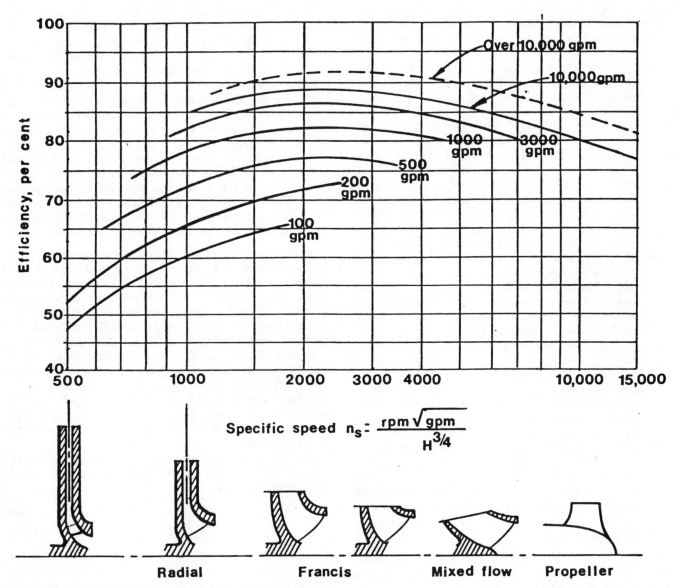

CENTRIFUGAL AND AXIAL FLOW PUMPS

Fig. 4.233 *Pump efficiency versus specific speed and pump size.*

design which will provide the optimum performance. Any deviations in the design from the optimum one result in higher loss and lower efficiency. The radial-type impellers can be designed up to a specific speed of about 1,500-1,800. The radial-type impellers of higher specific speeds have considerably lower efficiency as compared to mixed-flow impellers. Also a specific speed of about 4,000-4,500 may be the upper limit for the mixed-flow-type impellers.

If an impeller and diffuser are scaled up or down so the ratio of all homologous sections of both systems is the same, the following relations are satisfied at the same constant speed:

$$\frac{Q_1}{Q_2} = \left(\frac{D_1}{D_2}\right)^3 \qquad (4.262)$$

$$\frac{H_1}{H_2} = \left(\frac{D_1}{D_2}\right)^2 \qquad (4.263)$$

$$\frac{HP_1}{HP_2} = \left(\frac{D_1}{D_2}\right)^5 \qquad (4.264)$$

where:

Q_1 and Q_2 = the flowates
H_1 and H_2 = the heads produced
HP_1 and HP_2 = brake horsepowers corresponding to diameters D_1 and D_2.

In actual practice, it is difficult to maintain geometrical similarity and similarity of flow fields. Due to manufacturing reasons, the relative roughness of the walls in contact with the pumped fluid and relative running clearances usually are not the same and, consequently, the above relations are not satisfied. However, they can be used for making a preliminary estimate of the performance of the new pump.

4.246 Operating conditions and their effect on pump performance

The published pump performance curves usually refer to the performance of the pump at a fixed RPM and in clean water with specific gravity equal to 1 and of viscosity equal to 1 cp. However, in practice the pumps may be used to pump other liquids of different specific gravities and viscosities and may also operate at a different RPM. In such cases it is necessary to predict performance of the pump under actual operating conditions.

(a) *Effect of speed change on performance curves.* On changing the speed of a pump, the flow rate varies directly proportional to the speed. The head produced is proportional to the square of the speed while the brake horsepower is proportional to the cube of the speed. These relations are strictly followed only on neglecting the effect of speed on frictional and turbulence losses. However, these losses constitute only a small part of the total losses; therefore, these relations can be used for all practical purposes over a fairly wide range of speed changes. The efficiency of a pump remains constant during a speed change.

(b) *Effect of specific gravity.* The head produced by an impeller does not depend on specific gravity. Also, a centrifugal pump is a volumetric machine and the head produced is only a function of the volumetric flow rate. Consequently, the head-capacity curve does not depend on the specific gravity. The brake horsepower varies directly with the specific gravity and the pump efficiency remains constant irrespective of liquid density.

(c) *Effect of a small change of impeller diameter.* In many instances, the impeller diameter is slightly trimmed (turned down) to meet head-capacity requirements without resorting to a new pump design. This technique is also used if a pump develops too high a head at the design capacity. In this case, the capacity varies directly with diameter, the head varies with the square of the diameter, and the brake horsepower varies with the cube of the diameter. The pump efficiency does not change.

Combining the effect of all three variables—speed, specific gravity and diameter—the following relations are obtained (summarized in Table 4.21):

$$Q_2 = Q_1 \left(\frac{N_2}{N_1} \right) \left(\frac{D_2}{D_1} \right) \qquad (4.265)$$

$$H_2 = H_1 \left(\frac{N_2}{N_1} \right)^2 \left(\frac{D_2}{D_1} \right)^2 \qquad (4.266)$$

$$HP_2 = HP_1 \left(\frac{N_2}{N_1} \right)^3 \left(\frac{D_2}{D_1} \right)^3 \left(\frac{SG_2}{SG_1} \right) \qquad (4.267)$$

$$\eta_2 = \eta_1 \qquad (4.268)$$

where:

Q_1 H_1, HP_1, η_1, N_1, D_1 and SG_1 = Initial capacity, head, brake horsepower, efficiency, speed, diameter and specific gravity.

Q_2, H_2, HP_2, η_2, N_2, D_2 and SG_2 = new capacity, head, brake horsepower, efficiency, speed, diameter and specific gravity.

(d) *Effect of viscosity.* Most of the pumps handle water or other liquids of relatively low viscosity. However, in some instances pumps are built to handle liquids whose viscosity is greatly different from water. Viscous fluids have high internal resistance to flow. Consequently, the frictional losses and disk friction are increased, which results in low head and high brake horsepower. Viscosity also has an effect on leakage losses and it has been found that the viscosity reduces the capacity of a pump at the best efficiency point.

The complete effect of viscosity on centrifugal pump performance is not entirely known, although several papers have been published on this subject. The data in all the publications is based on laboratory tests. Of all the published works on this subject, the most popular and most frequently used data is the one published by the Hydraulic Institute (Figs. 4.234 and 4.235).

Figs. 4.234 and 4.235 provide a means of determining the performance of a centrifugal pump handling viscous fluids when its performance in water is known. They can also be used for selecting a pump for a given application. As these figures are based on experimental data, extrapolation beyond the limits shown is not recommended.

The following equations are used for determining the viscous performance when the water performance of the pump is known:

$$Q_{vis} = C_Q \times Q_W \qquad (4.269)$$

$$H_{vis} = C_H \times H_W \qquad (4.270)$$

$$E_{vis} = C_E \times E_W \qquad (4.271)$$

The following equations are used for approximating the water performance when the desired viscous capacity and head are given:

$$Q_W = \frac{Q_{vis}}{C_Q} \qquad (4.272)$$

$$H_W = \frac{H_{vis}}{C_H} \qquad (4.273)$$

The correction factors C_Q, C_H, and C_E are determined from Figures 4.234 and 4.235.

TABLE 4.21
SUMMARY OF AFFINITY LAWS

Diameter change only	Speed change only	Diameter & speed change
$Q_2 = Q_1 \left(\frac{D_2}{D_1} \right)$	$Q_2 = Q_1 \left(\frac{N_2}{N_1} \right)$	$Q_2 = Q_1 \left(\frac{D_2}{D_1} \times \frac{N_2}{N_1} \right)$
$H_2 = H_1 \left(\frac{D_2}{D_1} \right)^2$	$H_2 = H_1 \left(\frac{N_2}{N_1} \right)^2$	$H_2 = H_1 \left(\frac{D_2}{D_1} \times \frac{N_2}{N_1} \right)^2$
$HP_2 = HP_1 \left(\frac{D_2}{D_1} \right)^3$	$HP_2 = HP_1 \left(\frac{N_2}{N_1} \right)^3$	$HP_2 = HP_1 \left(\frac{D_2}{D_1} \times \frac{N_2}{N_1} \right)^3$

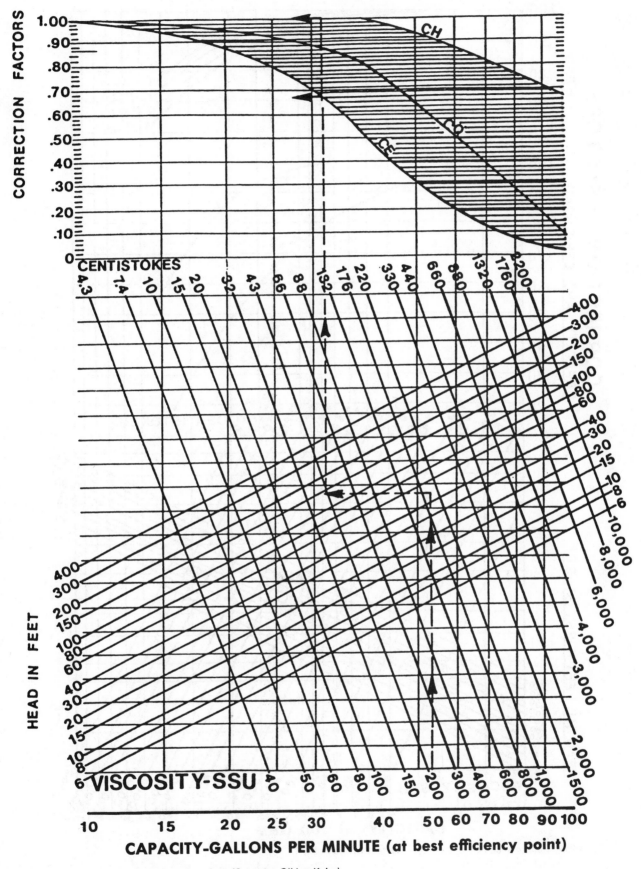

Fig. 4.234 *Performance correction chart (Courtesy OiLine-Kobe)*

In Figs. 4.234 and 4.235 and the preceding equations, the following designations are used:

Q_{vis} = capacity when pumping a viscous fluid.
H_{vis} = head when pumping a viscous fluid.

E_{vis} = efficiency when pumping a viscous fluid.
Q_w = water capacity.
H_w = water head.
E_w = water efficiency.

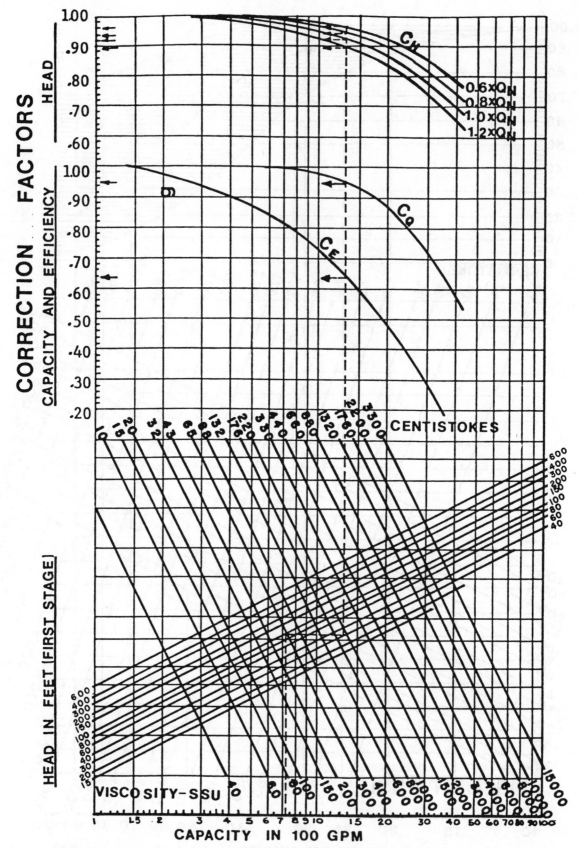

Fig. 4.235   *Performance correction chart (Courtesy OiLine-Kobe)*

The method of calculating the performance of a pump in a viscous liquid from water performance is given in Section 4.5311.

4.247   Cavitation and NPSH

If the absolute pressure of the liquid at any place inside a pump falls below the saturated vapor pressure corresponding to the operating temperature of the liquid, small vapor bubbles are formed and the dissolved gases are evolved. The vapor bubbles are carried by the flowing liquid to the regions of higher pressures where the bubbles condense or collapse.

Condensation of the bubbles is accompanied by a tremendous increase in pressure which is similar to water hammer or shock in nature. This phenomenon is known as cavitation.

Depending on the extent of cavitation, it can result in mechanical destruction, due to erosion and corrosion, and intense vibration.

Cavitation also has a significant effect on pump performance: the pump capacity as well as efficiency is reduced. Cavitation starts when the absolute pressure in a pump reaches vapor pressure and thus, to some extent, is related to the suction conditions. If the suction conditions are such that the pressure does not fall below the vapor pressure of the fluid at the operating temperature, cavitation will not set in. The minimum suction conditions required to prevent cavitation in a pump is known as the net positive suction head (NPSH). The minimum or required NPSH is measured experimentally.

The NPSH required by a vertical pump is equivalent to the hydraulic losses between the casing pressure opposite the first stage impeller and the entrance to the impeller vanes plus the difference in velocity head. This represents the net positive suction head required by the pump for proper operation. Any system must be so designed that the available NPSH of the system is equal to or exceeds the NPSH required by the pump.

Net positive suction head is the absolute pressure above the vapor pressure of the fluid pumped available in the casing opposite the first stage impeller to move and accelerate the fluid entering the impeller.

The available NPSH must be at least equal to the required NPSH if cavitation is to be prevented. An increase in the available NPSH provides a margin of safety against the onset of cavitation. It should be noted that the submersible pumps used for oil production usually do not experience cavitation because of sufficient submergence.

4.248 Axial thrust

Axial hydraulic thrust is the summation of unbalanced forces on an impeller acting in the axial direction. An ordinary single suction impeller with the shaft passing through the impeller eyes (Fig. 4.236) is subject to axial thrust because a portion of the front wall is exposed to the suction pressure, thus exposing relatively more back wall surface to the discharge pressure. If the discharge pressure were uniform over the entire impeller surface, the axial force acting toward suction

(downward direction) would be equal to the product of the net pressure generated by the impeller and the unbalanced annular area. However, that is not the case in actual practice. Due to rotation of the fluid in the impeller-diffuser gap, the pressure at the impeller periphery is much more than the pressure at the hub, as shown in Fig. 4.236. The actual pressure distribution in the impeller-diffuser gap cannot be determined from a rigorous analysis and it is estimated from the performance of similar previously built pumps. Whatever may be the pressure distribution, it always results in a force acting in the downward direction. For a given pump design, the magnitude of this force is inversely proportional to the flow rate, i.e., the higher the flow rate the smaller the force.

Another force acting on the impeller is the force due to the momentum of the fluid. The fluid enters the impeller axially and leaves in a radial or partly radial direction. The change in direction and velocity of the fluid results in another axial force which acts in the upward direction. The magnitude of this force is directly proportional to the flow rate squared.

The result of the two forces described above is the net hydraulic axial thrust developed by an impeller. Depending upon the pump design and flow rate, it can be in the downward direction, equal to zero, or in the upward direction. In a properly designed pump, thrust is maximum in the downward direction at shut-off (zero flow rate). Its magnitude gradually decreases on increasing the flow rate until it becomes equal to zero (usually near the best efficiency point). At higher flow rates, the thrust acts in the upward direction.

To eliminate the axial thrust of an impeller, a pump can be provided with both front and back wearing rings. The inner diameter of the back ring and the impeller skirt is made approximately the same to equalize thrust areas. Pressure approximately equal to the suction pressure is maintained on the back side of the impeller by drilling so-called balancing holes (Fig. 4.237) through the impeller wall near the inlet. The volumetric efficiency of such an impeller is slightly lower than that of an unbalanced impeller due to excess leakage past the wearing ring into the suction area through the balancing holes.

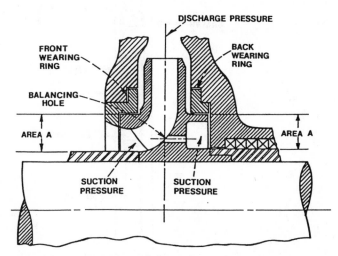

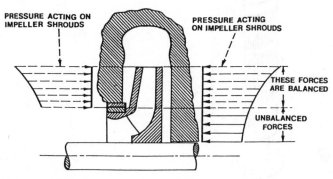

Fig. 4.236 Axial thrust on single suction impeller (Courtesy OilLine-Kobe)

Fig. 4.237 Balancing axial thrust of single-suction impeller by means of wear ring on back side and balancing holes. (Courtesy OilLine-Kobe)

Appropriate means should be provided in the design of a pump to handle the various axial forces which, in addition to the hydraulic thrust, include weight of the shaft, impellers and other parts attached to the shaft and the axial force due to action of the discharge pressure on the shaft cross sectional area at the discharge end.

A good reference on centrifugal pump principles is the work of Stepanoff.[3]

REFERENCES

1. Church, A. H., *"Centrifugal Pumps and Blowers,"* John Wiley & Sons, 1944.
2. Brown, et al., Unit Operations, John Wiley & Sons, 1951.
3. Stepanoff, A. J., *"Centrifugal and Axial Flow Pumps,"* John Wiley & Sons, Inc., May, 1967.
4. *Standard Handbook for Mechanical Engineers,* 7 ed. McGraw Hill Book Co.
5. Karossik et al., *Pump Handbook,* McGraw Hill Book Co., 1976.
6. *Hydraulic Institute Standards,* 13 ed.
7. *Basic Electricity for the Petroleum Industry,* 2 ed., Petroleum Extension Service, Univ. of Texas at Austin, 1977.
8. Brown, K. E., et al., *Technology of Artificial Lift Methods,* vol. 1, Pennwell Books, 1977.
9. Brill et al., *Two-Phase Flow in Pipes,* 1977.

4.3 COMPONENTS OF THE SUBMERSIBLE PUMPING SYSTEM

4.31 Introduction

A submersible pump system is comprised of an electric motor, a section to keep well fluid and motor oil separated (commonly called the seal or protector section), a gas separator, a multistage centrifugal pump, electric cable, switchboard and transformer. A complete range of accessories is available to supplement these principal elements.

In a typical application, "the pump" is run on tubing and submerged in the well fluid. Installations are applicable to crooked or directionally drilled holes.

In this section the various components of the pumping system are discussed. As mentioned previously, the

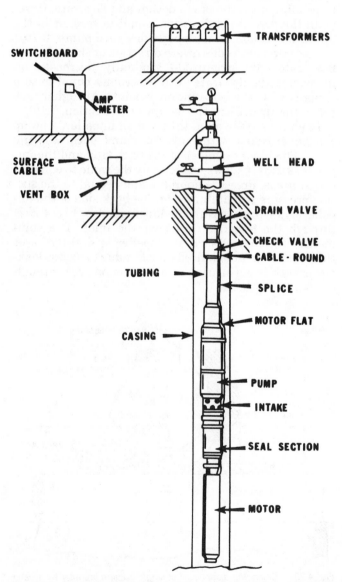

Fig. 4.31 Typical standard complete pumping system (Courtesy Bryon Jackson - Centrilift)

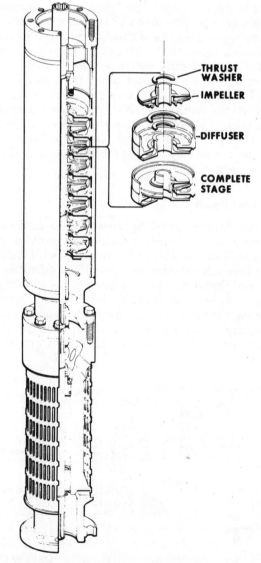

Fig. 4.32 Pump (Courtesy TRW-Reda)

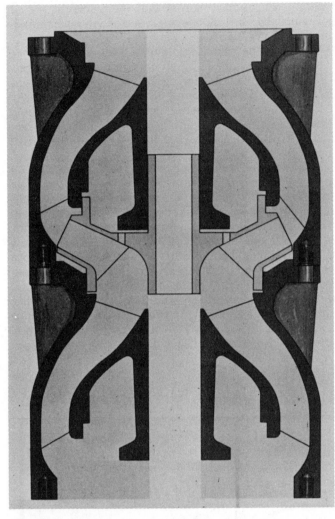

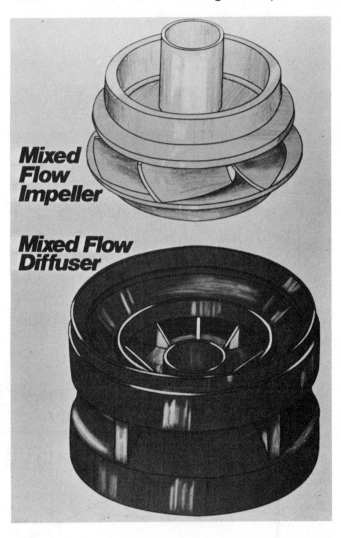

Mixed
Flow
Impeller

Mixed Flow
Diffuser

Fig. 4.33A Typical impeller and diffuser.

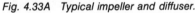

downhole pumping system consisting of the motor, protector and pump are called "the pump." It may also include a gas separator which is designed to separate free gas from the well fluid before the fluid enters the pump.

A typical complete system is shown in Fig. 4.31. Much of the descriptive material has been taken from references 1 through 5.

4.32 Submergible pump

4.321 Introduction

Submergible pumps are multi-staged centrifugal pumps. Each stage consists of a rotating impeller and a stationary diffuser. The type of stage used determines the volumes of fluid to be produced. The number of stages determine the total head generated and the horsepower required. Pumps are manufactured in a broad range of capacities for custom application to virtually all well conditions. Fig. 4.32 shows a typical single housing pump with standard intake.

The pressure energy change is accomplished as a pumped liquid surrounds the impeller; as the impeller rotates, the rotating motion of the impeller imparts a rotating motion to the liquid. The impeller imparts a motion tangential to the outer diameter of the impeller. This motion creates a centrifugal force which produces flow in a radial direction. Therefore the liquid is flow-

ing through the impeller with both a tangential and radial component. The result of these two components is the true direction of flow.

The diffuser changes some of the high velocity energy into relatively low velocity energy while directing the flow to the eye of the next impeller. For deeper applications, the impellers are of the floating or balanced type. In higher volume, larger units, a fixed type impeller is used. Refer to Fig. 4.33A and 4.33B for typical impellers.

In a floater pump, the impellers move axially along the shaft. In operating position, the impeller rests either on down-thrust pad or on up-thrust pad depending on the flow rate. In doing so, the axial thrust developed by each impeller is taken on individual up-thrust or down-thrust washers provided on each impeller and diffuser pads. The axial force due to pressure differential across the pump and acting at shaft end is taken by the thrust bearing in seal section. (Fig. 4.33C).

In a fixed pump, the impeller is fixed to the shaft and cannot move axially. Also it does not rest on any of the diffuser pads. The total axial thrust, developed by impellers as well as due to pressure difference, is taken on the external thrust bearing installed in the seal section.

In a combination pump, a certain percentage of the stages are floater and the remaining fixed. The pump is

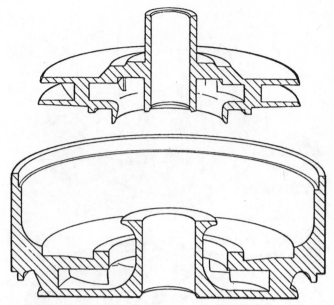

Fig. 4.33B Pancake stage impeller and diffuser (Courtesy TRW-Reda)

assembled in such a manner that the hydraulic thrust developed by the impellers is shared by each impeller.

However, the axial force due to pressure difference acting at shaft ends is taken only by the fixed stages. The external thrust bearing in such a design is not loaded at all during normal operation.

Each of the above-described pump designs have their own advantages, disadvantages, and limitations. They should be carefully analyzed for a particular application.

The axial thrust developed by an impeller depends on the hydraulic and mechanical design of the impeller and the operating point of the pump. A pump operating at a greater than design flow rate may produce excessive up-thrust. Inversely, a pump operating at less than design rate produces excessive downthrust as a result of which the external bearing and the downthrust washer may show signs of excessive wear. Due to these reasons, a centrifugal pump should be operated within a recommended capacity range. These recommended capacity ranges vary with different types of pump designs and are usually shown on pump performance curves. In the absence of such data, a range of 75-125%

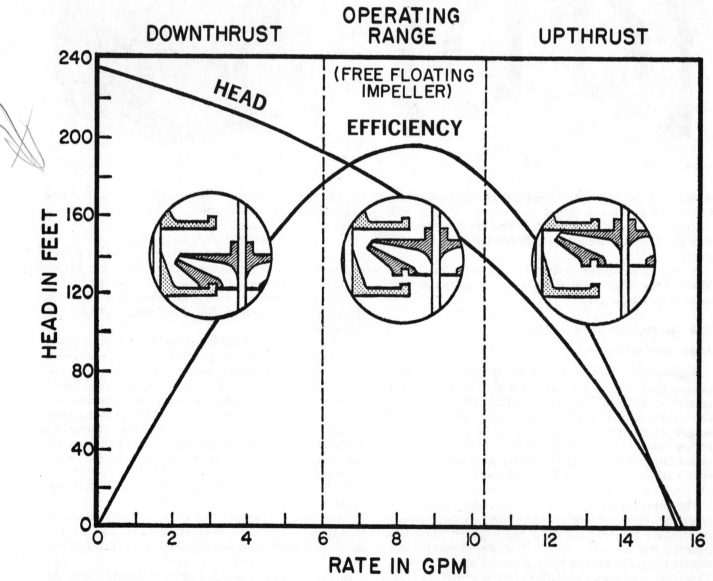

Fig. 4.33C Possible impeller positions

of the best-efficiency point may be taken as a good first approximation.

The overall length of a single section pump is limited to about 20-25 ft to facilitate proper assembly and handling. However, a number of pump sections can be joined in series to develop the required head. The maximum size (no. of stages) of a pump is determined based on one or more of the following limitations:

(a) horsepower rating of pump shaft
(b) pressure rating of pump housing
(c) load-carrying capacity of thrust bearing.

The longitudinal reactions or thrust on fixed impellers and shafts is transferred to the unit's thrust bearing, but for floating impellers is transferred to the pump housing.

The impellers are of a fully enclosed curved vane design, whose maximum efficiency is a function of impeller design and of a type whose operating efficiency is a function of the percent of design capacity at which the pump is operated. The mathematical relationship between head, capacity, efficiency, and brake horsepower is expressed as:

$$\text{bhp} = \frac{Q \times H \times \text{Sp. Gr.}}{E \times 3960}$$

where:

Q = volume, GPM
H = head, ft.
E = pump efficiency

Each stage of a submergible pump handles the same volume of fluid in the absence of free gas and produces a certain amount of head. This pressure head is additive; for example, if one stage produces 20 ft of head, 100 stages would produce 2,000 ft of head. The type of pump stage used determines the volume of fluid to be produced. The addition of more stages to a design will not increase volume unless the design did not have sufficient head initially.

Stages are manufactured of materials to provide optimum performance characteristics and maintain maximum corrosion and abrasion resistance. K-Monel shafts may be provided as standard equipment. Bolt-on heads and intakes make it possible to vary the capacity of a tandem pump in the field by using some or all of the pump sections.

4.322 Impeller diameters and casing sizes

The configuration and diameters of the pump impeller determines the amount of acceleration energy that is transmitted to the fluid. The impeller outside diameter is restricted by the internal diameter of the pump housing, which in turn is restricted by the well casing inside diameter. The internal diameter of the impeller is dependent on the outside diameter of the shaft, which must be strong enough to transmit power to all stages of the pump.

Submersible centrifugal pumps are manufactured for various sizes of well casings. The casing size ranges from 4½ in. and above. Table 4.31 illustrates approximate outside diameters of equipment available from submersible pump suppliers for the different casing O.D.'s. For any given diameter of pump housing and shaft, the impeller diameter will be constant.

TABLE 4.31
STANDARD PUMP DIAMETERS AVAILABLE

Casing O.D., in.	Weight per foot	Motor O.D., in.	Pump O.D., in.
4½	11.5	3¾	3⅜
	9.5		
5	All weights	3¾	3⅜
5½	20	4½	4.0
	17		
	15.5		
7	28	4½, 5½	4, 5⅜
	26		
	24		
	20		
8⅝ or greater	All weights	4½, 5½, 7⅜	4, 5⅛, 6¾

4.323 Thrust bearings

Thrust bearing designs are of the so-called solid shoe type or the pivot shoe type. The solid shoe is actually a misnomer. It implies that there is no flexibility in the individual shoes, but this is not the case. The bearing normally consists of six individual shoes which are mounted on a pedestal located at the center of the shoe. The pedestal is flexible enough so that it can deflect.

The operation of any bearing depends on the rotating member maintaining an oil film across the bearing surface. In the case of a pivot shoe or the so-called solid shoe bearing, the thrust runner pulls fluid with it across the pad. In either case, an oil film must be built and maintained or failure occurs.

Both the pivot shoe and pedestal-mounted shoe deflect at the entrance edge, thus allowing the oil film to be built. An oil film can be maintained for only a limited distance due to effects of viscosity, load, and temperature. Since the shoe must deflect, the oil film becomes wedge-shaped and the shoe must not be too long or it will allow the film to be squeezed out before it reaches the end of the shoe. For a given diameter there is an optimum number of shoes of a given width and length. To vary any one of the three requires a change in the other two.

The pedestal-mounted shoe is less tolerant of misalignment than the pivot shoe and its resistance to deflection requires that it be rated at less load.

One of the factors determining allowable bearing load is operating number, expressed symbolically:

$$\frac{\mu v}{5\ PL}$$

where:

μ = viscosity in reyns
v = velocity of thrust runner in feet per minute
P = load in psi
L = length of shoe in inches

An increase in loading decreases the operating number and lessens film thickness since P is in the denominator. An increase in temperature lessens viscosity and correspondingly decreases film thickness since μ is in the numerator. From this it can be seen that the viscosity of the oil used is an important factor to thrust bearing life and in keeping the assembling of parts as clean as possible.

In summary, the enemies of thrust bearing life are heat-reduced viscosity, misalignment, foreign particles, and vibration. Vibration and misalignment can cause destruction of the oil film which permits metal-metal contact. Dirt and/or foreign particles can cause scratching and contribute to loss of the oil film.

4.324 Thrust range

Impellers are designed so the net force is small, but downward at peak efficiency. (Fig. 4.33C) An impeller operating at a significantly greater than design capacity will show up-thrust wear since the force is less against the discharge side than against the suction side of the stage. Inversely, when the pump is operating at less than design rate, the force generated is greater on the discharge side of the stage and will result in down-thrust wear.

For these reasons, pumps should be operated within their capacity ranges for optimum impeller and thrust bearing wear. These capacity ranges vary with different type pumps. A good rule of thumb is that the capacity low range should not fall below 75% of the peak capacity (top pump efficiency) range and that the capacity high range should not exceed 25% of the peak capacity range. Figure 4.33C shows an example of this optimum range. The range is sometimes referred to as "thrust bearing range." By operating within this range, thrust bearing wear will be minimized.

The discharge rate and pressure of a submersible centrifugal pump depend on the rotational speed (rpm), size of impeller, impeller and diffuser design, number of stages, the dynamic head against which the pump is operating, and the physical properties of the fluid being pumped. The heads or pressures generated by each stage are additive. The total dynamic head of the pump is the product of the number of stages and the head generated by a single stage.

4.33 Protector or seal section

The seal section or the protector, in general, performs the following four basic functions:
  (a) connects the pump housing to the motor housing by connecting the drive shaft of the motor to the pump shaft;
  (b) houses the pump thrust bearing to carry the axial thrust developed by the pump;
  (c) prevents the entry of well fluid into the motor;
  (d) provides an oil reservoir to compensate for the expansion and contraction of motor oil due to heating and cooling of the motor when the unit is running or shut down.

The mechanical design and principle of operation of seal sections differs from one manufacturer to another. The main difference lies in the way the motor oil is isolated from the well fluid. Fig 4.34A shows the cross-section of one of the seal sections currently available. It is intended as an integral part of the motor which permits oil filling and sealing of motors at the factory rather than under adverse field conditions. In this design, the internal motor pressure is equalized to the wellbore pressure with the help of a metallic bellows installed at the bottom of the motor.

The motor oil is positively sealed from well fluid by means of an expansion bag which also compensates for expansion and contraction of motor oil. The entry of well fluid along the shaft is eliminated by the use of two mechanical shaft seals.

In the seal section shown in Fig. 4.34B, the motor oil is isolated from well fluid by using a high-specific-gravity liquid. Because of the direct communication between well fluid, barrier fluid and motor oil, the internal motor pressure is always equal to the well bore

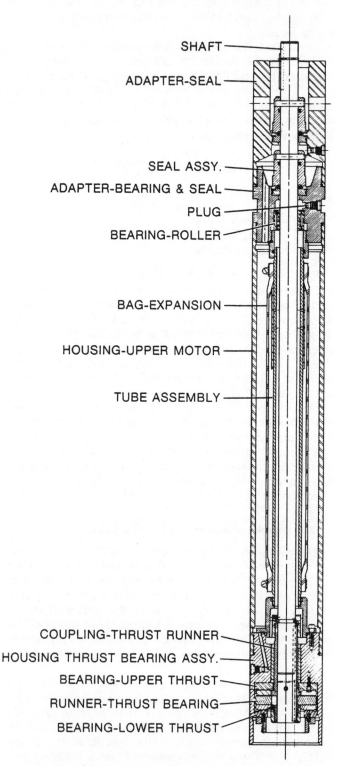

SHAFT

ADAPTER-SEAL

SEAL ASSY.

ADAPTER-BEARING & SEAL

PLUG

BEARING-ROLLER

BAG-EXPANSION

HOUSING-UPPER MOTOR

TUBE ASSEMBLY

COUPLING-THRUST RUNNER

HOUSING THRUST BEARING ASSY.

BEARING-UPPER THRUST

RUNNER-THRUST BEARING

BEARING-LOWER THRUST

Fig. 4.34A Motor seal section (Courtesy OiLine-Kobe)

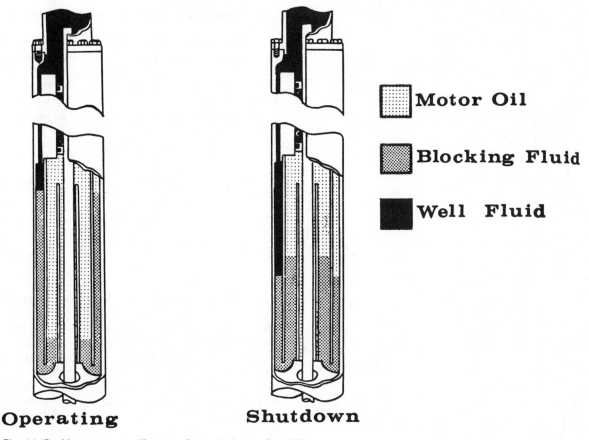

Operating   **Shutdown**

Motor Oil

Blocking Fluid

Well Fluid

Fig. 4.34B   Motor protector (Courtesy Byron Jackson - Centrilift)

pressure. Variation in the height of well fluid compensates for the expansion or contraction of motor oil.

The well fluid in the seal section shown in Fig. 4.34C communicates with the oil through a labyrinth path and thereby protects the motor from the adverse effects of well fluid entry.

All seal sections are equipped with high-capacity sliding-type thrust bearings capable of carrying all the axial loads developed by the pump. Depending on the design, these bearings can have fixed shoes or self-aligning tilting pads. Also, the pads may be pivoted centrally or slightly off-center.

All these design details may not be of much significance to a pump operator but do have an effect on pump operation. The centrally pivoted bearing can operate equally well in either direction of rotation. However, if the pads are pivoted off-center, severe damage may occur if rotated in the reverse direction. Also, the load-carrying capacity of the bearing can limit the pump size so as not to overload the bearing under given operating conditions.

Seal assemblies come in varying sizes to join different-sized motors to pumps. Sufficient room between the housing and well casing I.D. is of major importance when selecting pump and seal assemblies. This room allows passage between the housing and the ID of the well casing of the flat portion of the electric cable to the motor located on the bottom of the assembly. The drive shaft of the motor is connected to the pump shaft by the seal assembly shaft which may be splined on both ends. The lower end of the seal shaft fits on the motor shaft and is designed to allow for elongation of the motor

shaft due to temperature rises. The upper end of the seal shaft is engaged with the pump shaft in such a manner that the weight of the pump shaft, the longitudinal hydraulic load on the pump shaft, and in some cases the unbalanced longitudinal impeller loads are transmitted from the pump to the seal assembly shaft. These loads are in turn transferred to the thrust bearing.

4.34   Submergible motor

The motor is the driving force (prime mover) which turns the pump (Fig. 4.35A). The electrical motors used in submergible pump operations are two pole, three-phase, squirrel cage induction type. These motors run at a relatively constant speed of 3,500 rpm on 60 frequency and 2,915 on 50 Hz. The motors are filled with a highly refined mineral oil that must provide dielectric strength, lubrication for bearings, and good thermal conductivity. The thrust bearing of the motor carries the load of the motor's rotors. The nonconductive oil in the motor housing lubricates the motor bearings and transfers heat generated in the motor to the motor housing. Heat from the motor housing is in turn carried away by the well fluids moving past the exterior surface of the motor; therefore, a pumping unit's motor should never be set below the point of fluid entry unless some means of directing the fluid by the motor is utilized, such as a shroud over the motor, protector, and pump intake. A high starting torque enables it to reach operating speed in less than fifteen cycles.

The motor normally consists of a low-carbon steel housing with steel and brass laminations pressed

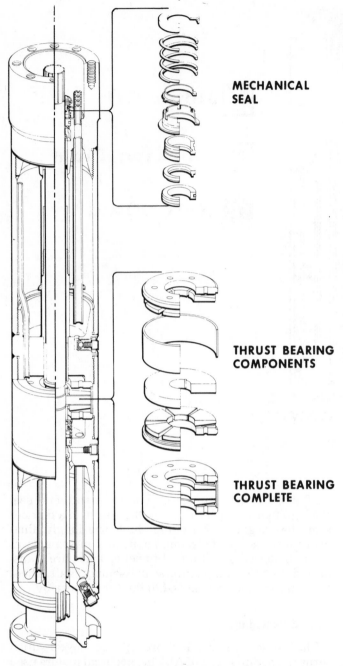

MECHANICAL SEAL

THRUST BEARING COMPONENTS

THRUST BEARING COMPLETE

Fig. 4.34C   Protector (Courtesy TRW-Reda)

inside. The steel laminations are aligned with the rotor sections and the brass laminations are aligned with the radial sleeve bearings. The rotor sections and bronze bearing sleeves are assembled on and keyed to a high strength steel shaft. The shaft is hollow to permit oil circulation. Various motor voltages are available to provide for optimum selection of motor, switchboard and cable combinations with minimum cost for voltage transformation.

The setting depth is a determining factor in motor voltage selection due to the voltage loss of a particular amperage and cable. When the voltage loss becomes too great, a higher voltage (lower amperage) motor is required. In deeper wells, economics becomes a factor; with a higher voltage motor, it is possible to use a smaller, less expensive cable. However, a higher volt-

age (more expensive) switchboard may have to be used.

The motor horsepower is calculated by multiplying the maximum horsepower per stage from the pump curve by the number of pump stages and correcting for the specific gravity of the fluid. Refer to design Section 4.5.

In general Table 4.32 shows the available motors:

TABLE 4.32
MOTORS AVAILABLE

Casing OD, in.	Horsepower (60 Hz)	Horsepower (50 Hz)
4½	up to 127.5	up to 106
5½	up to 240	up to 200
7	up to 600	up to 500
8⅝	up to 1020	up to 850

Amperage requirements may vary from 12 to 130 amps. The required horsepower is achieved by simply increasing the length of the motor section. The motor is made up of rotors usually 12-18 in. long that are mounted on a shaft and located in the electrical coils (stators) mounted within the steel motor housing. The larger single motor assemblies approach 30 ft in length and are rated up to 200 to 250 hp, while tandem motors approach 100 ft in length and have a rating up to 1,000 hp.

Fig. 4.35B shows motor characteristics curves for a typical submersible motor. These motor characteristics curves are based on output dynamometer measurements. Note that derating the motor 10% reduces the temperature rise only 4°F. from the normal temperature rise of 48°F. A derating of 20% reduces the temperature rise 8°F. In either case a 4 or 8° rise is insignificant and can be ignored. (The temperature rise for this curve is based on a cooling medium of water with a velocity of 1 ft/sec.)

Fig. 4.35C represents typical performance curves for an electrical motor showing change of speed, efficiency, power factor, amperage, and kilowatt input for a constant load with varying voltage. Operation at less than nameplate voltage results in lower speed, higher current (amperage) and higher kw. Lower speeds mean lower pump output, since the volume varies directly with the speed and pump head varies as the square of the speed. Operation at higher than nameplate voltage does not significantly affect amperage and kw but does result in a reduction in power factor. If there is a power factor penalty provision in the rate schedule, this is an important consideration. The ideal practice is to try and utilize 100% required surface voltage plus or minus 2%. This will provide the best overall performance. Greater than 2% variation is acceptable if it is necessary to operate from an existing power system.

Fig. 4.35D shows the temperature rise for 100% load in water (specific heat equals 1.0) and crude oil (specific heat equals 0.4) for a fully loaded motor. Note the higher temperature rise (92°F.) when crude oil is the cooling medium. If we derate 20% in this case, the heat rise is reduced by only 15°F. Again, 15° in relation to 300-350°F. is not a significant factor. From Fig. 4.35D, it is obvious that fluid velocity is just as important as fluid ambient temperature, if not more so.

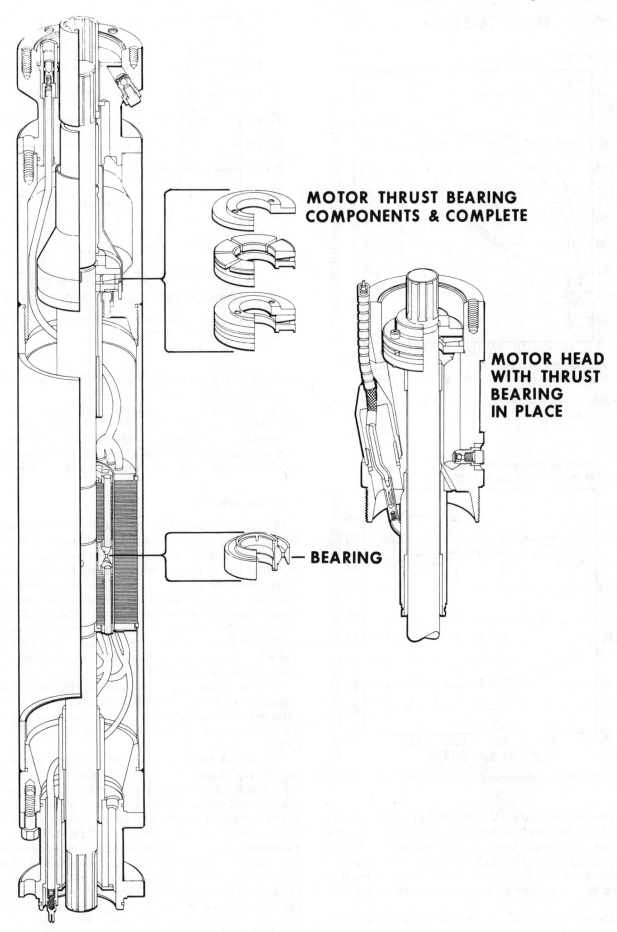

MOTOR THRUST BEARING COMPONENTS & COMPLETE

MOTOR HEAD WITH THRUST BEARING IN PLACE

—BEARING

Fig. 4.35A   Motor (Courtesy TRW-Reda)

Motor Test Data

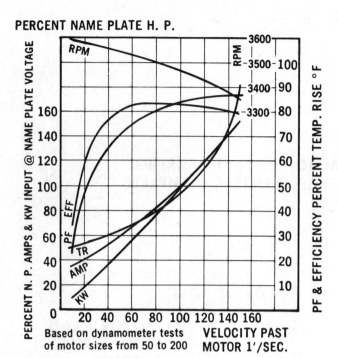

Fig. 4.35B   Typical motor test data (Courtesy Bryon Jackson - Centrilift)

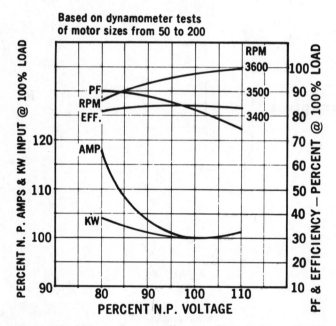

Fig. 4.35C   Motor performance (Courtesy Bryon Jackson - Centrilift)

The motor may have a thrust bearing which will carry all thrust loadings on the motor shaft. Similar to the seal assembly thrust bearing, this bearing is a Kingsberry sliding type. Although the motor rotors operate in either direction, the motor thrust bearing should be rotated in the proper direction to prevent early failure.

If the voltage of the installed motor is higher than nameplate voltage, the motor will run faster, will use more current (over excitation), have a lower power factor, and will develop more horsepower to supply the

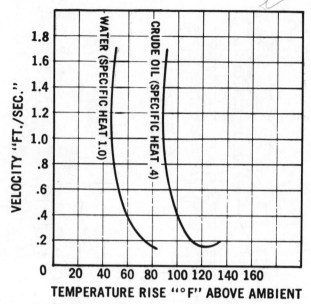

Fig. 4.35D   Temperature rise in pump motors (Courtesy Bryon-Jackson-Centrilift)

demand of the centrifugal pump. The opposite is the case if the voltage is low.

The materials for insulating motors are rated at approximately 180° C. (356°F). Some motors are impregnated by an epoxy method. This slot impregnation system is stated to improve heat transfer from the motor while bracing end turns against vibration.

4.35 Gas separators

The gas separator is normally a bolt-on section between the protector and the pump where it serves as the pump intake. The gas separator separates the free gas from the fluid and assists in routing the free gas away from the pump intake. Gas separators can be effective, but it is difficult to determine their exact efficiency. Numerous designs are available, and experimental work to determine their effectiveness is being conducted by research organizations. Sections 4.523 and 4.524 are on design. Note when and how much gas is desired to be vented.

Venting of the gas is not necessarily the optimum manner of pumping the well. Although the total intake volume is reduced, the discharge pressure is increased due to less gas in the column of fluids above the pump.

Figs. 4.36A and 4.36B illustrate typical gas separators. In Fig. 4.36A the well fluid enters the intake section. The fluid must reverse direction at this point so, due to a reduction of pressure, there is gas separation at the intake. The separated gas moves up the annulus and is vented out at the wellhead. Fluid still containing gas entering the gas separator intake moves downward and again reverses direction when fluid is picked up by the pickup impeller. This impeller causes a turbulence in the fluid that causes a vortex. This vortex causes the gas to move up along the shaft and the fluid to move on the outside in the inner annulus. This operation provides the first stage of the pump a higher density fluid as well as allowing the previous gas breakout to occur in this first stage. The gas separator is an aid in pre-

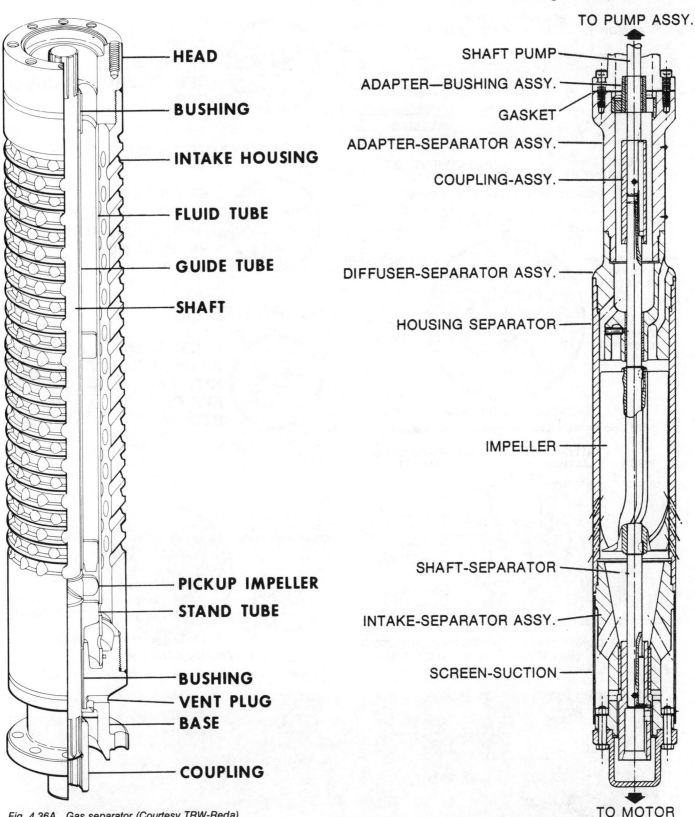

Fig. 4.36A *Gas separator (Courtesy TRW-Reda)*

Fig. 4.36B *Centrifugal gas separator (Courtesy OiLine-Kobe)*

venting gas lock and normally provides more efficient pumping of gassy wells.

Fig. 4.36B shows another gas separator. Its operation is based on the principle of separation of particles of different densities under the action of centrifugal forces. In the design under reference, the rotating impeller creates the field of centrifugal forces.

As the well fluid (consisting of free gas and liquid) passes through the impeller, it is subjected to the action of centrifugal forces. The liquid particles, being of higher density, are thrown towards the periphery of the impeller while gas particles form a core near the center. The gas is vented to the annular space while the gas-free liquid enters the pump impeller eye.

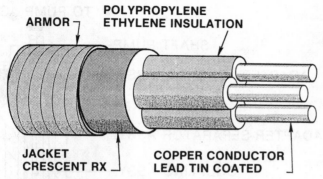

ARMOR

POLYPROPYLENE ETHYLENE INSULATION

JACKET CRESCENT RX

COPPER CONDUCTOR LEAD TIN COATED

Fig. 4.37A   Typical round cable (Courtesy TRW-Reda)

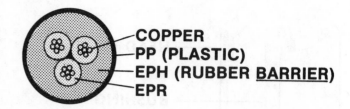

COPPER
PP (PLASTIC)
EPH (RUBBER <u>BARRIER</u>)
EPR

INSENSITIVE INSULATION

PP (PLASTIC)
NITRILE RUBBER
<u>HIGH TEMPERATURE</u>
PP (PLASTIC)
COPPER

G. P. NITRILE

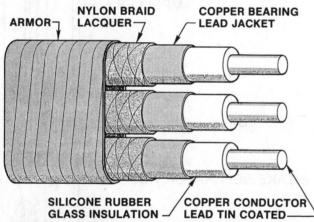

POLYPROPYLENE COPOLYMER INSULATION

TINNED COPPER STRAND (BLOCKED)

INTERLOCKING ARMOR

NITRILE JACKET

SOLID CONSTRUCTION

Fig. 4.37B   Round cable (Courtesy Bryon Jackson - Centrilift)

CURRENT

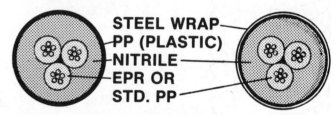

STEEL WRAP
PP (PLASTIC)
NITRILE
EPR OR
STD. PP

Fig. 4.37E   Typical cable (Courtesy Bryon Jackson - Centrilift)

ARMOR

NYLON BRAID
LACQUER

COPPER BEARING LEAD JACKET

SILICONE RUBBER GLASS INSULATION

COPPER CONDUCTOR LEAD TIN COATED

Fig. 4.37C   Typical flat cable (Courtesy TRW-Reda)

CL-81

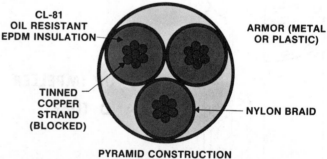

CL-81 OIL RESISTANT EPDM INSULATION

TINNED COPPER STRAND (BLOCKED)

ARMOR (METAL OR PLASTIC)

NYLON BRAID

PYRAMID CONSTRUCTION

Fig. 4.37F   Typical cable (Courtesy Bryon Jackson - Centrilift)

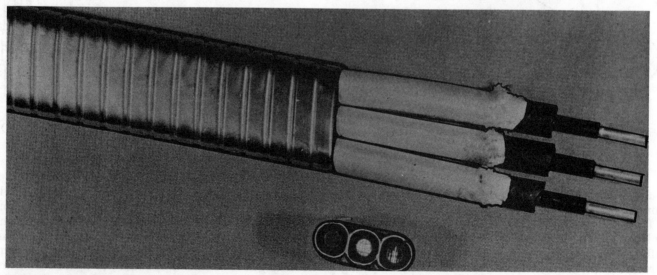

Fig. 4.37D   Flat cable (Courtesy Bryon Jackson - Centrilift)

4.36 Cables

4.361 Introduction

Power is supplied to the electric motor by an electric cable. A range of conductor sizes permits matching cables to motor requirements. Available in either round or flat styles, these insulated cables may be installed in well temperatures in excess of 300°F. The cable has an armor of steel, bronze or monel depending upon well conditions and requirements.

Much work continues in order to improve the various cables. Longer life is always being strived for and operation under conditions of higher temperatures is also desired.

Several cable configurations and designs are noted in Figs. 4.37A, B, C, D, E, and F. Pump companies provide both round and flat cable in conductor sizes from 2/0 through no. 6 in both copper or aluminum conductors. The proper cable size is governed by the amperage, voltage drop and space available between the tubing collar and casing. The best cable type is selected based on the bottomhole temperature and fluids encountered.

In electrical submersible cable the conductor elements used are copper and aluminum. Resistivity measurements on these two elements in terms of the standard ohm and taken at 20° C. are 10.37 for copper and 17.0 for aluminum.

Cable protectors must be provided and typical ones at the pumps are noted in Figs. 4.38A, B, and C.

Cable splicing takes place initially in going from flat cable by the pump to round cable by the tubing.

The resistivity of a conductor is inversely proportional to the number of free electrons in unit volume (Circular Mil-Foot), and this in turn depends upon the nature of the substance.

Considering the length of a conductor, for a given applied voltage, the volts per foot diminish as wire is lengthened; as a consequence the electron velocity diminishes and this in turn results in a reduction of current or in other words "resistance is directly proportional to conductor length."

Enlarging the cross section of a wire, on the other hand, has an inverse effect upon the resistance because the number of free electrons per unit length increases with the area. Under this condition the current will increase for a given applied emf since more electrons will be moved in unit time or, in other words, "resistance is inversely proportional to cross-sectioned area."

A formula for resistance in terms of resistivity ρ, length (L) in feet, and area (CM) in circular mils is:

$$R = \rho \frac{L}{CM}$$

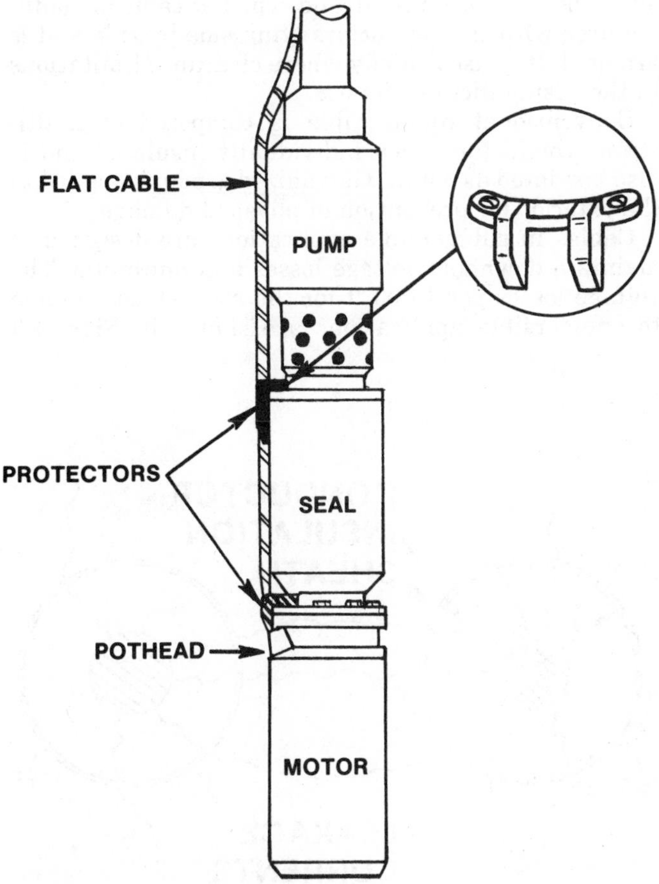

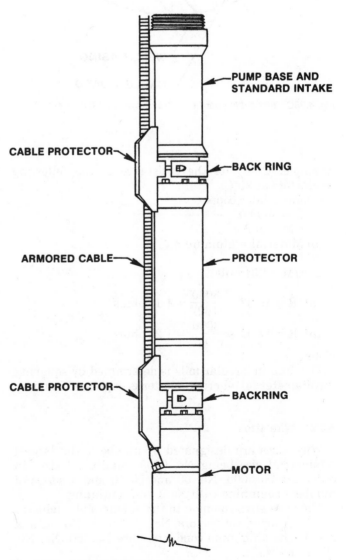

Fig. 4.38A Flat cable protector (Courtesy Bryon Jackson - Centrilift)

Fig. 4.38B Cable protectors

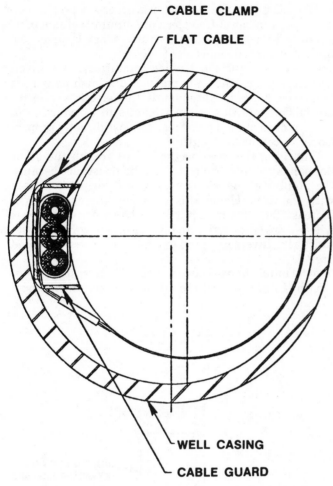

Fig. 4.38C   Flat cable guard (Courtesy Bryon Jackson - Centrilift)

Example: Calculate the resistance of the following conductors at 20°C.
- (a) material = copper
    L = 3000 ft
    CM = 162 mils
- (b) Material = aluminum
    L = 3000 ft
    CM = 204 mils

$$(a)\ \ R = 10.37 \times \frac{3000}{(162)^2} = 1.19\ \text{ohms}$$

$$(b)\ \ R = 17.00 \times \frac{3000}{(204)^2} = 1.22\ \text{ohms}$$

Note:

The area in circular mils is determined by squaring the diameter (d) where the latter is expressed in mils.

4.362   Wire sizes

Wire sizes are designated by numbers, the largest commercial size being No. 0000. Additional sizes in order are No. 000, No. 00 and No. 0 and a series of numbers beginning with No. 1 and continuing.

The wire sizes common to the submersible industry are for copper conductors No. 1, No. 2, No. 4, and No. 6. The aluminum conductors are No. 2/0, No. 1/0, No. 2 and No. 4.

Table 4.33 lists wire sizes from No. 1 to No. 8 with

TABLE 4.33
WIRE DATA (Single Round Wires, Standard Annealed Copper)

No.	Diam mils	Area cir mils	Ohms per 1,000 ft @ 20° C.	Lb. per 1,000 ft	Feet per lb.
1	289	83,690	0.1239	253.3	3.947
2	257	66,360	0.1563	200.9	4.977
4	204	41,740	0.2485	126.4	7.914
6	162	26,240	0.3951	79.46	12.58
8	128	16,510	0.6282	49.98	20.01

data on diameter in mils, area in circular mils, resistance in ohms per 1,000 ft at 20°C., lbs. per 1,000 ft, and number of feet per pound for single conductor standard annealed copper round wires.

4.363   Cable (submersible)

When cables are used in high voltage systems, each of the current-carrying conductors is often surrounded by a considerable thickness of high-grade insulation and sometimes a lead sheath. Although normal current is along the length of the conductor, there exist small crosswise currents—leakage currents—from one conductor to another (Fig. 4.37G).

The leakage resistance is generally small compared with the insulation resistance of cables and may be neglected in leakage-current computations.

Submersible pump cable is constructed either in a flat or a round configuration, and each conductor may either be solid or multi-strand. The flat cable is manufactured with the conductors lying side by side and is armored. It is used chiefly where clearance limitations in the casing dictates its use.

Conventional round cable is composed of multi-strand conductors, each individually insulated and is also insulated as a unit. Galvanized armor is applied to the exterior for prevention of physical damage.

Cables in submersible applications are designed to maintain downhole voltage losses to a minimum. The voltage losses per 1,000 ft for the size cables common to submersible applications are shown in Figs. 4.3

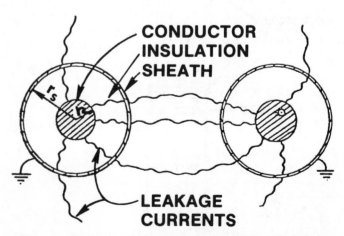

Fig. 4.37G   Cable current leakage (Courtesy Bryon Jackson - Centrilift)

TABLE 4.34
VOLTAGE LOSSES IN CABLES

Cable size	Voltage drop per amp. per 1000 ft @ 100% PF @ 20°
#12 Cu. or #10 AL	3.32
#10 Cu. or #8 AL	2.08
#8 Cu. or #6 AL	1.32
#6 Cu. or #4 AL	0.84
#4 Cu. or #2 AL	0.53
#2 Cu. or #1/0 AL	0.33
#1 Cu. or #2/0 AL	0.26

A (1 to 3) (Appendix 4.3). When sizing cable, the volt losses shown in Table 4.32 may be used. However, voltage loss curves are available from all submersible pump suppliers. (Appendix 4.3 A).

The insulation for these cables must withstand wellbore temperatures and pressures, and resist impregnation of well fluids. However, there are limitations for the cables used today due to the limitations of the materials used in their construction. Standard cables are generally rated for a 10-year life at a maximum temperature of 167° F. with life cut in half for each 15°F. above maximum. The environment the cable is operating under in the well will also directly affect its life.

Tijunelis and Wargin gave the following rules of thumb (Table 4.35) to be used in cable selection:[6]

TABLE 4.35
RULES OF THUMB IN OIL WELL SUBMERSIBLE CABLE SELECTION

ALL
cables and splices deteriorate with use in time!
                                    but . . . .
at rates depending on
              CONDITIONS + MATERIALS
When exposed to water, minerals and inorganics
    PLASTICS—Unaffected
    RUBBERS—Permeable
When exposed to oil
    PLASTICS—Only soften (unless highly plasticized)
    RUBBERS—Most soften
        Some harden at elevated temperatures
When exposed to gases
    PLASTICS—Permeable under pressure
    RUBBERS—Highly permeable
When BRIEFLY OVERHEATED
    PLASTICS—Melt
    RUBBER—Okay
When BRIEFLY OVERHEATED after prolonged
oil exposure
    PLASTICS—Melt quicker
    RUBBERS—Deform and crack.
LONG EXPOSURE to moderate heat
    PLASTICS—Slowly deform or embrittle.
    RUBBERS—Harden and crack.

They also listed the available commercial cables as noted in Table 4.36. The materials from which they are made are listed along with their features and limitations.

TABLE 4.36
OIL WELL CABLES IN USE TODAY

Configuration	Jacket	Insulation	Armor	Max.* Temp. Limit	QUALIFICATIONS
Flat	Lead	EPR Rubber	Metal	325°F	Lead splice to rubber difficult to make. Easily damaged, costly. Insulation has poor oil resistance.
Flat	Nitrile	EPR Rubber	Metal	250°F	Nitrile age hardens after prolonged use. Insulation has poor oil resistance.
Flat	None	Crosslinked Polyethylene	None	190°F	Crosslinked polyethylene softens and creeps at elevated temps and has poor oil resistance
Flat	None	PVC	None	200°F	PVC embrittles with age and has poor oil resistance
Flat	Neoprene	EPR Rubber	Metal	250°F	Neoprene has fair oil resistance and the EPR rubber insulation is poor in oil.
Round	Nitrile	Copolymer Polypropylene	Metal	230°F	Insulation limits use temp. because of its lower heat distortion temp.
Round	Epichlorohydrin rubber	Homopolymer Polypropylene	Metal	275°F	Insulation increases cables use temp. but decreases in low temp. usefulness.
Round	Acrylate rubber	EPR rubber	Metal	300°F	Limited by fair oil resistance and poor water resistance. Acrylate rubber hardens on aging.
Round	Nitrile	Crosslinked Polyethylene	None	190°F	Nitrile age hardens and crosslinked polyethylene has poor oil resistance.
Round	Polyethylene	Copolymer Polypropylene	None	150°F	Jacket has low heat distortion point and only fair oil resistance.
Round	EPR Rubber	EPR Rubber	Metal	325°F	Only fair oil resistance of jacket and insulation
Round	Nitrile	EPR Rubber	Metal	250°F	Nitrile hardens on aging. EPR rubber is not oil resistant

*This does not discount the temperature limit for effect of air, water, oil, long exposure time and resistance heating.

Tijunelis and Wargin discussed briefly those factors having detrimental effects on cables in centrifugal pump application.

They noted that the chemical environment had the most severe effect on cable life. It can change tensile properties, volume and hardness. One of the other factors having severe effects is gas permeation. Normally, this is noted when pulling the well due to depressurization of the cable. Severe swelling may occur and care should be taken in re-running any used cable.

Much research continues in the area of cable improvement and with the advent of new materials and construction procedures cable life should be much improved in the future.

4.364 Current-carrying capacity of cables

Current-carrying capacity is usually the major factor in determining the size conductor. The following are the recommended maximum amperes for each size cable.

No. 1 CU and 2/0 AL have a capacity of 115 amp maximum

No. 2 CU and 1/0 AL have a capacity of 95 amp maximum

No. 4 CU and 2 AL have a capacity of 70 amp maximum

No. 6 CU and 4 Al have a capacity of 55 amp maximum

Number 1 conductor is class A stranded which consists of 7 wires, each 0.1093 in. in diameter.

Number 2 conductor is class B stranded which consists of 7 wires each 0.0974 in. in diameter.

Number 4 conductor is class B stranded which consists of 7 wires each 0.0772 in. in diameter.

Number 6 conductor is solid and is 0.162 in. in diameter.

Table 4.37 lists data such as weights, diameters, and tensile strengths for copper conductor submersible cables.

TABLE 4.37
COPPER CONDUCTOR—3.0 KV CABLE

Type	Size	Weight (lbs/ft)	Diameter (in.)	Tensile strength
Round armored	No. 1	1.75	1.343	3800
	No. 2	1.65	1.265	3040
	No. 4	1.17	1.140	1930
	No. 6	0.88	0.970	1000
Round unarmored	No. 1	1.270	1.300	
	No. 2	1.039	1.210	
	No. 4	0.729	1.075	
	No. 6	0.523	0.980	
Flat armored	No. 1	1.42	0.69 × 1.83	
	No. 2	1.40	0.65 × 1.70	
	No. 4	1.39	0.59 × 1.51	
Lead sheathed	No. 1	2.43	0.79 × 1.90	
	No. 2	2.09	0.67 × 1.71	
	No. 4	1.75	0.57 × 1.53	

Table 4.38 lists data on types, weights, diameters and tensile strengths for aluminum conductor submersible cable.

TABLE 4.38
ALUMINUM CONDUCTOR—3.0 KV CABLE

Type	Size	Weight (lbs/ft)	Diameter (in.)	Tensile strength
Round armored	No. 2/0	1.39	1.425	2,085
	No. 1/0	1.16	1.335	1,660
	No. 2	0.87	1.195	1,042
	No. 4	0.74	1.080	657
Round unarmored	No. 2/0	0.76	1.310	---
	No. 1/0	0.65	1.200	---
	No. 2	0.51	1.12	---
	No. 4	0.39	1.00	---

3.365 Temperature effect on cables

For wells containing hydrocarbons and gas, polyethylene jacketed cable should be limited to 130° F. Polypropylene insulated with armor should be limited to 180°F. EPR lead sheath should be limited to 250°F.

Considerable thought should be given to the type and size cable selected for a submersible pump application, as it is a very important part of the installation and in many cases controls the life of the installation.

4.37 Switchboards

The standard switchboards are weatherproof and are available in a range of sizes and accompanying accessories to accommodate any pump installation. They range from simple units with pushbutton magnetic contactors with overload protection to more complex assemblies with fused disconnects, recording ammeter, under-voltage and overload protection, signal lights, timers for intermittent pumping, and instruments for automatic remote control operation. Refer to Figs. 4.39A and 4.39B for typical switchboards.

Switchboards are available for voltages from 440 through 4,800 v. Selection is based upon ratings of voltage, amperage, horsepower future requirements, and economics. Refer to the Section 4.5 for details on selection and to Appendix 4D for detailed information on swtichboards.

Underload or pump-off protection is necessary since low flow rates past the motor may not give adequate cooling. Automatic restart after shut-down and recording ammeters usually are furnished on all motor starter controllers.

Solid-state motor switchboards are available which replace conventional magnetic overload, underload, and phase failure mechanisms. The solid-state controllers offer instantaneous underload protection on all three-phase, time-delayed overload protections and automatic protection against single-phase conditions.

4.38 Transformers

Banks of three single-phase transformers, three-phase standard transformers, and three-phase auto transformers are manufactured for submersible use. These oil filled, self-cooling units are designed to convert primary line voltage to motor voltage requirements. They are equipped with taps to provide

Fig. 4.39A   Typical switchboard (Courtesy Bryon Jackson - Centrilift)

Fig. 4.39B   Typical switchboard (Courtesy Bryon Jackson - Centrilift)

maximum flexibility. Dry type transformers may be supplied for those locations which exclude the use of oil-filled transformers, such as offshore platforms. Where the voltage of the primary system is not compatible with the required surface voltage, a transformer is required. Auto transformers are available to increase a 440/480 line voltage to an 800 to 1,000 v range.

Three-phase transformers are available in any common range of primary and secondary voltages. However, where higher voltage primaries are available, it is more economical to use a bank of three single-phase transformers to step the primary voltage down to any voltage motor available (Figs. 4.310A and B). Additional information on transformers is found in Section 4.5 and Appendix 4E.

4.39   Junction box

A junction box is located between the wellhead and switchboard for safety reasons. Gas can travel up the cable and through the surface cable to the switchboard, causing a fire hazard or potential explosion. Hence, the junction box eliminates this gas travel. This is a vented weatherproof junction box.

Fig. 4.310A   Typical bank of three transformers (Courtesy Bryon Jackson - Centrilift)

TRANSFORMER CONNECTIONS

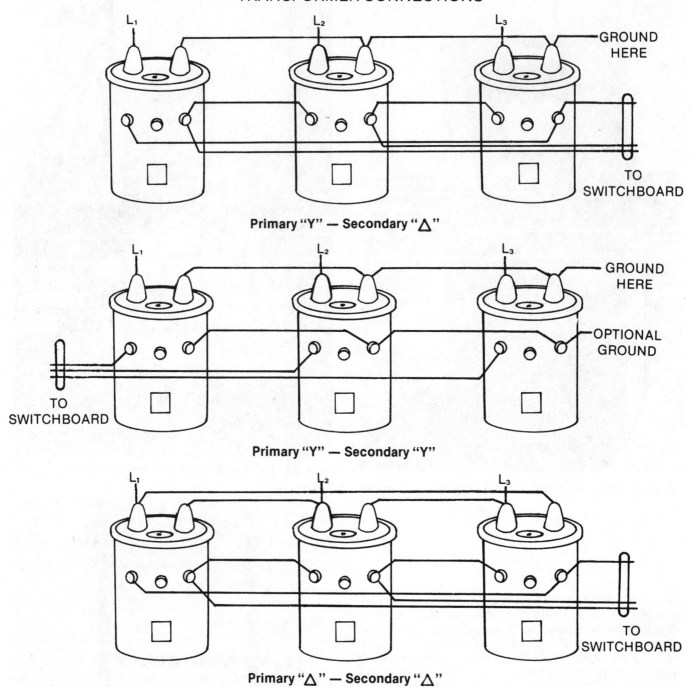

Fig. 4.310B   Transformer connections (Courtesy TRW-Reda)

A typical connection is noted in Fig. 4.311.

The junction box should be located at least 15 feet (minimum) from the wellhead and is normally mounted 2 to 3 ft above ground. Cables from the junction box to the wellhead and switchboard may be buried.

The junction box is normally constructed of 12 ga. plate and meets NEMA 3-R weatherproof specifications. It requires no cable splices or taping. Provision is made for grounding the power conductor and "I" wire. Armored cable clamps interlock and are adjustable. A lock tab provides security against unauthorized entry.

REFERENCES

1. Riling, Gene, "Submersible Pump Handbook," Byron Jackson Pump Division, Tulsa, Oklahoma.
2. "Reda Submergible Pump Catalog," TRW-Reda, Bartlesville, Oklahoma.
3. "Centrilift Submersible Pumps," Byron Jackson Pump Division, Tulsa, Oklahoma (catalog).
4. "Oil Dynamics Submersible Oil Well Pumps," Tulsa, Oklahoma (catalog).
5. "Kobe Electric Submersible Pumps Catalog," Kobe, Huntington Park, California.
6. Tijunelis, D. and Wargin, Robert, "Cable Selection Process for Submersible Oil Well Pumps," Southwestern Petroleum Short Course Association, Lubbock, Texas, 1977.

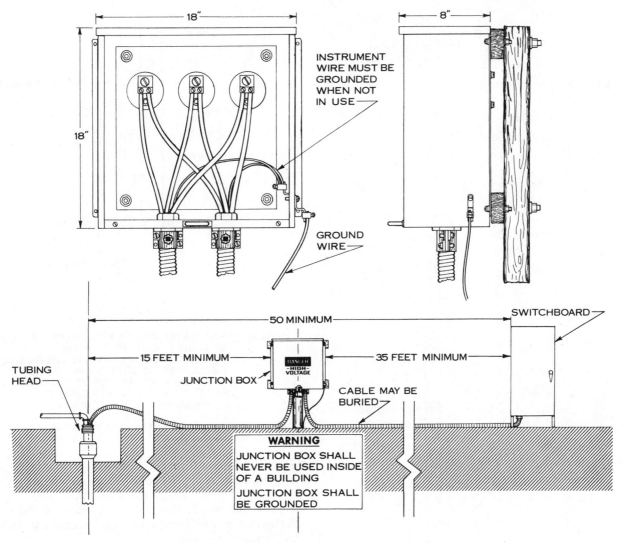

Fig. 4.311 Typical junction box connection (Courtesy TRW-Reda)

4.4 TYPES OF INSTALLATIONS

4.41 Introduction

Fig. 4.41 shows a complete pumping system including downhole and surface components. Much of the following descriptive material has been taken from references 1, 2, and 3.

A submergible electric pumping unit is composed of seven basic components: electric motor, multi-stage centrifugal pump, protector, power cable, motor flat cable, switchboard and/or an auto transformer, single three-phase transformer or a bank of three single-phase transformers.

All of the above items are available in numerous sizes and types to fit well specifications, such as casing size, desired producing volume, total lift, electrical power supply and environment.

In addition to these basic components, various auxiliary items are used. Some are required, while others are optional. The most common required items to complete an installation are cable clamps, cable reel, reel supports, shock absorber, tubing support and, in many

cases, a swage nipple. Other optional items are not required for an installation but recommended where applicable are flat cable guards, check valve, bleeder valve, centralizers, motor jackets, junction box, and downhole pressure sentries.

In some instances, usually in remote areas, engine generator sets are used instead of purchased utility power. Such generator sets may power multi-well installations or individual wells. In the latter case, transformers can usually be eliminated by supplying an alternator which is wound to supply the proper input voltage (required surface voltage).

This section discusses the various types of installations for both downhole and surface components. There are some special installations that can be made if desirable. Normally, the electrical pump installation is made without installing a packer on the tubing string. However, a packer can be installed if necessary. The pump can be run either below or above the packer but requires a special electrical cable by-pass through the packer if run below and should not be set in compression when run above.

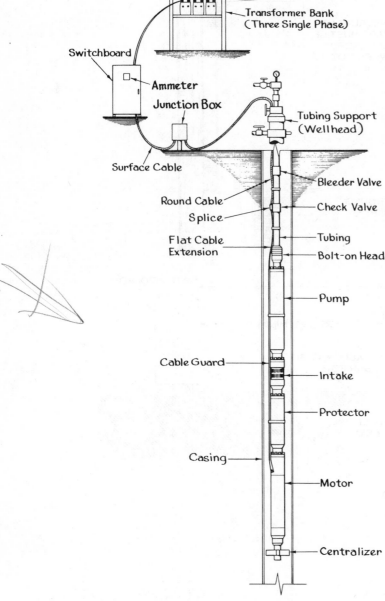

Fig. 4.41   Submersible centrifugal pumping unit (Courtesy TRW-Reda)

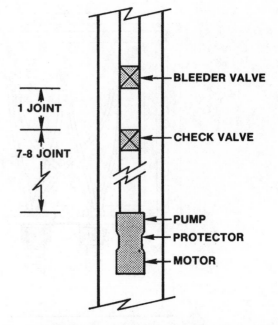

NOTES: (1) FOR WATER WELLS THE CHECK VALVE MAY BE ONLY 1 JOINT FROM PUMP
(2) FOR GASSY OIL WELLS THE CHECK VALVE SHOULD BE 7-8 JOINTS ABOVE PUMP
(3) THE CHECK VALVE CAN BE WIRELINE RETRIEVABLE IN WHICH CASE A BLEEDER VALVE IS NOT USED

Fig. 4.42   Detailed downhole equipment locations

4.42   Subsurface installations

4.421   Introduction

There are numerous downhole installations that can be made, and Fig. 4.42 shows a detailed standard installation including possible accessories such as check valves, and drain valves.

4.22   Standard installation

Fig. 4.43 represents how the majority of submergible pumping systems are presently applied.

Note that the unit is landed at a point above the perforations. Fluid entering the wellbore moves upward past the motor. This movement of well fluid cools the motor by dissipating heat generated by the motor. If the motor were set in the perforated area, the cooling effect would be decreased. Setting in the perforations could also result in housing damage due to jetting action of fluid entering the annulus through the perforations. Setting below the perforations would eliminate the cooling effect with a standard unit.

From the bottom, the first component of the system is the motor with the flat cable extension connected to the motor's pothead.

The next component is the protector section which functions as the equalizing chamber for the motor. As the motor operates, it expands and contracts oil in the protector section. The protector accommodates the motor by allowing the expanded motor oil to enter the protector and, conversely, by serving as a reservoir when the motor cools and requires additional oil. The protector seals the power end of the motor housing from the wellbore fluids while allowing pressure communications between the oil-filled unit and wellbore fluids.

The third component is the intake section where fluid enters the pump. Illustrated in Fig. 4.43 is the gas separator type intake, designed to help separate any free gas present in the fluid. It allows the separated gas to proceed up the casing annulus instead of entering the pump.

The next component is the centrifugal pump. Located at the top of the pump is the discharge head. The discharge head connects the tubing string, sometimes referred to as the discharge pipe, to the pump. Fluid enters the pump intake and is sent upwards through the pump stages by centrifugal force which builds pressure and simultaneously compresses then discharges the well fluid into the tubing string at a sufficient pressure to send the fluid to surface.

Above the pump is a check valve which allows the tubing to stand full of well fluid when the pump is shut down. The check valve prevents fluid in the tubing

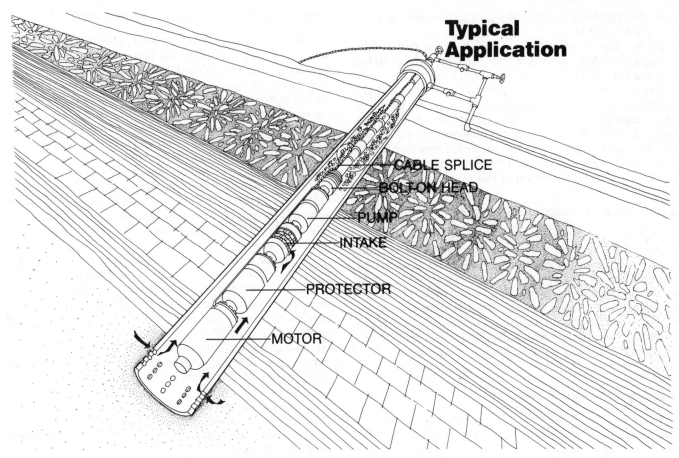

Typical Application

CABLE SPLICE
BOLT-ON HEAD
PUMP
INTAKE
PROTECTOR
MOTOR

Fig. 4.43   (Courtesy TRW-Reda)

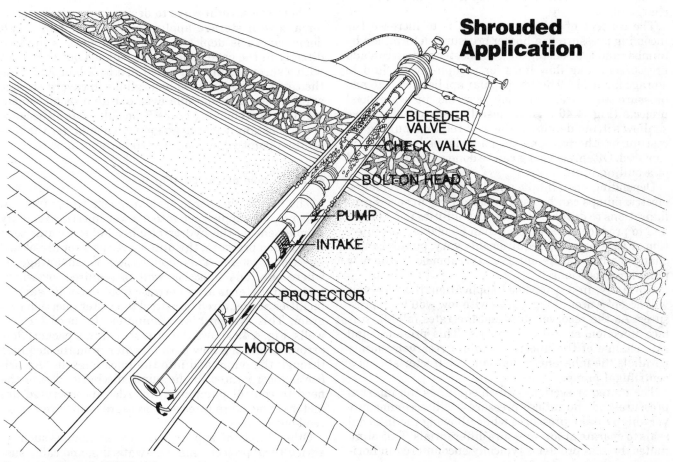

Shrouded Application

BLEEDER VALVE
CHECK VALVE
BOLT-ON HEAD
PUMP
INTAKE
PROTECTOR
MOTOR

Fig. 4.44   (Courtesy TRW-Reda)

from flowing back through the pump thus spinning the pump backwards.

Above the check valve is the bleeder valve. The bleeder provides a means of emptying the tubing string when a unit is pulled from the well.

The power cable runs parallel to the tubing. At the surface is a typical wellhead which seals the well and also suspends the tubing and other downhole equipment in the well. The electrical power cable runs from the wellhead to a junction box, from the junction box to a controller (switchboard). Power to the switchboard comes from transformation of the primary power.

4.423 Shrouded pumping system: (Fig. 4.44)

This system is essentially the same as the "typical standard installation." The shroud routes the well fluids by the motor for cooling purposes. With a shroud, the unit can be set in or below the perforations or, as an example, the shrouded unit can be set with the shroud in open ocean water and the fluid is directed by the motor at sufficient velocities to provide motor cooling. Note the flow arrows which depict the flow path of the incoming fluid and its travel to the pump intake.

4.424 Booster service: (Fig. 4.45)

In a booster application, the pumping system is installed inside a shallow set vertical section of casing, perhaps 100 ft long. An incoming line feeds fluid into the so-called can and the pump. The booster application also requires a shrouded motor for cooling. The pump is normally short coupled to the top of the can by the discharge pipe.

The concept of a booster system is to increase the incoming pressure to a higher pressure (determined by number and type of stages). The booster system can add pressure to long flow lines for pumping well fluid to storage facilities. Another popular application is to add pressure to water injection systems in waterflood projects (Fig. 4.46). These installations usually have shallow setting depths. If the well requires pulling for resizing or other reasons, the pump is easily pulled and replaced. Often, servicing can be done without the need of a pulling unit.

Depending upon intake and discharge requirements, electric pumps can be used in parallel or in series configurations to develop pressure capabilities to 5,000 psi and to produce rates greater than 60,000 b/d. As injection or booster pump requirements vary, pumps can be changed out to a different size to accommodate the new pumping conditions.

Normally, in a waterflood application, the reservoir will accept water at high rates with low well head pressures. As the reservoir fills and the rates fall off, surface injection pressure increases. This pressure and rate change of the injectivity into a reservoir closely parallels the inherent performance characteristics of a centrifugal pump.

The protector section—which equalizes the internal pressure and the intake pressure—will accommodate very high intake pressures while allowing the internal seals to operate at low pressure differentials. This eliminates the seal problems typically encountered in horizontal or shaft-driven high pressure pumps.

4.425 Production/injection system: (Fig. 4.47)

This system is referred to as a "closed injection system" for waterflood projects. It is possible to set a conventional submergible unit in a water supply well and inject the produced water directly into an injection well. From a typical water supply well, it is possible to develop pressures in excess of 5,000 psi at the surface for injection purposes. It is also possible to inject into more than one injection well at the same time.

Some significant benefits are available when injection requirements can be met with this system. The system requires no storage facilities for the produced water, no surface pumps, and no wellhouse and plant controls for injection regulation. All that is needed is monitoring the rate and pressure and changing the ammeter charts.

The natural rate/pressure characteristics of centrifugal pumps conform to injectivity performance. Capacity and pressure can be increased or decreased by changing out a pump section or even the motor, as reservoir conditions vary. Existing cable, transformers, and switchboard can often be used. The incremental cost of additional horsepower and pump capacity to meet high surface pressure requirements is nominal since many components of the high pressure system are identical.

The system requires no storage facilities, no foundation pads, housed structures, or source well/plant controls for volume regulation. Since the system is closed, corrosion control is simplified.

4.426 Cavern storage unit: (Fig. 4.48)

There are several ways to design a pumping system for a cavern storage application. In Fig. 4.48 a salt dome cavern is depicted where a product is dumped directly in the cavern and then pumped out as needed with a shrouded unit installed in a bore-hole adjoining the cavern. The use of submergibles in cavern storage applications is preferable over old fluid displacement methods for removing crude from salt caverns because of the constant leaching action which enlarges the cavern to a point where the cavern becomes unsafe.

4.427 Bottom discharge: (Fig. 4.49)

The bottom discharge system injects water from an upper water supply zone into a lower injection zone. By supplementing the hydrostatic head of the water zone with added pressure, the unit serves as a totally-closed injection system. This application is advantageous where only one injection well is required, or where injection patterns are so erratic that elaborate surface facilities are required to control the water injection program. This system also eliminates the need for two wells. Fig. 4.49 shows the unit seated in a packer; flow arrows depict the flow path of water from the acquifer through the pump and into the injection zone. An orifice is incorporated in the pump discharge due to the need to control pumping rate. Also incorporated are meters to measure rates and pressures which record on read-out devices.

Other kinds of applications are available, such as agricultural projects, mine de-watering, and cable suspended equipment. The applications shown provide a

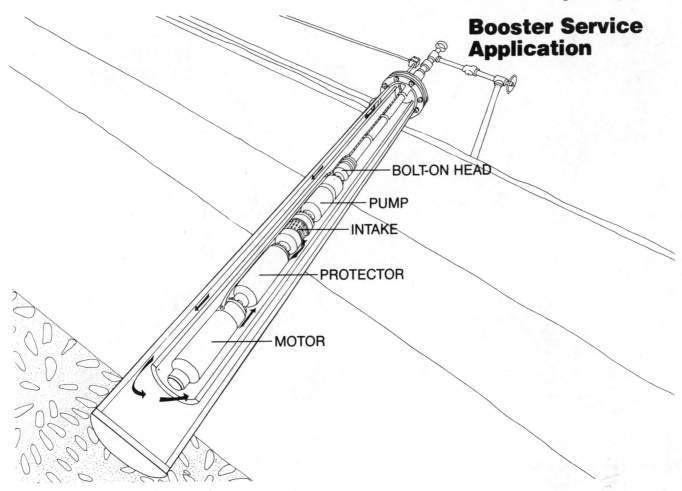

Booster Service Application

BOLT-ON HEAD

PUMP

INTAKE

PROTECTOR

MOTOR

Fig. 4.45   (Courtesy TRW-Reda)

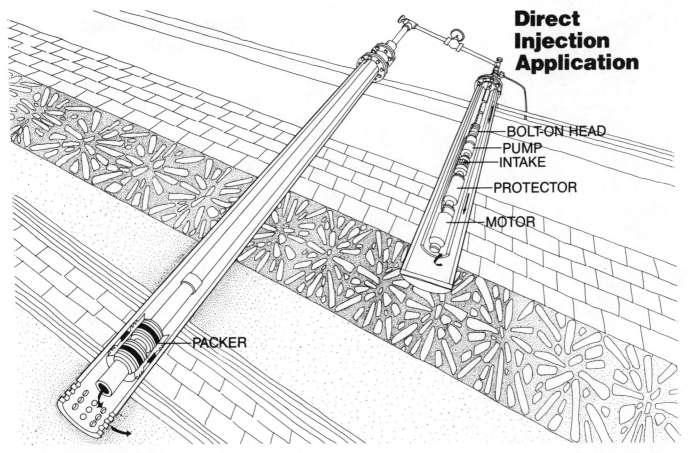

Direct Injection Application

BOLT-ON HEAD
PUMP
INTAKE

PROTECTOR

MOTOR

PACKER

Fig. 4.46   (Courtesy TRW-Reda)

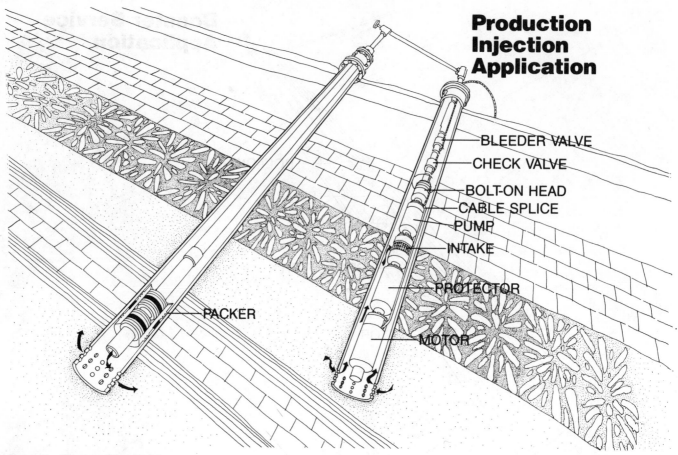

Production Injection Application

BLEEDER VALVE
CHECK VALVE
BOLT-ON HEAD
CABLE SPLICE
PUMP
INTAKE
PROTECTOR
MOTOR
PACKER

Fig. 4.47   *(Courtesy TRW-Reda)*

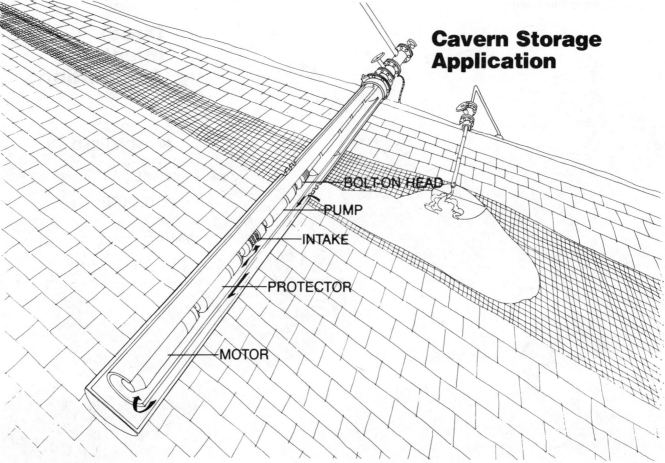

Cavern Storage Application

BOLT-ON HEAD
PUMP
INTAKE
PROTECTOR
MOTOR

Fig. 4.48   *(Courtesy TRW-Reda)*

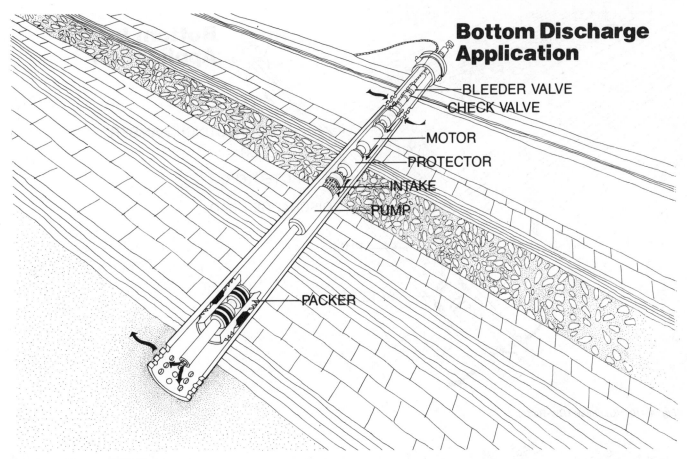

Bottom Discharge Application

— BLEEDER VALVE
— CHECK VALVE
— MOTOR
— PROTECTOR
— INTAKE
— PUMP
— PACKER

Fig. 4.49   (Courtesy TRW-Reda)

look at some significant ways in which submergible pumps are used.

The bottom discharge technique requires installation of the pump below the motor. The ends of the pump are reversed from normal operation, with the intake at the top allowing discharge of the media through a stinger in a permanent packer into the injection zone. Because the tubing string is utilized only as a means of suspending the unit, it can be used to inject chemicals into the water supply.

4.428   Bottom intake units (Fig. 4.410)

A bottom intake configuration is used in an application where casing clearance prohibits the desired production volume either due to tubing friction loss or the restrictions of a particular pump diameter. In bottom intake service, the pump and motor sections are reversed, with the pump intake located at the bottom of the unit. With a permanent packer installed downhole and a stinger in the pump intake, the well fluid is produced up the annulus. Because the complete system can be suspended by a small diameter, high tensile strength tubing string friction loss is minimized and the unit's efficiency and output are increased.

There are other applications which represent good bottom-intake applications, such as cavern storage projects. With a bottom-intake unit, a cavern could be pumped with the unit installed directly in the cavern. Fig. 4.410 shows a bottom-intake unit seated in a packer; fluid is coming into a stinger pipe going through the pump, then is discharged out the discharge head. After

fluid is discharged into the casing annulus, it passes by the protector and motor and goes to the surface. The tubing in this application is merely used to suspend the unit and the cable.

4.429   Cable suspended unit

The retrievable electrical pumping unit has been developed in recent years. It will undoubtedly receive more and more attention in the future. The idea of making the electrical pumping unit retrievable enhances its popularity especially in offshore wells that present very expensive tubing pulling jobs. Figure 4.411 shows this complete system with Figs. 4.412 and 4.413 showing two applications. The cable-suspended pump needs no tubing, but tubing can be used if desired.

In the operating position, a seating element, or "shoe," supports the pump and locking dogs engage it. The pump is then locked in position and, when not running, it cannot be pushed off its seat by bottom-hole-pressure buildup. Raising the motor by lifting on the power cable releases the dogs and the pump can be retrieved.

The seating element seals off the casing between the pump intake and discharge, permitting production up the casing without tubing. However, the cable-suspended pump can be run inside preinstalled tubing if desired. In this case, the seating element is on the bottom end of the tubing. Alternatively, the seating element may be installed within the tubing, installing a second "shoe" on the bottom of the tubing to receive a

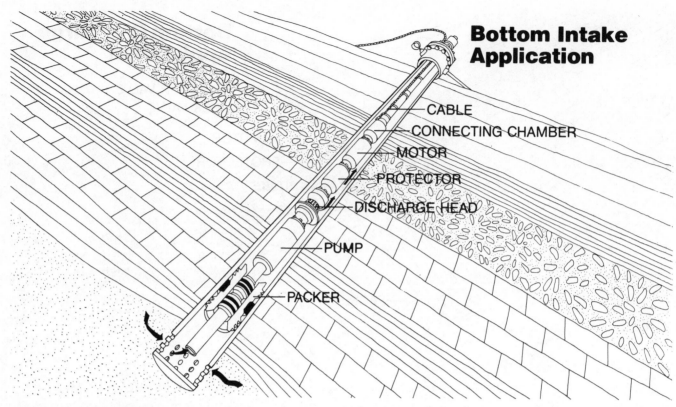

Bottom Intake Application

CABLE

CONNECTING CHAMBER

MOTOR

PROTECTOR

DISCHARGE HEAD

PUMP

PACKER

Fig. 4.410   (Courtesy TRW-Reda)

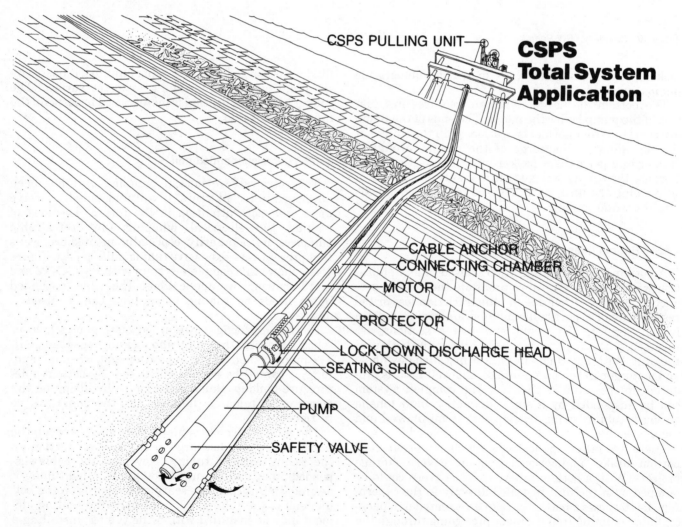

CSPS PULLING UNIT

CSPS Total System Application

CABLE ANCHOR

CONNECTING CHAMBER

MOTOR

PROTECTOR

LOCK-DOWN DISCHARGE HEAD

SEATING SHOE

PUMP

SAFETY VALVE

Fig. 4.411   Courtesy TRW-Reda)

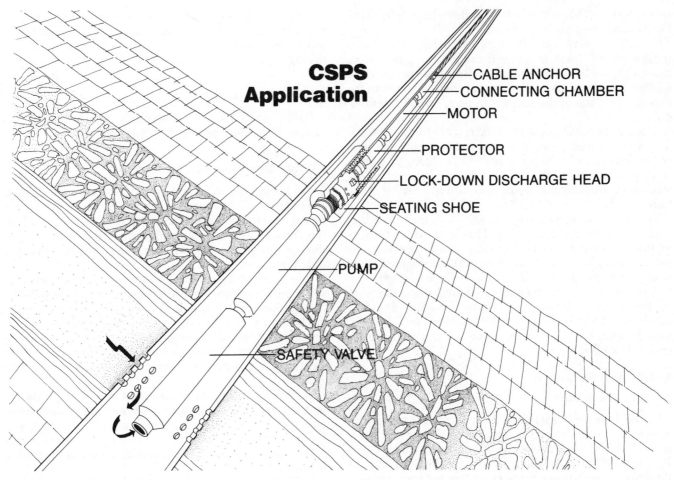

Fig. 4.412   (Courtesy TRW-Reda)

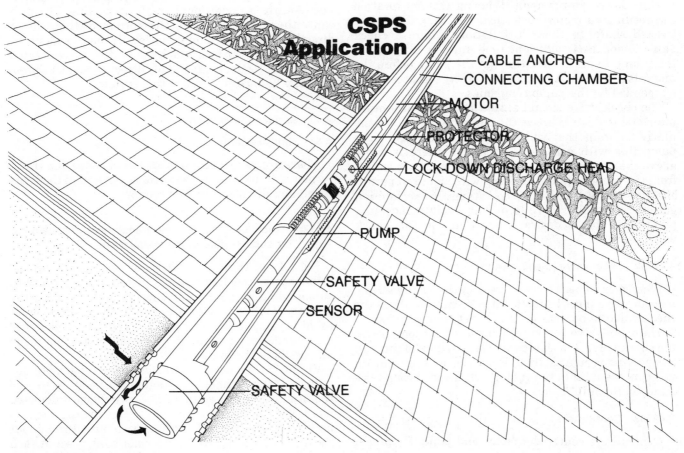

Fig. 4.413   (Courtesy TRW-Reda)

safety valve. This would afford a downhole lubricator for running and pulling the pump.

A safety valve can be installed below the pump intake so all fluid passing through the pump first passes through the safety valve. Because the valve is opened by pump discharge pressure, the valve will be closed when the pump is not running, providing shut-off.

The power cable must have extremely high strength for raising and lowering the pump, and present improvements are being made. This strength comes from armor which consists of a double layer of plow-steel wire for strength, counterwound to prevent rotation.

To avoid fishing jobs, a safety section is built into the cable at the unit's rope socket. This is to insure failure at that point if it is ever necessary to overstress the cable. No cable is left in the hole and the grooved neck of the unit is clear for the fishing tools to engage.

A pump change can be made without killing the well by the use of a lubricator system. By attaching the lubricator directly to the wellhead, the well is kept under control while the pumping unit is withdrawn and re-installed.

4.4210   Accessory items

4.42101   Check valve

A check valve, which is usually located 2 to 3 joints above the pump assembly, can be used to maintain a full column of fluid above the pump. If the check valve fails to hold or if the check is not installed, leakage of fluid from the tubing through the pump can cause a reverse rotation of the subsurface unit when the pump is shut down. Power applied during reverse rotation can result in a motor burn, thrust bearing failure, or twisted shaft. In those installations where a check valve is not used, sufficient time must be allowed for the tubing to drain back through the pump before the motor is restarted. A minimum of thirty minutes is recommended by the pump companies.

The check valve should always be placed at a minimum of 2 to 3 joints above the pump discharge, particularly for wells that will be pumping free gas. If the pump fills with gas and "gas locks," there is space above the pump for gas to rise, allowing gas to escape from the pump, filling the pump with liquid to permit effective start-up and pump operations. In special high gas-oil ratio wells, more than 2 or 3 joints above the pump may be a desirable location for the check valve.

4.42102   Drain valve (bleeder valve)

Whenever a check valve is used in the tubing string, a drain valve must be installed immediately above the check valve to prevent pulling a wet tubing string.

The drain valve is generally installed one joint above the check valve. If a check valve is not run and the well has no sand problems, there is no reason to run a drain valve as the fluid in the tubing will drain through the pump while pulling.

4.42103   Centralizers

Centralizers center the motor and pump for proper cooling and also, in some cases, to prevent cable dam-age due to abrasive rubbing. If centralizers are used in an installation, care should be taken to insure that the centralizers will not rotate or move up and down on the tubing.

4.42104   Cable bands

Cable bands are used to strap the power cable to the tubing. One band per 15 foot interval is normal.

4.42105   Downhole pressure monitors

Valuable reservoir and pump performance data is available with the use of downhole pressure monitors. By correlating reservoir pressure with the withdrawal rate, an operator can determine when it is necessary to change pump size, change injection rate, or consider well workover.

Two types of downhole pressure monitoring devices are available from the submersible pump suppliers. One system has the capability of continuously monitoring bottom hole pressure at the pump's setting depth, protecting the motor against overheating, detecting electrical failures, such as shorts to ground and giving a reading of the motor's operating temperature at any time.

The system requires no special wires. All signals are sent to the surface instruments over the regular power cable. (For operation on an ungrounded power system.) An optional, portable plug-in strip chart recorder can be provided to give a permanent record of pressure readings.

The other downhole pressure monitoring system works essentially the same with respect to pressure, but requires an additional small power-line. The unit can either be attached to the bottom of the unit or directly above the pump outlet. It has readout in impulses at the surface which can be directly converted to pressure readout. Means for direct readout in pressure can be incorporated in this system.

4.43   Surface installations

4.431   Introduction

There are various options for switchboard and transformer installations at the surface. Several of these are given and discussed in this section.

4.432   Typical switchboard and transformer installations for submergible pump installations

Figs. 4.414 through 4.420 show some of the typical switchboard and transformer installations used with submergible pump applications. In each instance the primary voltage is shown to be 12,500 v; however, it could be some other primary voltage. These sketches are intended to be a general guide only and do not cover all possible combinations.

Fig. 4.414 shows the transformer arrangement for reducing the primary voltage of 12,500 v to the required well-head voltage. The switchboard depicted has 1,500 v maximum capability. This arrangement is used where the 12,500 v primary is built throughout the field and a separate transformer bank is set at each well. This combination is most often used in sizes from

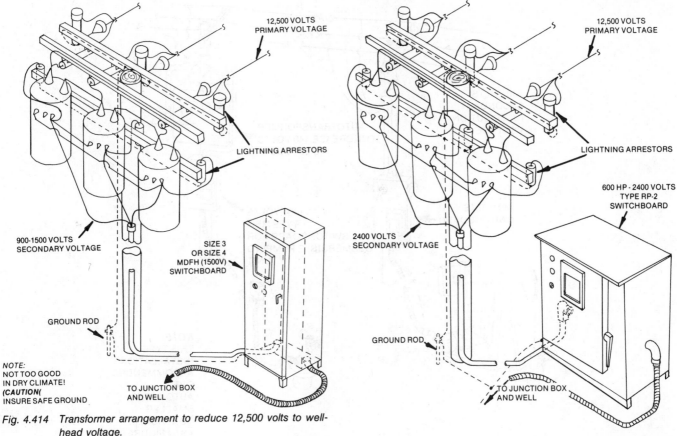

12,500 VOLTS
PRIMARY VOLTAGE

LIGHTNING ARRESTORS

900-1500 VOLTS
SECONDARY VOLTAGE

SIZE 3
OR SIZE 4
MDFH (1500V)
SWITCHBOARD

GROUND ROD

NOTE:
NOT TOO GOOD
IN DRY CLIMATE!
(CAUTION(
INSURE SAFE GROUND

TO JUNCTION BOX
AND WELL

Fig. 4.414   Transformer arrangement to reduce 12,500 volts to well-head voltage.

12,500 VOLTS
PRIMARY VOLTAGE

LIGHTNING ARRESTORS

600 HP - 2400 VOLTS
TYPE RP-2
SWITCHBOARD

2400 VOLTS
SECONDARY VOLTAGE

GROUND ROD

TO JUNCTION BOX
AND WELL

Fig. 4.415   Transformer arrangement to reduce 12,500 volts to 2,400 v.

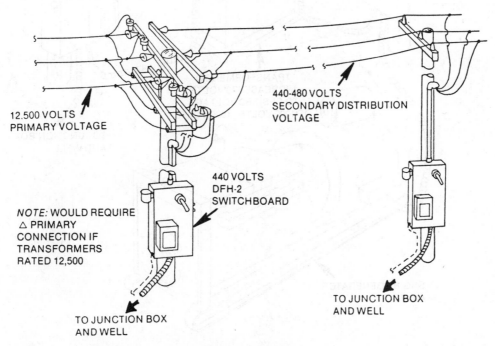

12,500 VOLTS
PRIMARY VOLTAGE

440-480 VOLTS
SECONDARY DISTRIBUTION
VOLTAGE

440 VOLTS
DFH-2
SWITCHBOARD

NOTE: WOULD REQUIRE
△ PRIMARY
CONNECTION IF
TRANSFORMERS
RATED 12,500

TO JUNCTION BOX
AND WELL

TO JUNCTION BOX
AND WELL

Fig. 4.416   Central transformer bank (12,500 v. to 762-830 v.)

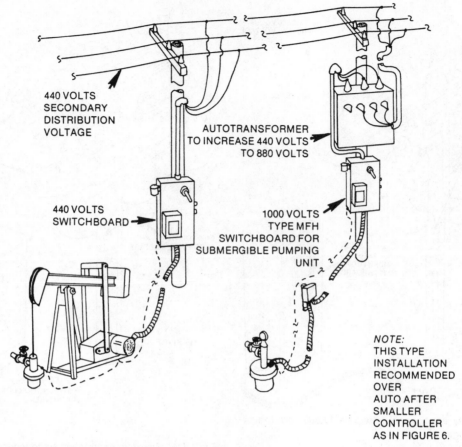

440 VOLTS
SECONDARY
DISTRIBUTION
VOLTAGE

AUTOTRANSFORMER
TO INCREASE 440 VOLTS
TO 880 VOLTS

440 VOLTS
SWITCHBOARD

1000 VOLTS
TYPE MFH
SWITCHBOARD FOR
SUBMERGIBLE PUMPING
UNIT

NOTE:
THIS TYPE
INSTALLATION
RECOMMENDED
OVER
AUTO AFTER
SMALLER
CONTROLLER
AS IN FIGURE 6.

Fig. 4.417   Electrical and beam pumps

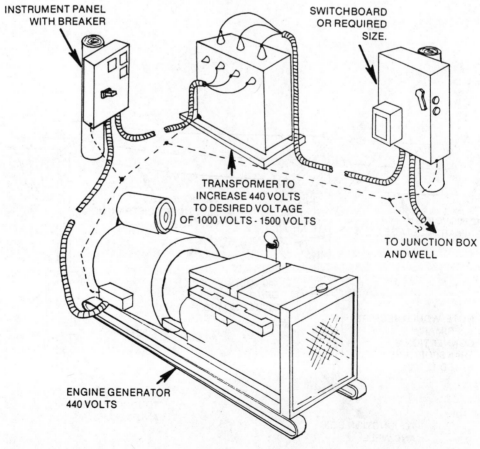

INSTRUMENT PANEL
WITH BREAKER

SWITCHBOARD
OR REQUIRED
SIZE.

TRANSFORMER TO
INCREASE 440 VOLTS
TO DESIRED VOLTAGE
OF 1000 VOLTS - 1500 VOLTS

TO JUNCTION BOX
AND WELL

ENGINE GENERATOR
440 VOLTS

Fig. 4.418   Engine generator with transformer

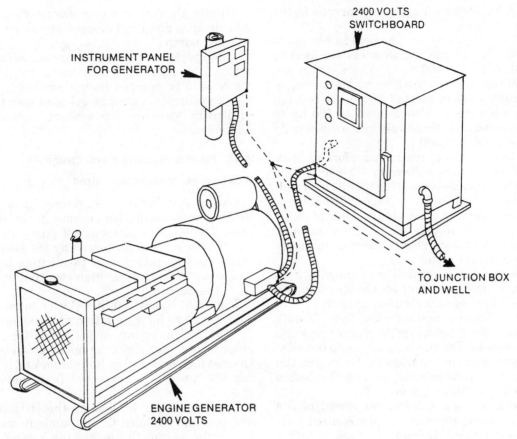

Fig. 4.419   Engine generator unit (2,400 v.)

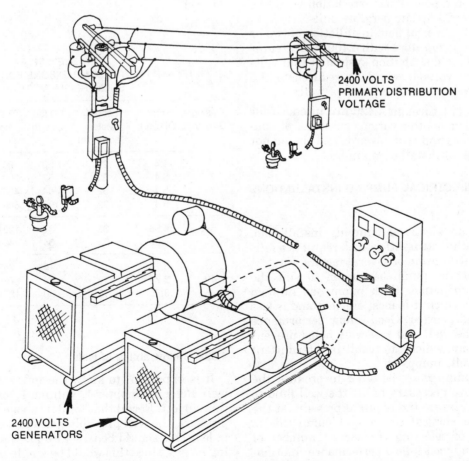

Fig. 4.420   Group of generators

50 hp up to 225 hp. Motor voltages are generally in the range of 900 to 1,400 v.

Fig. 4.415 shows the same arrangement as Fig. 4.414 with the exception that the transformers are used to reduce 12,500 volts to 2400 volts. The switchboard depicted is a 600 hp, 2,400 v switchboard. This arrangement is generally used in sizes of 200 hp to 600 hp; however, on deep wells it is used down to 100 hp to reduce cable size and cost. Motor voltages are generally in the range of 2,000 to 2,300 v.

Fig. 4.416 shows a typical central transformer bank reducing 12,500 v primary voltage to 762-830 v. This voltage is obtained by wye-connecting the 480 v secondary. The secondary distribution voltage of 762-830 v is built throughout a portion of the field or a line of wells. Usually the primary voltage of 12,500 v is built through the field with a number of these transformer banks serving different areas.

The maximum voltage capability of these switchboards is 1000 v. The wye point on the transformers and switchboard is to have a common ground.

Fig. 4.417 show beam pumps operating from the 440-v power system and a submergible pump operating from the same system. The submergible pump installation has a pole-mounted autotransformer to step up the 440 v to the required surface voltage of 880 v before going through a switchboard to the well.

Fig. 4.418 illustrates a transformer ahead of the switchboard increasing the 440 v to the required voltage. Various switchboard sizes may be used with this system.

Fig. 4.419 shows a 2400-v engine generator going directly to a 2400-v switchboard. The instrument panel is set at the side on a pole. This installation would be used for 200 hp to 520 hp submergible units.

Fig. 4.420 shows a typical field installation of two or more large engine generators with 2400-v or greater output serving a field distribution system at that voltage. This primary voltage would be reduced to the required voltage at each well or group of wells.

NOTE: Figures 4.414 through 4.420 are intended for general information on the various methods of connecting transformers and switchboards and should not be considered as final installation guides.

4.5   DESIGN OF ELECTRICAL PUMPING INSTALLATIONS

4.51   Introduction

The design of an electrical pumping installation, while normally rather straightforward, requires a fairly methodical consideration of several factors. First, it is important that the pump be chosen to suit the desired rate of production. Each pump has its own range of flow over which it is most efficient and is less subject to mechanical wear. Good inflow performance information for the well and reservoir helps prevent oversizing the pump, which can result in intermittent operation if the well "pumps off."

Secondly, the pump must be sized to produce the increase in pressure necessary to lift the well fluid to the surface and maintain the required pressure at the well head. In the vertical centrifugal pump this is merely a matter of selecting the correct number of stages. Here again, good inflow performance information is important.

Thirdly, the correct motor size can then be chosen to suit the flow and head along with the efficiency of the selected pump stage.

Pump performance and motor requirements are also affected by the characteristics of the fluid mixture which is to be pumped from a particular well. Therefore, consideration must be given to such factors as fluid density, viscosity, gas content, corrosiveness, and abrasiveness.

4.52   Factors affecting pump design

4.521   Flow configuration sizes

Casing size becomes extremely important in the design of an installation because it controls the maximum size (outside diameter) of pump and motor that can be run in the well. Generally the lowest cost, both initial and operating, will result from using a pump and motor of the largest diameter that will physically fit in the casing.

Table 4.51 shows that the relative submersible motor costs for a 120 hp installation are 1.00 for 7 in. casing, 1.44 for 5½ in. casing, and 2.30 for 4½ in. casing. In other words, the larger the motor diameter, the lower the cost for the same hp. Reliability and longer operating life can also be expected from motors of larger O.D.

The tubing size depends on the produced flow rate and is usually related to the pump diameter, i.e., the larger the pump, the larger the tubing. By far the majority of installations are for tubing flow since the pump is generally run on a tubing string and without a packer, although casing flow can be used in a well if desired.

TABLE 4.51
HP AND LENGTH COMPARISONS FOR ELECTRICAL
SUBMERSIBLE MOTORS

Casing Size, in.	Pump OD (in.)	Length Motor Section (ft)	Total Length	Total hp	hp per ft	Relative cost
7	5.4	20	20	120	6.25	1.00
	5.4	32	32	200	6.25	
	5.4	32	64	400	6.25	
	5.4	32	96	600	6.25	
5½	4.56	32	32	120	3.75	1.44
	4.56	32	64	240	3.75	
4½	3.75	18	18	25½	1.417	2.30
	3.75	18	44	76½	1.417	
	3.75	18	90	127½	1.417	

4.522   Inflow ability of well

It is important to know the inflow capability of the well. Refer to Chapter 1, Volume 1, for a discussion on this subject. Recall that for flow above the bubble point pressure, the well probably has a constant productivity index. This should be true also for water wells producing no gas, since this would be single phase liquid flow. For flow below the bubble point pressure, we may use

Vogel's procedure for flow efficiency equal to 1.00 and Standings extension of Vogel's work for flow efficiencies other than 1.00 (damaged or improved wells). For a more definite inflow curve we may refer to the work of Fetkovich requiring either a flow after flow or isochronal test on the well. An equation of the form $q_o = J_o^l (\overline{P}_R^2 - P_{wf}^2)^n$ will define the shape of the inflow curve for all ranges of flow.

If the inflow ability is known, we can design a pump for the maximum rate or any desired rate. This also assures us of selecting a pump to operate at near maximum efficiency.

For many wells the electrical pump will be capable of "complete pump-off," that is the pump's capacity will exceed the well's capacity. If this is done, care must be taken to insure that we do not "pump the well off."

It becomes a matter of the designer's judgement and his confidence in the inflow performance data as to how close to the pump intake the well should be drawn down. Most submersible centrifugal pumps will operate properly with 220 psig, or less, of pressure at the intake if the fluid being pumped is liquid. However, if there is free gas in the fluid approaching the pump's intake, the question becomes not one of minimum intake pressure but one of how much free gas can that particular pump handle without gas locking and/or how much of the free gas can be separated so that the pump can handle it.

In some cases we will use productivity index in length of head rather than pressure. For example in the English system we express PI in b/d/psi. For electrical pump design we may use b/d/ft of drawdown. In particular, this is quite common in water wells. The reason for this is to maintain continuity in design since the performance curves give pump ability in feet of head per stage.

4.523   Whether or not gas is to be pumped

As a general rule, most installations pump the production up the tubing without a packer in the well. This means that gas can be vented out the casing or routed through the pump.

If there is gas in the well there are, between the producing fluid level and the bottom of the well, a wide range of gas and liquid combinations that are significant to the pump's size and location in the well. It is impossible to say that any one criteria is always the best for selecting the pump and its location, since well and reservoir data available is not always of the same reliability, reservoir conditions may be changing with time, and other factors may be different from one well to another.

One possibility is to set the pump so that its intake pressure is above the bubble point pressure. There is then no free gas at the intake and the volume of flow that the pump must be sized for is simply the stock tank production rate times the formation volume factor. This can be done only if the bubble point pressure occurs above the end of the tubing string.

Another possibility is to set the pump so that its intake pressure is below the bubble point pressure. This has the advantage of shortening the tubing and cable lengths, but the pump must then handle a flow equal to the same stock tank production rate times a slightly smaller formation volume factor but plus the free gas that goes through the pump. As the pump is set higher in the well, the amount of free gas increases and care must be taken so that the flow entering the intake does not have a free gas to liquid ratio greater than the pump is capable of handling. Gas separators built integrally with the pump are a way of diverting free gas away from the pump so that it can be vented from the casing.

The pump and the motor are both affected by how much gas passes through the pump. Generally, gas will have a beneficial effect in the tubing and will reduce the horsepower required from the motor but will require the pump to handle a larger flow rate. The pump's ability to perform is greatly affected by the ratio of free gas to liquid that it must handle. As long as all the gas is in solution, the pump will perform normally as it would with a liquid of lower density and will continue to do so until the free gas to liquid ratio reaches about 0.1. Above about 0.1 the pump will likely begin to produce less than normal head and, as the free gas increases, will eventually gas lock and stop pumping any appreciable amount of fluid.

The calculations needed to give proper consideration to the effects of gas are quite lengthy and involved. Pump companies and others have developed computer programs for use in the proper sizing of pumping equipment in gassy wells.

4.524   Gas separation

One of the unsolved problems that we have today in electrical pumping is how to determine that volume of gas which it is possible to vent. In early years the gas may have been vented to the atmosphere, but present day practices prohibit this practice. The casing can be tied to the flow line and vented gas permitted to enter the flow line near the well head. In some cases a separate gas vent line may be necessary to prevent upping the well head pressure unnecessarily high. By controlling the back pressure on the casing, we can control, to some extent, the volume of gas vented.

However, we must be careful to leave the pump under a certain distance of liquid submergence. The blowing of casing gas around the bottom of the pump inlet is not a desirable practice and could lock or load the well, that is, with liquid in the tubing and gas in the pump we could reach a stymied condition. Various gas separators are available and some of these are discussed in Section 4.3.

We are slightly in the dark as to being able to say within 15-25% of how much gas we can vent. However, calculations should always be made for various values of GIP (gas ingestion percentage) to at least observe the effect on rates, hp, etc. A pump should then be selected to most efficiently produce the well based on our knowledge and experience in the area. We are fortunate in that the pump does cover a fairly wide range of rates; hence, we have some flexibility in being able to handle different gas volumes. Appendix 4K lists methods to estimate the percent of free gas that will go through the pump.

4.525   Deviated wells

The electrical submersible pump is designed to operate in a generally vertical position. It can, however, operate in deviated wells as long as it is not in a bind. The pump itself will, if necessary, function in a position approaching horizontal. The limit of deviation from the vertical is often determined by the unit's ability to maintain separation of its motor's oil and the well fluid. This is a matter of the manufacturer's design. For units designed with a flexible barrier between the motor oil and well fluid, this limit for deviated installations is removed.

4.526 Packers

The preferred manner to run an electrical submersible pump is without a packer so it is hanging free on the tubing string. It can be run beneath a packer, but this must be a special installation because the cable for power to the motor must by-pass through the packer. If a packer is required in the well, its selection should be made with care so the pump will have only a very little or no compressive weight on it. A permanent packer utilizing a long stinger type seal would work satisfactorily, keeping in mind that after the pump starts moving high volumes of hot fluids, the tubing will extend placing the pump in compression if allowance is not made.

In summary, where packers are required, use one that can be set without placing the tubing or pump in compression.

4.527 Viscosity effects

Viscosity affects the performance of centrifugal pumps by lowering the head-capacity curve, reducing the efficiency, and making the highest efficiency occur at a lower flow. For any one pump the effect on produced head is greater at higher flows and less at lower flows i.e., the head-capacity curves tend to rotate about the head at zero flow.

Since published pump performance curves are based on tests in which water (with a viscosity of about 30 S.S.U.) is the fluid, it is necessary to adjust the curves for fluids of higher viscosity. The amount of adjustment varies between pumps. Those with smaller flow passages will generally be affected more by high viscosity.

Deciding on the viscosity for which to adjust is also a factor. If gas is present then the live, or gas-saturated, viscosity of the oil must be used. If water is present along with the oil and an emulsion is formed, then the viscosity is several times greater than the average of the oil and water viscosity.

4.258 Temperature

The bottom hole temperature is important for installing an electrical submersible pump. It is necessary to know at what temperature the motor is going to operate. Also, the cable is selected with temperature being one of the controlling factors.

Even though the pump may not be set on bottom, a high rate of production will move the fluids rapidly up the tubing, bringing a much higher temperature to the pump than exists under static conditions. The higher the temperature at the pump, the shorter the motor life. For example, for each 18°F. rise in temperature above the motor insulation rating, the motor life is shortened by one-half. Cables are available that will operate successfully up to 350°F. but become more costly as the temperature becomes high.

Temperature must also be known to determine the total intake volume especially for handling gas.

4.529 Operating vs. unloading conditions

In the final horsepower selection of the motor, the operating hp requirements may be less than the unloading horsepower requirements. However, the unloading rate can be decreased to a value much less than the operating rate for unloading purposes. There will be instances when an oil well has been loaded with brine and the horsepower required for operation may be much less than the hp required for unloading. It may be necessary to compromise between the two hp requirements, keeping in mind that a motor can be overloaded as much as 20% for the short time period necessary to unload the well. This should always be checked in the final design to make certain that the well can be unloaded.

4.53 Detailed design of installations

4.531 Introduction

As far as design procedures are concerned, there are two types of wells: wells not producing gas and wells producing gas. If there is no gas in the well, the calculations for selecting the pumping equipment are relatively short and simple. The amount of head the pump must produce is simply a matter of (1) adding the feet of head required to lift the liquid to the surface, overcome the tubing friction, and provide the necessary well head pressure, then (2) subtracting the pressure caused by the height of fluid over the pump's intake. These are simple calculations because the specific gravity of the fluid is, for all practical purposes, the same throughout the well; therefore, the conversion between pressure in psi and pressure in feet of head is the same at all points in the well.

However, if the well is producing gas, the problem is much more complicated. Since the pressure and temperature are not the same at any two places in the well, the gas volume is also not the same and its proportion of the total mixture which is being removed is not the same. This results in a constantly changing density as the mixture of fluid and gas pass through the perforations, to the pump, and upward to the surface through the tubing. There is no constant conversion between pressure in psi and pressure in feet of head; therefore, it is necessary to make calculations at small intervals along the path of flow from the reservoir to the ground level. The pressure that the pump must produce can then be determined by a summation of the information acquired in the step by step calculations. Since these calculations are lengthy, voluminous, and relatively complex, they have been programmed for computer solution. Use of these programs is probably the only practical way to select or design a pump and motor in a gassy well.

Regardless of whether or not there is gas to be considered, there are sometimes special conditions that affect the pump and motor selection. If the well fluid is viscous then a pump of larger capacity and higher head will have to be selected and the motor size required will be increased. If there are corrosive, abrasive, or scaling conditions, then special consideration might need to be given to metallurgy or protective coatings.

The simplest installation from a design standpoint is a water well because no gas is being routed through the pump. The next simplest type of installation is a low gas-oil ratio oil well where the gas may or may not be routed through the pump. Third, is a well producing enough gas so that a portion or all of the gas must be pumped. Finally are special applications such as viscous crude oils.

4.532   General considerations in pump design

4.5321   Introduction

The sizing of a submersible pump, in most applications, is simple if the basic fundamentals of submersible equipment and well data are understood. Each application is an individual situation due to varying well conditions and type fluids that are to be pumped.

4.5322   Required well data

The initial data used for sizing a submersible unit is very important and must be reliable to insure the properly sized unit. The data required falls in the following four general categories:

(1) Inflow performance of the well and reservoir
(2) Physical dimensions of the well
(3) The makeup and characteristics of the well fluid
(4) The design objectives and any pre-set requirements of power supply, etc.

(1) The inflow performance of the well and reservoir establishes the maximum production capability of the well and also determines the suction side pressure for any flow less than maximum. Inflow performance is generally described as the static pressure at a known depth plus a flowing pressure at a minimum of one known flow rate. If there is no gas in the well, fluid levels instead of pressures are sufficient. The pressure for other production rates is determined by extending the inflow performance data in one of two generally accepted ways. The straight line productivity index (P.I.) is used if there is no gas or if all the gas is in solution. The downward curving inflow performance relationship (I.P.R.) is used when the reservoir pressure drops below bubble point pressure in the flow path to the well bore, thereby causing gas to come out of solution, and two phases flowing in the reservoir.

(2) Casing size and weight determine the pump and motor of maximum diameter that will fit in the well. This is important because generally the most efficient installation will result when the pump of largest diameter, having the proper flow range, is used.

(3) Bottom hole and perforated interval depths determine respectively the maximum possible pump setting depth and the maximum depth that the pump can be set without needing a motor jacket. If the perforations are above the motor, it is necessary to use a jacket or shroud to direct the flow past the motor thereby cooling it.

(4) The specific gravities and their percentages of the liquids and gas making up the mixture being pumped determine to a large extent the motor horsepower. Therefore, the specific gravity of the water and gas, oil API gravity, water cut and G.O.R. are needed.

(5) Viscosity, if available, is needed since published pump performance curves are based on water tests. If the viscosity is greater than that of water, a correction must be made to the head-capacity and horsepower curves. Standard correlations exist from which viscosity can be approximated from the oil's API gravity and temperature.

(6) Temperature of the well fluid near the bottom of the well and also at the well head are needed, particularly if there is gas present since the amount of gas in solution and the volume of any free gas are sensitive to the temperature and its change throughout the well and tubing. Also the selection of motor cable material is affected by the temperature of the liquid to which it is exposed.

(7) PVT data in the form of pressure, solution gas-oil ratio, and formation volume factor is needed if gas is present. If for any particular case the P.V.T. data is unknown, it can be approximated from standard correlations.

(8) Certain other information, not pertaining to the well and reservoir itself but more particularly to the pumping system, is also required. It is necessary to know how much pressure is required at the well head in order to deliver the production to its final destination. The power supply voltage available will determine the sizing of transformers and other electrical components. Whether it is 50 or 60 Hz will establish the speed and output of the pump. Tubing size is generally related to the pump's diameter and determines how much friction loss must be included in the total dynamic head. The tubing thread size and type must be known so check valves, bleeder valves, pump head and well head can be selected.

In order to become acquainted with the various performance curves and charts, this section covers the sizing procedures for a typical oil well. To demonstrate the differences in the two basic types of wells, i.e., those with and those without gas, the same well and reservoir data will be used for both with only the presence of gas being the difference. Understand that the two basic procedures can accommodate the range of variables such as differences in water cut, specific gravity, and viscosity experienced in selecting a pump and motor.

4.5323   Standard performance curves

Fig. 4.51 is a standard performance curve for a submersible pump. The head capacity curve is plotted with the head in feet and meters as ordinate (vertical) against capacity in barrels per day and M³/day as abscissae (horizontal). Fresh water (specific gravity 1.0) is the fluid used in rating submersible pumps. The head for a proposed application can be figured in feet, and the desired head and capacity can be read directly from the water curves without correction as long as the viscosity of the liquid is close to that of water.

The total stages required are found by the formula:

$$\text{Total stages} = \frac{\text{total dynamic head (ft)}}{\text{head (ft)/one stage}}$$

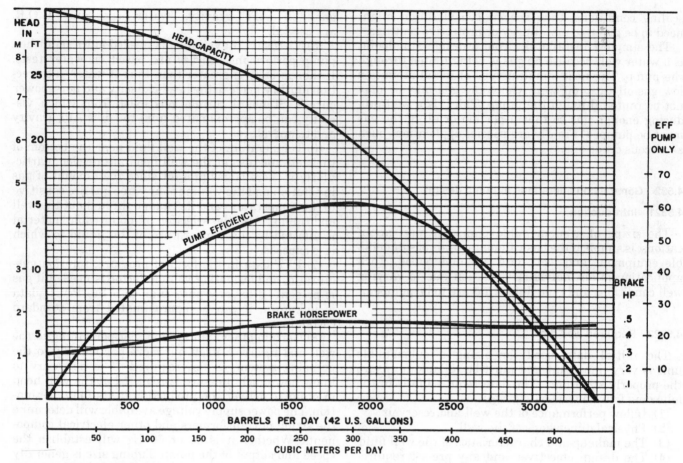

Fig. 4.51 Typical pump performance curve for 5½" O.D. well casing (Courtesy Bryon Jackson - Centrilift)

As an example and referring to Fig. 4.51 (standard performance curve), if the total calculated head were 5,000 ft and the volume required were 2,100 b/d, the number of stages would be found by entering the pump curve at the 2,100 b/d rate, moving up vertically to the head capacity curve and reading on the left side the head per stage as 19.7. Therefore, the required stages would be:

$$\text{Total stages} = \frac{5000 \text{ ft}}{19.7 \text{ ft/stage}} = 254 \text{ stages}$$

Other performance curves can be found in Appendix 4B.

The horsepower shown on the water curve will apply only to liquid with a specific gravity of 1.0. For other liquids, the water horsepower must be multiplied by the specific gravity of the liquid being pumped.

The required motor size is determined by multiplying the maximum horsepower per stage of the type pump selected by the number of stages by the specific gravity of the fluid:

$$\text{hp} = \text{hp/stage} \times \text{total stages} \times \text{specific gravity}$$

Again, using the standard performance curve and assuming a specific gravity (1.0) of fresh water, the hp/requirement of the 254 calculated stages would be found by taking the horsepower per stage from the pump curve. The horsepower requirement would be 0.435 hp/stage. The total horsepower requirement would be: hp = 0.435 hp/stage × 254 stages × 1.0 = 110 hp.

4.5324 Total dynamic head

Total dynamic head (TDH) is simply the total head required to be produced by the pump when it is pump-

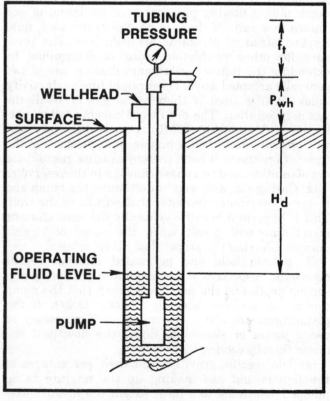

Fig. 4.52 Illustration of total dynamic head

ing at its desired rate. It is the difference between the head required at the pump discharge to deliver the flow to its final destination and any head that exists at the pump intake. (Fig. 4.52)

Some clarification may be needed here in describing total dynamic head. Many pump companies as well as design engineers refer to the total dynamic head as being the flowing wellhead pressure in feet plus friction to pump depth plus the effective lift. The effective lift depth is that depth at which the pump could be set to produce the desired flow rate. It would be total depth minus submergence or suction head. Designers may handle this in either way.

More specifically, when pumping a liquid without gas, the total dynamic head is the sum of (1) the friction losses in the tubing and surface flow line, (2) the difference in elevation between the final destination of the produced flow and the pump depth and (3) any significant losses in the discharge line due to valves, separator, etc., minus (4) the head that exists on the suction side of the pump's intake due to the column of liquid over the intake. These calculations can be made using head as the unit of pressure since the fluid density is the same throughout the pumping system.

However, if gas is present in the well, the density is not the same throughout the system and the calculations must be made in units of pounds per square inch

(psi) and then converted to head in order to use the standard performance pump curves.

Quite often, for design purposes, the losses and elevation difference in the surface flow line are replaced by a pressure at the well head which is sufficient to move the flow through and up the surface flow line.

EXAMPLE:

Data:

Required well head pressure	= 200 psig
Pump setting depth	= 10,570 ft
Tubing	= 2⅞″ EUE
Pumping rate	= 1600 b/d
Fluid pumped	= 70% API 40° oil
	30% water @ 1.05 S.G.
	Ave. density = 54.79
	lb_m/ft^3
Fluid over pump intake	= 650 ft

Well head pressure in ft of head =

$$200 \text{ psig} \times \frac{144}{54.79 \text{ #/ft}^3} = 526 \text{ ft.}$$

Friction loss in 10,570 ft of tubing =

$$10.57 \times 20.5 \ \frac{ft}{1000 \text{ ft}} = 217 \text{ ft}$$

(Fig. 4.53) Other friction loss curves are found in Appendix 4C.

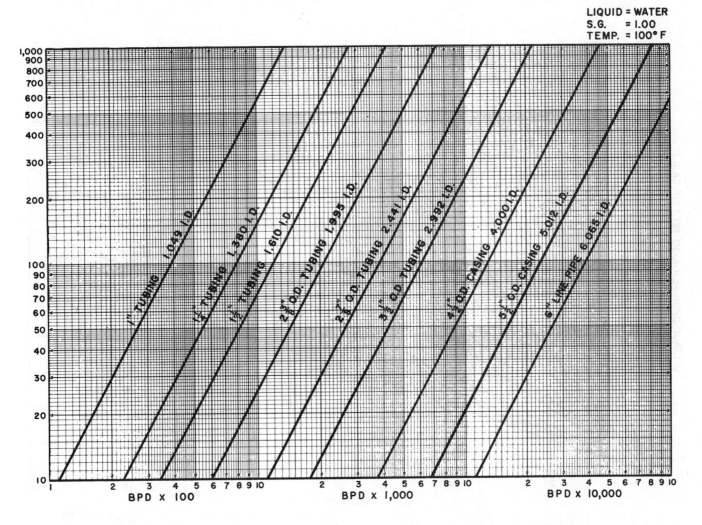

LIQUID = WATER
S.G.   = 1.00
TEMP. = 100° F

CALCULATED FROM HAZEN-WILLIAMS FORMULA $V = CR^{.63} S^{.54} .001^{-.04}$, WHERE C = 120

Fig. 4.53 Flow losses due to friction in A.P.I. pipe

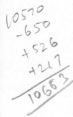

Elevation difference—pump to well head

$$= 10,570 \text{ ft}$$

Fluid over pump intake	= −650 ft
Total dynamic head (TDH)	= 10,633 ft

4.5325 Cable selection

The size and type cable selected for an application is determined by the current carrying capacity of the cable and by the environment (temperature and pressures) that the cable will be operating under. As a guide the current-carrying capacities of some of the cables are listed in Table 4.52.

TABLE 4.52
CABLE CURRENT-CARRYING CAPACITY

Cable No.	Max amp
1 CU	115
2/0 AL	115
2 CU	95
1/0 AL	95
4 CU	70
2 AL	70
6 CU	55
4 AL	55

Another chart used in sizing of submersible components is the cable voltage loss chart. (Fig. 4.54 and Appendix 4A).

Find the voltage loss per 1000 ft of cable by entering the chart at the amperage anticipated; move vertically up to the size cable to be used and at the temperature anticipated. Read the voltage loss at the left of the chart. On the example shown, the voltage loss would be 20/1,000 ft of cable at 58 amps and at a temperature of approximately 180°F. using number 2 cable.

EXAMPLE PROBLEM: TO DETERMINE "REQUIRED SURFACE VOLTAGE"

The definition of "required surface voltage" is the no load voltage required at the surface to accommodate the voltage of the motor used plus voltage losses due to cable size and other electrical components in the system.

Using the rated voltage and amperage of the motor, length and size of cable, plus cable loss chart (Fig. 4.54), calculate optimum surface voltage requirements for

Motor = 890 v, 58 amps

Cable = 3600 ft of No. 2 copper cable

The voltage loss for 58 amps and No. 2 copper cable was previously found to be 20/1,000 ft of cable. There-

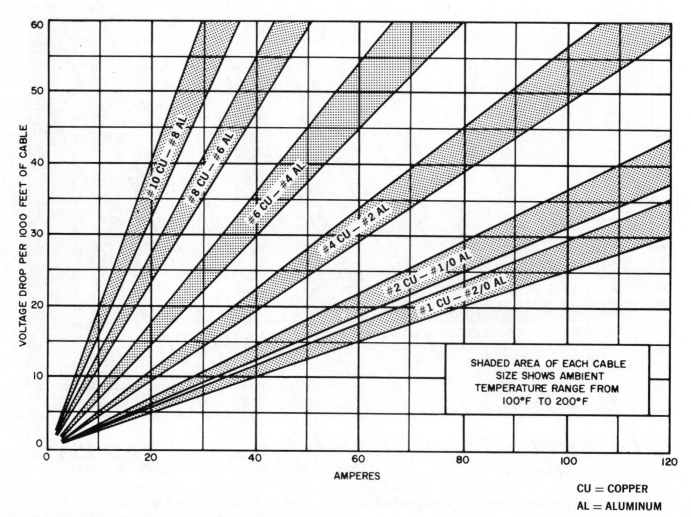

Fig. 4.54   Voltage loss chart (Courtesy Bryon Jackson - Centrilift)

CU = COPPER

AL = ALUMINUM

fore, the loss due to 3,600 ft of cable would be 20 v × 3.6 = 72 v. This added to the motor voltage would be 890 v + 72 v or 962 v. A good rule of thumb for voltage losses in a bank of three-phase transformers is 2.5% of the required voltage. In this case, it would be 962 v × 2.5% or 24 v. Total required voltage would be 962 v + 24 v or 986 or 990 v.

The amount of operating voltage is somewhat flexible within a 50-volt range. If the voltage cannot be exact, it should be slightly higher than required rather than lower. However, the voltage should be set as close to optimum as transformer settings will allow.

Some companies prefer to utilize higher voltage motors and larger switchboards in anticipation of increased production at a later date. Also, the higher voltage motors with lower amperage must be utilized in deeper wells where the casing size limits the size cable that can be used, therefore limiting amperage of motors.

If several options exist, then the determination of whether or not to use a 2,400 v system will depend upon an economic evaluation. The choice of motor voltage is a function of depth, casing size (may limit cable size), cable size, cable cost, switchboard cost, and cost of electric power. As a general rule we can use:

(A) Low hp shallow depths—440 v

(B) hp <70 intermediate depths —762-830 v

(C) 70 - 200 hp deep wells —1,500 v switchboard motors of 900 - 1,300 v

(D) >200 hp —Have a choice of 1,500 v or 2,400 v system. Depends on depth, economics of switchboards, cable costs and power costs

Based on cost, the following Example Problem is worked to determine whether or not to use a 2400 volt system.

EXAMPLE PROBLEM

Given:

Motor size = 150 hp
Depth = 6,000 ft
Have motor voltage choice
   (1) 2150 v—43 amp
   (2) 1150 v—80 amp

The following costs are relative and are subject to change, but should change relative to each other.

Costs: Switchboards, 2,400 v—$5300.00
                           1,500 v—$3000.00
             Cables, #2 Cable—$2.70/ft
                             —$16,200.00
                    #4 Cable—$1.85/ft
                            —$11,100.00
Total Costs: 2,400 v—5,300 + 11,100 = 16,400
             1,500 v—3,000 + 16,200 = 19,200

The savings with a 2,400 v system is $2800.00; therefore we would select a 2,400 v system. The following additional example shows how the final determination is based on operating costs:

Given: Same as previous example except
             depth = 3,000 ft
Costs: #4 cable = 1.85 × 3,000 = $5,550.00
             #2 cable = 2.70 × 3,000 = $8,100.00
             Savings #4 = $2,550.00
Total cost, 2,150 v = 5,300 + 5,550 = 10,850
Total cost, 1,500 v = 3,000 + 8,100 = 11,100
   Difference = $250.00

There is very little difference in costs; hence, we must make our decision based on power operating costs of both units.

43 amps in #4 cable—loss 23 v/1000
80 amps in #2 cable—loss 27 v/1000

For #4 cable (3,000)(23/1,000) = 69 v for motor of 2,150 v
Surface v = 2,150 + 69 = 2,219 v

$$kw = \frac{(v)(amps)(P.F.)\sqrt{3}}{1,000}$$
$$= \frac{(2,219)(43)(0.85)(1.73)}{1,000} = 140.31$$

For motor 1,150 v, 80 amps, #2 cable = (2,000)(27/1,000) = 81 v
Surface v = 1,150 + 81 = 1,231

$$kw = \frac{(1,231)(80)(0.85)(1.73)}{1,000} = 144.81$$

The difference is 4.5 kw higher for the 1,150 v motor.

$$(4.5)(24 \text{ hrs})(30 \text{ days}) = 3240 \frac{kwh}{mo}$$

The difference in costs = (3240)($0.01) = $32.40/month. Therefore our choice will be the 2,150 v system showing a savings of:
   Initial cost—$250.00
   Operating cost—$32.40/mo

4.5326   Transformer Sizing

In sizing the auto-transformer, three-phase transformer, or a bank of three single-phase transformers, the following equation is used:

$$kva = \frac{V_s \times A_m \times 1.73}{1,000} \qquad 4.51$$

where:
   kva = kilovolt amps
   V_s = required surface voltage
   A_m = motor nameplate amps or anticipated amperage that will be utilized

EXAMPLE PROBLEM:

required surface voltage = 990 v
amperage = 58 amps
$$kva = \frac{990 \text{ v} \times 58 \text{ amps} \times 1.73}{1,000} = 99.4$$

If three single-phase transformers are used, this 99 kva requirement would be divided by 3 to establish a value for each transformer. The auto-transformer, or three-phase transformer, would need to be of a minimum 100 kva size.

If it is anticipated that a well will require a larger pumping unit in the future, it would probably be economically feasible to install larger size transformers initially with appropriate top range.

Note: Transformers also must be selected for the

primary voltage available to accommodate the type hook-up required, ΔΔ, YΔ, or YY. Most submergible hook-ups are either ΔΔ or YΔ.

4.5327   Summary of pump sizing procedure

In sizing a submergible pumping unit, the following procedure should be followed in the sequence given:

(1) Collect and analyze the well, production, fluid, and electrical data.

(2) Determine the well's production capacity at the selected pump setting depth or determine pump setting depth at desired production rate. This includes determining what pump intake pressure (PIP) will be used for the design and the total volume to be pumped to achieve the required stock tank barrels.

(3) Calculate the total dynamic head (TDH) (Friction losses + system pressure + vertical lift).

(4) For the calculated capacity and total head, select from the various pump curves the pump type which will have the highest efficiency for the application. The selected pump must also be of the proper O.D. that will fit inside the casing of the well.

(5) Calculate from the type pump selected the number of stages required to supply the total required head at the required volume.

(6) Determine the motor horsepower required, using the highest specific gravity of the fluid that will be encountered for this calculation. Procedures of operation are sometimes used where the specific gravity of the live well fluid is used to calculate horsepower and steps are taken to unload kill fluid under minimum overload conditions. The protector type is usually determined from the series motor selected.

(7) Select the most economical cable size and type for the application from available technical data sheets.

(8) Determine voltage loss in the cable and the required surface voltage. The value of the required surface voltage will set the size of the switchboard.

(9) Calculate the KVA requirements in order to size the transformers.

(10) Select properly sized accessories, such as:
  (A) Tubing head, size, and type
  (B) Servicing equipment required for complete installation
  (C) Optional equipment

(11) Determine what other steps are required to ensure good operations such as:
  (A) Coat equipment for corrosion protection and use corrosive preventing materials
  (B) Use shroud, if required, etc.

4.533   Example problem #1A, water well

4.5331   Introduction

Electrical pumps are often utilized in all types of water wells including fresh water wells for domestic use for irrigation, fresh water wells for injection water, and salt water wells for injection purposes, and salt water wells for salt removal and possibly for gas dissolved in the salt water.

This represents the simplest design problem and will be covered first in this section. As in the case of an oil well, good well information is needed. Although we have some flexibility in the pump, it is preferable to have the pump operating at peak efficiency when producing the desired rate. If the maximum rate is desired refer to Section 4.54 for this procedure.

4.5332   Example problem #1A (no gas)

(1) Collect and analyze the data
  Data:
  Casing size = 8⅝ in. O.D.
  Tubing size (new) 5½ in. O.D.
  Depth = 2200 ft
  Perforations = 1900-2200 ft
  Power source = 12,500 v Primary
  Static fluid level = 500 ft from the surface
  Water specific gravity = 1.1
  Temperature = 120°F.
  Productivity index = 10 b/d/ft of Drawdown
  Desired rate = 10,000 b/d
  Surface flowline = 2,000 ft of 4 in. with an elevation rise of 30 ft
  (all pipe is new)
  Select a suitable electrical submersible pump and necessary related equipment.

(2) Determine well producing capacity
  In this problem a rate of 10,000 b/d is desired.

(3) Determine total dynamic head in feet required for 10,000 b/d

  (A) Drawdown in feet $= \dfrac{q}{PI \ (ft)} = \dfrac{10,000 \ b/d}{10 \ b/d/ft} =$ 1,000 ft

  Lift Head = 1,000 ft drawdown + 500 ft static fluid level = 1,500 ft (for safety set pump − 1600 ft)

  For no friction and no well head pressure, this would be the total head but we must account for both.

  (B) P_{wh} (wellhead pressure) in feet of head
  P_{wh} = Elevation Component + Friction Loss in Horizontal Line = 30 ft + (55 ft/1000)(2000) = 140 ft
  For friction loss see Appendix 4C (for new pipe).

  (C) Tubing friction loss, see Appendix 4C (new pipe)
  Friction loss for 10,000 b/d, 5 in. pipe = 18.5 ft/1000 ft (new) = (18.5)(1.6) = 29.6 ft
  Total dynamic head = lift + tubing friction + P_{wh} = 1,500 + 29.6 + 140 = 1670 ft

(4) Select a suitable pump
  Base selection on:
  (a) Casing size = 8⅝ in.
  (b) Rate = 10,000 b/d
  See Appendix 4G for casing data—Drift I.D. of 8⅝ in. 28 #/ft = 7.892 in. (I.D. = 8.017)
  See pump performance curves (Appendix 4B).
  Select I-300, capacity range = 8,000 to 11,500 b/d (Appendix 4BI-55)

(5) Determine number of stages required (see Appendix 4BI (55) pg. 206

Performance curves—8⅝ in. casing—3,500 rpm for I-300-60 Hz

At 10,000 b/d one stage develops 5950/100 = 59.5 of head

$$\text{No. of stages required} = \frac{1670 \text{ ft of head}}{59.5 \text{ ft/stage}} = 28 \text{ stages}$$

(6) Determine motor horsepower requirements

See performance curves for I-300, 60 Hz (Appendix 4BI (55) pg. 206

For 10,000 b/d = 5.85 hp/stage (based on γ_w = 1.00)

For a specific gravity of 1.1, hp =

$$(28 \text{ stages})\left(\frac{5.85 \text{ hp}}{\text{stage}}\right)(1.1) = 180 \text{ hp}$$

Note that maximum hp in operating range = 6.1 hp/stage

hp = (6.1)(28)(1.1) = 188 hp

Generally, select the maximum hp of 188 since additional hp may be needed for unloading purposes.

Select the motor (60 Hz)(refer to Appendix 4F). If the well has a temperature greater than 180°F., select a motor of sufficient hp such that 188 hp represents only 80% of the total hp. In this case the temperature is low; therefore, a 190 or 200 hp unit is sufficient. We find a 540 series motor of 200 hp of 1160 v and 105 amps.

(7) Check for Cable

Always use round cable if we have the clearance. See Section 4.5325 on cables. (Refer to Appendix 4A). The following cables are available from TRW-Reda.

(a) 3 KV – Redalene = standard (good for 180°F.) – GALV

(b) 3 KV – Redared – GALV (good for 300°F.)

(c) 3 KV – Polyethelene (good for corrosion at low temperatures (140°F.)

Select #1 CU (Appendix 4A)

(8) Determine cable voltage loss and select switch-board loss = 30 v/1,000 ft (Appendix 4A)

$$\frac{(1,670)}{1,000}(30) = 50 \text{ v}$$

Use polyethelene cable, select suitable switchboard.

(9) Select transformers

Refer to appendix for section on transformers.

Read surface voltage = 1,160 v + 50 v = 1210 v motor loss

$$\text{kva} = \frac{(v)(\text{Amp})\sqrt{3}}{1000} = \frac{(1210)(105)(1.73)}{1,000} = 220$$

Use bank of 3 transformers, 75 kva each = 225 kva total

(10) Select accessories

Leave selection of other equipment and parts to representative of manufacturer.

4.534 Example problem #1A for 50 Hz power

Check this same example for 50 Hz power. The I-300, 50 cycle pump is in the desired range (see Appendix 4BI (56) pg. 207. By way of comparison, for the same pump, the head developed is 10.6 meters. Converting to feet (3.2808 F/M), we have 34.77 ft/stage. The required number of stages = $\frac{1,670}{34.77}$ = 48 stages. This com-

pares to 28 stages needed for 60 Hz power. Hp for 50 Hz = 3.6 hp/stage. Total hp (50 Hz) = (3.6)(48)(1.1) = 190 hp. This compares to 188 hp for 60 Hz.

Submergible motors rotate at a speed of 3,450-3,500 rpm at a frequency of 60 Hz. At a frequency of 50 Hz, it rotates at a speed of 2,915 rpm. For each individual pump series, the diameter of the stages cannot be changed so the only variables are the different types of stages and the number of stages included in the pump. Therefore, any change in the performance of pumps is due to a speed change (rpm) only. The rotational speed can only be changed by varying the frequency. When this speed change occurs, it varies a centrifugal pump's operation according to affinity laws.

The affinity laws for a speed change only are:

$$Q_2 = Q_1\left(\frac{N_2}{N_1}\right), H_2 = H_1\left(\frac{N_2}{N_1}\right)^2, HP_2 = HP_1\left(\frac{N_2}{N_1}\right)^3$$

where:

Q_1, H_1, HP_1, N_1 = Initial capacity, head, brake horsepower and speed

Q_2, H_2, HP_2, N_2 = New capacity, head, brake horsepower and speed

Therefore, when a unit designed for 60 Hz operation is operated at 50 Hz, the 60 Hz performance is altered in the following percentages:

Capacity is 83.3% at 50 Hz

$$Q_2 = Q_1\left(\frac{N_2}{N_1}\right) = \frac{2915}{3500} = 83.3\%$$

Head is 69.4% at 50 Hz

$$H_2 = H_1\left(\frac{N_2}{N_1}\right)^2 = \left(\frac{2915}{3500}\right)^2 = 69.4\%$$

Brake horsepower is 57.8% at 50 Hz

$$HP_2 = HP_1\left(\frac{N_2}{N_1}\right)^3 = 57.8\%$$

4.535 Example problem #1B water well

An additional water well problem is worked to illustrate different conditions.

(1) Collect and Analyze the Data

Given:

Casing:	= 5½″ O.D.
Tubing	= 2⅜″ O.D.
Depth	= 5000′
Static liq. level	= 2500
γ_w	= 1.05 (No gas)
T	= 140°F.

A test on this producing water well shows it making 500 b/d with the liquid level at 3000 ft in the csg. At the surface there is 2000 ft of 2.0 in. flow line with an increase in elev. = 30 ft.

Desired rate = 1250 b/d. Use 100 ft safety in setting pump.

(It is quite common to use 200 to 300 ft safety.)

Our problem is to select an electrical submersible unit for:

(a) 60 cycle

(b) 50 cycle

(2) Production rate is given at 1,250 b/d desired.

(3) Determine total dynamic head

(A) Lift = 3750 ft

$$PI = \frac{500 \text{ b/d}}{3000 - 2500} = 1 \text{ b/d/ft}$$

Drawdown for 1,250 b = 1250 ft

Lift = 2,500 + 1,250 = 3,750 ft
Set pump at 3850 ft for S.F.

(B) Tubing friction (new pipe)(Appendix 4C) =

$$30 \text{ ft}/1,000 \frac{(3850)}{1,000} = 115.5 \text{ ft} \quad (\text{pg. 315})$$

(C) Surface head

Elevation + friction in surface flow line =
30 + (30)(2) = 90 ft (pg. 315)
Total head = 3,850 + 115.5 + 90 = 3955.5 ft

(4) Select pump (see Appendix 4B)
Use Reda—D-40 (Appendix 4BI-21) (pg. 172)
Centrilift—M-34 (Appendix 4BII-7)
Oil Dynamics RA-12 (Appendix 4BIII-4)
or Oiline-Kobe Model 13 SOF (Appendix 4B-IV)

(5) Stages required = $\frac{3955.5}{22.7}$ = 174 stages
(See Appendix 4BI-21)

(6) hp = 0.35 hp/stage
(.35)(174) = 61
(61)(1.05) = 65 hp

(7) Select motor
In cool wells motors can withstand 5-10% overload for a long period of time and up to 20% for unloading. Use 70 hp, 1170 v-38 amp; or 980 v-45 amp; or 785 v-57 amp. in 4½ in. motor series.

4.536 Comparative example problem #1C for water well production

In this example two problems are worked, one for 5½ in. casing × 2⅜ in. O.D. tubing and the other for 7 in. casing × 2⅞ in. tubing. Note that each example has the same well head pressure and the same depth of lift. However, less hp can be used in the 7 in. casing and 2⅞ in. tubing due to less frictional loss in the 2⅞ in. tubing and due to the fact that a larger O.D., more efficient pump, can be used in the 7 in.

EXAMPLE #1C (1)

Given:

5½ in. casing
2⅜ in. O.D. tubing
Perforations at 4000 ft
Static fluid level, 1000 ft from surface
100% water
BHT = 160°F.
PI = 1.0 b/d/ft drawdown
Wellhead pressure = 100 psig
Liquid specific gravity = 1.1
Desired production rate = 2,000 b/d
  Select pump for 2000 b/d
$$\frac{2000 \text{ b/d}}{1 \text{b/d/ft}} = 2000 \text{ ft drawdown}$$

1000 ft static fluid level + 2000 ft D.D. = 3000 ft lift
Set pump at 3,300 ft for safety.
For 2000 b/d friction = 72 ft/1000 ft in 2 in. with
  pump set at 3300 ft. (Appendix 4C)
Total friction = 72 ft/M ft (3.3 M') = 237.6 ft; 0.433
× 1.1 = 0.476 gradient for 1.1 SG water
100 psi wellhead pressure ÷ 0.476 = 210 ft head
Total head required to be developed
Lift   = 3000 ft
Tubing friction = 237.6
Surface head = 210
                 3447.6 ft to be dev. by pump

From Appendix 4BI-25 select D-55 pump for 5½ in. O.D. casing. From D-55 curve, pump develops 22 ft head/stage @ 2000 b/d; 3447.6 ft req. ÷ 22 ft/stage = 157 stages; 157 stages × 0.6 hp/stage = 94.2 hp for fresh water × (1.1 SG) = 103.6 hp

EXAMPLE PROBLEM #1-C (2)

Given:

Casing = 7 in. O.D.
Tubing = 2⅞ in. O.D.
Perf. = 4000 ft
Static bottom hole pressure = 1428 psig
100% water
BHT = 160°F.
PI = 2.10 b/d/psig
Treater pressure = 50 psig
Flowline size = 3 in., 2000 ft long, old pipe
Elevation change = 68 ft upwards
Liquid specific gravity = 1.1
  Select pump for 2000 b/d
  2,000 b/d ÷ 2.10 b/d/psi = 952.38 psi drawdown
  1,428 psi SIBHP − 952.38 = 475.62 FBHP
  0.433 × 1.1 = 0.476 psi/ft gradient
  475.62 fbhp ÷ (0.476) = 999.2 ft of water in hole
    pumping
  4,000 ft − 999.2 = 3,000.8 ft lift
  Set pump at 3,300 ft for safety. Tubing friction for
  2000 b/d = 30 ft/1000 ft in 2½ in.; 30 ft/M ft (3.3 M
  ft) = 99 ft total friction (Appendix 4C).
  Calculate head required at wellhead
  50 psig, 68 ft elevation and 2,000 ft of surface flow
    line
  50 psig ÷ (0.476) = 105.04 ft
  Elevation = 68 ft
  Friction in 2000 ft
  old 3 in.   = 37 ft
                210.04 ft.
Total dynamic head = Lift (ft)   = 3000.8
                     Tubing friction =   99
                     Surface head = 210.04
                     TDH   = 3309.84
                                   = 3310 ft

Note a choice of pumps, that is, G-52E or G-62E for 7 in. casing. Our choice would be G-52E if the PI data is weak or if it is anticipated that well static pressure may decline over a period of time (use G-52E pump) (Appendix 4BI-35).
  From G-52E curve the pump develops 39 ft/stage
    @ 2000 b/d.
  3310 ft head ÷ 39 ft/stage = 84.87 = 85 stages
  85 stages × 0.95 hp/stage = 80.75 hp
  80.75 hp fresh water × 1.1 SG = 88.82 hp

DISCUSSION OF PROBLEMS #1-C(1) and #1-C(2)

Note that each of the previous problems was for the same lift and same wellhead pressure. Due to having 7 in. casing, larger tubing was used to cut friction loss. With 7 in. casing we were able to use a larger diameter, more efficient, and more economical pump. The combination of lower tubing friction loss and more efficient pump allowed us to require only 88.82 hp motor load in 7 in. casing as opposed to 94.2 hp required in 5½ in. O.D. casing.

This emphasizes that initial completions are very important because casing sizes have a decided influence on the erfficiency of electrical pumping.

4.537   Class problems for water wells

CLASS PROBLEM #1-A

Given:

5½ in. casing × 2⅜ in. tubing
Perf 6000 ft
P_R = 2200 psi
Producing 100% water (γ_w = 1.07)
Temp bottom 160°F.
PI = 1 b/d/psi drawdown
P_{wh} = 120 psi
Desired flow rate = 1500 b/d
Select a complete pumping system.

CLASS PROBLEM #1-B

Given:

10¾ in. casing
4 in. tubing
Depth = 2,550 ft
P_R = 1200 psi
PI = 25 b/d/psi
Producing 100% water for flood purposes (γ_w = 1.10)
Flowing wellhead pressure = 140 psig
Desired flow rate = 20,000 b/d
Bottom hole temp = 150°F.
Surface flowing temp = 130°F.
Select a complete pumping system.

4.538   Example problem #2A, oil well pumping no free gas

This example problem is a low gas-oil ratio oil well making 15% water. It is assumed that no free gas comes through the pump for this example. The following simplified procedure is utilized. (Refer to Example #3 for procedures to handle gas.)

(1) Obtain and analyze data

Given:

Company, well name, number, field and location

Well data:

Casing = 5½ in. O.D., 17# to 6150 ft
Tubing = 2⅜ in. EUE 8RD
Perforations = 5900 to 5970; 6000 to 6030

Production data:

Static bottom hole pressure	2000 PSI @ 5950′ Datum
Flowing bottom hole pressure	1500 PSI
Present producing volume	475 bfpd (400 BO/D)
GOR	350 cubic feet per barrel
Bottom hole temperature	170°F.
Type reservoir	Solution gas drive
Bubble point pressure	2000 PSI

Fluid data:

A.P.I. gravity of oil	30°
Specific gravity of oil	0.876
Specific gravity of water	1.02
Specific gravity of gas	0.75

Power system:
Primary voltage—7,200/12,470

Other:

The well produces corrosive fluids.

It is desired to produce this well at the maximum rate possible, but maintaining a 300 psi intake pressure at the pump. This problem differs from the water well problem in that the intake volume at the pump is greater due to the formation volume factor for the oil.

Pump setting depth = 5,850 ft
Pump intake pressure = 300 PSI

It if further assumed that all gas is to be vented.

Referring to the procedures for pump selection of step one, note that the directions are to analyze the data received to ensure a properly sized unit. Start by checking well data, production data, fluid data and the power system. With the data provided, there will be no problems supplying a pump for the given information. The casing is large enough to accommodate conventional 400 series equipment and good production data has been provided.

(2) Determine Well's Productive Capacity

The decision has been made to try and create a 300 PSI PIP. A pump setting depth of 5,850 ft (50 ft above the perforations) has been selected. This is very reasonable as the 300 PSI pump intake pressure will probably be required due to the amount of gas to be vented. This is a practical design as, from experience with this type application, it has been found that approximately 300 PSI pump intake pressure for this type of well is required for good pumping conditions. The fifty-ft setting above the perforations is fine. This places the unit above the fluid entry point and fluid will pass the motor to carry away the heat.

The static bottom-hole pressure was taken at 5,950 ft. The pump will be set at 5,850 ft or 100 ft above the datum point. This will reduce the available drawdown pressure slightly. To find this decrease, we need to find the average specific gravity of the fluid below the pump.

Note that the well produces 475 b of total fluid with 400 b being oil. This is a 15% water cut and 85% oil cut. The fluid below the pump will be of the same oil-water ratio. The specific gravity of the gas-free oil is 0.876 and specific gravity of the water is 1.02. Average specific gravity is:

0.876 × 85% = 0.74
1.02 × 15% = 0.15
        Total = 0.89 ave.

This 100 ft will represent approximately

$$PSI = \frac{100 \text{ ft} \times 0.89}{2.31} \cong 40 \text{ PSI}$$

This neglects any free gas in the column. Therefore, the static bottom-hole pressure at the pump can be estimated to be $2000 - 40 = 1960$ PSI.

The production capacity can now be calculated. Since this is a solution gas drive reservoir, the general Vogel I.P.R. curve can be used to check the volume available for pumping.

The following illustrates the generalized I.P.R. curve of Vogel and how the calculated volume available for this application is found. (Fig. 4.55).[2]

Well test:

$$q_L = 475 \text{ BF/D} \qquad 475/q_m = 0.40$$
$$P_{wf} = 1500 \text{ psi} \qquad q_m = 1188 \text{ BF/D}$$
$$\overline{P}_R = 2000 \text{ psi} \qquad P_{wf} = 300 + 40 = 340 \text{ psi}$$
$$P_{wf}/\overline{P}_R = \frac{1500}{2000} = 0.75 \qquad P_{wf}/\overline{P}_R = 0.17$$
$$q/q_m = 0.94$$
$$q/q_{max} = 0.40$$
$$q = 0.94 \ (1188) = 1117 \text{ BF/D} = \text{flow rate for}$$
$$P_{wf} = 340 \text{ psi at well datum depth giving 300}$$
psi at pump intake.

Note from the I.P.R. calculation that the maximum rate possible (q_m) is approximately 1188 b/d and at 300 PSI operating PIP, the volume available (q) will be 1117 b/d or a reasonable rate to size for would be 1,125 bfpd of stock tank barrels.

As a comparison of the I.P.R. method vs. straight-line drawdown on a solution gas driven reservoir, calculate the well's capacity using the straight-line method.

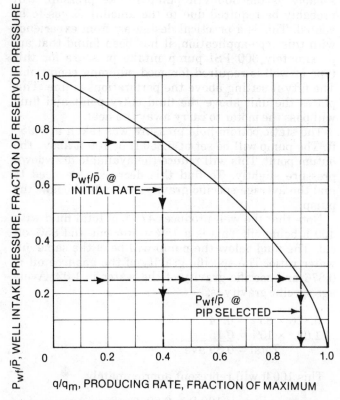

Fig. 4.55 Inflow performance relationship for wells in solution gas drive reservoirs

Calculation:

$$PI = \frac{b/d}{SIBHP - FBHP}$$
$$= \frac{475}{2000 - 1500} = 0.95 \text{ B/PSI}$$

Available drawdown to 300 PSI, PIP is
$$2000 - 40 - 300 = 1,660 \text{ PSI}$$
$$1,660 \text{ PSI} \times 0.95 \text{ B/PSI} = 1577 \text{ b}$$

If the straight line method had been used, the capacity of the well would be highly optimistic and possibly result in an oversized unit. The well's productivity should be checked after the unit has operated for some period of time. Then, if the well's capacity is greater than anticipated, a higher volume unit can be sized and installed.

Note that the volume to size for is 1,125 BF/D of stock tank barrels. The volume of reservoir barrels that must be pumped to provide these stock tank barrels must now be determined. Sufficient data is available to determine a formation volume factor of 1.075 at $P_{wf} = 300$ psi (standings correlation).

The reservoir liquid intake rate will be

$$1125 \text{ b/d} \times 85\% \text{ oil} \times 1.075 = 1028 \text{ BO/D}$$
$$1125 \text{ b/d} - 15\% \text{ water} = \underline{199} \text{ BW/D}$$
$$\text{TOTAL} = 1227 \text{ BF/D}$$

This assumes that no free gas is going through the pump. (All vented out the casing.)

A later example will show how to include gas pumped in determining total intake volume.

(3) Determine total dynamic head

The total dynamic head is composed of (1) the vertical lift, (2) friction loss in the tubing, and (3) surface wellhead pressure. Other examples show how to include the horizontal flow line and differences in elevation.

(A) Vertical lift

The vertical lift is determined by calculating where the operating fluid level will be with the 300 PSI pump intake pressure. The 300 PSI must be converted to feet and subtracted from the pump setting depth.

Assume this 300 PSI will be an oil gradient. Gas will be neglected. Therefore, the 300 PSI represents:

$$\text{Head feet} = \frac{300 \text{ PSI} \times 2.31 \text{ ft/PSI}}{0.876 \text{ specific gravity}} = 790 \text{ ft}$$

Vertical lift is equal to:
$$H_d = 5850 \text{ ft (PSD)} - 790 \text{ ft (FOP)} = 5060 \text{ ft}$$

where:

PSD = pump setting depth
FOP = fluid operating level in feet (fluid over pump-submergence-FOP)

(B) Friction loss
Given:

5850 ft of 2⅜ in.
1227 BFPD rate

Friction loss from the friction loss chart for this rate in 2⅜ in. tubing is 40 ft/1000 ft. (Refer to Appendix 4C)

$$F_t = 40 \text{ ft} \times 5.85 = 234 \text{ ft}$$

Use 250 ft for friction losses so that it includes check and bleeder losses.

Note: The friction loss through the check and bleeder are small compared to the total head and can be neglected, however a figure of 5-10 ft each may be used.

(C) Surface wellhead pressure

Given:

200 PSI system (tubing) pressure = P_{wh} = wellhead pressure (use a 0.890 average specific gravity)

Converting to feet of head

$$\text{Feet of head} = \frac{200\ \text{PSI} \times 2.31\ \text{ft/PSI}}{0.89\ \text{Specific Gravity}} = 520\ \text{ft}$$

Total Dynamic Head

$$\text{TDH} = 5060\ \text{ft}\ (H_d) + 250\ \text{ft}\ (F_t) + 520\ \text{ft}\ (P_{wh}) = 5830\ \text{ft}$$

(4) Select the type and size of pump

Since this unit is to be installed in 5½ in. O.D. casing, a 400 series (4 in. O.D.) pump should be selected. At the 1227 BF/D rate the D-40 is the most efficient of the TRW Reda 400 series pumps, and the M-34 or G-48 pumps are the most efficient of the centrilift pumps (Appendix 4B).

(5) Determine the number of stages necessary

At 1227 B/D, the head per stage on the D-40 Reda pump is approximately 23 ft. Therefore, the required staging is:

$$\text{No. of Stages} = \frac{\text{TDH}}{\text{ft/stage}} = \frac{5830}{23} = 254\ \text{stages}$$

(refer to Appendix 4B)

A D-40 pump with 254 stages is available in a 150 housing. (Refer to various catalogs.)

(6) Determine motor hp

hp = no. of stages × (hp/stage) × specific gravity

hp = 254 stages × 0.35 hp/stage × 0.890 (using max hp)

hp = 79

Note: This horsepower requirement is for the well under operating conditions.

A 456 series Reda motor (4.56 in. O.D.) can be used due to casing I.D. limitations. From Appendix 4F a 90 hp Reda motor is available in the 456 series. This would be a good selection for this application.

Note: If the well is killed with brine or heavier fluid, the 90 hp will be approximately 10% overloaded while the kill fluid is pumped out. This will have to be taken into account on the well's initial start-up.

(7) Select the cable

The size cable and motor voltage and amperage for most economical operations must be determined. At this point there is an economical choice of what equipment to select and still provide equipment that will do the job. We do not want a marginal design where the equipment is operating at its limit.

To select the cable, note 170°F. well temperature and 5950 ft of cable required. This allows 100 ft more than pump setting depth to be used for surface cable.

5850 + 100 = 5950 ft of cable

Refer to Appendix 4F where there are four 90 hp

motors of different voltages and amperage to choose from.

For this installation and for a Reda selection the 1,260 v, 45 amp motor and No. 4 copper cable should be selected. The 45 amps fits the range of No. 4 cable's (largest size that can be used in 5½ in. casing) current carrying capabilities. Redalene cable is the best cable selection for 170°F. operations. Also, a 1,500 v switchboard can be used.

If the 57-amp motor had been selected, then the current carrying capacity of the No. 4 cable would be approaching its limits. If the 1,500 v motor were selected, then a 2,400 v switchboard would be required at a higher cost. The 2,000 v motor could have been selected and No. 6 cable used with the 29 amp motor, but a 2,400 v switchboard would have to be used and the No. 6 cable would possibly need to be changed out at a later date.

(8) Determine voltage loss in cable and surface voltage

Given:

5950 ft of #4 copper cable
45 amp 1,260 v motor

From the voltage loss chart of Appendix 4A it is found that the voltage loss for 45 amps with #4 cable at 170°F. is 24 v per 1000 ft of cable. The required surface voltage is:

$$V_s = [(24\ v \times 5.95) + 1260] \times 102.5\% = 1438\ v$$

where the 2.5% allows for transformer loss.

Surface voltage of 1425 to 1450 v will be proper voltage for this application.

(9) Calculate required kva

Given:

1450 v surface voltage
45 amps (normal operation)

where:

$$\text{kva} = \frac{1450\ v \times 45\ \text{amps} \times 1.73}{1,000} = 113$$

Three single-phase transformers will be used (standard domestic operation). The value required per single transformer is:

$$\frac{113\ \text{required kva}}{3\ \text{single-phase transformers}} = 37.67\ \text{each}$$

Note: 37.5 kva transformers could be used because the amperage, once the kill fluid has been unloaded, will be less than motor nameplate amps (45) (Appendix 4E). The calculated horsepower requirement for the installation is 86 hp and the initial horsepower requirement did not include gas in the specific gravity of the fluid. However, due to the slight difference in cost between 37.5 kva and 50 kva transformers and the future flexibility with the 50 kva transformers, recommendations would be to use three 50 kva, single-phase transformers.

(10) Select proper accessories

The tubing is 2⅜ in. EUE 8RD so no swage to the pump will be required as the pump head is 2⅜ in. EUE 8RD thread. The check valve and bleeder valve must also be ordered with 2⅜ in. EUE 8RD threads. The proper tubing head must be selected for casing pressures anticipated and for the location of the well.

(11) Determine what steps are required to ensure good operations.

This well is corrosive, so precautions should be taken to combat this corrosive environment, such steps as: (1) Use plastic coated equipment, (2) Use stainless steel or monel bands for strapping the cable to the tubing, (3) Use a flat cable with corrosive resistant pothead, (4) Use corrosive resistant bolts and vent plugs throughout equipment, if necessary. Other pump selections are possible by referring to Appendix 4B.

4.539   Example problem #2B, oil well producing no gas

In this example the significant thing is that no gas is produced or at least the amount of gas is so small that it can be ignored. The procedure of Section 4.5327 will be followed.

(1) Collect and analyze the data.

General: List the owner of the well, its identification by field name and well number and its geographical location.

Well data:

Casing—7 in. O.D., 23 #/ft
Tubing—2⅞ in., EUE, 8 round thread
Depth to bottom of casing—11,000 ft
Perforated interval—10,600 to 10,650 ft

Reservoir data:

Static pressure—2900 psig at 10,000 ft
Flowing pressure—2540 psig @ 10,000 ft
        for 1000 STB/D (Total
                  liquid
Water cut—30%
Oil API gravity—40°
Water specific gravity—1.05
Dead oil viscosity—3.6 cp @ 100°F.
                 1.6 cp @ 200°F.
Well temperature—225°F @ 10,000 ft
Temperature at well head—160°F.

Pumping Data:

Pressure required at well head—200 psig
Power supply—60 Hz volts
Desired production rate—1600 STB/D (Total
                      liquid)
Desired pump setting depth as deep as necessary.

Special Problems: No sand, scale, corrosion or paraffin problems.

A quick analysis of the above data shows that 7 in. casing will accommodate a pump having a capacity equal to 1600 STB/D. No other conditions appear to prohibit the selection of pumping equipment for this well.

(2) Determine the well and reservoir's inflow performance at the desired production rate

The ability of the well and reservoir to deliver the desired production rate to the pump will determine the pump's intake pressure. In the reservoir data the static pressure is given as 2,900 psig at 10,000 ft and a flowing pressure of 2,540 psig at the same depth is given for a rate of 1000 STB/day.

$$J = \frac{q}{P - P_{wf}} \text{ (assume constant)}$$

where:

J = productivity index
q = production rate (STB/D)
P = static pressure (psig)
P_{wf} = flowing pressure (psig)

In this problem $J = \dfrac{1000 \text{ STB/D}}{2900 \text{ psig} - 2540 \text{ psig}}$
$\qquad\qquad = 2.78.$

The flowing pressure at the desired flow rate of 1600 STB/D will then be

$$2900 \text{ psig} = \frac{1600 \text{ STB/D}}{2.78} = 2325 \text{ psig at a}$$
$$\text{depth of 10,000 ft}$$

(3) Determine total dynamic head (TDH)

In Section 4.5324 Total Dynamic Head was defined as the sum of (1) the friction losses in the tubing and surface flow line, (2) the difference in elevation between the final destination of the produced flow and the pump, and (3) any significant losses in the discharge line due to values, separator, etc. minus (4) the head that exists on the suction side of the pump's intake due to the column of liquid over the intake.

Much of the well data used to calculate total dynamic head is in units of psig, while the units of friction loss and pump performance are feet of head. Therefore it is necessary to convert the well data to feet of head according to the specific gravity of the fluid. Since the specific gravities given are for a 60°F. temperature they should be adjusted for the higher temperature of the well. In this case it will be close enough to correct to a temperature of 190°F. which is approximately the average of the bottom hole and well head temperatures.

Reference material shows that the specific gravity of water at 190°F. is 96.7% of what it is at 60°F. and for 40° API oil it is 93.4% as much. The average specific gravity is then calculated as follows:

Water = 1.05 × 0.967 × 0.3 = 0.30
Oil = 0.825 × 0.934 × 0.7 = 0.54
                          0.84 ave. @
                           190°F.

The conversion from psig to feet for this fluid is then $\dfrac{144 \text{ sq in.}}{62.4 \text{ #/°F.} \times 0.84} = 2.75$ which is the number of feet of head that is equivalent to one psi of pressure.

Flowing pressure = 2325 psi = (2.75)(2325)
                     = 6394 ft
Effective lift = 10000 − 6394 = 3606 ft

For safety set pump at 4000 ft from surface.

In this example friction losses and elevation change in the surface flow line and also all other significant losses in the discharge line are provided for in the 200 psig pressure required at the well head.

The total dynamic head then computes as follows:

Tubing friction loss = 4 ×
20.5 ft/1000 = 102 ft
Elevation difference between pump and well head = 4000 ft
Required head at well head = 200 psig × 2.75 = 550 ft
Suction pressure at pump = − 394 ft (minus)
TDH = 4258 ft

(4) Select pump type

The 7 in. casing allows a 513 Centrilift series pump to be selected. This is the largest pump diameter that will go in a 7 in. casing and, therefore, the best selection for size. Within the 513 series the 1600 STB/day producing rate fits the optimum range of the I-42B (Centrilift) stage type and it is therefore selected. (Refer to Appendix 4B-II).

(5) Determine number of pump stages

At 1600 B/D the I-42B pump produces 38.3 ft per stage (Appendix 4B-II). The total number of stages required is then $\frac{4258}{38.3}$ TDH = 111.2 (use 112) stages.

(6) Determine motor horsepower

At 1600 B/D the I-42B pump requires 0.69 horsepower per stage (Appendix 4B-II) when pumping fluid of 1.0 specific gravity. The motor horsepower required at operating conditions is then 112 stages × 0.69 HP/stg × 0.84 S.G. = 65 Horsepower. A 75 hp, 1350 v, 35 amp, 544 series motor is selected because it is the next larger sized motor than the 65 horsepower required at operating conditions and the 544 series motor is designed for use with the 513 series pump and will go in the 7 in. casing. The extra horsepower is enough to allow for increased horsepower that may be required at start-up.

(7) Select cable

The cable size is selected to keep the voltage drop in the cable to an optimum amount when considered with the other electrical equipment, and it must also be able to fit in the well casing.

For the combination of 1350 v, 35 amp motor and pump setting depth of 4000 ft, a No. 4 copper cable is selected (Appendix 4A). The voltage drop is 18 ft per 1000 ft of cable and is 1.3% of the nameplate voltage. The length of cable needed will be the pump setting depth (4000 ft) plus 100 ft allowance to make switchboard connections for a total of 4100 ft of cable. The type of cable should be of a type which is good for the 225°F. well temperature.

More than one cable size can often be selected and the size selected will have an effect on the switchboard and power costs. Therefore, economic comparisons should generally be made using current cable, switchboard and power costs.

(8) Determine required surface voltage

The 4,000 ft of No. 4 copper cable will have a total voltage drop of 18 ft/1000 ft × 4 = 72 v. Therefore, the required surface voltage will be 1,350 at the motor plus 72 v drop in the cable for a total of 1422 v at the surface.

(9) Determine size of transformer

Transformer size is designated by kva and is determined by multiplying the surface voltage by the amps by 1.73 and dividing by 1,000. In this case kva = $\frac{1.73 \times 1422 \text{ v} \times 35 \text{ amps}}{1,000}$ = 86 kva total. For a bank of three single-phase transformers the kva for each will then be 29 kva each and the next commercial size of 37.5 kva transformer would be selected.

(10) Select proper accessories

The 2⅜ EUE 8RD tubing will connect directly to the pump head, so no adapting connector is necessary. The check and bleeder valve must have the same threads as the tubing and the tubing head must be suitable for the tubing, casing, cable size, pressure and vertical load.

4.5391   Class problems for oil wells pumping no gas

CLASS PROBLEM #2-A

Given:

Casing 5½ in. (17#/ft)
Tubing 2⅜ in.
Well depth = 4234 ft
Performations = 4148 to 4158 ft
$\overline{P}_R$ = 900 psig
For P_{wf} = 550 psi, q_L = 625 b/d (30 oil)
Gas-liquid ratio = 15 scf/b
Bottom hole temp = 98°F.
Well head flowing pressure = 50 psig
Desired rate = 800 b/d total
°API of oil − 36°
Water specific gravity = 1.02

Select a complete pumping system.

CLASS PROBLEM 2-B

Given:

3 in. tubing
7 in. casing
Depth = 10,000 ft
Well produces 25% oil, 75% water
$\overline{P}_R$ = 2800 psi
PI = 10 b/d/psi
°API of oil = 36
γ_w = 1.07
Gas-oil ratio = 100 scf/B (assume free gas is vented)
Separator pressure = 100 psi
Flow line = 3000 ft of 4 in. with an increase in elevation of 200 ft

Design a complete pumping system neglecting effect of free gas. (Gas-liquid ratio is very low at 25 scf/B).

4.5310 Design example problems #3, wells with medium to high gas-oil ratios

4.53101 Introduction

Wells that pump gas offer a more difficult design problem. Refer to Sections 4.523 and 4.524 for detailed discussions on this topic. In this section a procedure and example problem will be given for handling gas. For our purposes assume that all or a certain percentage of the gas is to be pumped.

In general, the solution is firstly to determine the intake pressure (flowing bottom hole pressure if pump is set on bottom) and secondly to determine the pump discharge pressure from an appropriate multiphase flow correlation. This determines the differential between the pump intake and pump discharge which represents the "ΔP" that the pump must develop. By starting with the intake pressure we assume pressure increments and determine he average pressure gradient in increments from which the average head developed can be determined. From these head increments the stages required between pressure increments can be determined. The total number of stages can be determined by summing up the incremental stages. Finally, hp requirements can be determined and the design finalized.

Section 4.53102 represents a procedure assuming a homogeneous mixture in the pump and further assuming that the pump can develop a head according to a no-slip density calculation. The temperature across the pump can be assumed to be constant due to the pump's short length. The various fluid properties needed, such as B_o, R_s, etc., should be obtained from PVT analysis of the particular crude. If this is not available we can use general correlations such as those of Standing which can be found in Volume I of this series.

4.53102 Design procedure for submersible pumps for wells producing gas through pump

(1) Determine pump intake pressure
(2) Determine pump discharge pressure—from multiphase flow correlation
(3) Determine "ΔP" between intake and discharge pressures
(4) Starting with intake pressure, select pressure increments so that $(P_{intake} = P_{wf}) + \Sigma$ "$\Delta P's$" = $P_{discharge}$
(5) Determine density at each pressure chosen by:
  (A) Find volume of oil, gas and water at each pressure
  (B) Find mass of oil, gas, and water at each pressure
  (C) ρ_{mix} = mass/volume
(6) Find pressure gradient at each pressure point
(7) Find average pressure gradient between pressure points
(8) Convert average pressure gradients to feet of head between pressure increments
(9) Find volume flow rate in b/d at each pressure point
(10) Determine average volume flow rates between pressure points
(11) Select pump for each average volume flow rate and obtain feet of head developed for each case

(12) Determine pressure developed per stage by multiplying (avg. gradient) × (ft/stage)
(13) Determine number of stages needed per pressure increment by dividing press. incre. by PSI/Stage
(14) Determine total number of stages by Σ (Step 13)
(15) Determine HP/Stage from performance curves
(16) Determine total HP by Σ (Step 15)

A combination pump will likely occur, that is, 120 stages of X-50 and 60 stages of Z-60.

Note: This will be a pump for operating conditions and more stages may be required for unloading.

4.53103 Detailed procedure for determining the number of stages needed between the 500 and 700 PSI increments of example #3A

To illustrate these calculations, complete detailed calculations are given for one increment, that is, between the intakes of 500 psi and 700 psi. The following step-wise procedure is given.

EXAMPLE PROBLEM #3-A; PUMPING GAS

Given:

Depth = 7000 ft
Tubing = 2⅜ in. O.D.
5½ in. casing
$\overline{P}_R$ = 1000 PSI
P_{wf} = 500 PSI
P_{wh} = 200 PSI
G/O = 500 scf/b   (GLR = 500/2 = 250 scf/B)
γ_g = 0.65
q_L = 500 b/d
$q_o + q_w$ = 500 b/d (50% water)
γ_w = 1.07
γ_o = 35° API
Temp. surf. = 120°F.
Temp. bottom = 160°F.
Pumping 100% of gas (assumed)

Find number of stages needed between the pump intake pressure of 500 psi and 700 psi. We will show the procedure for this 200 psi increment of pressure.

(1) Find volume of oil, gas and water associated with one stock tank BBL of oil at 500 PSI and 700 PSI. (Assume 200 psi increment.)

$Vol_{500 \text{ psi}}$ = Vol oil + vol gas + vol water   (4.52)
Basis = 1 stock tank bbl
Vol oil = (1) B_o = (1)(1.08) = 1.08
Vol oil (ft³) = (1.08)(5.61) = 6.059
Vol gas = 500 scf/stock tank bbl

$$B_g = \left(\frac{1}{5.615}\right)\left(\frac{14.7}{514.7}\right)$$

$$\left(\frac{620}{520}\right)\left(\frac{.97}{1}\right) = .00577 \text{ B/scf}$$

or $B_g = .00504 \dfrac{TZ}{P}$ b/scf   (4.53)

$B_{g700 \text{ psi}}$ = 0.00404 bbl/scf
Vol water = (1)(B_w) = 1.0 B/B
(assume no change in volume for the water.)
$B_{o \, (500)}$ = 1.08 R_s = 80 SCF/B
$B_{o(700)}$ = 1.094 R_s = 120 SCF/B
Free gas$_{(500)}$ = 500 − 80 = 420 SCF/B
Free gas$_{(700)}$ = 500 − 120 = 380 SCF/B
Volume of oil, water, and free gas at 500 psig associated with 1 STB = Vol oil + vol water +

vol gas $= (1.08) + (1.0) + 420(.00577) = 4.5034$

Total volume produced $= (250)(4.5034) =$ 1125.85 b

For 700 psig we have $(1.094) + (1.0) + 380(.00404) = 3.6292$ b/STB

Total volume$_{700} = (3.6292)(230) = 907.3$ b

(2) Find total mass flow rate in lb$_{mass}$/d

Find mass per STB

(A) Mass oil $\gamma_o(350) = (0.8498)(350) = 297.3$ lbs

$\hspace{10em}(4.54)$

(B) Mass water $\gamma_w(350) = (1.07)(350) = 374.5$ lbs

(C) Mass gas γ_g (GOR) $(.0764) = (0.65)(0.764)(500) = 24.83$ lbs

Mass per STB of oil, water and gas $= 297.3 + 374.5 + 24.83 = 696.63$ lbs

Total mass flow rate $= (696.63)(250) = 174,157.5$ lb$_m$/day $=$ constant value for no slippage at each pressure point

(3) Find pressure gradient in psi/ft at 500 psi and 700 psi

(A) Find $\rho_{mix} = \dfrac{mass}{vol}$

$\rho_{500} = \dfrac{696.63}{(4.5034)(5.61)} = 27.55$ lb$_m$/ft^3

$Grad_{500} = \dfrac{\rho}{144} = \dfrac{27.55}{144} = 0.1913$ psi/ft

$\rho_{700} = \dfrac{696.63}{(3.6292)(5.61)} = 34.185$ lb$_m$/ft^3

$Grad_{700} = \dfrac{34.185}{144} = 0.2474$ psi/ft

(4) Find average gradient

Avg gradient $= \dfrac{0.1913 + 0.2374}{2} = 0.2143$

(5) Find average volume

Flow rate $= \dfrac{Vol_{(500)} + Vol_{(700)}}{2} = \dfrac{1125.85 + 907.3}{2}$

$\hspace{4em}= 1017$ b/d

(6) Select pump for this average flow rate. Select Reda D-40, Centrilift M-34 (see Appendix 4B) ODI R9, or Kobe Model II SOF.

(7) Convert to psi/stage

Reda $= 25$ ft/stage

Avg. gradient $= 0.2143$ psi/ft

$(25)(0.2143) = 5.36$ psi/stage

(8) Find stages per increment of pressure

$\dfrac{200 \text{ psi}}{5.36 \text{ psi/stage}} = 37.3$ stages (Use 38) (Reda D-40)

4.53104 Example problem #3 B, detailed procedure for determining total stages, hp, etc.

In order to determine the pump discharge pressure required in a well producing a large amount of gas we can use any one of several multiphase flow correlations. Several methods are listed below:

(a) Hagedorn & Brown

(b) Orkiszewski

(c) Dun & Ros

(d) Beggs and Brill

(e) Poettmann & Carpenter

The following data is available for the design of this installation and is the same as the previous example except we are pumping 100% oil. (Refer to Figs. 4.56 A and B)

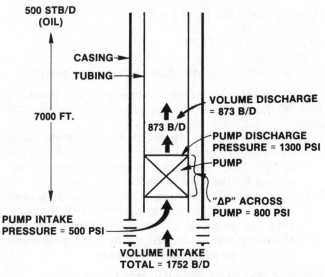

Fig. 4.56A Illustration of Example Problem 3-B for handling gas

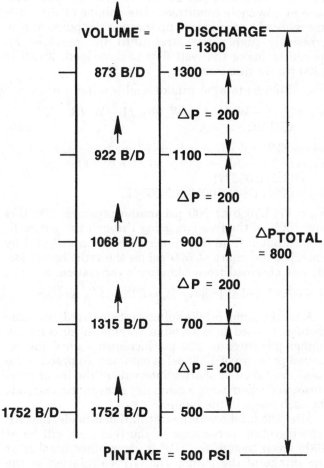

Fig. 4.56B Illustration of Increments of Pressure and Flow Rates for Example Problem 3-B

Given:

Well depth $= 7000$ ft

Tubing size $= 2\frac{3}{8}$ in. OD

$\overline{P}_R = 1000$ psi

$P_{wf} = 500$ psi

$P_{wh} = 200$ psi

G/O $= 500$ scf/bbl

$\gamma_g = 0.65$

$q = 500$ b/d (all oil)

Casing size = 5½ in. (17 #/ft)
BHT = 160°F.
Surface Temp. = 120°F.
°API = 35
GIP = 100% (percent of free gas entering pump)

In the first calculation assume you are pumping 100% of the gas. Further assume no slippage in the pump.
(1) Determine pump discharge requirements using method of Hagedorn & Brown. Refer to Appendix C Vol. 2-A. The pump discharge pressure required for the well to flow is 1300 psig.
(2) Inasmuch as the volume of the 500 b of stock tank oil and its related gas is variable with temperature and pressure, a calculation of the volume at the pump intake and the volume as it progresses through the pump must be made.

Method of calculating volume in pump

The total volume going through the pump consists of the oil, water, and gas at the temperature and pressure at downhole conditions. The volume of the three are reduced as they move up through the pump at progressively higher pressures until the pressure required to make the well flow is developed, which is 1300 psi for this example.

Volume total at intake = oil + water + gas

$$V_{in} = q_o (1 - W_c)B_o + q_o(W_c) + q_o(1 - W_c)(R - R_s)(GIP)B_g \quad (4.55)$$

where GIP = fraction volume of gas being pumped.
$$V_{in} = 500(1 - 0)B_o + q(0) + 500(1 - 0)(500 - 80)(1.0)(0.00577)$$
$$V_{in} = 500(1.08) + 500(420)0.00577$$

V_{in} = 1751.70 b at 500 psi intake pressure, 500 B/O and 100% of the free gas going through the pump. R_s was obtained from Standing's correlation in Vol. I by entering the chart at 500 psi on the right hand side. B_o was obtained from Standing's correlation, also. B_g was calculated as follows: $B_g = 0.00504 \dfrac{ZT}{P}$ bbl/scf.

After the pump intake volume is calculated, the same method is used to determine volume at successively higher pressures in 200 psi increments until the required pressure of 1300 psi is obtained. In practice the steps are taken at shorter intervals at the lower pressures, and when using a computer a calculation is made for each stage in the pump.

After the intake volume calculation is made, the GIP (gas ingestion percentage of the free gas) will be at 100% even though the GIP is some other number at the intake of the pump. The GIP calculation at the intake pressure determines the amount of total gas that will go through the pump and up the annulus. As pointed out previously, the volume of gas pumped and the volume vented is not easily determined.

The specific gravity of the liquid at the various pressures can be easily calculated by considering that a constant mass is moving through the pump.

$$\rho_g = \frac{28.97 \times \gamma_g \times P}{Z R T} \quad (4.56)$$

$$\rho_g = \frac{28.97 \times .65 \times 14.7}{1 \times 10.73 \times 520} = 0.0496 \text{ lb/cu ft at standard conditions}$$

SG 35° API oil = 0.8489
Wt. gas = .0496 lb/cu ft × 500 G/O = 24.8 lbs
Wt. oil/bbl = .8489 × 350 lb/bbl = 297 lbs
Wt. one bbl oil + related gas = 321.8 lbs

The weight of 500 b/d of stock tank oil and its related gas would be 160,900 lbs/d. The intake volume of 1751.70 b/d × 350 lb/bbl would give a weight of 613,095 lbs if it were 1.0 SG. 160,900 ÷ 613,095 = .262 S.G. and a gradient of 0.113. The S.G. and gradient will be required in the pump calculations. Additional calculations for pressures of 700, 900, 1100 and 1300 psi will be made as follows:
(a) $V_{700} = 500(1 - W_c)B_o + q(W_c) + q(1 - W_c)(R - R_s)(GIP)B_g$
$= 500(1)1.094 + 500(0) + 500(1)(500 - 120)(1).00404$
$= 1314.60$ b/d
$SG_{700} = 160,900 ÷ (1314.6 × 350) = .3497$
Gradient at 700 = .151 psi/ft
(b) $V_{900} = 500(1)1.11 + 500(0) + 500(1)(500 - 170)(1).00311$
$= 1068$ b/d
$SG_{900} = 160,900 ÷ (1068 × 350) = .430$
Gradient at 900 = 0.186 psi/ft
(c) $V_{1100} = 500(1.16) + 500(500 - 225)0.00249 = 922$ b/d
$SG_{1000} = 160,900 ÷ (922 × 350) = 0.4986$
Gradient at 1100 psi = 0.218 psi/ft
(d) $V_{1300} = 500(1.163) + 500(500 - 240)0.00224 = 873$ B/D
$SG_{1300} = 0.527$
Gradient at 1300 psi = 0.228 psi/ft.

Pump selection:

The pump curve selected for the range of 1751.70 b/d intake volume and 873 b/d discharge volume is the D-40 pump (Appendix 4B-I). The method used will introduce some error due to the fact that 200 psi steps are used instead of small pressure increments.
V_{500} = 1751.70 SG_{500} = 0.262 Gr_{500} = 0.113
V_{700} = 1314.60 SG_{700} = 0.3497 Gr_{700} = 0.151
V_{avg} = 1533 B/D SG_{avg} = 0.305 Gr_{avg} = 0.132

From the D-40 curve, one stage will develop 18.4 ft head at 1533 b/d. The average gradient of 0.132 psi/ft × 18.4 ft/stage = 2.42 psi/stage; 200 psi ÷ 2.42 psi/stage = 83 stages.
The horsepower per stage when pumping 1.0 SG water is 0.35 hp/stage; therefore, the horsepower per stage pumping a mixture with 0.305 SG would be 0.35 × 0.305 or 0.107 hp/stage and 0.107 × 83 = 8.88 hp for 84 stages. The same calculations are made for each of the 200 psi pressure increments:

V_{avg} for V_{700} and V_{900} = 1191 b/d
SG_{avg} for V_{700} and V_{900} = 0.389
Gr_{avg} for V_{700} and V_{900} = 0.169 psi/ft
Head/Stage at 1191 b/d = 23.3 ft
23.3 ft × .169 = 3.93 psi/stage
200 psi ÷ 3.93 = 51 stages
0.34 hp/stage × .389 SG = 0.132 hp/stage
51 × .132 = 6.745 HP for 51 stages

For the first two calculations we have a total number of stages of 134 (83 + 51) and a horsepower requirement of 15.625 (8.88 + 6.745).

Similar calculations are made for the other two increments and we get 40 stages and 35 stages with horsepower requirements of 5.94 and 5.57 respectively. The total pump consists of 209 stages (83 + 51 + 40 + 35) and a horsepower requirement of 27 horsepower (8.88 + 6.745 + 5.94 + 5.57).

Final Summary (Figs. 4.56 A and B):

Pump size = 209 stages (D-40, Reda)
Horsepower = 27
Intake volume = 1751.7 b/d
Discharge volume = 897 b/d
Discharge pressure = 1300 psi
Intake pressure = 500 psi
Pressure developed = 800 psi
(Refer to Figs. 4.56A & B. Other selections can be made from Appendix 4B)

This pump is what is required at this set of conditions; however, it may take more stages to unload the well if it has been killed with a heavy fluid. Also the horsepower requirements while pumping the heavy fluid will be greater.

4.53105 Computer design for example problem #3 B, no slippage—no pump deterioration (computer solution by TRW-Reda)

This program is entitled Old Kermit and makes most of the same assumptions as we did in Example 3 of Seciton 4.53104 for long-hand calculations. One change is that the Orkiszewski correlation was used and much closer increments were taken, that is for every stage instead of 200 psi increments. The main assumption is that no slippage is involved and that the pump is capable of developing a head arrived at from the density of the mixture of gas and liquid.

For clarification the data is given again:

Depth = 7000 ft
q = 500 B/D oil
2⅜" O.D. Tubing
5½" casing
γ_g = 0.65
°API = 35
Pumping 100% of Gas

Surf. Temp. = 120°F.
P_{wf} = 500 psi
G/O = 500 scf/B
$\overline{P}_R$ = 1000 psi
P_{wh} = 200 psi
BHT = 160°F.

Example 3-B (no slippage) was run by computer; an explanation of the various tables and print-outs are given as follows:

(1) Table 4.53 shows a printout of values of Z, Free gas, B_o and solution gas versus pressure. These values were used in the calculations.

(2) Table 4.54 shows the results and the values of pressure, the corresponding total volume flow rate, the gradient of the mixture and the specific gravity of the mixture. Note that the discharge pressure is 1214 psi by the Orkiszewski correlation as compared to 1300 psi for the Hagedorn and Brown correlation. All present day used multiphase flow correlations will give different answers. In some cases the results will check very closely whereas in others they may differ quite a bit.

In addition it shows a summary of the pump design results and it is noted that 21 stages of D-55 (Reda) and 173 stages of D-40 (Reda) stages are needed. This gives a total of 194 stages required. Note that this compares to 209 stages of D-40 pump by long-hand calculations. Horsepower requirements are essentially the same being 27.84 by computer as compared to 27 by long hand.

(3) Table 4.55 shows the same calculations as Table 4.54 except the program was limited to a D-40 pump. Note that 206 stages were needed with 28.29 hp. This compares very closely with the long hand calculations showing 209 stages of D-40.

(4) Table 4.56 shows the results if only 50% of the gas is pumped and the other 50% being vented. The total volume flow rate through the pump is reduced starting with an intake volume of 1144.77 b/d as compared to 1750.6 b/d for pumping 100% gas.

Table 4.56 also shows the corresponding pressure, volume flow rates, gradients and specific gravities for pumping 50% gas.

Interestingly, a total of 211 stages is needed to handle 50% gas (less volume) as compared to 194 stages

TABLE 4.53
VARIABLES USED IN CALCULATIONS

	Pressure	Z	Barrels free	SCF Free	Pressure	Solution gas	Oil formation
1	15.0000	.9886	51.8742	498.8273	15.0000	1.1727	1.0468
2	100.0000	.9820	13.0656	488.4691	100.0000	11.5309	1.0507
3	200.0000	.9742	6.7116	473.4204	200.0000	26.5796	1.0566
4	300.0000	.9665	4.3819	456.6782	300.0000	43.3219	1.0631
5	400.0000	.9587	3.1689	438.7317	400.0000	61.2683	1.0702
6	500.0000	.9510	2.4235	419.8332	500.0000	80.1668	1.0778
7	600.0000	.9416	1.9150	400.1395	600.0000	99.8605	1.0857
8	700.0000	.9322	1.5476	379.7590	700.0000	120.2410	1.0941
9	800.0000	.9240	1.2713	358.7715	800.0000	141.2285	1.1027
10	845.0000	.9203	1.1677	349.1460	845.0000	150.6541	1.1068
11	1000.0000	.9080	.8812	315.2087	1000.0000	184.7913	1.1210
12	1500.0000	.8755	.3590	198.8107	1500.000	301.1893	1.1714
13	2000.0000	.8532	.0980	74.0406	2000.0000	425.9594	1.2277
14	3000.0000	.8157	.0000	.0000	3000.0000	694.2666	1.3548
15	4000.0000	.7782	.0000	.0000	4000.0000	981.8721	1.4985

TABLE 4.54
RESERVOIR DATA AND PUMP SELECTION RESULTS
(100% of Gas—No Slip Case)

Pressure	Flow	GR	Spec. grav.
500.00	1750.64	.1128	.2603
552.63	1618.93	.1219	.2813
555.58	1611.55	.1224	.2825
556.54	1604.13	.1230	.2838
598.82	1504.57	.1309	.3021
600.73	1499.07	.1314	.3032
649.82	1410.93	.1395	.3219
652.59	1405.96	.1400	.3231
709.29	1308.41	.1502	.3465
712.49	1304.13	.1507	.3477
777.13	1217.61	.1611	.3718
780.75	1212.76	.1618	.3733
853.90	1129.41	.1734	.4001
857.97	1125.83	.1739	.4014
939.16	1054.55	.1854	.4279
943.65	1050.60	.1861	.4294
1033.49	985.33	.1981	.4571
1038.41	983.00	.1986	.4582
1135.07	937.40	.2082	.4804
1140.33	934.92	.2087	.4817
1210.37	901.87	.2162	.4989

Pump selection process selected 21 D-55 series pumps.

Total dynamic head 3302.1528

Minimum pump discharge pressure 1213.874

Pressure drop across pump 713.8740

    Well name   1   60 cycle power

Fluid characteristics		Wellbore conditions	
GOR	500.00 SCF/BBL	Measured depth	7000.00 ft
Oil gravity	35.0 API	SIBHP	1000.00 psi
Gas gravity	.65	Intake pressure	500.00 psi
Water cut	.00 percent	Wellhead pressure	200.00 psi
Viscosity	8.00 CP at 100.00 F	Perforations	7000.00 ft
Viscosity	2.40 CP at 200.0 F	Bottom hole temperature	160.00
Mean temperature	140.65 F	Free gas at intake	100.0 percent
PI --------	BPD/PSI	Tubing size	1.99 inches
		Vertical pump depth	7000.00
		Wellhead temperature	120.00

Pump configuration			Pump performance	
			Intake volume	1750.64 BPD
			Discharge volume	901.87 BPD
			Discharge pressure	1213.87 psi
			Drawdown	.00 psi
Stages		Type	FBHP	1000.00 psi
21		D-55	Desired flow	500.00 BPD
173		D-40	Horsepower	27.84

for 100% gas. However 31.56 hp is required compared to 27.84 hp for 100% gas. The principal reason for this is that the discharge pressure increased to 1832 psi for 50% gas as compared to 1214 psi for 100% gas. By venting 50% of the gas we lose its benefit in lightening the flowing gradient in the tubing string and hence have a higher discharge pressure.

(5) Table 4.57 shows the effect of having to pump 50% gas. This in return reduces the gas liquid ratio to 250 scf/B. In turn, Table 4.57 assumes that we are pumping 50% water at intake conditions. Table 4.57 also shows the corresponding pressures, volume flow rates, gradients and specific gravities.

For 50% water and GIP = 50%, we need additional hp (41.25) and 231 pump stages.

4.53106 Computer calculations for example problem #3B (adjustments made for pump deterioration) (Solution by TRW-Reda)

In this calculation an adjustment has been made for the true performance of the pump. Therefore, we would expect more stages and more horsepower as compared to the no-slip calculations. These calculations were made on the computer using the Orkiszewski correlation and certain pump performance modification fac-

TABLE 4.55
RESERVOIR DATA AND PUMP SELECTION RESULTS
(100% Gas—No Slip Case)

Pressure	Flow	GR	Spec. grav.
500.00	1750.64	.1128	.2603
531.79	1671.08	.1181	.2726
533.61	1666.54	.1184	.2733
571.24	1572.35	.1254	.2894
573.40	1566.95	.1256	.2904
618.23	1467.64	.1342	.3097
620.76	1463.07	.1346	.3106
672.75	1369.77	.1436	.3314
675.68	1364.49	.1441	.3326
735.65	1273.12	.1543	.3560
729.01	1268.63	.1548	.3572
807.01	1179.25	.1662	.3836
810.62	1175.04	.1668	.3849
897.28	1100.10	.1779	.4106
891.50	1096.39	.1785	.4120
976.67	1022.14	.1911	.4410
980.75	1018.02	.1918	.4427
1072.65	966.38	.2020	.4661
1078.71	968.99	.2025	.4673
1178.00	917.15	.2127	.4908
1188.41	914.59	.2133	.4921
1210.73	901.70	.2162	.4990

Total dynamic head   3301.5504
Minimum pump discharge pressure   1213.874
Pressure drop across pump   713.8740
     Well name   1   60 cycle power

Fluid characteristics		Wellbore conditions	
GOR	500.00 SCF/BBL	Measured depth	7000.00 ft
Oil gravity	35.0 API	SIBHP	1000.00 psi
Gas gravity	.65	Intake pressure	500.00 psi
Water cut	.000 percent	Wellhead pressure	200.00 psi
Viscosity	8.00 CP at 100.00 F	Perforations	7000.00 feet
Viscosity	2.40 CP at 200.00 F	Bottom hole temperature	160.00
Mean temperature	140.65 F	Free gas at intake	100.0 percent
PI ------	BPD/PSI	Tubing size	1.99 inches
		Vertical pump depth	7000.00
		Wellhead temperature	120.00

Pump configuration		Pump performance	
		Intake volume	1750.64 BPD
		Discharge volume	901.70 BPD
		Discharge pressure	1213.87 psi
		Drawdown	.00 psi
		FBHP	1000.00 psi
Stages	Type	Desired flow	500.00 BPD
206	D-40	Horsepower	28.29

(Pump type held constant)

tors. The pump performance deteriorates to some extent as more and more gas is routed through the pump. As mentioned previously, the various companies hold this information confidential, but some deterioration probably starts when the in situ free gas to liquid volume exceeds 0.1 and shows complete deterioration at values of in situ free gas to liquid volume exceeds 3 to 1. Additional research is being conducted all the time and better means of gas separation are needed so that the gas can be vented. The venting of gas probably varies to a great extent depending upon the flow pattern. Refer to Sections 4.523 and 4.524 on gas venting and separation.

Another rule of thumb that sometimes assures good operations is if the intake volume is less than two times the stock tank volume then the calculations based on no slippage are reasonably accurate.

Solution to problem #3-B (pump deterioration included)

(6) Table 4.58 shows the calculations for pumping 100% gas and pump deterioration considered. Note that

TABLE 4.56
RESERVOIR DATA AND PUMP SELECTION RESULTS
(50% Gas—100% Oil) No Slip Case

Pressure	Flow	GR	Spec. grav.
500.60	1144.77	.1668	.3850
579.54	1086.07	.1855	.4280
564.09	1021.40	.1866	.4307
677.64	916.58	.2075	.4789
682.93	911.02	.2087	.4817
784.62	827.09	.2293	.5293
789.09	828.61	.2303	.5314
783.59	820.10	.2312	.5836
888.83	765.01	.2474	.5709
692.11	762.34	.2482	.5728
1000.20	707.65	.2667	.6154
1006.32	706.15	.2672	.6166
1167.30	676.66	.2786	.6430
1163.31	675.05	.2793	.6445
1264.61	643.14	.2926	.6752
1271.77	641.40	.2934	.6770
1413.88	606.73	.8091	.7133
1421.68	604.83	.3100	.7154
1575.99	583.82	.3208	.7403
1584.24	583.82	.3208	.7404
1741.06	583.32	.3214	.7416
1749.82	563.32	.3214	.7417
1831.97	583.32	.3217	.7424

Total dynamic head   4141.5068
Minimum pump discharge pressure   1832.258
Pressure drop across pump   1332.2578
    Well name   1   60 cycle power

Fluid characteristics		Wellbore conditions	
GOR	500.0 SCF/BBL	Measured depth	7000.00 ft
Oil gravity	35.0 API	SIBHP	1000.00 psi
Gas gravity	.65	Intake pressure	500.00 psi
Water cut	.00 percent	Wellhead pressure	200.00 psi
Viscosity	6.00 CP at 100.00 F	Perforations	7000.00 feet
Viscosity	2.40 CP at 200.00 F	Bottom hole temperature	160.00
Mean temperature	140.53 °F	Free gas at intake	50.0 percent
		Tubing size	1.99 inches
		Vertical pump depth	7000.00
		Wellhead temperature	120.00

Pump configuration			Pump performance	
			Intake volume	1144.77 BPD
			Discharge volume	583.32 BPD
			Discharge pressure	1832.26 psi
			Drawdown	.00 psi
Stages		Type	FBHP	1000.00 psi
58		D-40	Desired flow	500.00 BPD
153		D-20	Horsepower	31.56

263 stages are required as compared to 194 stages for a no-slip calculation.

(7) Table 4.59 shows the calculations for pumping 50% of the gas and pump deterioration considered. Note that 213 stages are required compared to 211 for no pump deterioration. HP's are practically the same. This shows that for low gas volumes the results are practically the same.

(8) Table 4.510 shows the calculations for the well producing 50% water and pumping 50% of the gas. Note that 231 stages requiring 41.25 hp are needed. This checks exactly with the no-slip calculation indicating that no pump deterioration occurs for the well making 50% water and venting 50% of the free gas at intake.

(9) Table 4.511 A&B summarizes all the results of these two calculations. The greatest difference occurs when we are pumping all the gas and the well is producing 100% oil. The more water we produce, the lower the gas to liquid ratio and hence the more efficient the pump. If the pump is assumed to be operating based on the no-slip density it requires 194 mixed stages as compared to 263 mixed stages if we consider pump deterioration. Horsepower requirements are essentially the same. Table 4.512 shows the difference in stages

TABLE 4.57
RESERVOIR DATA AND PUMP SELECTION RESULTS
(50% Gas—50% Water) No Slip Case

Pressure	Flow	GR	Spec. grav.
500.00	822.38	.2610	.6024
607.52	745.08	.2878	.6641
613.83	741.77	.2891	.6670
744.95	679.02	.3153	.7277
752.42	676.11	.3167	.7308
904.27	626.09	.3403	.7854
912.72	625.95	.3415	.7880
1051.11	593.95	.3594	.8294
1090.28	592.84	.3601	.8310
1269.23	571.01	.3735	.8618
1278.91	569.83	.3742	.8635
1468.08	546.36	.3894	.8987
1478.33	544.85	.3904	.9010
1675.23	541.66	.3929	.9068
1685.61	541.66	.3929	.9068
1882.83	541.66	.3933	.9077
1893.21	541.66	.3933	.9077
2090.63	541.66	.3937	.9086
2101.03	541.66	.3937	.9086
2298.65	541.66	.3941	.9095
2309.05	541.66	.3942	.9096
2506.88	541.66	.3945	.9105
2517.30	541.66	.3946	.9105
2621.50	541.66	.3948	.9110

Total dynamic head   5391.1217
Minimum pump discharge pressure   2628.298
Pressure drop across pump   2128.2975
    Well name   1   60 cycle power

Fluid characteristics		Wellbore conditions	
GOP	500.0 SCF/BBL	Measured depth	7000.00 ft
Oil gravity	35.0 API	SIBHP	1000.00 PSI
Gas gravity	.65	Intake pressure	500.00 psi
Water cut	50.00 percent	Wellhead pressure	200.00 psi
Viscosity	8.00 CP at 100.00 F	Perforations	7000.00 feet
Viscosity	2.40 CP at 200.00 F	Bottom hole temperature	160.00
Mean temperature	140.10 F	Free gas at intake	50.0 percent
PI -------	BPD/PSI	Tubing size	1.99 inches
		Vertical pump depth	7000.00
		Wellhead temperature	120.00

Pump configuration		Pump performance	
Stages	Type	Intake volume	822.38 BPD
231	D-20	Discharge volume	541.66 BPD
		Discharge pressure	2628.30 psi
H.P. = 41.25		Drawdown	.00 psi
		FBHP	1000.00 psi

and horsepower for a no-slip calculation and for the pump deteriorating.

Table 4.512 shows that at the higher gas volumes, the pump requires considerably more stages (69). At the lower gas volumes, there is no difference in the number of stages required.

4.53107   Summary of example problems #3A and #3B

In summary we can make the following comments:
(1) The purpose of these example problems has been to show a more precise method to use when all the P.V.T. data on a specific crude oil is available. The amount of gas that is in solution is quite variable with different oils from different areas and can affect the volume b calculations, etc., quite a bit.

(2) We use general correlations, such as Standing's or Lasater's, if P.V.T. data is not available. Using the correct P.V.T. data can make a difference in the pump calculations as well as the tubing flow calculations.

(3) Be aware that after the calculation of the volume for the first stage, the total gas in the system is constant and you use a GIP of 100 even though the GIP for the first stage can be set at less than 100. This

TABLE 4.58
RESERVOIR DATA AND PUMP SELECTION RESULTS
(100% Gas—100% Oil) (Pump Deterioration Included)

Pressure	Flow	GR	Spec. grav.
500.00	1750.64	.0361	.0833
517.17	1707.68	.0416	.0960
518.15	1705.22	.0419	.0967
538.47	1654.37	.0487	.1125
539.63	1651.45	.0492	.1134
557.16	1607.59	.0554	.1278
561.83	1595.90	.0571	.1318
562.79	1593.50	.0575	.1326
582.76	1543.58	.0651	.1502
583.91	1540.64	.0655	.1512
608.12	1485.80	.0745	.1719
609.51	1483.30	.0749	.1729
638.23	1431.74	.0841	.1940
639.87	1428.80	.0846	.1952
674.06	1367.41	.0963	.2223
676.03	1368.87	.0970	.2239
717.46	1297.47	.1110	.2561
719.85	1294.28	.1117	.2577
789.22	1228.20	.1271	.2933
772.08	1224.40	.1280	.2955
831.36	1152.28	.1469	.3389
834.78	1148.51	.1479	.3414
905.07	1084.48	.1669	.3851
809.08	1080.97	.1680	.3876
991.62	1008.49	.1921	.4433
996.35	1004.33	.1936	.4467
1090.31	958.52	.2036	.4699
1085.48	958.11	.2041	.4711
1195.85	808.78	.2146	.4952
1201.30	906.15	.2152	.4966
1212.30	900.96	.2164	.4994

Total dynamic head   3298.9224
Minimum pump discharge pressure   1213.874
Pressure drop across pump   713.8740
   Well name   1   60 cycle power

Fluid characteristics		Wellbore conditions	
GOR	500.0 SCF/BBL	Measured depth	7000.00 ft
Oil gravity	35.0 API	SIBHP	1000.00 psi
Gas gravity	.65	Intake pressure	500.00 psi
Water cut	.00 percent	Wellhead pressure	200.00 psi
Viscosity	8.00 CP at 100.00 F	Perforations	7000.00 feet
Viscosity	2.40 CP at 200.00 F	Bottom hole temperature	160.00
Mean temperature	140.65 F	Free gas at intake	100.0 percent
		Tubing size	1.99 inches
		Vertical pump depth	7000.00
		Wellhead temperature	120.00

will determine the amount of gas going through the pump and up the casing.

(4) In actual practice the performance of the pump head capacity curve is diminished by a factor input into the computer. This is shown by the two computer printouts for the same problem and is summarized in Table 4.511.

4.53108   Example problem #3C, well pumping gas (computer solution by Centrilift)

Given:

Casing = 7 in.
Depth = 11,000 ft
   (bottom of casing)

Dead oil viscosity = 3.6 cp @ 100°F.
Dead oil viscosity = 1.6

Tubing = 2⅞ in. Perforations, 10,600 to 10,650 ft
$\overline{P}_R$ = 2900 psi at 10,000 ft
P_{wf} = 2540 psi for q = 1000 b/d
Water cut 30%
Oil API gravity—40
γ_w = 1.05

cp @ 200°F.
Temp = 225°F. at 10,000 ft
Well head flowing temp = 160°F.
P_{wh} = 200 psig
Power supply = 60 Hz
Desired Rate = 1600 STB/D
Pump Depth = 10,570 ft

(1) Collect and analyze data:
Because of the gas, the additional data required is:
Producing GOR—500 scf/STB
Gas specific gravity—0.75

TABLE 4.59
RESERVOIR DATA AND PUMP SELECTION RESULTS

(50% Gas—100% Oil) (Pump Deterioration Included)

Pressure	Flow	GR	Spec. grav.
500.00	1144.77	.1442	.3328
571.82	1039.40	.1767	.4079
576.14	1033.07	.1789	.4128
667.97	926.73	.2053	.4788
673.19	921.25	.2065	.4765
779.55	831.03	.2283	.5268
785.55	826.37	.2295	.5297
785.55	826.37	.2295	.5297
877.44	769.77	.2459	.5675
882.66	767.12	.2467	.5694
989.33	713.10	.2647	.6109
995.37	710.04	.2656	.6134
1115.41	679.54	.2775	.6404
1121.97	677.94	.2781	.6418
1251.82	646.26	.2912	.6721
1258.94	644.53	.2920	.6738
1399.98	610.13	.3075	.7096
1407.71	608.24	.3084	.7117
1561.20	563.32	.3207	.7402
1569.45	563.32	.3208	.7402
1726.24	583.32	.3213	.7415
1734.50	583.32	.3214	.7416
1825.41	583.32	.3217	.7423

Total dynamic head   4141.6048
Minimum pump discharge pressure   1832.258
Pressure drop across pump 1332.2578

Well name   1   60 cycle power

Fluid characteristics		Wellbore conditions	
GOR	500.0 SCF/BBL	Measured depth	7000.00 ft
Oil gravity	35.0 API	SIBHP	1000.00 psi
Gas gravity	.65	Intake pressure	500.00 psi
Water cut	.00 percent	Wellhead pressure	200.00 psi
Viscosity	8.00 CP at 100.00 F	Perforations	7000.00 feet
Viscosity	2.40 CP at 200.00 F	Bottom hole temperature	160.00
Mean temperature	140.53 F	Free gas at intake	50.0 percent
PI -------	BPD/PSI	Tubing size	1.99 inches
		Vertical pump depth	7000.00
		Wellhead temperature	120.00

Pump configuration		Pump performance	
		Intake volume	1144.77 BPD
		Discharge volume	583.32 BPD
		Discharge pressure	1832.26 psi
		Drawdown	.00 psi
Stages	Type	FBHP	1000.00 psi
60	D-40	Desired flow	500.00 BPD
153	D-20	Horsepower	31.44

PVT data:

Pressure	Solution GOR	Formation Volume Factor
0	0	1.00
300	110	1.07
600	250	1.14
900	340	1.19
1,200	420	1.24
1,500	500	1.29
13,000	500	1.26

(2 to 6) Determine inflow performance, TDH, pump selection and motor horsepower

All of these items will be determined and printed out by the computer program. Since there is gas being produced and flow is below the bubble point pressure the program determines inflow performance according to the inflow performance relationship (IPR) (Vogel Solution).

The data is listed on an input data sheet, Table 4.513. In addition to the data already discussed,

TABLE 4.510
RESERVOIR DATA AND PUMP SELECTION RESULTS
(50% Gas—50% Water) (Pump Deterioration Included)

Pressure	Flow	GR	Spec. grav.
500.00	822.38	.2610	.6024
607.52	745.08	.2878	.6641
618.83	741.77	.2891	.6670
744.95	679.02	.3153	.7277
752.42	676.11	.3167	.7308
904.27	628.09	.3403	.7854
912.72	625.95	.3415	.7880
1081.11	593.95	.3594	.8294
1090.28	592.84	.3601	.8310
1269.23	571.01	.3735	.8618
1278.91	569.83	.3742	.8635
1468.08	546.36	.3894	.8987
1478.33	544.85	.3904	.9010
1675.23	541.66	.3929	.9066
1685.61	541.66	.3929	.9068
1882.83	541.66	.3933	.9077
1893.21	541.66	.3933	.9077
2090.63	541.66	.3937	.9086
2101.03	541.66	.3937	.9086
2298.65	541.66	.3941	.9095
2309.05	541.66	.3942	.9096
2506.88	541.66	.3945	.9105
2517.30	541.66	.3946	.9105
2621.50	541.66	.3948	.9110

Total dynamic head   5391.1217
Minimum pump discharge pressure   2628.298
Pressure drop across pump   2128.2975
      Well name   1   60 cycle power

Fluid characteristics		Wellbore Conditions	
GOR	500.0 SCF/BBL	Measured depth	7000.00 ft
Oil gravity	35.0 API	SIBHP	1000.00 psi
Gas gravity	.65	Intake pressure	500.00 psi
Water cut	50.00 percent	Wellhead pressure	200.00 psi
Viscosity	8.00 CP at 100.00 F	Perforations	7000.00 feet
Viscosity	2.40 CP at 200.00 F	Bottom hole temperature	160.00
Mean temperature	140.10 F	Free gas at intake	50.0 percent
PI ------	BPD/PSI	Tubing size	1.99 inches
		Vertical pump depth	7000.00
		Wellhead temperature	120.00

Pump configuration		Pump performance	
		Intake volume	822.38 BPD
		Discharge volume	541.66 BPD
		Discharge pressure	2628.30 psi
		Drawdown	.00 psi
		FBHP/	1000.00 psi
Stages	Type	Desired flow	500.00 BPD
231	D-20	Horsepower	41.25

TABLE 4.511-A
SUMMARY TABULATION—ASSUMING NO PUMP DETERIORATION

Correlation	GIP (%)	Water (%)	Intake vol.	Discharge vol.	Discharge press.	TDH (ft)	ΔP across pump	Stages	Type pump	hp
Hagedorn Long-hand	100	0	1752	897	1300		800	209	D-40	27
Orkiszewski computer	100	0	1750.6	902	1214	3302	714	194	D-55 D-40	27.8
Orkiszewski computer	100	0	1750.6	902	1214	3302	714	206	D-40	28.2
Orkiszewski computer	50	0	1145	583	1832	4142	1332	211	D-40 D-20	31.56
Orkiszewski computer	50	50	822	542	2628	5391	2128	231	D-20	41.25

TABLE 4.511-B
SUMMARY TABULATION ACCOUNTING FOR TRUE PUMP PERFORMANCE (According to Reda)

Correlation	GIP (%)	Water (%)	Intake vol. (b/d)	Discharge vol. (b/d)	Discharge press.	TDH (ft)	ΔP across pump	Stages	Pump	hp
Orkiszewski computer	100	0	1750.6	901	1214	3299	714	263	D-55(34) D-40(229)	28
Orkiszewski computer	50	0	1145	583	1832	4142	1332	213	D-40(60) D-20(152)	31.4
Orkiszewski computer	50	50	822	542	2628	5391	2128	231	D-20	41.25

TABLE 4.512
DIFFERENCE IN REQUIRED STAGES

Pump condition	STAGES		Additional stages for actual condition	*HP
	No Slip	True		
100% gas—no water	194	263	69	
50% gas—no water	211	213	2	
50% gas—50% water	231	231	0	

*There is no change in hp requirements for all cases.

the input data sheet provides for a statement of how much of the free gas at the pump intake will go through the pump. This is a statement defining the efficiency of any gas separator used along with how the casing may or may not be vented. Also, there is an entry that states the maximum capability of the pump to handle gas and still produce the head shown on published curves. This prevents the selection of a pump at a setting depth at which there is too much free gas.

Tables 4.514 through 4.519 are the printout of this computer-assisted selection. They show that if the pump setting depth is 10,570 ft, all the gas is in solution and that the other essentials are:

Required pump
discharge pressure = 3006 psig

Suction pressure = 2492 psig
Total dynamic head = 514 psig or 1453 ft
Intake flow rate = 1921 b/d
Pump selection = 32 stages of Y-62B
Horsepower at
operating condition = 25 hp
Horsepower required for 32
stages of Y-62B if it were
pumping a liquid of 1.0
specific gravity = 32 hp
Alternate pump selections = 70 stages of Z-69
or 64 stages of
N-80

The printouts of Tables 4.514 through 4.519 also show that:

(1) A setting depth of 4000 ft or where the intake pressure is 20 psig for the pump would be too shallow because the free gas is too much for the pump to handle.

(2) The bubble point is 7656 ft deep at which 34 stages of a Y-62B pump could handle the required flow using 27 hp.

(3) The highest point in the well that the pump can be located and not exceed its free gas to liquid capability is a setting depth of 5226 ft at which point it would require 63 stages of Y-62B and 49 horsepower.

(7 through 11) Select cable and determine surface voltage, transformer size, accessories and other requirements.

TABLE 4.513
INPUT DATA SHEET FOR EXAMPLE 3-C

1.1 Well identification			70
01 DEMONSTRATION WELL WITH GAS			
1.2 Date of request (Month/Day/Year)	71	01 16 76	76
1.3 P.V.T. data to be read (Enter Yes or No)	78	YES	80
2.1 Datum point—vertical depth (ft)	01	10000.	10
2.2 Datum point—static pressure (PSIG)	11	2900.	20
2.3 Datum point—flowing pressure (PSIG) at rate in 2.4	21	2540.	30
2.4 Datum point—total liquid flow rate (STB/D)	31	1000.	40
2.5 Datum point—liquid temperature (°F.)	41	225.	50
2.6 Desired total liquid flow rate (STB/D)	51	1600.	60
2.7 Water cut percentage (0/0)	61	30.	70
2.8 Producing gas-oil ratio (SCF/STB)	71	500.	80
3.1 Depth to top of perforations (ft)	01	10600.	10
3.2 Bottom hole depth (ft)	11	11000.	20
3.3 Tubing dia. (in.)	21	2.875	30
3.4 Casing O.D. (in.)	31	7.	40
3.5 Oil API gravity	41	40.	50
3.6 Water specific gravity	51	1.05	60
3.7 Gas specific gravity	61	0.75	70
3.8 Liquid temperature at wellhead (°F.)	71	160.	80
4.1 Viscosity data—units	01	CP	10
4.2 temperature A (°F.)	11	100.	20
4.3 viscosity A	21	3.6	30
4.4 temperature B (°F.)	31	200.	40
4.5 viscosity B	41	1.6	50
4.6 Percentage of free gas through pump (0/0)	51	15.	60
4.7 Max. allowable free gas-liquid ratio through pump (CF/CF)	61	0.1	70
4.8 Required wellhead pressure (PSIG)	71	200.	80
Depth pressure profile requested _____			
Pump setting locations requested _____			
5.1 depth no. 1 (Feet)	01	Y 10570.	10
5.2 depth no. 2 (Feet)	11	Y 4000.	20
5.3 free gas-liquid ratio no. 1 (CF/CF)	21	Y 0.	30
5.4 free gas-liquid ratio no. 2 (CF/CF)	31	Y 0.667	40
5.5 pressure no. 1 (PSIG)	41	Y 20.	50
5.6 pressure no. 2 (PSIG)	51	Y 2000.	60
5.7 Electric power supply available (cycles)	61	60.	65
5.8 Sales requested series number	66	400.	70
5.9 Sales requested stage type	71	N-80	80

	Pressure (PSIG)			Pressure volume temperature data Solution GOR (SCF/BBL)			Formation volume factor		
6.1	01	0.	07	08	0.	14	15	1.	20
6.2	21	300.	27	28	110.	34	35	1.07	40
6.3	41	600.	47	48	250.	54	55	1.14	60
6.4	61	900.	67	68	340.	74	75	1.19	80
7.1	01	1200.	07	08	420.	14	15	1.24	20
7.2	21	1500.	27	28	500.	34	35	1.29	40
7.3	41	13000.	47	48	500.	54	55	1.26	60
7.4	61	0.	67	68		74	75		80

TABLE 4.514
SUBMERSIBLE PUMP SELECTION AND EVALUATION

Well program demonstration well
Original input data   Date 1-16-76

Datum point—vertical depth	10000.0 feet
Datum point—static pressure	2900.0 psig
Datum point—flowing pressure	2540.0 psig
Datum point—total liquid flow rate	1000.0 STB/DAY
Datum point—liquid temperature	225.0 deg-F
Desired liquid flow rate	1600.0 STB/DAY
Water cut percentage	30.0 percent
Producing gas-oil ratio	500.0 SCF/STB
Depth to top of perforations	10600.0 feet
Bottom hole depth	11000.0 feet
Tubing O.D.	2.8750 inches
Casing O.D.	7.0000 inches
Oil API gravity	40.0
Water specific gravity	1.0500
Gas specific gravity	0.7500
Liquid temperature at wellhead	160.0 deg-F
Oil viscosity at 100.0 degrees Fahrenheit	3.6000 CP
Oil viscosity at 200.0 degrees Fahrenheit	1.6000 CP
Percentage of free gas through the pump	15.0 percent

Max. allowed free gas/liquid ratio through pump		0.1000 CF/CF
Required wellhead pressure		200.0 psig
Requested depth pressure profile (sizing requested)		
A) Depth number 1	(Yes)	10570.00 feet
B) Depth number 2	(Yes)	4000.00 feet
C) Free gas-liquid ratio number 1	(Yes)	0.00 CF/CF
D) Free gas-liquid ratio number 2	(Yes)	0.67 CF/CF
E) Pressure number 1	(Yes)	20.00 PSIG
F) Pressure number 2	(Yes)	2000.00 PSIG
Electric power available		60. cycles
Sales requested series number	(Yes)	400.
Sales requested stage type	(Yes)	N-80

PRESSURE VOLUME TEMPERATURE DATA

Pressure (PSIG)	Solution G.O.R. (SCF/BBL)	Formation volume factor
0.0	0.0	1.0000
300.0	110.00	1.0700
600.0	250.00	1.1400
900.0	340.00	1.1900
1200.0	420.00	1.2400
1500.0	500.00	1.2900
13000.0	500.00	1.2600

TABLE 4.515
SUBMERSIBLE PUMP SELECTION AND EVALUATION

Well program demonstration well
Pump selections at flow rate requested   Date   1-16-76
Depth to producing level   2962. feet   Bottom hole flowing pressure   2638. PSIG
Depth pressure profile

	Vertical setting depth (feet)	Pump suction pressure (PSIG)	Free gas to liquid ratio (CF/CF)	Reqd. pump discharge pressure (PSIG)	
Size	10570.	2491.8	0.0	3006.2	A) Depth no. 1
Size	4000.	247.8	3.2900	1508.6	B) Depth no. 2
Size	7656.	1500.0	0.0	2037.3	C) Free GLR no. 1
Size	5226.	662.7	0.6600	1683.6	D) Free GLR no. 2
Size	3026.	20.0	59.9000	1253.7	E) Pressure no. 1
Size	9125.	2000.0	0.0	2536.5	F) Pressure no. 2

Pump selection A) requested depth no. 1

Pump setting depth	10570.0 feet	Discharge pressure	3006.2 psig
Oil flow rate	1120.0 STB/DAY	Water flow rate	480.0 STB/DAY
Intake conditions			
Pressure	2491.8 PSIG	Temperature	228.7 Deg-F
Free gas/liquid ratio	0.0000 CF/CF	Pumping viscosity	29.2 SSU
Liquid flow rate	1920.9 BBL/DAY	Liquid density	48.5939 LB/CF
Free gas flow rate	0.0 BBL/DAY	Gas density	0.0041 LB/CF
Total flow rate	1920.9 BBL/DAY	Flowing fluid density	48.5934 LB/CF

Pump specs. Series	Type	No. of Stages	Selection range & min flow (BPD)			Pressure PSIG	Output feet	Discharge volume	Horsepower Actual-Max.	
Fixed series selection										
513	Y-62B	32	1700	2200	1700	515.	1453.	1919.	25.	32.
	Totals	32				515.	1453.		25.	32.
Fixed type selection										
			(see fixed series selection)							
Sales requested series										
400	Z-69	70	1700	2600	1700	518.	1462.	1919.	24.	32.
	Totals	70				518.	1462.		24.	32.
Sales requested type										
400	N-80	64	1590	3520	1590	522.	1474.	1919.	29.	38.
	Totals	64				522.	1474.		29.	38.

TABLE 4.516
SUBMERSIBLE PUMP SELECTION AND EVALUATION

Well program demonstration well
Pump selections at flow rate requested
Pump selection C) requested free GLR no. 1   Date   1-16-76

Pump setting depth	7656.0 feet	Discharge pressure	2037.3 PSIG
Oil flow rate	1120.0 STB/DAY	Water flow rate	480.0 STB/DAY

Intake conditions

Pressure	1500.0 PSIG	Temperature	209.8 Deg-F
Free gas/liquid ratio	0.0000 CF/CF	Pumping viscosity	29.3 SSU
Liquid flow rate	1923.8 BBL/DAY	Liquid density	48.6414 LB/CF
Free gas flow rate	0.0 BBL/DAY	Gas density	0.0041 LB/CF
Total flow rate	1923.8 BBL/DAY	Flowing fluid density	48.6409 LB/CF

Pump specs		No. of	Selection range			Pressure	Output	Discharge	Horsepower
Series	Type	Stages	& min flow (BPD)			PSIG	feet	Volume	Actual-Max.
Fixed series selection									
513	Y-62B	34	1700	2200	1700	543.	1543.	1922.	27. 34.
	Totals	34				543.	1543.		27. 34.
Fixed type selection									
			(see fixed series selection)						
Sales requested series									
400	Z-69	74	1700	2600	1700	543.	1544.	1922.	26. 34.
	Totals	74				543.	1544.		26. 34.
Sales requested type									
400	N-80	67	1590	3520	1590	542.	1542.	1922	30. 40.
	Totals	67				542.	1542.		30. 40.

TABLE 4.517
SUBMERSIBLE PUMP SELECTION AND EVALUATION

Well program demonstration well
Pump selections at flow rate requested
Pump selection D) requested free GLR no. 2   Date   1-16-76

Pump setting depth	5226.4 feet	Discharge pressure	1683.6 PSIG
Oil flow rate	1120.0 STB/DAY	Water flow rate	480.0 STB/DAY

Intake conditions

Pressure	662.7 PSIG	Temperature	194.0 Deg-F
Free gas/liquid ratio	0.0992 CF/CF	Pumping viscosity	30.1 SSU
Liquid flow rate	1767.5 BBL/DAY	Liquid density	51.5489 LB/CF
Free gas flow rate	175.4 BBL/DAY	Gas density	2.2641 LB/CF
Total flow rate	1942.9 BBL/DAY	Flowing fluid density	47.8122 LB/CF

Pump specs		No. of	Selection range			Pressure	Output	Discharge	Horsepower
Series	Type	Stages	& min flow (BPD)			PSIG	feet	Volume	Actual-Max.
Fixed series selection									
513	Y-62B	63	1700	2200	1700	1023.	2907.	1923.	49. 64.
	Totals	63				1023.	2907.		49. 64.
Fixed type selection									
			(see fixed series selection)						
Sales requested series									
400	Z-69	137	1700	2600	1700	1024.	2907.	1923.	47. 62.
	Totals	137				1024.	2907.		47. 62.
Sales requested type									
400	N-80	125	1590	3520	1590	1025.	2912.	1923.	56. 74.
	Totals	125				1025.	2912.		56. 74.

TABLE 4.518
SUBMERSIBLE PUMP SELECTION AND EVALUATION

Well program demonstration well
Pump selections at flow rate requested
Pump selection F) requested pressure no. 2   Date 1-16-76

Pump setting depth	9124.7 feet	Discharge pressure	2536.5 PSIG
Oil flow rate	1120.0 STB/DAY	Water flow rate	480.0 STB/DAY
Intake conditions			
Pressure	2000.0 PSIG	Temperature	219.3 Deg-F
Free gas/liquid ratio	0.0000 CF/CF	Pumping viscosity	29.3 SSU
Liquid flow rate	1922.3 BBL/DAY	Liquid density	48.6179 LB/CF
Free gas flow rate	0.0 BBL/DAY	Gas density	0.0041 LB/CF
Total flow rate	1922.3 BBL/DAY	Flowing fluid density	48.6174 LB/CF

Pump specs		No. of	Selection range			Pressure	Output	Discharge	Horsepower
Series	Type	Stages	& min flow (BPD)			PSIG	feet	Volume	Actual-Max.
Fixed series selection									
513	Y-62B	34	1700	2200	1700	545.	1544.	1921.	27. 34.
	Totals	34				545.	1544.		27. 34.
Fixed type selection									
			(see fixed series selection)						
Sales requested series									
400	Z-69	73	1700	2600	1700	538.	1524.	1921.	25. 33.
	Totals	73				538.	1524.		25. 33.
Sales requested type									
400	N-80	67	1590	3520	1590	544.	1543.	1921.	31. 40.
	Totals	67				544.	1543.		31. 40.

TABLE 4.519
SUBMERSIBLE PUMP SELECTION AND EVALUATION

Well program demonstration well
Pump selections at flow rate unresolved   Date 1-16-76
*****Last flow rate adjustment resulted in

Oil flow rate	368. STB/DAY
Water flow rate	158. STB/DAY
Total flow rate	526. STB/DAY

*****This rate is unacceptable for requests

B) Depth no. 2	4000.00	Because GLR above maximum limit
E) Pressure no. 1	20.00	Because GLR above maximum limit

CLASS PROBLEM #3-A (WELL PUMPING GAS)

Given:

7 in. - 23#/ft	GOR = 500 scf/B
3 in. tubing	$\gamma_g = 0.65$
Depth = 8000 ft	Desired rate = 2000 b/d
Perforations = 7930 – 7960	°API = 40
	Temp bottom = 180°F
$\overline{P}_R = 2600$ psi	Wellhead temp = 120°F
$P_{wf} = 2000$ psi for 1000 b/d	$P_{wh} = 200$ psi

(Assume behavior according to Vogel's reference curve)

Design a complete pumping system.
  (a) Pumping all the gas
  (b) Venting 50% of the gas

CLASS PROBLEM #3-B (WELL PUMPING GAS)

Given:

Depth = 8000 ft	$P_{wh} = 180$ psi
Tubing = 2⅜" O.D.	GOR = 600 scf/B
Casing 5½ in.	$\gamma_g = 0.7$
$\overline{P}_R = 1200$ psi	$q_L = 800$ b/d for $P_{wf} =$
$P_{wf} = 400$ psi	400 psi
	$\gamma_o = 36°$ API

Find the number of stages needed between the pump intake pressure of 400 psi and 600 psi for temp = 170°F.
  (a) Producing all oil
  (b) Producing 50% water ($\gamma_w = 1.12$)

4.5311 Example problem #4, pumping viscous crudes

4.53111 Introduction

The performance of centrifugal submersible pumps changes when pumping viscous crude oils or emulsions. This is noted by an increase in brake horsepower with reductions in efficiency, head and rates. Various viscosity conversions can be found in Appendix 4L (1) through 4L (5). Additional discussions on viscosity can be found in Section 4.235. Riling presents a design procedure for handling viscous crudes and his procedure is followed in this section.[1]

Fig. 4.57 shows an example of how a centrifugal pump's performance curves are altered due to the effects of moderate to high viscosities. The exact effects of viscosity cannot be predicted and additional research is needed in this area. Of course, we cannot predict the true effects of viscosity in multiphase flow

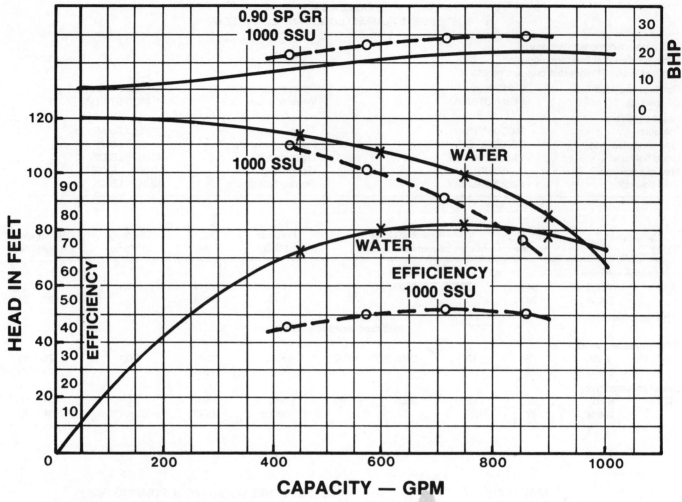

Fig. 4.57 *Performance curves showing Effect of Viscosity*

in pipes, and the passage of the fluids through the stages of a centrifugal pump become much more complex. Emulsions are a separate and, additionally, much more complex problem. Various assumptions are presently used to correct for water percentages, such as doubling the crude viscosity. As a general rule the mixture viscosity of oil and water can be expected to increase for water cuts greater than 20%, reaching a maximum between 55-75% and reducing back to the water viscosity for water cuts greater than 80%. Of course this varies depending upon the composition of the crude and water. Mixtures of different crudes and different water will not necessarily form emulsions at the same water cuts. Riling noted from experience that for water-cuts between 20% and 40% the mixture viscosity increased by a factor of 2 to 3 over the crude viscosity. For cases where emulsions are known to exist such as 55-75% water cut a factor or of 5 to 6 may be required.

Various additional figures can be found in the Appendix 4L to assist in viscosity calculations:

Tables 4.520 and 4.521 provide a means of determining the performance of a conventional-sized submersible pump handling a viscous liquid when its performance with water is known. They also can be used as an approximate method in the selection of a pump and motor for a given application.

4.53112 Design procedure

Riling gave the following procedure in selecting a submergible pumping unit for handling viscous crudes:

(1) Determine the total dynamic head that would be required for pumping water of a 1.0 specific gravity (using the procedures described earlier in this section.)

(2) Determine from tests or from Fig. 4L.1 (Appendix 4L) the gas-free crude viscosity at reservoir temperature.

(3) Correct the gas-free crude viscosity of step 2 for gas saturation by use of Fig. 4L.2 of Appendix 4L or by actual tests.

(4) Convert the viscosity units to SSU units (Fig. 4L.3—Appendix 4L).

(5) Correct the mixture viscosity for water cut if data is available.

(6) Refer to Tables 4.520 or 4.521 and apply the correction factors to the capacity and head so the proper size pump and horsepower motor can be determined.

(7) Select the proper pump and motor.

(8) Size additional subsurface and surface equipment to compliment corrected design as described in previous procedures.

TABLE 4.520
PERFORMANCE CORRECTION CHART
APPROXIMATE CHANGES DUE TO VISCOSITY AND
SPECIFIC GRAVITY
60% maximum efficiency pumps

Viscosity S.S.U. at pumping temperatures	Performances as percentages of performance with water, all at maximum efficiency			
	Capacity factor	Head factor	New efficiency	Horsepower factor
50	100.0	99.5	57.5	104.0 × 80
80	98.5	98.5	54.0	107.8 × 80
100	98.0	98.0	52.0	110.8 × 80
150	96.0	96.0	47.5	116.3 × 80
200	94.0	94.0	44.5	119.1 × 80
300	91.0	91.0	40.0	124.2 × 80
400	88.0	89.0	37.0	127.0 × 80
500	85.0	86.0	34.0	128.0 × 80
600	83.0	84.5	32.5	129.5 × 80
700	80.5	82.5	30.5	130.0 × 80
800	79.0	81.0	29.0	131.0 × 80
900	77.0	80.0	28.0	132.0 × 80
1,000	75.5	78.0	26.5	133.0 × 80
1,500	69.0	72.5	22.0	136.5 × 80
2,000	63.0	67.5	19.5	131.0 × 80
2,500	58.0	63.0	17.0	129.0 × 80
3,000	54.0	59.0	15.0	127.5 × 80
4,000	47.0	50.5	12.0	118.7 × 80
5,000	41.0	45.0	10.0	111.0 × 80

γ_o = Specific gravity of oil at pumping temperature

TABLE 4.521
PERFORMANCE CORRECTION CHART
APPROXIMATE CHANGES DUE TO VISCOSITY AND
SPECIFIC GRAVITY
70% maximum efficiency pumps

Viscosity S.S.U. at pumping temperatures	Performances as percentages of performance with water, all at maximum efficiency			
	Capacity factor	Head factor	New efficiency	Horsepower factor
50	100.0	100.0	69.5	100.8 × 80
80	99.5	100.0	67.5	103.1 × 80
100	99.0	99.5	65.5	105.3 × 80
150	98.0	98.5	62.0	109.0 × 80
200	96.5	97.0	59.0	111.0 × 80
300	94.0	94.5	55.0	113.1 × 80
400	92.0	92.0	51.5	115.0 × 80
500	90.5	90.5	49.0	117.0 × 80
600	89.0	89.5	47.5	117.3 × 80
700	87.0	88.0	45.7	117.3 × 80
800	85.5	86.0	44.0	117.0 × 80
900	84.0	85.0	43.0	116.2 × 80
1,000	83.0	84.0	42.0	116.0 × 80
1,500	78.0	79.0	37.0	113.5 × 80
2,000	74.0	75.5	34.7	112.6 × 80
2,500	70.5	72.5	32.0	111.6 × 80
3,000	67.0	69.5	30.0	108.8 × 80
4,000	61.0	64.0	27.0	101.2 × 80
5,000	55.0	60.0	25.0	92.4 × 80

γ_o = Specific gravity of oil at pumping temperature.

4.53113 Example problem for viscous crudes

For a detailed illustration of the correct procedure to follow in sizing a submersible centrifugal pumping unit for a viscous liquid where corrected pump curves are not available, the following example is given:

Given:

Casing—7 in. O.D., (23 #/ft)
Tubing—5,200 ft of 2⅞ in. O.D. EUE 8rd
Perforations—5,300 ft to 5,400 ft
Pump setting depth—5,200 ft (100 ft above top perforations)
Operating fluid level—200 ft fluid over pump

Production data:

Discharge pressure—50 PSIG
GOR—50 scf/B
Bottom hole temperature—130° F.
Desired volume—1,700 b/d
Total head (water calculations)—5,215 ft (200 ft FOP)

Fluid data:

A.P.I. gravity of oil—16
Water cut—30%
Specific gravity of water—1.02

(1) Total dynamic head for water is found to be 5215 ft giving 200 ft of fluid over the pump.
(2) From Figure 4L.2 (Appendix 4L), find the gas-free crude viscosity for 16° API crude at 130°F to be approximately 140 centipoises (cp).
(3) From Figure 4L.1 (Appendix 4L), find the gas-saturated crude oil viscosity for 140 cp at a GOR of 50 scf/B to be approximately 65 cp.
(4) From Figure 4L.3 (Appendix 4L), convert 65 cp to SSU units using an average specific gravity of 1.0. Read 400 SSU.
(5) Assume the 30% water cut increases the actual viscosity by a ratio of 2:1. Corrected viscosity would then be 2 × 400 = 800 SSU.
(6) Assume a 70% efficient pump and use Table 4.521 to obtain the following correction factors:

$$\text{Capacity factor } (Q_{vis}) = 85.5\%$$
$$\text{Head factor } (H_{vis}) = 86.0\%$$
$$\text{Horsepower factor } (bhp_{vis}) = 117.0 \ (1.02) = 119.3$$
$$\text{Corrected total head} = \frac{5215}{0.86} = 6064 \text{ ft}$$
$$\text{Corrected capacity} = \frac{1700}{0.855} = 1990 \text{ b/d}$$

(7) Select a suitable pump and motor from the standard performance curves of Appendix 4BII using the corrected values of Step 6.

For example one pump satisfying these requirements for a corrected rate of 1990 b/d is from Appendix 4BII - (19) (Y-62B-Centrilift).

The number of stages will be $\frac{6064}{44.5} = 136$ stages

Horsepower requirements are 136 × 1.0 hp/stage × 119.3 = 162 hp

(8) Additional equipment is selected as previous examples.

CLASS PROBLEM #4-A (PUMPING VISCOUS CRUDES)
Given:

5½ in. casing × 2⅜ in. O.D. tubing
Perforations = 6400 − 6450 ft
Tubing length = 6300 ft
$\underline{PI} = 5$
$\overline{P}_R = 1750$ psia
Water cut = 50% ($\gamma_w = 1.07$)
Desired rate = 1500 b/d (total liquid)
$P_{wh} = 120$ psi
°API = 14
Bottom hole temp = 160°F.

Design a complete pumping system accounting for viscous effects and assume no gas is to be pumped.

CLASS PROBLEM #4-B (PUMPING VISCOUS CRUDES)
Given:

7 in. casing
3 in. tubing
Tubing length = 8000 ft
Perforations = 8030 − 8090
$\underline{PI} = 2$
$\overline{P}_R = 2600$ psi
Water cut 40% ($\gamma_w = 1.10$)
Desired rate = 1800 b/d (total liquid)
$P_{wh} = 160$ psig
°API = 12
Temp @ 8000 ft = 200°F.
GOR = 100 scf/b

Design a complete pumping system (assume pumping no gas).

4.54 Operational approaches to pump design

4.541 Introduction

The design of oil well subsurface pumping installations involves the solution of four major problems. First, what is the minimum production flow rate obtainable from a well? Second, what is the maximum production flow rate obtainable from the well? Third, what is the most economical installation capable of handling a desired production flow rate? Finally, what is the most profitable production flow rate and its corresponding installation?

When attempting to pump gassy oil wells, some critical questions arise, regardless of the particular problem which is being solved. Is it necessary to vent gas? How much gas is it necessary to vent? Is it permissible to vent gas? How much gas is it permissible to vent? How shallow is it possible to set the pump? How deep is it necessary to set the pump? What is the size of the required pump? What is the pressure increase that the pump has to develop?

Since the oil industry started using subsurface pumps to produce gassy oil wells, there have been no clear answers for these questions.

4.542 Basic pump variables and characteristics

Following are listed the basic variables in the pumping problem.

Pumping problem basic variables

q_L' = Stock tank flow rate
D_{PM} = Pump setting depth

f_{gPM} = Free gas pumped fraction
P_{PMI} = Pump inlet pressure
P_{PMO} = Pump outlet pressure
T_S = Sink temperature (casing temperature just below the pump inlet)
T_{PM} = Pump temperature
P_{PML} = Pump low pressure
P_{PMH} = Pump high pressure
q_{PMLP} = Pump low pressure volumetric flow rate
q_{PMHP} = Pump high pressure volumetric flow rate
ΔP_{Ext} = External differential pressure existing across the pump, from the inlet to the outlet
ΔP_{Inl} = Pressure loss in the pump inlet
ΔT_{PM} = Temperature increase due to the pump operation
ΔP_{Out} = Pressure loss in the pump outlet
ΔP_{Int} = Internal differential pressure across the pump, from the low to the high pressure

Operational variables—q_L', D_{PM}, f_{gPM}
Environmental variables—
    External—P_{PMI}, P_{PMO}, T_{PM}
    Internal—T_{PM}, P_{PML}, P_{PMH}
Pump configuration variables—q_{PMLP}, q_{PMHP}
Differential variables
    $\Delta P_{Ext} = P_{PMO} - P_{PMI}$
    $\Delta P_{Inl} = P_{PMI} - P_{PML}$
    $\Delta T_{PM} = T_{PM} - T_S$
    $\Delta P_{Out} = P_{PMH} - P_{PMO}$
    $\Delta P_{Int} = P_{PMH} - P_{PML}$

A complete nomenclature is at the end of this section.

Basic pump characteristics:
(1) Pump flow rate range upper limit (q_{PMUL})—The maximum volumetric flow rate the pump can handle or displace.
(2) Pump flow rate range lower limit (q_{PMLL})—The minimum volumetric flow rate the pump can handle or displace.
    Note: for an electric motor-powered centrifugal pump the operating speed is normally constant. The q_{PMUL} and q_{PMLL} values may be read from the pump performance curve. For reciprocating pumps the operation speed may vary.
(3) Permissible pump temperature (T_{PMLIM})
(4) Permissible internal differential pressure (ΔP_{INHL}) – The excess internal pressure the pump can undergo without bursting.
(5) Permissible external differential pressure (ΔP_{EXHL}) – The excess external pressure the pump can undergo without collapsing.
(6) The pump inlet flowing pattern configuration, a geometrical characteristic which will influence the value of ΔP_{Inl}.
(7) The pump outlet flowing pattern configuration, a geometrical characteristic which will influence the value of ΔP_{Out}.

4.543 Solution model and graphical representations

4.5431 Introduction

The pump is the interface between the natural flow of fluids coming from the reservoir, through the casing, into the pump inlet, and the artificially induced flow going from the pump outlet, up through the tubing string, through the flowline, and into the gas-liquid separator. The pump environment is characterized principally by the pump inlet pressure, the pump outlet pressure, the pump temperature, the pump low pressure, and the pump high pressure.

Knowing the reservoir static pressure and the separator operating pressure, analysis determines how the pump environment changes when the operational variables—stock tank flow rate, free gas pumped fraction, pump setting depth—change. The model accounts for reservoir inflow performance, the composition of the flowing mixture, the properties of the fluids, and the flowing system configuration.

For any pump internal environment point (P_{PML}, T_{PM}, P_{PMH}) the pump volumetric input and the pump volumetric output are calculated. These are, respectively, the pump low pressure volumetric flow rate (q_{PMLP}) and the pump high pressure volumetric flow rate (q_{PMHP}).

Knowing the values of the environmental variables (P_{PMI}, P_{PMO}, T_{PM}, P_{PML}, P_{PMH}, q_{PMLP}, q_{PMHP}) it is possible to determine the pump horsepower requirement, using any existing or improved calculation method, when designing a new installation. It is also possible to determine what can improve the installation performance when analyzing the well operation, knowing the pump characteristics and the installation power capacity.

The interaction between the variables of the pump problem and the pump characteristic variables are quite complex. Because of this, a graphical approach helps to solve the problem.

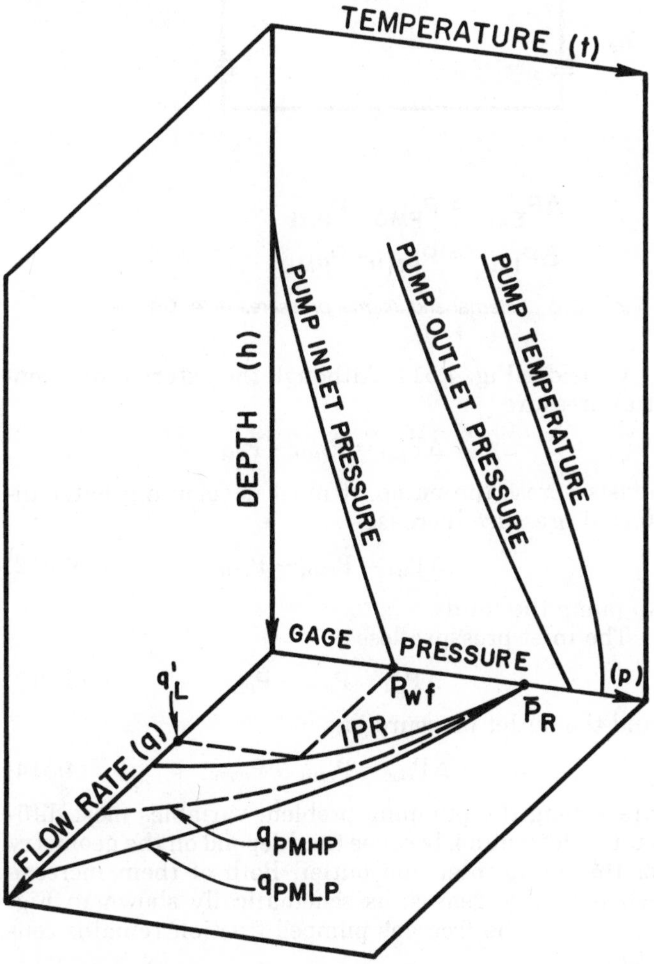

Since the pumping problem is principally concerned with volumetric flow rates, pressures and setting depths, a tri-dimensional Cartesian system may be defined, where these quantities may be represented on axes q, p, h, shown in Fig. 4.58. Temperatures are also represented along the p axis on the plane p/t, h.

Many calculations are involved in simulating the flow of fluids through the complete flowing system required for determinating P_{PML}, P_{PMH}, q_{PMLP}, and q_{PMHP}. Therefore, computer programs should handle them.

4.5432 External and internal environments

Suppose it is desired to produce the stock tank flow rate

$$q'_L = q'_o + q'_w \qquad (4.57)$$

from a well with static pressure $\overline{P}_R$ and gas-oil ratio GOR. A pump set at the depth D_{PM} will also pump a fraction of the gree gas, f_{gPM} (Fig. 4.59). From the well inflow performance relationship, calculate the bottomhole flowing pressure P_{wf}.

Assume that a sink exists just below the pump inlet. Fluids enter the well perforations, flow through the casing, and rise to the sink. Estimate the sink temperature (T_S) using Shiu's seventh flowing temperature correlation:[8]

$$T_S = T_{BH} - g_T[Z - A(1 - e^{-Z/A})]$$

where:

T_S is the sink temperature, °F
T_{BH} is the bottomhole temperature, °F
g_T is the geothermal gradient, °F/ft

Fig. 4.58 Graphic representation framework for the pumping problem (after Couto)

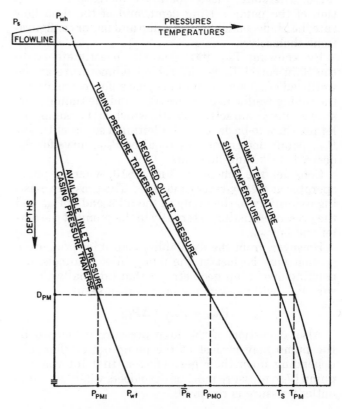

Fig. 4.59 Pump external environment (after Couto)

Z is the distance between the source (perforated interval) and the sink (pump inlet), ft

A is the relaxation distance, ft

given by:

$$A = W_t^{0.5253} \times \rho_L^{2.9303} \times d^{-0.2904} \times \gamma_o^{0.2608} \times \gamma_g^{4.4146} \times e^{-4.2051}$$

where:

W_t is the total mass flow rate, lb_m/sec

$\rho_L = \rho_o(1 - f'_w) + \rho_w \times f'_w$ is the stock tank liquid density, lb_m/ft^3

ρ_o is the oil stock tank density, lb_m/ft^3

ρ_w is the water stock tank density, lb_m/ft^3

d is the inside diameter of the pipe (casing), in.

γ_o is the API gravity of the oil

γ_g is the specific gravity of the water

The sink pressure may be evaluated using a multiphase flow correlation, calculating up the casing pressure traverse since the bottomhole pressure is known. The sink pressure becomes the pump inlet pressure, and the locus of the inlet pressure points is the available P_{PMI} curve plotted in Fig. 4.59 (also the casing pressure traverse for the desired flow rate).

The pump temperature is estimated from

$$T_{PM} = T_S + \Delta T_{PM} \qquad (4.58)$$

The quantity ΔT_{PM} represents the local temperature increase due to the pump operation. It is negligible for sucker rod and hydraulic pumps and may be estimated as a function of the motor heat loss for an electric motor-powered centrifugal pump. Since the efficiencies of subsurface centrifugal pumps and electric motors vary in narrow ranges, they are invariant with the pump and motor sizes, for the purpose of calculating ΔT_{PM}. Therefore, the temperature increase is a function of the pump setting depth and of the fluid flow rate, i.e., independent of the pump and motor configurations themselves.

By knowing T_{PM}, it is possible to estimate—with the Shiu correlation—the tubing temperature at the wellhead (T_{wh}) and from multi-phase flow correlations, the tubing wellhead pressure (P_{wh}) and the tubing pressure at any pump setting depth, which is the pump outlet pressure to be developed. Plotting a series of D_{PM} vs P_{PMO} points determines the required P_{PMO} curve for the desired stock tank flow rate.

Consider the scheme of Fig. 4.510, where a pump operates at steady-state conditions. The pressure existing externally to the pump at the inlet end is P_{PMI} and the pressure existing externally to the pump at the outlet end is P_{PMO}.

However, from the available pump inlet pressure a portion may be lost at the pump inlet channels, depending upon their geometry, so that internally a pressure P_{PML} will exist as the pump low pressure.

$$P_{PML} = P_{PMI} - \Delta P_{Inl} \qquad (4.59)$$

Also, a portion of the high pressure developed by the pump may be lost at the pump outlet channels, depending upon their geometry, so that internally a pump high pressure must be developed, when the pump outlet pressure is required:

$$P_{PMH} = P_{PMO} + \Delta P_{Out} \qquad (4.510)$$

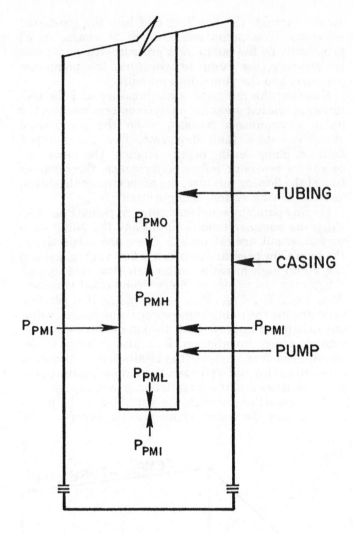

$$\Delta P_{Ext} = P_{PMO} - P_{PMI}$$

$$\Delta P_{Int} = P_{PMH} - P_{PML}$$

Fig. 4.510 External and internal pressures (after Couto)

Consider Fig. 4.511. Although the external differential pressure

$$\Delta P_{Ext} = P_{PMO} - P_{PMI} \qquad (4.511)$$

exists across the pump, it must develop a greater internal pressure increase

$$\Delta P_{Int} = P_{PMH} - P_{PML} \qquad (4.512)$$

to pump the fluids.

The inlet pressure loss

$$\Delta P_{Inl} = P_{PMI} - P_{PML} \qquad (4.513)$$

and the outlet pressure loss

$$\Delta P_{Out} = P_{PMH} - P_{PMO} \qquad (4.514)$$

are among the pumping problem variables most difficult to determine, because they depend on the geometry of the pump inlet and outlet. Both of them increase when D_{PM} decreases, as schematically shown in Fig. 4.512, and the free gas pumped fraction remains constant.

Both ΔP_{Inl} and ΔP_{Out} seem to have minor importance for sucker rod and centrifugal pumps. They may be

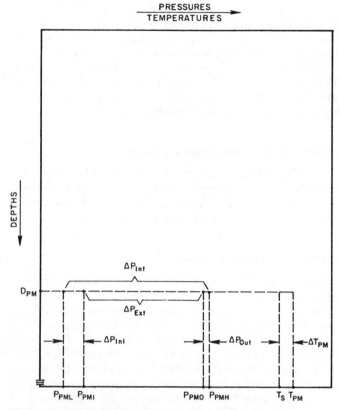

Fig. 4.511   *Pumping problem differential variables (after Couto)*

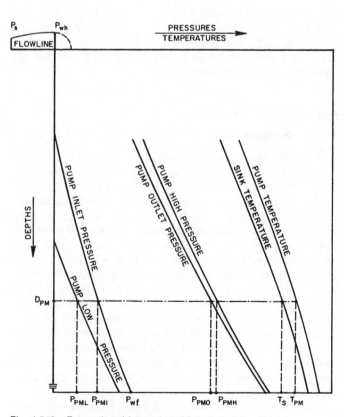

Fig. 4.512   *External and internal environments (after Couto)*

specially important in hydraulic pumps. Therefore, in general $P_{PML} = P_{PMI} - \Delta P_{Inl}$ and $P_{PMH} = P_{PMO} + \Delta P_{Out}$.

When ΔP_{Inl} is negligible $P_{PML} = P_{PMI}$ (4.515)

When ΔP_{Out} is negligible $P_{PMH} = P_{PMO}$ (4.516)

4.5433   In situ volumetric flow rates

To size a pumping installation correctly, determine the in situ volumetric flow rates. Assume that the pump is set at the depth D_{PM}, where the pump temperature is T_{PM}. At the low pressure and pump temperature conditions it is possible to calculate, through appropriate correlations, the oil, water, and gas formation volume factors (B_o, B_w and B_g) and the gas solubility in the oil, R_s, and the water, R_{sw}.

With this information, calculate the low pressure in situ liquid flow rate:

$$q_L = q'_o \times B_o + q'_w \times B_w \qquad (4.517)$$

and the low pressure in situ gas flow rate:

$$q_g = [q'_o(GOR - R_s) - q'_w \times R_{sw}] f_{gPM} \times B_g. \qquad (4.518)$$

The total in situ volumetric flow rate, or the pump low pressure volumetric flow rate, is then:

$$q_{PMLP} = \{q'_o \times B_o + q'_w \times B_w + [q'_o(GOR - R_s) - q'_w \times R_{sw}] f_{gPM} \times B_g\}_{(P_{PML}, T_{PM})} \qquad (4.519)$$

or

$$q_{PMLP} = q'_L \{f'_o \times B_o + f'_w \times B_w + [f'_o(GOR - R_s) - f'_w \times R_w] f_{gPM} \times B_g\}_{(P_{PML}, T_{PM})} \qquad (4.520)$$

At the high pressure and pump temperature conditions it is possible to calculate the total in situ volumetric flow rate:

$$q_{PMHP} = \{q'_o \times B_o + q'_w \times B_w + [q'_o(GOR - R_s) - q'_w \times R_{sw}] f_{gPM} \times B_g\}_{(P_{PMH}, T_{PM})} \qquad (4.521)$$

or

$$q_{PMHP} = q'_L \{f'_o \times B_o + f'_w \times B_w + [f'_o(GOR - R_s) - f'_w \times R_{sw}] f_{gPM} \times B_g\}_{(P_{PMH}, T_{PM})} \qquad (4.522)$$

Equations 4.520 and 4.522 show how the pump volumetric input and output flow rates may be calculated. Notice from the equations that the connective element between the environment and the pump volumetric input/output is the free gas pumped fraction.

Centrifugal pumps will always require that the low pressure flow rate be equal to or less than the highest rate recommended by the manufacturer. Also, the high pressure flow rate should be equal to or greater than the lowest recommended rate. However, the actual flow rate inside the pump, q_{PM}, varies along the pump axis. Therefore, it may be written $q_{PMUL} \geqslant q_{PMLP} \geqslant q_{PM} \geqslant q_{PMHP} \geqslant q_{PMLL}$.

As stated, a reciprocating pump may have its operation speed varied, so that its net displacement may vary between q_{PMUL} and q_{PMLL}. If D_{Act} is the pump displacement corresponding to the actual pump operating speed, it may be written $q_{PMUL} \geqslant D_{MXG} \geqslant D_{Act} \geqslant D_{MNG} \geqslant q_{PMLL}$ where the variables D_{MXG} and D_{MNG} are the net displacements necessary to handle respectively, the maximum and the minimum amount of free gas that can be pumped: $D_{MXG} = q_{PMLP}$ for $f_{gPM} = (f_{gPM})_{Max}$ and $D_{MNG} = q_{PMLP}$ for $f_{gPM} = (f_{gPM})_{Min}$.

4.5434   Differential pressure constraints

Refer to Fig. 4.513. The maximum internal differential pressure the pump housing will undergo is $\Delta P_{INH} = P_{PMH} - P_{PMI}$. It is always required that $\Delta P_{INH} \leq \Delta P_{INHL}$ to prevent pump internal rupture. The maximum permissible pump high pressure is therefore:

$$P_{PMHMX} = P_{PMI} + \Delta P_{INHL} \qquad (4.523)$$

As seen from Fig. 4.513, the pump must be set above point B to avoid bursting. The maximum external differential pressure the pump housing will face is $\Delta P_{EXH} = P_{PMI} - P_{PML} = \Delta P_{Inl}$.

It is required that $\Delta P_{EXH} \leq \Delta P_{EXHL}$. The minimum permissible pump low presure is:

$$P_{PMLMN} = P_{PMI} - \Delta P_{EXHL} \qquad (4.524)$$

Figure 4.513 infers that the pump may be set at any depth without risk of collapsing.

4.5435   Inlet pressure requirements

Refer to Fig. 4.513. The pump inlet pressure required to drive the fluids to be pumped into the pump is

$$P_{PMID} = \Delta P_{Inl} = P_{PMI} - P_{PML} = \Delta P_{EXH} \qquad (4.525)$$

where P_{PMID} stands for required (for driving) pump inlet pressure.

Therefore, it is always required that $P_{PMI} \geq P_{PMID}$. If a fraction $f_v = (1 - f_{gPM})$ of the free gas is to be vented, there must exist a pressure at the gas-liquid interface within the annulus sufficient to drive the gas up through the annulus and into the flowline. This re-quired (for venting) pump inlet pressure (P_{PMIV}) must offset the flowline inlet pressure (P_{wh}) and the dynamic gas column. This assumes that gas must be vented into the surface flow line with wellhead pressure of P_{wh}.

Therefore, the Required (for Venting) Pump Inlet Pressure is $P_{PMIV} = P_{wh} +$ dynamic gas column and it is also required that $P_{PMI} \geq P_{PMIV}$. The pump may not be set above the level corresponding to point A.

4.5436   Pump temperature constraint

In Fig. 4.513 a straight line representing the permissible pump temperature (T_{PMLIM}) is shown. Since it is required that $T_{PM} \leq T_{PMLIM}$, the pump should not be set below the depth equivalent to point C.

4.5437   Final considerations

Equations 4.520 and 4.522 refer to a particular pump whose ΔP_{Inl} and ΔP_{Out} characteristics must be known. The use of such equations depends on the availability of information from the pump manufacturer. Using these equations will allow a complete analysis of the well production capability for the pump under consideration.

By neglecting any pressure drop at either the inlet or outlet, the pump low pressure becomes equal to the pump inlet pressure and the pump high pressure equal to the pump outlet pressure. Then, Equations 4.520 and 4.522 assume the forms:

$$q_{PMLP} = q'_L \{f'_o \times B_o + f'_w \times B_w + [f'_o(GOR - R_s) - f'_w \times R_{sw}] f_{gPM} \times B_g\}_{(P_{PMI}, T_{PM})} \qquad (4.526)$$

and

$$q_{PMHP} = q'_L \{f'_o \times B_o + f'_w \times B_w + [f'_o(GOR - R_s) - f'_w \times R_{sw}] f_{gPM} \times B_g\}_{(P_{PMO}, T_{PM})} \qquad (4.527)$$

Equations 4.526 and 4.527 no longer depend on the pump configuration. Normally, centrifugal pumps ΔP_{Inl} and ΔP_{Out} are neglected and it is possible to perform one well production capability analysis for all available configurations.

4.544   Solution approaches

4.5441   Introduction

All three operational variables (q'_L, f_{gPM}, D_{PM}) directly or indirectly determine the pump low pressure volumetric flow rate and the pump high pressure volumetric flow rate, according to Equations 4.520 and 4.522. Although the pump setting depth does not directly appear in any of those equations, it must be assumed to determine the pump low pressure (given stock tank flow rate). Also, the pump high pressure is a function of the pump setting depth. Furthermore, the pump low pressure and the pump high pressure are directly related one to the other, given the free gas pumped fraction, the stock tank flow rate and the pump setting depth.

Because of this special feature of the pumping problem and because of the tri-dimensional Cartesian system (q, p/t, h) chosen as a medium to represent the pumping problem variables, different approaches may solve the problem.

For instance, each time that one of the operational variables of Equations 4.520 and 4.522 is held constant,

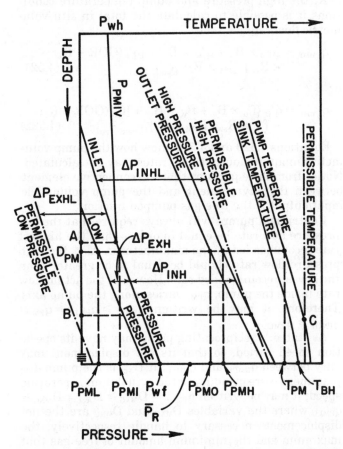

Fig. 4.513 Pumping problem requirements and constraints (after Couto)

a graphical solution approach results. These operational approaches include systems approach, flow rate approach, and free gas approach.

4.5442 The systems approach

4.54421 Introduction

In the systems approach, the pump is set at a particular setting depth. A flowing system configuration results, characterized by the tubing string length and the casing component length. By choosing another pump setting depth, another flowing system configuration is defined, and the process may be repeated until all the possible flowing systems are analyzed.

The Systems Approach determines the variations in the environment caused by changes in the stock tank flow rate and the free gas pumped fraction. After that, the volumetric flow rates at the inlet and outlet of the pump are calculated.

Two cases must be considered. In case A, the pressure losses across the inlet and outlet may be neglected. As noted earlier, the low and high pressure flow rates do not depend on the pump configuration itself (equations 4.526 and 4.527).

When the two in situ flow rates are plotted against q'_L, using f_{gPM} as a parameter, two types of analysis are possible: (a) Pump versatility analysis, when a particular pump configuration is chosen and the objective is to determine the minimum and the maximum stock tank flow rates obtainable from the system and the corresponding needs for gas venting. If the pump configuration is varied, all system production possibilities may be analyzed; (b) Pump selection analysis, when a particular stock tank flow rate is chosen. For this desired stock tank flow rate, the objective is to determine what pump configurations may be used and the corresponding needs for gas venting.

In case B, the pressure losses ΔP_{Inl} and ΔP_{Out} may not be neglected, as when hydraulic pumps pump high flow rates. In this case, the pump volumetric flow rates depend on the pump configuration from equations 4.520 and 4.522. This means that the pump configuration is determined before using the equations. Therefore, only the pump versatility analysis may be performed after a plot of q_{PMLP} and q_{PMHP} against q'_L. The pump selection has already been made.

The complete analysis of the production possibilities requires a plot for each usable pump.

4.54422 Level equivalent master region

The clue for the understanding of the flowing system production capability analysis is the definition of the level equivalent master region. This is shown on a graph which displays the problem boundaries and shows the system behavior, for different stock tank flow rates and free gas pumped fractions, but for a particular pump setting depth.

To begin, assume that the pump will be set at the well perforated interval. Refer to Fig. 4.514, where gauge pressures (p) are plotted on the abscissa and volumetric flow rates (q) on the ordinate. On this plane, p × q will be defined as the master region equivalent to this particular level of operation, the perforated interval depth.

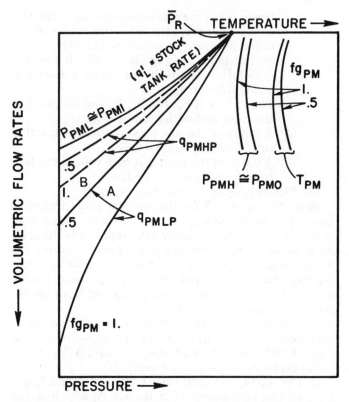

Fig. 4.514 Level equivalent master region (after Couto)

First, the Inflow Performance Relationship of the well is plotted. Either a linear or a curved relationship may be used. In the following sections a Vogel type inflow curve is used. Notice that with the pump set at the same depth as the perforated interval, the inflow performance curve gives for any rate the pressure existing at the pump inlet. Thus, it is possible to call that curve the available pump inlet pressure curve.

Using Equations 4.526 and 4.527, and the free gas pumped fraction as a parameter, a family of q'_L vs. q_{PMLP} curves and a family of q'_L vs. q_{PMHP} curves are plotted using the pump inlet pressure curve as a reference. Since every stock tank flow rate value corresponds with one pump inlet pressure value, the curves P_{PMI} vs. q_{PMLP} and P_{PMI} vs. q_{PMHP} are plotted instead to represent the q'_L vs. q_{PMLP} and q'_L vs. q_{PMHP} curves (Fig. 4.514).

The following procedure may be used when plotting in situ flow rates calculated from Equations 4.526 and 4.527 for any pump depth:

(1) Assume a pump setting depth (D_{PM})
(2) Assume several free gas pumped fractions (f_{gPM})
(3) Assume several stock tank flow rates (q'_L)
(4) For each combination of free gas pumped fraction and stock tank flow rate, calculate:
  (a) The sink temperature and the available pump inlet pressure. The sink temperature may be evaluated using the Shiu flowing temperature correlation. The inlet pressure is calculated from the bottomhole up to the sink, using a multiphase flow correlation.
  (b) The required pump outlet pressure and the pump temperature $T_{PM} = T_S + \Delta T_{PM}$.
      The process is iterative, since ΔT_{PM} depends on the pump outlet pressure and vice versa. The outlet pressure is calculated from the separator to the tubing extremity, that is, the pump outlet, using multiphase flow correla-

tions. ΔT_{PM} may be estimated as a function of the flow rate, the inlet pressure and the outlet pressure, as mentioned earlier, or may be considered negligible, in which case the calculation of the pump outlet pressure is straightforward.

(c) The values of B_o, B_w, B_g, R_s and R_{sw}, both at the inlet and outlet conditions.

(d) A value of the low pressure volumetric flow rate, from Equation 4.526.

(e) A value of the high pressure volumetric flow rate, from Equation 4.527.

(5) Plot P_{PMI} vs. q_{PMLP} and P_{PMI} vs. q_{PMHP}.

Refer to Fig. 4.515 for the various plots, including the inflow performance, i.e., the available pump inlet pressure curve. The quantity ΔP_{Int} represents the pressure increase the pump has to develop to deliver the stock tank flow rate q'_L at the particular free gas pumped fraction. The quantity Δq_{PM} represents the decrease in volume of the fluids, occurring from the inlet to the outlet of the pump.

Notice, in Fig. 4.516, that there is no difficulty in representing P_{PMI} vs. q_{PMLP} and P_{PMI} vs. q_{PMHP} from Equations 4.520 and 4.522 and also ΔP_{Inl} and ΔP_{Out}, when these pressure losses are not negligible.

In Fig. 4.514, the solid lines for $f_{gPM} = 1$ and $f_{gPM} = 0.5$ are the boundaries of a Region A, which is the locus of the (q'_L vs q_{PMLP}) points. Similarly the dashed lines for $f_{gPM} = 1$ and $f_{gPM} = 0.5$ are the boundaries of a Region B, which is the locus of the (q'_L vs q_{PMHP}) points. Regions A and B normally overlap themselves. However, the points of both regions correspond. Thus the name level equivalent master region will apply to the set of both regions.

The master region is a geometrical representation of

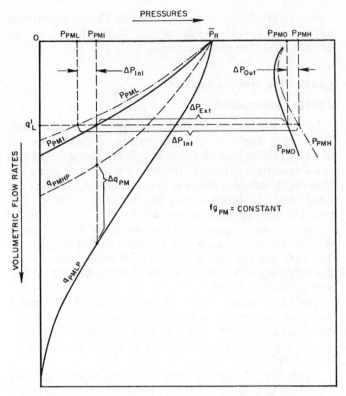

Fig. 4.516 *External and internal pressure increase (after Couta)*

the flowing system behavior which does not account for the restrictions that the pump itself imposes. Using a pump as part of the flowing system entails a series of *interferences* in the master region, reducing its dimensions and limiting the ability of the system to deliver fluids.

4.54423 Interferences

Interference is any reduction in the system's ability to deliver fluids, caused by any operational, technical, or economical requirements or constraints.

The available pump inlet pressure curve is represented in Fig. 4.517. Also represented are three required (for venting) pump inlet pressure curves, calculated ($P_{PMIV} = P_{wh} +$ dynamic gas column). Notice that when the free gas pumped fraction equals 1, no gas is being vented and P_{PMIV} is zero for any stock tank flow rate values. If 50% of the free gas is to be pumped, the maximum stock tank flow rate obtainable from the system is limited to q'_{L_1}. For any stock tank flow rate greater than q'_{L_1} the available pump inlet pressure becomes less than the pressure required for venting. As a consequence, for $f_{gPM} = 0.5$ that portion of the master region for which the pump inlet pressure is less than A becomes inaccessible.

Since ΔP_{Inl} is considered negligible, it follows that the required (for driving) pump inlet pressure is always zero, for any stock tank flow rates and free gas pumped fractions $P_{PMID} = \Delta P_{INL} = 0$.

The pump inlet pressure should be at least 300 psig for centrifugal pumps operating in high GOR wells. It is not clear, however, if 300 psig is required for venting the gas, for driving the fluids into the pump, or to avoid excessive amounts of in situ gas volumes at the pump inlet which might cause pump gas lock. In any case, the required pump inlet pressure depends

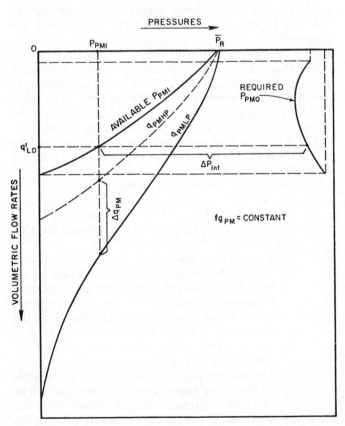

Fig. 4.515 *Pressure increase and volume reduction (after Couto)*

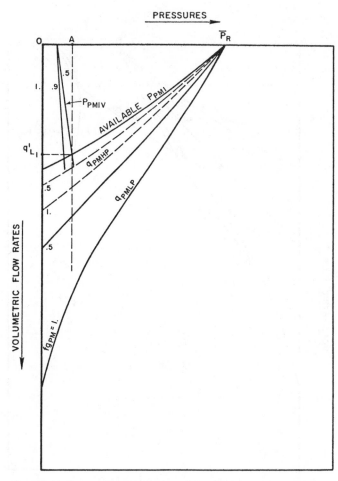

Fig. 4.517 Gas venting interference (after Couto)

values of the pump inlet and pump outlet pressures are known.

From Fig. 4.519, this new interference causes the loss of a portion of the master region equivalent to the triangle ABC. This means that any stock tank flow rate greater than q'_{L_3} will require some gas venting to meet criterion. Even if the bottomhole gas-liquid separation is 100% efficient and all free gas can be vented, the maximum stock tank flow rate obtainable from the system is limited to q'_{L_4}. Assuming, however, that it is impossible to vent more than 50% of the free gas so that f_{gPM} is at least 0.5, the maximum stock tank flow rate is reduced to q'_{L_5}. This interference is attributed to separation.

Another interference is from the pump internal differential pressure limitation. In Fig. 4.521 the permissible pump high pressure curve was derived by adding to each point of the available pump inlet pressure curve the quantity ΔP_{INHL}, i.e., the maximum permissible excess differential pressure inside the pump housing:

$$P_{PMHMX} = P_{PMI} + \Delta P_{INHL}.$$

Also represented in Fig. 4.521 are two pump high pressure curves. The maximum stock tank flow rate is limited to q'_{L_6}, as a consequence of the pump internal differential pressure limitation. For any stock tank flow rate greater than q'_{L_6} the actual pump high pressure would be greater than the permissible pump high pressure, with the risk of bursting. Venting some gas would make this interference more critical.

Fig. 4.522 shows how to account for a non-negligible

on both stock tank flow rate and free gas pumped fraction.

Fig. 4.518 shows how to account for non-negligible inlet pressure losses for hydraulic pumps. Two pump low pressure curves were plotted. For example, when the free gas pumped fraction equals one, the maximum stock tank flow rate obtainable from the system is approximately limited to q'_{L_2}. For any stock tank flow rate slightly greater than q'_{L_2} the gauge P_{PML} would be negative.

The limiting pump low pressure volumetric flow rate curve is plotted in Fig. 4.519. The limiting pump low pressure volumetric flow rate (q_{PMLIM}) is that rate beyond which the in situ input volumetric flow rate is so large that the pump operation becomes inefficient, impossible, or risky because of gas lock.

Some manufacturers have suggested that the pump inlet volumetric flow rate should not be greater than twice the pump outlet volumetric flow rate. More consistently, O'Neil suggested that the pump inlet volumetric flow rate should not exceed twice the stock tank flow rate. A general criterion suggests that the pump low pressure volumetric flow rate should not exceed the product of the in situ low pressure liquid flow rate by a gas lock factor, GLF: $q_{PMLIM} = GLF (q'_o \times B_o + q'_w \times B_w)$ and $q_{PML} \leq q_{PMLIM}$.

In this work a gas lock factor of 2 was used as recommended by O'Neil. Notice that for other types of pumps, such as sucker rod and hydraulic, the GLF may be calculated once the pump geometry and the

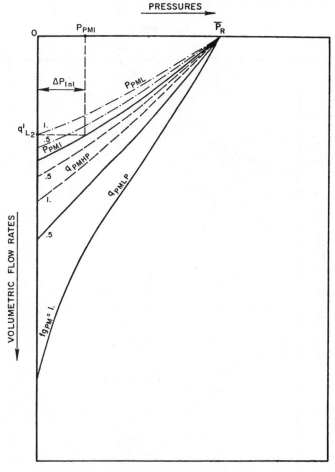

Fig. 4.518 Inlet pressure loss interference (after Couto)

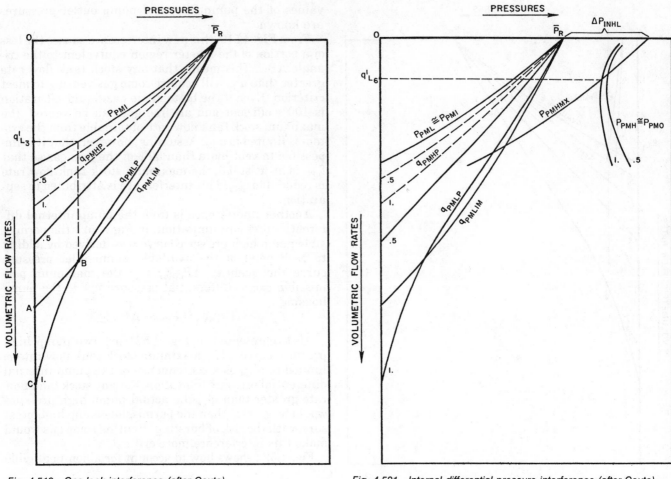

Fig. 4.519   Gas lock interference (after Couto)

Fig. 4.521   Internal differential pressure interference (after Couto)

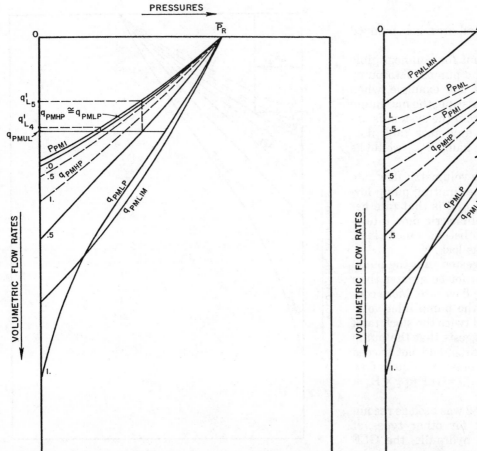

Fig. 4.520   Capacity range upper limit interference (after Couto)

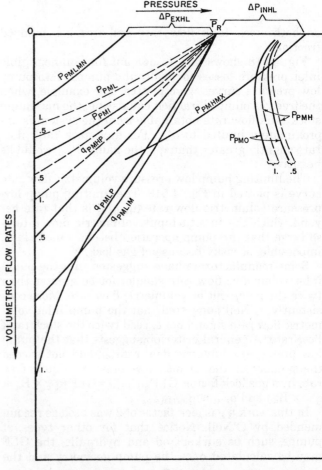

Fig. 4.522   External differential pressure interference (after Couto)

ΔP_{out} which will alter the pump high pressure from the pump outlet pressure. The permissible pump low pressure curve is also plotted in Fig. 4.522, derived by $P_{PMLMN} = P_{PMI} - \Delta P_{EXHL}$. The figure illustrates how to analyze the risk of pump collapse. If the permissible pump low pressure curve cuts any pump low pressure curve there will be pump external differential pressure interference.

Temperature interference analysis is shown in Fig. 4.523. The motor heat loss is proportional to the internal pressure differential $\Delta P_{Int} = P_{PMH} - P_{PML}$, and so will be temperature increase, ΔT_{PM}. Consequently, the pump temperature curve will look like Fig. 4.523 where a straight line represents the permissible pump temperature. Thus, very high and very low stock tank flow rates will tend to cause pump temperature interferences.

Finally, a straight line is plotted in Figure 4.524, representing the pump flow rate range lower limit of the pump. If all the free gas is admitted into the pump, the stock tank flow rate will have a value of q'_{L7}. Any attempt to produce less than q'_{L7} will make the pump high pressure flow rate less than q_{PMLL}. Also, when producing the flow rate q'_{L7}, any attempt to vent gas will cause the same problem. Therefore, q'_{L7} represents a minimum limit for the stock tank flow rate obtainable from the system.

4.54424 Dominance

Several interferences cause limitations to the maximum stock tank flow rate, and some interferences cause limitations to the minimum stock tank flow rate. When determining the maximum and minimum stock tank flow rate possible from a system, the problem is to find what interference or what combination of interferences dominates the other interferences, either from the maximum or minimum side of the question. One interference dominates when it plays any role in determining the final form of the master region. Different combinations of dominances may determine a number of different master region configurations.

4.54425 Pump incompatibility

A pump may be inadequate to handle a desired stock tank flow rate in the context of a particular flowing system. Furthermore, a pump may be totally imcompatible with the flowing system, in which case it will be unable to handle any stock tank flow rate in the context of that system.

For example, assume a system where the maximum gas venting is limited to 70% (Fig. 4.525). If the pump in this system has the flow rate range lower limit indicated in the figure, it must have upper limit equal to or greater than X. Otherwise it would be incompatible with the system, as it may be seen. Notice, however, that $q_{PMUL} \geq X$ is a necessary, but not sufficient, condition.

Fig. 4.526 shows a pump which is compatible with the previous system and can handle any stock tank flow rate within the interval $q'_{LMN} - q'_{LMX}$. As a matter of interest, notice that q'_{LX} is the only stock tank flow rate that can be obtained with any free gas pumped fraction. As the flow rate approaches q'_{LMN}, more free gas

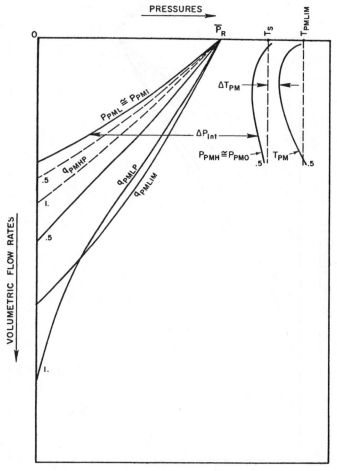

Fig. 4.523   Temperature interference (after Couto)

must be admitted to the pump. As the flow rate approaches q'_{LMX}, more free gas must be vented.

4.54426 Pump versatility analysis

The level equivalent master region is a graphic representation of the system behavior which permits a quick analysis of the versatility of the pump in the context of that system. As an illustration, consider Fig. 4.527, from which several conclusions may be drawn:

(1) The maximum stock tank flow rate obtainable from the system is q'_{LMX}. A greater production will cause $P_{PMI} < P_{PMIV}$ and/or $q_{PMLP} > q_{PMUL}$.

(2) In order to obtain q'_{LMX} it is necessary to pump exactly 50% of the free gas. An attempt to pump more than 50% of the free gas will cause $q_{PMLP} > q_{PMUL}$. An attempt to pump less than 50% will cause $P_{PMI} < P_{PMIV}$.

(3) The maximum stock tank flow rate obtainable from the system without venting gas is q'_{LMXNV}. A greater production rate, without venting gas, would cause $q_{PMLP} > q_{PMUL}$. Gas lock would occur beyond point A.

(4) It is possible to produce q'_{LMXNV} venting any amount of free gas. But the pump high pressure will probably increase because less gas is being used for lifting purposes in the tubing string. Therefore, the operation would have to be justified on an economical basis.

(5) The minimum stock tank flow rate obtainable

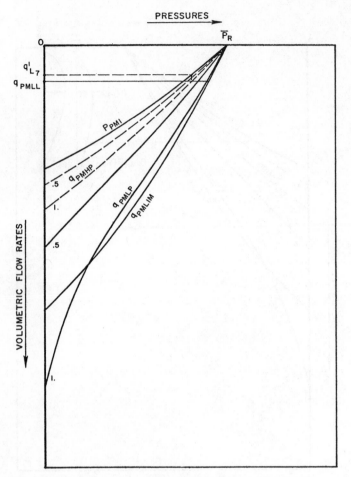

Fig. 4.524   Capacity range lower limit interference (after Couto)

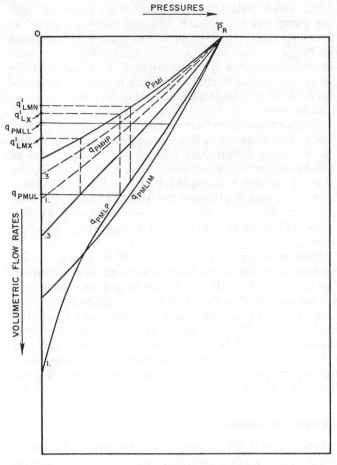

Fig. 4.526   Pump compatibility analysis (after Couto)

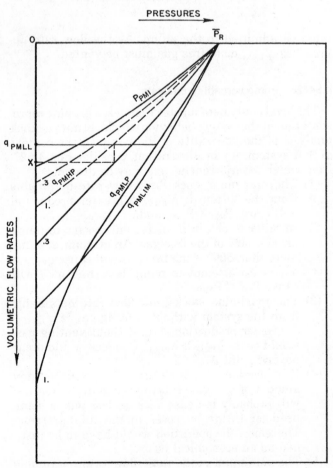

Fig. 4.525   Pump incompatibility analysis (after Couto)

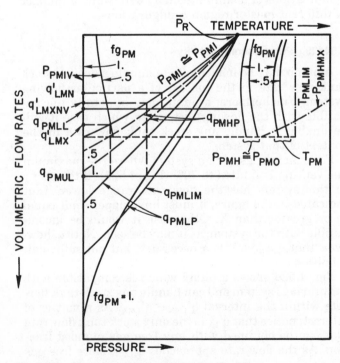

Fig. 4.527   Pump versatility analysis (after Couto)

from the system is q'_{LMN}. A smaller production will cause q_{PMHP} to be less than q_{PMLL}.

(6) In order to obtain q'_{LMN} it is necessary to pump all the free gas. Any attempt to vent gas will also cause q_{PMHP} to be less than q_{PMLL}.

(7) Any desired stock tank flow rate q'_L such that $q'_{LMN} < q'_L < q'_{LMX}$ is possible.

Similar analyses may be made for any master region configurations.

4.54427   Pump selection analysis

If the desired stock tank flow rate q'_{LD} is predetermined, the pump configuration may be immediately selected (Fig. 4.528).

If there is a pump (or a set of interconnectable pumps) whose recommended flow rate range covers the interval (A-B), the desired flow rate is feasible without gas venting. If 50% of the free gas is to be vented, then another pump (or set of pumps) whose operational rate range covers the interval (C-D) will be sufficient.

A pump covering the interval (C-B) will be flexible enough to handle q'_{LD} with the free gas pumped fraction varying between 0.5 and 1.

4.54428   Further interferences

Further interferences may cause the master region to shrink even more. The maximum brake horsepower with which the pump can operate, and any sort of economic interference should be considered. These should be checked after a pump sizing procedure has been applied to the problem data and before a final decision is made.

4.54429   Generalizations

The preceding analysis assumed a system with the tubing string extended to the well perforations, i.e., a system where no casing component was present. For such a system, a level equivalent master region was defined, based on the well inflow performance relationship.

As indicated previously, if any casing portion is introduced into the system, a corresponding portion of the tubing string must be eliminated from the system. This will define a new pump setting depth, or a new operational level within the well. Or, conversely, if a different pump setting depth is defined, a portion of the tubing string must be cut and an equivalent portion of casing must be introduced.

For any other pump setting depth a master region may be derived, based on a shifted inflow performance curve. Refer to Fig. 4.529, where the casing pressure traverse is represented for a number of stock tank flow rates, as calculated by multiphase flow correlations, starting from the true IPR curve. Also represented are the corresponding pump outlet pressure curves (for a particular free gas pumped fraction).

At any level in the well it is possible to know (for any feasible stock tank flow rate) the available pump inlet pressure, the required pump outlet pressure, and the pump temperature for any value of free gas pumped fraction. Therefore, all the elements necessary for defining a level equivalent master region may be calculated.

The level equivalent master region may be defined when the pump is set below the well perforations. The difficulty is that a downward flow will exist in the casing-tubing annulus for which there is not yet a specific multiphase flow correlation available.

This process may be repeated for different separation pressures, flowline configurations (diameters and lengths), and tubing string diameters, making it possible to cover all the dimensions of the pumping problem. Final decisions concerning a new installation may then be based upon economical considerations.

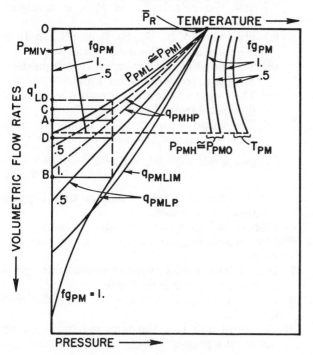

Fig. 4.528   *Pump selection analysis (after Couto)*

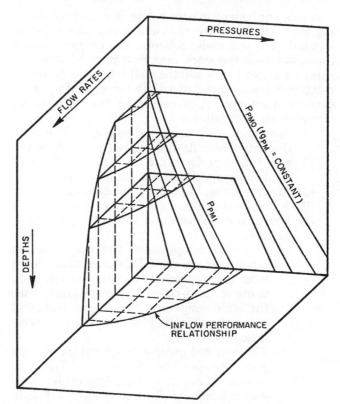

Fig. 4.529   *Shifted inflow performance curves (after Couto)*

4.5443 The flow rate approach

4.54431 Introduction

In the flow rate approach, the stock tank flow rate is set at a particular value, whereas the pump setting depth and the free gas pumped fraction may vary. The entire range of flow rates the well can deliver may be investigated by applying the flow rate approach to a number of different stock tank flow rates. The relationships between the operational variables and the environmental variables, and between these variables and the in situ low/high pressure volumetric flow rates, may be represented in a rate equivalent master region. The same conceptual framework previously developed applies to this approach, *mutatis mutandi.*

If case A is considered (no pressure loss at the inlet or outlet of the pump), Equations 4.526 and 4.527 apply; since these equations do not depend on the pump, two analyses may be performed. First is pump versatility analysis. If the pump configuration is particularized, an analysis may determine the minimum and the maximum free gas pumped fraction with which the stock tank flow rate under consideration may be obtained, and the minimum and maximum pump setting depth.

The second is pump selection analysis. When the pump configuration varies, it is possible to determine the pump which is best suited to handle the flow rate under consideration, as well as the pump setting depth and the free gas pumped fraction permissible variation ranges. Also, if any restriction in the casing internal diameter limits the pump setting depth, this type of interference may quickly be analyzed and its effect easily accounted.

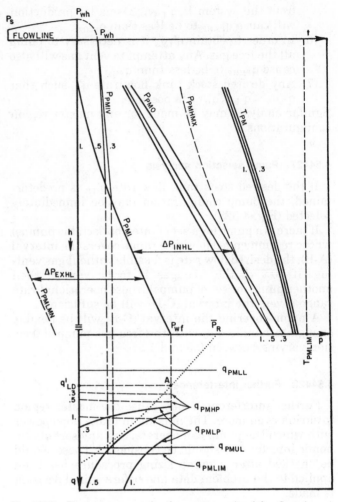

Fig. 4.530   Flow rate approach - the master region (after Couto)

4.54432 Rate equivalent master region

A rate equivalent master region is a graph displaying the well flow problem boundaries under different pump depths and under different gas pumping conditions, but with the stock tank flow rate set at a constant value. In Fig. 4.530 the well inflow performance relationship was plotted at a depth corresponding to the perforated interval. A procedure for the construction of the graph is as follows:

(1) Assume a stock tank flow rate q'_{LD}. Calculate the bottomhole flowing pressure, P_{wf}, from the well inflow performance relationship. The bottomhole temperature is assumed known.

(2) Assume several pump setting depths.

(3) Assume several free gas pumped fractions.

(4) For each combination of D_{PM} and f_{gPM} values, calculate:

(a) The sink temperature and the available pump inlet pressure. This later is calculated from the bottomhole upwards to the sink, using a multi-phase flow correlation, after the sink temperature has been evaluated using the Shiu flowing temperature correlation.

(b) The required pump outlet pressure and the pump temperature $T_{PM} = T_S + \Delta T_{PM}$. The process is iterative: the pump outlet pressure depends on ΔT_{PM} and vice versa. P_{PMO} is calculated from the separator to the tubing

extremity, i.e., the pump outlet, using multiphase flow correlations. ΔT_{PM} may be estimated as a function of q'_{LD}, P_{PMI}, and P_{PMO}, or may be considered negligible, in which case the calculation of P_{PMO} is straightforward.

(c) The values of B_o, B_w, B_g, R_s and R_{sw}, both at the inlet and at the outlet conditions.

(d) The value of the low pressure volumetric flow rate, from Equation 4.526.

(e) The value of the high pressure volumetric flow rate, from Equation 4.527.

(f) The value of the limiting pump low pressure volumetric flow rate, as follows:

$$q_{PMLIM} = 2 \times q'_{LD}(f'_o \times B_o + f'_w \times B_w)$$
$$(P_{PMI}, T_{PM})$$

A value other than 2 may be used if pump tests are available.

(5) Plot pump depth vs pump inlet pressure.

(6) Plot D_{PM} vs P_{PMO} and D_{PM} vs T_{PM} for each f_{gPM} value.

(7) Plot the pump inlet pressure vs the limiting low pressure volumetric flow rate.

(8) Plot P_{PMI} vs q_{PMLP} and P_{PMI} vs q_{PMHP} for each f_{gPM} value.

(9) Calculate and plot the required (for venting) pump inlet pressure: $P_{PMIV} = P_{wh} +$ dynamic gas column.

(10) For any particular pump to be analyzed plot the following information:

(a) The limiting temperature as a straight line.

(b) The maximum permissible pump high pressure:

$$P_{PMHMX} = P_{PMI} + \Delta P_{INHL}$$

and the minimum permissible pump low pressure:

$$P_{PMLMN} = P_{PMI} - \Delta P_{EXHL}$$

(c) The upper and lower permissible flow rates, q_{PMUL} and q_{PMLL}.

4.54433 Interferences

Flow rate master region interferences are quickly analyzed with Fig. 4.530. Pump temperature interferences never occur for this illustration. Pump internal differential pressure interferences may occur for f_{gPM} less than 0.50. Pump external differential pressure interference is impossible because ΔP_{EXHL} is greater than P_{wf}. The limiting pump low pressure flow rate interferes but is dominated by the flow rate range

upper limit. The flow rate range lower limit does not interfere. The pressure required to vent gas interferes for any f_{gPM} different from 1.

4.54434 Pump incompatibility

Pump incompatibility is easily detected with the help of Fig. 4.531. For instance, any pump for which $q_{PMLL} >$ X, or $T_{PMLIM} < Y$, or $\Delta P_{INHL} < Z$ will be incompatible with the stock tank flow rate q'_{LD}.

4.54435 Pump versatility analysis

In Figure 4.532 the pump versatility analysis was performed and the pump can be set at any depth in the interval $D_{PMMIN} - D_{PMMAX}$. To set the pump at D_{PMMIN}, pump exactly 50% of the free gas. Pumping less gas would make the pump inlet pressure less than the pressure required for venting. Pumping more gas will force the pump out of its flow rate range lower limit. If all the free gas is to be pumped, the pump setting depth must be restricted to the interval $A - D_{PMMAX}$.

4.54436 Pump selection analysis

Pump selection analysis is performed with the help of the flow rate master region in Fig. 4.533. Assuming that it is impossible to vent gas and supposing that there is a liner at the depth A which prevents

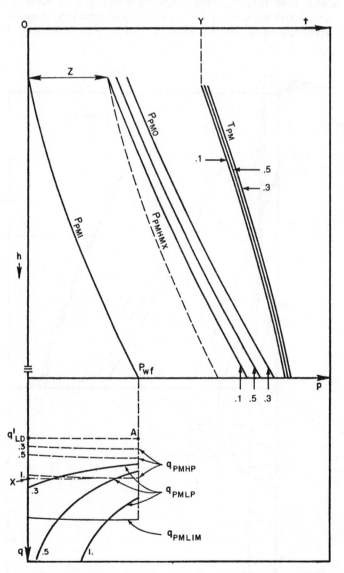

Fig. 4.531   *Flow rate approach - pump incompatibility analysis (after Couto)*

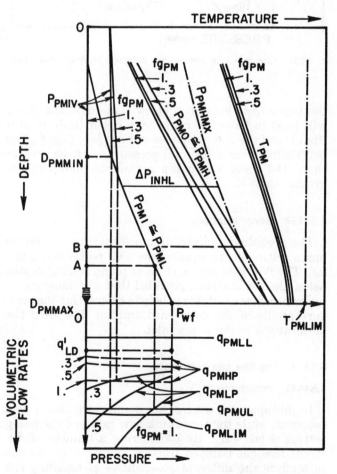

Fig. 4.532   *Flow rate approacch - pump versatility analysis (after Couto)*

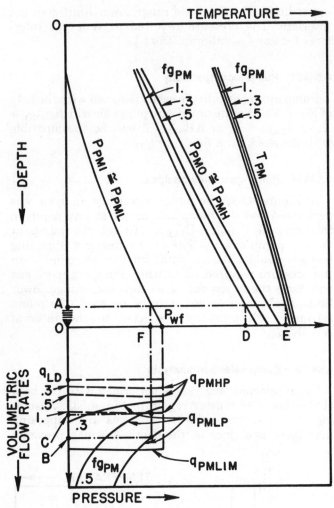

Fig. 4.533   *Flow rate approach - pump selection analysis (after Couto)*

the pump from being run below that point, the pump will need to have the rate range upper limit greater than B and the rate range lower limit less than C. The permissible excess internal pressure must be greater than (D–F) and the pump permissible temperature greater than E.

4.54437   Generalizations

The preceding analysis assumed that the maximum pump setting depth equaled the well perforated interval. The flow rate approach finds pump setting depths below the perforations, provided that a reliable correlation becomes available to help in calculating the pressure profile of the downward annular flow from the perforations to the pump inlet.

4.5444   The free gas approach

4.54441   Introduction

In this approach the free gas pumped fraction is held constant, while the stock tank flow rate and the pump setting depth vary. By considering a number of different free gas pumped fraction values, it is possible to analyze the different possibilities of handling the free gas when producing the well.

Master region—The free gas master region shown

in Fig. 4.534 was derived using a procedure similar to those in the previous approaches: A free gas pumped fraction value is assumed, several stock tank flow rates and pump setting depths are assumed, and the necessary variables are calculated for each combination of rate and depth values.

Interference—If the pump characteristics are imposed on the master region, as in Fig. 4.535, interference analyses may be performed.

Incompatibility—Pump incompatibility appears to be difficult to analyze using this approach, although the analysis may be performed.

Pump versatility analysis—The ability of a pump to handle different stock tank rates at different depths, with the particular f_{gPM} under study, is analyzed in the same manner as previously explained. For instance, the maximum flow rate obtainable with the pump in Fig. 4.535 is q'_{L2}, and the pump must be set at a depth corresponding to the point A to make that production. A greater flow rate will make the inlet pressure insufficient to vent the gas (with the pump set at point A) or the pump temperature greater than the permissible pump temperature (if the pump is set below point A).

Pump selection analysis—Once a particular stock tank flow rate is chosen, the best pump to handle it, under the particular free gas pumped fraction, may be determined using the master region. For instance, if

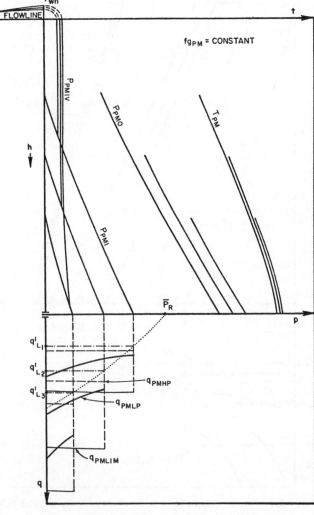

Fig. 4.534   *Free gas approach - the master region (after Couto)*

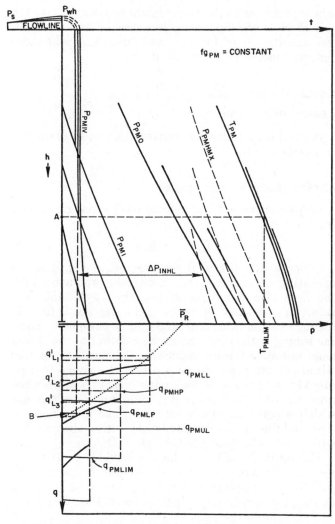

Fig. 4.535 Free gas approach - pump versatility and selection analyses (after Couto)

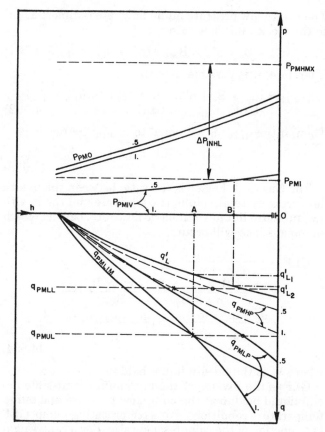

Fig. 4.536 Inlet pressure approach (after Couto)

a pump exists whose rate range upper limit is equal to B (Fig. 4.535) this pump will be better suited to handle the flow rate q'_{L_2} than the one in the figure.

4.54442 Generalizations

It is also possible to extend this approach to solve problems of pumping below the casing perforations, depending on the availability of downward annular multiphase flow correlation.

4.5445 Other solution approaches

Equations 4.520 and 4.522 offer other graphical approaches to solve the pumping problem. One example is shown in Fig. 4.536. There, the constant variable is the pump inlet pressure.

Fig. 4.536 shows how the stock tank flow rate varies with the depth for a particular pump inlet pressure value. It also shows the influence of the free gas pumped fraction. Since it represents all three operational variables, this approach appears promising. Using the pump characterized in Fig. 4.536, the maximum stock tank rate obtainable when pumping all the free gas is q'_{L_1} and the pump must be set at A. The maximum stock tank rate obtainable when $f_{gPM} = 0.5$ is q'_{L_2} and the pump will be set at B.

4.545 Special topics

4.5451 Pump design criterion

When observed from a capacity range standpoint, the pumps on the market do not seem to follow any apparent sequential criterion. From the preceding study a criterion is proposed which may help establish upper and lower flow rate range limits to which the pumps of a series should be designed and manufactured.

First, the minimum volumetric output that a pump must handle is the in situ high pressure liquid volumetric flow rate:

$$(q_{PMHP})_{f_{gPM}=0} = q_o + q_w = q'_o \times B_o + q'_w \times B_w \tag{4.528}$$

This is slightly greater than the stock tank flow rate. Therefore, if for a particular pump the rate range lower limit is set equal to the stock tank flow rate $q_{PMLL} = q'_L = q'_o + q'_w$, the pump will actually see a volumetric output always greater than the range lower limit:

$$q_{PMHP} = q_o + q_w + q_g > q_{PMLL} = q'_o + q'_w \tag{4.529}$$

Secondly, the limiting pump low pressure volumetric flow rate criterion should be examined. It says that the pump low pressure volumetric flow rate must be equal to or less than the in situ liquid flow rate multiplied by a gas lock factor:

$$q_{PMLIM} = (q'_o \times B_o + q'_w \times B_w)(GLF) \tag{4.530}$$

and

$$q_{PMLP} \leqslant q_{PMLIM}$$

The in situ low pressure liquid flow rate is almost equal to the stock tank flow rate:

$$(q'_o \times B_o + q'_w \times B_w) \cong (q'_o + q'_w) = q'_L \quad (4.531)$$

Therefore, it is possible to write:

$$q_{PMLIM} = (q'_o \times B_o + q'_w \times B_w)(GLF) \cong (q'_o + q'_w)$$
$$(GLF) = q_{PMLL} \times GLF \quad (4.532)$$

Then, if the q_{PMUL} is set equal to q_{PMLIM} the result is:

$$q_{PMUL} = GLF \times q_{PMLL} \quad (4.533)$$

The Gas Lock Factor is the ratio between the in situ low pressure total volumetric flow rate and the in situ low pressure liquid volumetric flow rate, beyond which pump gas lock will occur:

$$GLF = \frac{q_{PMLIM}}{q'_o \times B_o + q'_w \times B_w}$$
$$= \frac{(q'_o \times B_o + q'_w \times B_w + q'_g \times B_g)_{Max}}{q'_o \times B_o + q'_w \times B_w}$$
$$= \frac{q_o + q_w + (q_g)_{Max}}{q_o + q_w} = \frac{q_L + (q_g)_{Max}}{q_L} = \frac{1}{(\lambda_L)_{Min}} \quad (4.534)$$

where λ_L is the no-slip liquid holdup

GLF is the inverse of the minimum permissible no-slip liquid holdup at the pump low pressure and pump temperature conditions. For a reciprocating pump, GLF is a function of the compression ratio. For a centrifugal pump, GLF is a function of the stage head capacity.

Let the overlap factor be the ratio beween the rate range lower limit of the $(i + 1)^{th}$ pump and the rate range upper limit of the ith pump in a series of interconnectable pumps:

$$\text{overlap factor} = OF = \frac{(q_{PMLL})_{i+1}}{(q_{PMUL})_i} \quad (4.535)$$

The GLF is, by nature, greater than 1, whereas the OF must be less than 1 to cause the overlap between the capacity range of the pumps. Therefore, the pumps of a series of interconnectable pumps should satisfy the following relations:

$$(q_{PMUL})_i = (q_{PMLL})_i \times GLF \quad (4.536)$$

$$(q_{PMLL})_{i+1} = (q_{PMUL})_i \times OF \quad (4.537)$$

For instance, if GLF = 1.67 and OF = 0.90 a series of pumps like the one that is listed below will result:

Pump #	q_{PMLL}	q_{PMUL}
1	1000	1670
2	1500	2500
3	2250	3750
4	3375	5625
5	5063	8438
6	7594	12,656

It will be always possible to select single or tandem pumps in this series to operate in gassy oil wells and handle any stock tank flow rates from slightly less than 1,000 to slightly less than 12,656 if f_{gPM} may be varied between 1.0 and 0.0.

This reasoning suggests an approach to use pumps in sequence and should be the object of further study. Of course, neither GLF nor OF must be constant for the whole pump series. It is convenient to always make OF less than 1, principally to take care of any errors in the calculation of $(q_{PMHP})_i$ and $(q_{PMLP})_{i+1}$, when designing tapered pump installations.

4.5452 Control of gas

4.54521 Introduction

Several techniques may control the transit of the gas in pumped wells.

4.54522 Surface choke control

Let the variable f_{gv} be defined as the free gas vented fraction:

$$f_{gv} = 1 - f_{gPM} \quad (4.538)$$

Suppose that the well represented in Fig. 4.537 is operating at steady-state conditions, producing with the help of a pump set at D_{PM} the stock tank flow rate of q'_L. Assume that 60% of the free gas is flowing through the annulus path and 40% is pumped through the tubing path, when the choke existent at the casing head outlet is totally opened to the flowline. In this situation, the pressure drop across the choke will be nil ($\Delta P_C = \Delta P_{CMin} = 0$), the casing head pressure will be minimum ($P_{ch} = (P_{ch})_{Min} = P_{wh}$), the fraction of free gas which is being naturally vented will be $f_{gv} = (f_{gv})_{Nat} = 0.6$), and the gas/liquid "interface" within the annulus will be in its highest position ($D_I = (D_I)_{Min}$).

The natural ability of the gas bubbles to discriminate at the pump inlet level whether to go through path A or through the path B appears to be governed by the geometry at the neighborhood of the pump inlet,

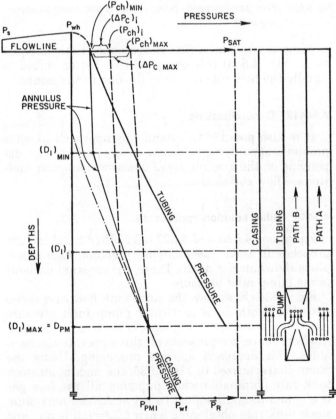

Fig. 4.537 Gas control: surface choke control (after Couto)

the properties of the fluids at the pump inlet, the buoyancy force acting upwards on the bubbles, and the dragging force attracting the bubbles to the axis of the pump, caused by the liquid entering the pump.

The buoyancy force is principally a function of the pressure at the pump inlet, which is a major factor in the control of the size of the bubbles and, therefore, of the axial upward bubble velocity. The dragging force depends principally on the liquid flow rate going into the pump, which is a major factor controlling the radial, centripetal bubble velocity. Therefore, if the liquid flow rate and the pressure at the pump inlet remain constant, the amount of gas being naturally vented should also remain constant.

If the casing choke is adjusted to position i, which introduces a pressure drop across the choke $(\Delta P_c)_i$, the casing head pressure will increase to P_{ch_i}. The gas/liquid interface will be forced down to a position D_{I_i}. However, the liquid flow rate and the pressure at the pump inlet level do not change. Consequently the bubble velocities do not change. Therefore, the discriminating ability of the bubbles does not change, and the free gas vented fraction remains unchanged: $f_{gv} = (f_{gv})_{Nat} = 0.6$.

Subsequent reductions in the size of the casing choke will increase ΔP_c and P_{ch}; the interface level will lower, with the f_{gv} remaining constant, until a choke position is reached where the gas/liquid interface is forced to its lowest level $D_I = (D_I)_{Max} = D_{PM}$, the casing head pressure reaches its maximum value $P_{ch} = (P_{ch})_{Max}$, and the pressure drop across the choke also reaches its maximum value $\Delta P_c = (\Delta P_c)_{Max}$. The free gas vented fraction is still $f_{gv} = (f_{gv})_{Nat} = 0.6$.

When the travelling gas bubbles reach the pump inlet, they immediately incorporate themselves into the continuous phase of gas existing there. Hence, there is no further need to consider buoyancy and dragging forces acting upon the bubbles, since there are no more bubbles. What actually happens is that the control of the venting process is suddenly transferred to the surface casing choke. And since no more than the present flow rate of gas can go through the present choke opening under the present (maximum) ΔP_c, the f_{gv} will be at its natural value $f_{gv} = (f_{gv})_{Nat} = 0.6$.

Up to this point, any choke reduction was offset by an increase in ΔP_c, caused by a downward shifting of the gas/liquid interface, so that the flow of gas through the choke remained constant. From this critical reduction point onwards, the gas/liquid interface can not be further lowered, since it has reached the pump inlet. Therefore, a subsequent reduction in size of the choke can not be counterbalanced by a corresponding increase in ΔP_c. It will cause the gas flow rate through the choke to decrease under a constant choke pressure drop $\Delta P_c = (\Delta P_c)_{Max}$.

Consequently, the free gas vented fraction begins to decrease, departing from the natural free gas vented fraction. The more the choke is reduced, the more the f_{gv} approaches zero. When the choke is completely closed the free gas vented fraction becomes zero.

This controlling procedure assumes that a flow pattern exists near the pump outlet in the tubing string, such that no slugs are formed there, and the pump outlet pressure remains practically free of oscillations after the gas/liquid interface reaches the pump inlet

and more gas is forced into the pump so that no heading will occur.

Under such conditions, the fraction of free gas to be vented may be controlled between $f_{gv} = (f_{gv})_{Nat}$ and $f_{gv} = 0$. Notice that during the process the amount of free gas will remain constant and the amount of vented gas will vary from a maximum to zero.

4.54523 Shifting the pump within the well

For a particular stock tank flow rate, the natural free gas vented fraction will vary with the pump setting depth from a minimum value—when the pump is set opposite the perforations, to a maximum value—with the pump set at the gas/liquid interface level where it will approach 1. In other words, for any particular stock tank flow rate, the value of the natural free gas vented fraction is a function of the pump submergence.

If the pump inlet pressure is greater than the saturation pressure and this is greater than the tubing wellhead pressure, the value of the natural free gas vented fraction may be varied between 0 and 1 by setting the pump somewhere between the level where the saturation pressure is reached within the annulus and the gas/liquid interface level with the casing wide open. Therefore, the first manner of increasing the gas vented fraction is by varying the pump setting depth from the perforation level to the open casing gas/liquid interface level.

Notice that the amount of free gas increases as the pump is raised in the well; so will the amount of vented gas and, of course, the value of $(f_{gv})_{Nat}$. The limitations to this process are those imposed by the different types of interferences which may reduce the permissible pump setting depth interval.

4.54524 Bottomhole "separation"

Another way of increasing the natural free gas vented fraction without increasing the amount of free gas is by using a bottomhole gas/liquid separator. A bottomhole gas/liquid separator is any device designed to alter the natural ability of the gas bubbles to discriminate between paths A and B in favor of the annular path and in detriment to the tubing path.

To accomplish that objective, a bottomhole separator works to increase the axial upward bubble velocity and/or to decrease the radial, centripetal bubble velocity. Because of this, geometry is one of the main factors to be considered when designing bottomhole separators. Hopefully, a bottomhole separator can increase the free gas vented fraction from $f_{gv} = (f_{gv})_{Nat}$ to $f_{gv} \cong 1.0$.

4.54525 Recycling gas

The value of the fraction of free gas pumped may be made greater than 1 by recycling some gas downwards through the annulus into the pump. Assume that it is desired to have $f_{gPM} = 1.5$. Then, $f_{gv} = 1 - f_{gPM} = 1 - 1.5 = -.5$, with the negative sign indicating downward injection of gas through the annulus, rather than upward gas venting.

Refer to Fig. 4.538. If P_{wh}, the tubing pressure at the well head, is greater than P_{ch}, the casing head gas pressure corresponding to the pump inlet pressure, it is easy to separate and recycle gas without having to compress it at the surface.

The tubing pressure P_{wh} may be naturally greater than P_{ch} or a pressure drop may be induced at the flowline inlet to make P_{wh} greater than P_{ch}.

Recycling gas in pumped wells might accomplish one or more objectives. First, it might diminish the internal pressure drop across the pump and, consequently, the pump horsepower consumption. Second, it might increase the pump high pressure volumetric flow rate and make it greater than the pump rate lower limit without making the pump low pressure volumetric flow rate greater than the pump rate upper limit. Finally, it might carry down atomized chemicals to be injected into the tubing string through the pump. Finally, even when pump inlet pressure is greater than the saturation pressure—implying no free gas at the pump inlet—it is possible to recycle gas.

4.54526   Summary

The different possibilities of controlling the gas in a pumped oil well are schematically shown in Fig. 4.539. Fig. 4.539 represents a well producing a stock tank flow rate q'_{LD} with the pump set at the depth D_{PM} where the free gas fraction being naturally pumped is $(f_{gPM})_{Nat} = 0.6$. The pump low pressure volumetric flow rate curves for different values of f_{gPM} are shown. The scheme shows clearly the effect of the different gas controlling procedures. The use of an efficient bottomhole separator supplemented with choke control and recycling tech-

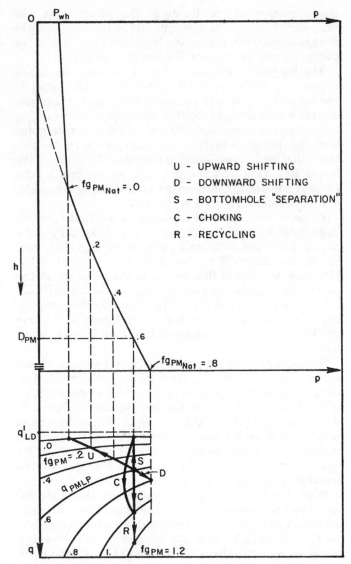

Fig. 4.539   Effect of different gas control techniques (after Couto)

niques should enable the operator to exercise almost perfect control over the gas.

4.5453   Surface determination of the pump inlet pressure

The previous discussions suggest a procedure to determine the pump inlet pressure, from the surface. There are three cases to be considered.

In case 1 the pump inlet pressure is less than the saturation pressure and no gas is being vented. Take note of the casing head gas pressure. The pump inlet pressure equals the casing head gas pressure plus the static gas column down to the pump inlet:

$$P_{PMI} = P_{ch} + \Delta P_{SG}, \qquad (4.539)$$

where ΔP_{SG} denotes the pressure due to the static gas column.

In case 2 the pump inlet pressure is less than the saturation pressure and some gas is being vented. Reduce the casing choke gradually. The casing head gas pressure will gradually build up, whereas the gas flow rate through the choke will remain essentially constant. When the casing head gas pressure stops building up (and the gas flow rate through the choke begins to diminish), the value of the maximum casing head

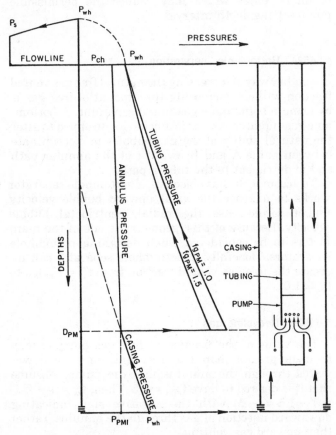

Fig. 4.538   Gas control: recycling gas (after Couto)

gas pressure is noted. The pump inlet pressure is equal to the maximum casing head gas pressure plus the dynamic gas column in the annulus down to the pump inlet:

$$P_{PMI} = (P_{ch})_{Max} + \Delta P_{DG} \qquad (4.540)$$

where ΔP_{DG} is the pressure due to the dynamic gas column.

In case 3, the pump inlet pressure is greater than the saturation pressure. In this case, no gas can be vented. Note the casing head gas pressure and determine the depth of the gas/liquid interface within the annulus. The pump inlet pressure equals the casing head gas pressure plus the static gas column from the surface down to the gas/liquid interface depth plus the liquid static column from the interface level down to the pump inlet: $P_{PMI} = P_{ch} +$ static gas column + static liquid column.

4.5454 Pumping water wells

The model in this study applies equally to water wells, irrespective of presence or absence of gas. Equations 4.520 and 4.522 or 4.526 and 4.527 when applied to water wells degenerate to:

$$q_{PMLP} = (q'_w \times B_w)_{(P_{PML}, T_{PM})} \cong q'_w \qquad (4.541)$$

and

$$q_{PMHP} = (q'_w \times B_w)_{(P_{PMH}, T_{PM})} \cong q'_w \qquad (4.542)$$

Special care must be taken when using centrifugal pumps to produce water wells to prevent the pump low pressure from being less than the water saturation pressure at the pump temperature. If water vaporization takes place inside the pump, cavitation may occur and damage the pump.

4.5455 Pumping below the perforations

A belief exists that pumping a gassy oil well below the casing perforations will allow venting 100% of the free gas, and that this will permit a liquid production greater than the maximum liquid production that can be made with the pump above the perforations. As a matter of fact, venting 100% of the free gas is *possible* whenever, without restriction at the casing outlet, the pump is set just at the gas/liquid interface level. Therefore, the maximum production a well can make, with gas venting, is limited to the stock tank flow rate $q'_L{}^*$ shown in Fig. 4.540. It will occur when $P_{wf} = P_{PMIV}$ and will be obviously limited by P_{wh}. The pump, of course, must be set opposite the perforations so that P_{PMI} will also be equal to P_{PMIV}.

Under such conditions, if no restriction is introduced at the casing head, f_{gv} may be made equal to 1, and the maximum amount of gas that can be vented from the well will be vented. The necessity of using $f_{gv} = 1$ or the convenience of reducing its value, by choking at the surface, depends on q_{PMLP} and q_{PMHP} and on the magnitude of the inlet pressure loss, ΔP_{Inl}. If the free gas is being totally vented ($f_{gv} = 1$), lowering the pump will not significantly alter the values of the low and high volumetric flow rates. However, the pump inlet pressure may be increased to meet the requirement $P_{PMI} \geq \Delta P_{Inl}$ if the inlet pressure loss is not negligible.

If the free gas is not being totally vented ($f_{gv} < 1$),

lowering the pump will also significantly decrease the value of q_{PMLP} and slightly decrease the value of q_{PMHP}. On the other hand, if *no gas is being vented at all*, the production is no more limited by P_{wh}, and there is a possibility of producing the well's potential stock tank flow rate, $(q'_L)_{Max}$, depending on the practicability of accomplishing $P_{PMI} \geq \Delta P_{Inl}$, $q_{PMLP} \leq q_{PMLL}$, and $q_{PMHP} \geq q_{PMLL}$, assuming that all the remaining requirements and constraints are also being met.

4.5456 The temperature problem

There are three important temperatures to be evaluated when solving pumping problems: tubing wellhead temperature (T_{wh}), pump temperature (T_{PM}), and sink temperature (T_s).

Sink temperature—For the highest flow rates and for the lowest flow rates a well can deliver, according to its inflow performance, the sink temperature will be approximately equal to the geothermal temperature at the sink depth. However, for intermediate flow rates the sink temperature may diverge considerably from the geothermal temperature.

For very high flow rates the bottomhole flowing pressure is very low. The sink must be set close to the reservoir level and its temperature will have to be approximately equal to the reservoir temperature.

For very low flow rates the dynamic temperature profile along the casing will be approximately equal

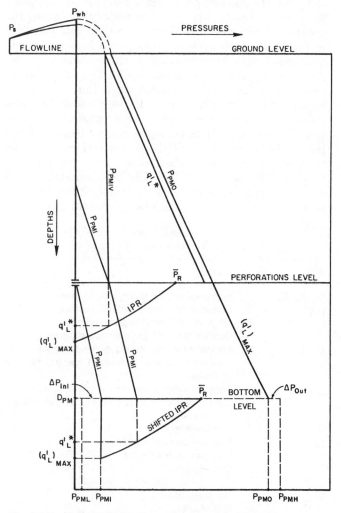

Fig. 4.540 Pumping below the perforations (after Couto)

to the geothermal profile, and the sink temperature will be approximately equal to the geothermal temperature, no matter the level the sink occupies within the casing.

For intermediate flow rates the sink temperature may diverge considerably from the geothermal temperature, if the pump is set far from the perforated interval. There is sufficient mass flow rate to carry upwards large amounts of heat and there is enough bottomhole flowing pressure to permit setting the pump away from the reservoir level.

The Shiu seventh flowing temperature correlation may be used to evaluate the sink temperature. However, this correlation was developed for tubing flowing temperatures. The sink temperature determination should be, therefore, the object of further study.

Pump temperature—Once the sink temperature has been evaluated, the pump temperature may be estimated through:

$$T_{PM} = T_s + \Delta T_{PM}. \qquad (4.58)$$

Determining ΔT_{PM}, the temperature rise due to the pump operation, is complex. Fortunately, in most of the cases the quantity has diminutive importance when compared to T_s, and may be neglected.

For centrifugal pumps, in wells where ΔT_{PM} might have significance, the temperature rise may be estimated assuming that the flowing fluids will dissipate all the heat loss occurring in the motor. If the work to compress the gas being pumped is considered negligible in comparison to the work necessary to raise the pumped liquid pressure from the pump low pressure to the pump high pressure, the pump horsepower requirement may be estimated as the hydraulic horsepower necessary to pump the liquid. The input horsepower may then be calculated as a function of the motor and pump efficiencies. If the product of these efficiencies is considered practically constant, as they actually are, the input horsepower may be considered relatively independent of the pump and motor types themselves. However, when a computer approach is used, a more accurate ΔT_{PM} may be estimated, based on the overall horsepower consumption. From this, the heat loss—the part of the energy consumption cor-

TABLE 4.546
NOMENCLATURE

Symbol	Quantity
B_g, B_o, B_w	Formation volume factor, gas, oil, water
D_{Act}	Reciprocating pump actual volumetric displacement, variable with the pump speed.
D_I	Depth of the gas/liquid interface within the annulus
D_{MNG}	Alternative pump net displacement necessary to handle the minimum amount of free gas that can be pumped
D_{MXG}	Alternative pump net displacement necessary to handle the maximum amount of free gas that can be pumped
D_{PM}	Pump setting depth
f_{gPM}	Free gas pumped fraction
f_{gv}	Free gas vented fraction
FO	Overlap factor
f'_o, f'_w	Stock tank liquid fraction, oil, water
GLF	Gas lock factor
GOR	Gas/oil ratio
h	Depth

Symbol	Quantity
p	Gauge pressure
P_{ch}	Casing pressure at the wellhead
P_{PMH}	Pump high pressure
P_{PMHMX}	Permissible pump high pressure
P_{PMI}	Pump inlet pressure
P_{PMID}	Pump inlet pressure required for driving the fluids into the pump
P_{PMIV}	Pump inlet pressure required for venting the gas
P_{PML}	Pump low pressure
P_{PMLMN}	Permissible pump low pressure
P_{PMO}	Pump outlet pressure
$\overline{P}_R$	Reservoir static pressure
P_s	Separator pressure
P_{sat}	Saturation pressure
P_{wf}	Bottom hole pressure
P_{wh}	Tubing pressure at the wellhead
q	In situ volumetric flow rate
q'	Stock tank flow rate
q_{Act}	Centrifugal pump actual volumetric flow rate, variable along the pump axis
q'_L, q'_o, q'_w	Volumetric flow rate at surface conditions, liquid, oil, water
q'_{LD}	Desired liquid flow rate
q'_{LMN}	Minimum stock tank flow rate that the pump can handle
q'_{LMX}	Maximum stock tank flow rate that the pump can handle
q'_{LMXNV}	Maximum stock tank flow rate without venting gas
q_{PMHP}	Pump high pressure volumetric flow rate
q_{PMLL}	Pump flow rate range lower limit
q_{PMLP}	Pump low pressure volumetric flow rate
q_{PMLIM}	Limiting pump low pressure volumetric flow rate
q_{PMUL}	Pump flow rate range upper limit
R_s	Solubility of gas in oil
R_{sw}	Solubility of gas in water
t	Temperature
T_{BH}	Bottomhole temperature
T_{PM}	Pump temperature
T_{PMLIM}	Permissible pump temperature
T_s	Sink temperature
T_{wh}	Tubing wellhead temperature
ΔP_c	Differential pressure across the choke
ΔP_{EXH}	External differential pressure across the pump housing
ΔP_{EXHL}	Permissible ΔP_{EXH}
ΔP_{Ext}	Differential pressure existing externally to the pump
ΔP_{INH}	Internal differential pressure across the pump housing
ΔP_{INHL}	Permissible ΔP_{INH}
ΔP_{Inl}	Pressure loss at the pump inlet channels
ΔP_{Int}	Differential pressure existing internally at the pump
ΔP_{Out}	Pressure loss at the pump outlet channels
Δq_{PM}	Volumetric flow rate reduction across the pump
Δt_{PM}	Pump temperature increase due to the pump operation itself
λ_L	No-slip liquid holdup

Subscripts	
Ext	External
g	Gas
Inl	Inlet
Int	Internal
L	Liquid
Max	Maximum
Min	Minimum
Nat	Natural
o	Oil
Out	Outlet
w	Water

responding to the motor and pump inefficiencies—is converted to temperature rise by taking into account the mass flow rate and the liquid specific heat.

Tubing wellhead temperature—Once the T_{PM} has been evaluated, the Shiu correlation may be again applied to estimate the tubing wellhead temperature.

REFERENCES

1. Couto, L. E. D., "Selection, Positioning and Operation of Centrifugal, Hydraulic and Sucker Rod Pumps," M.S. Thesis, University of Tulsa, (1978).
2. Brill, J. P., and H. D. Beggs, *Two-Phase Flow in Pipes,* 3rd edition (1977).
3. Couto, L. E. D., and M. Goland, "General Inflow Performance Relationship for Solution-Gas Reservoir Wells," preprint SPE-9765 (1981).
4. O'Neil, R. K., "Application and Selection of Electrical Submergible Pumps," preprint SPE-5907 (1976).
5. *Reda Submergible Pump Catalog,* published by Reda Pump Company.
6. *Submersible Pump Handbook,* published by Centrilift Inc. (1975).
7. Shiu, K. C., and H. D. Beggs, "Predicting Temperatures in Flowing Oil Wells," presented at the Energy Technology Conference and Exhibition, (Houston; November 5-9, 1978).
8. Shiu, K. C., "An Empirical Method of Predicting Temperatures in Flowing Wells," M.S. Thesis, The University of Tulsa Petroleum Engineering Department (1976).
9. Vogel, J. V., "Inflow Performance Relationships for Solution-Gas Drive Wells," *Journal of Petroleum Technology,* (January 1968) 83-92.

4.6   INSTALLATION AND OPERATION OF ELECTRICAL CENTRIFUGAL PUMPING SYSTEMS

4.61   Introduction

There are several factors that should be emphasized concerning the installation and operation of electrical pump systems.

Once a submersible is correctly sized and its operation is properly monitored, the installation should become a relatively trouble-free, economical operation.

4.62   Equiment dimensions

Fig. 4.61 illustrates a submergible pumping unit installed inside either casing, liner, or open hole and the related unit part dimensions that must be used to calculate the clearance available on the flat cable side.

The following equation can be used to find the minimum clearance that could possibly be available or the maximum O.D. of the total unit.

The illustration and equation assume that the protector used will be the same size as the pump series and that no flat guards will be used. (In tight applications, flat guards are not used.) If flat guards are used, then this dimension must be added. The O.D. of the protector must be used in place of the pump's O.D. in the equation if a larger O.D. protector than pump O.D. is used.

Note: On all applications where the clearance is very close, it is recommended that a full gauge tool (slightly larger than equipment O.D.) should be run prior to the installation of equipment. This is to insure that there are no burrs or tight spots in the hole. This operation will prevent many expensive mistakes when installing submergibles in tight or heavy walled casing. The following equation can be used with Fig. 4.61.

$$\frac{H}{2} - \left[\frac{P}{2} + C + B - \left(\frac{H - M}{2} \right) \right] = \text{Pump clearance on flat cable side}$$

where:

  H = I.D. of hole
  M = largest diameter at any point on motor
  P = pump or protector diameter
  C = thickness of armored flat cable
  B = thickness of band

EXAMPLE:

Given:

  5½ in. API 8RD casing, 17# (4.892 in. I.D.; 4.767 in. drift diameter)
  400 series pump (4.0 in. O.D.)
  456 series motor (4.56 in. O.D.)
  Band thickness 0.030 in.
  3 kv high temp flat cable—size 6
  Conductor (0.505 in. × 1.20 in.)

Note: On flat cable construction, the thickness can vary from a maximum to a minimum due to manufacturing tolerances. The maximum thickness for #6 3 kv HT flat is 0.55 in. The maximum should be used when calculating maximum dimension of equipment.

Drift diameter—the minimum inside diameter that the supplier (mill) must provide for API standard casing, drill pipe and tubing.

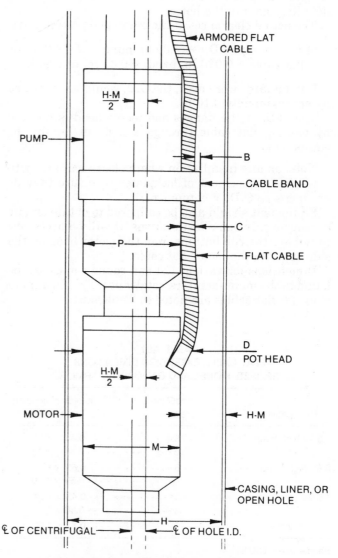

Fig. 4.61 *Equipment dimensions (Courtesy TRW-Reda)*

For casing, the pipe must pass a drift mandrel of the following size to meet API standards:

Casing	Diameter
Size up to 8⅝ in. D, inclusive	⅛ in. smaller than tabulated
Sized over 8⅝ in. D to 13⅜ in. D, inclusive	5/32 in. smaller than tabulated
Sizes over 13⅜ in. D	3/16 in. smaller than tabulated

The pipe I.D. should never be less than the drift diameter if there is no casing damage and, in many cases, will be larger than drift. This is why it is important to run a gauge before installing equipment. The I.D. of pipe can be used for calculations:

Calculation:

$$\frac{H}{2} - \left[\frac{P}{2} + C + B - \left(\frac{H - M}{2}\right)\right] = \text{clearance}$$

$$\frac{4.892}{2} - \left[\frac{4.00 \text{ in.}}{2} + 0.55 \text{ in.} + 0.030 \text{ in.} - \left(\frac{4.892 \text{ in.} - 4.56 \text{ in.}}{2}\right)\right] = 2.446 \text{ in.} - 2.00 \text{ in.} + 0.55 \text{ in.} + 0.030 \text{ in.} - 0.166 = .032 \text{ in.}$$

This maximum dimension (clearance) takes into consideration the total unit with the centerline dimensions of both pump and motor. This indicates the unit will definitely go into the hole.

The actual clearances at the pump and protector are:

(4.892 in. pipe I.D.) − (4.00 in. pump O.D. + 0.55 in. flat cable + 0.030 in. band) = 0.31 in. or (5/16 in.).

If flat guards were used, the O.D. would be increased by approximately 0.16 in.

The O.D. at the flanges and motor head is reduced by passing flat cable through a slot cut out in the equipment.

Note: on any installation where clearances are critical—check placement of lockplates to insure they do not increase O.D. of equipment.

Equipment should also be calipered to obtain on site dimensions. In some installations, it will be necessary to remove the rub buttons from the motor head on the side opposite the motor flat cable.

Dimensions for various flat and round cables can be found in the manufacturers' catalogues. Typical dimensions for flat cables are noted in Table 4.61.

TABLE 4.61
FLAT CABLE EXTENSION DIMENSIONS
BRONZE, MONEL AND GALVANIZED ARMOR

Type cable	Conductor size	Outside dimensions maximum
1.5 kv high temp.	8	0.44 × 1.02
	6	0.48 × 1.12
3 kv high temp.	6	0.55 × 1.32
	4	0.59 × 1.50
1.5 kv standard	8	0.43 × 1.01
	6	0.46 × 1.12
3 kv standard	5	0.62 × 1.53

Band—use 0.030 in.
Flat guard—use 0.16 in.

4.63 Banding material application

Banding material applications are generally accepted as follows:

Type	Code	Use
Galvanized steel	(Painted black)	For normal installations where corrosion is not a problem.
Stainless steel	(Coded with green paint)	Used for most oil-well applications with mild corrosive environments.

Caution: Certain concentrations of H_2S will cause embrittlement and failure of SS bands.

Monel	(Coded with red paint)	For installations suspected of being highly corrosive with high concentration of H_2S.

Cable clamps come in several lengths to accommodate different O.D. equipment. The lengths and types are:

22 in. Black Acme Cable Clamp
32 in. Black Acme Cable Clamp
22 in. Monel Acme Cable Clamp
32 in. Monel Acme Cable Clamp
40 in. Monel Acme Cable Clamp
22 in. Stainless Acme Cable Clamp
32 in. Stainless Acme Cable Clamp
42 in. Stainless Acme Cable Clamp

Note: where unarmored cable is used, cable clamp saddles must be used to prevent damage to cable by metal bands.

4.64 Installation practices

There are also several installation practices that should be observed such as location and proper installations of the check and drain valves.

4.641 Running and pulling

Riling noted the following good practices in running or pulling an electrical pump:[3]

(1) Insure rig is aligned over well.
(2) The cable sheave should be no higher than 30 ft from the wellhead with 45 feet as a maximum. The proper O.D. sheave should be used for the cable size with 48 in. O.D. as a recommended minimum.
(3) Do not hurry operations.
(4) Insure that the equipment is not handled in a rough manner.
(5) Insure that excessive strain is kept off the cable by placing an additional person at the spooler to prevent tension on the cable.
(6) Insure that the proper equipment, such as slips or backup, are of the type that prevent the cable from being damaged or wrapped around the tubing. A backup should always be used when tightening tubing.
(7) Run or pull the tubing slowly (1000-1500 ft/hr) to insure that the cable is not damaged. (75% of cable damage is due to mishandling.)
(8) Make certain that the cable is banded to the tubing correctly and that care is being taken not to damage the cable with the slips.

(9) A junction box is always recommended; but if, by chance, it is not used, insure that the cable is vented between the wellhead and the switchboard. (Note: Gas can still migrate through and under the insulation on a vented cable.) Insure that the switchboard is located a safe distance from the well (50 ft minimum).

(10) Never handle a reel of cable by making a sling or using chains which touch the cable. The proper way to move a reel of cable is to insert a piece of pipe through the reel to serve as an axle and lift with wire line or chain sling with a spreader bar attached to axle. Never allow reel of cable to roll against objects which might crush or damage cable or reel.

(11) A pick-up nipple (lifting sub) should always be used in the pump discharge head when suspending total assembly unit.

(12) Locate the switchboard a safe distance from the transformers for safety.

4.642   Equipment handling

The following procedures in the handling of equipment and cables are recommended to properly install or pull a unit:

(1) The pump, motor, and cable must be assembled and handled during installation or removal according to the manufacturer's instructions. The manufacturer's field representative should be on all jobs. He should be allowed sufficient time to check out the equipment.

 The pump serviceman's job at the installation is a mechanical one with set procedures and one in which he has been well trained. Assembly of the unit must be done as carefully and as cleanly as possible.

(2) The cable reel should be 75-100 ft away from the installation rig, and the cable guide wheel over which this cable passes should not be more than 30 ft from the ground. At all times there should be slack between the cable reel and the cable guide wheel.

(3) Feeding the cable from top of the reel reduces the tension on the cable. One drawback to this method is if the unit is dropped, the reel will be pulled towards the rig. A cable cutter is recommended to insure against this.

(4) If the cable is coiled on the ground, it is extremely important that the person coiling the cable remains outside the coils for safety reasons.

(5) Tension applied to the cable can result in a slight elongation which, in turn, could result in additional heat build-up in the cable. An elongation can also weaken the armor protection so that the jacket and/or insulation ruptures and becomes a cable failure. Proper procedures in the care of the cable can and do reduce cable failure. Careful handling is imperative if cable life is to be prolonged.

(6) The flat cable must be protected by a steel guard and the round cable must be securely clamped to the tubing. It is very important that the cable be run straight up the tubing. Rotation of the tubing should not be allowed while running the pump.

(7) If directional wells are encountered, it may be necessary to run centralizers to provide additional protection for the cable and keep the motor centered in the casing. This will eliminate any hot spots that could result if the motor is lying against the low side of the well casing. The centralizers will also minimize the amount of wear on the cable as it is being run in the hole.

(8) When a pump is pulled, the cable should be carefully checked both visually and with a ohmmeter. An insulation tester can also be utilized. Any questionable cable should not be run back in a well.

 Bad spots can be removed and the cable repaired by splicing in new sections. Extreme care is necessary when making such repairs, as a poor splice will result in a short circuit. Occasionally, it is possible to extend cable life by swapping ends.

(9) A temporary heated shelter should be erected around the cable spool when the ambient temperature falls below 40° F. if the cable is new and at 0° F. if the cable is used.

(10) Flexure can be reduced by running the cable over the largest possible sheave during installation or pulling. A 54 in. O.D. sheave is recommended.

(11) When pulling the tubing the bleeder plug in the drain valve should be sheared as soon as fluid is observed in the tubing.

 This shearing of the plug is best accomplished by using a sinker bar on a wire line; however, a go-devil or breakoff bar (short length of sucker rod) may be used.

 IT IS EMPHASIZED THAT THIS LATTER METHOD SHOULD <u>NOT</u> TAKE PLACE UNTIL FLUID IS OBSERVED AT THE SURFACE, SINCE SEVERE DAMAGE TO THE CHECK VALVE AND PUMP CAN RESULT IF THE BAR IS DROPPED IN A DRY STRING.

 Damage can also result due to impact loading from the bar hitting a low fluid level in the tubing. Also the bar would have impact on the downhole equipment should the tubing be free of fluid due to a tubing leak or faulty check valve.

4.643   Motor controller settings, electro-mechanical

(1) The proper settings for the motor controller are as follows:

(a) The overload current relay should be set at a maximum of 120% of the motor's nameplate amperage. To avoid tripping the relay on slight overloads, the settting should be not less than 110% of the motor's nameplate amperage.

 This relay is a hand reset type. If tripped, the surface and subsurface equipment should be completely checked out before restarting.

(b) The undercurrent relay should be set at 80-85% of the nameplate amperage. Eighty % is normally considered the standard setting. This will give maximum protection under pumpoff, gas locking or pump intake plugging conditions. In some instances, if the fluid gradient is very light, the relay may be adjusted lower if there is adequate fluid production to permit reasonable satisfactory pump operations and cooling of the motor.

The undercurrent setting should never be set below the no load motor current.

(c) The time delay relay should be set from 1 to 5 sec to prevent nuisance tripping during momentary power interruptions or gas unloading.

(d) The time delay on the automatic restart timer should be set for a minimum of 30 min. This is to assure that a unit is not started while backspinning.

4.644   Cable handling and banding

Since handling of equipment is such an important part of insuring successful and economical submersible pump operations, the following cable handling and banding procedures should be closely followed on all installations.

(1) Correctly center the rig over the hole.

(2) Set cable reel 75 ft to 100 ft from wellhead if possible. Use bands in the middle of each joint and 18 in. above each coupling.

(3) Use proper tension when handling the cable.

(4) Band squarely across the cable and tubing. The band should be at right angles to the tubing. The bottom and top edges of the band should be flush with the tubing. If not, remove the band and install properly.

(5) Band the flat cable and flat guards in a straight line up the side of the protector and pump. Start immediately above the pothead with a section of flat guard which has the bottom end slightly tapered. Continue with the flat guard until reaching a position slightly below the splice of the flat cable and round cable.

(6) Arrange the flat guards in such a way as to avoid banding over voids, screens, and changes of sections. Where necessary, cut the box end of flat guard squarely with a hacksaw for proper alignment. File the edge and corners round and smooth.

(7) Before setting slips, place the cable in the slip door guide slot.

(8) Band the round cable in a straight line up the tubing.

(9) Use a backup on every joint where there is any chance of the tongs turning the pipe in the slips.

(10) Avoid loose bands. If a band is too loose, take it off and install correctly.

(11) Keep slips in good condition using sharp dies of the non-rotating type.

(12) Be sure that the swivel lock on the hook is latched and that the hook is not free to swivel.

(13) Avoid spots in the flat guard where an end or edge can dig into the flat cable.

(14) If the cable is crushed and/or the armor is broken sufficiently to where insulation damage is suspected—STOP! Call for approval from the pump serviceman before proceeding.

(15) A spreader bar must be used over all submersible cable reels at all times.

At any time when the tubing is being pulled and the cable is not following exactly, stop and contact the pump company representative for directions.

4.645   Surface equipment

4.6451   Grounded vs. ungrounded systems

The literature generally states that distribution systems are better protected if the system is grounded. Fast-acting relays can be applied to detect grounds on a live conductor, disconnect it, and prevent excessive damage to the system and to human life. Where the system with a ground is available to endanger people and their well being, this is a necessary and proper reason for a grounded system.

In a submersible pump installation, the motor and all of the cable (except a few feet which could be in conduit) should be underground and inaccessible. In this case, an ungrounded system is better because if the power cable is damaged, and it does have maximum exposure to the possibility of physical damage, a single line ground will not prevent successful operation. This case would be the equivalent of operating with one corner of the secondary delta grounded.

In those installations with an individual isolating transformer for each pump, an ungrounded secondary gives the best overall service. Multiple units operating from a large single substation should be considered only when the logistics leave little if any choice. Individual units operating through autotransformers attached to an existing 440 v system are acceptable when the economics of another method is unacceptable.

4.6452   Transformers

Since most submersible pumps are designed to have low current (approximately 100 amps or less), the voltage might be any value from 400 to 2,300 v.

EXAMPLE:

A 975 v, 92 amp, 150 hp motor located at a depth of 100 ft would require 1005 v, but at 5000 ft it would require 1125 v, and at 8000 ft it would require 1215 v. The manufacturer offers and can give delivery on a transformer with taps for these ranges of voltage requirements.

The most trouble-free installation has one pumping unit per isolating transfer or a bank of three single phase transformers. Multiple units operating from a large substation without individual isolating transformers will be more costly to maintain than single unit installations. Single unit undergrounded installations can be operated with a line to ground cable fault whereas substations with multiple units connected should be a grounded system.

In this case, without isolating transformers for each unit, a cable ground means an inoperative unit, and rework is necessary at that time.

4.646   Location of downhole equipment

4.6461   Check valve

The check valve should always be located at least 2 joints above the top of the pump and, if high gas-oil ratio conditions are known to exist, it could be located at a higher position above the pump. Accordingly, if gas locking occurs there is more free space for the gas to rise above the pump. This allows liquid entry and, in turn, allows the pump to be alleviated from gas locking.

4.6462 Bleeder or drain valve

The bleeder valve is run in conjunction with a check valve and is usually installed one to two joints above the check valve. The bleeder valve's function is to prevent pulling a wet string. There are several types of drain valves on the market. The most simple type utilizes a hollow pin that is punctured by dropping a bar. With the plug sheared off there is communication between tubing ID and annulus. Some drain valves require pressure build-up in the tubing that shears a pin located externally on the tubing. Other drain valves are available that are removable or can be punctured by use of a wireline.

Drain or bleeder valves also come as a combination unit when larger size casing is used to suspend the pumping unit.

4.6463 Motor on top of pump

The motor can be run on top of the pump on bottom intake type units. This will allow a larger O.D. pump to be run on bottom since the cable does not have to bypass the pump.

4.65 Operating practices

4.651 Introduction

There are several operating practices that should be followed in running an electrical pump. This should not imply that this system is so complicated that a trained specialist should be there all the time. However, each operator should be familiar with the pump and certain basic concepts concerning electrical principles.

4.652 Choke operation

Normally, the pump should operate against the lowest back pressure possible at the surface. This reduces the total dynamic head to a minimum. However, perhaps due to lack of good inflow data, an incorrectly-sized pump may have been installed or its setting depth may not be correct for the well conditions. A choke at the surface can be installed which, in turn, creates additional back pressure and controls the rate which can place the pump in its allowable operating range. All submergible installations should have a choke installed at the wellhead.

4.653 Start-up of the pump

When a submersible unit is started, the load voltage should be no less than 95% of no load voltage. If it is less than 95%, this could mean that there is inadequate electrical capacity available which may be insufficient transformation and/or insufficient conductor size. With adequate capacity, the starting time is less than 20 Hz. The current inrush is 450% in the first cycle and decreases immediately. The average inrush over the 20 Hz starting time is 250%.

A phase rotation meter should be used in order to verify the correct direction of rotation. This eliminates the need to stop and change rotation in those cases where the initial production may be carrying fracture sand which could result in a locked pump.

To assure correct hook-up so that the pump rotates correctly, the electric leads should be marked. Some pumps will pull more amperage rotating in one direction as compared to the other, giving a means of detecting correct hook-up. The pump may actually move some fluids even though rotating in the wrong direction such as 250 b/d in the wrong direction and 600 b/d in the correct direction.

As a precaution any time a pump has been shut down it should not be immediately restarted. In particular if there is no check valve in the tubing or if the check valve is leaking, the pump is probably reverse rotating due to fluids running back through the pump from the high hydrostatic head in the tubing. If power to start the pump is applied during this time, a tremendous torque is applied to the pump shaft and failure may occur.

4.654 Casing pressure

Some controversy exists about the proper methods to produce wells in regard to casing pressure. The migration rates of oil and gas are dependent upon the difference in pressure between the well bore and the reservoir pressure; therefore, the produced gas/oil ratio may be the same for any oil production rate, regardless of whether the gas is freely produced from the casing or whether an attempt is made to restrict gas production by holding casing pressure. With a stable casing pressure the balance of the gas is produced through the pump. Therefore, the gas production from the normal well will not be increased or decreased by the opening or closing of surface casing valves for a constant oil producing rate. It is advantageous in some gassy applications of submergibles to hold a certain amount of casing pressure. This casing pressure helps to compress the fluid in the annulus and provide the pump with a denser fluid to pump, but probably lowers the production rate depending upon flowing pressure.

In the case where casing pressure is used for better producing conditions, a casing pressure regulator must be provided to insure stable casing pressures and to insure that the casing pressure does not continue to rise, thereby restricting the well capabilities or possibly pushing the fluid level below the pump. In a well of this type, there is usually a specific casing pressure where the pump and the well produce at maximum efficiency. Different casing pressures, along with well tests, must be conducted in order to determine the proper casing pressure. In most cases, it is recommended that submergible operations be conducted with the casing vented. A well may pump off due to maintaining an excessive casing pressure.

4.655 Testing

A means of determining the pumping rate should be provided on start-up. The unit should be closely observed for two or three days and good initial start-up data should be obtained.

Periodic tests and equipment analysis are required to obtain the most efficient service from any artificial lift system. The subsurface electric pump is no exception.

Tests are recommended on the following schedule:
(1) Upon initial start-up
(2) 5 to 7 days after start-up

(3) On new installations, every 2 weeks until the well stabilizes

(4) Monthly thereafter

The test data should include a minimum of:

1. Running amperage
2. Pumping rate
3. BHP
4. Tubing wellhead pressure and casing pressure
5. Gas (MCFPD) (Both vented and pumped)

A careful study should be undertaken on any pump installation that does not produce as originally designed, since the problems may be either with the pump and motor assembly or a wellbore deficiency. A liquid level in the casing can be obtained with acoustic devices. A working fluid level should be a direct indication of the pump suction pressure. With poor suction pressure data, pump performance cannot be accurately established.

The use of downhole-pressure measurement devices is recommended.

4.7   ANALYSIS AND TROUBLESHOOTING

4.71   Introduction

Once the unit is in the well and operating, it should be analyzed to determine if it is functioning properly. Even though the pump is operating satisfactorily, we need data from this operation in order to compare with later malfunctions.

The importance of collecting good production and operating data on an installation cannot be overstressed. This data can be used as follows:

(1) Checking the present design
(2) Resizing if necessary
(3) Detecting and preventing possible problems that could cause a failure
(4) Economic evaluation
(5) As an aid in analyzing and determining the reasons for any failures

It is important to have good production information throughout the operating life of a submergible pump. If a pump fails, past operating data should be evaluated to determine the most desirable pump that should be rerun in the well.

This section is a guideline for understanding applications. All situations cannot be covered, and pumping operations must be learned from experience.

The following simple rules should be followed in analyzing well applications:

(1) Use all information available
(2) Do not make assumptions without the facts
(3) Communicate with the original designer for complete details whenever possible
(4) Be sure the data is accurate; if the data is questionable, analysis will be questionable

4.72   Determining productivity of wells

For test purposes, a submersible centrifugal pump can be operated with the discharge valve completely or partially closed for short periods of time. The head pressure generated under such conditions remains constant. This characteristic permits a method for determining the productivity of a well produced by a submersible centrifugal pump.

The only necessary equipment is a pressure gauge and a valve. The pressure gauge is installed in the discharge tubing at the well head between the valve and the pump.

The following procedure can be used for determining PI:

(1) Shut the well in for sufficient time to allow the static fluid level to stabilize. (If the well is not being produced, the discharge tubing should be filled prior to the test.)
(2) With the tubing full, close the valve.
(3) Start the pump
(4) Record the surface pressure. This reading represents the pressure for zero rate of production.
(5) Open the valve.
(6) Check the rate of production until it becomes constant.
(7) Close the valve.
(8) Record the pressure at the instant of closing. (If gas is present, pressure build up will be slow). This reading represents the pressure for the constant rate of production measured in Step #6.

From these two points, the productivity of the well can be determined (Fig. 4.71).

In this diagram the head created by the pump, running with the discharge valve closed under conditions of the static fluid level, is represented by column H. The pressure read at the surface gauge represents the P_1 part of this column, with the static fluid level being the distance FL_1 below the surface. With the pump running again and with the valve closed at a given producing rate, the head generated by the pump will again be H. But the pressure read at the surface gauge is only the P_2 part of this fluid column, the working level being located the distance L_2 from the surface. The drawdown of the fluid in feet is represented by $L_2 - L_1$ and equals the head-feet equivalent of differ-

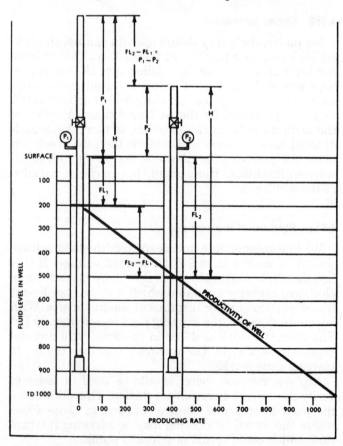

Fig. 4.71   Determination of well productivity

ence in pressures of $P_1 - P_2$ as noted on the pressure gauges. This relation can be written in the following form:

$$L_2 - L_1 = \frac{P_1 - P_2}{G_s} \qquad (4.71)$$

where:

$L_2 - L_1$ = drawdown in feet for a given rate or production.

$P_1 - P_2$ = difference in the observed surface pressures in PSI

G_s = a constant representing the fluid gradient in lbs/sq in./ft due to the weight of a 1-ft column of fluid being pumped. The usual values for G_s are as follows: For fresh water, 0.434; for saltwater, 0.45-0.50; for 40° A.P.I. oil 0.36. Other values can be determined depending upon fluid density.

On the diagram of Fig. 4.71 the drawdown is shown for a 400 b/d rate of production and is projected to the 1,000 ft fluid level, indicating the theoretical maximum output of the well. Only two points are used on the chart to determine the drawdown curve. A number of readings can be taken at different rates of production to check the correctness of the curve.

4.73   Analysis of existing installations

As more knowledge is acquired of submergible pumping installations and well operations, there is a large amount of information that can be obtained from an existing installation.

4.731   Example problem #1

As an example, the following would represent the analysis of an existing installation and illustrates a portion of the information that can be obtained.

Note: If a well's fluid is near the specific gravity of water the pump's operation will follow the pump's performance curve very closely.

Analysis:

Situation: Check out of an existing installation.

Data from original installation report:
(1) Pump—159 stages (G-52)—REDA
(2) Motor—60 hp (540 series—990 v 39 amps)
(3) Casing—7 in. 23#
(4) Tubing—2⅞ in. EUE 8RD 5900 ft

Data from well site:
(1) Tubing pressure—170 PSI
(2) Casing pressure—no gauge
(3) Unit running continuous (pulling 39 amps)

Data from production foreman or pumper:
Note: This is information that must be obtained from well test data.
(1) Production—2250 bf/d
(2) Water Cut—80%
(3) Gas—Too small to measure

Enough data is now available to analyze the installation by following the procedures:
(1) Refer to the pump curve for the G-52 pump (Fig. 4.72).
(2) Note where the production rate falls on the pump performance curve, and if it is in the recommended range of the installed pump.

(3) Determine the feet of head per stage on the curve where the pump is operating.
(4) Multiply this by the number of stages in the installed pump to determine what head the pump is working at.
(5) With this data, you can determine very closely where the fluid level in the well is by determining what vertical lift the pump is working against.

The pump is producing 2250 BF/D and the pump is operating to the right of the curve. By reading the head capacity curve at the 2250 b/d rate, we find the feet of head per stage is 34. The total head generated by the pump would be:

TDH = 159 stage × 34 ft/stage = 5406 ft

Included in the total head is the friction loss in the tubing at a 2250 b/d rate and the 170 PSI tubing pressure. These values must be backed out of the total head to determine the vertical lift and thereby determine the operating fluid level. Friction loss from the friction loss chart for 2⅞ in. tubing at 2250 b/d would be:

F_t = 45 ft/1000 ft × 5900 ft = 265 ft   (see Appendix 4C)

The head due to tubing pressure is:

$$\text{feet of head} = \frac{170 \times 2.31}{1.0 \, (\text{water gradient assumed due to water cut})} = 393 \text{ ft}$$

The vertical lift will be the total head minus the friction loss and tubing pressure:

V_1 = 5406 ft − 265 F_t − 393 F_t = 4750 ft (rounded off)

The fluid level is, therefore, at ±4750 ft or ±1150 ft over the pump setting depth. The calculation will not be far off in a well that makes very little gas.

This well is producing just inside the recommended capacity range. The well would also support a higher volume unit if surface facilities were available to handle the increased water production.

Initial design:

An initial estimation of the well's design could be estimated from the information available from the pull and run. This can be determined very quickly with two quick calculations. The initial sizing of a pump is always at the highest efficiency of the pump if it is possible or, in other words, a pump is available for the required volume at peak efficiency. The design will enver be far from peak efficiency.

In this case, the volume could be assumed to be 2000 bf/d. From the curve, this is 39 ft per stage. For the 159-stage pump, this would represent 159 stages multiplied by 39 ft per stage or 6150 ft of total dynamic head. The friction loss at 2000 bf/d rate for 2⅞ in. tubing is 40 ft or 1000 ft of tubing or 240 total head ft. Subtracting this from the total head leaves 5910 ft. Most pump designs are for some submergence, even in a high water cut. One hundred PSI or approximately 100 ft over the pump would be realistic with 50 PSI of system pressure.

4.732   Example problem #2

Situation: Check out and analyze an existing installation

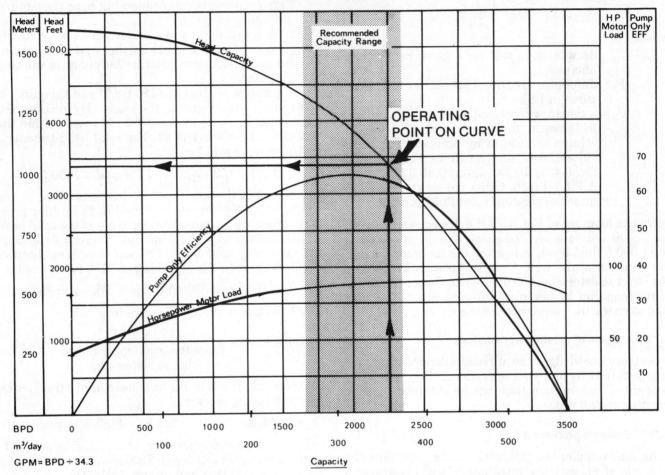

Fig. 4.72 Performance curves for G. 52 pump

Given:

Pump—159 Stages of G-52
Motor—60 hp (540 series, 990 v 39 amps)
Casing—7 in 23#
Tubing—2⅞ in. EUE 8RD
Tubing pressure—200 PSI
Casing pressure—vented and hooked to flow line
   with check valve in line
Amperage—34-35 amps and fluctuating slightly
Production—1600 to 1700 bf/d
Water cut—40%
Gas—350 to 450 Mcfd

This type of well is more difficult to analyze than those with higher water cut. Proper data for complete analysis is generally not available in the field for this type well. This is the type of well where the field foreman or engineer will often ask the following question:

This well was designed for 2000 BF/D and we have a high fluid level (about 1,500 ft of fluid over the pump). We are only getting 1600 to 1700 BF/D. Why?

(An analysis of the pumping well conditions with the data that is available will provide an answer).

Analysis:

(1) The well is producing 350 to 450 MCF of gas per day and 1600 to 1700 BF/D with a 40% water cut. This represents approximately 950 to 1000 BO/D. Using the 1000 BO/D figure and the 450 MCF gas production, this represents a GOR of:

$$GOR = \frac{450,000}{1000} = 450 \text{ cu ft/bbl}$$

(2) The casing pressure can be assumed to be 200 PSI as the casing is tied into the flow line and pressure must exceed 200 psi before gas is vented into the system.

(3) The 1500 ft of fluid over the pump cannot be evaluated unless the gradient of fluid is known. This 1500 ft of fluid could represent more capacity available, but it could also represent a foam column where very little increased production would be realized with a pump change. For example, if the gradient of the fluid in the annulus was 0.20 psi, it would represent 300 psi of fluid.

We now have enough information to explain what has probably occurred. From what we have learned previously, it can be assumed that the well was initially designed for 2000 bf/d and in reality is handling this volume of fluid or even a greater volume at reservoir conditions to produce the 1600 plus bf/d in stock tank barrels. In checking the pump performance curve of the G-52 pump it appears that the application is not good since the unit is operating out of the recommended capacity range. This would not be the case in this application as the pump is operating in the peak capacity range at the pump sub-surface setting depth and we are observing surface data.

The reduced amperage of the motor also provides a clue that the unit is pumping a light fluid and the ammeter chart verifies this. The production of this well could possibly be increased with a change in casing pressure.

Note: This should not be attempted without the use of a casing regulator and controlled test.

4.74   Analysis of well producing viscous crude

The performance of centrifugal pumps are affected when handling viscous fluids. The effects are a marked increase in brake horsepower, a reduction in head, reduction in efficiency, and a reduction in capacity.

A clue for pinpointing a viscosity problem is when we have an application that is sized correctly, but the rate is below expectations and the horsepower is above what it should be. Another clue is that viscosity problems occur in fields of low gravity crude, low bottomhole temperatures and water cuts in the 30% and 70% range. Many times in a field such as this, the problem will be analyzed as a pump problem when in reality viscosity is the problem. Also emulsions may be formed, and the tubing pressure loss becomes excessive.

If a viscosity problem is suspected, a quick way to check this is to open a bleeder valve on the tubing wellhead and observe the type crude and flow from this bleeder. Heavy viscous fluid will not separate readily as it comes out of the bleeder valve. For water percentages between 30-70%, emulsions are quite likely to occur.

4.75   Monitoring operations

4.751   Introduction

The performance of a submergible pump may be monitored using a recording ammeter since the pump loading of the motor is sensitive to the specific gravity of produced fluid and rate. The ammeter chart also indicate voltage variations when the pump is operating in a stabilized condition. Ammeter charts may be used as diagnostic tools to locate pump, motor, and wellbore problems. Typical ammeter performance charts are noted in Figures 4.73(A) through 4.73(O).

Unfortunately the ammeter is one of the least understood tools. The ammeter chart, much like a physician's electro-cardiogram, is a recording of the heartbeat of the submersible electrical motor. The analysis of amp charts can provide valuable information for the detection and correction of minor operational problems before they become costly major ones.

An electrical submersible pump package is typically comprised of six major components:

(1) One or more transformers providing the proper surface power (transformer bank).
(2) One motor control panel housing the necessary switchgear and surface controls necessary for operation (switchboard).
(3) A length of three conductor special power transmission cables to link power from the switchboard to the motor.
(4) One or more (in tandem) electrical submersible three-phase two pole constant speed induction motors (motor).

(5) One motor to bore fluid isolation section (seal or protector).
(6) One or more (in tandem) submersible centrifugal pump(s).

A properly designed combination of these components will provide the desired fluid production. When a change in equipment operation or well characteristics occurs, the interaction between these components will be upset. In these cases, the imbalance is reflected on the recording ammeter chart.

Switchboard

The typical switchboard is constructed in two major sections: the high voltage compartment, and the low voltage compartment.

The high voltage side is comprised of four basic elements:

(A) surface power input cable
(B) the contactor
(C) a reducing current transformer
(D) downhole power output cable

The input power circuit is opened or closed by the contactor, thereby supplying the power to the motor via the downhole cable. The current transformer supplies low voltage, usually 110-120 to the low voltage controls in the board.

The low voltage side provides the necessary controls to activate the contactor. These controls monitor one essential variable, the amperage. The controls consist of three manually reset current-sensitive circuit breakers, called overload relays (one on each leg), activated by current over a preset level, and one automatically resetting current-sensitive circuit breaker, called the under current relay (one on one leg), activated by current under a preset level.

Activation of any of the overloads causes briefly delayed deactivation of the contactor, resulting in complete unit shut down. No automatic restart circuits are activated.

Activation of the undercurrent relay causes deactivation of the contactor and instant shutdown of the unit. An automatic restart sequence then begins through a timing relay and restart occurs. During restart, the undercurrent relay is blocked from the circuit for a brief period by a second timing relay. The automatic restart sequence may be bypassed by use of the hand-off auto switch located on the switchboard.

The amperage is recorded from one power leg and displayed on the ammeter.

4.752   Recording ammeter

The recording ammeter is located visibly on the switchboard. Its function is to record the input amperage of the motor. This is accomplished by the use of a reducing current transformer coupled to one leg of the cable inside the switchboard. The linearly reduced amperage is then plotted on a circular chart whose grid carries the proper abscissa multiplier to indicate the actual cable amperage.

4.753   Ammeter chart analysis

Assuming that the recording ammeter is functioning properly, a number of changes in operating conditions may be defined by proper interpretation of the amp chart. Some of the potentially damaging conditions are:

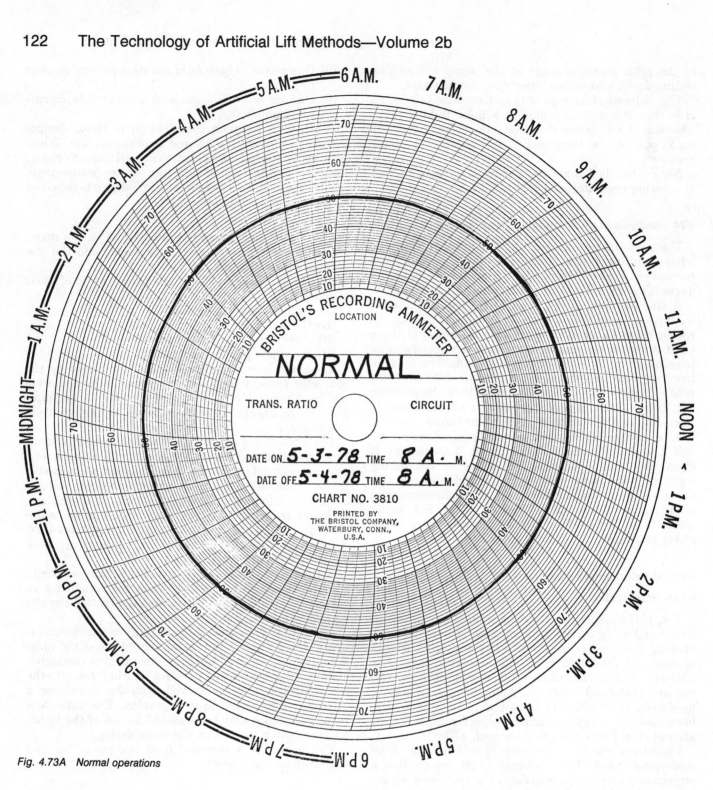

Fig. 4.73A   Normal operations

(1) Primary power line voltage fluctuations
(2) Low amperage operation
(3) High amperage operation
(4) Erratic amperage operation

The following examples [Fig. 4.73(A) through 4.73(O)] deal with the proper interpretation of ammeter charts and their inter-relationship with other guides in the troubleshooting and preventative maintenance of electrical submergible pumps.

(1) Normal operation [Fig. 4.73(A)]

A characteristic of three-phase two pole constant speed induction motors under a non-varying load is the constant amperage drawn. An ideal submergible installation is designed such that the actual horsepower to be used is within approximately 10% of the rated nameplate horsepower, and such that the total dynamic head and producing rate vary from actual to design by approximately 5%. Under these conditions, the ammeter should draw a smooth symmetrical curve at an amperage near nameplate.

Figure 4.73(A) illustrates ideal conditions. Standard operations may produce a curve above or below nameplate amperage, but it should be a smooth symmetrical one to be considered ideal. A well may not produce a smooth constant value curve, but constant variations from day to day are probably normal with operations for that well. Any deviations from a well's normal operation is a clue to possible problems or changing well conditions.

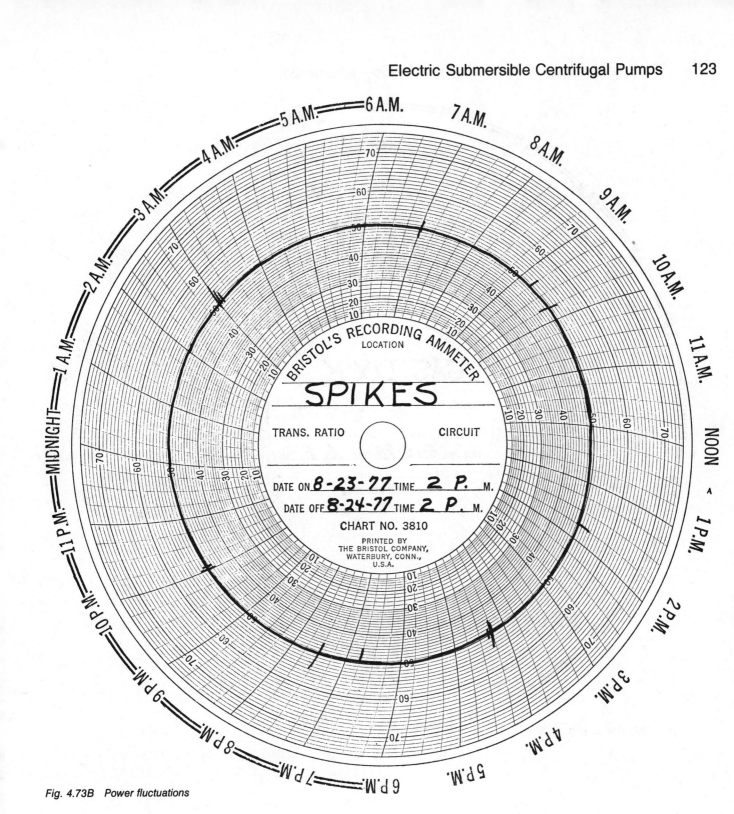

Fig. 4.73B   Power fluctuations

(2) Power fluctuations [Fig. 4.73(B)]

Under continuous normal operations, the power output by the motor remains relatively constant. Under this condition the amperage varies inversely with the voltage. Consequently, if the primary power supply voltage fluctuates, the amperage will fluctuate in an attempt to retain constant horsepower output. The fluctuations will be reflected on the amp chart as in Fig. 4.73(B). The most common cause of power fluctuation is periodic heavy drain loading of the primary power system. For example, such a drain, or sag, may be caused by the start-up of a high horsepower injection pump. Occasionally, it may be a combination of smaller simultaneous drains. If this is the case, some effort must be made to respace these drains such that their combined impact is small. By correlating the fluctuations with time, it may be possible to determine the exact cause. This type of operation should not be detrimental to equipment as long as the spikes are not severe. The same type spikes will be observed during an electrical disturbance such as a lightning storm.

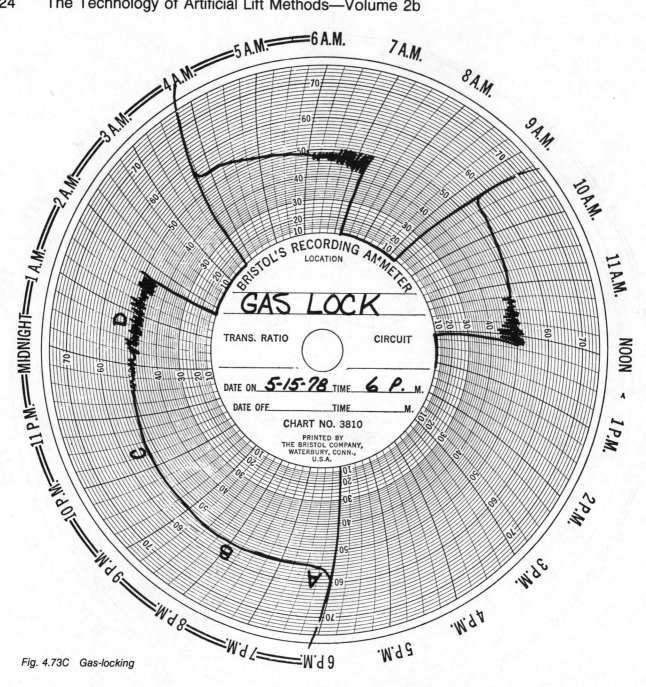

Fig. 4.73C Gas-locking

(3) Gas locking [Fig. 4.73(C)]

Fig. 4.73(C) shows the chart of a pump which has gas locked and shut down. Section A shows start-up. At this time, the annular fluid level is high; thus, the production rate and amperage are accelerated slightly due to the reduced total dynamic fluid head. Section B shows a normal operating curve as the fluid level nears the design value. Section C shows a decrease in amperage as the fluid level falls below design and fluctuation as gas begins to break out near the pump. Finally, Section D shows erratic low amperage as the fluid level nears the pump's intake. Cyclic loadings of free gas and slug fluid eventually cause undercurrent shutdown of the unit.

It is possible to remedy this situation by lowering the pump to a point where gas breakout is low enough to permit continuous operation. If lowering the pump is not feasible, it may be possible (depending on the unit configuration) to choke production back until a suitable fluid level is established. If neither of these possibilities exist, a system of programmed downtime cycling should be designed for the maximum fluid withdrawal, using the least number of cycles. The pump should be resized on the next pump changeout.

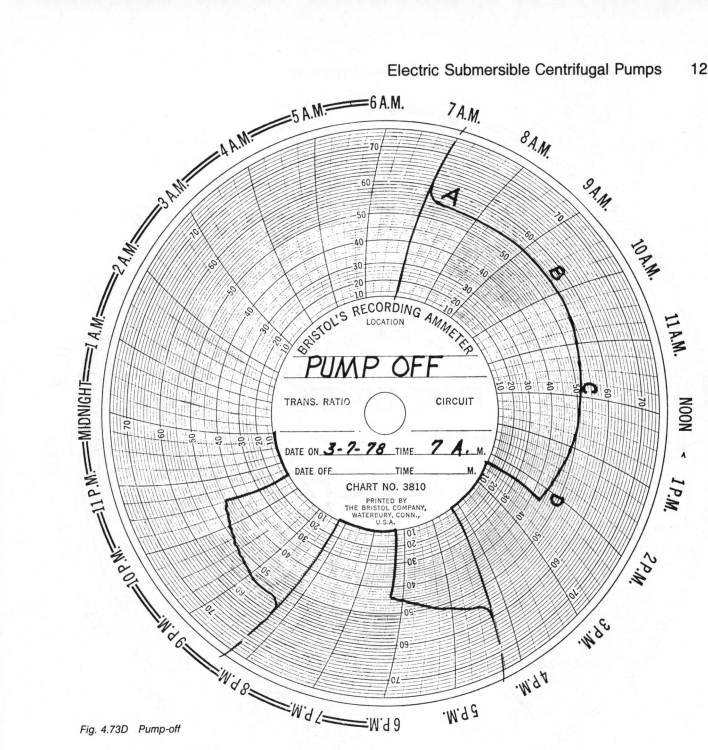

Fig. 4.73D   Pump-off

(4) Fluid pump off conditions [Fig. 4.73(D)]

Fig. 4.73(D) shows the chart of a unit which has pumped off and shut down on undercurrent, restarted automatically and shut down for the same reason.

Analysis of Sections A, B, and C are identical to that for gas locking except no free gas breakout fluctuations are evident due to the assumption of no gas present. In Section D, the fluid level approaches the pump intake, and the rate and amperage decline. Finally, the preset undercurrent level is reached, and the unit drops off the line.

When a unit drops off line due to undercurrent, an automatic restart sequence is triggered. As is shown, the unit restarted automatically after the preset time delay. During shut down, the fluid rose slightly. When the unit restarted, the fluid level had not reached static level. Thus, the pump off cycle began somewhere in Section C. The problem is that the unit is too large for the application. Remedial action is the same as that for gas locking.

If the unit is to be lowered, care should be taken to insure that the unit will not be underpowered due to the depressed fluid level and resultant increased total dynamic head.

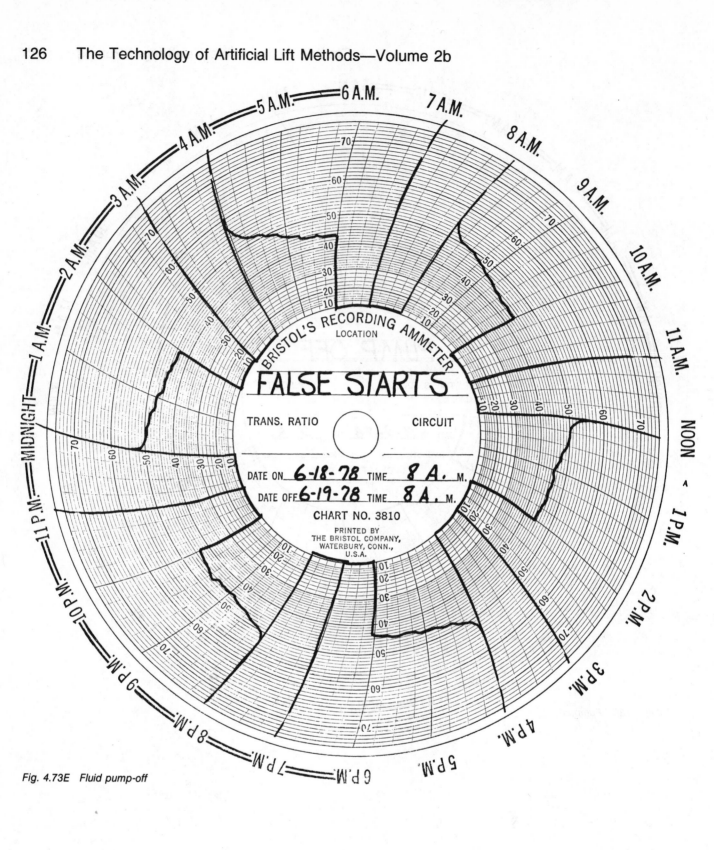

Fig. 4.73E   Fluid pump-off

(5) Fluid pump off conditions—restart failure [Fig. 4.73(E)]

Fig. 4.73(E) shows a chart from a unit which has shut down on underload, failed in an attempt to restart automatically, timed out, and restarted beginning the cycle again.

Analysis of this plot is similar to that for pump-off of fluid conditions, except that the auto-restart delay is not of sufficient length to allow adequate annular fluid build-up for loading the pump on alternate restart attempts. This unit is considerably oversized. If the unit is cycled, the downtime would probably be the maximum available on the 5 hour timer. The pump should be resized on the next changeout. Stimulation for increased production should possibly be considered.

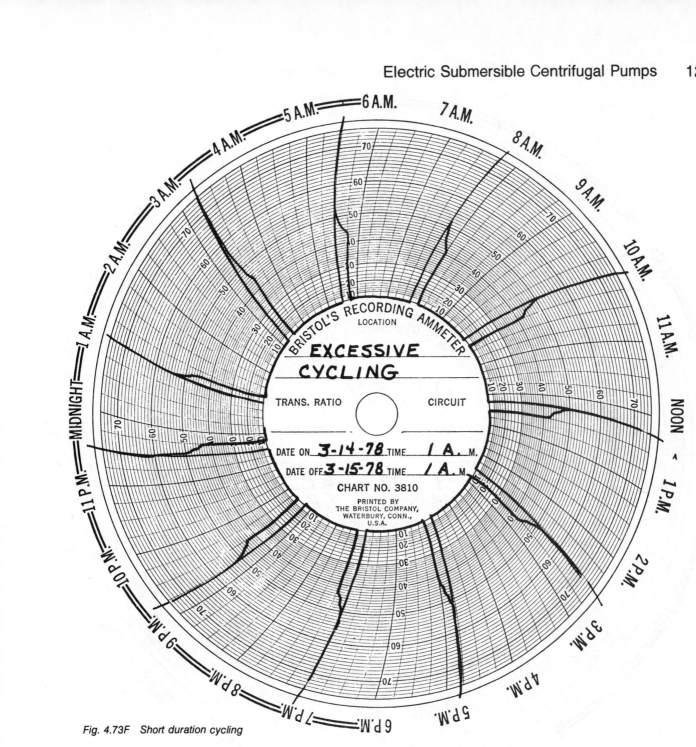

Fig. 4.73F   Short duration cycling

(6) Frequent, short duration cycling [Fig. 4.73(F)]

Fig. 4.73(F) shows a chart similar to that of fluid pump off conditions except that the running times are more brief and the cycles more frequent. This configuration chart usually applies to a unit which is too large for the application. It would be unusual to find a submergible unit missized this badly. If the productivity of the well appears to be compatible with the unit, other problems probably exist.

An immediate suggestion would be to run a fluid level determination immediately after unit shutdown. If the sounding survey shows fluid over the pump, a check for high tubing pressure should be made. If the discharge line is plugged or a valve is closed against the flow, a reduction in fluid production should occur, accompanied by a drop in amperage. If the discharge pressure is reasonable or low, check for low fluid production rate immediately after pump up. An abnormally low rate may be caused by a tubing leak. Generally, a tubing leak near the surface will result in reduce fluid production and accompanying erratic amperage. This type of operation is extremely detrimental to submergible motors and should be corrected immediately.

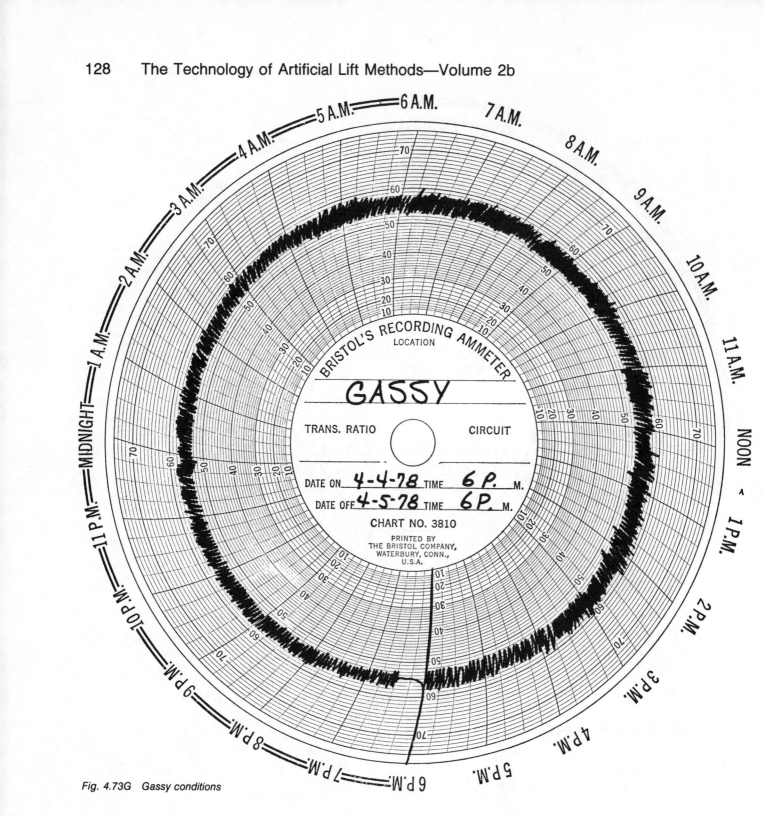

Fig. 4.73G   Gassy conditions

(7) Gassy conditions [Fig. 4.73(G)]

Fig. 4.73(G) shows the chart of a unit which is operating near designed levels, but is handling light gassy fluids.

The fluctuations are caused by entrained and free gas penetrating heavier fluid production. This condition is usually accompanied by a reduction in total fluid production (stock tank barrels). A submergible pump will attempt to pump whatever is present at the pump intake. It will attempt to pump the predesigned number of barrels or whatever fluid is available, including gas. One barrel of gas may represent a very small stock tank contribution, but represents a substantial volume through the pump.

This type of chart may also exist when pumping an emulsified fluid where the intake is being plugged momentarily by the emulsion. On an emulsion block, the spikes will usually drop below the normal amperage line.

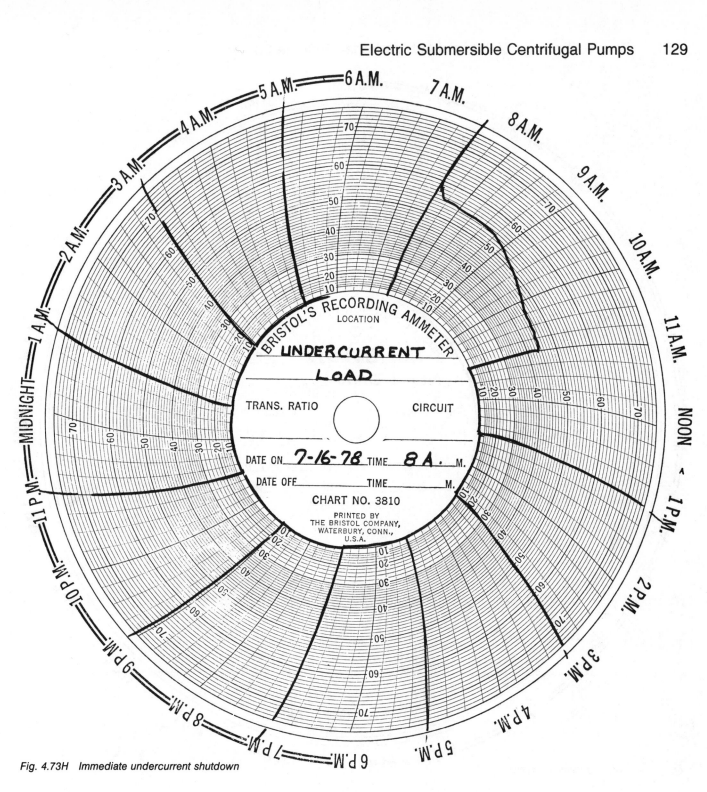

Fig. 4.73H   Immediate undercurrent shutdown

(8) Immediate undercurrent shutdown [Fig. 4.73(H)]

Fig. 4.73(H) shows the chart of a unit which was starting, running a very short time, and shutting down due to undercurrent. This cycle is repeated by the automatic restart sequence. Generally, this type recording is caused by fluid lacking sufficient density or volume to load the motor to an amperage above the undercurrent setting. If productivity tests show fluid available at the pump intake, it is possible to solve this problem by lowering the undercurrent shutdown amperage. This change should be done after consulting the pump company representative. Another cause of this type of recording is failure of the timing relay used to block the undercurrent relay from the control circuit during the automatic restart sequence. This problem is also best solved after consulting the pump company representative. Several areas should be checked to pinpoint the problem.

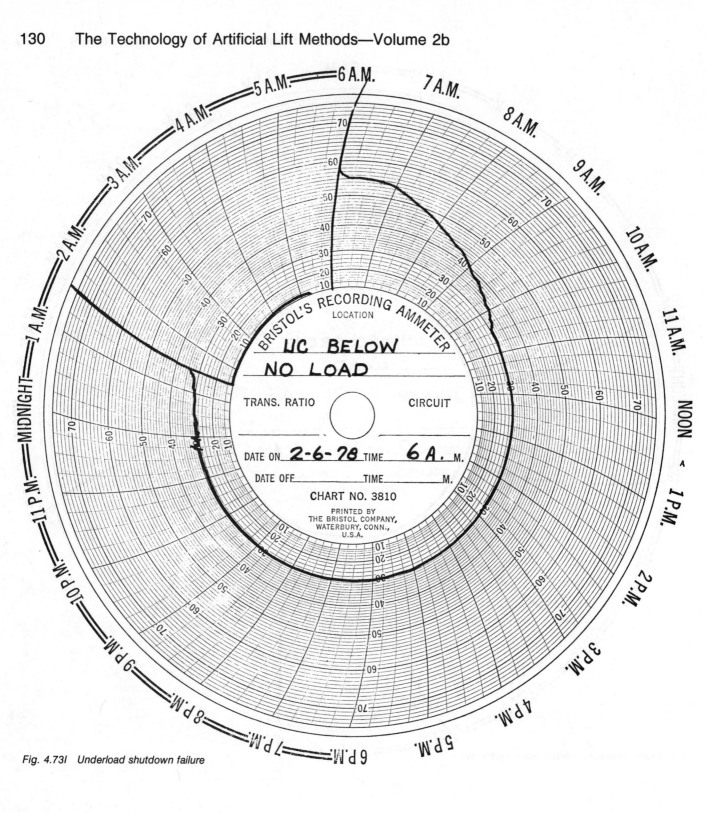

Fig. 4.73I   Underload shutdown failure

(9) Underload shutdown failure [Fig. 4.73(I)]

Fig. 4.73(I) shows a normal start-up followed by a decline in amperage due to the no-load idle amperage of the motor. Finally, after a period of loadless operation, the unit faults on overload.

This recording is typical of a unit oversized for the application and poor relay protection set points. The unit eventually pumps the well down to a point where the undercurrent relay should drop the unit off line. In this case, however, the undercurrent relay failed or was preset below idle amperage of the motor. With the fluid production retarded, the motor ran at idle until heat build-up caused a motor or cable burn. Fluid passage by the motor provides the cooling mandatory for a submergible operation.

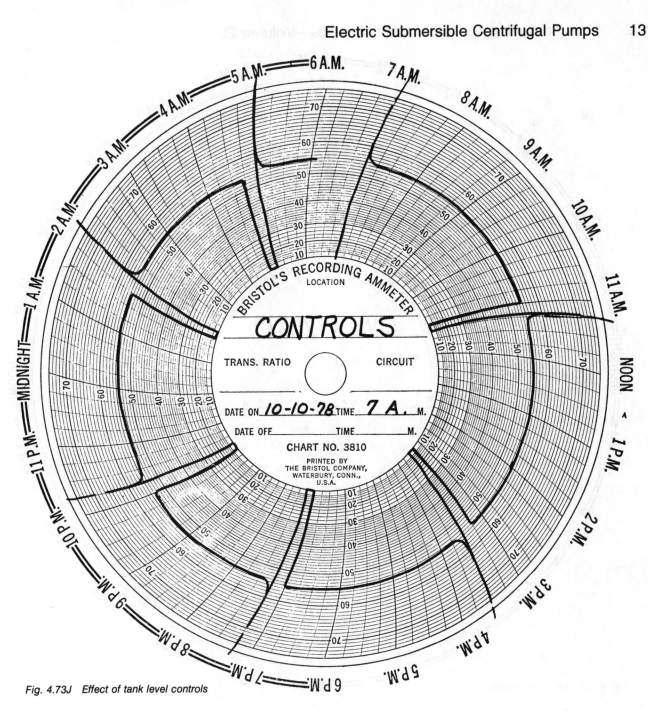

Fig. 4.73J   Effect of tank level controls

(10) Tank level controls [Fig. 4.73(J)]

Fig. 4.73(J) shows an amperage recording of a unit which is being controlled by a tank switch. The switch drops the unit off line and starts the auto-restart sequence. This type of operation is normal, but the restart delay is far too short. In almost all cases, when a unit is shut down, fluid will tend to fall back through the pump, spinning the unit backwards (backspin). Attempting to restart any submergible pump in a backspin mode may result in damaged equipment such as twisted or broken shafts. A minimum of thirty minutes is required to ensure against backspin by allowing all fluid levels to stabilize. Minimum downtime can be determined by checking voltage generated by the backspin and determining how long it actually takes the well to stabilize. A convenient way to ensure against this is to set the auto-restart delay timer above thirty minutes and to use the H-O-A switch on the switchboard for an automatically delayed start.

It is risky to depend on a check valve unless it has been determined that the check valve is holding properly.

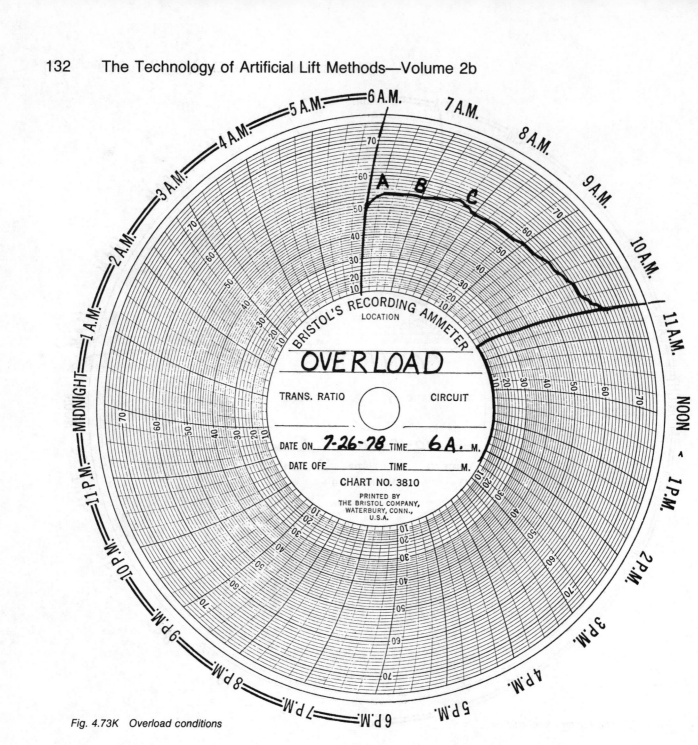

Fig. 4.73K   Overload conditions

(11) Normal overload conditions [Fig. 4.73(K)]

Fig. 4.73(K) shows a recording chart of a unit which has shut down due to overload (high current) conditions.

Section A of the curve shows start up at some amperage below nameplate (normal for some unit configuration) and gradually rising to normal. Section B shows the unit running normally. Section C shows a gradual rise in amperage until the unit finally drops off line due to overload. Until the cause of this overload has been corrected, restart should not be attempted.

Automatic restart sequences are not instigated due to the manual reset required by the overload relays. The complete installation should be checked out before a restart of the unit is attempted.

Common causes of this type shutdown are increases in fluid specific gravity or viscosity (such as heavy brines or muds), sand production, emulsions, or mechanical problems such as lightning, motor overheat, or wearing equipment.

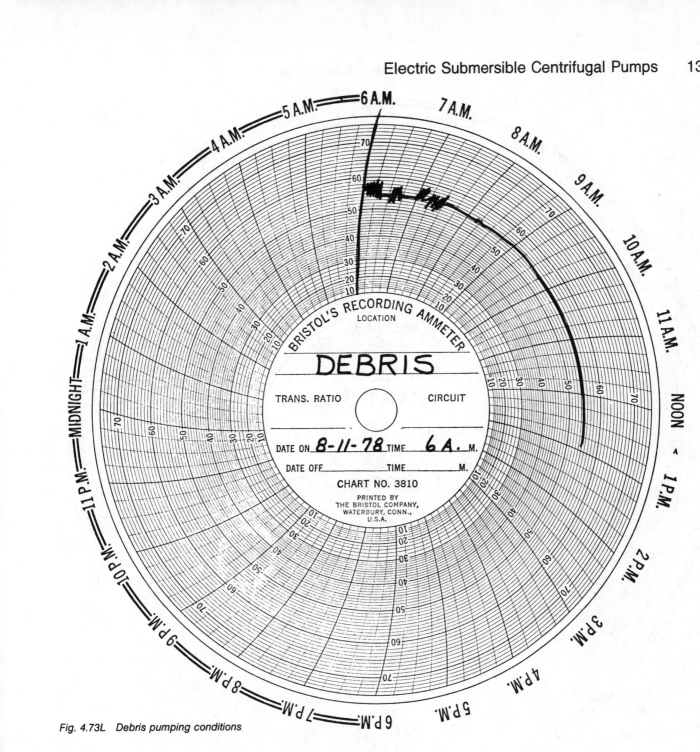

Fig. 4.73L Debris pumping conditions

(12) Debris pumping conditions [Fig. 4.73(L)]

Fig. 4.73(L) shows the recording of a unit which started, pumped erratically for a short period, and then proceeded under normal conditions.

This type operation is expected when cleaning a well of such debris as scale, loose sand and weighted muds or brines. This type operation is not unusual but is not recommended where avoidable.

The horsepower required is a function of the specific gravity of the fluid. Therefore, if it is necessary to kill a well, use the lightest possible kill fluids and consult the pump representative on start up horsepower. He can determine if the motor is of sufficient size to "clean up" the kill fluid. Under certain circumstances, it may be necessary to hold back pressure on the well to prevent excess amperage. If a well produces loose sand initially, the well should be produced slowly at a reduced capacity initially to provide a slow drawdown on the formation.

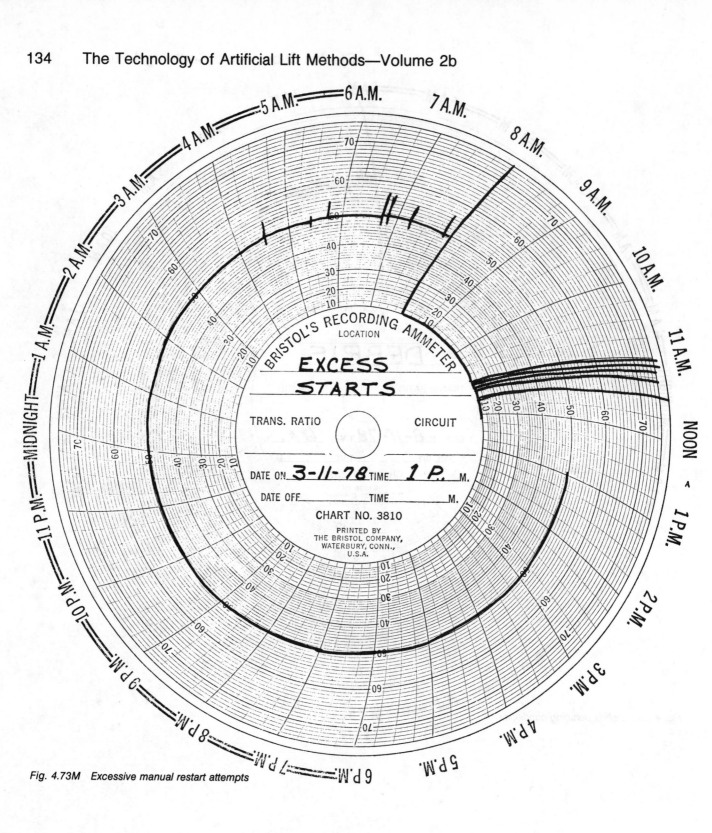

Fig. 4.73M   Excessive manual restart attempts

(13) Excessive manual restart attempts [Fig. 4.73(M)]

Fig. 4.73(M) shows a relatively normal recording chart until power fluctuation kicks were noticed. Finally, the unit dropped off line due to overload. It is also evident that manual restarts were attempted. If a single manual restart attempt fails under these conditions, the unit should be checked by a pump company representative. In this case power fluctuations, such as a lightning storm, caused the unit to shut down. When the unit did not start, problems should have been looked for elsewhere. If, for example, a primary line disconnect had been burned, the unit would attempt to restart under single phase conditions, immediately shutting down. This type restart attempt would eventually cause equipment failure.

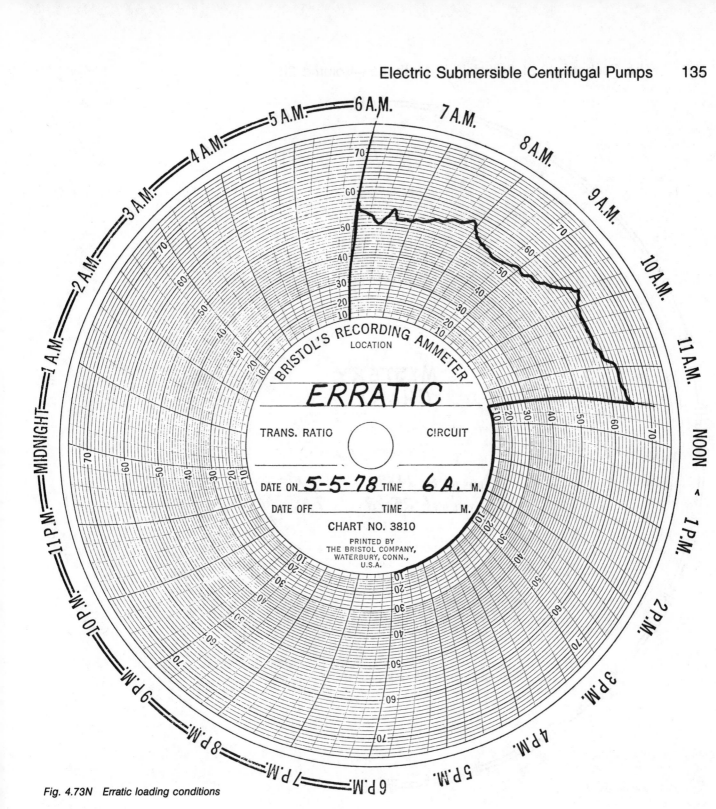

Fig. 4.73N *Erratic loading conditions*

(14) Erratic loading conditions [Fig. 4.73(N)]

Fig. 4.73(N) exhibits an unpredictable varying recording. This type recording is usually produced by fluctuations in fluid specific gravity, or large changes in surface pressure. The unit finally drops off line due to overload and will not automatically restart. Manual restart should not be attempted until the unit is thoroughly checked by a pump serviceman and the cause of the problem determined.

Some typical results or simultaneous causes for overload failure of this nature are a frozen pump, burned motor, burned cable and blown fuses (primary and/or secondary).

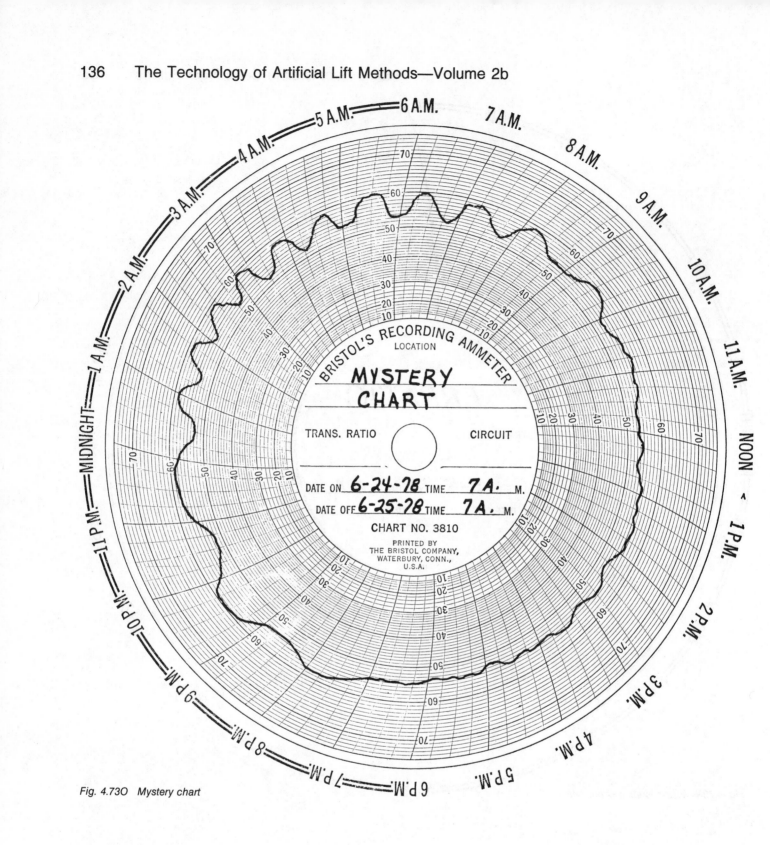

Fig. 4.73O   Mystery chart

(15) Mystery loading [Figure 4.73(O)]

Fig. 4.73(O) exhibits a recording chart that can be called the mystery chart. The motor appears to load during period A and lose some load during B. The use of all data available is necessary to analyze charts of this type.

This recording could be the result of a normal well producing characteristics, or it could be something as simple as the chart paper deforming due to the change in temperature from day to night.

4.76   Failure analysis

4.761   Introduction

This section will aid the engineer, technician, or fieldman involved with submersible operations to become more knowledgeable concerning the causes of equipment failures and will provide recommendations for reducing or preventing them.

This discussion of failures analysis is divided into three sections: possible causes, recommendations for reducing failures by becoming familiar with these causes, and trouble shooting.

4.762 Causes of failures

(1) Excessive overload (OL) for an extended period of time
(2) Seal or protector section leak
(3) Well conditions—insufficient fluid movement, high temperature, corrosion, and abrasives in the fluid stream
(4) Bad or faulty installation
(5) Switchboard troubles
(6) Faulty equipment
(7) Worn out pump
(8) Lightning
(9) Bad electrical system

4.7621 Causes of motor failures

(1) Excessive motor overload. This can be due to any one reason or combination of the following reasons.

The reason should be determined and corrected as soon as possible if a motor is running in a highly overloaded condition.

(A) Abnormally high specific gravity of the well fluid
(B) Bad design (undersized motor) due to poor data
(C) Worn out pump
(D) Low or unbalanced voltage

(2) Seal or protector section leak. A leaking seal section allows well fluids to enter the motor and usually results in a motor failure. Possible causes for a seal section leak are:

(A) Worn-out pump causing seal-damaging vibrations
(B) Broken mechanical seals from rough handling
(C) Defective seal section construction
(D) Bad installation methods and/or procedure

(3) Well conditions—insufficient fluid movements

(A) This occurs when the production volume is not sufficient to cool the motor. The recommended fluid velocity is ¾ to 1 ft/sec. with an absolute minimun of ½ ft/sec.
(B) This also occurs when a unit is set below the perforations in a well and a cooling jacket is not installed to direct the fluid by the motor.

(4) Corrosion. The deterioration of metal due to corrosion can result in holes in the housings. These holes will allow well fluids to enter the motor or cause loss of pressure in the pump.

(5) Faulty installation. This possible cause is the result of a faulty original installation, faulty equipment, or the existence of bad electrical conditions (insufficient voltage).
Note: providing adequate voltage is very important

(6) Motor switchboard. Motor switchboards do not normally suffer a great deal of component failures. However, the presence of dirt or moisture can cause electrical devices to malfunction; therefore, the cabinets should be gasketed.

High voltage surges can cause the protective device's failure. Some overload relays will not operate if the ambient temperature falls far below freezing. Heating devices should be utilized in sub-zero environments. Extremely high ambient temperatures will also lower the amperage loading required to actuate a relay. Shading or shelters can be provided to prevent direct exposure to sun or other heat sources, or ambient temperature compensated relays can be used to insure proper production.

(7) Faulty equipment. Occasionally a manufacturing defect will go undetected at the plant and in the field. It is very likely that such an equipment defect will result in a short run.

(8) Worn out pump. Since the longitudinal reactions or thrust on centrifugal pump impellers and shafts is transferred to the unit's thrust bearing, a sustained pump overload or underload condition will generally result in failure of the thrust bearing rather than the pump. Pumps will normally fail because of wear or will become locked because of scale, sand, paraffin, or deposition build-up. The degree of wear may be greatly accelerated by the presence of entrained abrasives such as sand in the pumped fluids.

(9) Lightning. Damage to transformers, switchboards and motors can result from lightning striking at or near the surface equipment.

(10) Electrical system. Low or unbalanced voltage and current can be detrimental to good submersible operation and will result in equipment failures.

Note: If there is any three-phase unbalance in the system, it should be corrected before attempting to operate the unit. Current unbalance between phases should not exceed 5%. The percent of current unbalance is defined and determined as follows:

$$\% \text{ I unbalanced} = \frac{\text{max. I difference from ave. I}}{\text{ave. I}} \times 100$$

Example: phase #1 = 50 amps
phase #2 = 48 amps
phase #3 = 52 amps

Add the three readings of the phases and divide the total to obtain the average current.

50 + 48 + 52 = 150 amps ÷ 3 = 50 amp average
Calculate the greatest amp difference from the average and divide this difference by the average to obtain the percentage of unbalance.
50 amps (average) − 48 amps = 2 amps ÷ 50 = 4%

In this example, the percent of unbalance would be satisfactory since the percentage does not exceed 5%.

4.7622 Causes of pump failures

A pump failure is usually the result of one of the following reasons:

(1) Downthrust wear, due to producing below peak efficiency

(2) Upthrust wear, due to producing above peak efficiency

(3) Wear due to producing abrasives

(4) Plugged or locked stages, due to deposition build up.

(5) Longevity wear

(6) Twisted shaft due to locked pump or starting before fluid columns in tubing and annulus have equalized after a shut down

(7) Corrosion

On initial start-up, the formation may sometimes tend to produce large slugs of sand. This is especially true when the producing zone is an unconsolidated sand formation. This problem can be minimized by either maintaining back pressure on the tubing or by cycling lifted fluids from which the produced abrasives have been removed back into the well bore and thereby minimizing the amount of fluid removed from the formation. The well should be swabbed for a sufficient time before the pump is installed to remove unconsolidated sand. If fluid is cycled back into the annulus, care must be taken to assure that sufficient fluid from the formation is passing the motor to achieve the proper cooling of the motor.

Note: If a well produces sand initially, do not shut the unit off until the sand has cleared the system.

To protect the external housings of the motor, pump and seal sections, various types of coatings are available. For severe corrosive environments, a non-metal coating is available that has been used with excellent results for several years. There are also several effective metal coatings.

4.7623 Protector or seal failures

Protector or seal failures can occur for the following reasons: (1) a worn pump's vibration, resulting in leakage through the mechanical seals, (2) bad handling procedures, resulting in a protector failure due to breaking or cracking ceramic parts of the mechanical seals, (3) improperly serviced unit, and (4) numerous cycles.

4.7624 Causes of cable failures

(1) Mechanical damage during running or pulling operations can be caused by
   (A) Crushing
   (B) Stretching
   (C) Crimping
   (D) Cutting

(2) Cable deterioration is due to
   (A) High temperatures
   (B) High pressure gas
   (C) Corrosion
   (D) Normal aging

(3) Excessive amperage load creating high conductor temperature is capable of breaking down the insulation.

4.76241 Cable (round)

Initial installation

The cable should be properly sized to reduce the possibility of cable failure. Good handling procedure during running operations is essential to minimize the chances of cable failure.

Re-installation of round cable

If it has received proper care, the round cable can usually be rerun. However, cable being pulled from a gassy well will usually be damaged, especially in the lower portion, due to gas which has penetrated the jacket. As the cable is exposed to the atmosphere (pressure reduction), this gas escapes and results in the formation of blisters which eventually pop and damage the insulation. In some cases, the cable can be reversed, that is, the top portion is run as the bottom portion and reused. In some cases a portion of the cable should be replaced.

4.76242 Cable (flat)

The flat cable must withstand the most severe conditions of the power cable string. It is subjected to the most severe abrasion when running and pulling than any other portion of the cable string. Furthermore, the flat cable is located where pressures and temperatures are usually quite high. The heat generated by the normal operating current in the motor flat is 1.9 to 2.3 times greater than the heat generated in the primary cable. Any pump operations creating higher operating currents can result in temperature rises within the flat cable of 100°F. or greater.

Should a well be pulled for any reason, the flat cable by the pump should be changed before rerunning the equipment. Once a unit is pulled and exposed to the atmosphere, cable insulation deterioration is accelerated and failures often occur when the unit is rerun and returned to production.

4.763 Failure analysis and prevention

4.7631 Motor failures

The possible causes of motor failures have been previously discussed. This portion will list recommendations or practical precautionary steps that should be taken to eliminate unnecessary early failures. These recommendations are listed under two categories: (1) initial installation and (2) re-installation.

Initial installation:
   (1) First, the motor should be sized correctly for the installation
   (2) A sufficient volume of fluid for cooling purposes should pass the motor.
   (3) Surface voltage should be correct and balanced.
   (4) Underload and overload devices should be set properly for best protection.
   (5) Operating data should be monitored periodically.

Reinstallation:

If a motor is pulled and checks good electrically, there is no reason it cannot be flushed, refilled with new oil and rerun. However, the decision to rerun a motor should also take into account the prior operating conditions and the age of the motor. It is usually a worn out pump or contamination of motor oil that causes motor problems after rerun.

When a unit is pulled (if the motor is rerun) the following precautions should be taken:
   (1) Flush and refill the motor.

(2) Change out the protector or seal section.

(3) If there is any doubt as to the pump's condition, it should be changed out.

Pump conditions can ordinarily be determined by an experienced serviceman. Pump wear is a function of:

(A) Amount of fluid pumped

(B) Nature of the fluid

(C) How much abrasive material has passed through the pump

(D) Length of the pump's run

4.7632 Field checkout

4.76321 Introduction

There are several checks to be made which can solve some equipment problems or prevent the unit from being damaged. These checks do not involve contact with any parts of the electrical system. The checks are divided into two conditions which cover most situations. The operator should become familiar with the well's normal operation so that he will immediately detect anything unusual.

4.76322 Situation #1

Note that the unit is running, but running erratically. The amp chart shows repeated starts and stops, and/or the amperage has dropped sharply. The following sequence of analysis is suggested:

If the amps have dropped sharply, this can indicate a hole in tubing, twisted pump shaft gas lock, or a change in well conditions. Any of these will result in a severely reduced volume of fluid to the surface. If the unit is running and little or no fluid is being produced and ample time shows on the ammeter for fluid to have surfaced, immediate shut down of the unit is recommended.

Note: Always shut unit off with the small on-and-off control on the switchboard before throwing the main on-off switch.

If a normal volume of fluid is being produced and low amperage is indicated, check the volume of gas production and the ammeter calibration. If the unit is starting and going down immediately, this may indicate pumping against a closed or plugged valve, excess casing pressure build-up holding fluid below the pump intake, defective underload (UL), or malfunctioning or improperly set timer in the switchboard

Check to see if the casing pressure and tubing pressure are normal. Calibration of gauges may be necessary. Check for a closed or plugged valve in the system. If the trouble appears to be electrical, notify company electrician or serviceman.

4.76323 Situation #2

If upon arrival the well site is found to be down, check in this order:

(A) Check that the ammeter is operating properly. If the ammeter is operating properly the recording chart should give a good indication of what has happened. The amp chart should indicate whether the unit went down on UL or OL. If a unit goes down on UL, it will not re-start automatically. The OL dash-pots must be manually reset before the unit will re-start.

(B) If the unit is down on OL, first check the primary fuses at the transformers to see if they are disconnected. If so, report this to your electrician or call a serviceman to check out the unit.

(C) If everything checks properly on the surface and subsurface and a sufficient time has elapsed (30 min.) since shutdown, reset OL and try *one* start. Watch the amperage very closely and check fluid production. If the unit again goes down on OL, do not try to restart it. Report the malfunction so that a serviceman can be called.

(D) If the unit is down on UL and has not restarted after normal time out, try one start. Watch the unit very closely, as the UL setting may require adjustment (lowering from present setting) or there may be a bad relay in the switchboard.

The following are some things that the field operator should never do:

(A) Do not reset UL or OL settings without prior approval. Do not modify switchboard circuit.

(B) Do not attempt repeated starts on the unit; they can damage equipment that might otherwise be okay

(C) Do not attempt to restart a unit until 30 minutes have elapsed since shutdown to insure that the unit is not backspinning.

4.76324 Situation #3

Figures 4.74(A) and 4.74(B) show pressure and amperage recordings of a field case. Note that the well is being produced in cycles and that additional changes in amperage is occurring during each producing cycle. The following data is known concerning this well:

Depth of pump—2095 ft

Tubing—2⅞ in.

Pump—Reda D-31 with 109 stages

Motor—40 hp, 900 v, 28.5 amps

Recommended range—147 to 213 m³/day (925-1340 b/d).

Figure 4.74(B) shows the pump stopping seven times per day on undercurrent shut-off and remaining off for two hours each time it shuts down. Its running time varies somewhat indicating erratic and probably gassy conditions.

The decision was made to try and utilize a choke on the production string at the surface to see if the pump could run continuously within its allowable range.

Figures 4.74(C) and 4.74(D) show the first attempts in trying first a 48/64 in. choke, then a 24/64" choke. Note that 4 hours of continuous running occurred with the 48/64 in. choke and that 10½ hours occurred with the 24/64 in. choke. The wellhead pressure is more stable with the 24/64 in. choke and the amperage has gradually decreased to 24 amps. However it eventually (after 10½ hours) shuts down.

Finally [as shown in Figures 4.74(E) and 4.74(F)] the well runs continuously with a 20/64 in. choke at the surface and produces 148 m³/day (931 b/d), which is within the lower recommended range of the pump. Note that the amperage is now constant at 23 amps and the surface wellhead pressure constant at 110 psig. Normal pump life can be expected under these conditions.

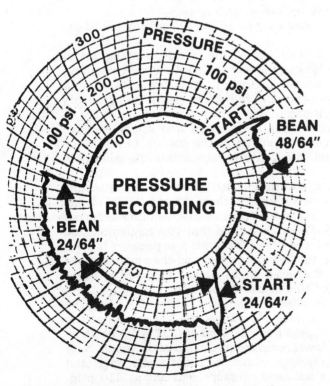

Fig. 4.74A *Surface wellhead pressure recording (Courtesy Petrobras)*

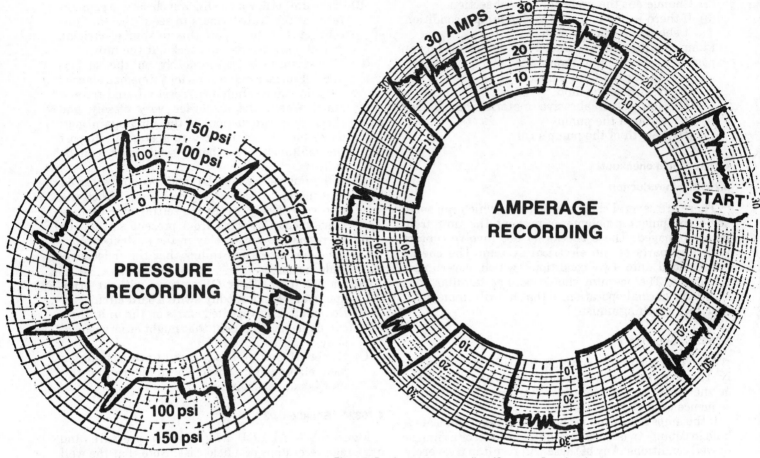

Fig. 4.74B   *Amperage recording (Courtesy Petrobras)*

Fig. 4.74C *Surface wellhead pressure recording (Courtesy Petrobras)*

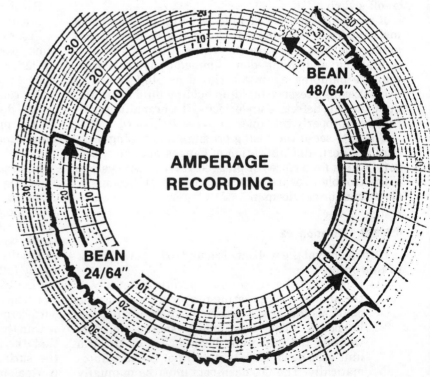

Fig. 4.74D *Amperage recording (Courtesy Petrobras)*

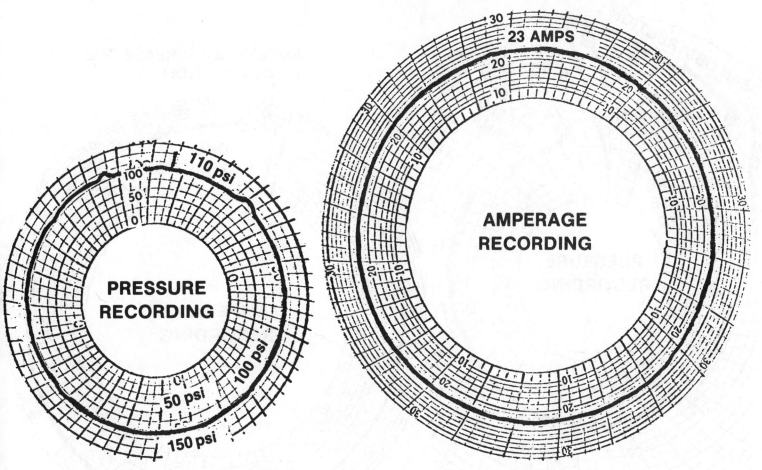

Fig. 4.74E Surface wellhead pressure recording (Courtesy   Fig. 4.74F Amperage recording (Courtesy Petrobras)
Petrobras)

A careful analysis and repeated trials on different choke sizes resulted in a good operation and eliminated having to pull the pump.

One other thing that can be determined during any part of the test is the motor's no load amperage. By blocking out the underload protection (jumper), the motor would continue to operate after dropping off line and this would indicate no-load amperage. This would indicate the amperage where the UL setting must be set above to insure unit dropping out.

4.76325 Situation #4

An analysis of the surface wellhead pressure recording, along with the amperage chart [Fig. 4.75(A) and 4.75(B)], shows an abnormal increase in surface operating pressure. In checking the surface wellhead connections, it was found that iron sulfate had formed a restriction in the choke body. After removal of this restriction, normal operations were resumed.

4.764 Troubleshooting

4.7641 Unit has stopped and will not restart

(1) Make certain main disconnect switch is open.
(2) Check overload relays and reset if open. Check continuity across O.L. contacts. Clean or replace NC top block if necessary.
(3) Check the resistance phase to ground and phase to phase with ohmmeter at cable connections

going to well. Make sure your 0 to 0 reading indicates a perfect balance. Check 0 to 0 resistance very closely. A burned motor can be very slightly unbalanced and yet not grounded. *Any* unbalance indicates a defective motor or possibly a defective cable. Calculate the additive resistance of the motor and cable string from cable and motor resistance charts. A resistance that is much lower than what calculations show may indicate the cable insulation has been subjected to excess heat and has allowed the three conductors to pull together in a dead short.

(4) If downhole equipment checks O.K., check all fuses in the switchboard. If the ohmmeter shows zero with the leads on each end of a fuse the fuse is fine (Ohmmeter set on RX1 scale.) Check the external automatic shutdown circuit for continuity.
(5) Check all switchboard terminal connections for possible loose, open, or dirty contacts.
(6) Investigate the line fuses or transformer fuses. Usually you can see if a fuse is open when dropout fuses are used.
(7) Check switchboard control voltage at no-load conditions; determine if the voltage is normal. If all the above check O.K. or the problem has been found and remedied, try to start the unit *one time*. During the attempted start, leave a voltmeter connected to observe voltage drop. Clip on the ammeter and be ready to observe the current. If observed voltage on the attempted start drops

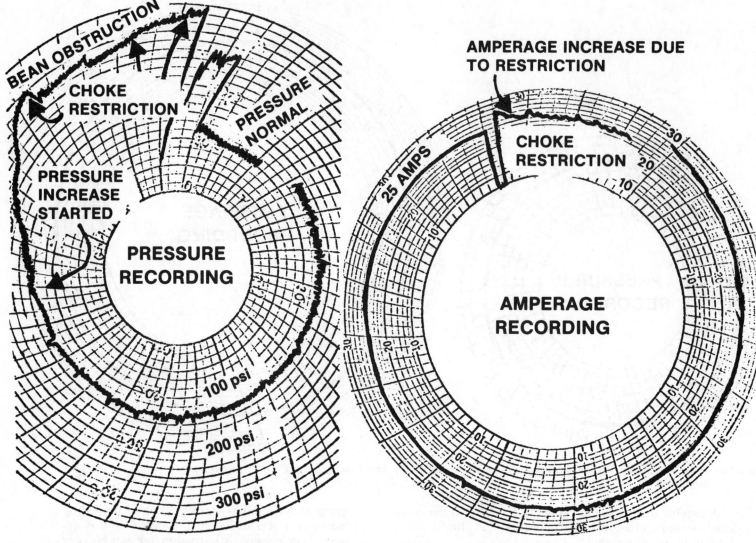

Fig. 4.75A Surface wellhead pressure recording (Courtesy Petrobras)

Fig. 4.75B Amperage recording (Courtesy Petrobras)

below 60% of no-load voltage, then power system trouble is indicated.

For a starting voltage requirement, consider optimum to be 70% of the switchboard voltage. This means that at the instant the start button is engaged, the voltage will dip approximately 30% and may take several seconds to come back to normal. This dip in voltage on starting will also be reflected in the starting current. In all cases, the high starting current should diminish to normal or near-normal amperage in a matter of seconds. Usually, it will reduce very rapidly after start to a little above normal, then gradually decrease to normal as the well produces. If the current remains extremely high for four or five seconds, it may indicate one of the following: a stuck or tight pump, single phase condition, a pump in a bind in crooked well. Stop the pump, and determine the cause of trouble.

Use the following as a guide to start amperage:

(1) If the amperage is 3½ times the nameplate amperage, the unit is probably running. If so, it indicates a very easy start with sufficient voltage.

(2) If the amperage is 4 times nameplate amperage, a normal start. The unit should be running. This indicates sufficient voltage.

(3) If the amperage is 4½ times nameplate amperage, the unit is probably running. If so, it indicates the unit is a little tight, but voltage is still sufficient.

(4) If the amperage is 5 to 6 times nameplate amperage (all three phases), this indicates a locked rotor condition.

Several factors will affect starting current and voltage. For example, a new pump is usually tighter. An old pump with a deposition of iron sulfide or calcium is usually harder to start, especially if it has been down some length of time. An old pump which is clean and well broken in will generally start with less amperage and with less voltage dip.

Another important factor is the torque available at the motor. The torque value will vary with the different voltage and amperage of units of the same horsepower. The following example shows how to determine the torque:

Data:

    70 hp motor, 980 v 45 amps
    6500 ft setting depth
    #4 cable

Calculations:

    Surface voltage = 980v × 160v loss = 1140 v

Assume 3.0 nameplate current inrush or 135 amps on start-up. The cable voltage loss at start-up is now:

$$160v \times 3.0 = 480 \text{ v}$$

Voltage reaching the motor is 1140-480 or 660 v. This is 67% of NP voltage.

There is 1.5 ft-lbs torque/hp @100% voltage or $1.5 \times 70 = 105$ ft-lbs

The torque varies as the square of the voltage; therefore:

$$(.67)^2 \times 105 = 47 \text{ ft-lbs at the unit on start-up.}$$

This unit would probably have difficulty starting if any wear or binding effect was present.

A higher voltage motor of the same horsepower would provide better torque characteristics.

Example: 70 hp motor, 1,170 v 38 amps
6,500 ft setting depth
#4 cable

Calculations:

Surface voltage = 1170v + 134v loss = 1304 v

Assume 3.0 nameplate current inrush or 114 amps on start-up. The cable voltage loss at start-up is now:

$$134 \text{ v} \times 3.0 = 402$$

Voltage reaching the motor is 1304v − 402v or 902 v. This is 77% of NP voltage.
Torque calculations are:

$$(0.77)^2 \times 105 = 62.25 \text{ ft-lbs at unit on start-up.}$$

The torque available at a unit can be calculated closely, provided good data on inrush current and voltage loss is known.

(8) If one phase indicates very low or no current, the line is single-phased. Unbalanced current on each 0 also indicates power problems. This must be repaired by the power company or qualified electricians. In the process of checking for a single 0 condition, energizing the unit for two seconds should be sufficient to observe and note the current of each separate phase.

(9) If all electrical checks on cable and motor downhole are O.K. and all three phases still carry abnormal high current the pump is probably stuck. A stuck pump may be broken loose by reversing and attempting to start, then returning to correct rotation to make the attempt.
Note: A maximum of three attempts is sufficient. If the unit does not break clear, it will require pulling or another method, such as acidizing.

If there is no check valve, acid may be spotted down the tubing into the pump to help clear scale buildup. If this fails, it must be pulled and sent to the factory for repairs.

(10) If the pump is stuck and the voltage does not drop below the required starting voltage, amperage will remain high. However, if the voltage drops too low to start the unit, the amperage will also be lower than normal starting amperage.

(11) Note: Always investigate all possibilities of trouble on the surface before pulling the tubing to check out the unit.

4.7642 Troubleshooting procedures

Table 4.71 is presented as a guide to help in troubleshooting an electrical pumping system.

TABLE 4.71
TROUBLESHOOTING PROCEDURES

Symptoms	Possible Causes	Corrective Measures
1. Switchboard will not operate	No voltage to switchboard	Check fuses on primary system, transformer, and main switch. Check for voltage at potential transformer. Check switchboard fuse.
	Loose or open contacts	See that overload relay contacts are clean and closed—may be checked with ohmmeter to determine if contacts are solidly closed. Check all other relay contacts and door interlock switches to determine if damage has occurred.
	Loose or open terminals	With screwdriver, check all terminal screws at relays, door switches, and terminal strips. Due to vibration, screws will often loosen during shipment of equipment.
	Open circuit on remote control, float switches, or pressure switches	Check continuity on all such circuits. If such remote control has been used, and later disconnected, make certain that float switch jumper is in place in the switchboard.
	Defective unit	Test unit per instructions.
2. Switchboard is operating, but blows fuses, or opens overload relays—high current	Cable damaged during installation or shipment	Open main switch—Caution: Always make visual inspection, making certain all three switch contacts are open. If in doubt, check with instruments. Using ohmmeter or megger on appropriate scale, check each conductor, going to well, reading from conductor to proven ground. A zero or very low reading indicates that cable is grounded. Read also from conductor to conductor 1-2, 2-3 and 1-3. Figure what the resistance should be from your motor and

Symptoms	Possible Causes	Corrective Measures
2. Switchboard is operating, but blows fuses, or opens overload relays—high current	Cable damaged during installation or shipment	cable resistance charts. These three readings should be near to zero. On lower scales, they will read 1 to 6 ohms depending on size of motor. A very high reading from conductor to conductor will indicate an open circuit. This could be in the cable or motor. Caution: Before pulling, make this test again at the nearest open end of cable going to well, junction box, open splice, extra fuse panel, etc. Make certain trouble is not on surface before proceeding to pull unit.
	Fuses too small or overload relays set too high on scale (too low amperage). Dashpot overloads may have wrong type fluid or delay ports open too wide.	Check setting of overload relays and fuse sizes, replace or adjust as necessary—considering type of fuse (superlag, fusetron, or other) and starting current of pumping unit. Refer to fuse chart, also to motor starting current.
	Low voltage, high voltage, single ϕ power, unbalanced voltage	Using voltmeter, check voltage at switchboard controls and multiply by potential transformer ratio (See section on transformers). If power transformer taps need to be adjusted, it is necessary to use power company personnel or electricians, depending on ownership and location.
	Pump stuck	If the foregoing measures have indicated that voltage, cable, and motor are O.K., it is possible that the pump is stuck from sand, deposition, or other causes. This can often be remedied by reversing any two (2) leads at switchboard and bumping pump in reverse rotation two or three times. If this fails, unit must be pulled to determine and correct fault. Acid may be spotted in pump if there is no check valve.

Symptoms	Possible Causes	Corrective Measures
2. Switchboard is operating, but blows fuses, or opens overload relays—high current	Unit in bind due to crooked place in well bore. (This is not a common problem.)	Usually, raising or lowering unit two or three joints in well will relieve this condition.
	Check valve leaking	If check valve is leaking at time of attempted start, and well is not stabilized, pump will be running backwards. This can be determined with voltmeter or triplett ohmmeter on 3 to 12 volt AC scale. A voltage reading will indicate pump spinning backwards. Caution: Attempting to start in this condition will result in very high current and possible damage to pump and motor. Wait for well to stabilize. In automatic operation, be sure timer relay is set for sufficient delay to permit stabilization of well. Make notation on installation report to change check valve on next pull.
3. Pump has been operating normally, but goes off and will not restart due to high current	Weather conditions	Determine if there has been lightning or high wind in the area of power system. If so, check for blown fuses. Make all checks set forth in preceding discussions.
	Power system surges or low power capacity and low voltage	Find out if any unusually heavy electrical loads have been added to or taken from the power system which might affect the voltage at pump installation. Possibly larger wire, larger transformers, or surge capacitors are needed. Obtain all possible information.
	Well conditions	Check flow line or gathering system to determine if any unusual amount of sand or mud is in evidence. Possible well clean-out is indicated.
	Pump condition	Consider past running time of pump, well history, sand, mud, etc. Possibly thrust wash-

Symptoms	Possible Causes	Corrective Measures
		ers and bearings worn causing undue friction. Unit should be pulled and changed out to avoid burned motor.
4. Unit stops running —low current	Pump gas-locked. Underload may be set too low.	Relieving pressure on casing will sometimes remedy this. May be necessary to load casing with water or other fluid. In this gassy condition, a gas separator should be installed, and check valve in-installed several joints above pump. This unit should be operated with casing open. This condition can usually be relieved by lowering pump two (2) or three (3) joints. Casing may be closed temporarily to increase pressure forcing more gas in solution. Open casing after gas-lock is broken.
	Well pumped off	May be necessary to acidize or frac well or sand pump to clean out at perforations or open hole. If well will produce well above the pump's minimum range, it may be possible to choke back produced fluid for continuous operation. Refer to pump data chart to determine minimum-maximum recommended pumping range. Also refer to ammeter chart for analysis.
	Undercurrent relay needs adjustment	In cases of light fluid load but sufficient fluid to meet minimum pumping requirements, undercurrent relay may be adjusted to operate at lower current. In this case, the production rate should be closely observed for several days. A timing delay relay may be added to the circuit to delay the drop-out time by the undercurrent relay.
	Where generator is used instead of established power system.	When a generator slows down, the frequency (cycles), volt-

Symptoms	Possible Causes	Corrective Measures
	Lowered speed of generator.	age, current, and power consumption all go correspondingly lower. Speed generator up to normal speed to raise current.
5. Pump runs but produces little or no fluid.	Tubing leak	Make pressure test at tubing head to determine if a leak exists. If so, tubing must be pulled and faulty joint or joints removed. Occasionally, a high or low current may be noted, depending on location of the leak, working fluid level, size of unit, etc. However, low or high current does not always indicate a tubing leak.
	Plugged pump intake	Will need to be pulled and cleaned. Can sometimes be cleaned by reverse rotation.
	Obstruction in flow line —sanded up, closed valve, etc.	Check pressure on flow line at wellhead. If abnormally high, take appropriate measures to clear line.
	Total head in pump not sufficient for application.	Check design of pump in connection with operating fluid level per application section.
	Broken pump shaft, protector shaft, or motor shaft.	Unit will need to be pulled and faulty piece of equipment replaced. Where undercurrent relay is employed, this and the last two conditions will usually stop pump on undercurrent.
	Reverse rotation	Reverse any two of the three conductors at well cable connections and try in the opposite direction.
	Well pumping off or very low working fluid level	Determine, if possible, working fluid level and refer to "Pump Gas-Locked" and "Well Pumped Off" in the section "Unit Stops Running—Low Current."

4.77 Maintenance program

When the motor controller becomes dusty and dirty, or if routine maintenance is due, the following procedures are recommended:

(A) Caution: make certain that a ground wire is connected to the controller enclosure and that it is properly grounded.

(B) Turn H-O-A selector switch to "off" position.

(C) Turn disconnect switch to "off" position.

(D) Pull each of the three fuses on the primary side of the transformer bank.

(E) Using hot gloves (tested and certified "safe") and a voltmeter, verify that there is no voltage on the incoming lines or on the line side of the disconnect switch. The phase-to-phase and phase-to-ground on all *three* phases should show zero volts.

(F) Use compressed air to remove all dust and dirt from enclosures and components.

(G) If required, replace the door gasket material.

(H) Replace the louver filter material if equipped.

(I) Inspect the components for loose connections—electrical and physical—broken parts, burned (dark brown) coils, components that have deteriorated, or any component which you suspect as being or might become defective.

(J) Almost all control circuit contacts are fixed inside a relay housing, thereby reducing the dust and dirt problem. Therefore, when the contacts of a relay are burned completely up or very dirty and corroded, the complete relay should be replaced.

(K) Dirty or burned contacts on the OL relays and contactor should be replaced.

(L) Check to make sure that the disconnect switch safety devices will prevent entry into the high voltage compartment when the disconnect switch is in the "on" position. Also check door latches, locks, and hinges.

(M) Service and set the OL and UC relays per the previous recommendations.

(N) Annually check the contactor's stationary and movable contacts.

(O) Disconnect power cable marking phases 1, 2, and 3.

(P) Make a visual inspection of the transformer bank to determine that the transformers do not have broken insulators or loose wires (internal or external), that all gasket material is in good condition and will not allow water to enter the transformers, to see that the oil is not contaminated, and that fluid is at the proper level.

(Q) Re-energize the transformer bank and cycle motor controller in both hand and auto positions.

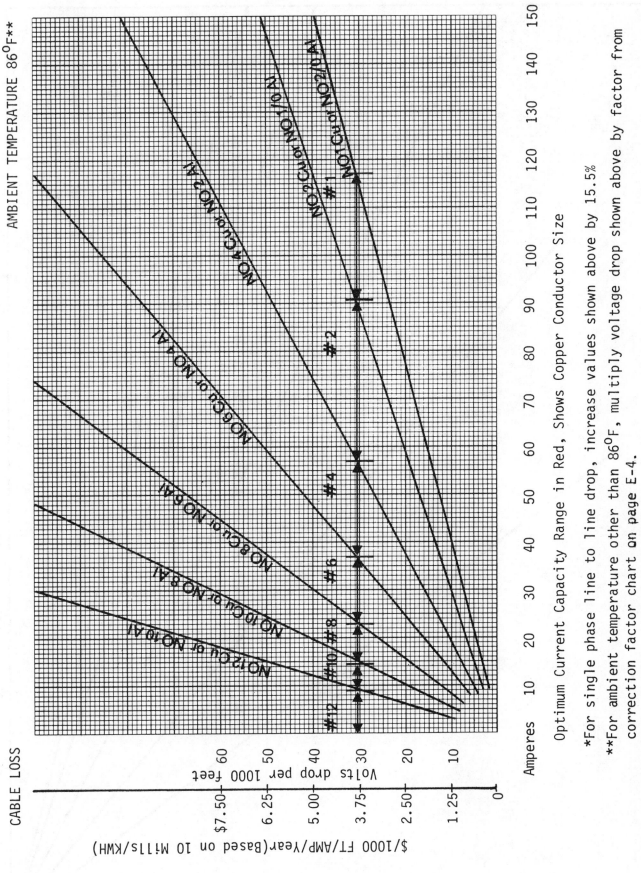

CABLE LOSS CHART
3 PHASE*

AMBIENT TEMPERATURE 86°F**

CABLE LOSS

Optimum Current Capacity Range in Red, Shows Copper Conductor Size

*For single phase line to line drop, increase values shown above by 15.5%

**For ambient temperature other than 86°F, multiply voltage drop shown above by factor from correction factor chart on page E-4.

Fig. 4A (1) Voltage drop chart (Courtesy TRW-Reda)

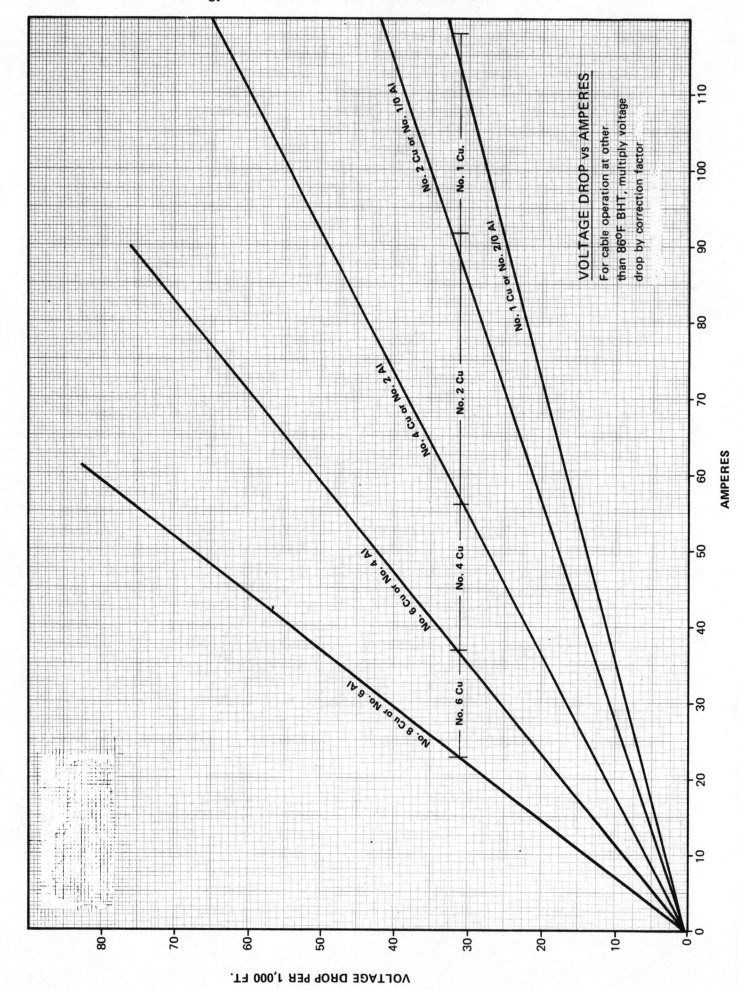

VOLTAGE DROP vs AMPERES

For cable operation at other than 86°F BHT, multiply voltage drop by correction factor

Fig. 4A (2)   Volt drop vs. amperes at 86°F (Courtesy OilLine-Kobe)

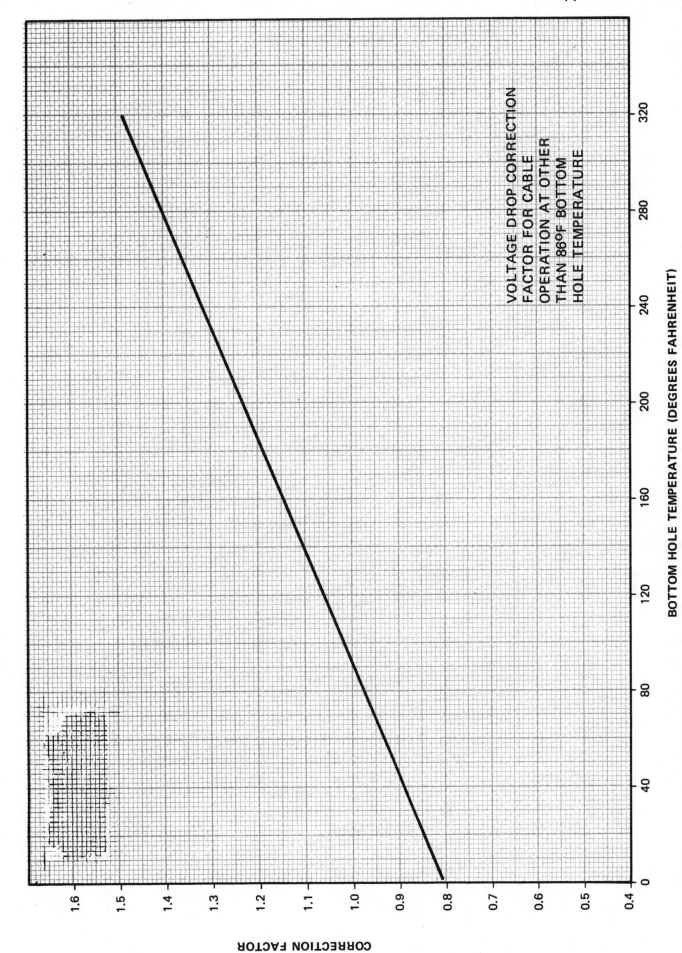

VOLTAGE DROP CORRECTION
FACTOR FOR CABLE
OPERATION AT OTHER
THAN 86°F BOTTOM
HOLE TEMPERATURE

BOTTOM HOLE TEMPERATURE (DEGREES FAHRENHEIT)

CORRECTION FACTOR

Fig. 4A (3)   Voltage drop correction factor for cable operation at other than 86°F bottom-hole temperature. (Courtesy OilLine-Kobe)

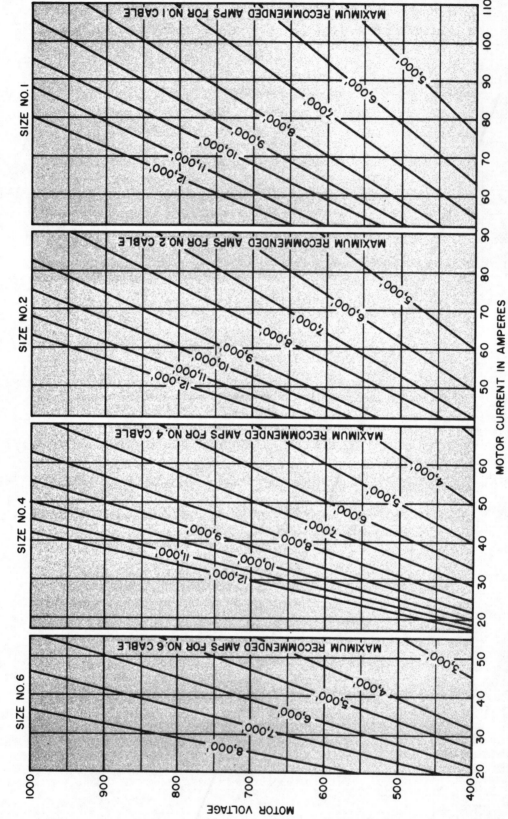

Fig. 4A (4) Maximum recommended cable lengths for various motor volt/amp ratings (Courtesy Bryon Jackson - Centrilift)

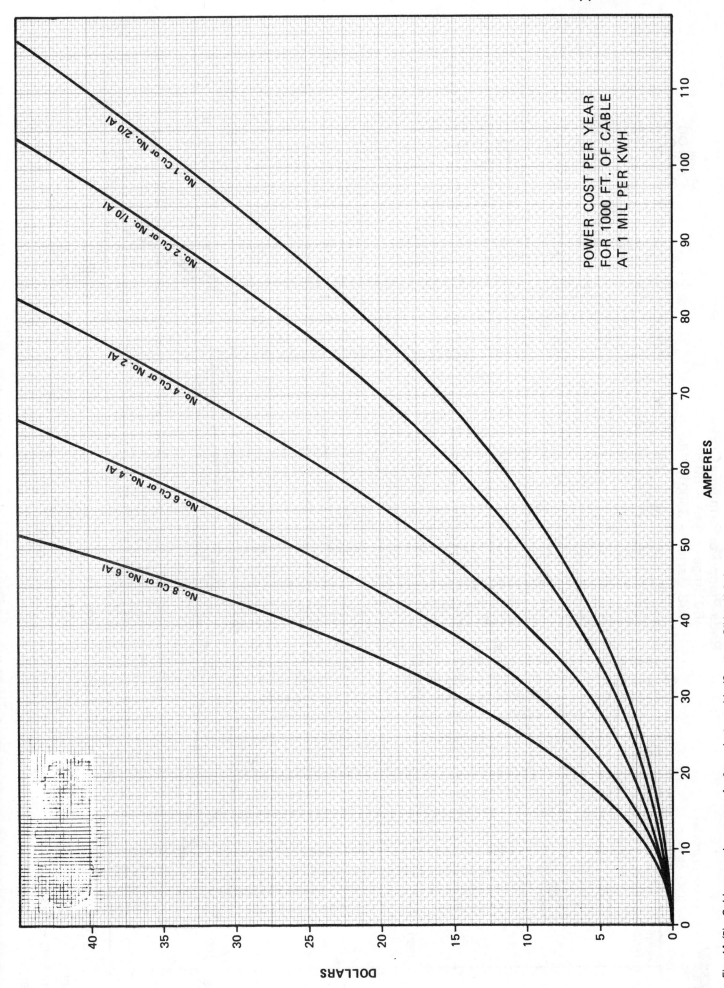

POWER COST PER YEAR
FOR 1000 FT. OF CABLE
AT 1 MIL PER KWH

No. 1 Cu or No. 2/0 Al

No. 2 Cu or No. 1/0 Al

No. 4 Cu or No. 2 Al

No. 6 Cu or No. 4 Al

No. 8 Cu or No. 6 Al

AMPERES

DOLLARS

Fig. 4A (5)   Cable power loss curves for 3-conductor cable (Courtesy OilLine-Kobe)

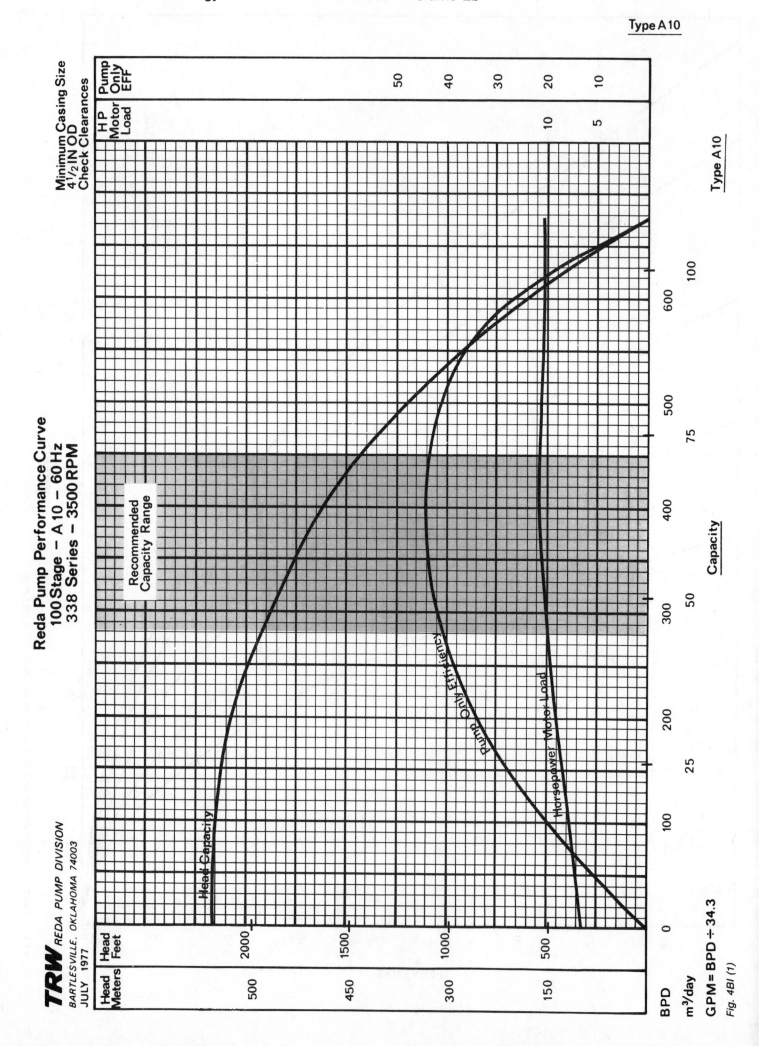

Fig. 4BI (1)

Type A10

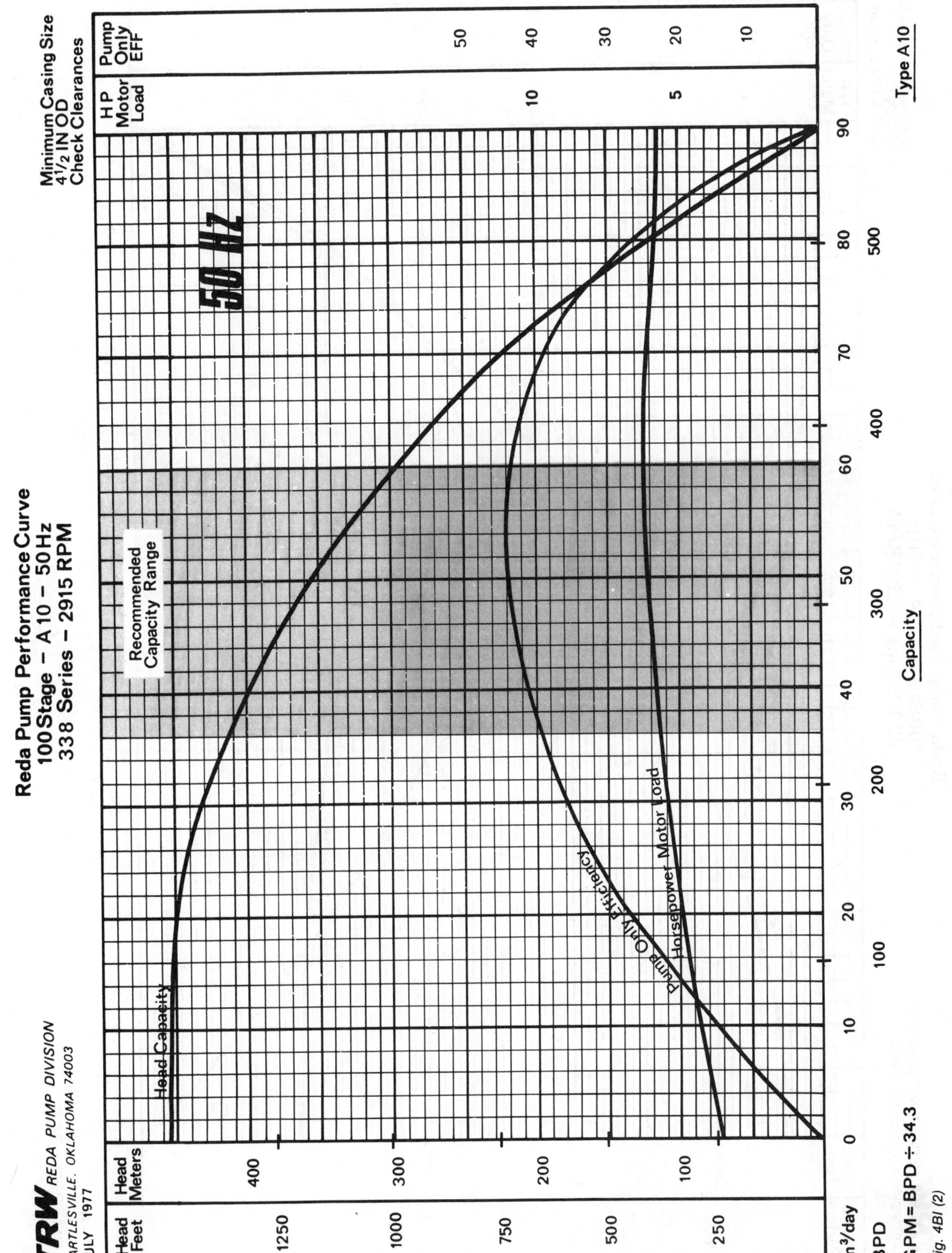

Reda Pump Performance Curve
100 Stage – A10 – 50 Hz
338 Series – 2915 RPM

TRW REDA PUMP DIVISION
BARTLESVILLE. OKLAHOMA 74003
JULY 1977

Minimum Casing Size
4½ IN OD
Check Clearances

50 Hz

Recommended
Capacity Range

Head Capacity

Pump Only Efficiency

Horsepower Motor Load

	HP Motor Load	Pump Only EFF
		50
	10	40
		30
	5	20
		10

Type A10

Capacity

Head Feet	Head Meters									
1250	400									
1000	300									
750	200									
500										
250	100									

m³/day

BPD

GPM = BPD ÷ 34.3

Fig. 4BI (2)

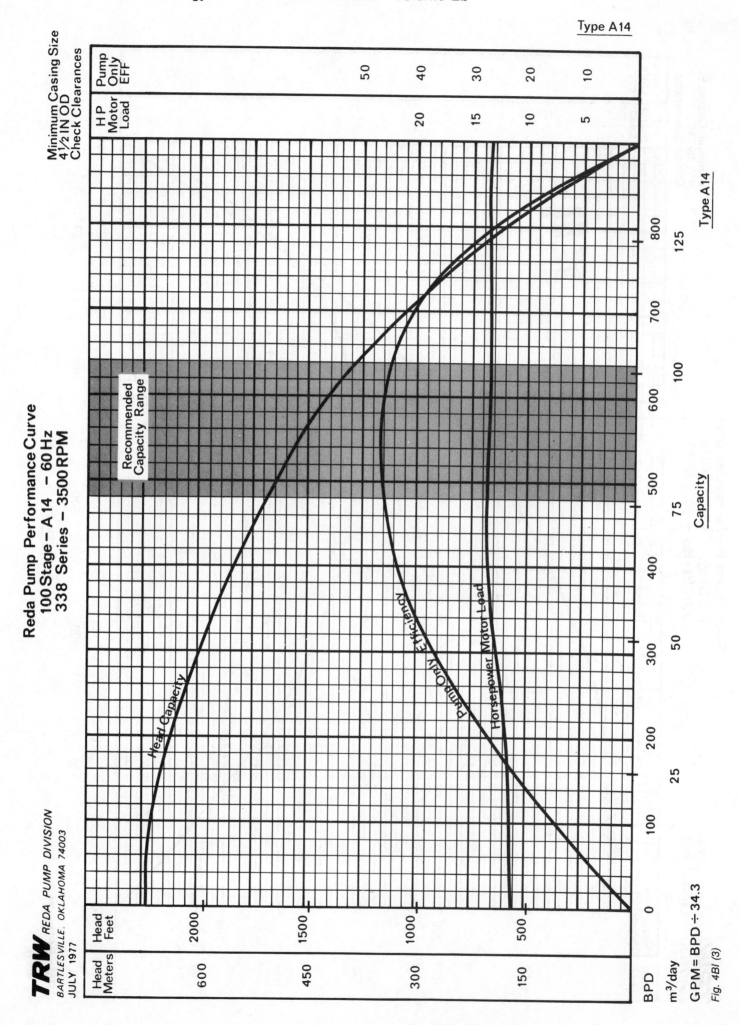

Type A14

Reda Pump Performance Curve
100 Stage – A14 – 60 Hz
338 Series – 3500 RPM

TRW REDA PUMP DIVISION
BARTLESVILLE, OKLAHOMA 74003
JULY 1977

Minimum Casing Size
4½ IN OD
Check Clearances

GPM = BPD ÷ 34.3

Fig. 4BI (3)

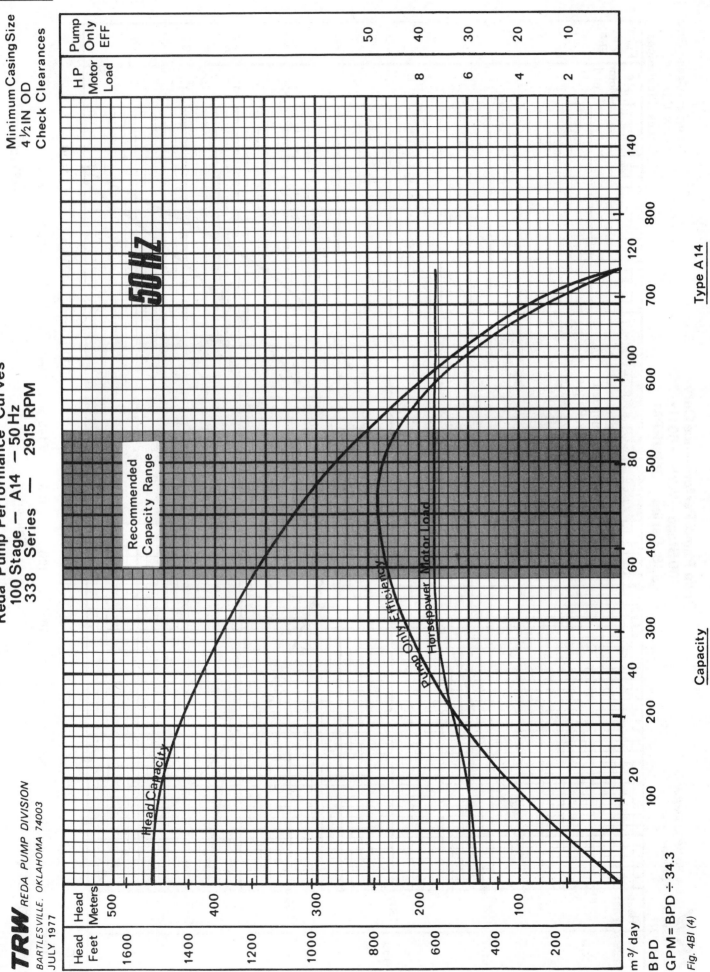

Type A14

Reda Pump Performance Curves
100 Stage — A14 — 50 Hz
338 Series — 2915 RPM

TRW REDA PUMP DIVISION
BARTLESVILLE. OKLAHOMA 74003
JULY 1977

Minimum Casing Size
4½ IN OD
Check Clearances

50 HZ

Recommended
Capacity Range

Head Capacity

Pump Only Efficiency

Horsepower Motor Load

Head Feet	Head Meters
1600	500
1400	400
1200	300
1000	
800	200
600	
400	100
200	

HP Motor Load	Pump Only EFF
	50
8	40
6	30
4	20
2	10

m³/ day
BPD
GPM = BPD ÷ 34.3

Capacity

Type A14

Fig. 4Bl (4)

Type A25

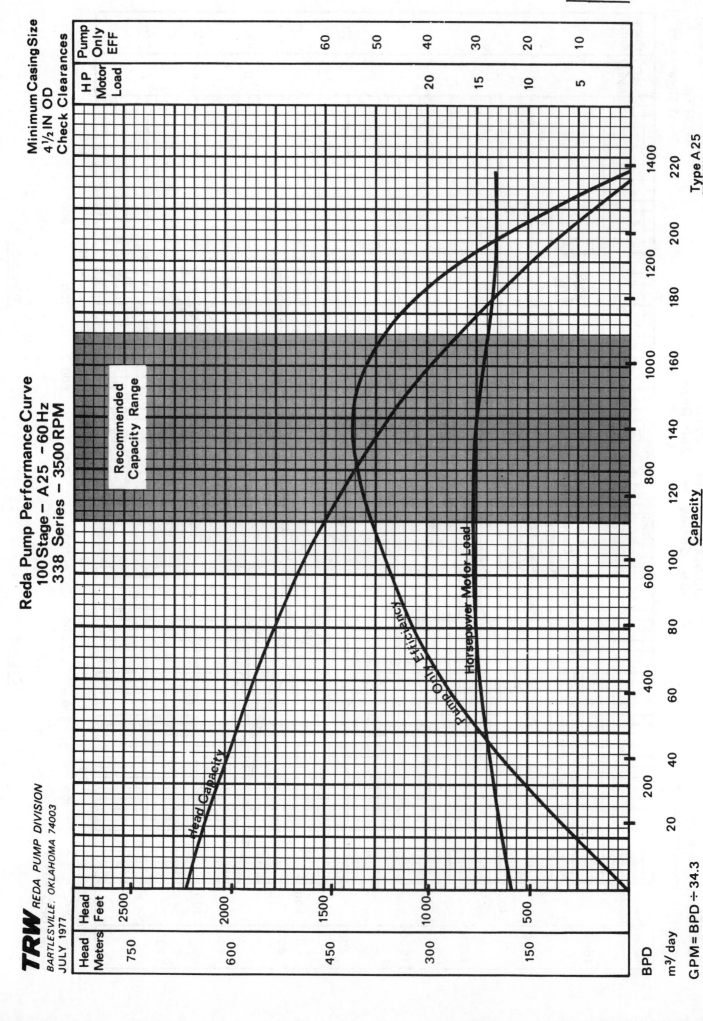

Reda Pump Performance Curve
100 Stage – A25 – 60 Hz
338 Series – 3500 RPM

TRW REDA PUMP DIVISION
BARTLESVILLE, OKLAHOMA 74003
JULY 1977

Minimum Casing Size
4½ IN OD
Check Clearances

Recommended Capacity Range

Head Capacity

Pump Only Efficiency

Horsepower Motor Load

Type A25

BPD

GPM = BPD ÷ 34.3

Fig. 4BI (5)

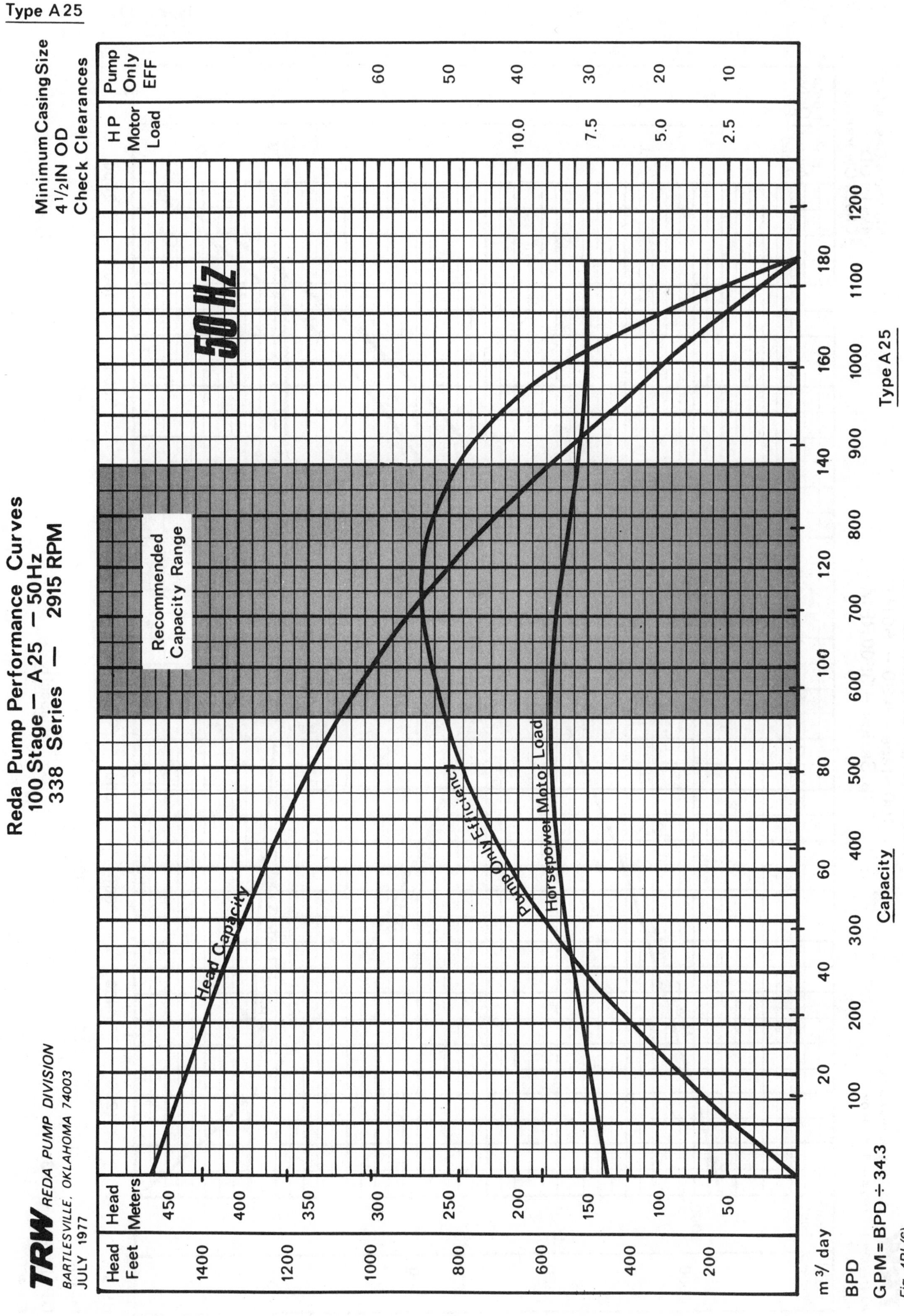

Type A 25

Reda Pump Performance Curves
100 Stage – A 25 – 50 Hz
338 Series –   2915 RPM

TRW REDA PUMP DIVISION
BARTLESVILLE, OKLAHOMA 74003
JULY 1977

Minimum Casing Size
4½ IN OD
Check Clearances

50 Hz

Recommended
Capacity Range

Head Capacity

Pump Only Efficiency

Horsepower Motor Load

Type A 25

Capacity

Pump Only EFF	HP Motor Load
60	
50	
40	10.0
30	7.5
20	5.0
10	2.5

Head Feet	Head Meters
1400	450
	400
1200	350
1000	300
800	250
	200
600	150
400	100
200	50

m³/ day

BPD

GPM = BPD ÷ 34.3

Fig. 4BI (6)

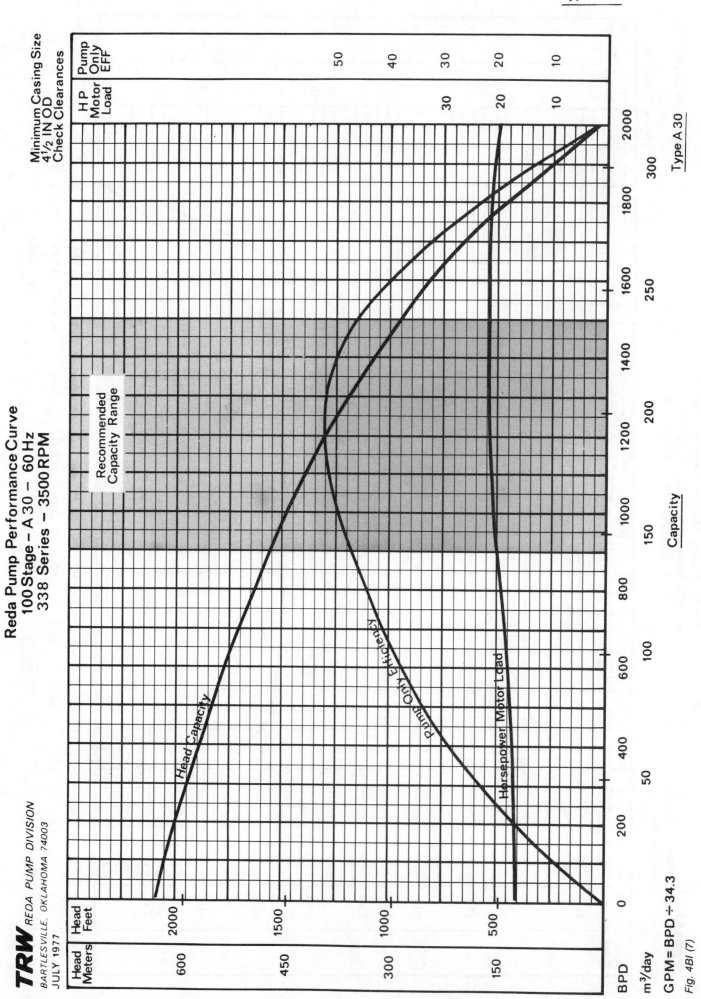

Type A 30

Fig. 4Bl (7)

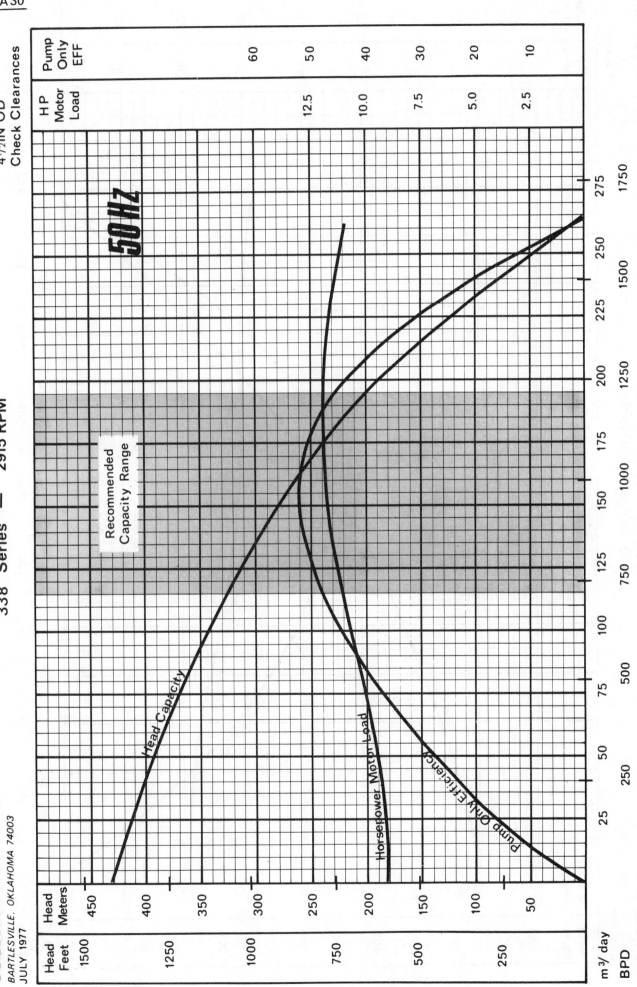

Type A 30

Reda Pump Performance Curves
100 Stage — A 30 — 50 Hz
338 Series — 2915 RPM

TRW REDA PUMP DIVISION
BARTLESVILLE, OKLAHOMA 74003
JULY 1977

Minimum Casing Size
4¹/₂ IN OD
Check Clearances

50Hz

Recommended Capacity Range

Head Capacity

Horsepower Motor Load

Pump Only Efficiency

Type A 30

Capacity

m³/day

BPD

GPM = BPD ÷ 34.3

Fig. 4BI (8)

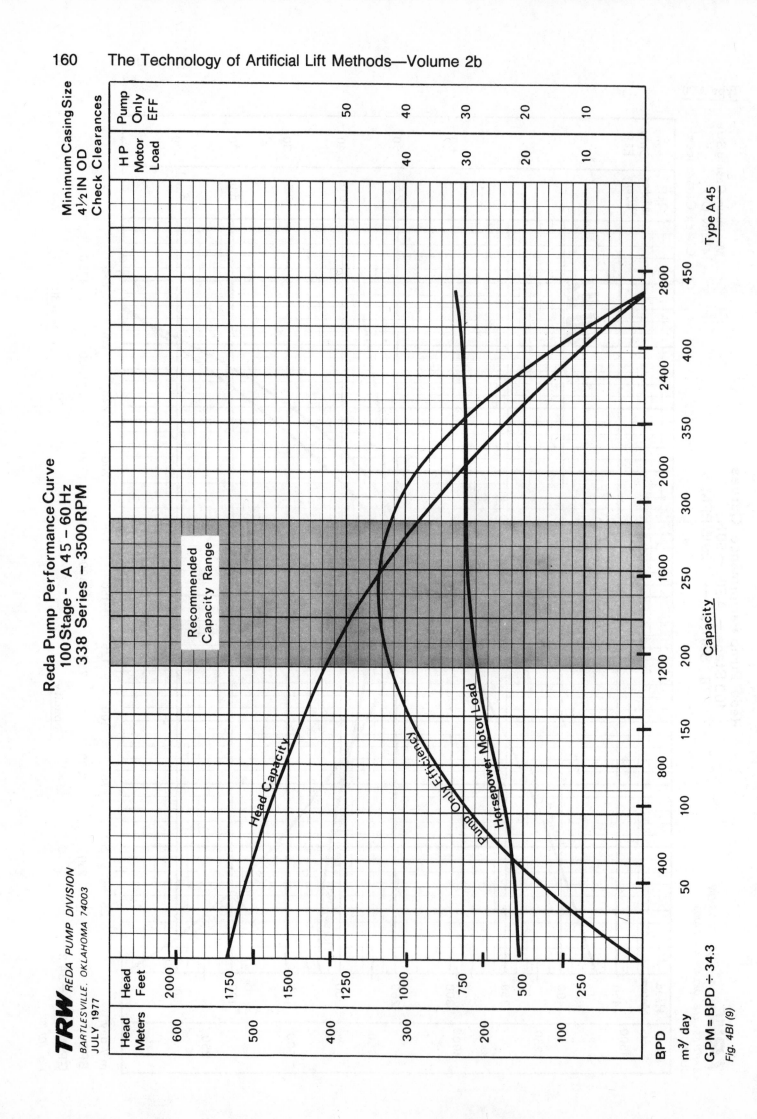

Reda Pump Performance Curve
100 Stage – A 45 – 60 Hz
338 Series – 3500 RPM

TRW REDA PUMP DIVISION
BARTLESVILLE. OKLAHOMA 74003
JULY 1977

Minimum Casing Size
4½ IN OD
Check Clearances

Type A45

GPM = BPD ÷ 34.3

Fig. 4Bl (9)

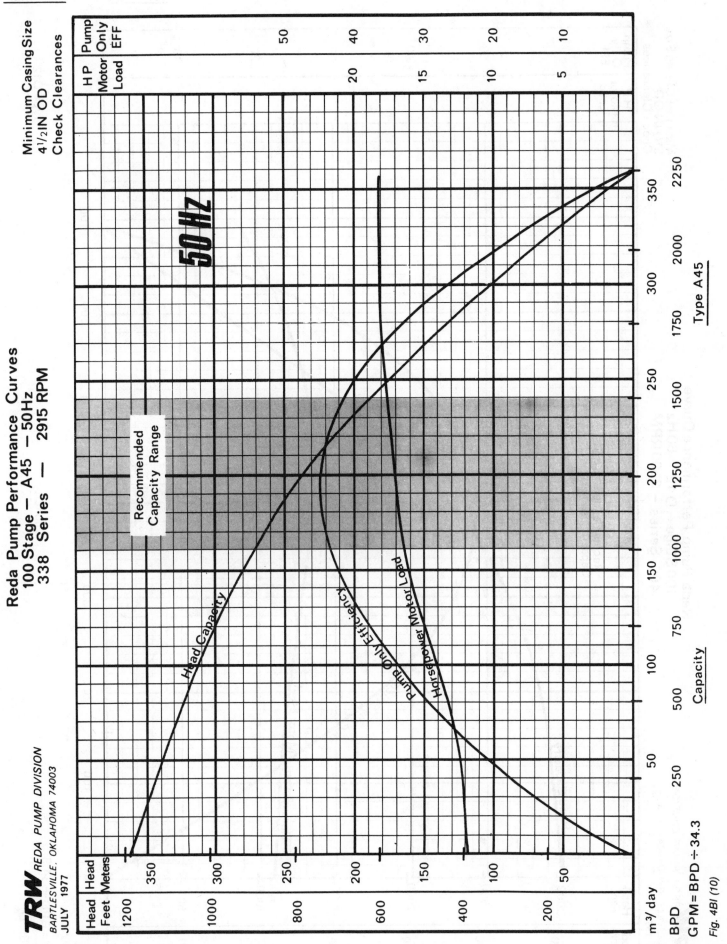

Reda Pump Performance Curves
100 Stage — A45 — 50 Hz
338 Series — 2915 RPM

Minimum Casing Size
4½ IN OD
Check Clearances

50 HZ

Recommended
Capacity Range

Head Capacity

Pump Only Efficiency

Horsepower Motor Load

Capacity

Type A45

TRW REDA PUMP DIVISION
BARTLESVILLE. OKLAHOMA 74003
JULY 1977

GPM = BPD ÷ 34.3

Fig. 4BI (10)

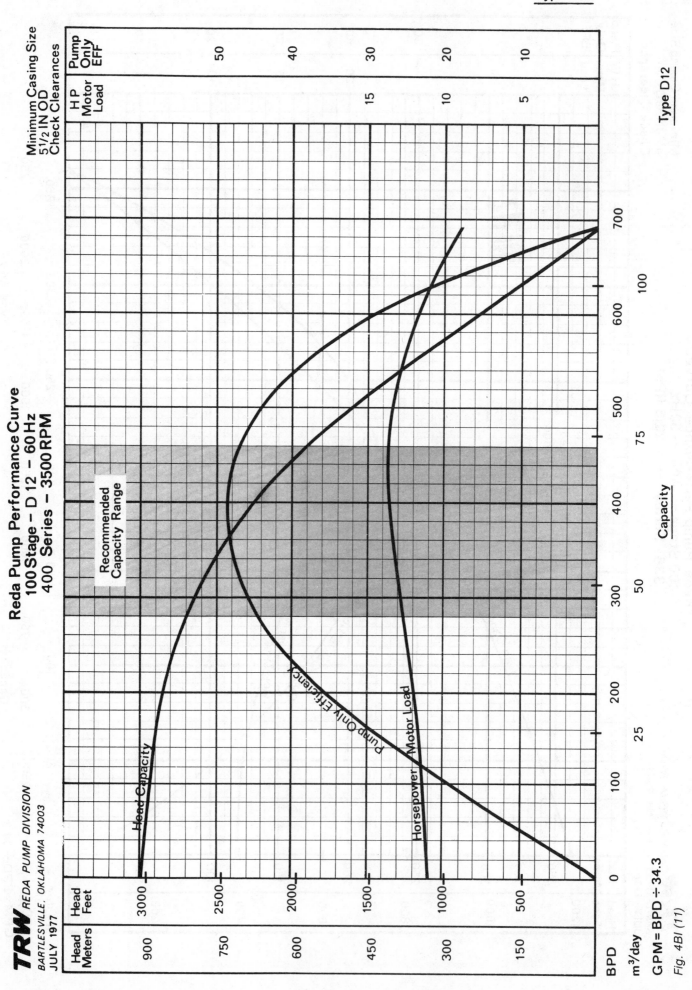

Fig. 4BI (11)

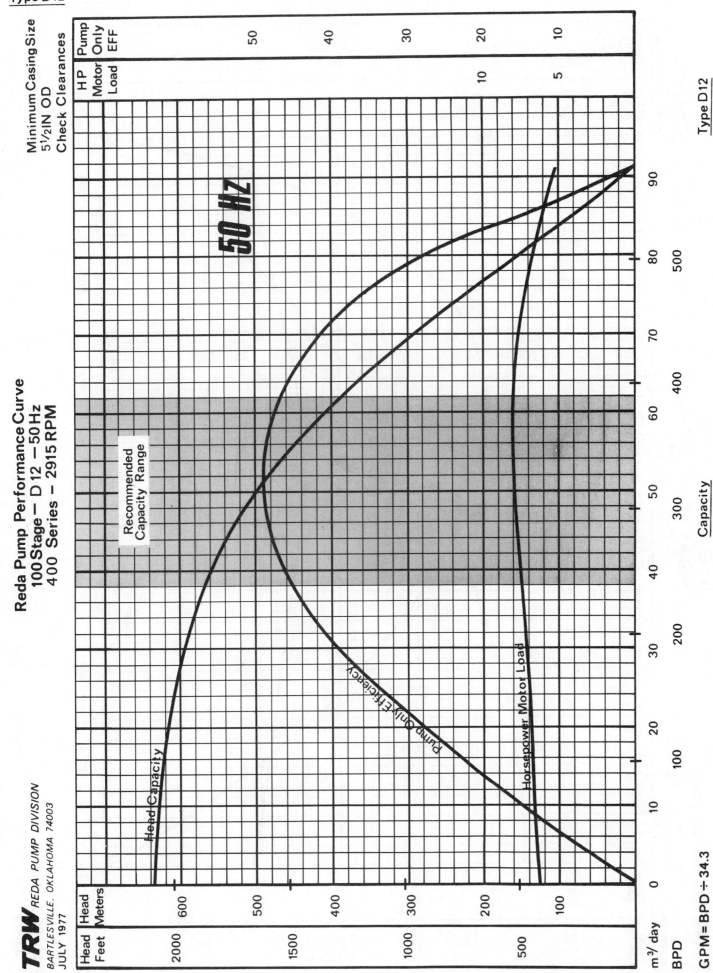

Type D12

Reda Pump Performance Curve
100 Stage – D 12 – 50 Hz
400 Series – 2915 RPM

Minimum Casing Size
5½ IN OD
Check Clearances

50 HZ

Recommended Capacity Range

Head Capacity

Pump Only Efficiency

Horsepower Motor Load

TRW REDA PUMP DIVISION
BARTLESVILLE, OKLAHOMA 74003
JULY 1977

GPM = BPD ÷ 34.3

Fig. 4BI (12)

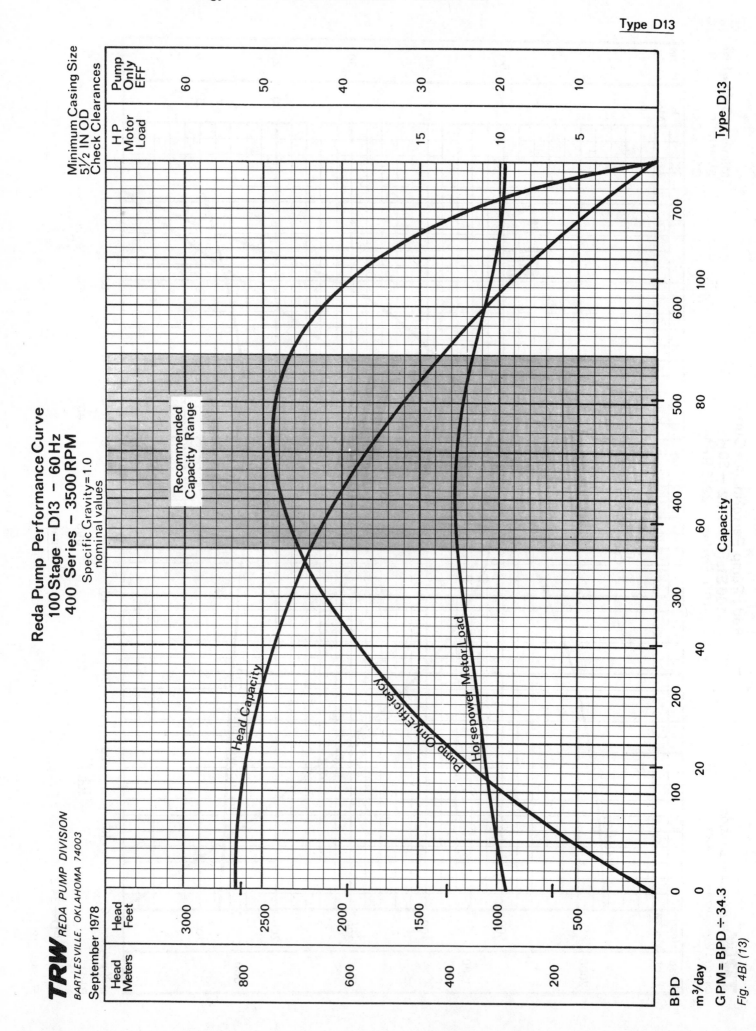

Fig. 4BI (13)

Type D13

Minimum Casing Size
5½ IN OD
Check Clearances

Reda Pump Performance Curves
100 Stage – D13 – 50 Hz
400 Series — 2915 RPM

Specific Gravity = 1.0
nominal values

TRW REDA PUMP DIVISION
BARTLESVILLE, OKLAHOMA 74003
September 1978

GPM = BPD ÷ 34.3

Fig. 4BI (14)

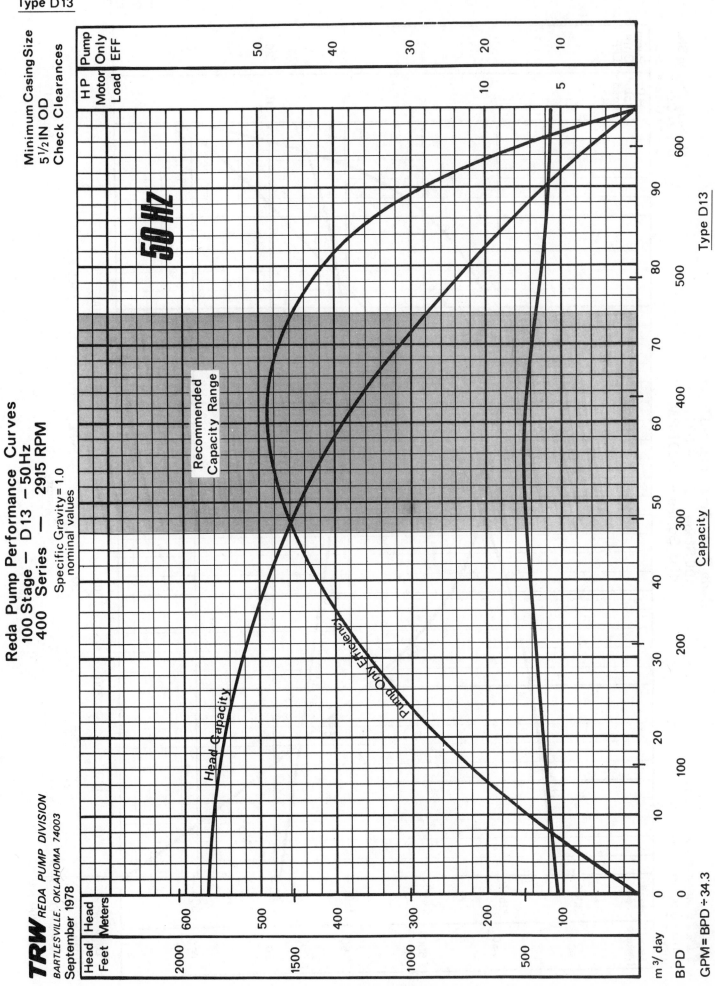

50 Hz

Recommended
Capacity Range

Head Capacity

Pump Only Efficiency

Type D13

Capacity

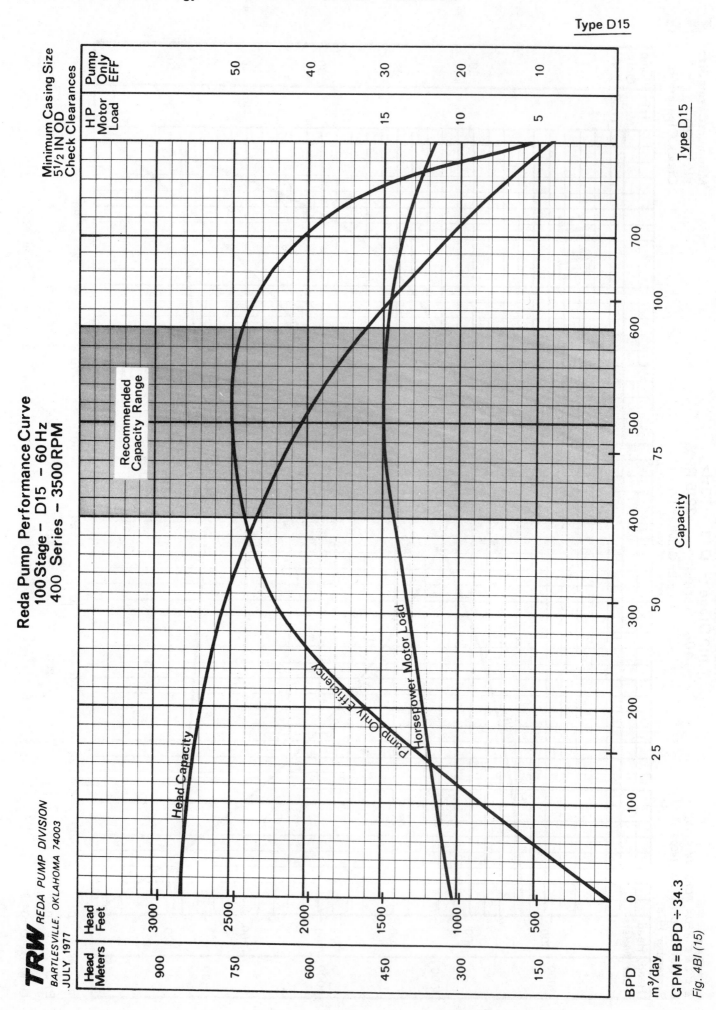

Reda Pump Performance Curve
100 Stage – D15 – 60 Hz
400 Series – 3500 RPM

TRW REDA PUMP DIVISION
BARTLESVILLE, OKLAHOMA 74003
JULY 1977

Type D15

Minimum Casing Size
5½ IN OD
Check Clearances

Recommended Capacity Range

Head Capacity

Pump Only Efficiency

Horsepower Motor Load

Capacity

BPD = BPD ÷ 34.3
GPM = BPD ÷ 34.3

Fig. 4Bl (15)

Type D15

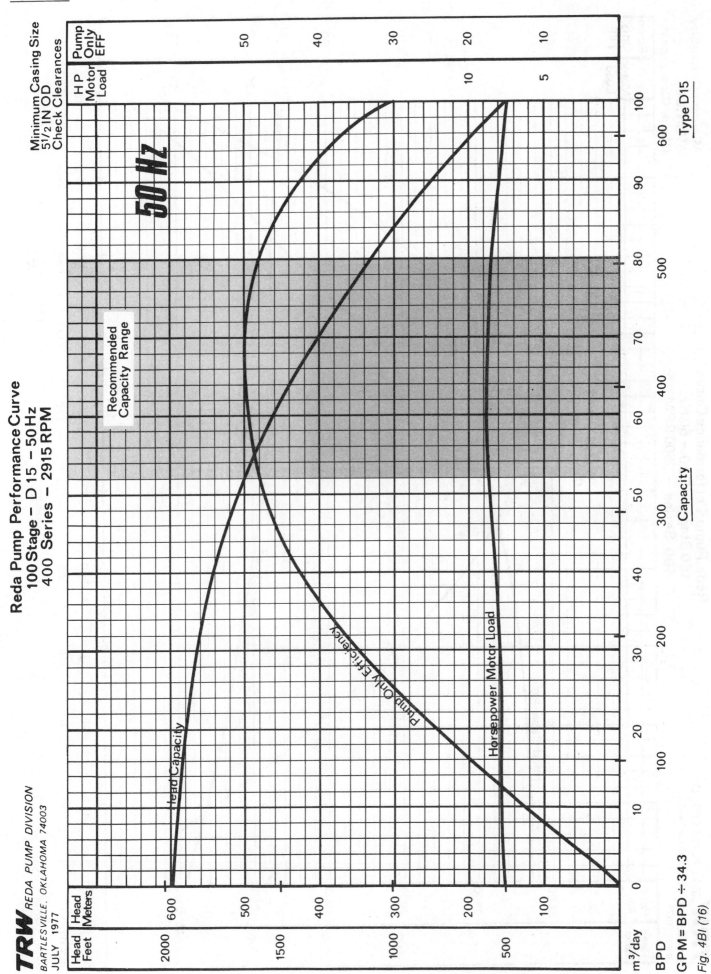

Reda Pump Performance Curve
100 Stage – D 15 – 50 Hz
400 Series – 2915 RPM

TRW REDA PUMP DIVISION
BARTLESVILLE. OKLAHOMA 74003
JULY 1977

GPM = BPD ÷ 34.3
Fig. 4BI (16)

Type D15

Minimum Casing Size
5½ IN OD
Check Clearances

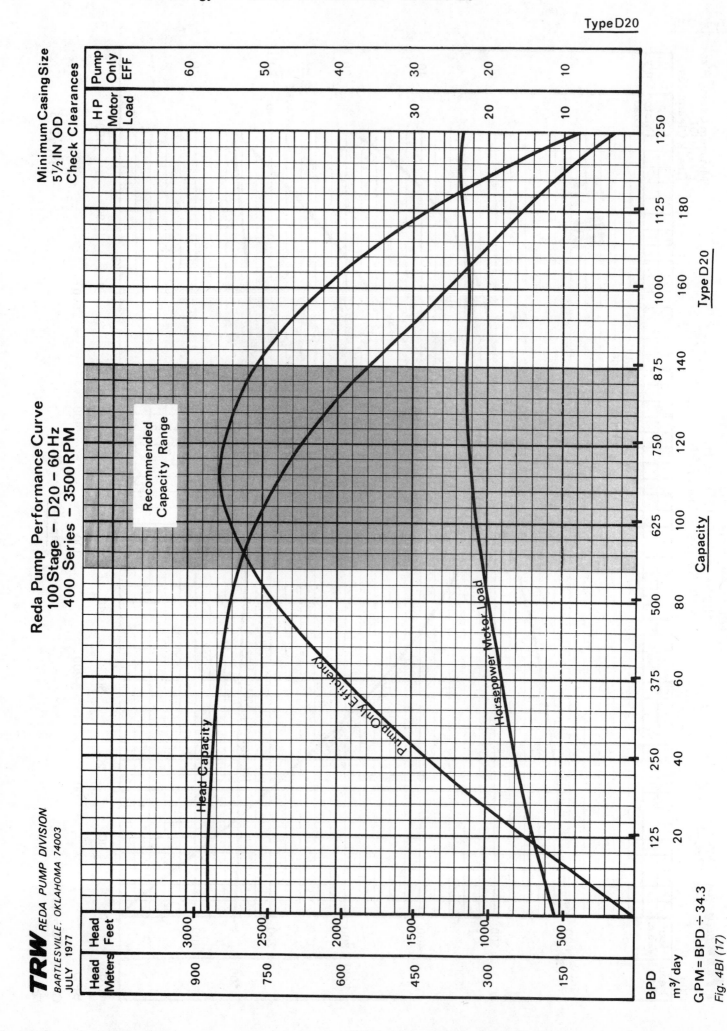

Reda Pump Performance Curve
100 Stage – D20 – 60 Hz
400 Series – 3500 RPM

Type D20

Minimum Casing Size
5½ IN OD
Check Clearances

Recommended Capacity Range

Head Capacity

Pump Only Efficiency

Horsepower Motor Load

TRW REDA PUMP DIVISION
BARTLESVILLE, OKLAHOMA 74003
JULY 1977

GPM = BPD ÷ 34.3

Fig. 4BI (17)

Type D20

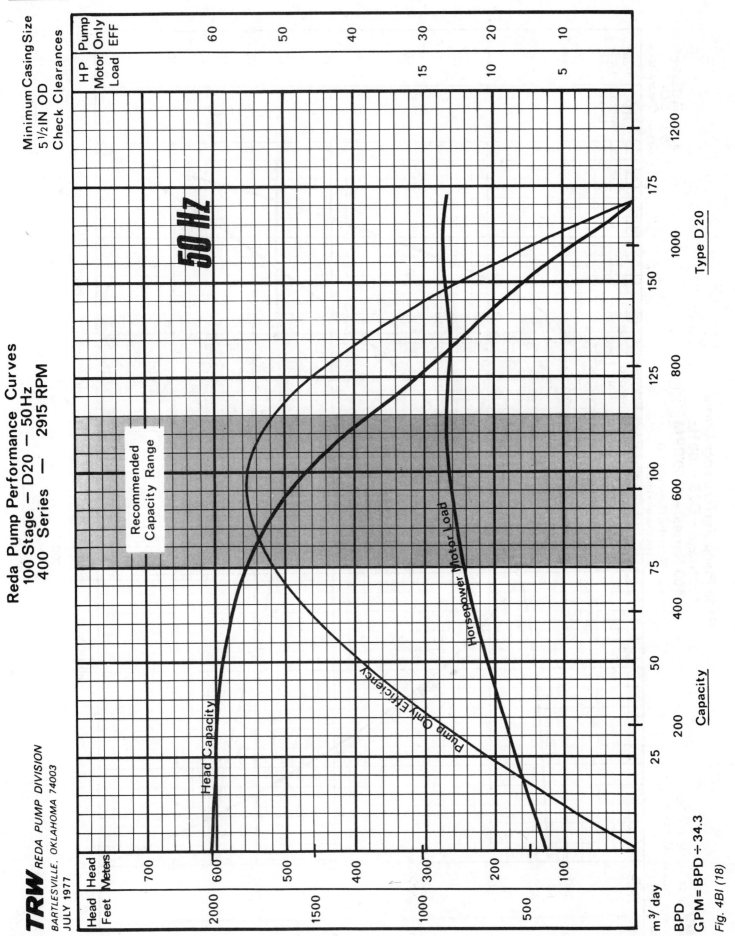

Reda Pump Performance Curves
100 Stage — D20 — 50 Hz
400 Series — 2915 RPM

Minimum Casing Size
5½ IN OD
Check Clearances

TRW *REDA PUMP DIVISION*
BARTLESVILLE, OKLAHOMA 74003
JULY 1977

GPM = BPD ÷ 34.3

Fig. 4BI (18)

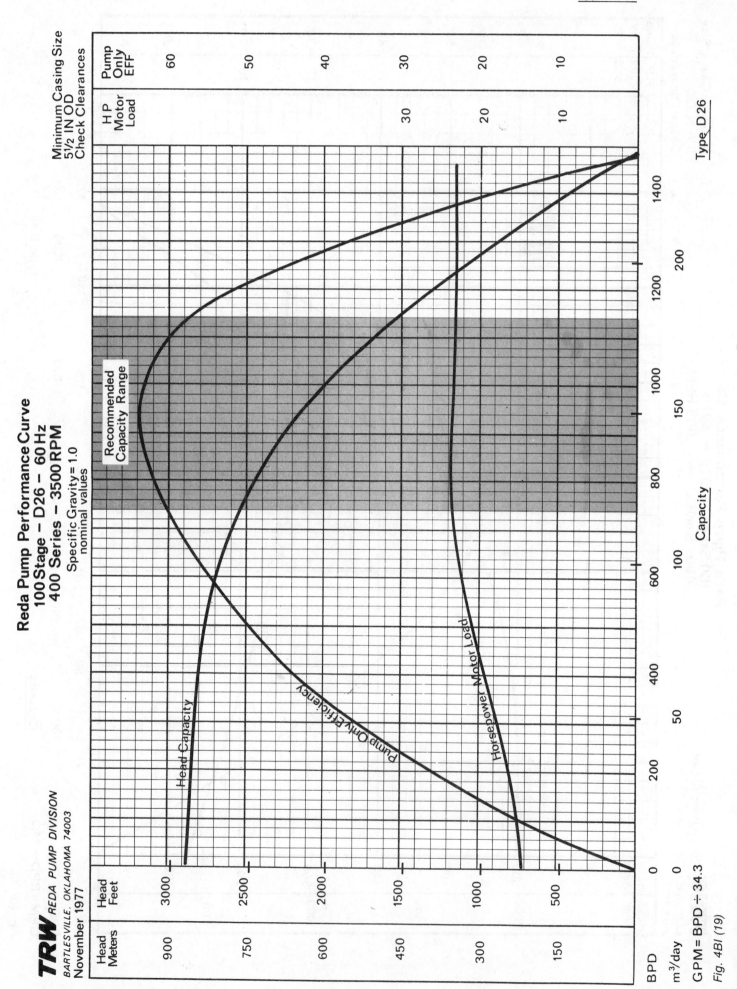

Reda Pump Performance Curve
100 Stage – D26 – 60 Hz
400 Series – 3500 RPM
Specific Gravity = 1.0
nominal values

Type D26

Fig. 4BI (19)

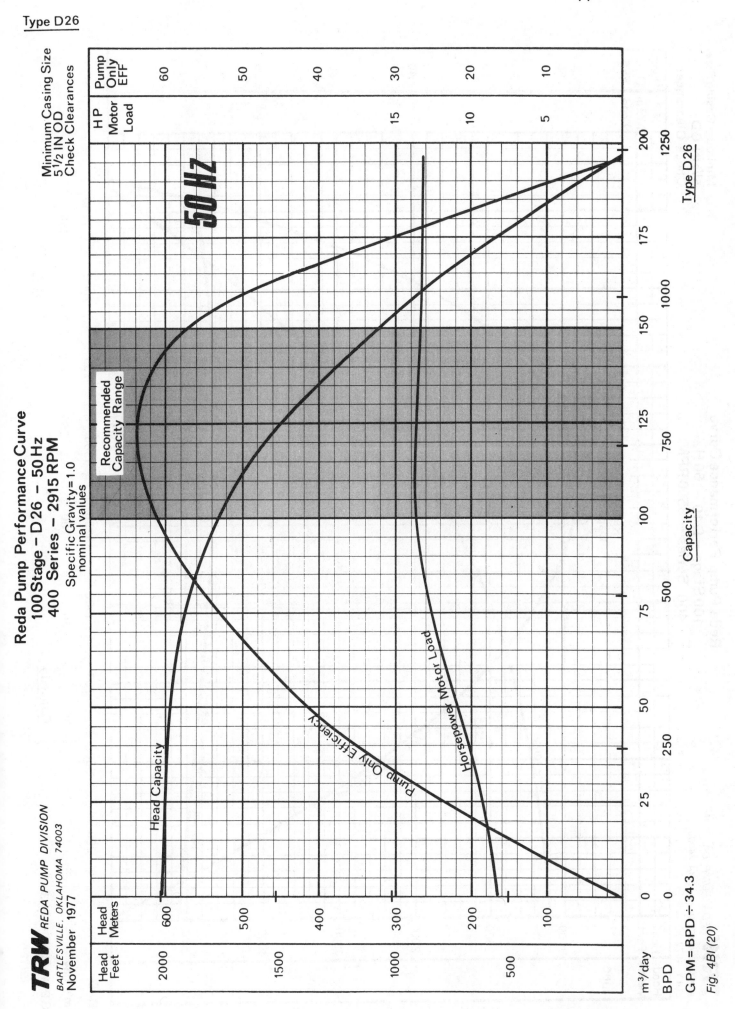

Type D26

Minimum Casing Size
5½ IN OD
Check Clearances

Reda Pump Performance Curve
100 Stage – D26 – 50 Hz
400 Series – 2915 RPM
Specific Gravity = 1.0
nominal values

50 Hz

Recommended
Capacity Range

Head Capacity

Pump Only Efficiency

Horsepower Motor Load

Capacity

Type D26

TRW *REDA PUMP DIVISION*
BARTLESVILLE, OKLAHOMA 74003
November 1977

GPM = BPD ÷ 34.3

Fig. 4BI (20)

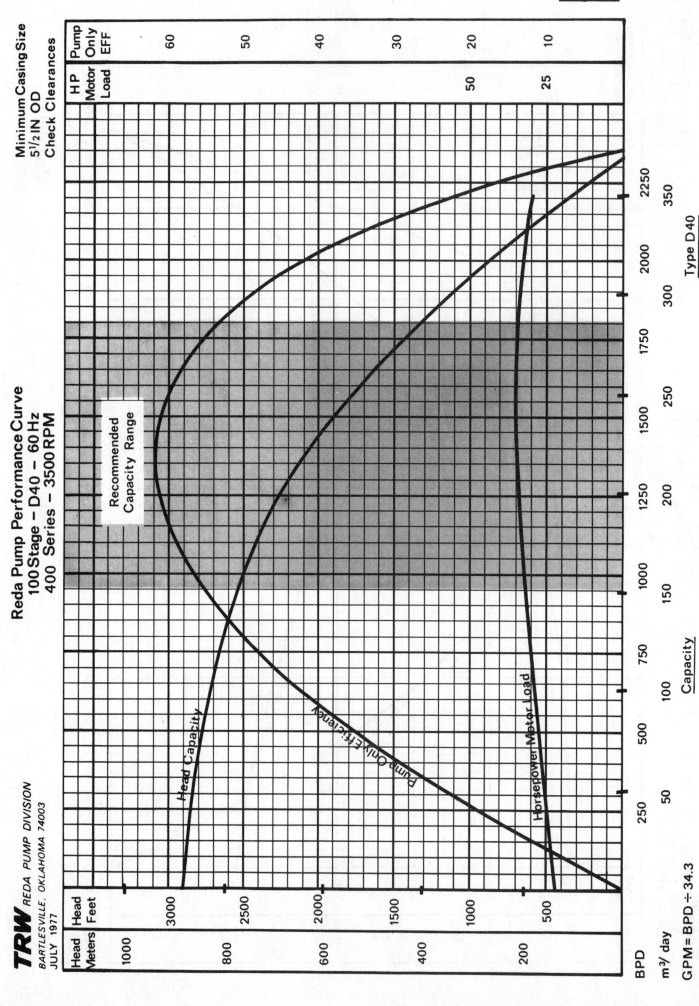

Type D 40

Fig. 4BI (21)

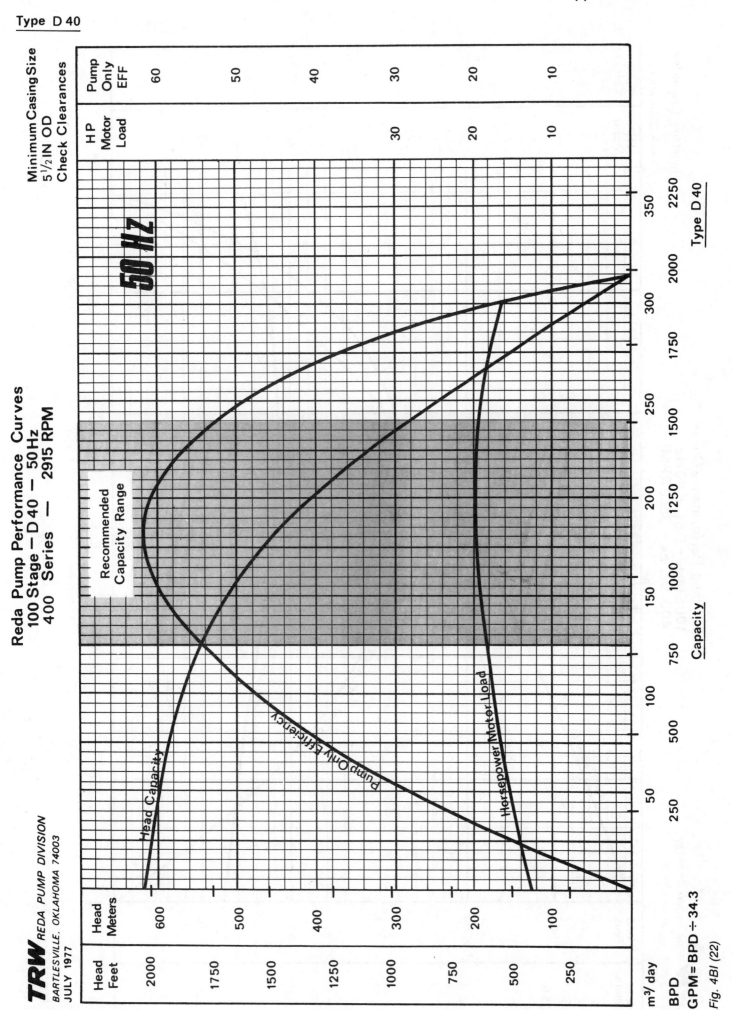

Type D 40

Fig. 4BI (22)

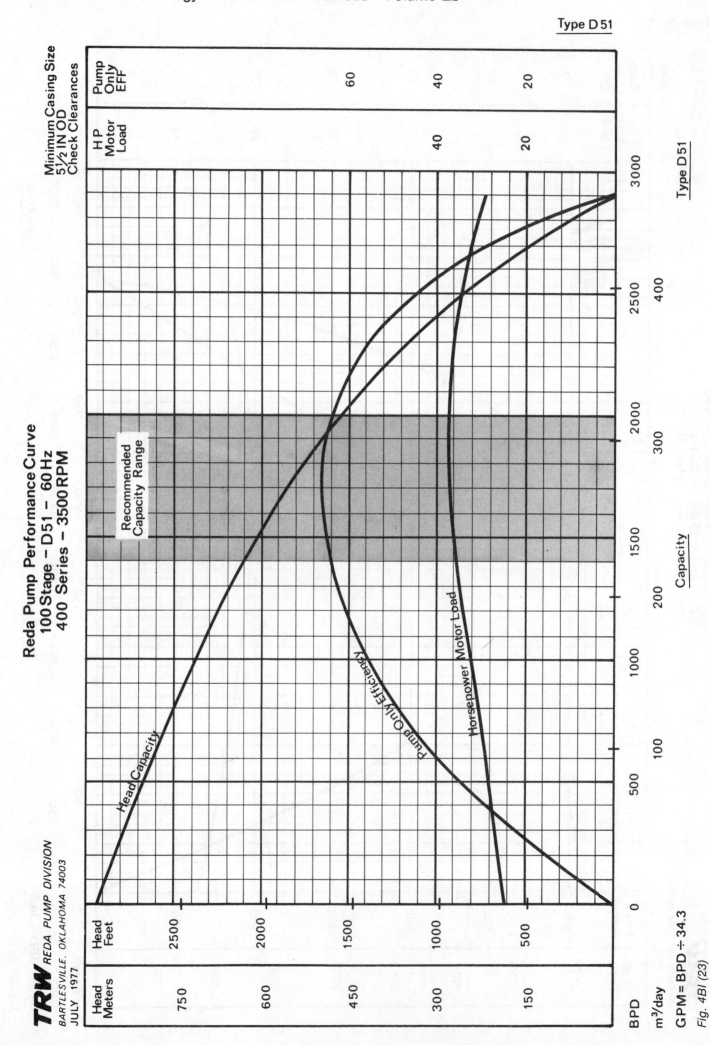

Reda Pump Performance Curve
100 Stage – D 51 – 60 Hz
400 Series – 3500 RPM

TRW REDA PUMP DIVISION
BARTLESVILLE, OKLAHOMA 74003
JULY 1977

Minimum Casing Size
5½ IN OD
Check Clearances

Type D 51

GPM = BPD ÷ 34.3

Fig. 4Bl (23)

Type D 51

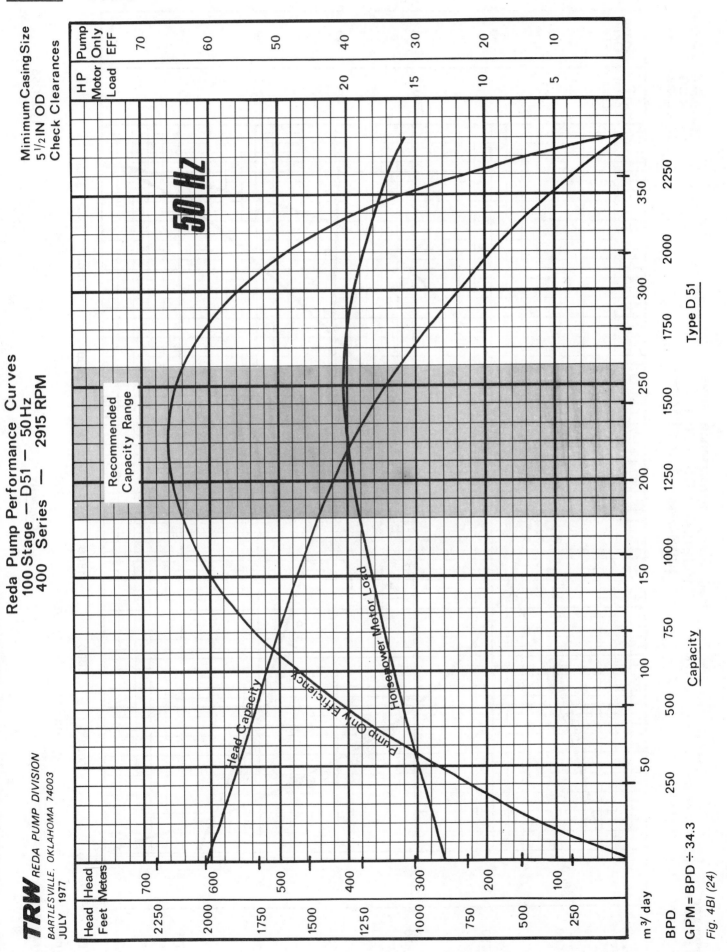

Reda Pump Performance Curves
100 Stage — D51 — 50 Hz
400 Series — 2915 RPM

50 Hz

Recommended Capacity Range

Minimum Casing Size
5½ IN OD
Check Clearances

Head Capacity

Pump Only Efficiency

Horsepower Motor Load

H P Motor Load	Pump Only EFF
	70
	60
	50
20	40
15	30
10	20
5	10

Capacity

Type D 51

TRW REDA PUMP DIVISION
BARTLESVILLE, OKLAHOMA 74003
JULY 1977

Head Feet	Head Meters
2250	700
2000	600
1750	500
1500	400
1250	
1000	300
750	200
500	
250	100

m³/ day   250   500   750   1000   1250   1500   1750   2000   2250

BPD   50   100   150   200   250   300   350

GPM = BPD ÷ 34.3

Fig. 4Bl (24)

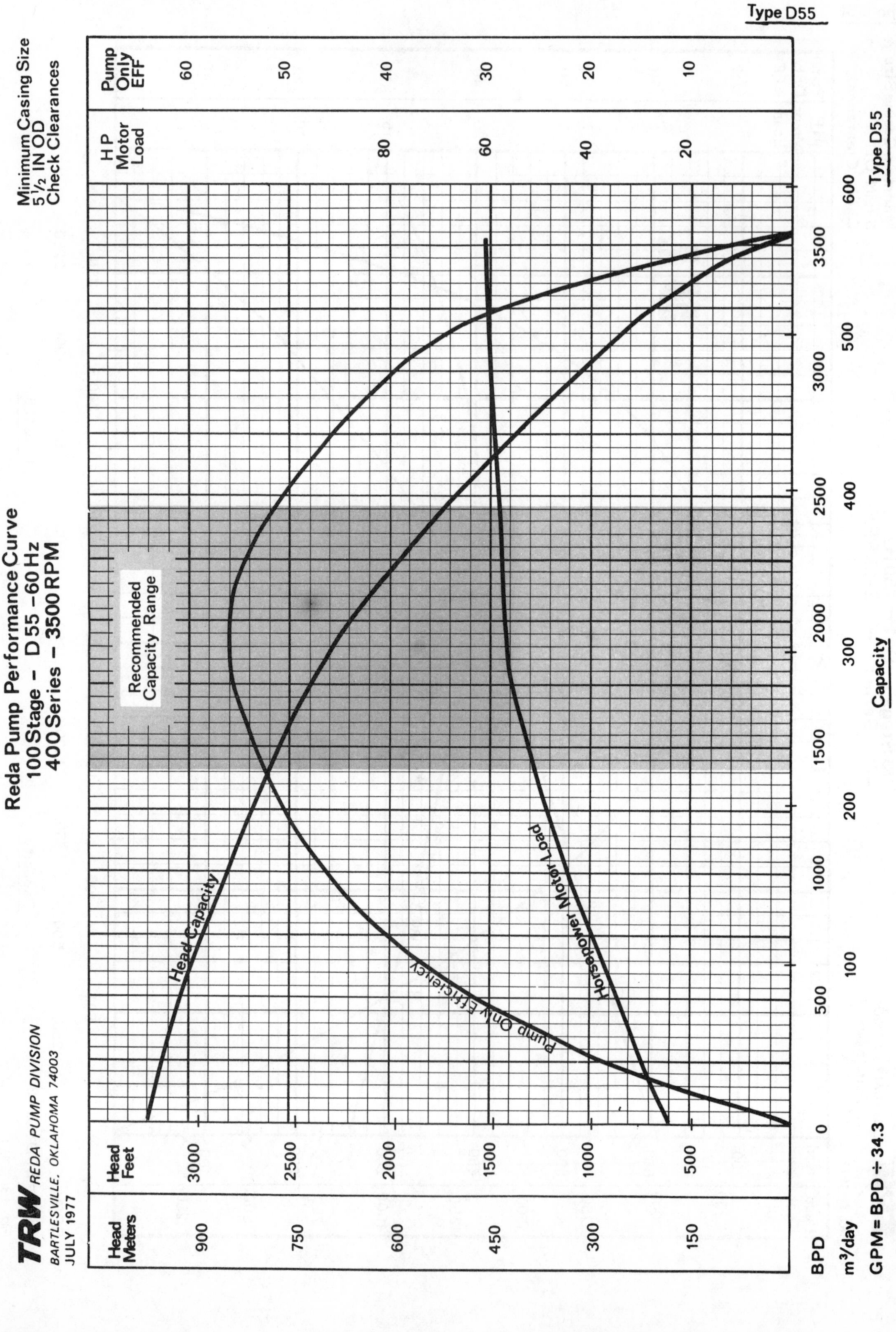

Type D55

Reda Pump Performance Curve
100 Stage – D 55 – 60 Hz
400 Series – 3500 RPM

Minimum Casing Size
5½ IN OD
Check Clearances

TRW REDA PUMP DIVISION
BARTLESVILLE, OKLAHOMA 74003
JULY 1977

GPM = BPD ÷ 34.3

Fig. 4BI (25)

Type D55

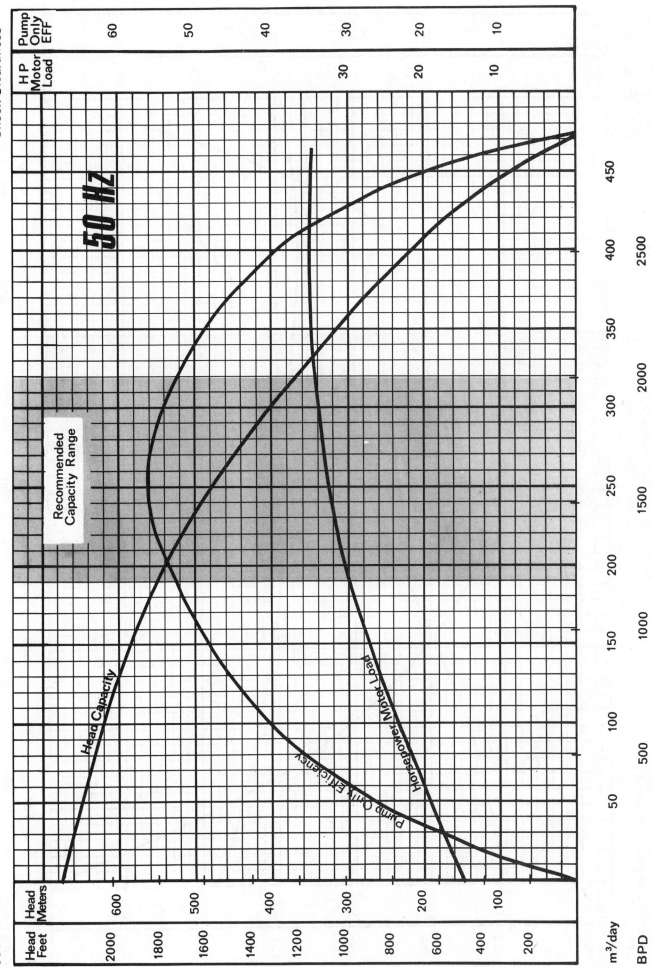

Type D55

Minimum Casing Size
5½ IN OD
Check Clearances

Reda Pump Performance Curve
100 Stage — D55 — 50 Hz
400 Series – 2915 RPM

50 Hz

Recommended
Capacity Range

Head Capacity

Pump Only Efficiency

Horsepower Motor Load

TRW REDA PUMP DIVISION
BARTLESVILLE, OKLAHOMA 74003
JULY 1977

Capacity

Type D55

m³/day

BPD

GPM = BPD ÷ 34.3

Fig. 4BI (26)

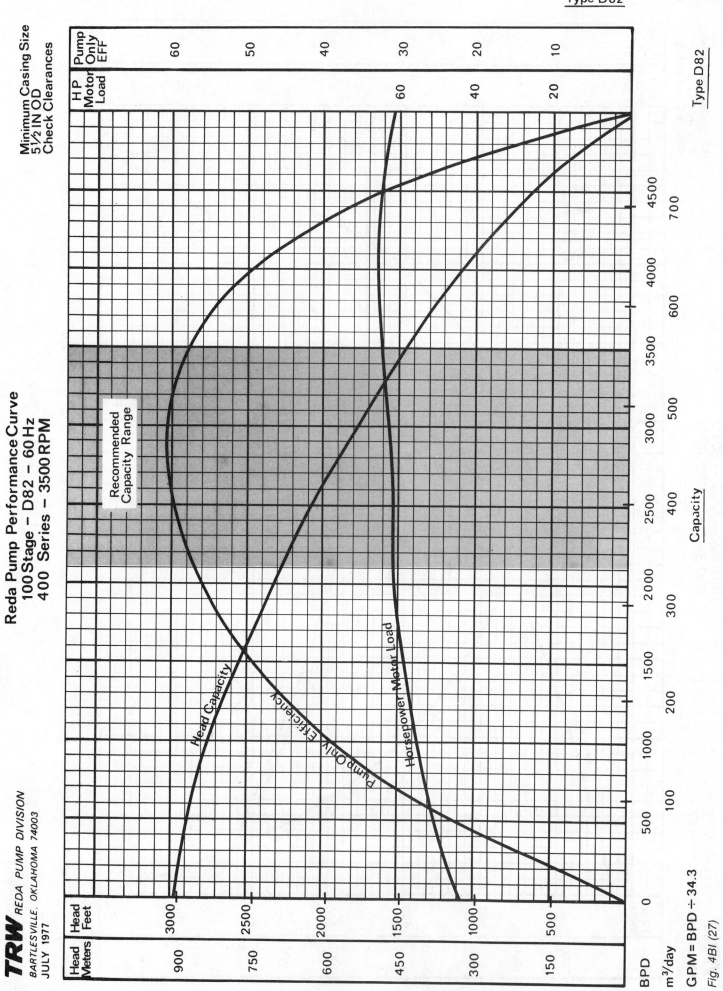

GPM = BPD ÷ 34.3

Fig. 4BI (27)

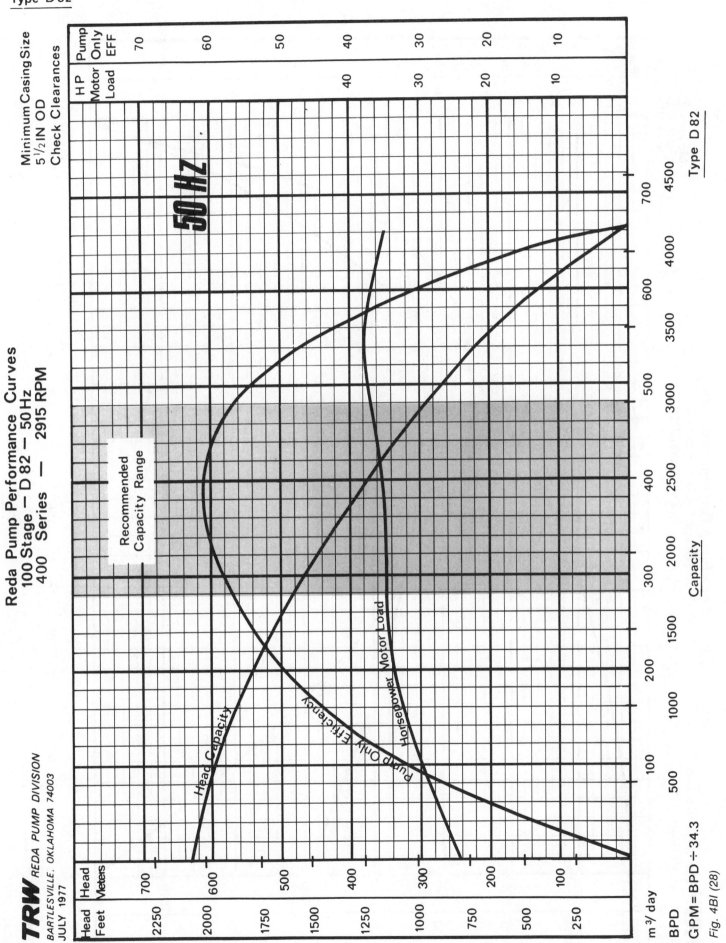

Reda Pump Performance Curves
100 Stage – D 82 – 50 Hz
400 Series — 2915 RPM

Type D 82

50 Hz

Recommended Capacity Range

Head Capacity

Pump Only Efficiency

Horsepower Motor Load

Minimum Casing Size 5½ IN OD
Check Clearances

TRW REDA PUMP DIVISION
BARTLESVILLE, OKLAHOMA 74003
JULY 1977

GPM = BPD ÷ 34.3

Fig. 4BI (28)

Type D 82

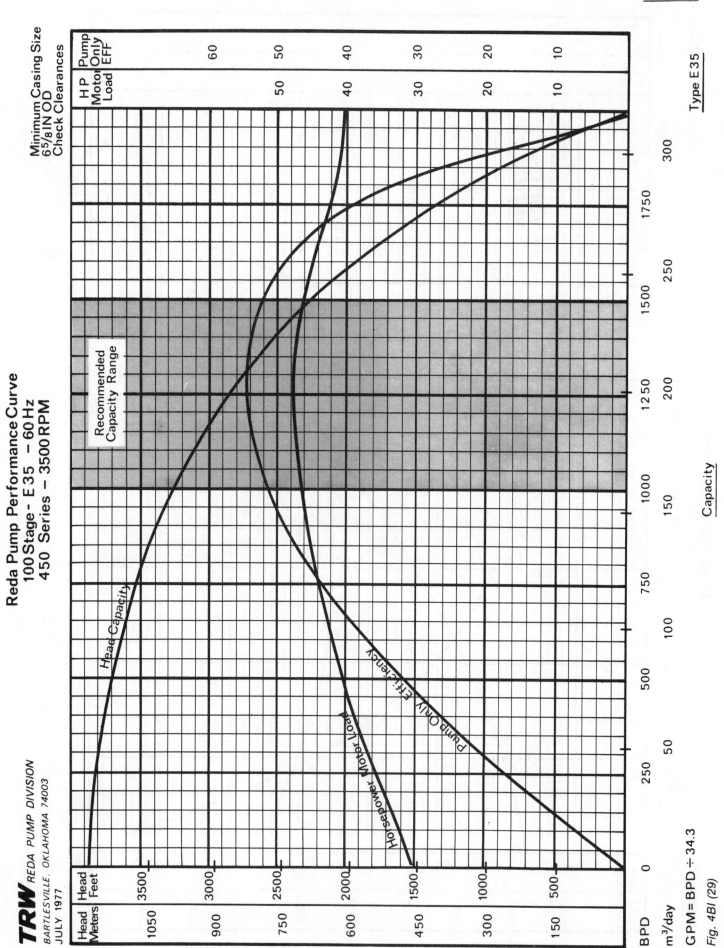

Fig. 4BI (29)

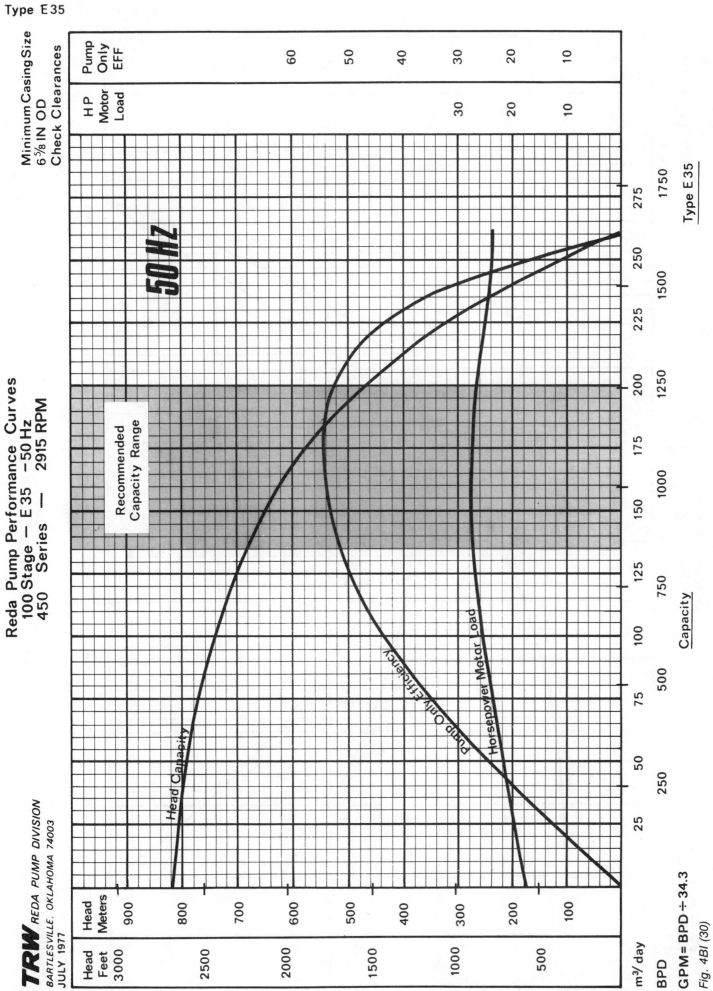

Type E 35

Minimum Casing Size
6⅝ IN OD
Check Clearances

HP Motor Load	Pump Only EFF
	60
	50
	40
30	30
20	20
10	10

Reda Pump Performance Curves
100 Stage — E 35 – 50 Hz
450   Series —   2915 RPM

50 HZ

Recommended
Capacity Range

Type E 35

Capacity

Head Capacity

Pump Only Efficiency

Horsepower Motor Load

TRW REDA PUMP DIVISION
BARTLESVILLE, OKLAHOMA 74003
JULY 1977

Head Feet	Head Meters
3000	900
	800
2500	700
2000	600
	500
1500	400
	300
1000	
	200
500	
	100

m³/ day

BPD

GPM = BPD ÷ 34.3

Fig. 4BI (30)

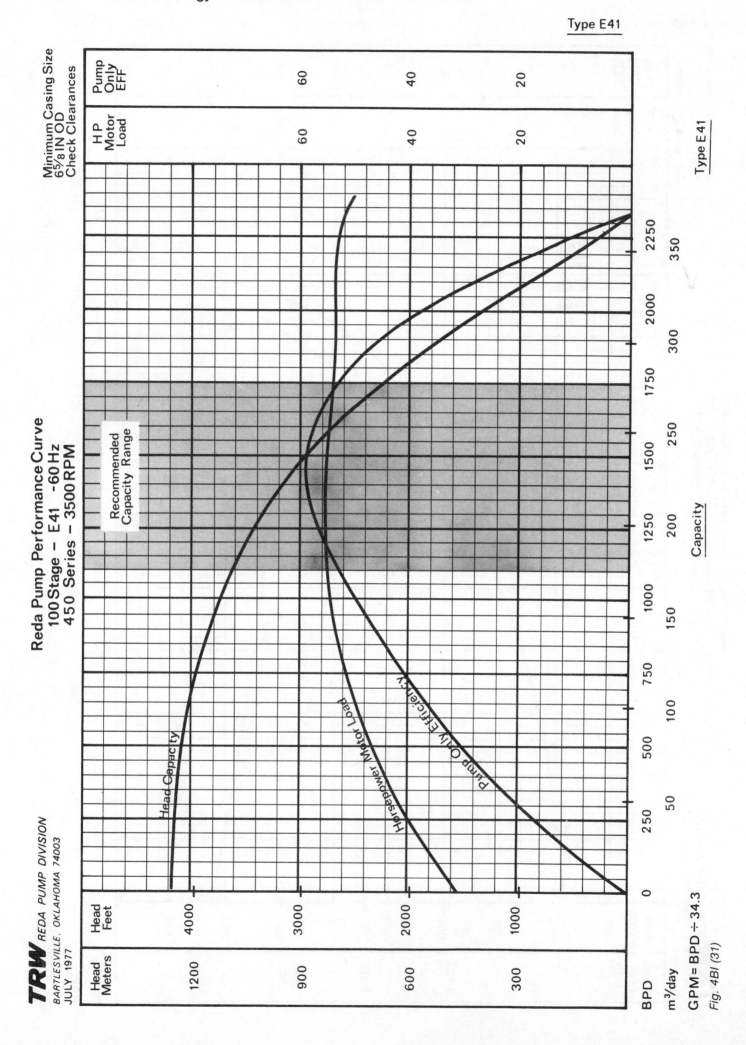

Reda Pump Performance Curve
100 Stage – E41 –60 Hz
450 Series – 3500 RPM

Minimum Casing Size
6⅝ IN OD
Check Clearances

Type E41

TRW REDA PUMP DIVISION
BARTLESVILLE, OKLAHOMA 74003
JULY 1977

GPM = BPD ÷ 34.3

Fig. 4Bl (31)

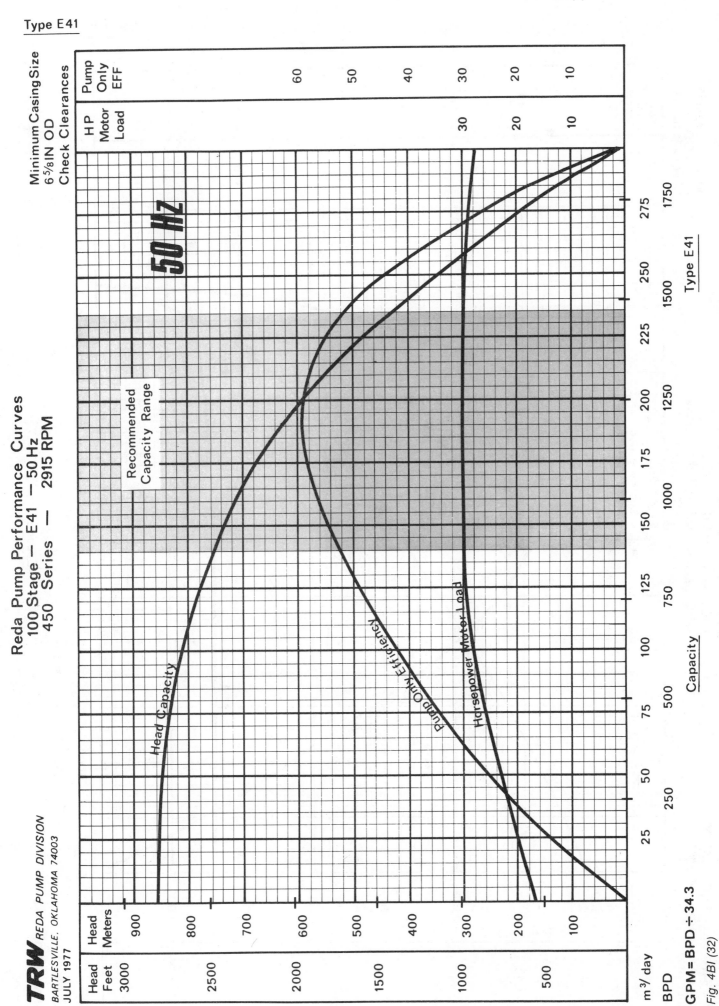

Reda Pump Performance Curves
100 Stage – E41 – 50 Hz
450 Series — 2915 RPM

Type E41

Minimum Casing Size
6⅝ IN OD
Check Clearances

TRW REDA PUMP DIVISION
BARTLESVILLE. OKLAHOMA 74003
JULY 1977

GPM = BPD ÷ 34.3

Fig. 4Bl (32)

Type E 100

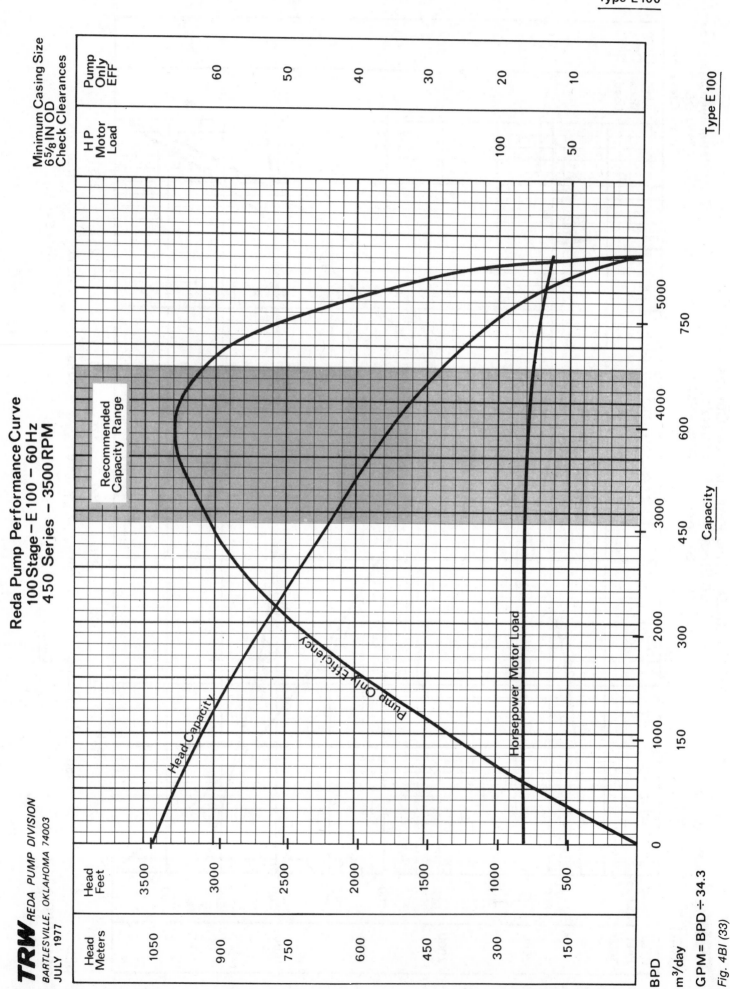

TRW REDA PUMP DIVISION
BARTLESVILLE, OKLAHOMA 74003
JULY 1977

Reda Pump Performance Curve
100 Stage – E 100 – 60 Hz
450 Series – 3500 RPM

Minimum Casing Size
6⅝ IN OD
Check Clearances

Recommended Capacity Range

Head Capacity

Pump Only Efficiency

Horsepower Motor Load

Type E 100

Capacity

GPM = BPD ÷ 34.3

Fig. 4BI (33)

Reda Pump Performance Curves
100 Stage — E 100 — 50 Hz
450 Series — 2915 RPM

Type E 100

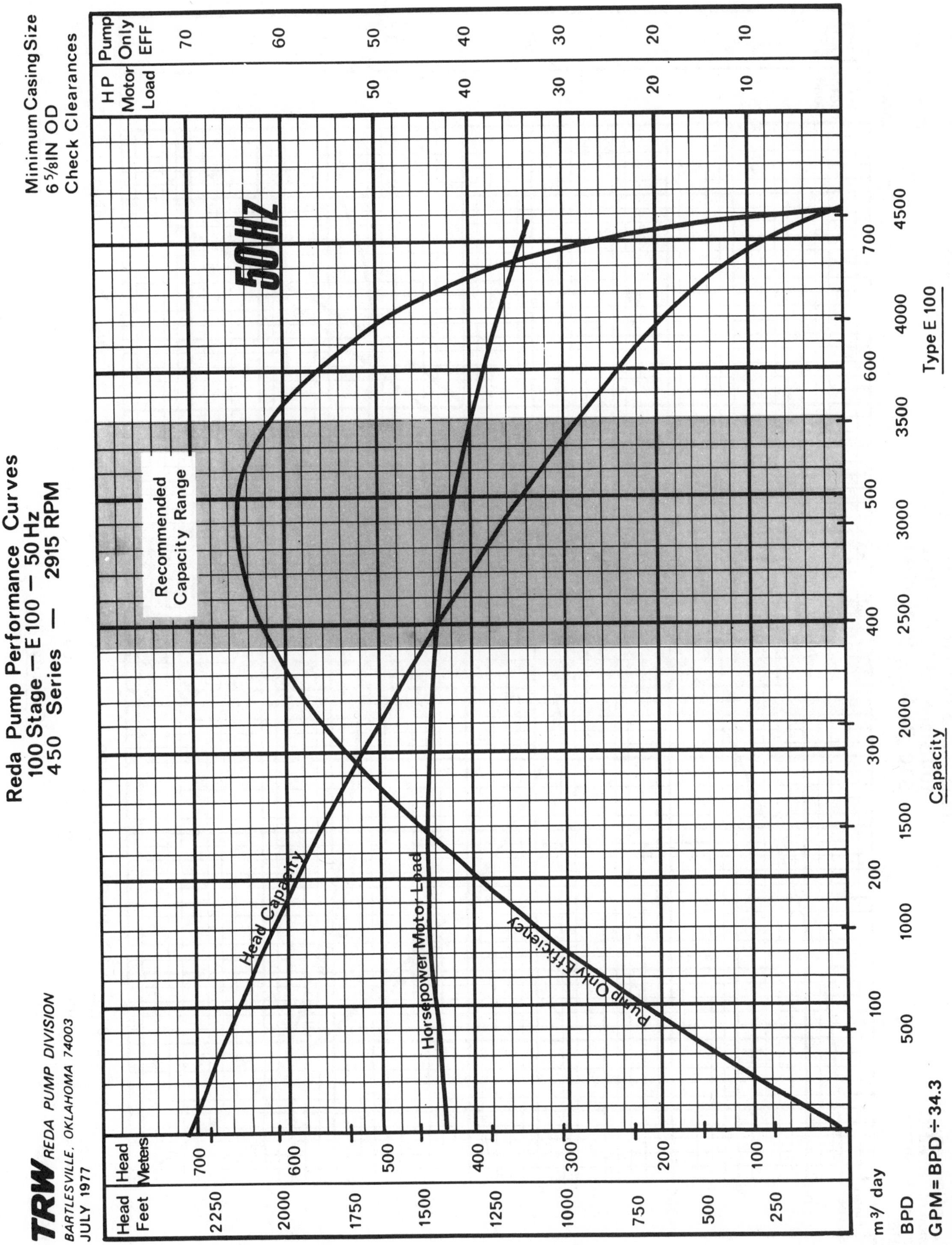

Minimum Casing Size
6 5/8 IN OD
Check Clearances

TRW REDA PUMP DIVISION
BARTLESVILLE, OKLAHOMA 74003
JULY 1977

GPM = BPD ÷ 34.3

Fig. 4BI (34)

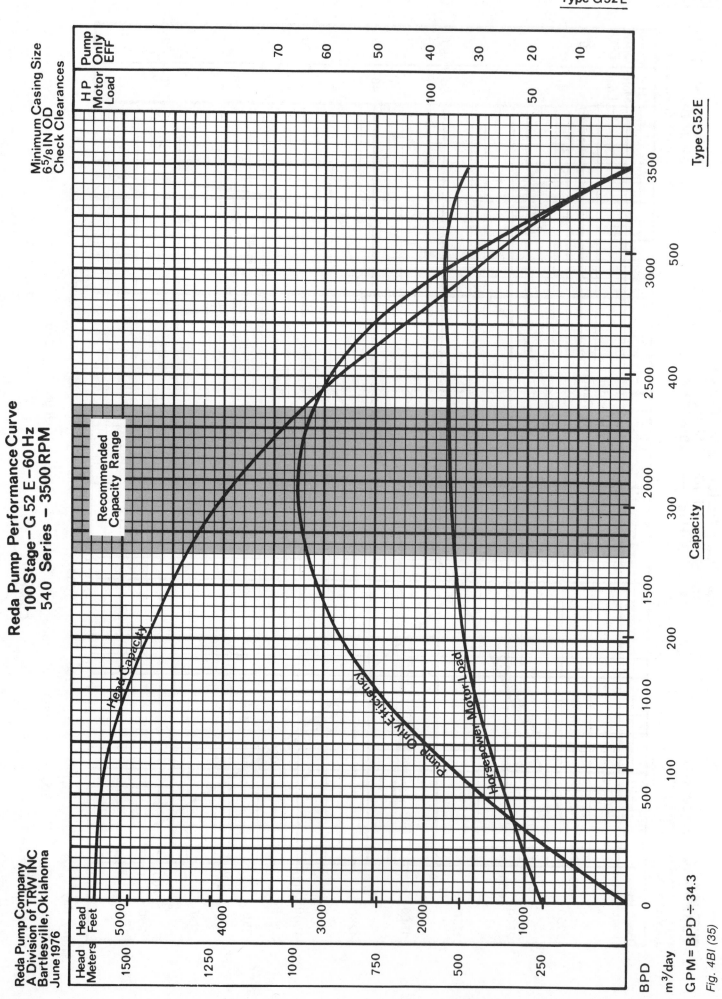

Reda Pump Company
A Division of TRW INC
Bartlesville, Oklahoma
June 1976

Reda Pump Performance Curve
100 Stage – G 52 E – 60 Hz
540 Series – 3500 RPM

Minimum Casing Size
6⁵/8 IN OD
Check Clearances

Type G52E

Recommended
Capacity Range

Head Capacity

Pump Only Efficiency

Horsepower Motor Load

GPM = BPD ÷ 34.3

Fig. 4BI (35)

Type G52E

Type G52E

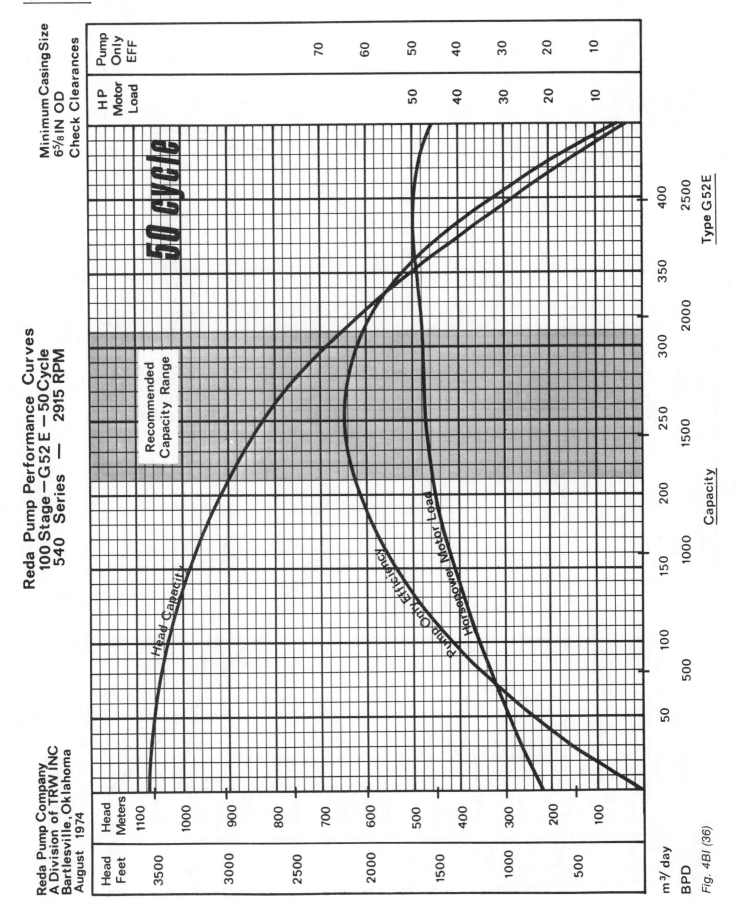

Reda Pump Performance Curves
100 Stage—G52 E—50 Cycle
540 Series —   2915 RPM

Reda Pump Company
A Division of TRW INC
Bartlesville, Oklahoma
August 1974

Minimum Casing Size
6⅝ IN OD
Check Clearances

50 cycle

Recommended Capacity Range

Head Capacity

Pump Only Efficiency

Horsepower Motor Load

Type G52E

Capacity

HP Motor Load	Pump Only EFF
	70
	60
50	50
40	40
30	30
20	20
10	10

Head Feet	Head Meters
3500	1100
	1000
3000	900
2500	800
	700
2000	600
1500	500
	400
1000	300
	200
500	100

m³/day
BPD

Fig. 4BI (36)

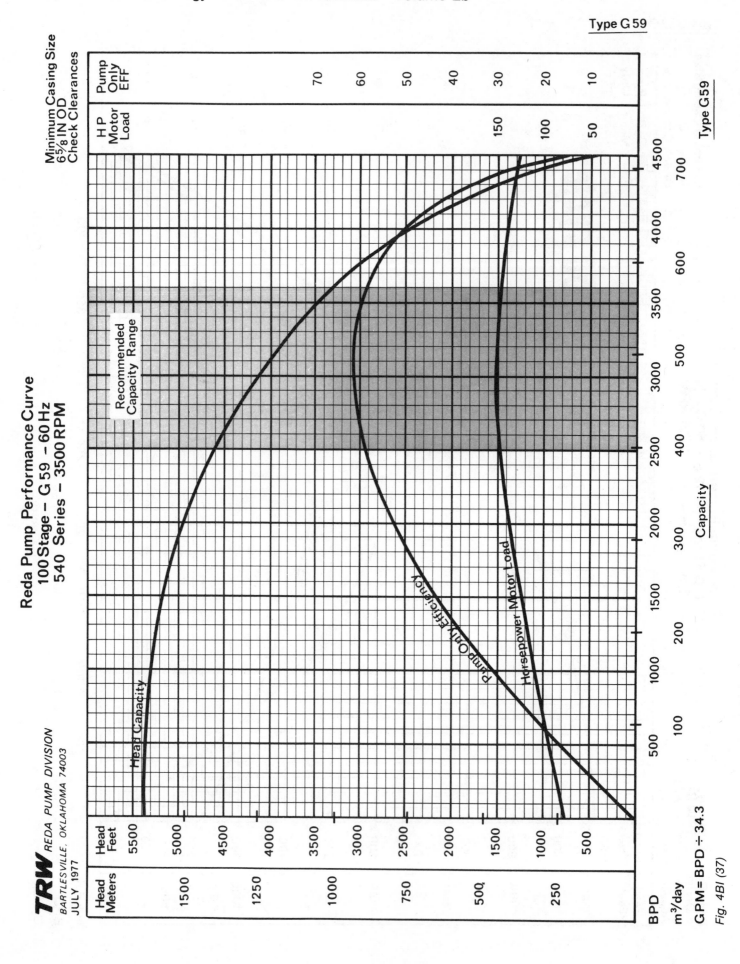

Reda Pump Performance Curve
100 Stage – G 59 – 60 Hz
540 Series – 3500 RPM

TRW REDA PUMP DIVISION
BARTLESVILLE, OKLAHOMA 74003
JULY 1977

Type G 59

Minimum Casing Size
6⅝ IN OD
Check Clearances

Recommended Capacity Range

Head Capacity

Pump Only Efficiency

Horsepower Motor Load

GPM = BPD ÷ 34.3

Fig. 4Bl (37)

Type G59

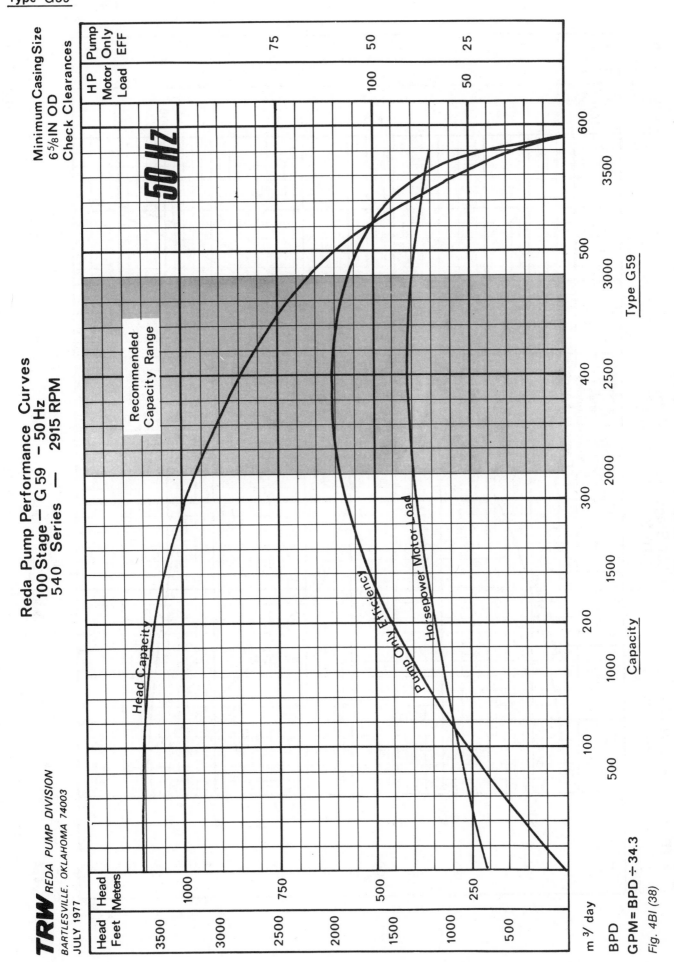

Reda Pump Performance Curves
100 Stage — G 59 — 50 Hz
540 Series — 2915 RPM

Minimum Casing Size
6⅝ IN OD
Check Clearances

TRW REDA PUMP DIVISION
BARTLESVILLE, OKLAHOMA 74003
JULY 1977

GPM = BPD ÷ 34.3

Type G59

Fig. 4BI (38)

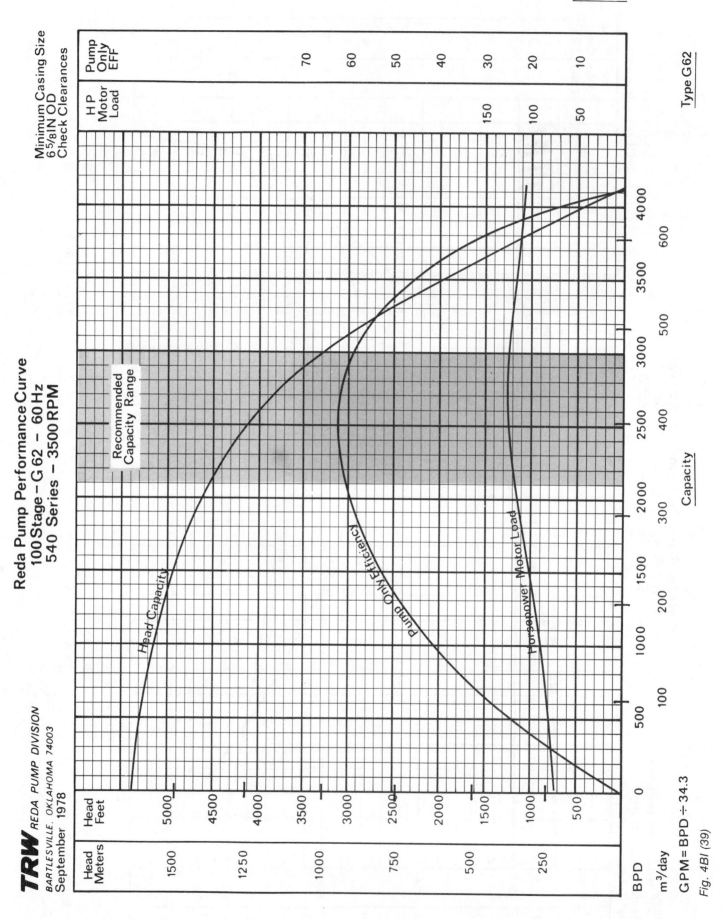

Reda Pump Performance Curve
100 Stage – G 62 – 60 Hz
540 Series – 3500 RPM

TRW REDA PUMP DIVISION
BARTLESVILLE, OKLAHOMA 74003
September 1978

GPM = BPD ÷ 34.3

Fig. 4BI (39)

Type G62

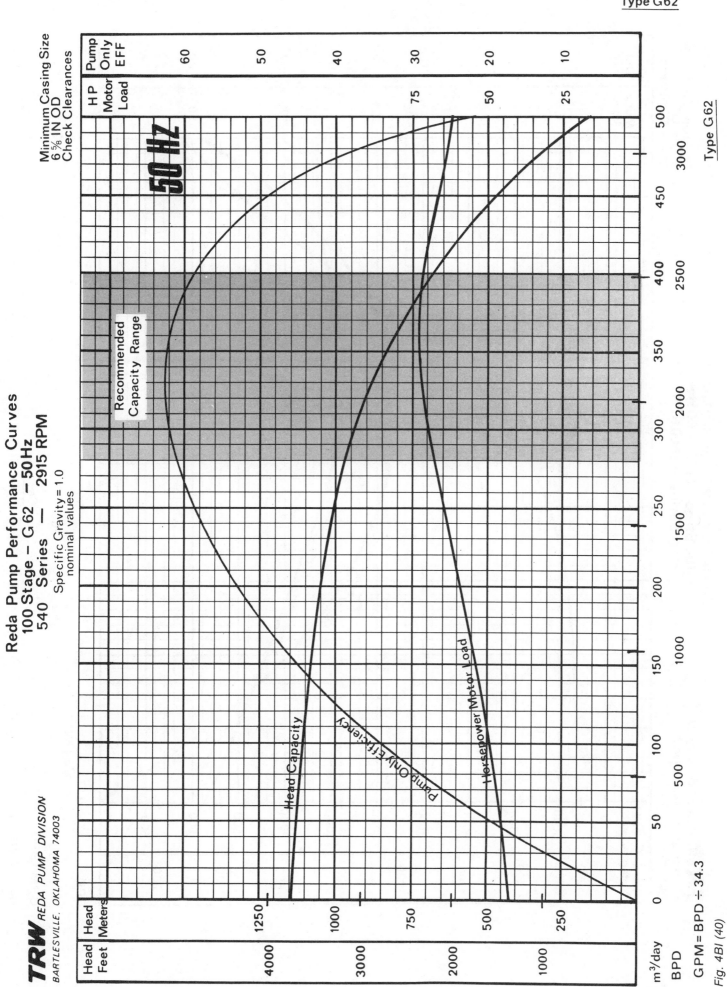

Reda Pump Performance Curves
100 Stage – G 62 – 50 Hz
540   Series   —   2915 RPM

Specific Gravity = 1.0
nominal values

Minimum Casing Size
6 5/8 IN OD
Check Clearances

TRW REDA PUMP DIVISION
BARTLESVILLE, OKLAHOMA 74003

GPM = BPD ÷ 34.3

Fig. 4BI (40)

Type G62

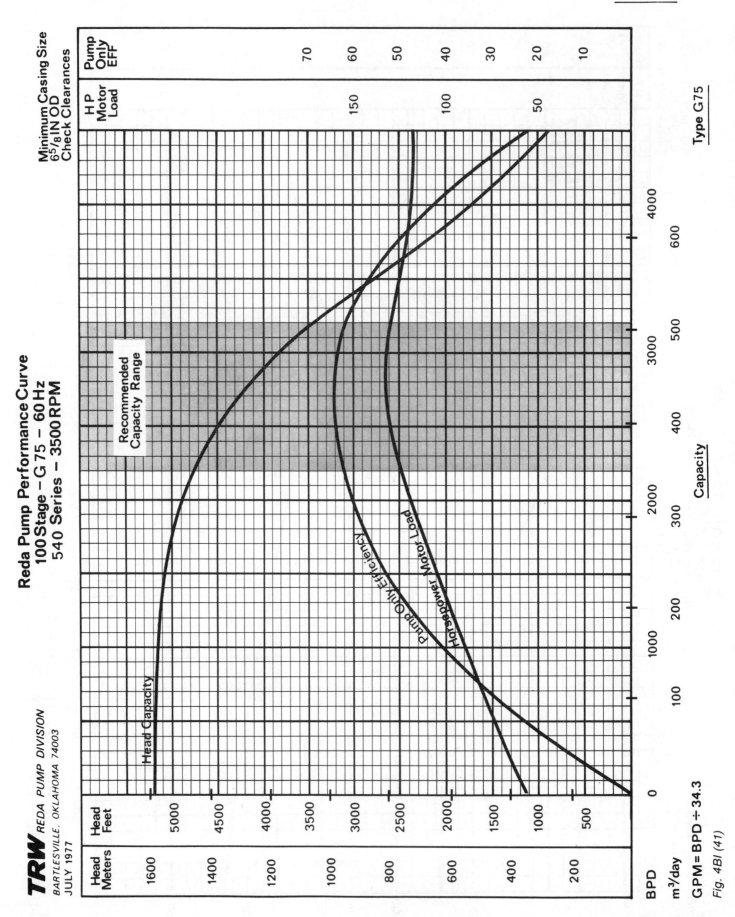

Fig. 4BI (41)

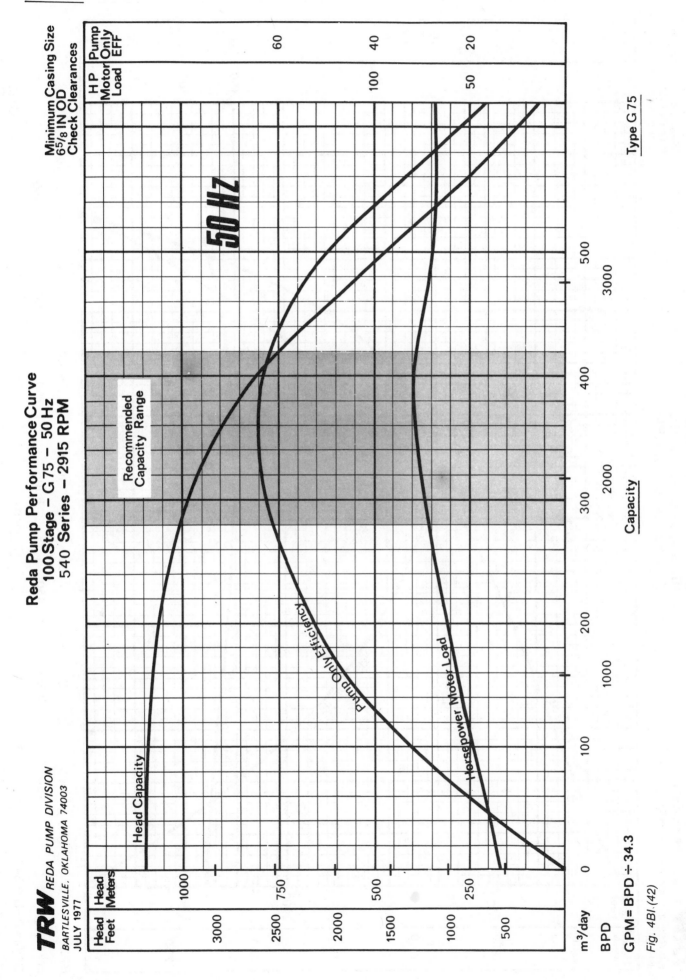

Type G 75

Minimum Casing Size
6⁵/₈ IN OD
Check Clearances

Reda Pump Performance Curve
100 Stage – G 75 – 50 Hz
540 Series – 2915 RPM

TRW REDA PUMP DIVISION
BARTLESVILLE, OKLAHOMA 74003
JULY 1977

50 Hz

Recommended Capacity Range

Head Capacity

Pump Only Efficiency

Horsepower Motor Load

Type G 75

GPM = BPD ÷ 34.3

Fig. 4Bl (42)

Type G 110

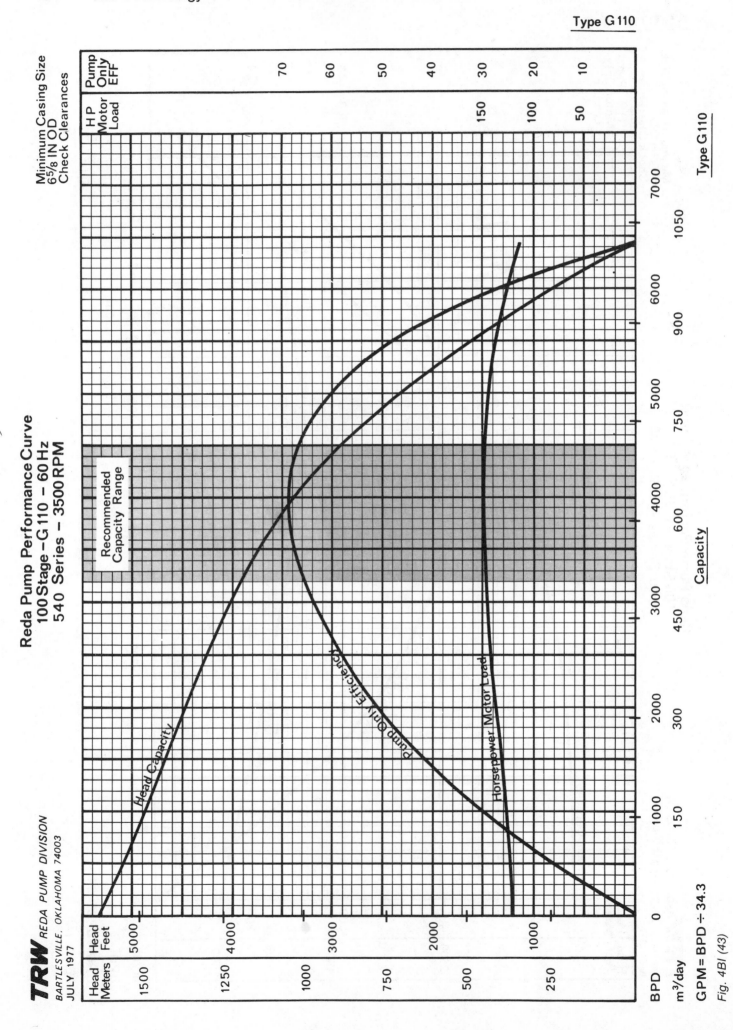

Reda Pump Performance Curve
100 Stage – G 110 – 60 Hz
540 Series – 3500 RPM

Minimum Casing Size
6⅝ IN OD
Check Clearances

TRW REDA PUMP DIVISION
BARTLESVILLE, OKLAHOMA 74003
JULY 1977

GPM = BPD ÷ 34.3

Fig. 4BI (43)

Type G 110

Type G110

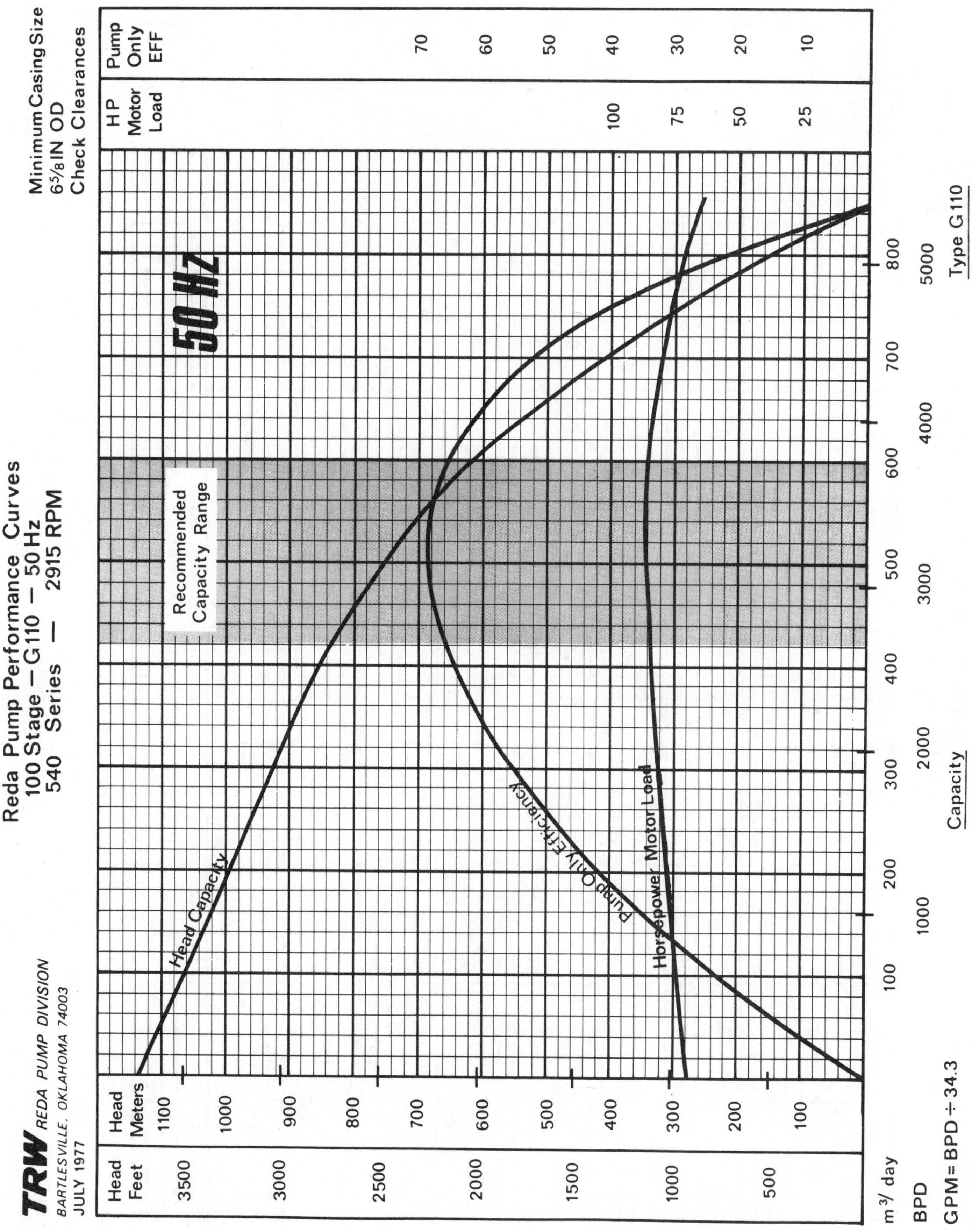

Reda Pump Performance Curves
100 Stage – G110 – 50 Hz
540 Series – 2915 RPM

TRW REDA PUMP DIVISION
BARTLESVILLE, OKLAHOMA 74003
JULY 1977

Minimum Casing Size
6⁵/₈ IN OD
Check Clearances

	HP Motor Load	Pump Only EFF
		70
		60
		50
	100	40
	75	30
	50	20
	25	10

50 Hz

Type G110

Recommended Capacity Range

Head Capacity

Pump Only Efficiency

Horsepower Motor Load

Capacity

Head Feet	Head Meters
3500	1100
	1000
3000	900
2500	800
	700
2000	600
1500	500
	400
1000	300
	200
500	100

m³/ day   100 200 300 400 500 600 700 800

BPD   1000 2000 3000 4000 5000

GPM = BPD ÷ 34.3

Fig. 4BI (44)

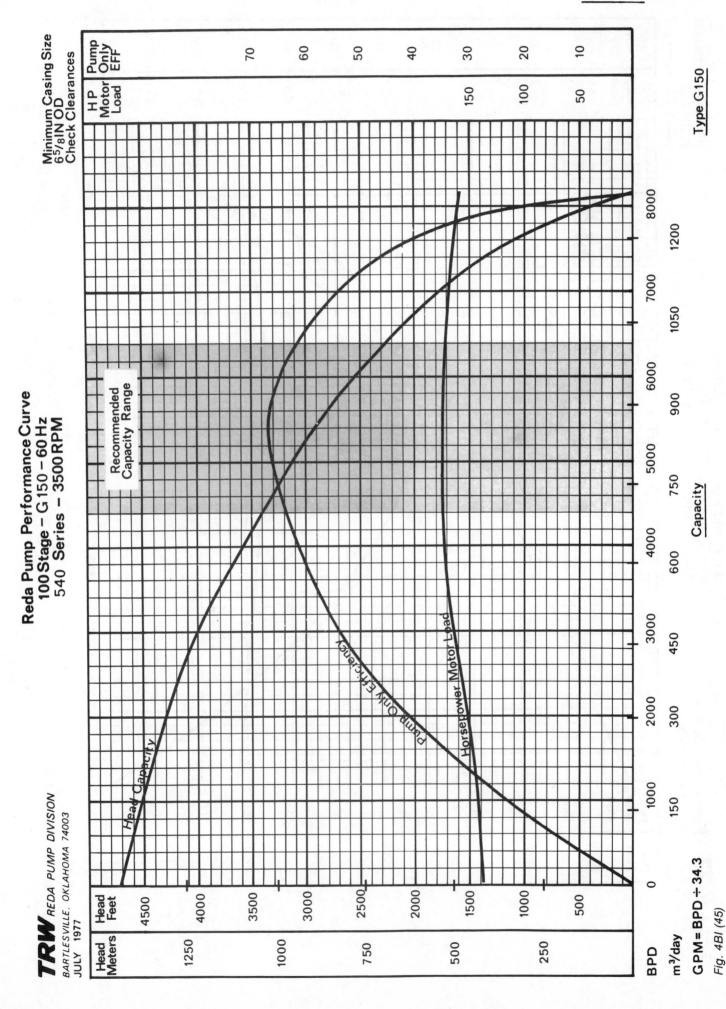

TRW REDA PUMP DIVISION
BARTLESVILLE, OKLAHOMA 74003
JULY 1977

Reda Pump Performance Curve
100 Stage – G 150 – 60 Hz
540 Series – 3500 RPM

Minimum Casing Size
6⁵/₈IN OD
Check Clearances

Type G150

Head Capacity

Pump Only Efficiency

Horsepower Motor Load

Recommended Capacity Range

Capacity

GPM = BPD ÷ 34.3

Fig. 4BI (45)

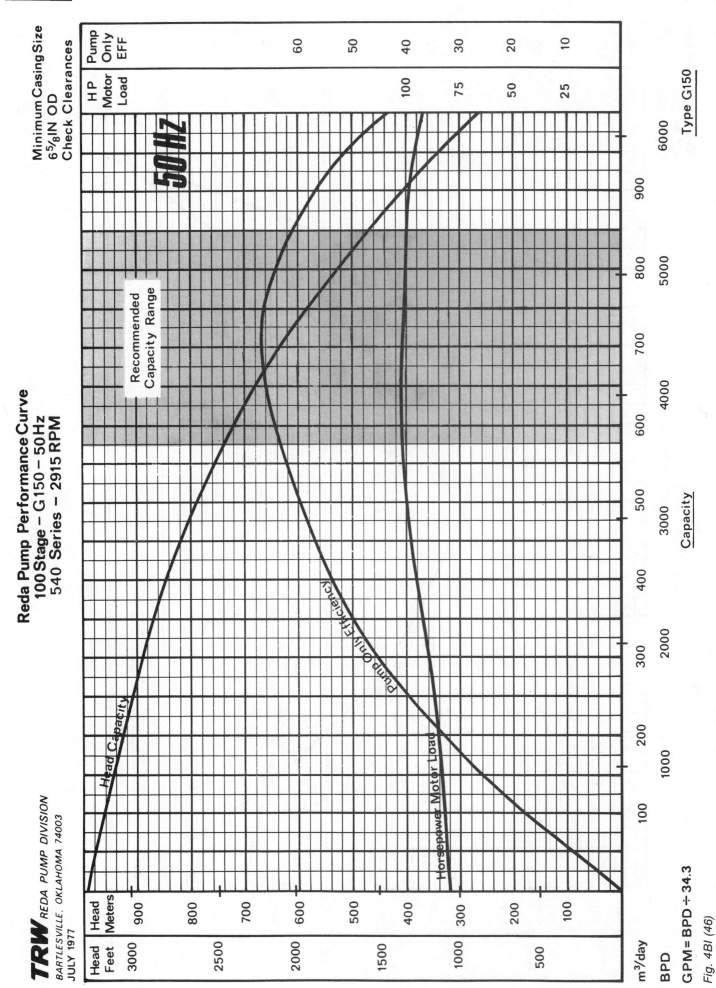

Reda Pump Performance Curve
100 Stage – G150 – 50 Hz
540 Series – 2915 RPM

TRW *REDA PUMP DIVISION*
BARTLESVILLE, OKLAHOMA 74003
JULY 1977

Type G150

Minimum Casing Size
6⁵⁄₈ IN OD
Check Clearances

50 Hz

Recommended
Capacity Range

Head Capacity

Pump Only Efficiency

Horsepower Motor Load

Type G150

Capacity

GPM = BPD ÷ 34.3

Fig. 4Bl (46)

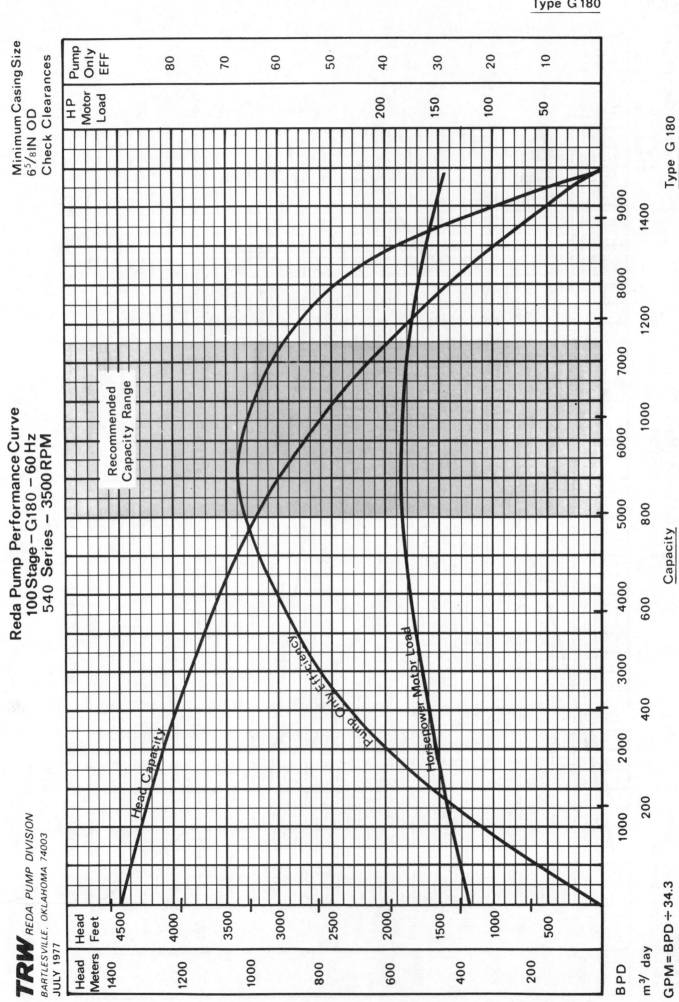

Type G 180

Reda Pump Performance Curve
100 Stage – G180 – 60 Hz
540 Series – 3500 RPM

Minimum Casing Size
6⁵/₈ IN OD
Check Clearances

TRW REDA PUMP DIVISION
BARTLESVILLE, OKLAHOMA 74003
JULY 1977

Recommended Capacity Range

Head Capacity

Pump Only Efficiency

Horsepower Motor Load

Type G 180

Capacity

GPM = BPD ÷ 34.3

Fig. 4BI (47)

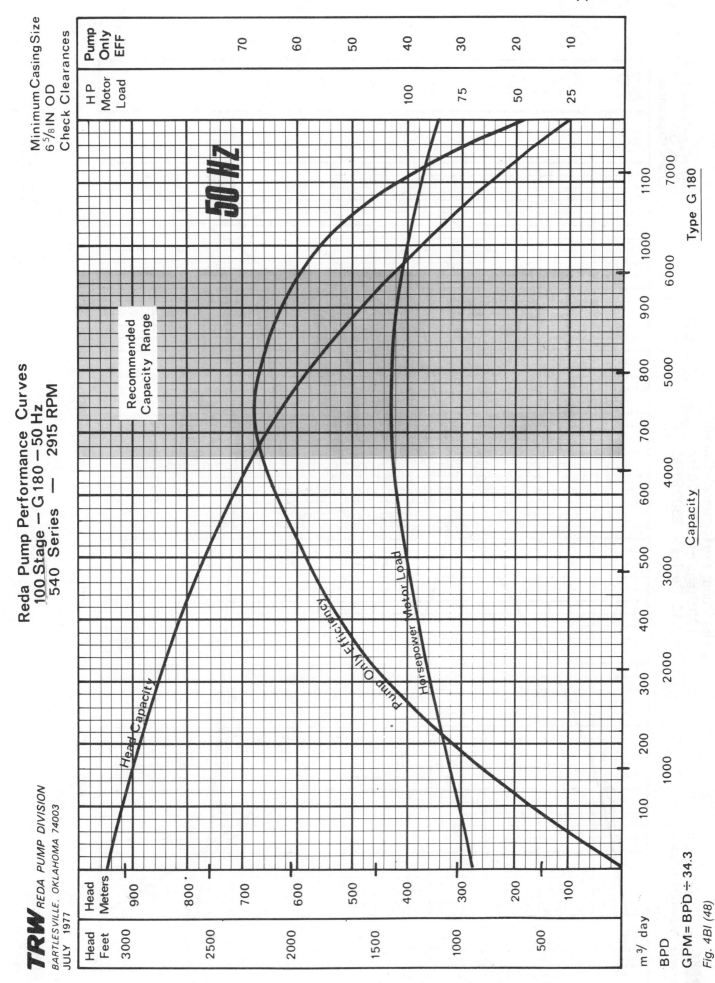

Reda Pump Performance Curves
100 Stage – G 180 – 50 Hz
540 Series — 2915 RPM

TRW REDA PUMP DIVISION
BARTLESVILLE, OKLAHOMA 74003
JULY 1977

Minimum Casing Size
6 5/8 IN OD
Check Clearances

50 Hz

Recommended
Capacity Range

Type G 180

Capacity

Head Capacity

Pump Only Efficiency

Horsepower Motor Load

	Pump Only EFF	HP Motor Load
	70	
	60	
	50	
	40	100
	30	75
	20	50
	10	25

Head Feet	Head Meters
3000	900
	800
2500	700
2000	600
	500
1500	400
1000	300
	200
500	100

m³/ day

BPD

GPM = BPD ÷ 34.3

Fig. 4BI (48)

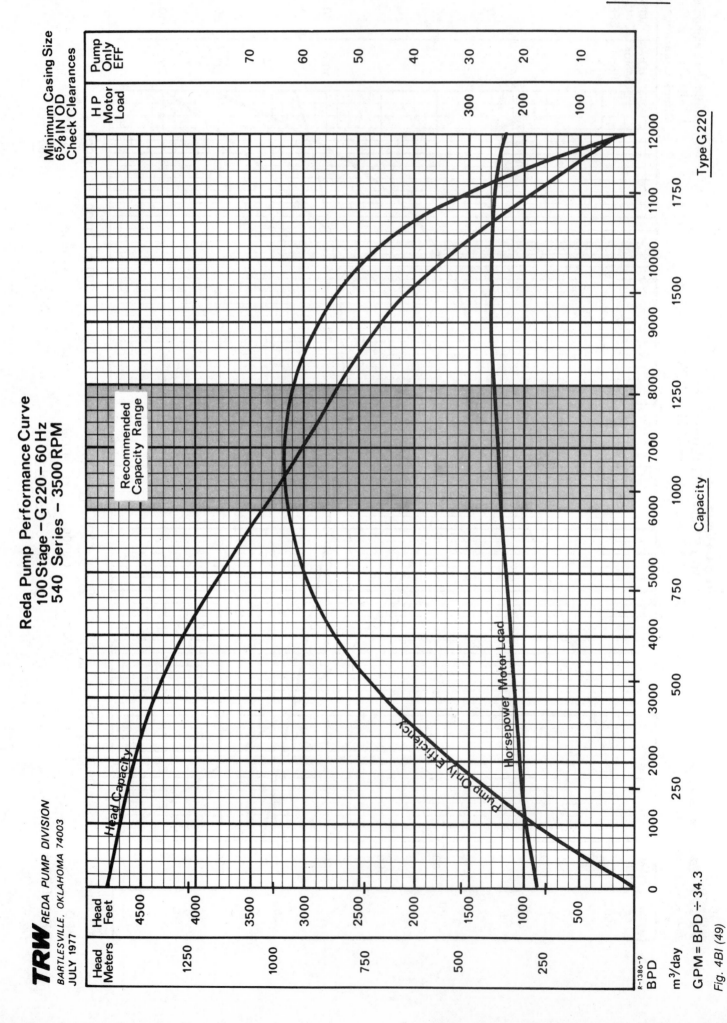

Type G220

Reda Pump Performance Curve
100 Stage – G 220 – 60 Hz
540 Series – 3500 RPM

TRW REDA PUMP DIVISION
BARTLESVILLE, OKLAHOMA 74003
JULY 1977

Minimum Casing Size
6⅝ IN OD
Check Clearances

GPM = BPD ÷ 34.3

Fig. 4BI (49)

Type G 220

Reda Pump Performance Curves
100 Stage—G 220 – 50 Hz
540 Series — 2915 RPM

TRW REDA PUMP DIVISION
BARTLESVILLE, OKLAHOMA 74003
JULY 1977

Minimum Casing Size
6⅝ IN OD
Check Clearances

50 HZ

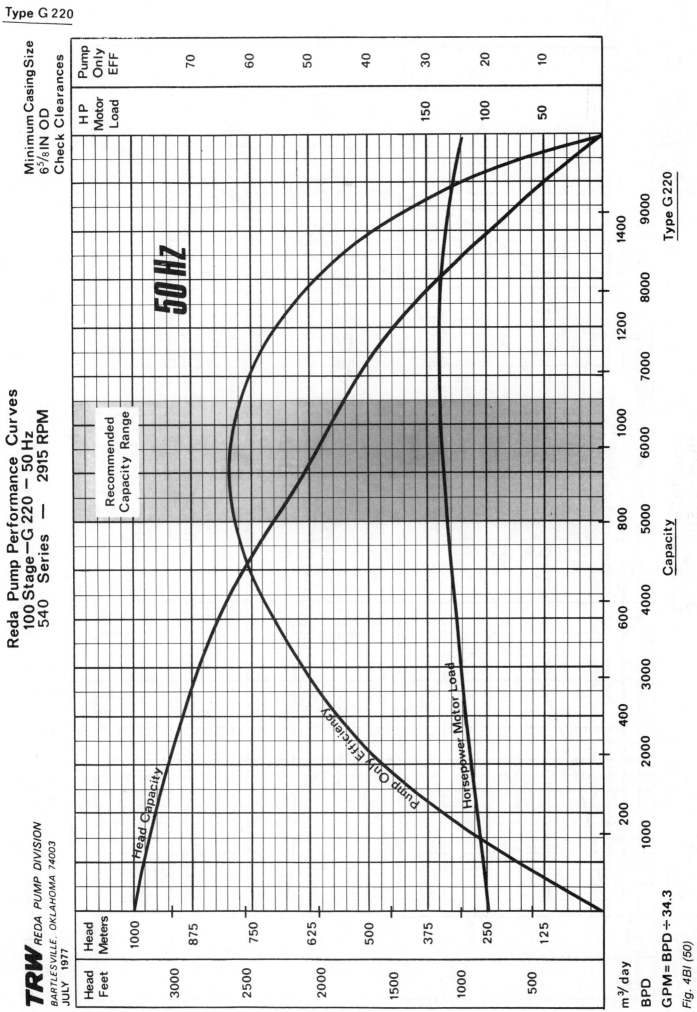

GPM = BPD ÷ 34.3

Fig. 4BI (50)

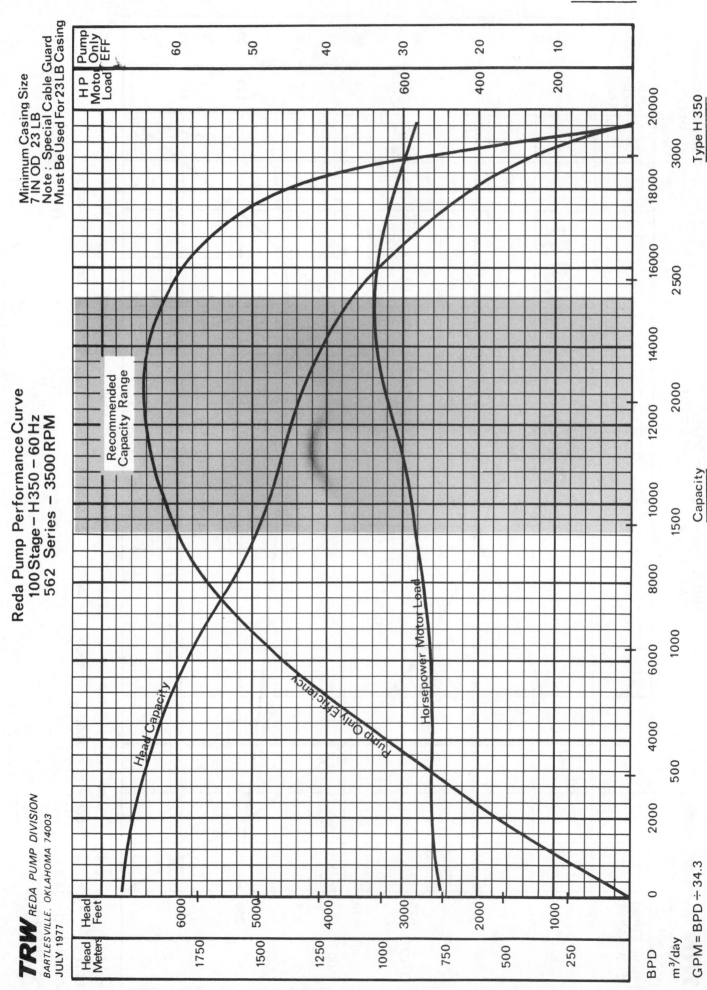

Reda Pump Performance Curve
100 Stage – H350 – 60 Hz
562 Series – 3500 RPM

Type H 350

Minimum Casing Size
7 IN OD 23 LB
Note : Special Cable Guard
Must Be Used For 23LB Casing

TRW REDA PUMP DIVISION
BARTLESVILLE. OKLAHOMA 74003
JULY 1977

GPM = BPD ÷ 34.3

Fig. 4BI (51)

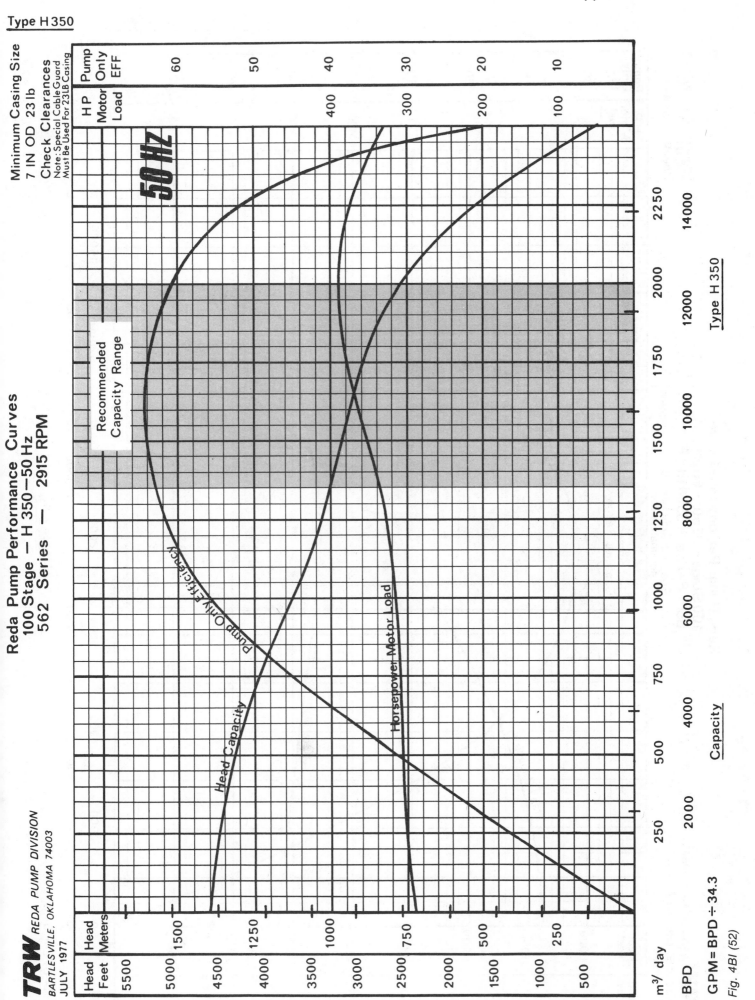

Type H 350

Minimum Casing Size
7 IN OD 23 lb

Check Clearances
Note: Special Cable Guard
Must Be Used For 23 LB Casing

50 Hz

Recommended
Capacity Range

Pump Only Efficiency

Head Capacity

Horsepower Motor Load

Reda Pump Performance Curves
100 Stage – H 350 – 50 Hz
562 Series – 2915 RPM

TRW REDA PUMP DIVISION
BARTLESVILLE, OKLAHOMA 74003
JULY 1977

GPM = BPD ÷ 34.3

Type H 350

Capacity

Fig. 4BI (52)

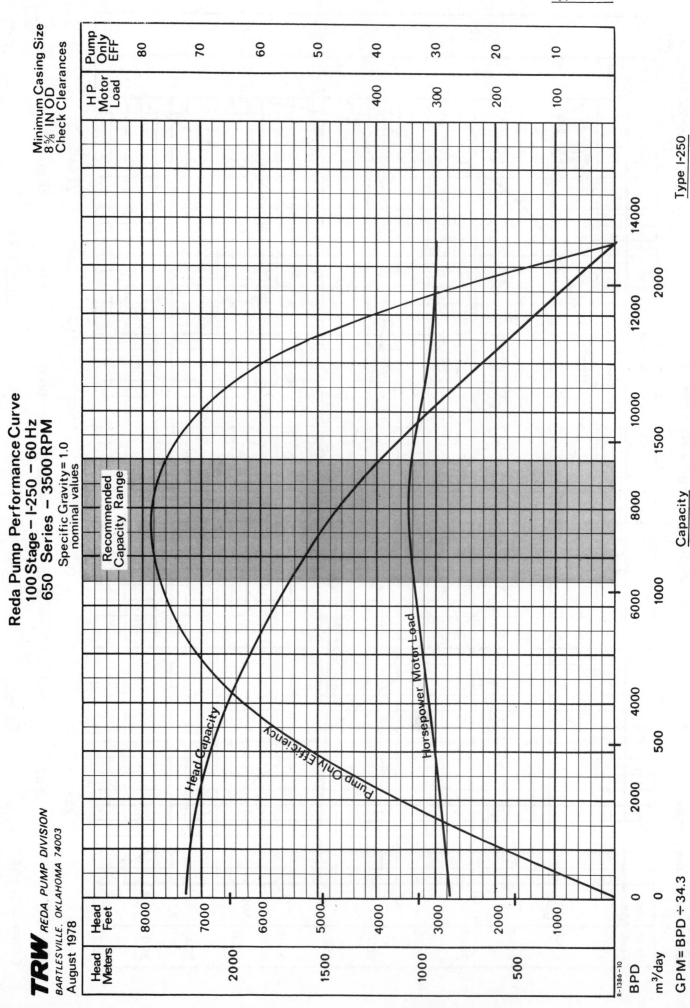

Type I-250

Reda Pump Performance Curve
100 Stage – I-250 – 60 Hz
650 Series – 3500 RPM
Specific Gravity = 1.0
nominal values

TRW REDA PUMP DIVISION
BARTLESVILLE, OKLAHOMA 74003
August 1978

Minimum Casing Size
8⅝ IN OD
Check Clearances

Capacity

Type I-250

GPM = BPD ÷ 34.3

Fig. 4BI (53)

Type I-250

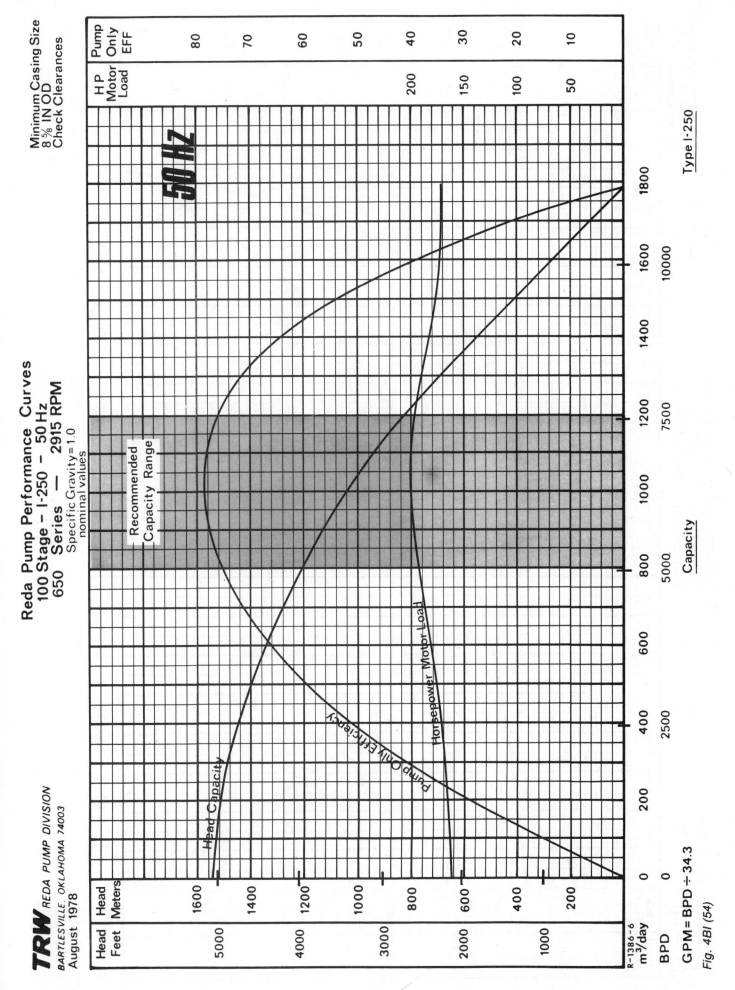

Reda Pump Performance Curves
100 Stage – I-250 – 50 Hz
650 Series – 2915 RPM
Specific Gravity=1.0
nominal values

TRW REDA PUMP DIVISION
BARTLESVILLE, OKLAHOMA 74003
August 1978

Minimum Casing Size
8⅝ IN OD
Check Clearances

50 HZ

Type I-250

Recommended
Capacity Range

Capacity

Head Capacity

Pump Only Efficiency

Horsepower Motor Load

Head Feet	Head Meters		Pump Only EFF
1600			80
1400			70
1200			60
1000			50
800			40
600			30
400			20
200			10

HP Motor Load
200
150
100
50

R-1386-6
m³/day   0   200   400   600   800   1000   1200   1400   1600   1800

BPD   0   2500   5000   7500   10000

GPM= BPD ÷ 34.3

Fig. 4BI (54)

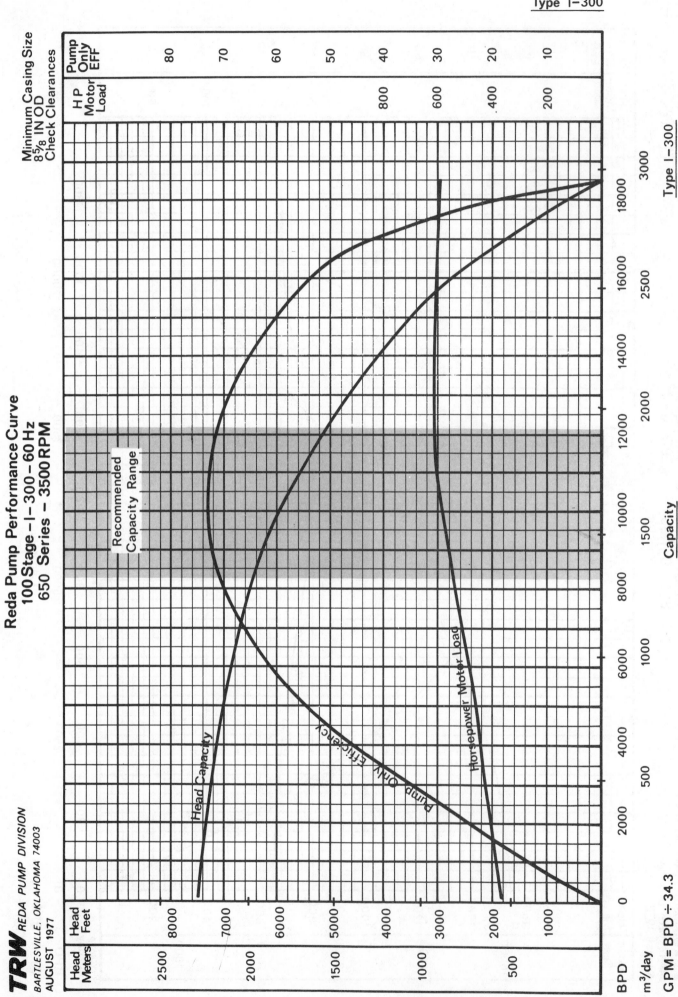

Type I–300

Reda Pump Performance Curve
100 Stage – I – 300 – 60 Hz
650 Series – 3500 RPM

TRW REDA PUMP DIVISION
BARTLESVILLE. OKLAHOMA 74003
AUGUST 1977

Minimum Casing Size
8⅝ IN OD
Check Clearances

Recommended Capacity Range

Head Capacity

Pump Only Efficiency

Horsepower Motor Load

Capacity

Type I–300

BPD
m³/day

GPM = BPD ÷ 34.3

Fig. 4Bl (55)

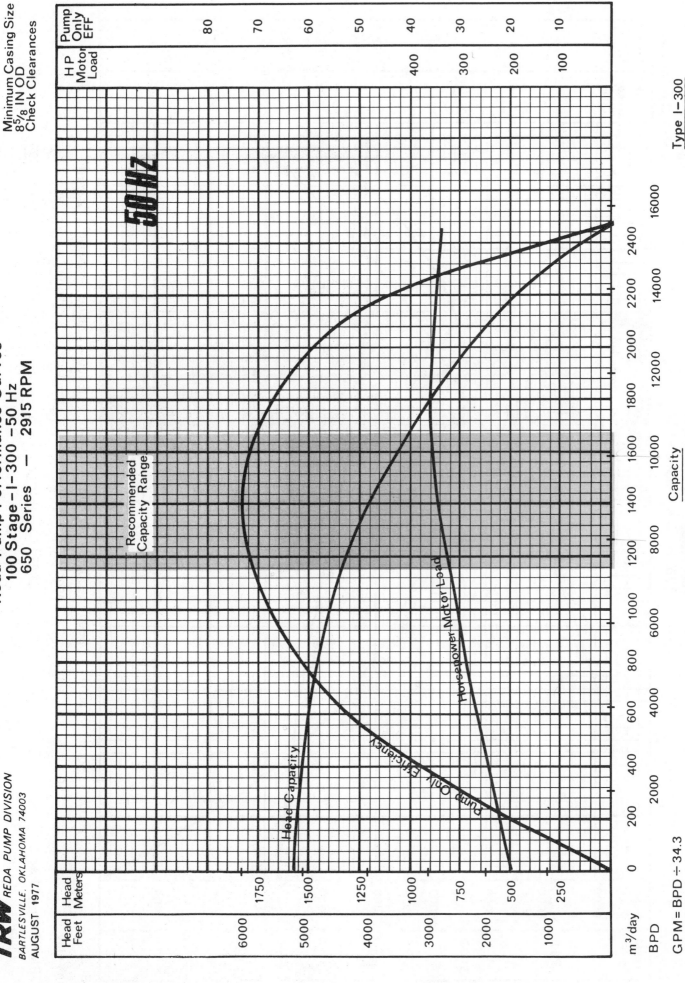

Type I–300

Reda Pump Performance Curves
100 Stage–I–300 –50 Hz
650   Series   —   2915 RPM

TRW REDA PUMP DIVISION
BARTLESVILLE, OKLAHOMA 74003
AUGUST 1977

GPM = BPD ÷ 34.3

Fig. 4BI (56)

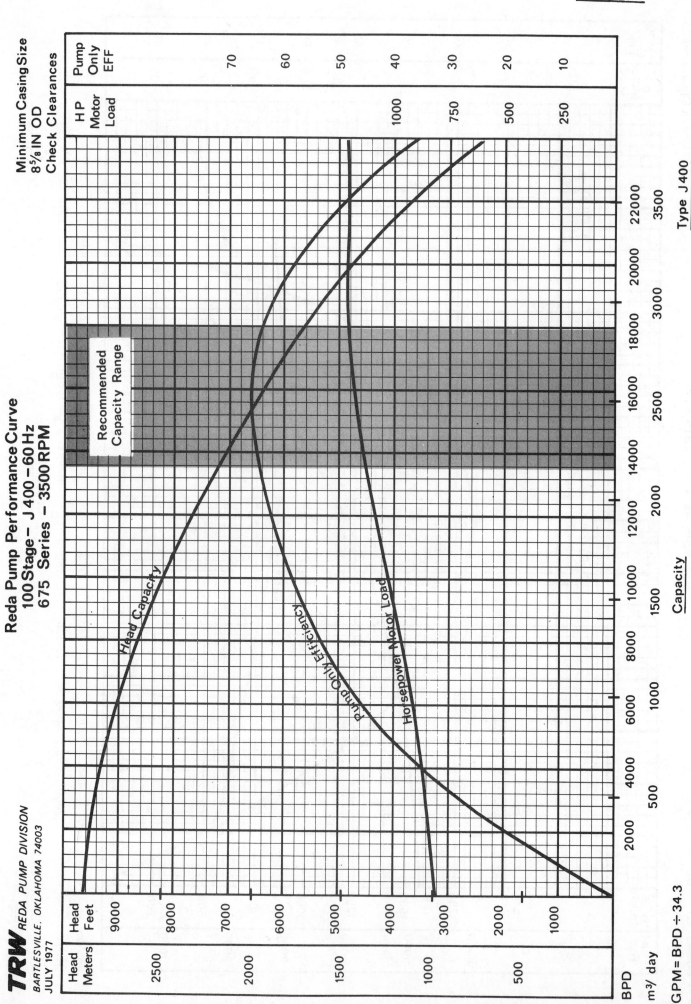

Type J 400

Reda Pump Performance Curve
100 Stage – J 400 – 60 Hz
675 Series – 3500 RPM

Minimum Casing Size
8⅝ IN OD
Check Clearances

TRW REDA PUMP DIVISION
BARTLESVILLE, OKLAHOMA 74003
JULY 1977

GPM = BPD ÷ 34.3

Fig. 4Bl (57)

Type J 400

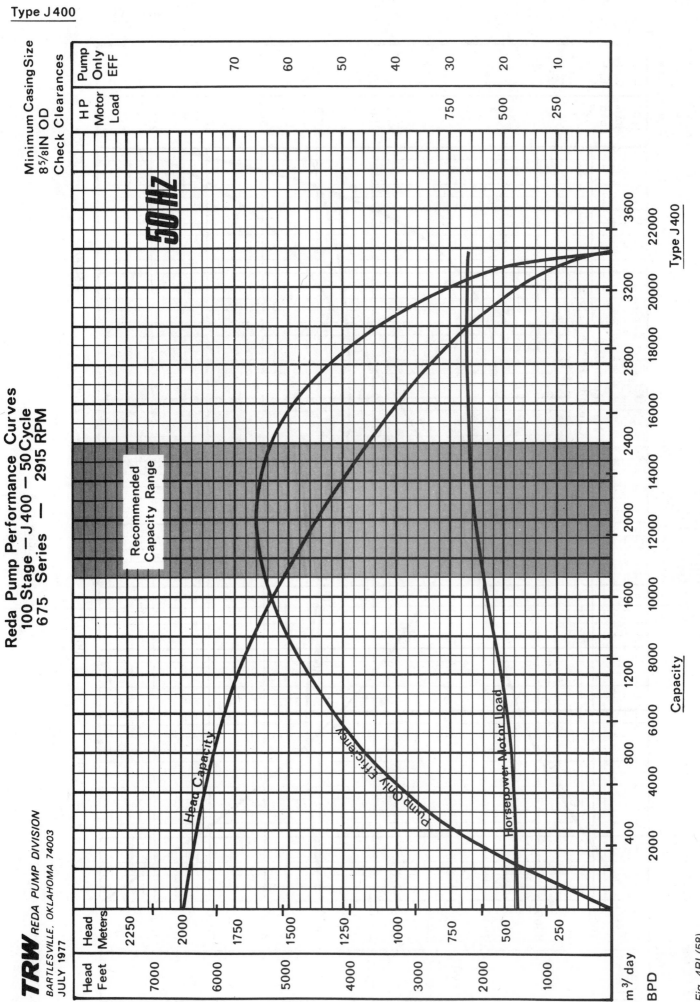

Fig. 4Bl (58)

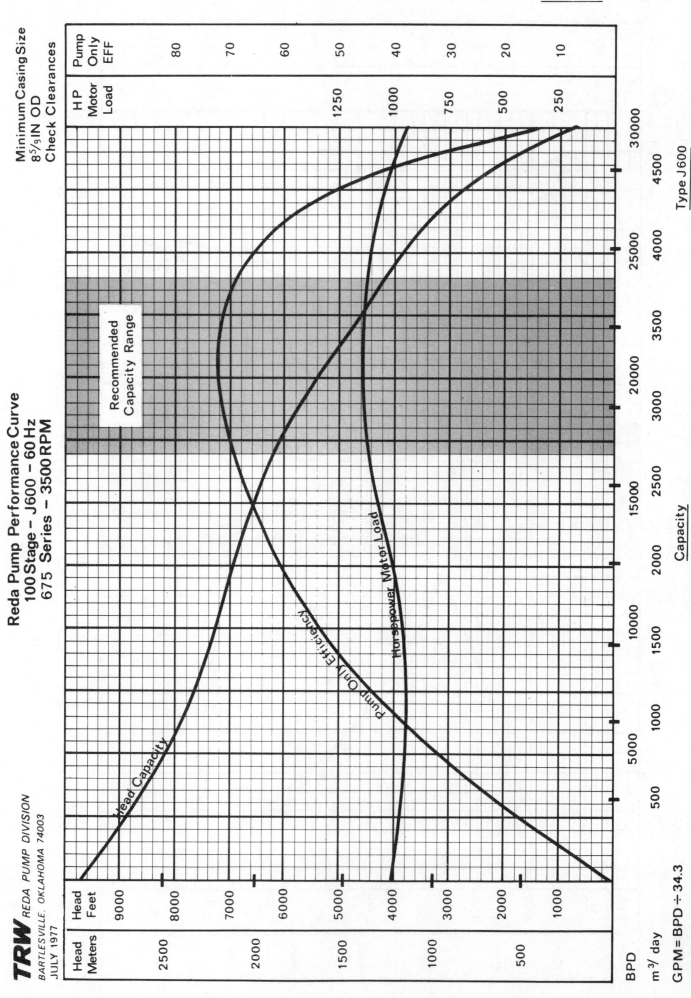

Reda Pump Performance Curve
100 Stage – J600 – 60 Hz
675 Series – 3500 RPM

Type J600

Minimum Casing Size
8 5/8 IN OD
Check Clearances

TRW REDA PUMP DIVISION
BARTLESVILLE, OKLAHOMA 74003
JULY 1977

GPM = BPD ÷ 34.3

Fig. 4BI (59)

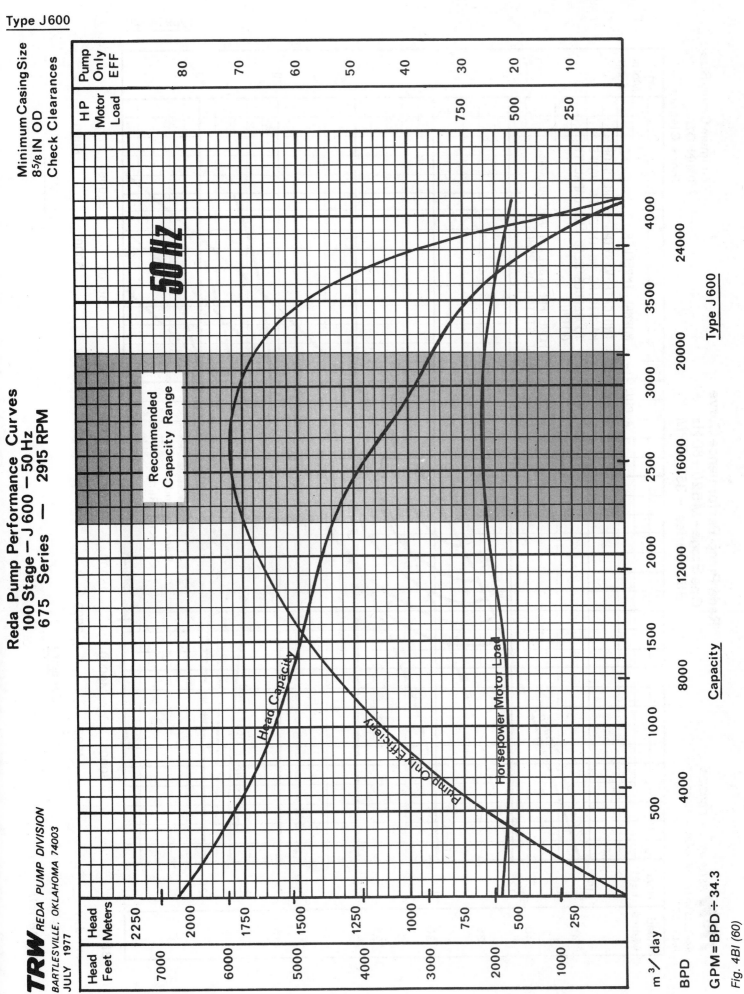

Type J600

Minimum Casing Size
8⅝ IN OD
Check Clearances

Reda Pump Performance Curves
100 Stage – J 600 – 50 Hz
675 Series – 2915 RPM

TRW REDA PUMP DIVISION
BARTLESVILLE, OKLAHOMA 74003
JULY 1977

GPM = BPD ÷ 34.3

Fig. 4BI (60)

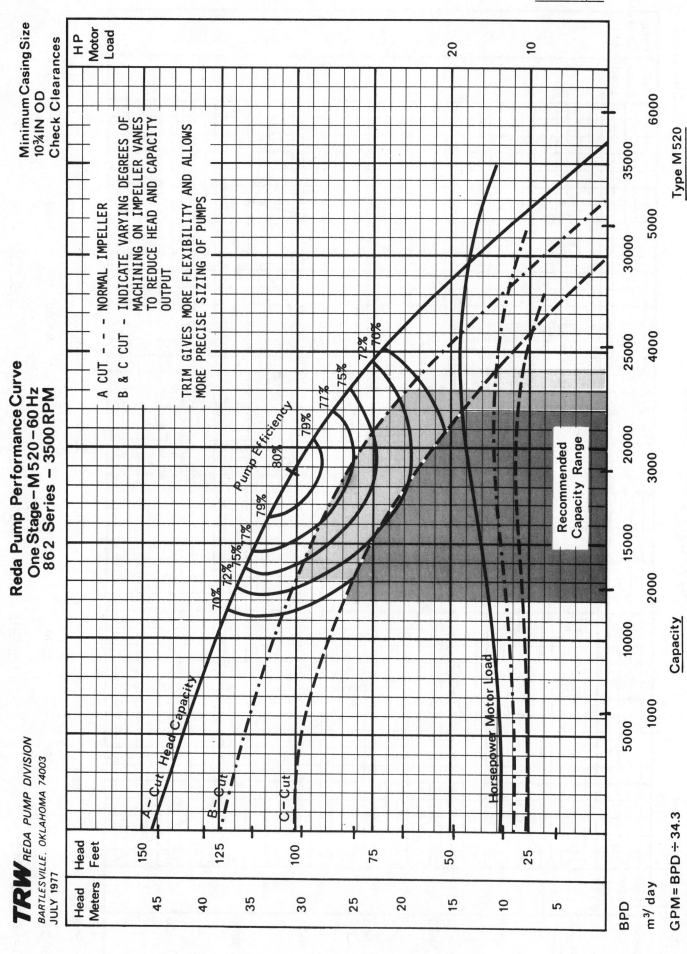

Fig. 4Bl (61)

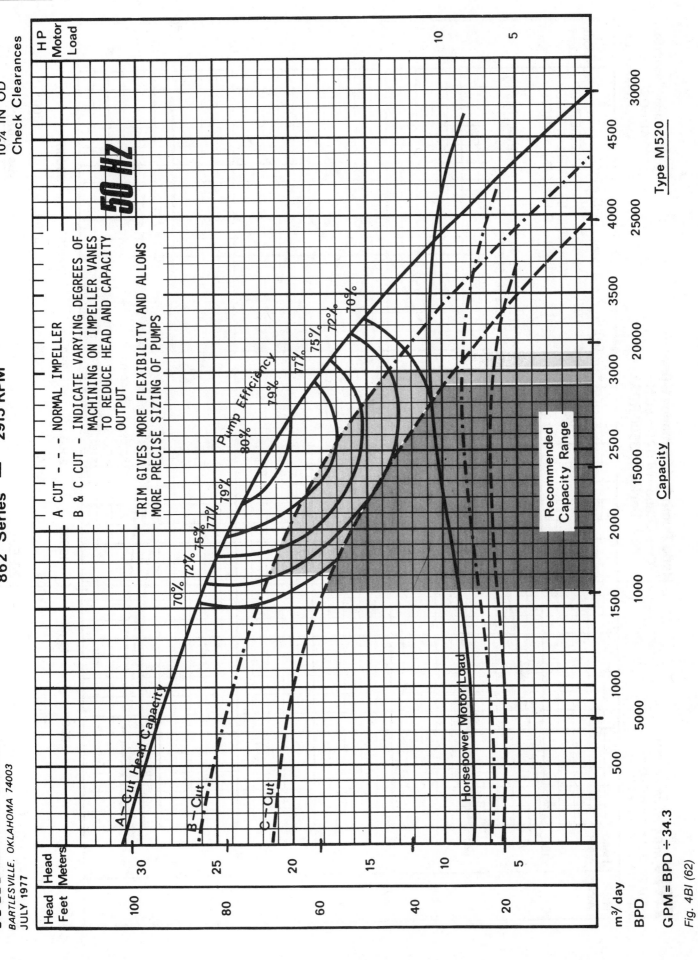

Reda Pump Performance Curves
One Stage—M520—50 Hz
862 Series — 2915 RPM

Type M520

Minimum Casing Size
10¾ IN OD
Check Clearances

TRW REDA PUMP DIVISION
BARTLESVILLE, OKLAHOMA 74003
JULY 1977

GPM=BPD÷34.3

Fig. 4BI (62)

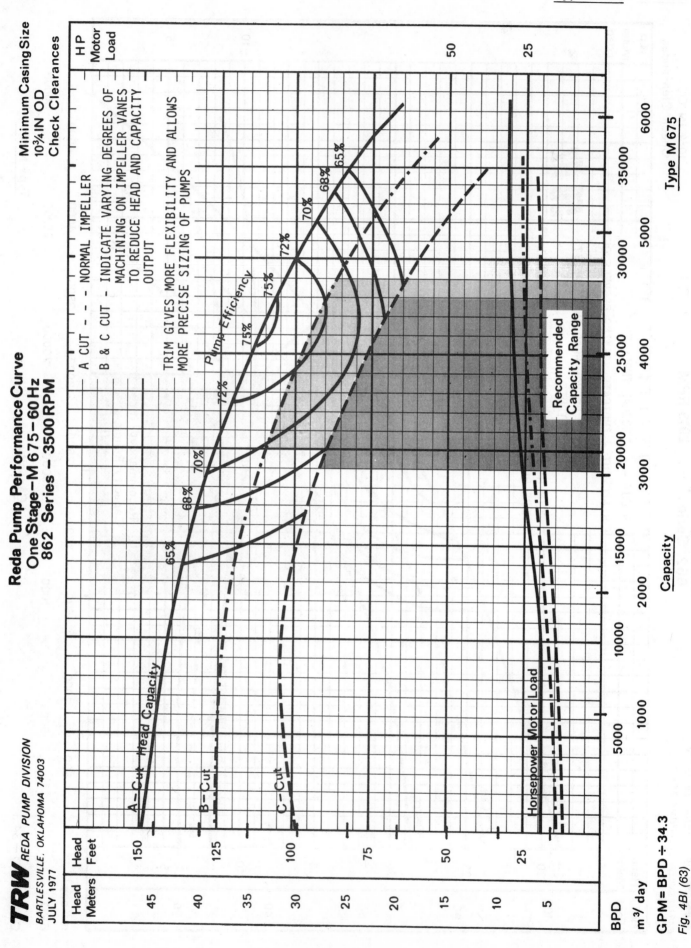

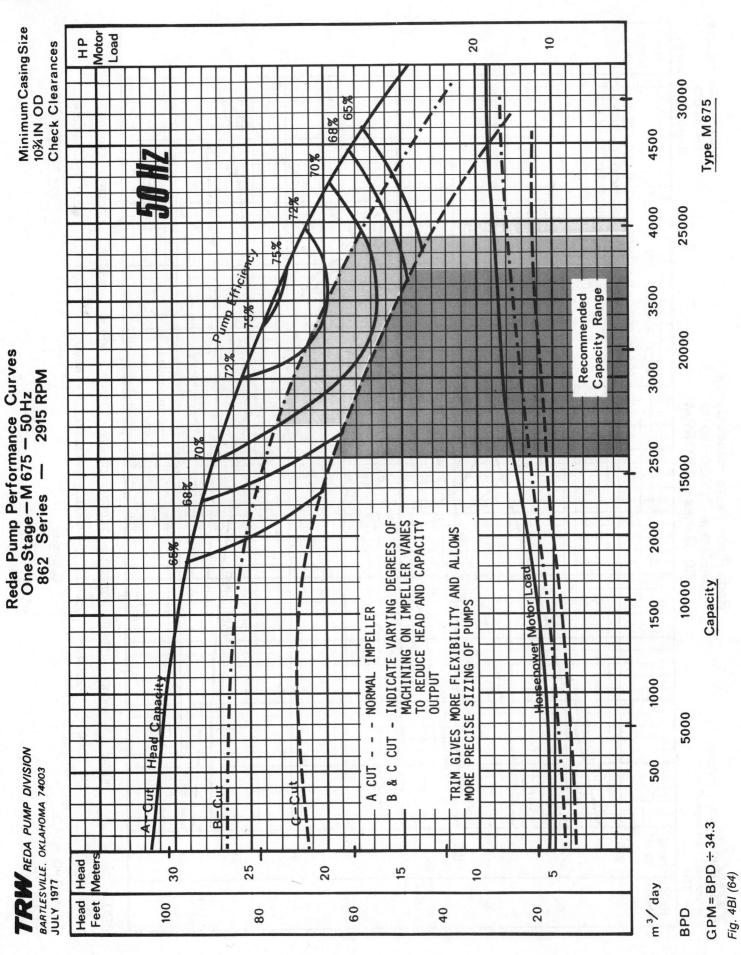

Reda Pump Performance Curves
One Stage—M 675 – 50 Hz
862 Series — 2915 RPM

Type M675

Minimum Casing Size
10¾ IN OD
Check Clearances

HP Motor Load

50 HZ

65%
68%
70%
72%
75%
75%
72%
70%
68%
65%

Pump Efficiency

Recommended Capacity Range

A–Cut   Head Capacity
B–Cut
C–Cut

A CUT - - - NORMAL IMPELLER
B & C CUT - INDICATE VARYING DEGREES OF
           MACHINING ON IMPELLER VANES
           TO REDUCE HEAD AND CAPACITY
           OUTPUT

TRIM GIVES MORE FLEXIBILITY AND ALLOWS
MORE PRECISE SIZING OF PUMPS

Horsepower Motor Load

20
10

TRW REDA PUMP DIVISION
BARTLESVILLE, OKLAHOMA 74003
JULY 1977

Head Feet   Head Meters

100   30
        25
80
        20
60
        15
40
        10
20   5

m³/ day
BPD
GPM= BPD ÷ 34.3
Fig. 4BI (64)

Type M675

Capacity

500   1000   1500   2000   2500   3000   3500   4000   4500
5000   10000   15000   20000   25000   30000

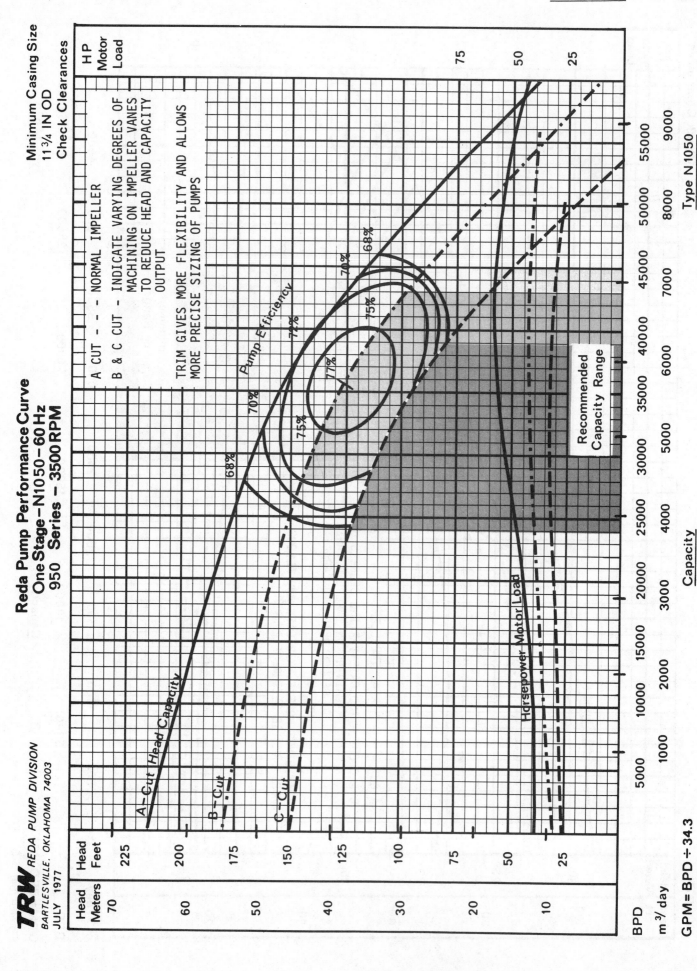

Reda Pump Performance Curve
One Stage—N1050—60 Hz
950 Series — 3500 RPM

TRW REDA PUMP DIVISION
BARTLESVILLE, OKLAHOMA 74003
JULY 1977

Type N1050

Minimum Casing Size
11¾ IN OD
Check Clearances

A CUT — — - NORMAL IMPELLER
B & C CUT - INDICATE VARYING DEGREES OF
MACHINING ON IMPELLER VANES
TO REDUCE HEAD AND CAPACITY
OUTPUT

TRIM GIVES MORE FLEXIBILITY AND ALLOWS
MORE PRECISE SIZING OF PUMPS

GPM = BPD ÷ 34.3

Fig. 4BI (65)

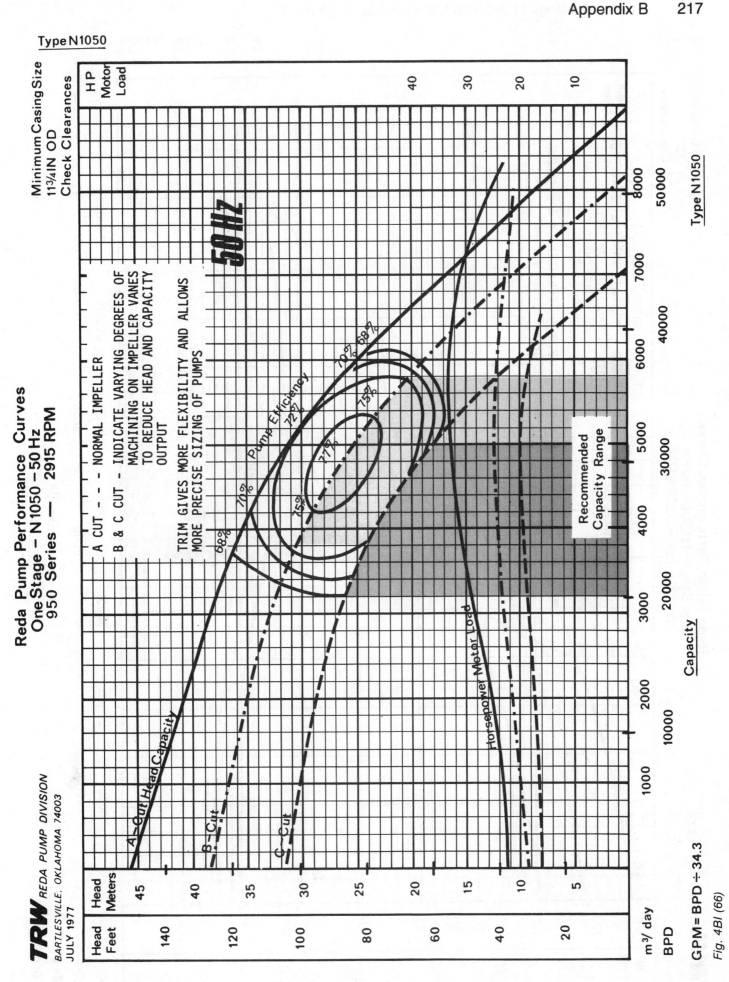

Reda Pump Performance Curves
One Stage – N1050 – 50 Hz
950 Series — 2915 RPM

Type N1050

Minimum Casing Size
11¾ IN OD
Check Clearances

TRW REDA PUMP DIVISION
BARTLESVILLE, OKLAHOMA 74003
JULY 1977

A CUT – – – NORMAL IMPELLER

B & C CUT – INDICATE VARYING DEGREES OF
MACHINING ON IMPELLER VANES
TO REDUCE HEAD AND CAPACITY
OUTPUT

TRIM GIVES MORE FLEXIBILITY AND ALLOWS
MORE PRECISE SIZING OF PUMPS

50 Hz

Recommended
Capacity Range

GPM = BPD ÷ 34.3

Fig. 4BI (66)

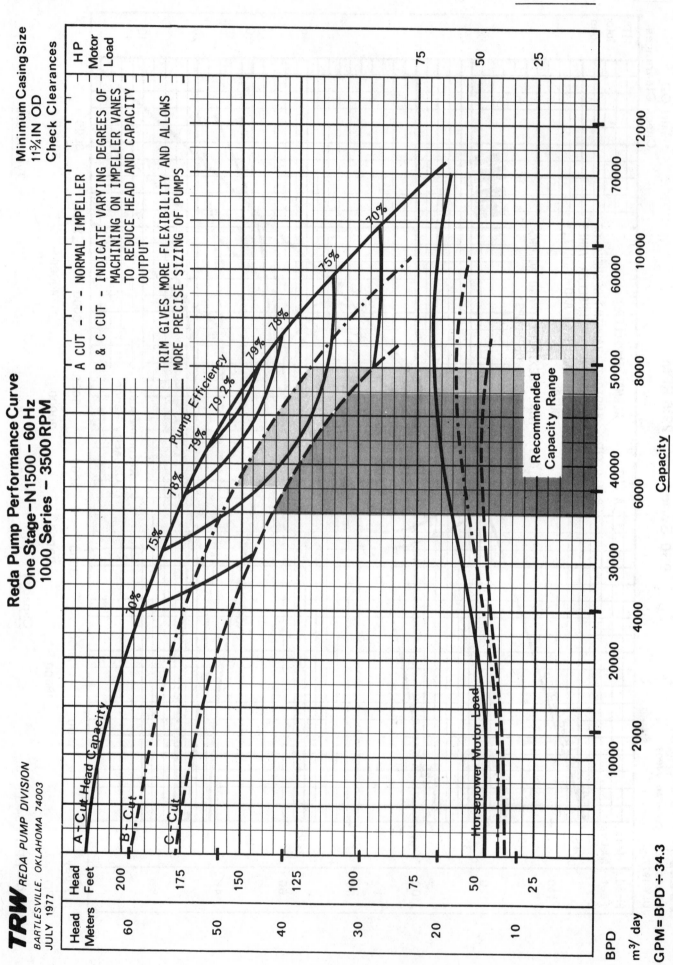

Type N1500

Reda Pump Performance Curve
One Stage—N1500 – 60 Hz
1000 Series – 3500 RPM

TRW REDA PUMP DIVISION
BARTLESVILLE. OKLAHOMA 74003
JULY 1977

Minimum Casing Size
11¾ IN OD
Check Clearances

A CUT – – NORMAL IMPELLER
B & C CUT – INDICATE VARYING DEGREES OF
MACHINING ON IMPELLER VANES
TO REDUCE HEAD AND CAPACITY
OUTPUT

TRIM GIVES MORE FLEXIBILITY AND ALLOWS
MORE PRECISE SIZING OF PUMPS

Recommended
Capacity Range

GPM = BPD ÷ 34.3

Fig. 4BI (67)

Type N 1500

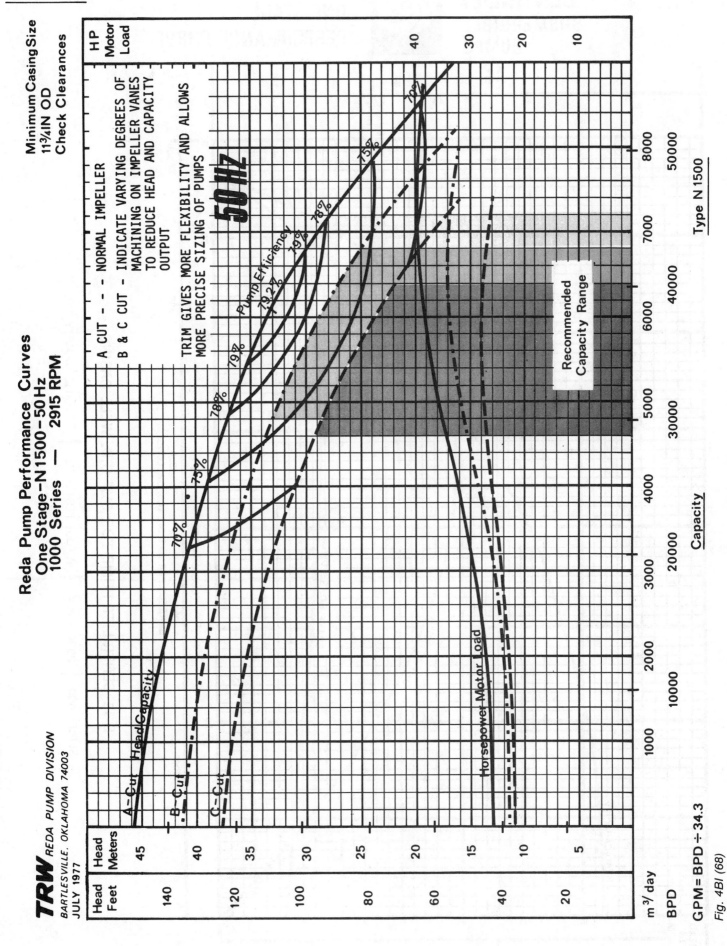

Fig. 4BI (68)

CENTRILIFT®
SUBMERSIBLE PUMPS

B-11
ONE STAGE
PERFORMANCE CURVE

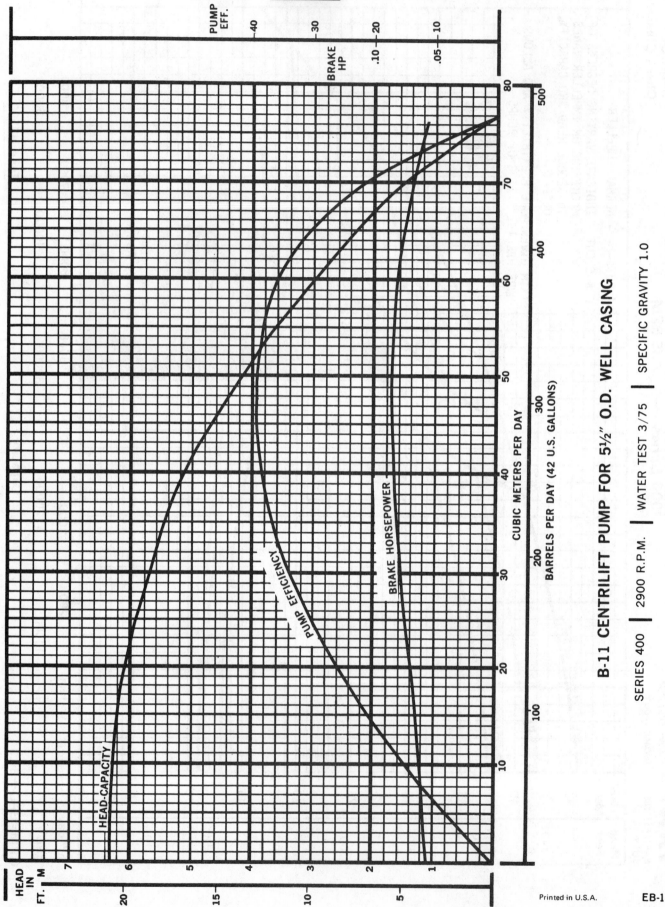

B-11 CENTRILIFT PUMP FOR 5½" O.D. WELL CASING

| SERIES 400 | 2900 R.P.M. | WATER TEST 3/75 | SPECIFIC GRAVITY 1.0 |

HEAD-CAPACITY

PUMP EFFICIENCY

BRAKE HORSEPOWER

PUMP EFF.

BRAKE HP

CUBIC METERS PER DAY

BARRELS PER DAY (42 U.S. GALLONS)

HEAD IN FT. M

Fig. 4BII (1)

Printed in U.S.A.

EB-1

"60 HERTZ"

CENTRILIFT® SUBMERSIBLE PUMPS

B-11 ONE STAGE PERFORMANCE CURVE

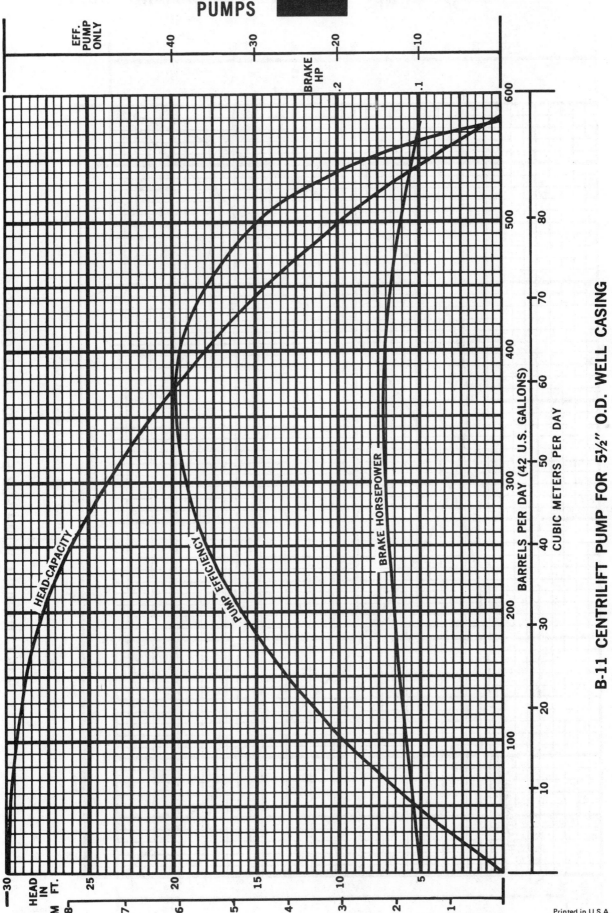

B-11 CENTRILIFT PUMP FOR 5½" O.D. WELL CASING

| SERIES 400 | 3475 R.P.M. | WATER TEST 3/75 | SPECIFIC GRAVITY 1.0 |

EB-1.1

Fig. 4BII (2)

Printed in U.S.A.

5/78

CENTRILIFT SUBMERSIBLE PUMPS

W-18 ONE STAGE PERFORMANCE CURVE

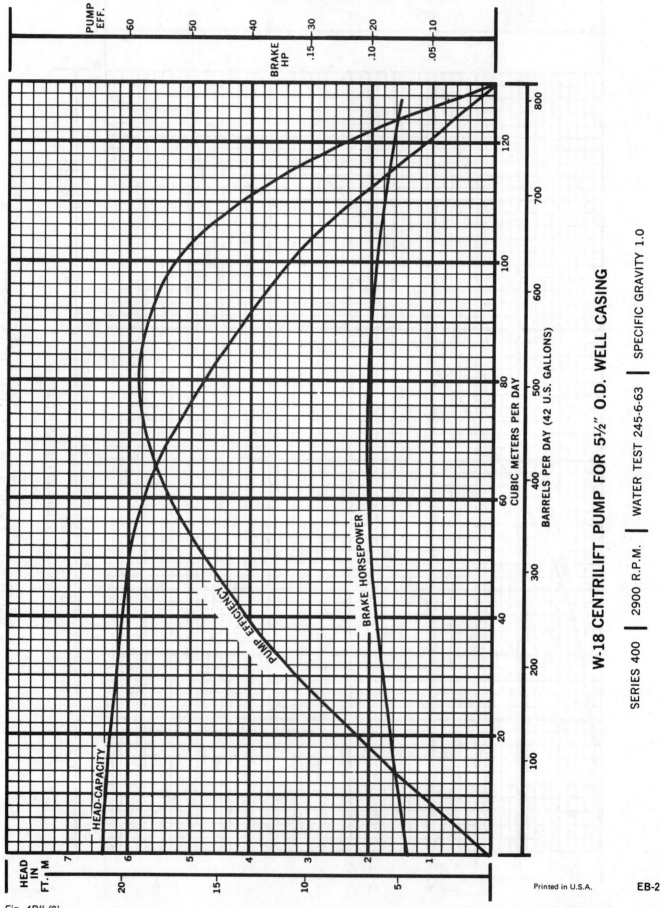

W-18 CENTRILIFT PUMP FOR 5½" O.D. WELL CASING

SERIES 400 | 2900 R.P.M. | WATER TEST 245-6-63 | SPECIFIC GRAVITY 1.0

5/78

Fig. 4BII (3)

Printed in U.S.A.   EB-2

CENTRILIFT® SUBMERSIBLE PUMPS

W-18 ONE STAGE PERFORMANCE CURVE

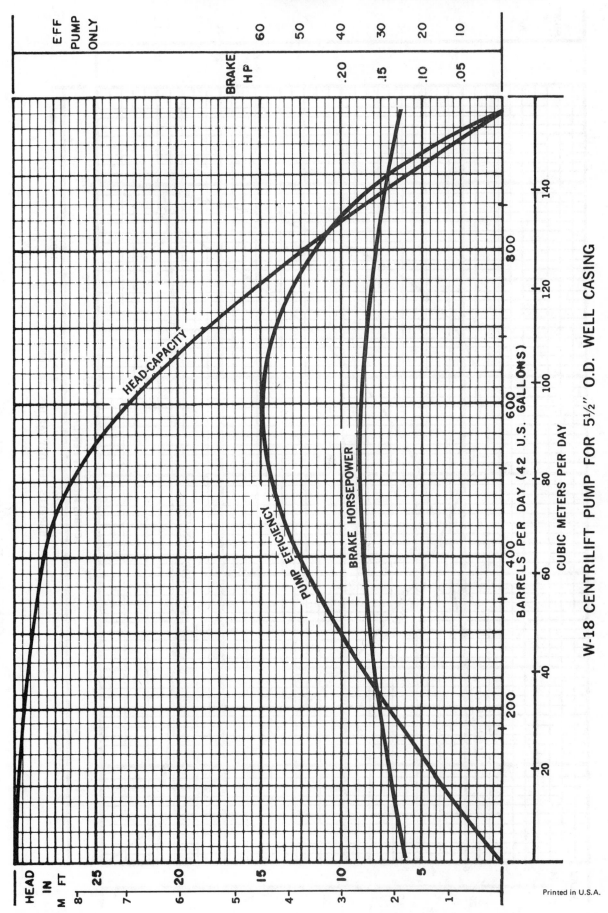

W-18 CENTRILIFT PUMP FOR 5½" O.D. WELL CASING

SERIES 400 | 3475 R.P.M. | WATER TEST NO. 245-6-63 | SPECIFIC GRAVITY 1.0

Fig. 4BII (4)

CENTRILIFT®
SUBMERSIBLE PUMPS

H-27
ONE STAGE
PERFORMANCE CURVE

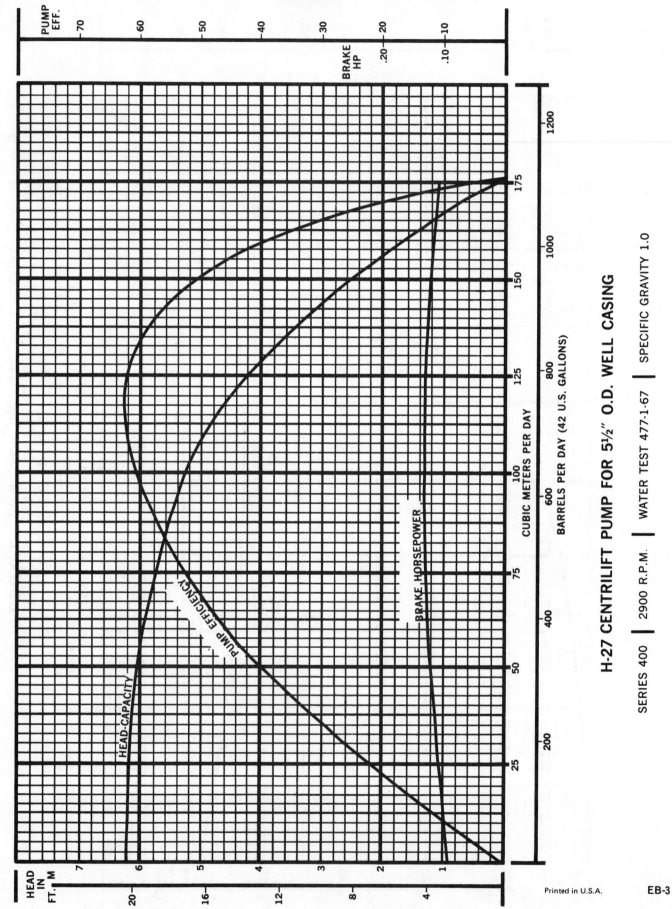

H-27 CENTRILIFT PUMP FOR 5½" O.D. WELL CASING

| SERIES 400 | 2900 R.P.M. | WATER TEST 477-1-67 | SPECIFIC GRAVITY 1.0 |

5/78

Fig. 4BII (5)

Printed in U.S.A.   EB-3

CENTRILIFT® SUBMERSIBLE PUMPS

H-27 ONE STAGE PERFORMANCE CURVE

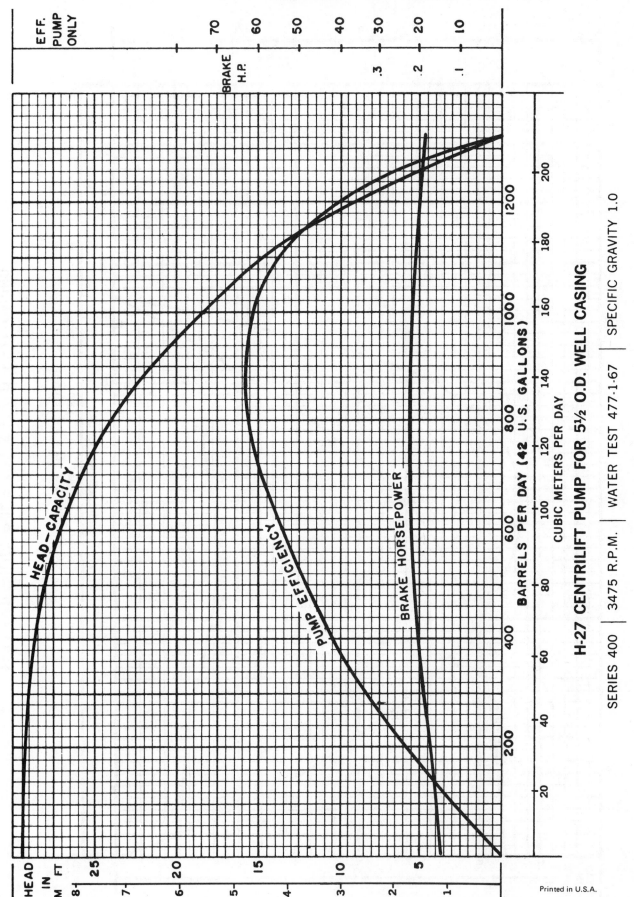

H-27 CENTRILIFT PUMP FOR 5½ O.D. WELL CASING

SERIES 400 | 3475 R.P.M. | WATER TEST 477-1-67 | SPECIFIC GRAVITY 1.0

EB-3.1

Fig. 4BII (6)

CENTRILIFT® SUBMERSIBLE PUMPS

M-34
ONE STAGE
PERFORMANCE CURVE

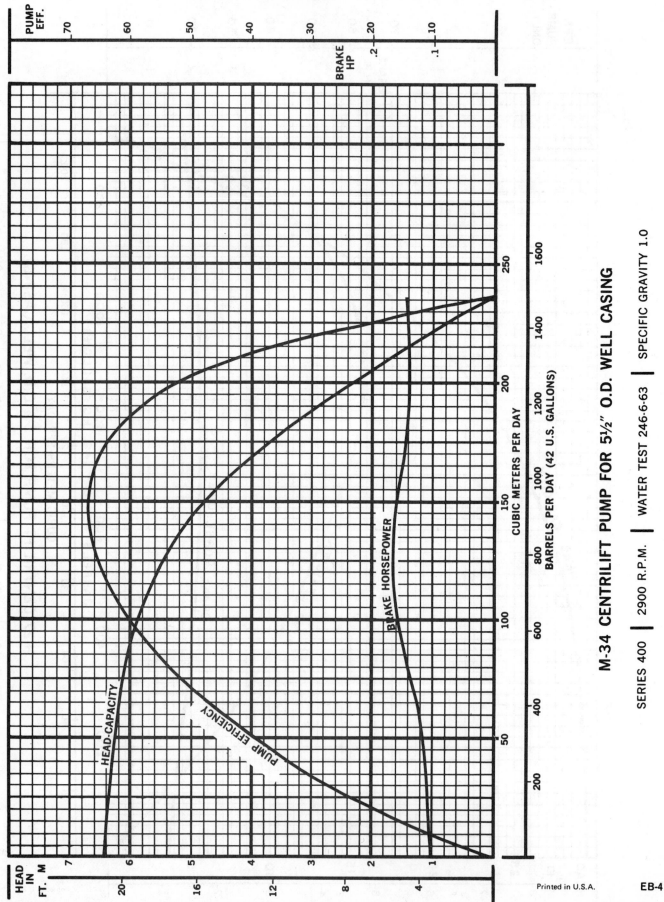

M-34 CENTRILIFT PUMP FOR 5½" O.D. WELL CASING

| SERIES 400 | 2900 R.P.M. | WATER TEST 246-6-63 | SPECIFIC GRAVITY 1.0 |

5/78

Fig. 4BII (7)

Printed in U.S.A.

EB-4

CENTRILIFT® SUBMERSIBLE PUMPS

M-34 ONE STAGE PERFORMANCE CURVE

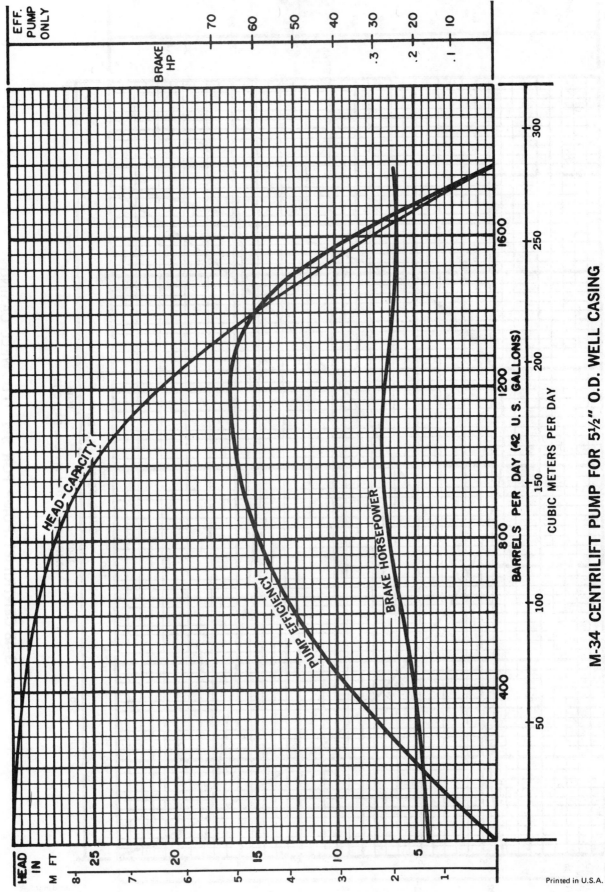

M-34 CENTRILIFT PUMP FOR 5½" O.D. WELL CASING

SERIES 400 | 3475 R.P.M. | WATER TEST NO. 246-6-63 | SPECIFIC GRAVITY 1.0

EB-4.1

Fig. 4BII (8)

CENTRILIFT® SUBMERSIBLE PUMPS

G-48 ONE STAGE PERFORMANCE CURVE

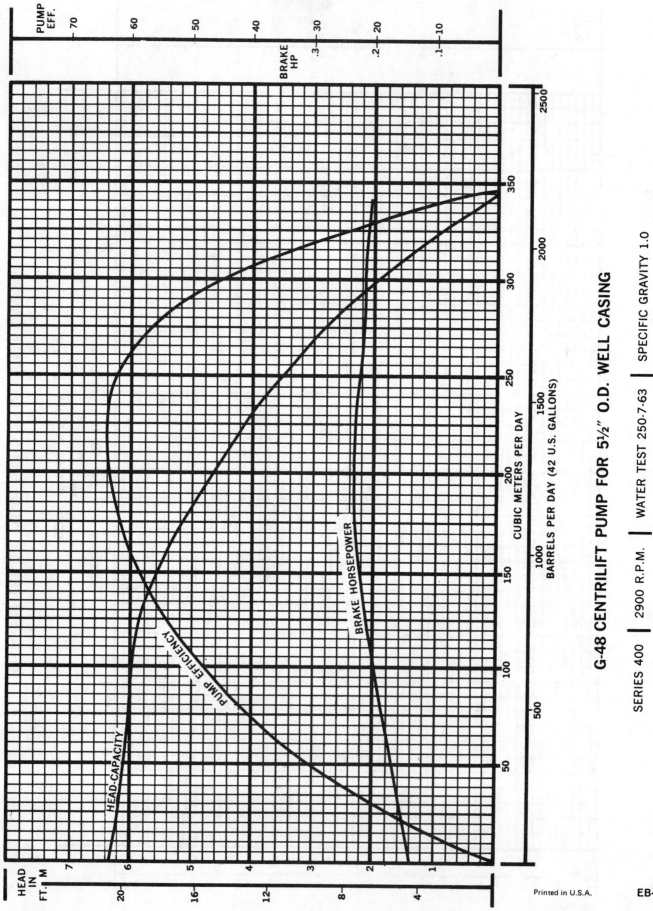

G-48 CENTRILIFT PUMP FOR 5½" O.D. WELL CASING

| SERIES 400 | 2900 R.P.M. | WATER TEST 250-7-63 | SPECIFIC GRAVITY 1.0 |

Fig. 4BII (9)

CENTRILIFT®
SUBMERSIBLE PUMPS

G-48 ONE STAGE PERFORMANCE CURVE

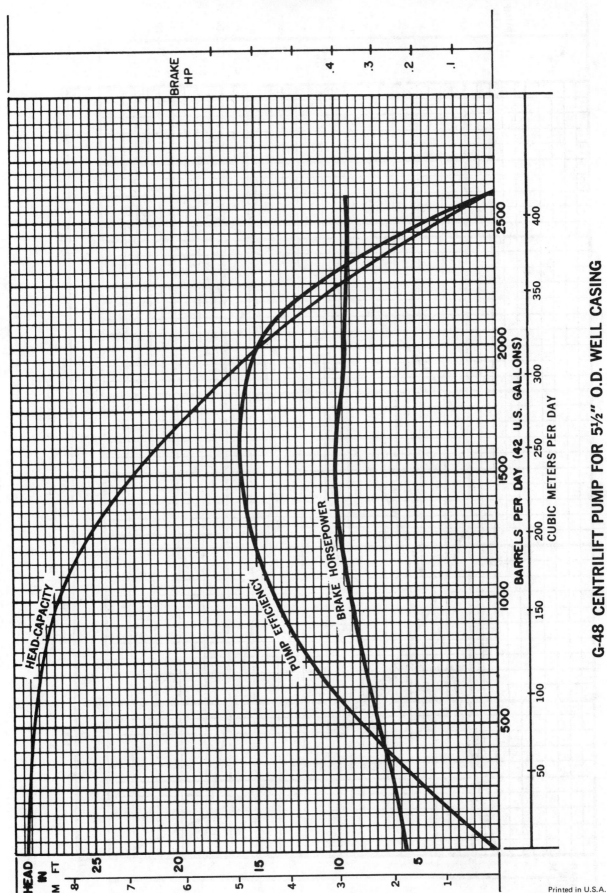

G-48 CENTRILIFT PUMP FOR 5½" O.D. WELL CASING

SERIES 400 | 3475 R.P.M. | WATER TEST NO. 250-7-63 | SPECIFIC GRAVITY 1.0

EB-5.1

Fig. 4BII (10)

CENTRILIFT® SUBMERSIBLE PUMPS

J-61 ONE STAGE PERFORMANCE CURVE

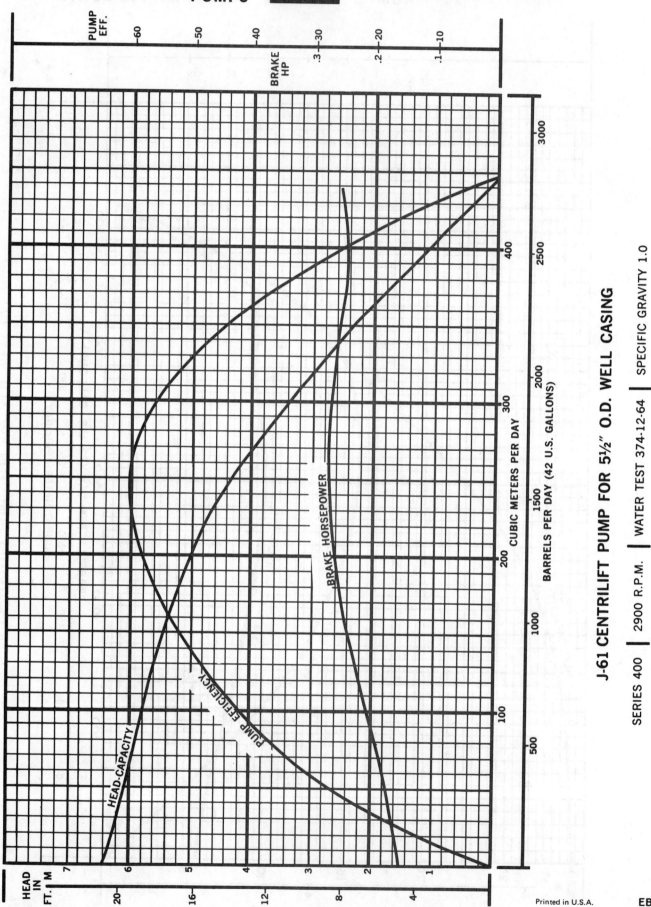

J-61 CENTRILIFT PUMP FOR 5½" O.D. WELL CASING

SERIES 400 | 2900 R.P.M. | WATER TEST 374-12-64 | SPECIFIC GRAVITY 1.0

5/78

Fig. 4BII (11)

Printed in U.S.A.

EB-6

CENTRILIFT® SUBMERSIBLE PUMPS

J-61 ONE STAGE PERFORMANCE CURVE

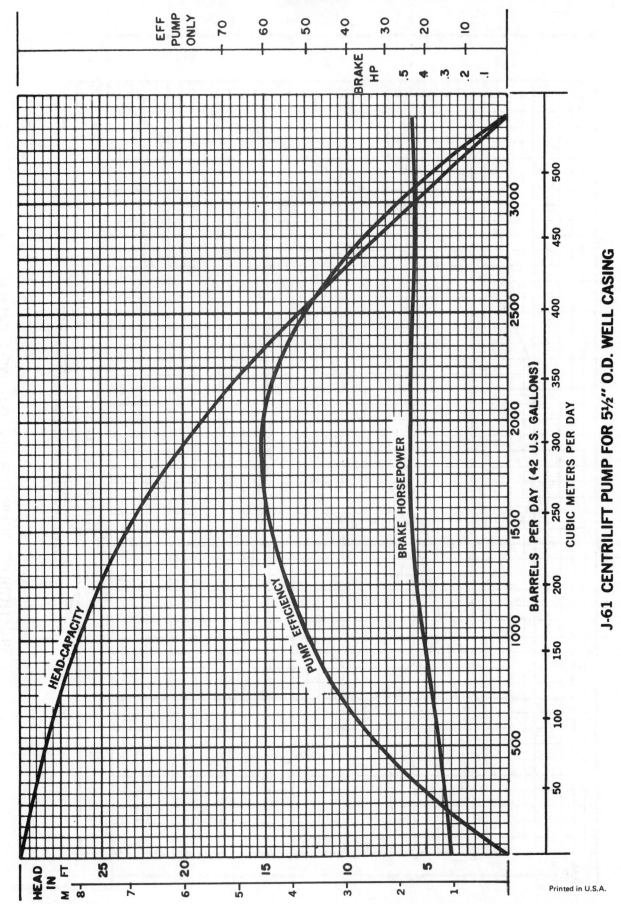

J-61 CENTRILIFT PUMP FOR 5½" O.D. WELL CASING

SERIES 400 | 3475 R.P.M. | WATER TEST NO. 565-4-68 | SPECIFIC GRAVITY 1.0

EB-6.1

Fig. 4BII (12)

CENTRILIFT® SUBMERSIBLE PUMPS

Z-69 ONE STAGE PERFORMANCE CURVE

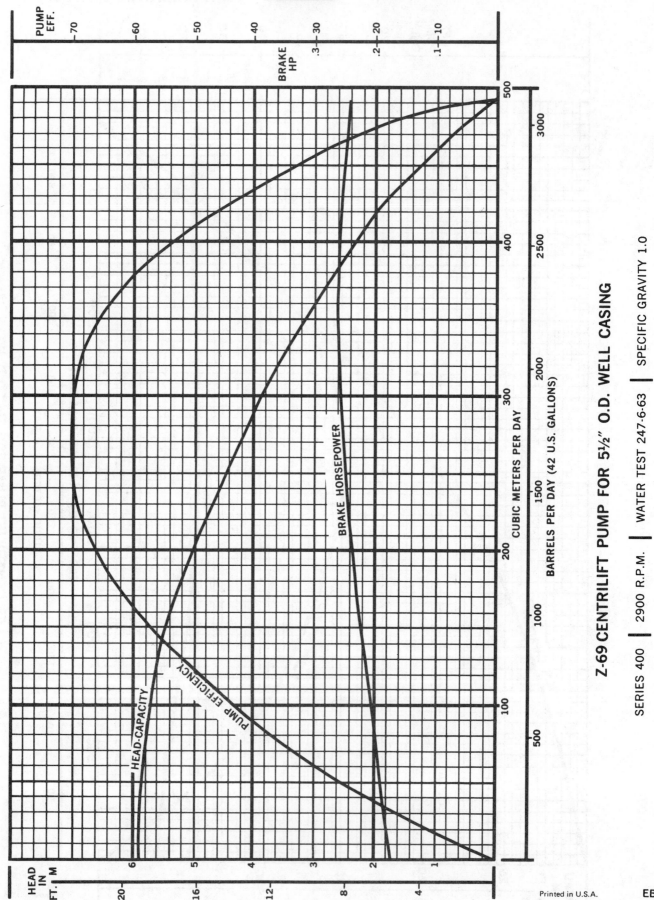

Z-69 CENTRILIFT PUMP FOR 5½" O.D. WELL CASING

| SERIES 400 | 2900 R.P.M. | WATER TEST 247-6-63 | SPECIFIC GRAVITY 1.0 |

CENTRILIFT®
SUBMERSIBLE PUMPS
Z-69 ONE STAGE PERFORMANCE CURVE

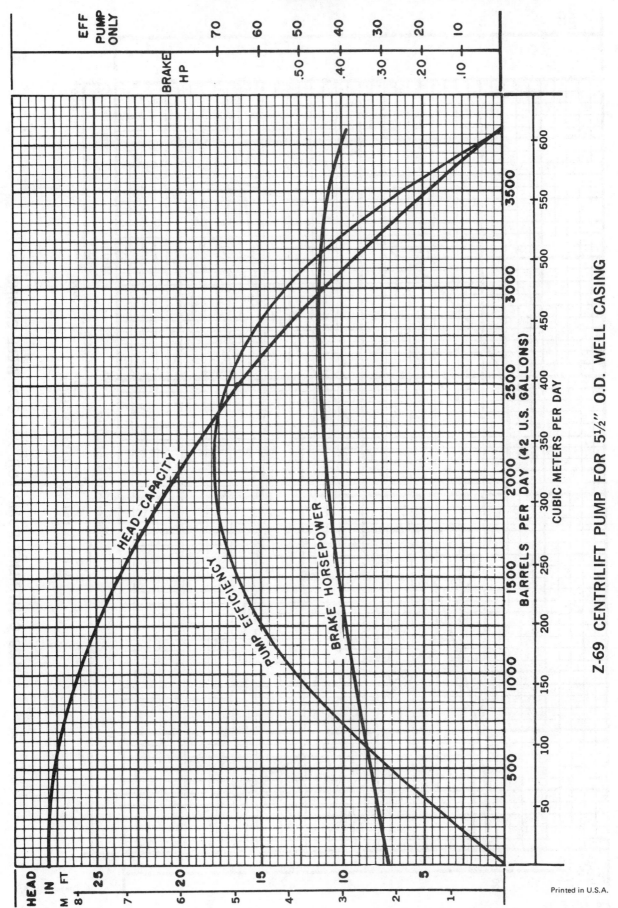

Z-69 CENTRILIFT PUMP FOR 5½" O.D. WELL CASING

SERIES 400 | 3475 R.P.M | WATER TEST NO. 247-6-63 | SPECIFIC GRAVITY 1.0

EB-7.1

Fig. 4BII (14)

Printed in U.S.A.   5/78

CENTRILIFT® SUBMERSIBLE PUMPS

N-80 ONE STAGE PERFORMANCE CURVE

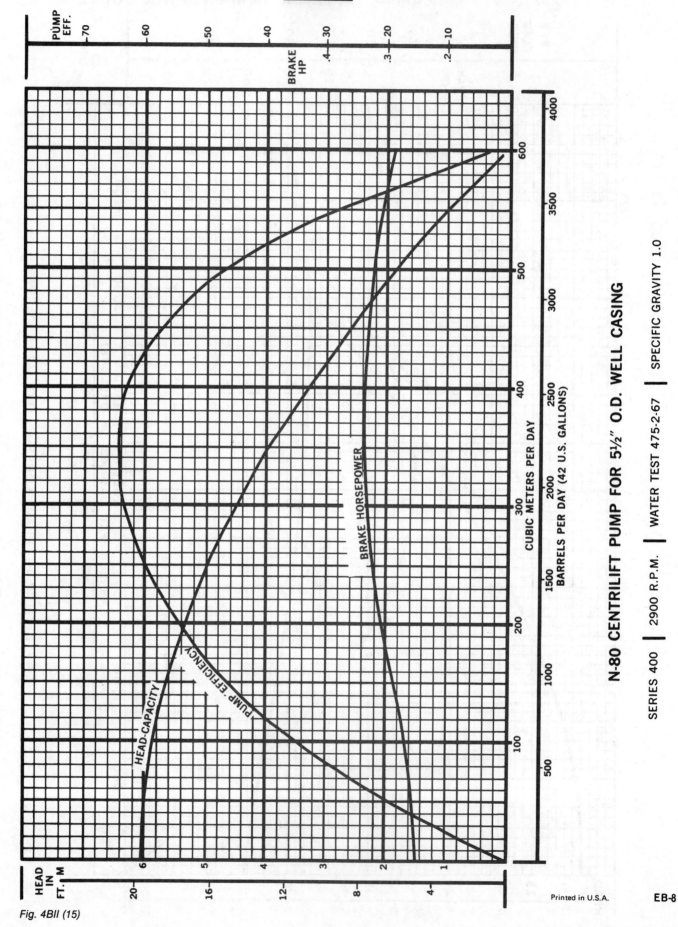

N-80 CENTRILIFT PUMP FOR 5½" O.D. WELL CASING

SERIES 400 | 2900 R.P.M. | WATER TEST 475-2-67 | SPECIFIC GRAVITY 1.0

HEAD-CAPACITY

PUMP EFFICIENCY

BRAKE HORSEPOWER

PUMP EFF.

BRAKE HP

HEAD IN FT. M

BARRELS PER DAY (42 U.S. GALLONS)

CUBIC METERS PER DAY

5/78

Fig. 4BII (15)

CENTRILIFT® SUBMERSIBLE PUMPS

N-80 ONE STAGE PERFORMANCE CURVE

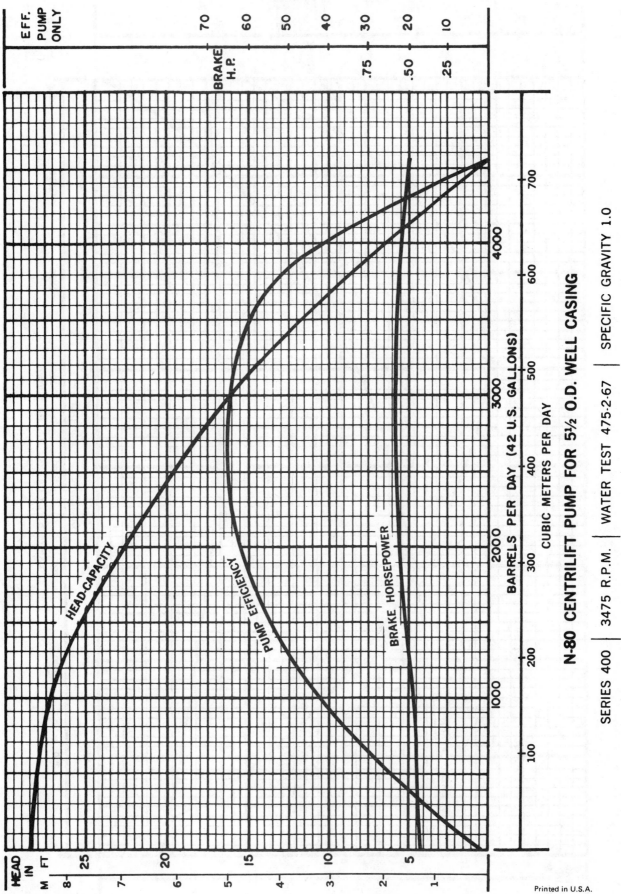

N-80 CENTRILIFT PUMP FOR 5½ O.D. WELL CASING

SERIES 400 | 3475 R.P.M. | WATER TEST 475-2-67 | SPECIFIC GRAVITY 1.0

EB-8.1

Fig. 4BII (16)

CENTRILIFT® SUBMERSIBLE PUMPS

I-42B ONE STAGE PERFORMANCE CURVE

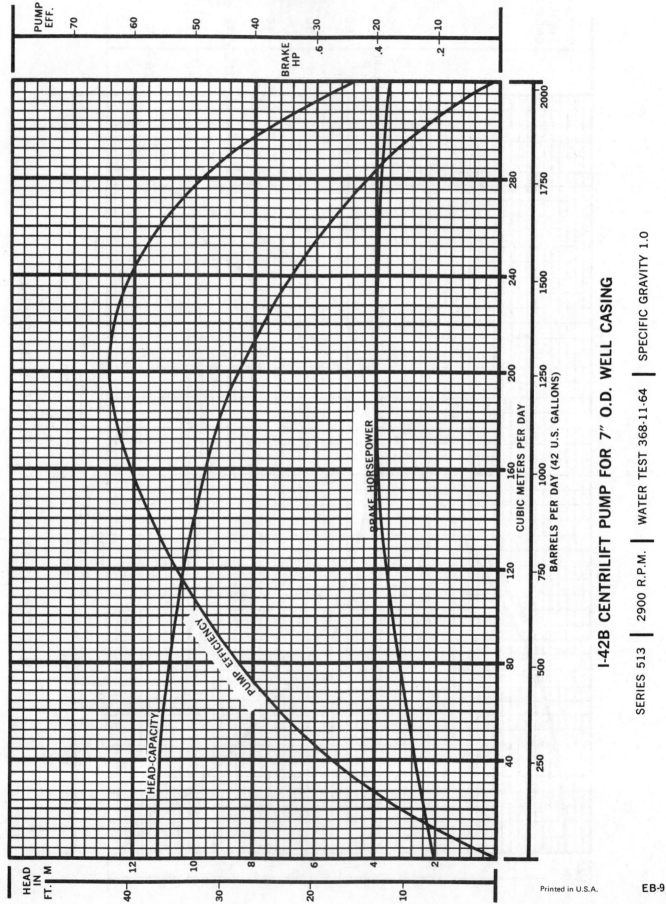

I-42B CENTRILIFT PUMP FOR 7" O.D. WELL CASING

| SERIES 513 | 2900 R.P.M. | WATER TEST 368-11-64 | SPECIFIC GRAVITY 1.0 |

5/78

Fig. 4BII (17)

Printed in U.S.A.

EB-9

CENTRILIFT®
SUBMERSIBLE PUMPS

I-42B ONE STAGE PERFORMANCE CURVE

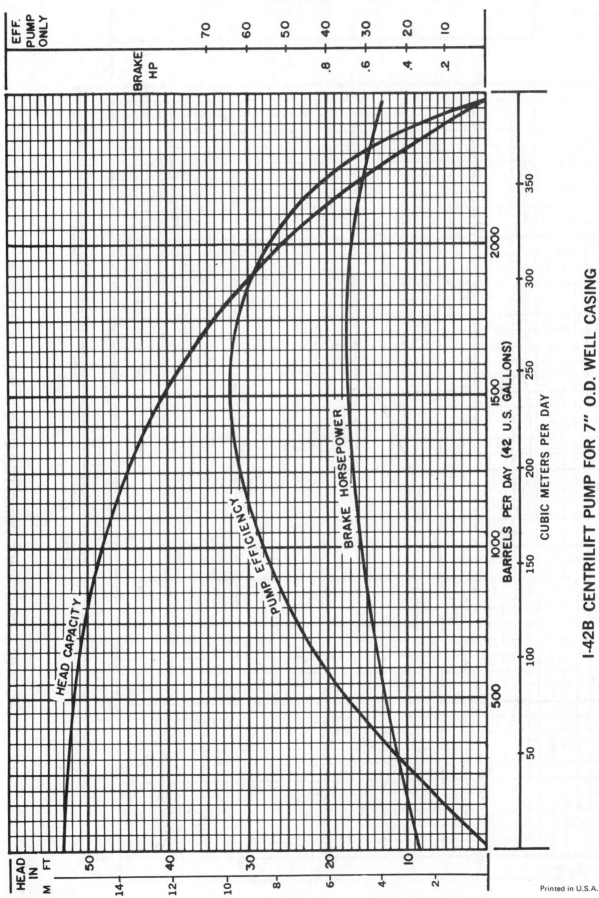

I-42B CENTRILIFT PUMP FOR 7" O.D. WELL CASING

SERIES 513 | 3475 R.P.M. | WATER TEST NO. 368-11-64 | SPECIFIC GRAVITY 1.0

EB-9.1

Fig. 4BII (18)

CENTRILIFT®
SUBMERSIBLE PUMPS

Y-62 B
ONE STAGE
PERFORMANCE CURVE

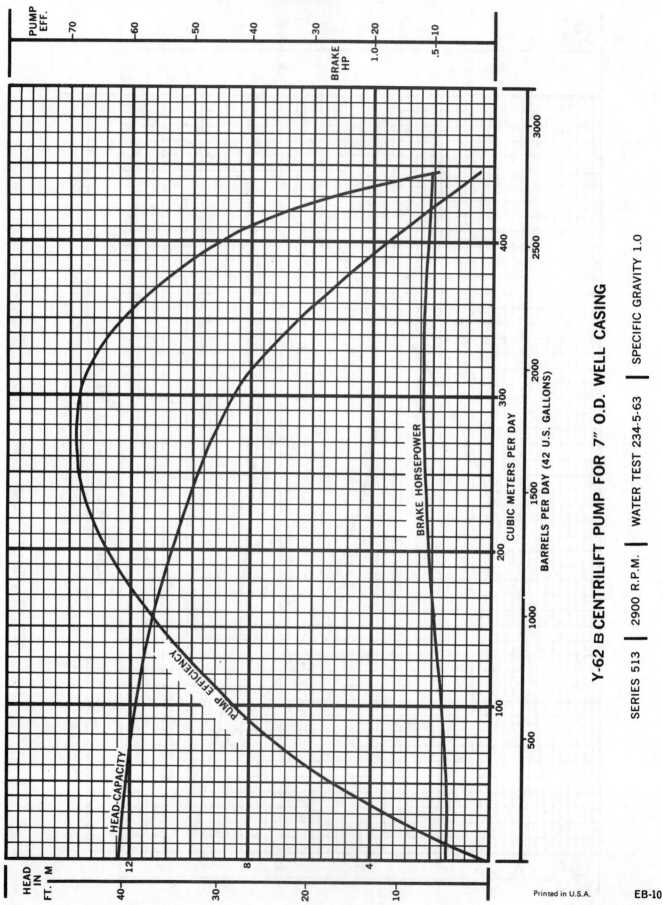

Y-62 B CENTRILIFT PUMP FOR 7" O.D. WELL CASING

| SERIES 513 | 2900 R.P.M. | WATER TEST 234-5-63 | SPECIFIC GRAVITY 1.0 |

5/78

Fig. 4BII (19)

CENTRILIFT® SUBMERSIBLE PUMPS

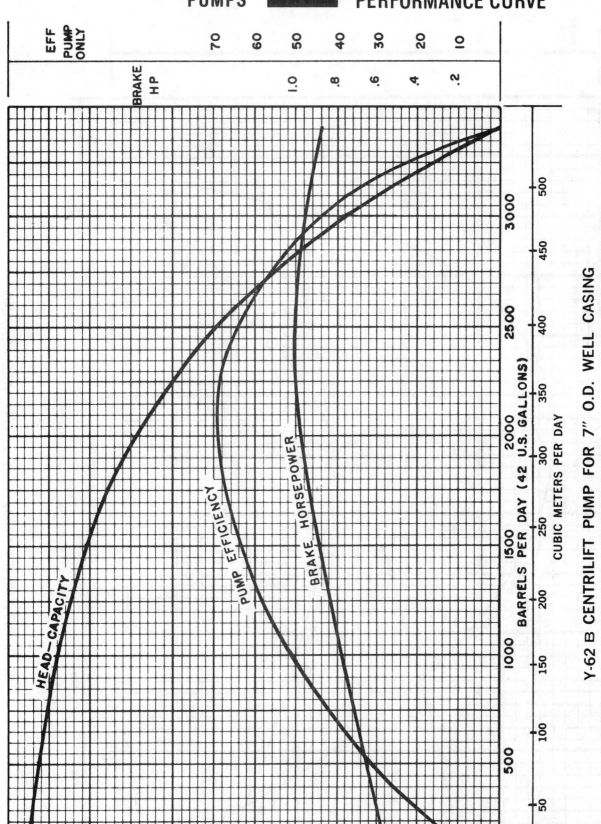

Y-62 B ONE STAGE PERFORMANCE CURVE

Y-62 B CENTRILIFT PUMP FOR 7" O.D. WELL CASING

SERIES 513 | 3475 R.P.M. | WATER TEST NO. 234-5-63 | SPECIFIC GRAVITY 1.0

EB-10.1

Fig. 4BII (20)

CENTRILIFT® SUBMERSIBLE PUMPS

K-70 ONE STAGE PERFORMANCE CURVE

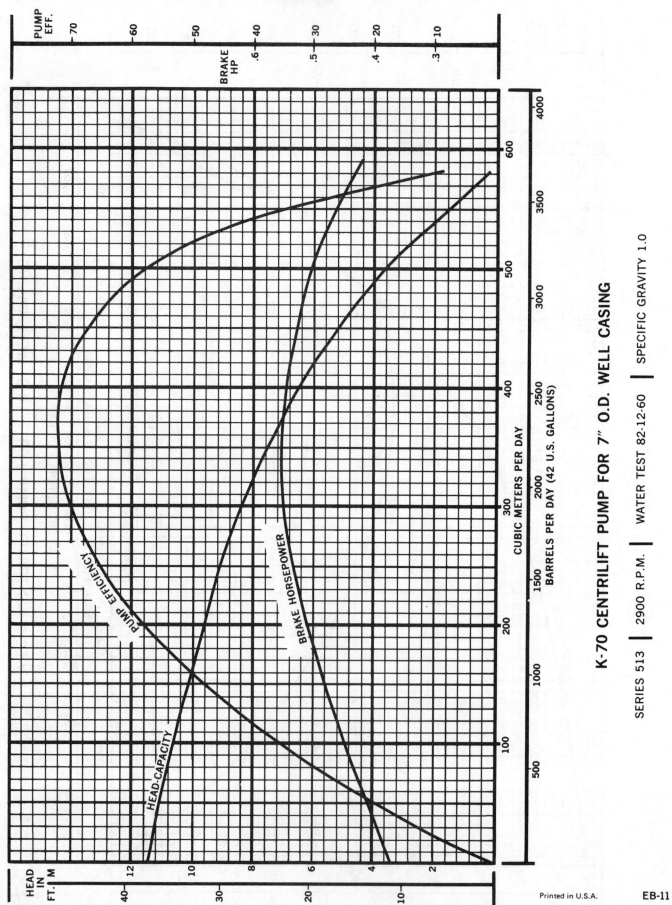

K-70 CENTRILIFT PUMP FOR 7" O.D. WELL CASING

SERIES 513 | 2900 R.P.M. | WATER TEST 82-12-60 | SPECIFIC GRAVITY 1.0

CENTRILIFT® SUBMERSIBLE PUMPS

K-70 ONE STAGE PERFORMANCE CURVE

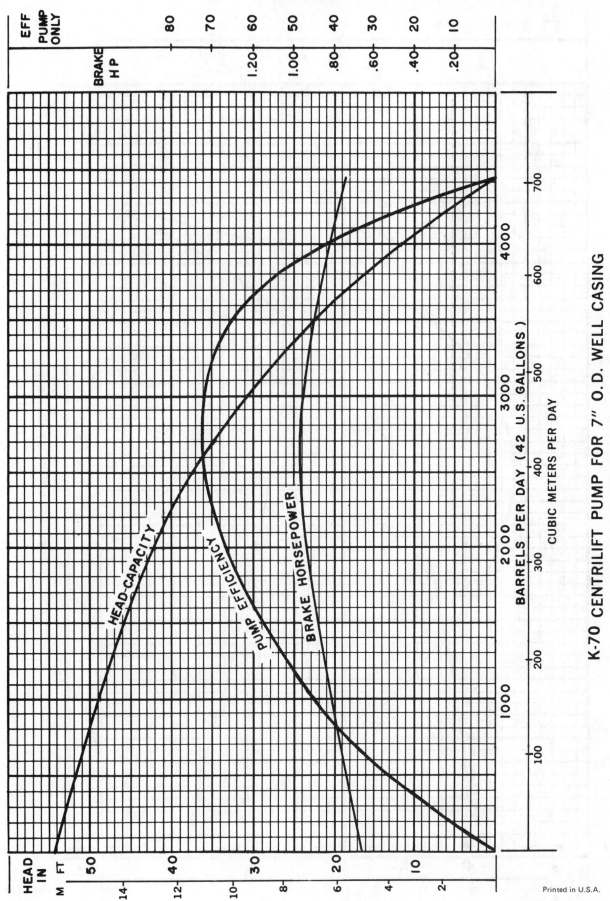

K-70 CENTRILIFT PUMP FOR 7" O.D. WELL CASING

SERIES 513 | 3475 R.P.M. | WATER TEST NO. 82-12-60 | SPECIFIC GRAVITY 1.0

EB-11.1

Fig. 4BII (22)

Printed in U.S.A.   5/78

CENTRILIFT®
SUBMERSIBLE PUMPS

C-72
ONE STAGE
PERFORMANCE CURVE

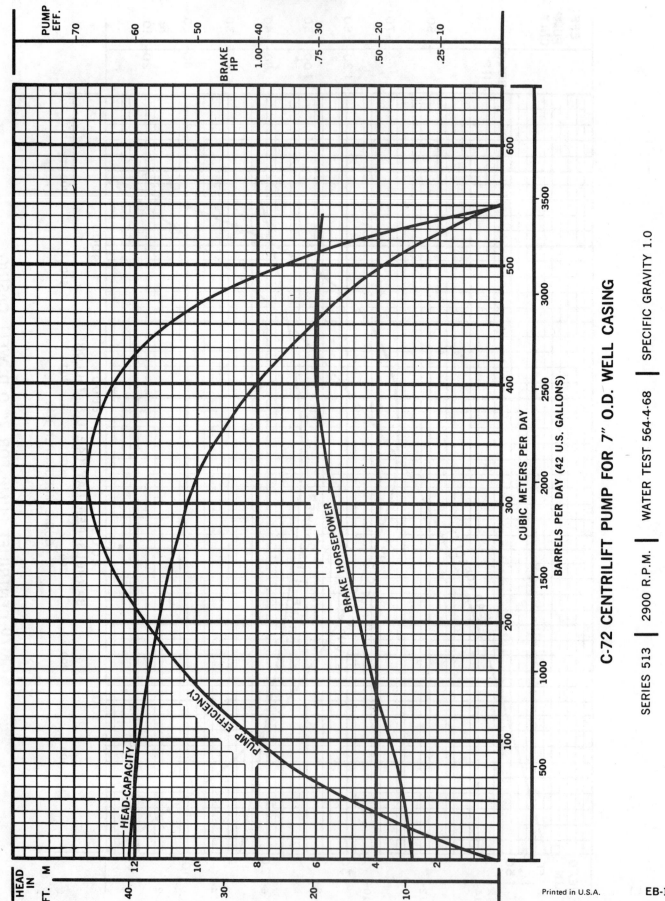

C-72 CENTRILIFT PUMP FOR 7" O.D. WELL CASING

SERIES 513 | 2900 R.P.M. | WATER TEST 564.4-68 | SPECIFIC GRAVITY 1.0

5/78

Fig. 4BII (23)

Printed in U.S.A.

EB-12

CENTRILIFT® SUBMERSIBLE PUMPS

C-72 ONE STAGE PERFORMANCE CURVE

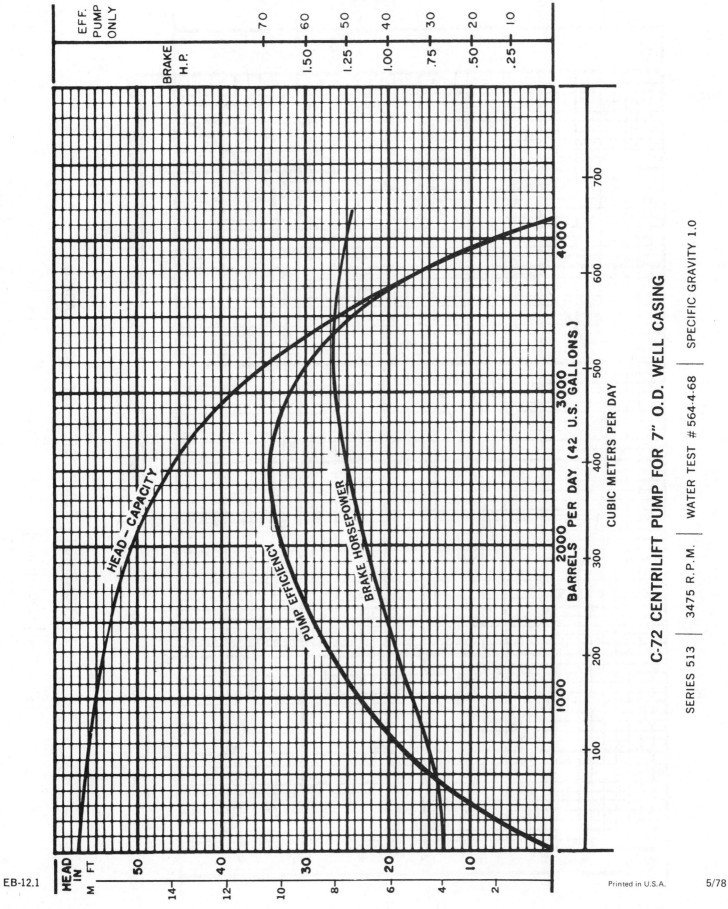

C-72 CENTRILIFT PUMP FOR 7" O.D. WELL CASING

| SERIES 513 | 3475 R.P.M. | WATER TEST # 564-4-68 | SPECIFIC GRAVITY 1.0 |

EB-12.1

Fig. 4BII (24)

CENTRILIFT® SUBMERSIBLE PUMPS

KA-100 ONE STAGE PERFORMANCE CURVE

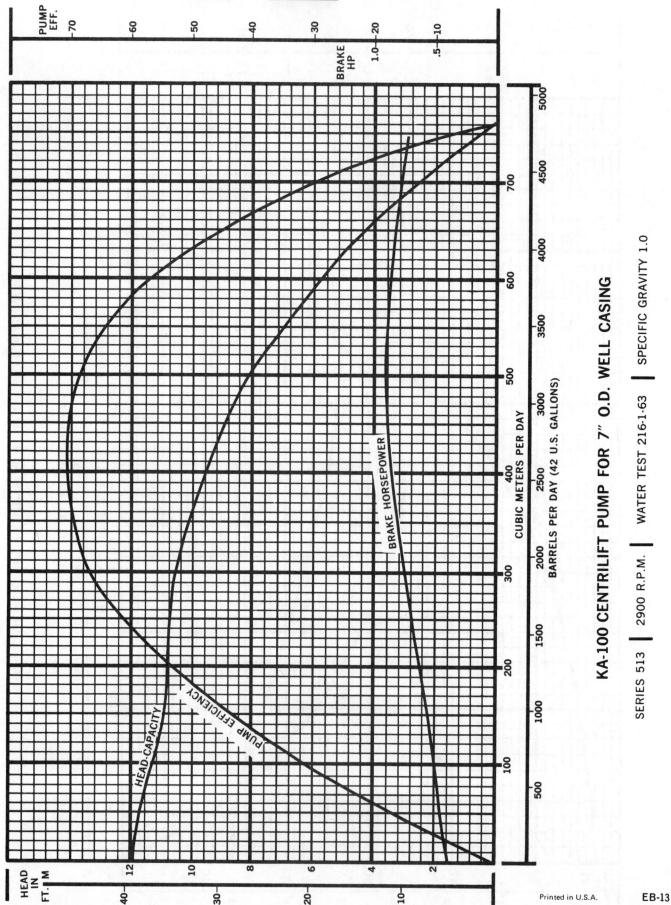

KA-100 CENTRILIFT PUMP FOR 7" O.D. WELL CASING

SERIES 513 | 2900 R.P.M. | WATER TEST 216-1-63 | SPECIFIC GRAVITY 1.0

5/78

Fig. 4BII (25)

Printed in U.S.A.　EB-13

"60 HERTZ"

CENTRILIFT® SUBMERSIBLE PUMPS
KA-100 ONE STAGE PERFORMANCE CURVE

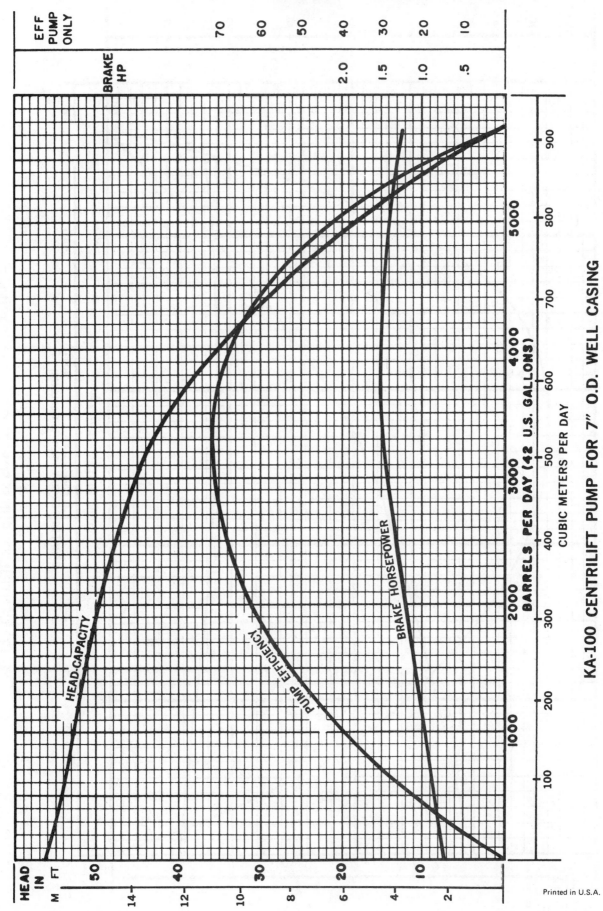

KA-100 CENTRILIFT PUMP FOR 7" O.D. WELL CASING

SERIES 513 | 3475 R.P.M. | WATER TEST NO. 216-1-63 | SPECIFIC GRAVITY 1.0

Printed in U.S.A.   5/78

Fig. 4BII (26)

CENTRILIFT® SUBMERSIBLE PUMPS

E-127 ONE STAGE PERFORMANCE CURVE

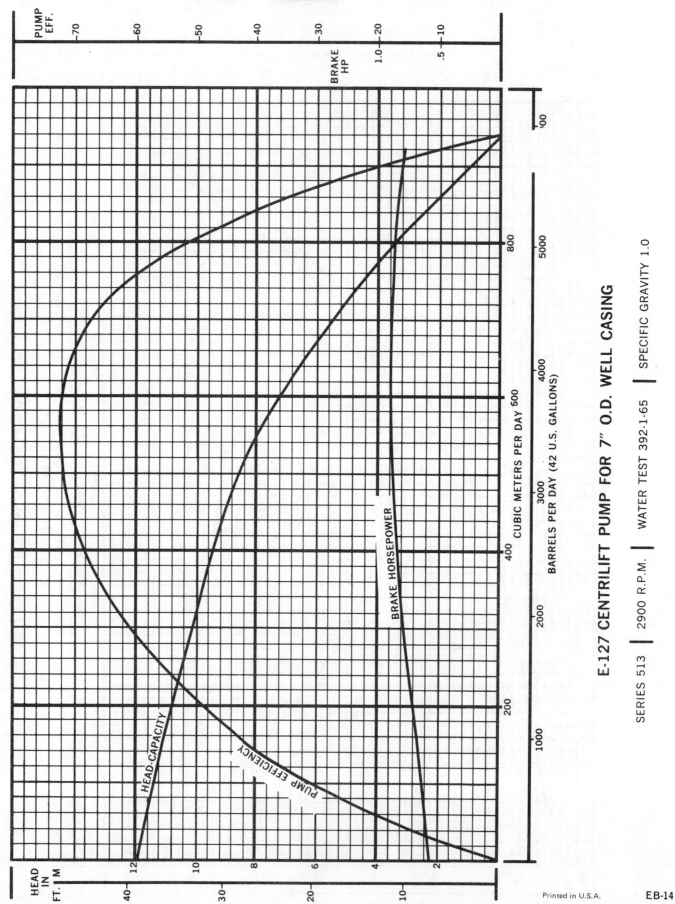

E-127 CENTRILIFT PUMP FOR 7" O.D. WELL CASING

SERIES 513 | 2900 R.P.M. | WATER TEST 392-1-65 | SPECIFIC GRAVITY 1.0

HEAD-CAPACITY

PUMP EFFICIENCY

BRAKE HORSEPOWER

5/78

Fig. 4BII (27)

Printed in U.S.A. EB-14

"60 HERTZ"

CENTRILIFT® SUBMERSIBLE PUMPS

E-127 ONE STAGE PERFORMANCE CURVE

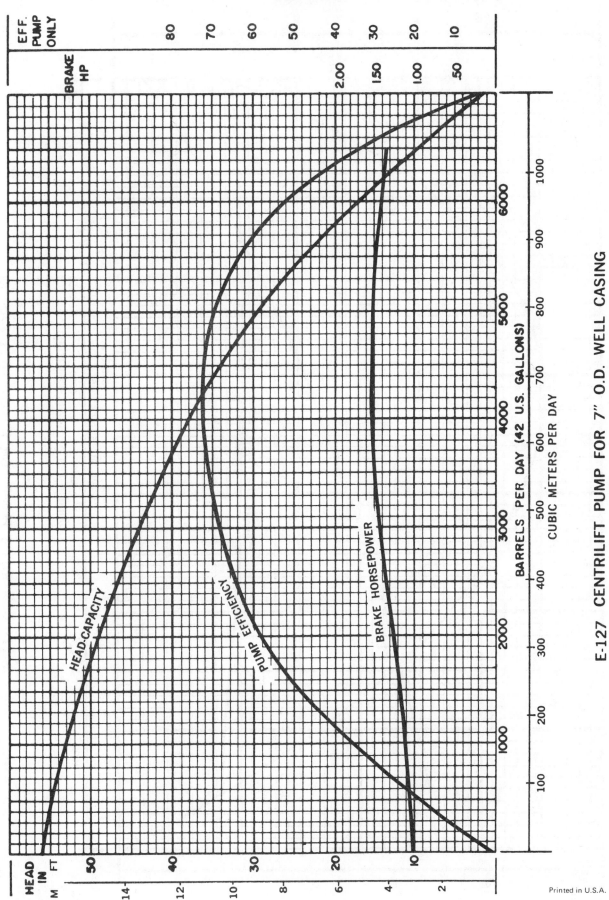

E-127 CENTRILIFT PUMP FOR 7" O.D. WELL CASING

| SERIES 513 | WATER TEST | 3475 R.P.M. | NO. 392-1-65 | SPECIFIC GRAVITY 1.0 |

CENTRILIFT SUBMERSIBLE PUMPS

S-175 ONE STAGE PERFORMANCE CURVE

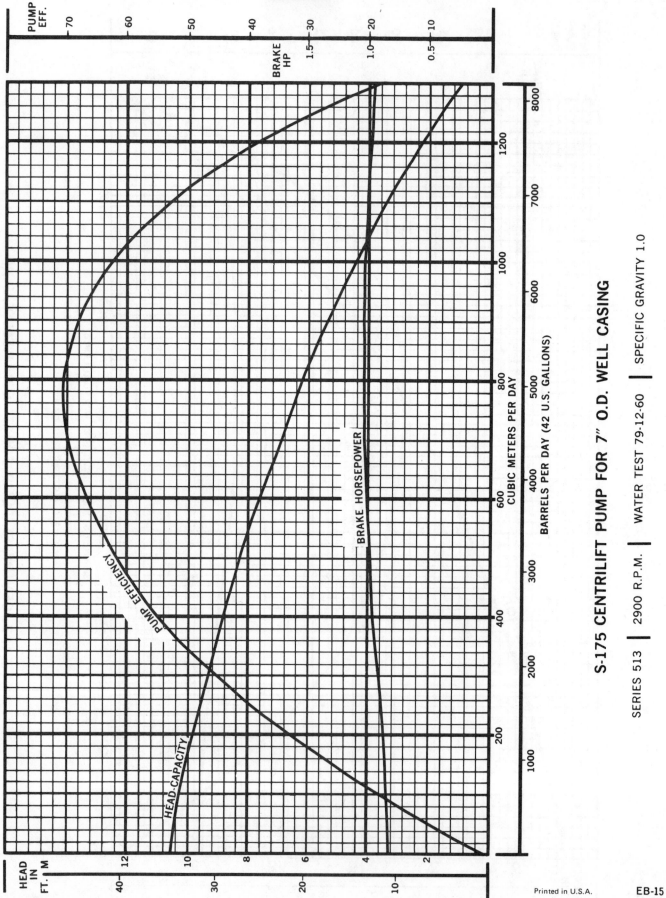

S-175 CENTRILIFT PUMP FOR 7" O.D. WELL CASING

SERIES 513 | 2900 R.P.M. | WATER TEST 79-12-60 | SPECIFIC GRAVITY 1.0

5/78

Fig. 4BII (29)

Printed in U.S.A. EB-15

CENTRILIFT® SUBMERSIBLE PUMPS

S-175 ONE STAGE PERFORMANCE CURVE

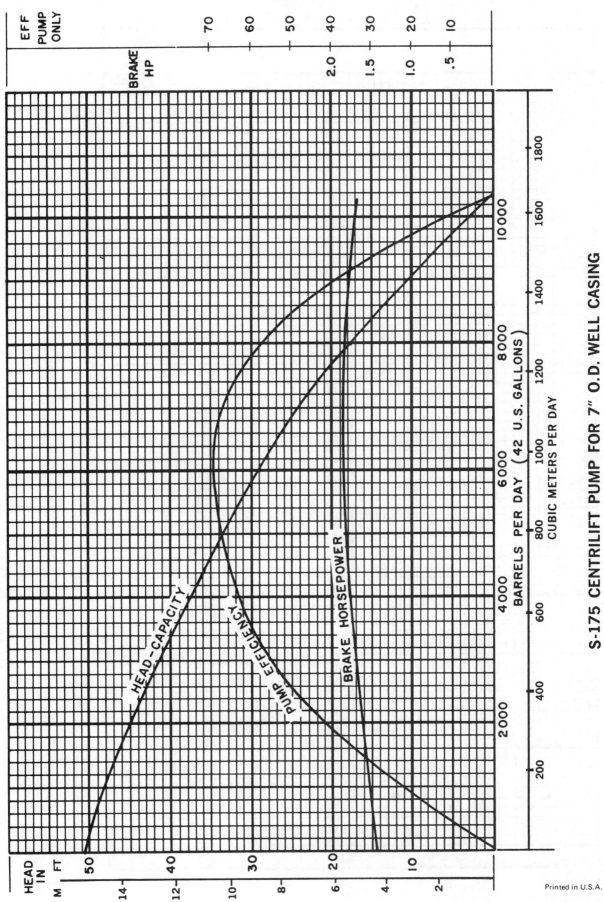

S-175 CENTRILIFT PUMP FOR 7" O.D. WELL CASING

| SERIES 513 | 3475 R.P.M. | WATER TEST NO. 79-12-60 | SPECIFIC GRAVITY 1.0 |

EB-15.1

Fig. 4BII (30)

CENTRILIFT®
SUBMERSIBLE PUMPS

D-225B
ONE STAGE
PERFORMANCE CURVE

"50 HERTZ"

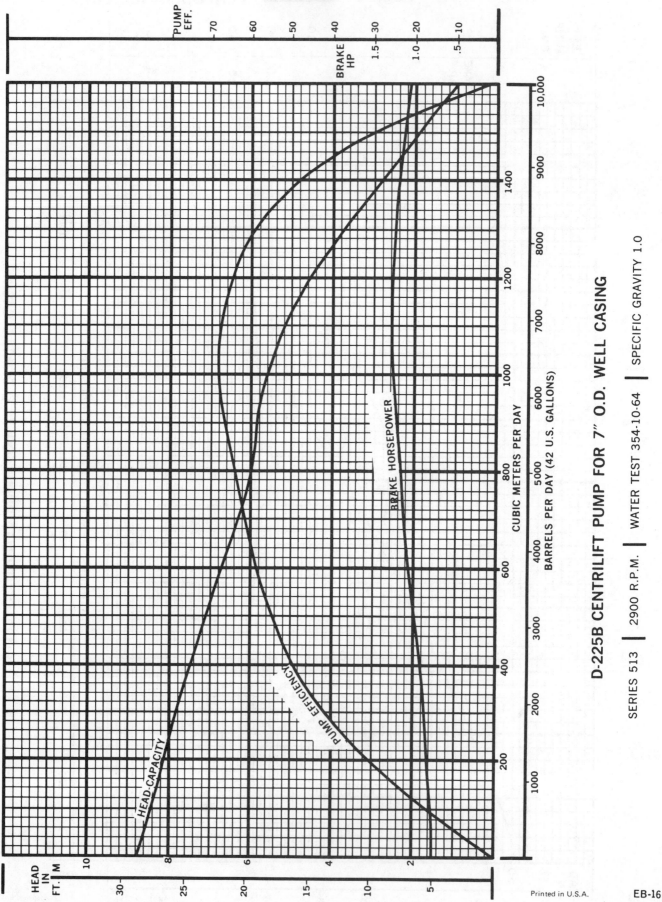

D-225B CENTRILIFT PUMP FOR 7" O.D. WELL CASING

| SERIES 513 | 2900 R.P.M. | WATER TEST 354-10-64 | SPECIFIC GRAVITY 1.0 |

5/78

Fig. 4BII (31)

Printed in U.S.A.

EB-16

CENTRILIFT® SUBMERSIBLE PUMPS

D-225B ONE STAGE PERFORMANCE CURVE

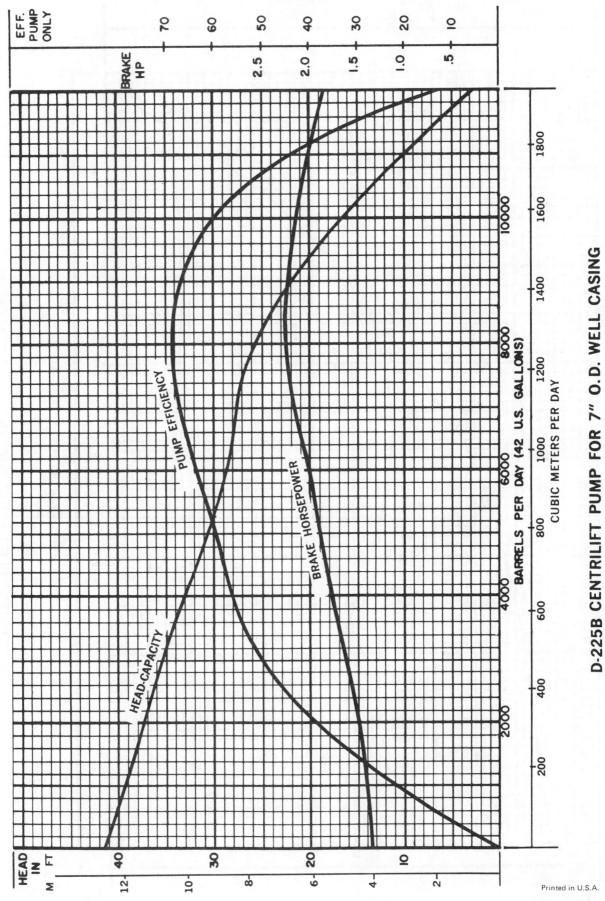

D-225B CENTRILIFT PUMP FOR 7" O.D. WELL CASING

SERIES 513 | 3475 R.P.M. | WATER TEST NO. 354-10-64 | SPECIFIC GRAVITY 1.0

EB-16.1

Fig. 4BII (32)

Printed in U.S.A.   5/78

CENTRILIFT® SUBMERSIBLE PUMPS

A-177 ONE STAGE PERFORMANCE CURVE

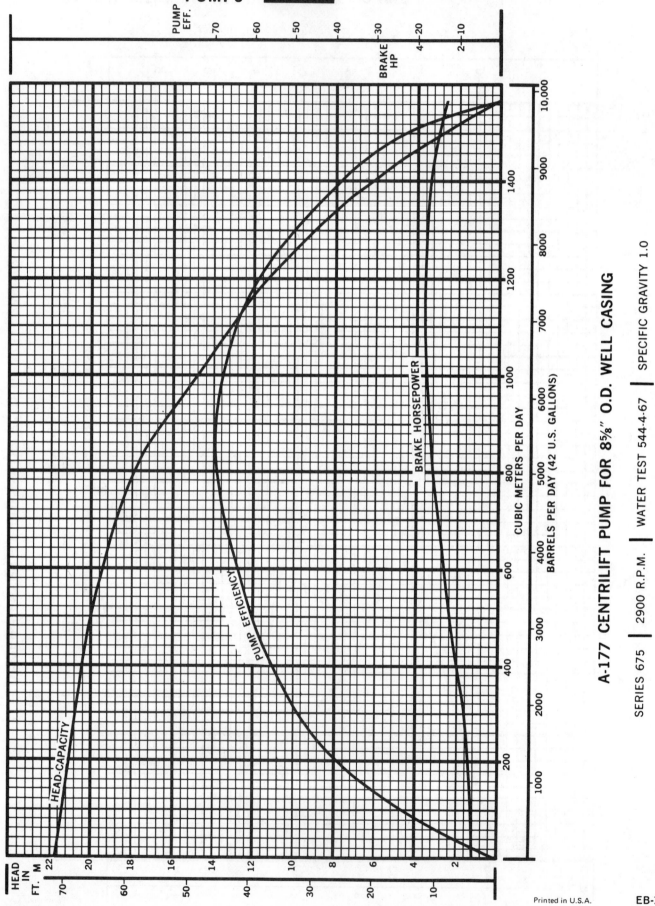

A-177 CENTRILIFT PUMP FOR 8⅝" O.D. WELL CASING

| SERIES 675 | 2900 R.P.M. | WATER TEST 544-4-67 | SPECIFIC GRAVITY 1.0 |

5/78

Fig. 4BII (33)

Printed in U.S.A.　　EB-17

CENTRILIFT®
SUBMERSIBLE PUMPS

A-177 ONE STAGE
PERFORMANCE CURVE

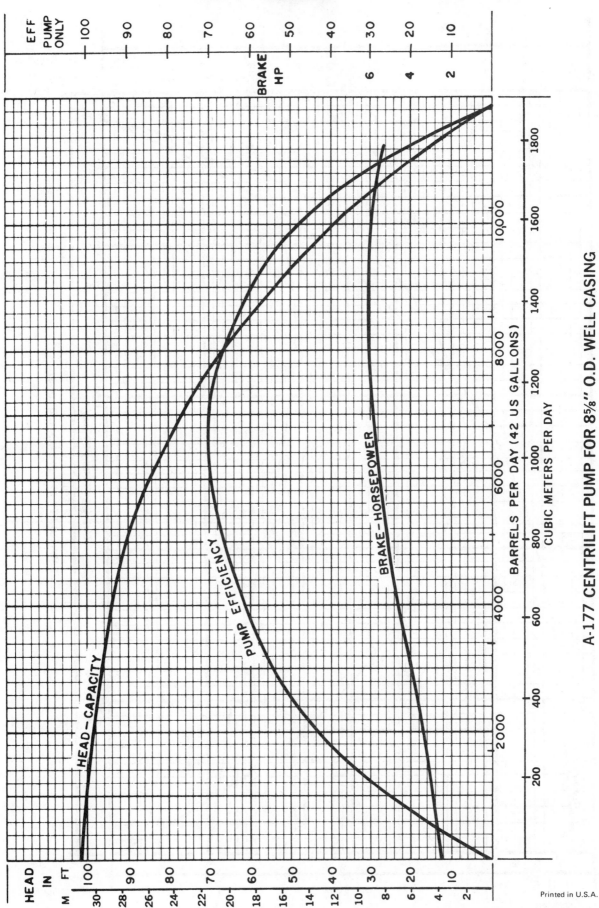

A-177 CENTRILIFT PUMP FOR 8⅝" O.D. WELL CASING

SERIES 675 | 3475 R.P.M. | WATER TEST NO. 544-4-67 | SPECIFIC GRAVITY 1.0

EB-17.1

Fig. 4BII (34)

CENTRILIFT® SUBMERSIBLE PUMPS

P-320A ONE STAGE PERFORMANCE CURVE

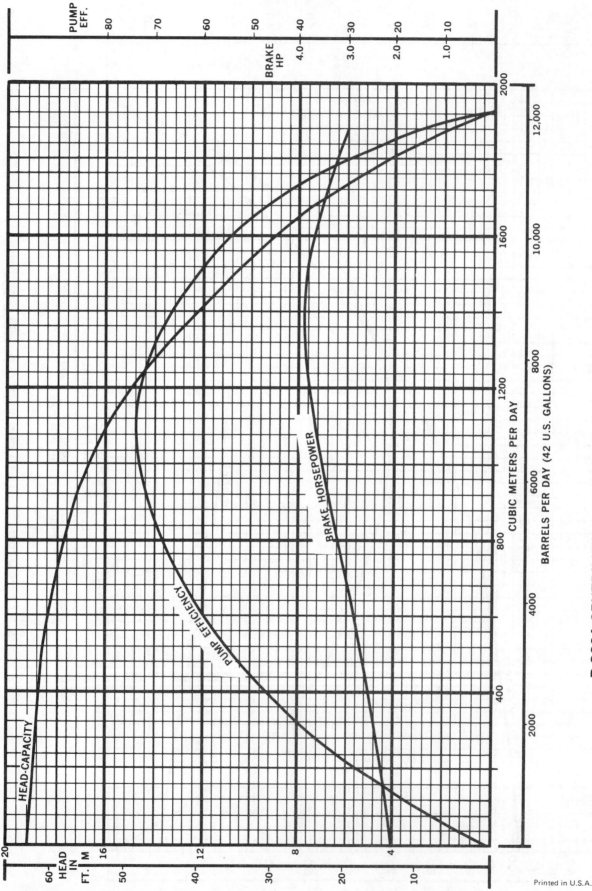

P-320A CENTRILIFT PUMP FOR 8⅝" O.D. WELL CASING

| SERIES 675 | 2900 R.P.M. | WATER TEST 381-1-65 | SPECIFIC GRAVITY 1.0 |

Fig. 4BII (35)

5/78

Printed in U.S.A.   EB-18

CENTRILIFT® SUBMERSIBLE PUMPS

P-320A ONE STAGE PERFORMANCE CURVE

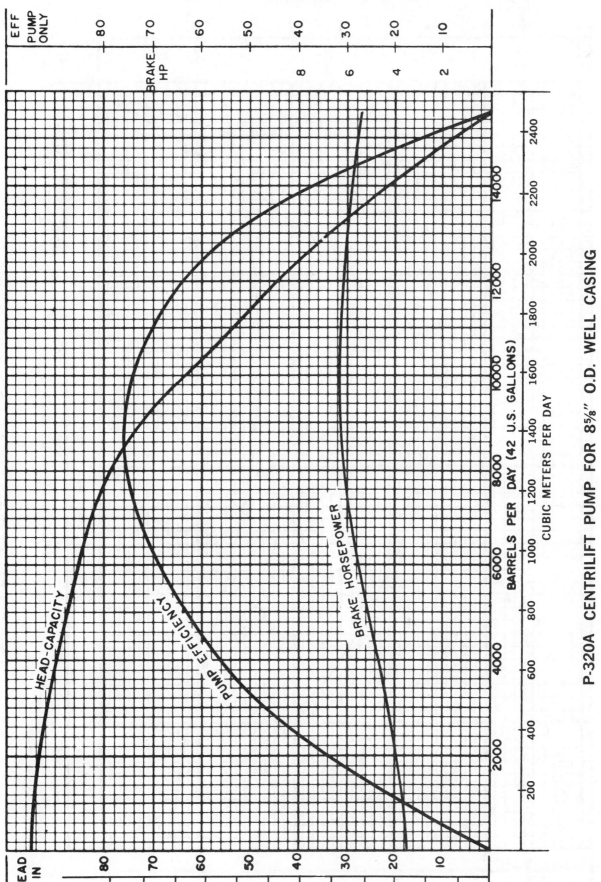

P-320A CENTRILIFT PUMP FOR 8⅝" O.D. WELL CASING

| SERIES 675 | 3475 R.P.M. | WATER TEST NO. 385-1-65 | SPECIFIC GRAVITY 1.0 |

EB-18.1

Fig. 4BII (36)

CENTRILIFT®
SUBMERSIBLE PUMPS

R-330
ONE STAGE
PERFORMANCE CURVE

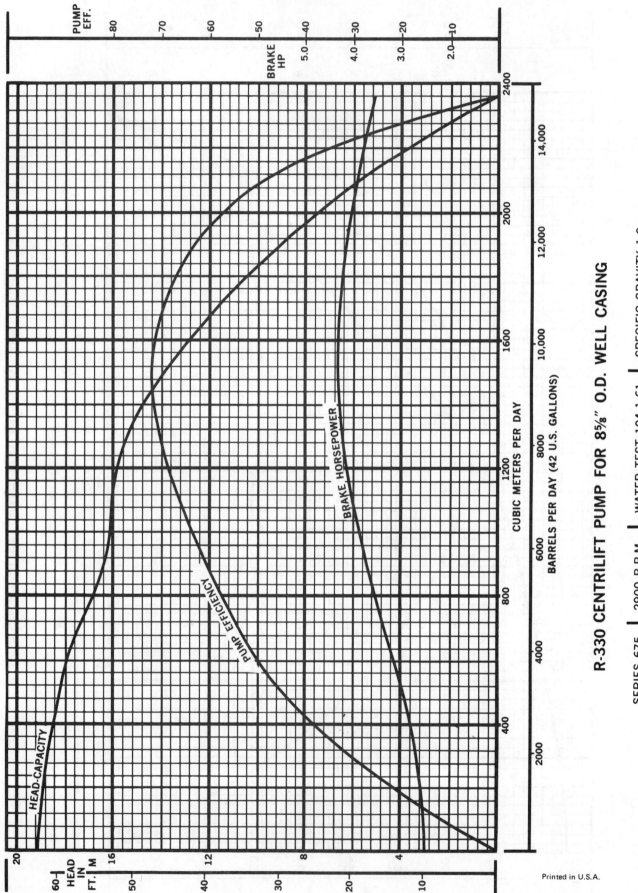

R-330 CENTRILIFT PUMP FOR 8⅝" O.D. WELL CASING

| SERIES 675 | 2900 R.P.M. | WATER TEST 104-1-61 | SPECIFIC GRAVITY 1.0 |

HEAD-CAPACITY

PUMP EFFICIENCY

BRAKE HORSEPOWER

CENTRILIFT® SUBMERSIBLE PUMPS

R-330 ONE STAGE PERFORMANCE CURVE

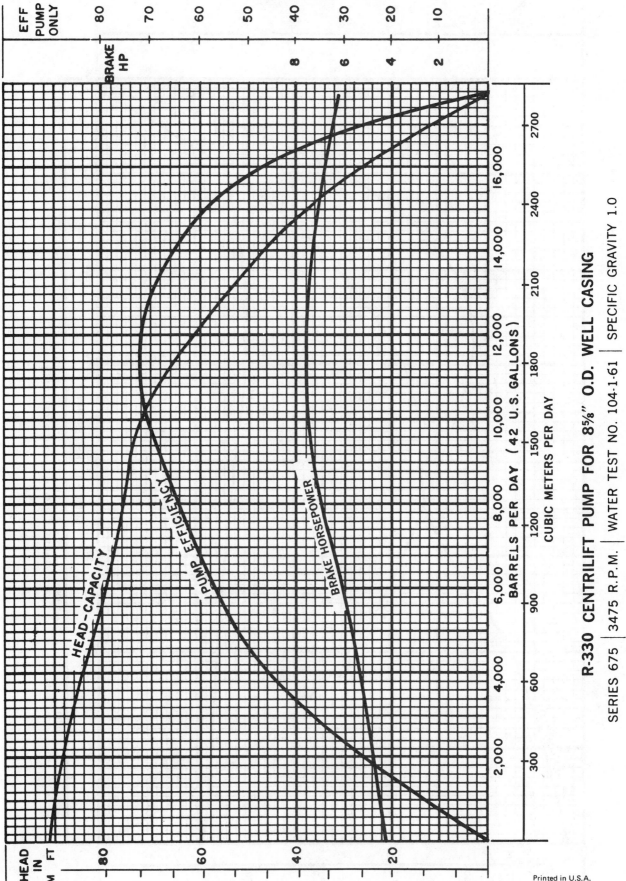

R-330 CENTRILIFT PUMP FOR 8⅝" O.D. WELL CASING

SERIES 675 | 3475 R.P.M. | WATER TEST NO. 104-1-61 | SPECIFIC GRAVITY 1.0

EB-19.1

Fig. 4BII (38)

CENTRILIFT® SUBMERSIBLE PUMPS

L-500 ONE STAGE PERFORMANCE CURVE

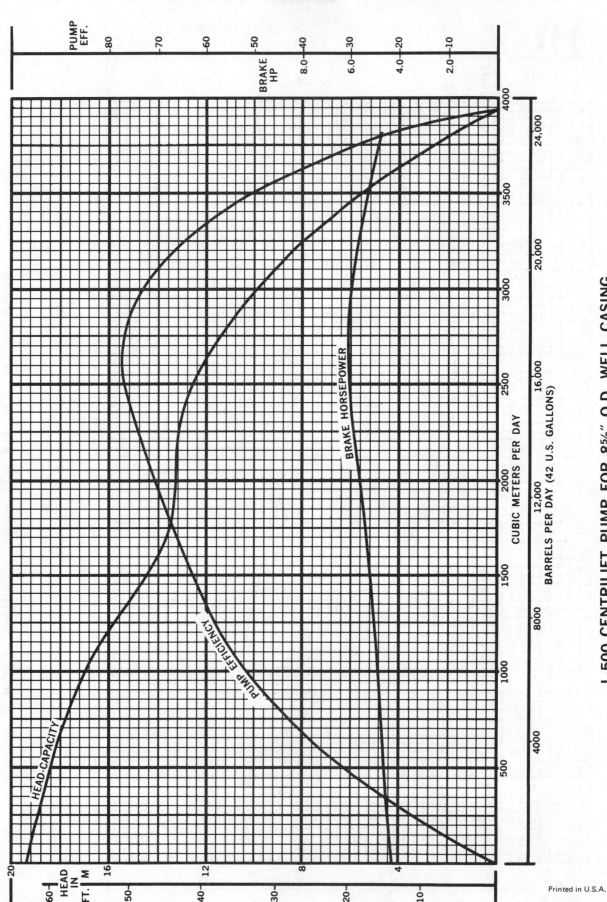

L-500 CENTRILIFT PUMP FOR 8⅝" O.D. WELL CASING

| SERIES 675 | 2900 R.P.M. | WATER TEST 390-1-65 | SPECIFIC GRAVITY 1.0 |

5/78

Fig. 4BII (39)

Printed in U.S.A.   EB-20

CENTRILIFT® SUBMERSIBLE PUMPS

L-500 ONE STAGE PERFORMANCE CURVE

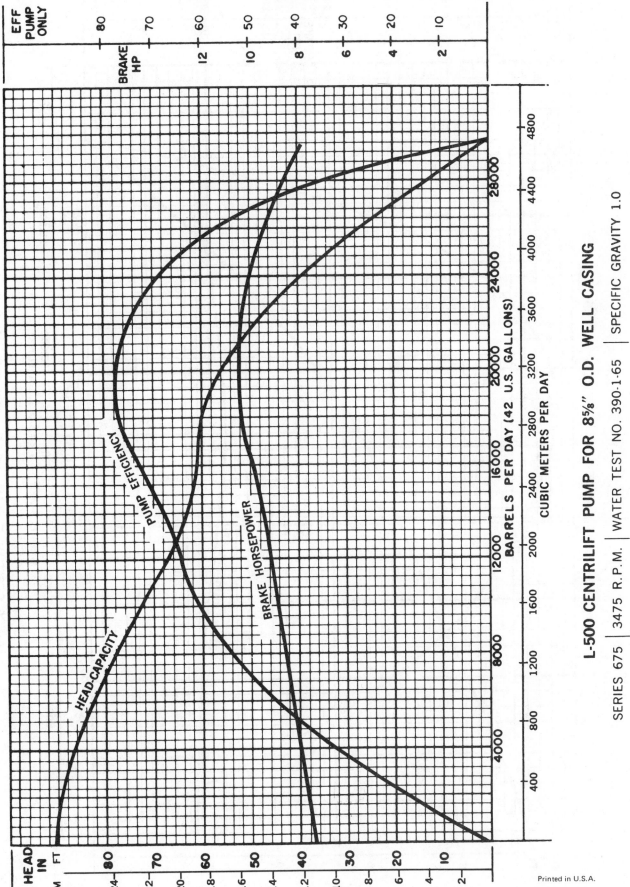

L-500 CENTRILIFT PUMP FOR 8⅝" O.D. WELL CASING

SERIES 675 | 3475 R.P.M. | WATER TEST NO. 390-1-65 | SPECIFIC GRAVITY 1.0

EFF PUMP ONLY

BRAKE HP

BARRELS PER DAY (42 U.S. GALLONS)

CUBIC METERS PER DAY

PUMP EFFICIENCY

HEAD-CAPACITY

BRAKE HORSEPOWER

HEAD IN FT / M

Printed in U.S.A.   5/78

Fig. 4BII (40)

CENTRILIFT® SUBMERSIBLE PUMPS

IA-600 ONE STAGE PERFORMANCE CURVE

"50 HERTZ"

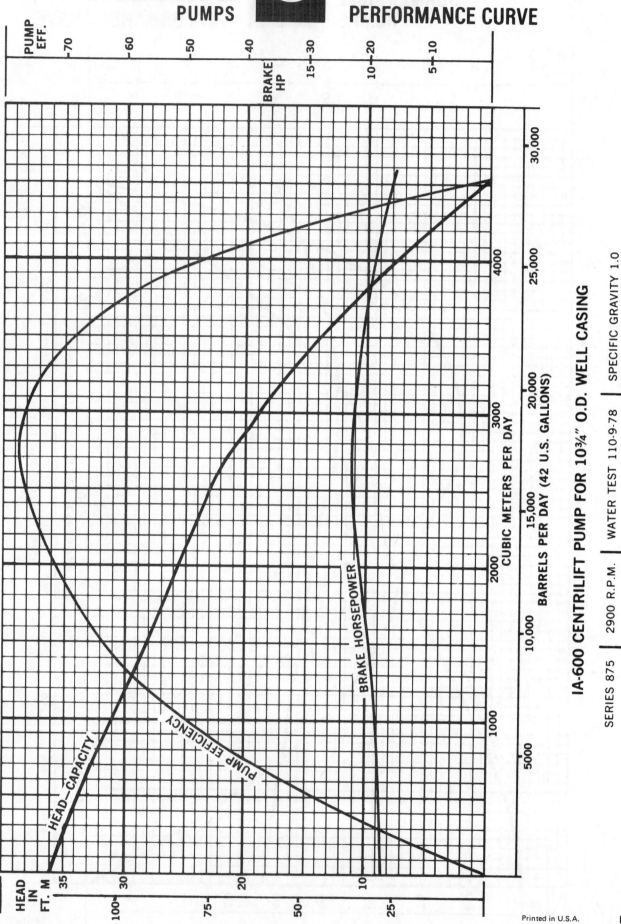

IA-600 CENTRILIFT PUMP FOR 10¾" O.D. WELL CASING

| SERIES 875 | 2900 R.P.M. | WATER TEST 110-9-78 | SPECIFIC GRAVITY 1.0 |

5/78

Fig. 4BII (41)

Printed in U.S.A.

EB-21

CENTRILIFT® SUBMERSIBLE PUMPS

IA-600 ONE STAGE PERFORMANCE CURVE

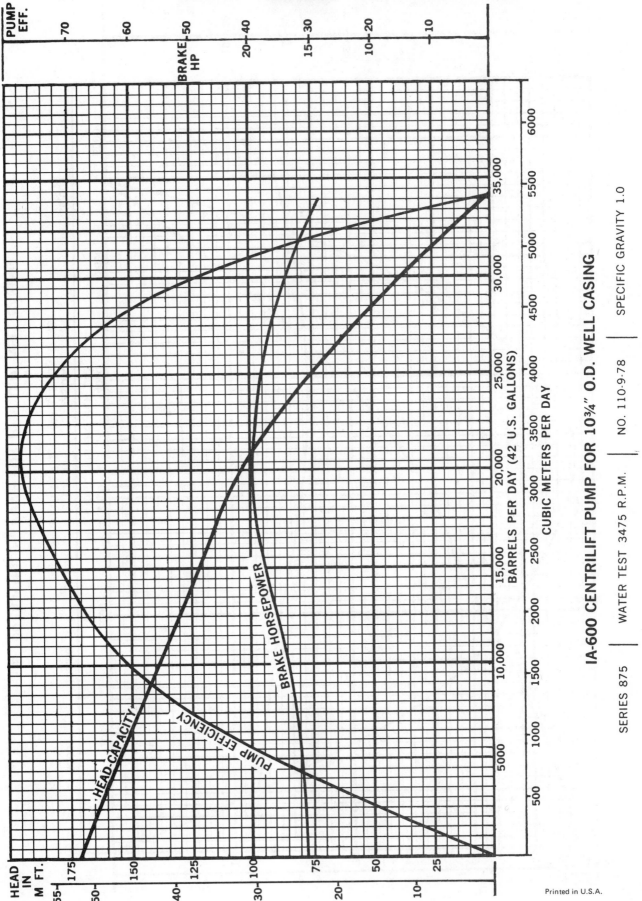

IA-600 CENTRILIFT PUMP FOR 10¾" O.D. WELL CASING

SERIES 875 | WATER TEST 3475 R.P.M. | NO. 110-9-78 | SPECIFIC GRAVITY 1.0

EB-21.1

Fig. 4BII (42)

Printed in U.S.A.   5/78

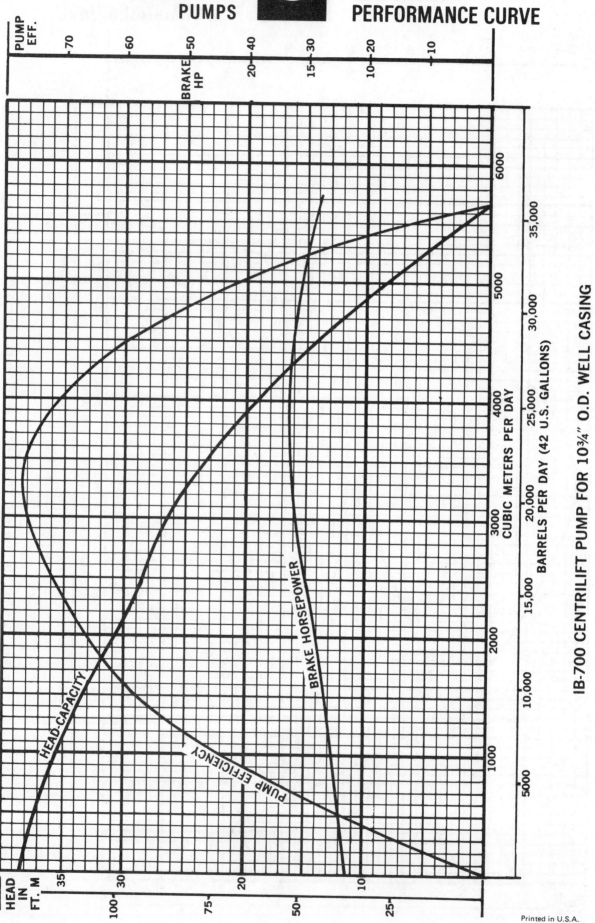

CENTRILIFT® SUBMERSIBLE PUMPS

IB-700 ONE STAGE PERFORMANCE CURVE

PUMP EFF.

— 70
— 60
BRAKE — 50
HP
— 40
— 30
— 20
— 10

— 20
— 15
— 10

HEAD-CAPACITY

BRAKE HORSEPOWER

PUMP EFFICIENCY

CUBIC METERS PER DAY

1000   2000   3000   4000   5000   6000

BARRELS PER DAY (42 U.S. GALLONS)

5000   10,000   15,000   20,000   25,000   30,000   35,000

IB-700 CENTRILIFT PUMP FOR 10¾" O.D. WELL CASING

| SERIES 875 | 2900 R.P.M. | WATER TEST 116-10-78 | SPECIFIC GRAVITY 1.0 |

HEAD IN FT. M

35
30
20
10

100
75
50
25

5/78

Fig. 4BII (43)

Printed in U.S.A.

EB-22

CENTRILIFT®
SUBMERSIBLE PUMPS

IB-700 ONE STAGE PERFORMANCE CURVE

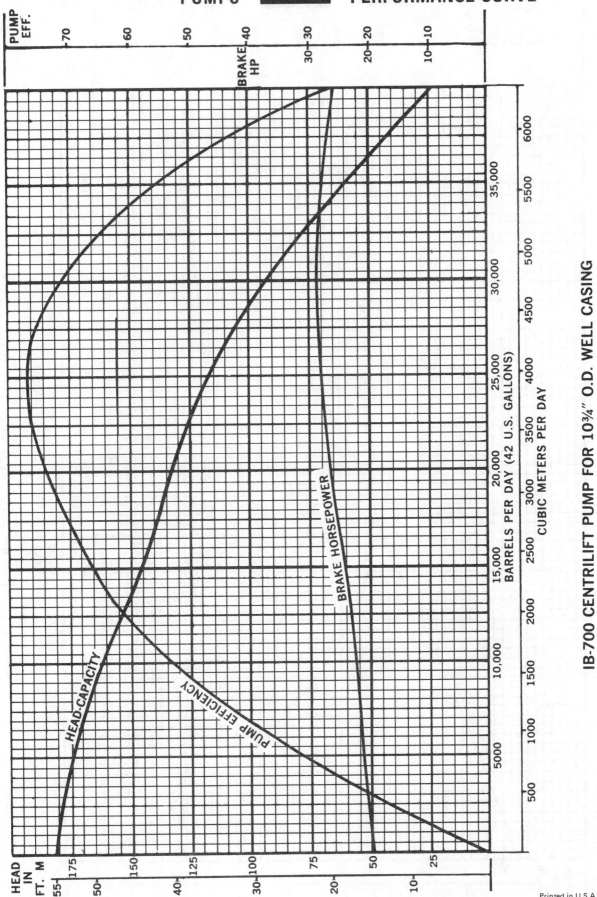

IB-700 CENTRILIFT PUMP FOR 10¾" O.D. WELL CASING

| SERIES 875 | WATER TEST 3475 R.P.M. | NO. 116-10-78 | SPECIFIC GRAVITY 1.0 |

EB-22.1

Fig. 4BII (44)

Printed in U.S.A.   5/78

CENTRILIFT® SUBMERSIBLE PUMPS

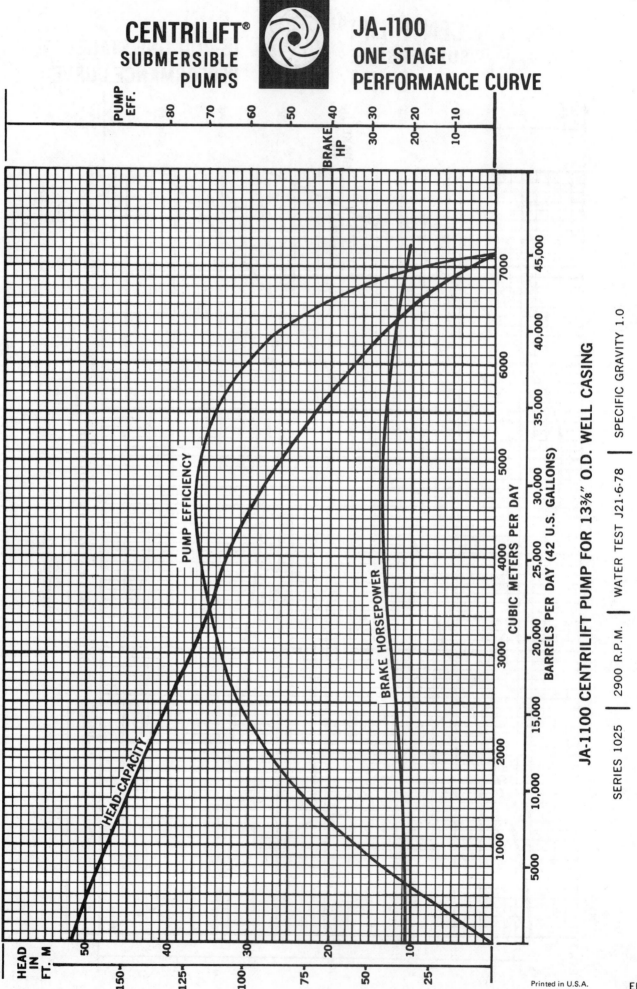

JA-1100 ONE STAGE PERFORMANCE CURVE

JA-1100 CENTRILIFT PUMP FOR 13⅜" O.D. WELL CASING

SERIES 1025 | 2900 R.P.M. | WATER TEST J21-6-78 | SPECIFIC GRAVITY 1.0

5/78

Fig. 4BII (45)

Printed in U.S.A.

EB-23

CENTRILIFT® SUBMERSIBLE PUMPS

JA-1100 ONE STAGE PERFORMANCE CURVE

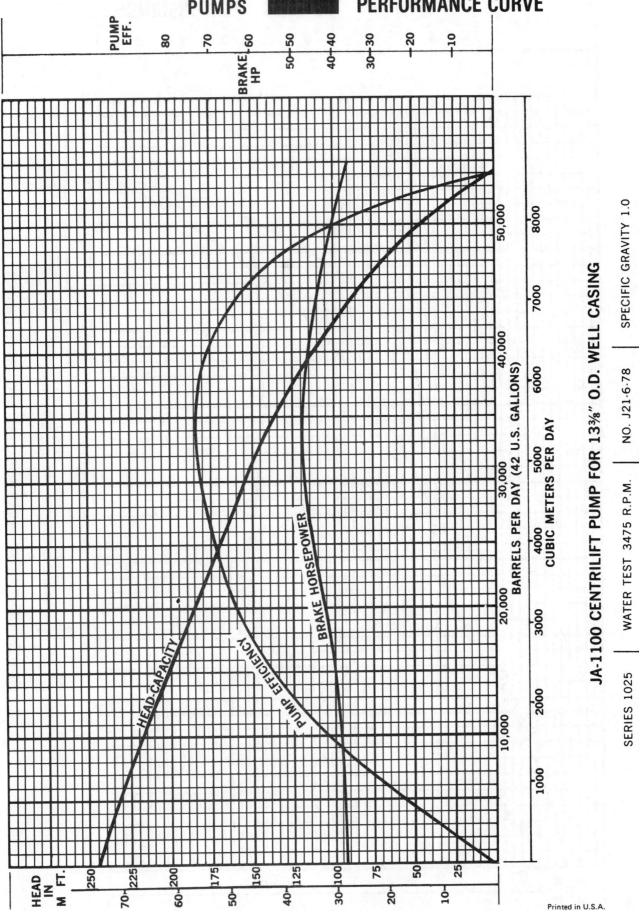

JA-1100 CENTRILIFT PUMP FOR 13³⁄₈" O.D. WELL CASING

SERIES 1025 | WATER TEST 3475 R.P.M. | NO. J21-6-78 | SPECIFIC GRAVITY 1.0

BARRELS PER DAY (42 U.S. GALLONS)

CUBIC METERS PER DAY

HEAD-CAPACITY

PUMP EFFICIENCY

BRAKE HORSEPOWER

PUMP EFF.

BRAKE HP

HEAD IN M FT.

CENTRILIFT® SUBMERSIBLE PUMPS

JB-1300 ONE STAGE PERFORMANCE CURVE

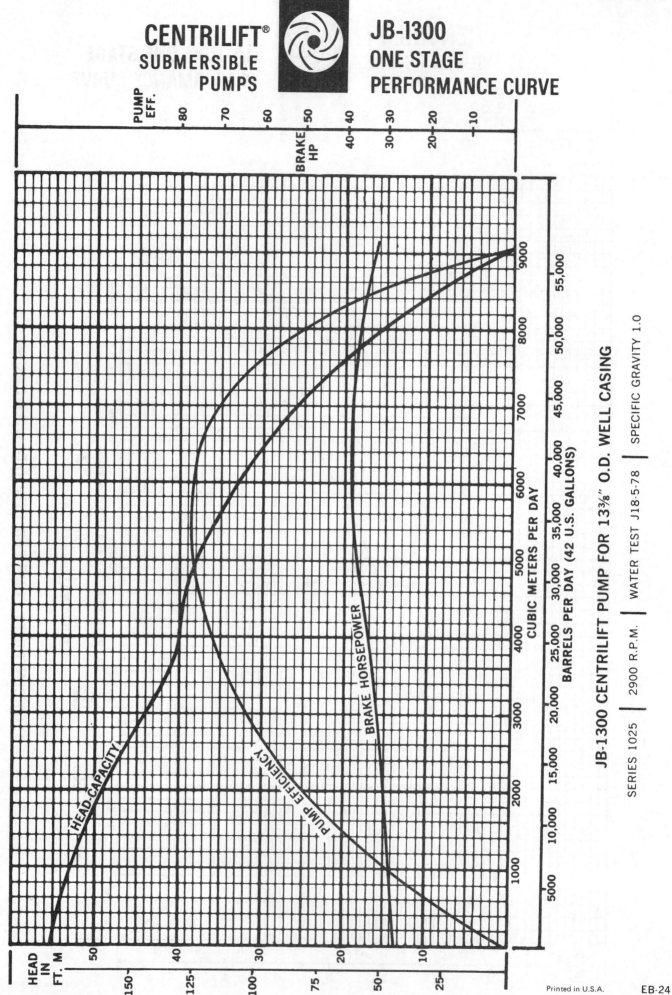

JB-1300 CENTRILIFT PUMP FOR 13³⁄₈" O.D. WELL CASING

| SERIES 1025 | 2900 R.P.M. | WATER TEST J18-5-78 | SPECIFIC GRAVITY 1.0 |

PUMP EFF.

BRAKE HP

HEAD-CAPACITY

PUMP EFFICIENCY

BRAKE HORSEPOWER

CUBIC METERS PER DAY

BARRELS PER DAY (42 U.S. GALLONS)

HEAD IN FT. M

CENTRILIFT® SUBMERSIBLE PUMPS

JB-1300 ONE STAGE PERFORMANCE CURVE

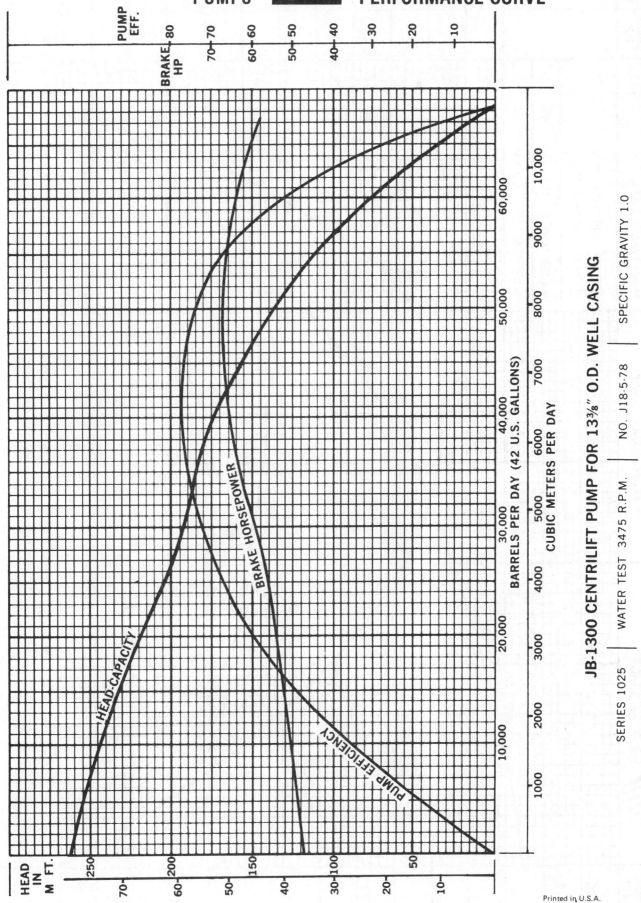

JB-1300 CENTRILIFT PUMP FOR 13⅜" O.D. WELL CASING

| SERIES 1025 | WATER TEST 3475 R.P.M. | NO. J18-5-78 | SPECIFIC GRAVITY 1.0 |

EB-24.1

Fig. 4BII (48)

Printed in U.S.A.   5/78

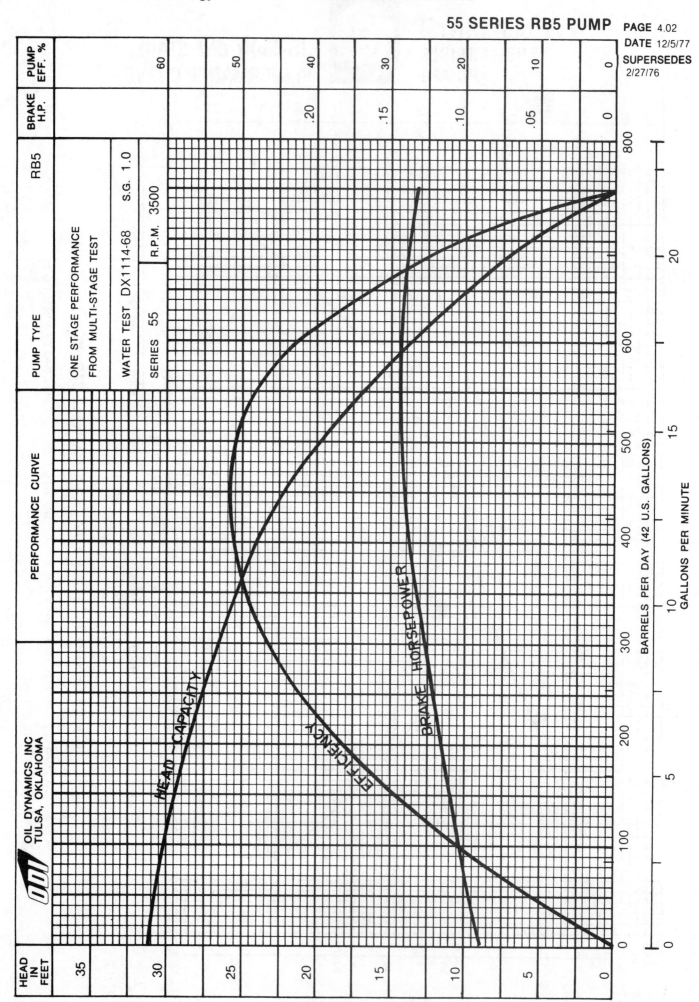

55 SERIES RB5 PUMP

Fig. 4BIII (1)

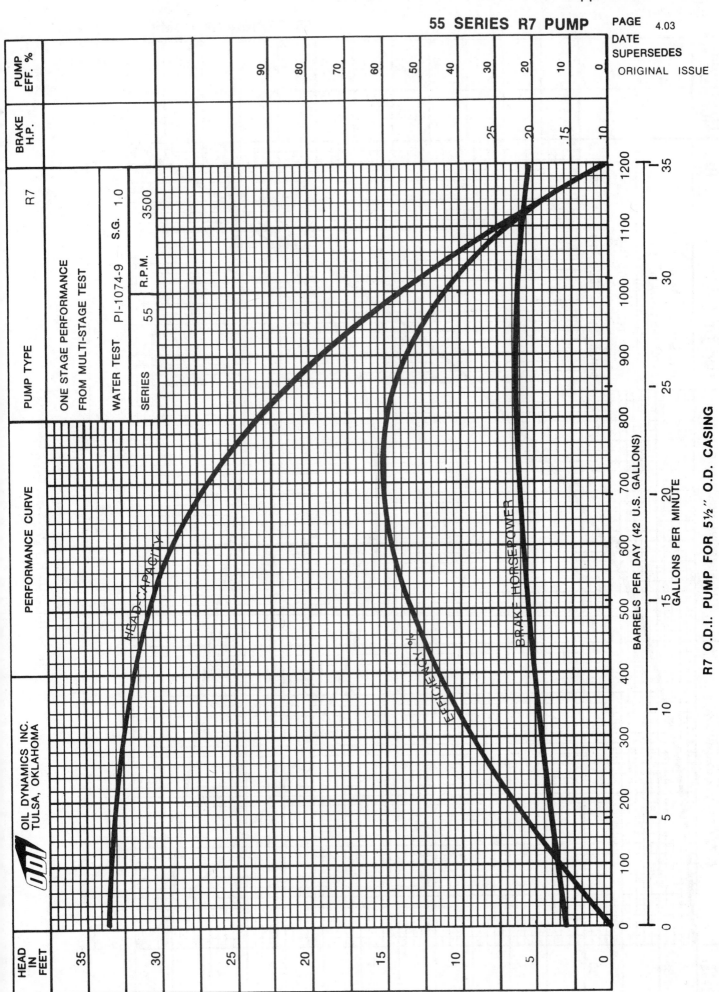

55 SERIES R7 PUMP

PAGE 4.03
DATE
SUPERSEDES
ORIGINAL ISSUE

| PUMP EFF. % | | 90 | 80 | 70 | 60 | 50 | 40 | 30 | 20 | 10 | 0 |

| BRAKE H.P. | | | | | | | 25 | | 20 | .15 | .10 |

OIL DYNAMICS INC.
TULSA, OKLAHOMA

PUMP TYPE — R7

PERFORMANCE CURVE

ONE STAGE PERFORMANCE
FROM MULTI-STAGE TEST

WATER TEST   PI-1074-9   S.G. 1.0

SERIES   55   R.P.M.   3500

HEAD-CAPACITY

EFFICIENCY

BRAKE-HORSEPOWER

BARRELS PER DAY (42 U.S. GALLONS)

GALLONS PER MINUTE

R7 O.D.I. PUMP FOR 5½" O.D. CASING

HEAD IN FEET: 35, 30, 25, 20, 15, 10, 5, 0

Fig. 4BIII (2)

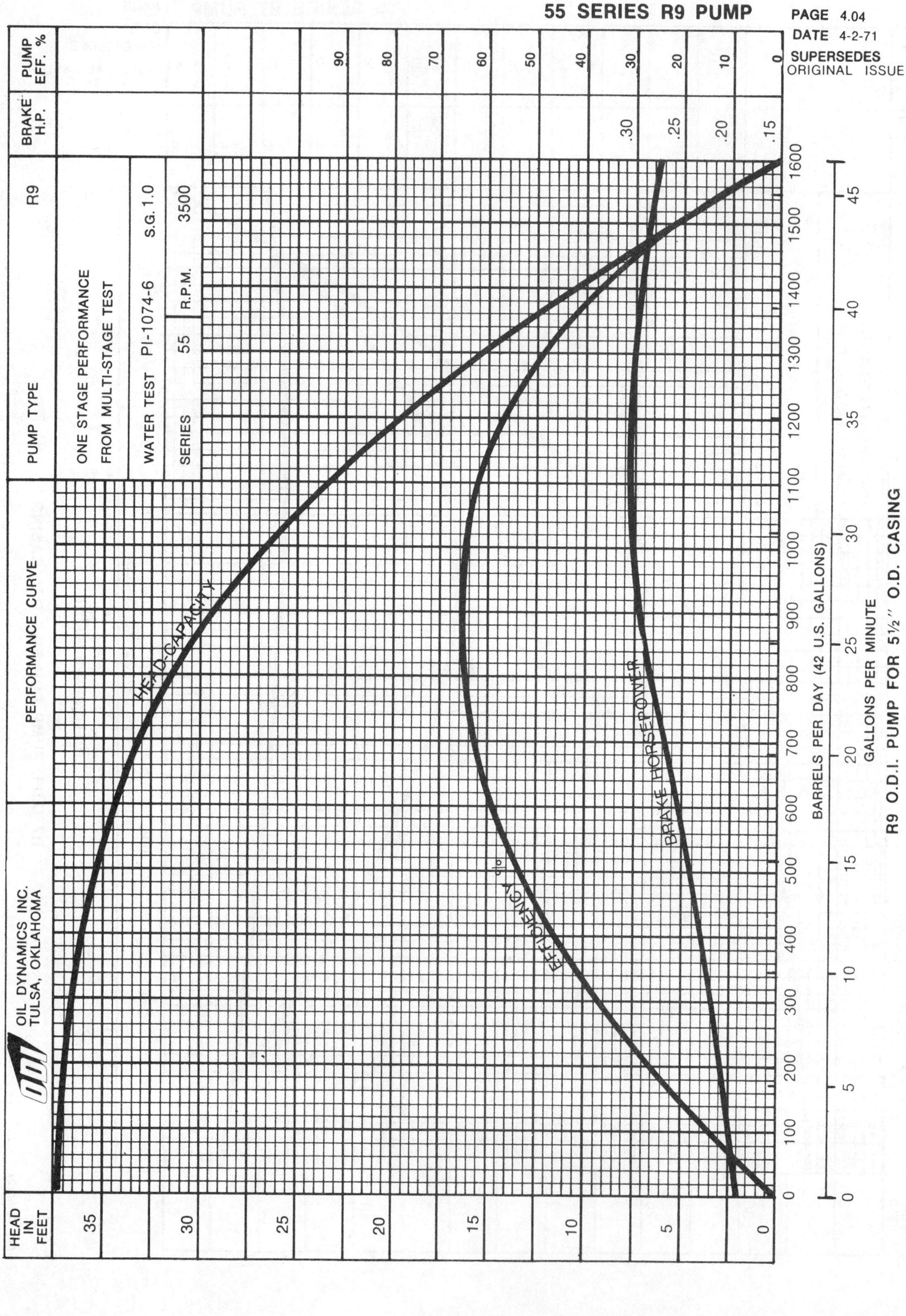

55 SERIES R9 PUMP

PAGE 4.04
DATE 4-2-71
SUPERSEDES
ORIGINAL ISSUE

OIL DYNAMICS INC.
TULSA, OKLAHOMA

PUMP TYPE	R9	BRAKE H.P.	PUMP EFF. %

ONE STAGE PERFORMANCE FROM MULTI-STAGE TEST

WATER TEST   PI-1074-6   S.G. 1.0

SERIES 55   R.P.M. 3500

PERFORMANCE CURVE

HEAD-CAPACITY

EFFICIENCY

BRAKE HORSEPOWER

BARRELS PER DAY (42 U.S. GALLONS)

GALLONS PER MINUTE

R9 O.D.I. PUMP FOR 5½" O.D. CASING

HEAD IN FEET

Fig. 4BIII (3)

55 SERIES RA12 PUMP

PAGE 4.05
DATE 4-2-71
SUPERSEDES
ORIGINAL ISSUE

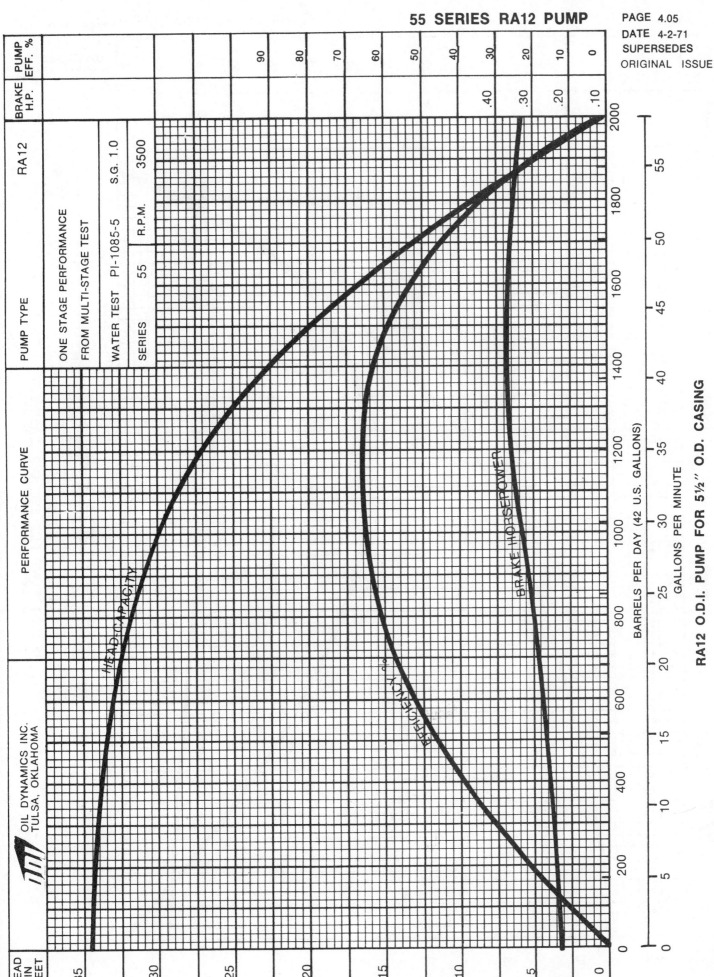

OIL DYNAMICS INC.
TULSA, OKLAHOMA

PERFORMANCE CURVE

PUMP TYPE	RA12
ONE STAGE PERFORMANCE FROM MULTI-STAGE TEST	
WATER TEST PI-1085-5	S.G. 1.0
SERIES 55	R.P.M. 3500

HEAD CAPACITY

EFFICIENCY %

BRAKE HORSEPOWER

RA12 O.D.I. PUMP FOR 5½" O.D. CASING

BARRELS PER DAY (42 U.S. GALLONS)

GALLONS PER MINUTE

Fig. 4BIII (4)

55 SERIES RA16 PUMP

PAGE 4.06
DATE 4-2-71
SUPERSEDES
ORIGINAL ISSUE

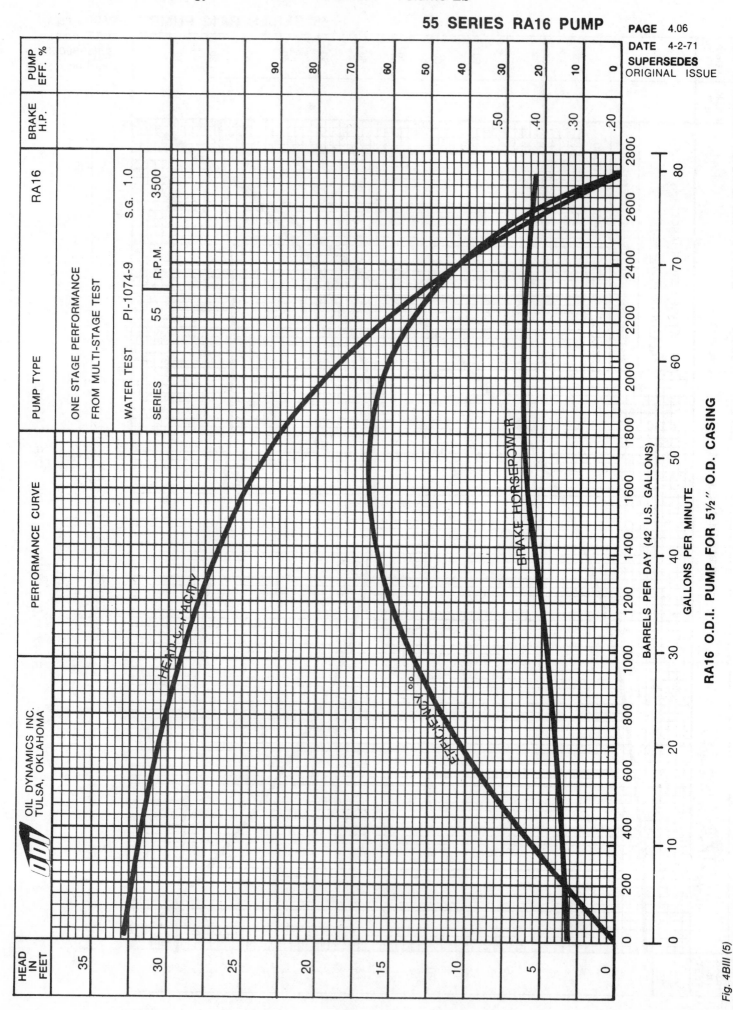

RA16 O.D.I. PUMP FOR 5½" O.D. CASING

Fig. 4BIII (5)

55 SERIES RA22 PUMP

PAGE 4.07.1
DATE 1-11-79
SUPERSEDES
5-12-71

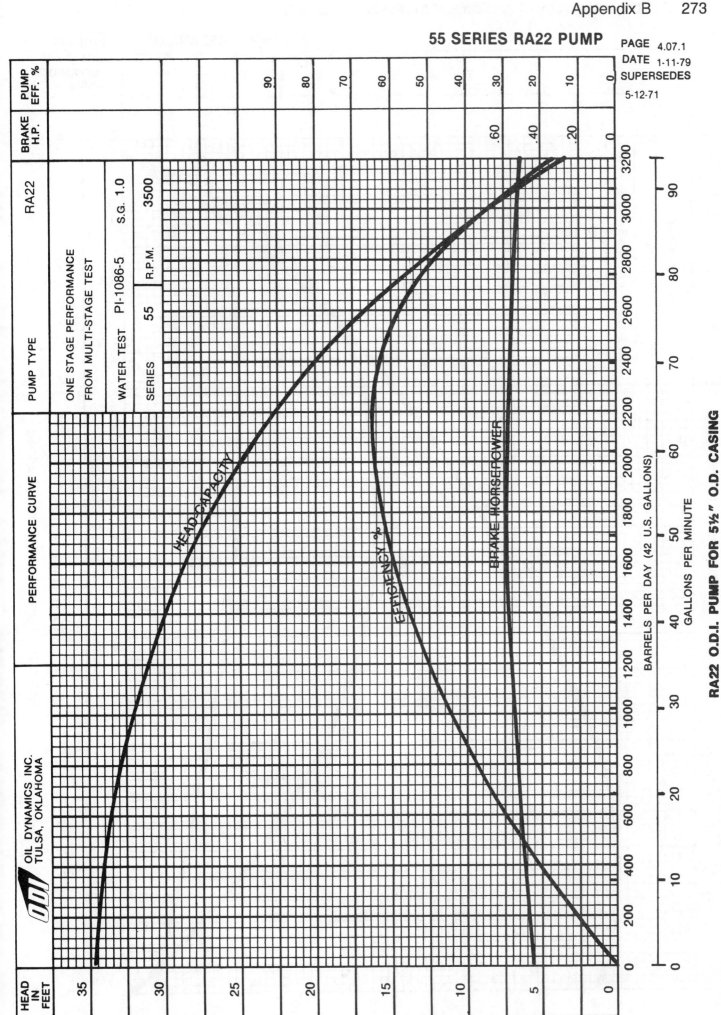

RA22 O.D.I. PUMP FOR 5½" O.D. CASING

Fig. 4BIII (6)

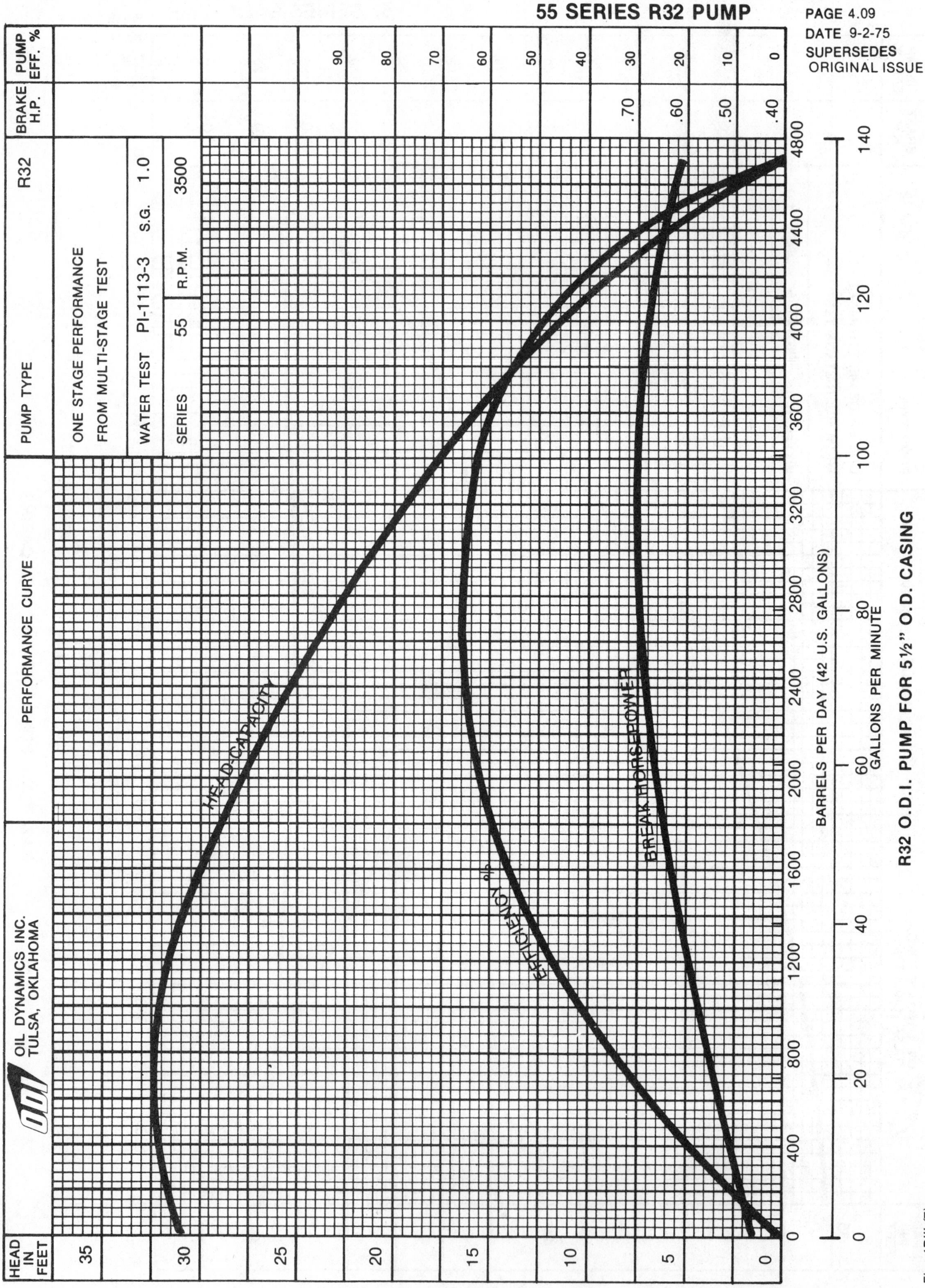

55 SERIES R32 PUMP
PAGE 4.09
DATE 9-2-75
SUPERSEDES
ORIGINAL ISSUE
OIL DYNAMICS INC.
TULSA, OKLAHOMA
PERFORMANCE CURVE
PUMP TYPE
R32
ONE STAGE PERFORMANCE
FROM MULTI-STAGE TEST
WATER TEST
PI-1113-3
S.G.
1.0
SERIES
55
R.P.M.
3500
HEAD IN FEET
BRAKE H.P.
PUMP EFF. %
HEAD-CAPACITY
EFFICIENCY %
BREAK HORSEPOWER
BARRELS PER DAY (42 U.S. GALLONS)
GALLONS PER MINUTE
R32 O.D.I. PUMP FOR 5½" O.D. CASING

Fig. 4BIII (7)

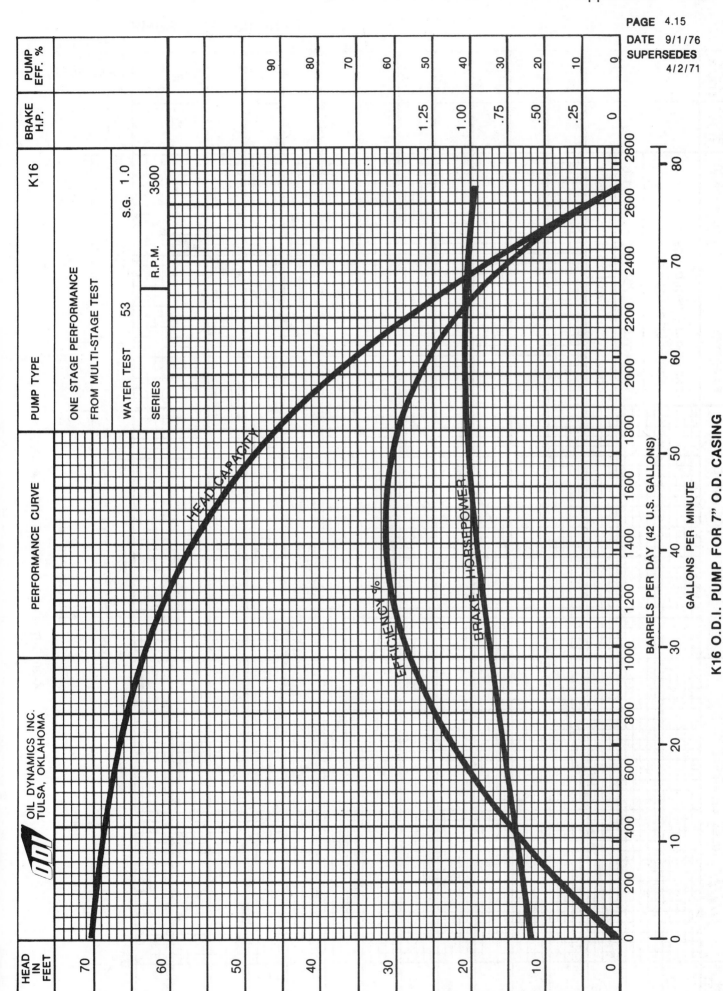

PAGE 4.15
DATE 9/1/76
SUPERSEDES
4/2/71

K16 O.D.I. PUMP FOR 7" O.D. CASING

Fig. 4BIII (8)

70 SERIES K-20 PUMP

PAGE 4.16
DATE 4-29-77
SUPERSEDES 4-2-71
(K20A)

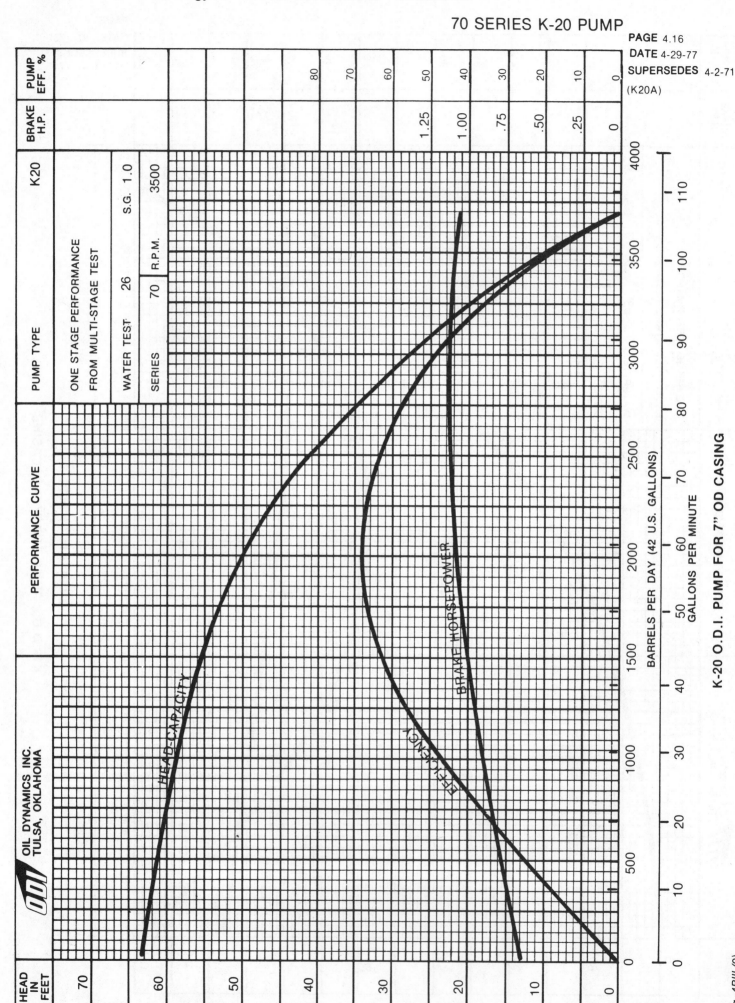

OIL DYNAMICS INC.
TULSA, OKLAHOMA

PERFORMANCE CURVE

PUMP TYPE

ONE STAGE PERFORMANCE
FROM MULTI-STAGE TEST

WATER TEST   26   S.G.   1.0

SERIES   70   R.P.M.   3500

K20

HEAD-CAPACITY

BRAKE HORSEPOWER

EFFICIENCY

BARRELS PER DAY (42 U.S. GALLONS)

GALLONS PER MINUTE

K-20 O.D.I. PUMP FOR 7" OD CASING

Fig. 4BIII (9)

70 SERIES K28 PUMP

PAGE 4.17.1
DATE 8-1-78
SUPERSEDES
ORIGINAL ISSUE

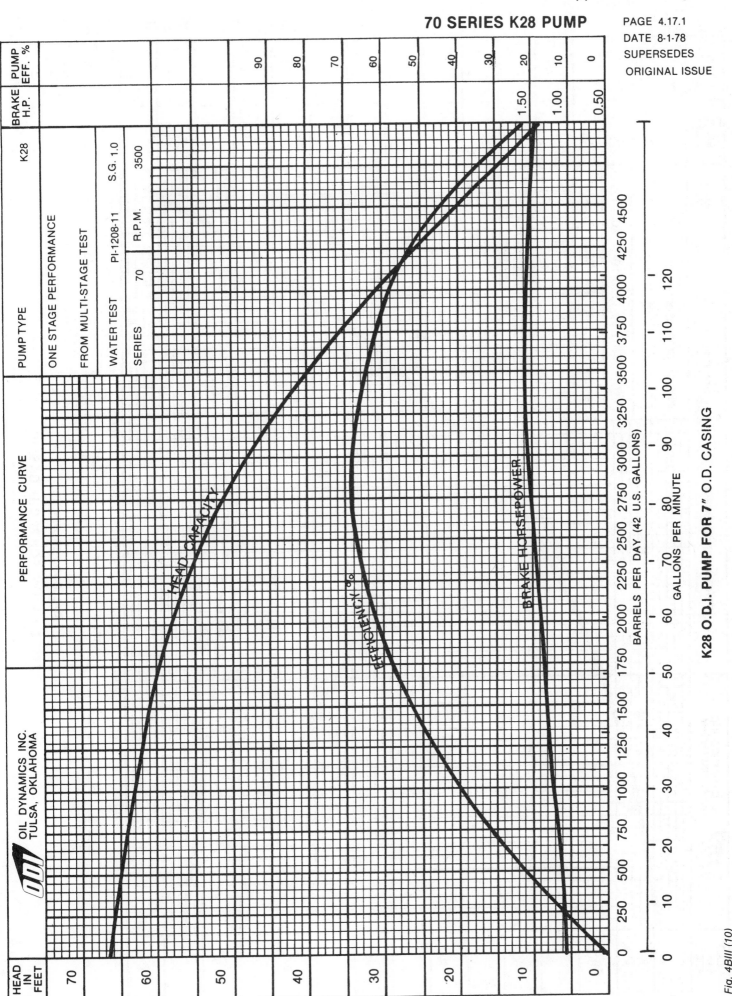

OIL DYNAMICS INC.
TULSA, OKLAHOMA

PERFORMANCE CURVE

PUMP TYPE

ONE STAGE PERFORMANCE

FROM MULTI-STAGE TEST

| WATER TEST | PI-1208-11 | S.G. 1.0 |
| SERIES | 70 | R.P.M. 3500 |

K28

HEAD CAPACITY

EFFICIENCY %

BRAKE HORSEPOWER

K28 O.D.I. PUMP FOR 7" O.D. CASING

BARRELS PER DAY (42 U.S. GALLONS)

GALLONS PER MINUTE

Fig. 4BIII (10)

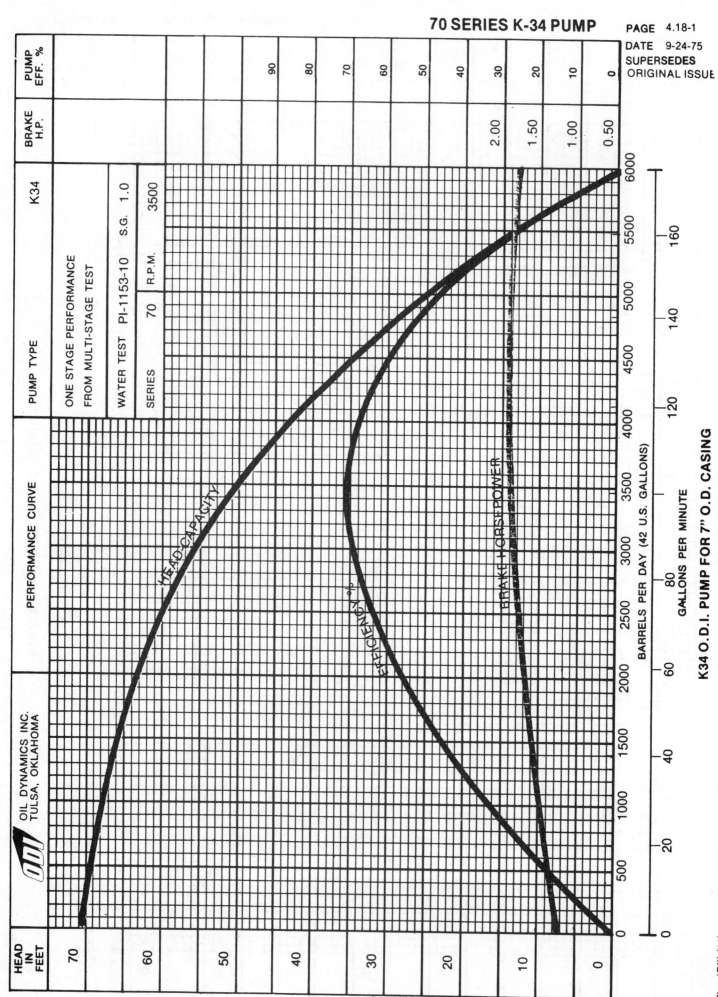

70 SERIES K-34 PUMP

PAGE 4.18-1
DATE 9-24-75
SUPERSEDES
ORIGINAL ISSUE

K34 O.D.I. PUMP FOR 7" O.D. CASING

Fig. 4BIII (11)

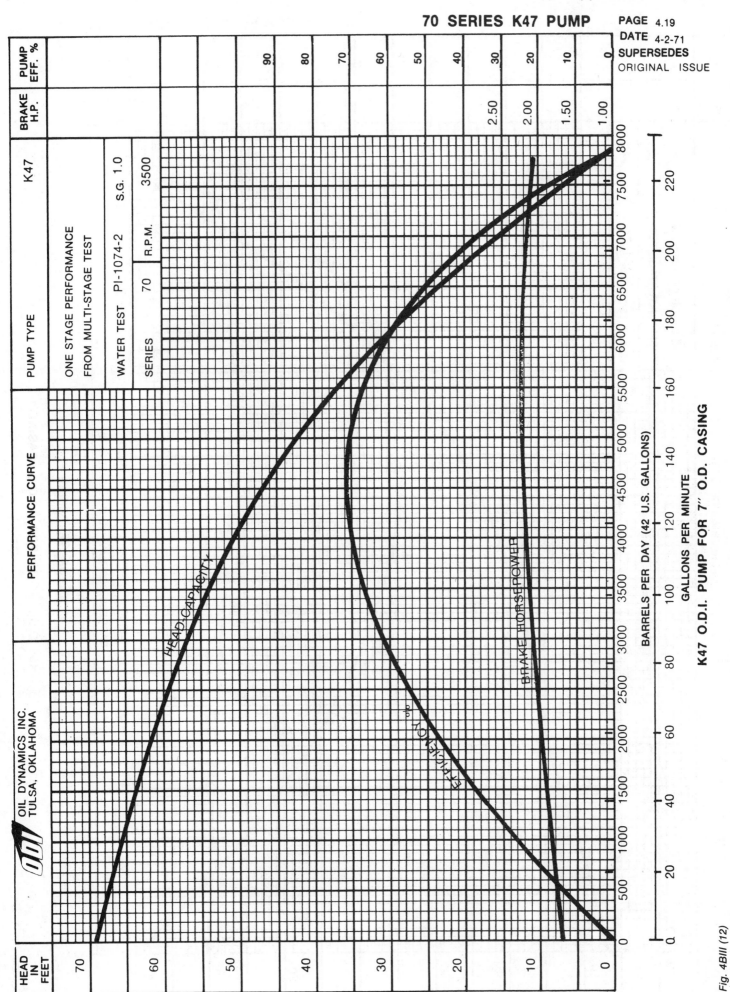

70 SERIES K47 PUMP

PAGE 4.19
DATE 4-2-71
SUPERSEDES
ORIGINAL ISSUE

K47 O.D.I. PUMP FOR 7" O.D. CASING

Fig. 4BIII (12)

70 SERIES K60 PUMP

PAGE   4.20
DATE   4-2-71
SUPERSEDES
ORIGINAL ISSUE

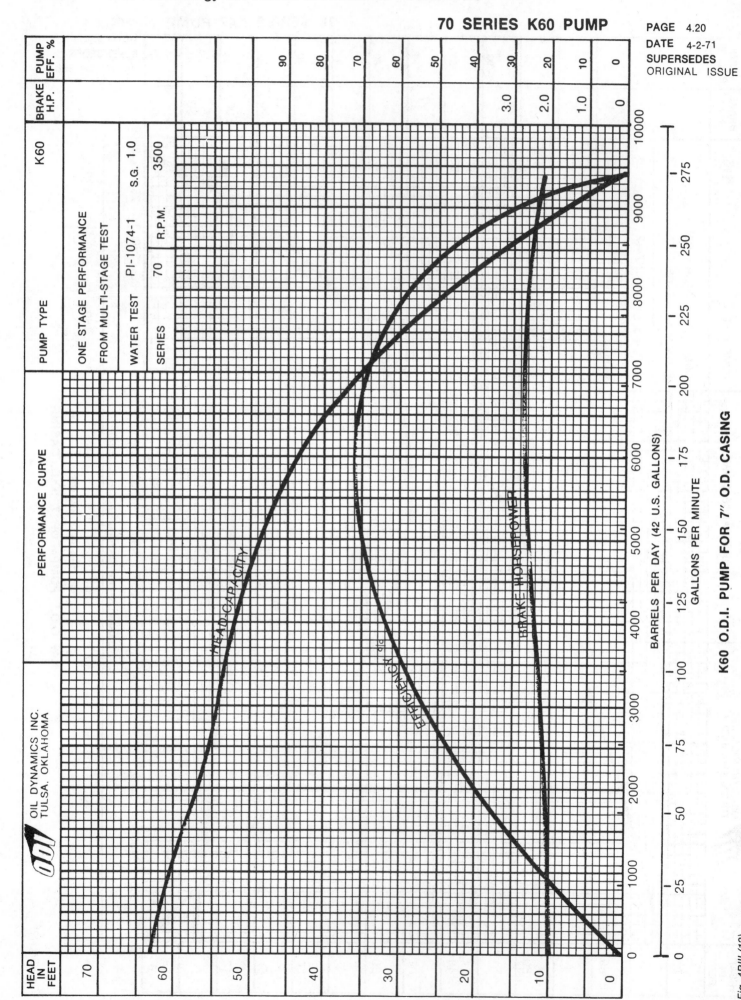

Fig. 4BIII (13)

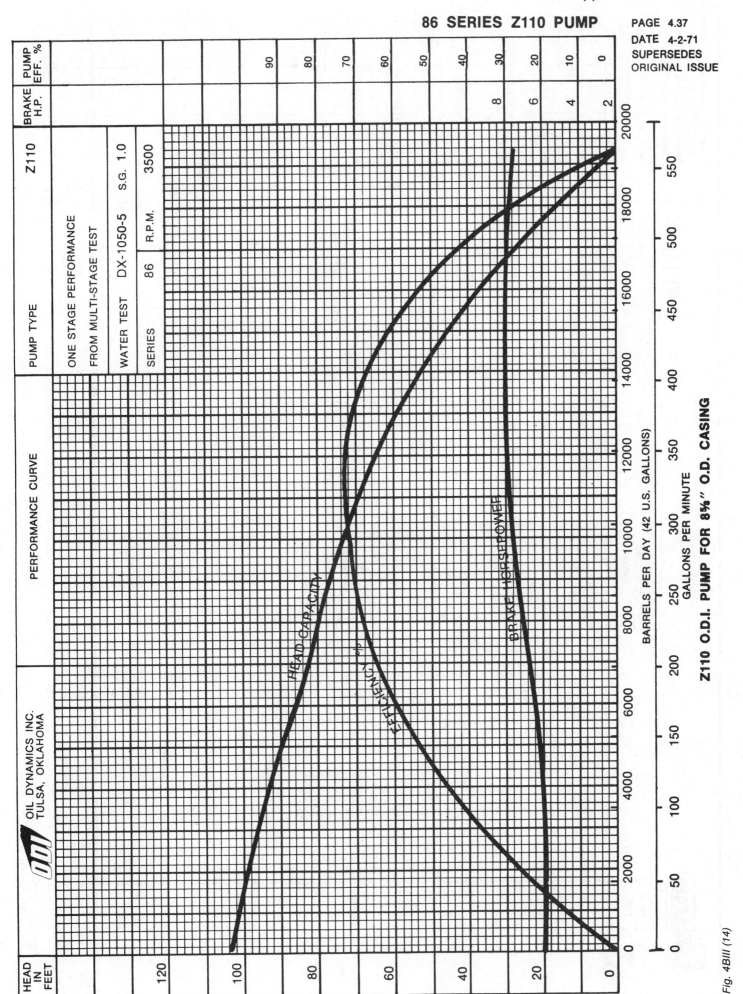

86 SERIES Z110 PUMP

PAGE 4.37
DATE 4-2-71
SUPERSEDES
ORIGINAL ISSUE

OIL DYNAMICS INC.
TULSA, OKLAHOMA

	Z110
PUMP TYPE	
ONE STAGE PERFORMANCE	
FROM MULTI-STAGE TEST	
WATER TEST	DX-1050-5	S.G. 1.0
SERIES	86	R.P.M.	3500

PERFORMANCE CURVE

HEAD CAPACITY

EFFICIENCY

BRAKE HORSEPOWER

BARRELS PER DAY (42 U.S. GALLONS)

GALLONS PER MINUTE

Z110 O.D.I. PUMP FOR 8⅝" O.D. CASING

Fig. 4BIII (14)

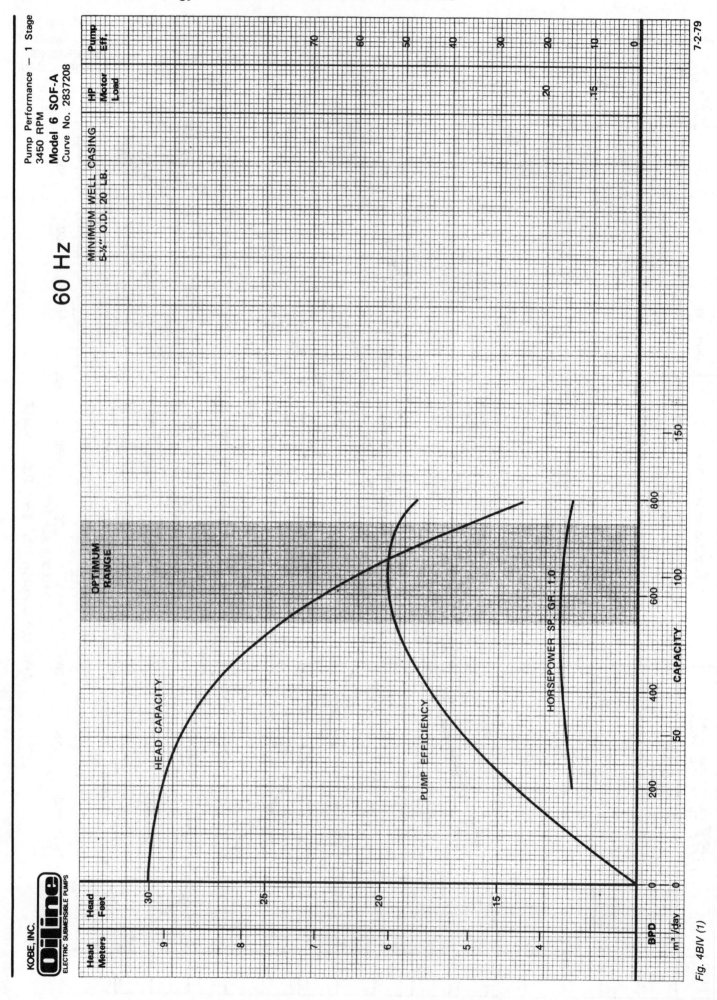

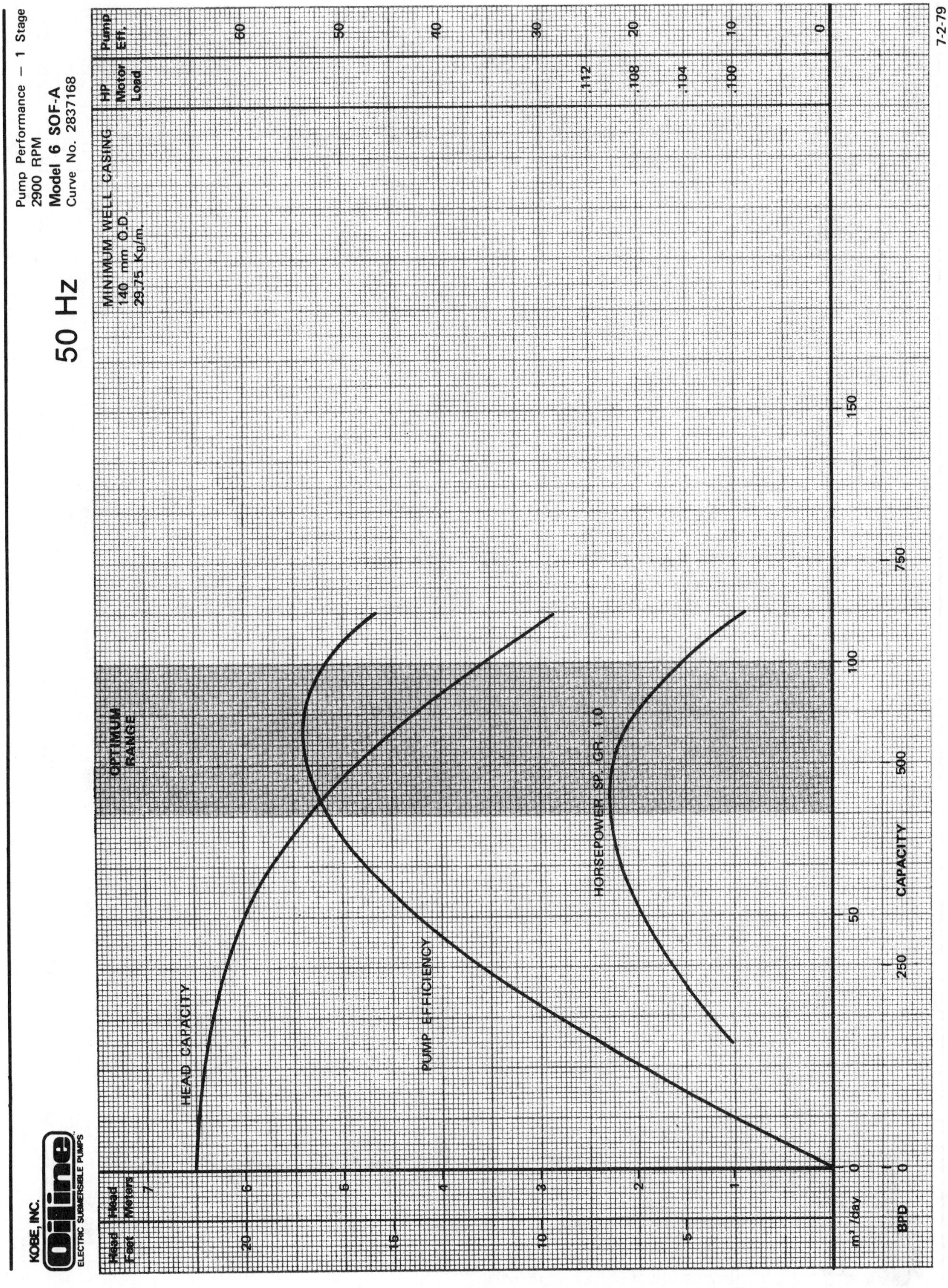
KOBE, INC.
Oilline
ELECTRIC SUBMERSIBLE PUMPS
Pump Performance – 1 Stage
2900 RPM
Model 6 SOF-A
Curve No. 2837168
50 Hz
MINIMUM WELL CASING
140 mm O.D.
29.75 Kg/m.
Head Feet
Head Meters
HP Motor Load
Pump Eff.
OPTIMUM RANGE
HEAD CAPACITY
PUMP EFFICIENCY
HORSEPOWER SP. GR. 1.0
CAPACITY
m³/day
BPD
7-2-79

Fig. 4BIV (2)

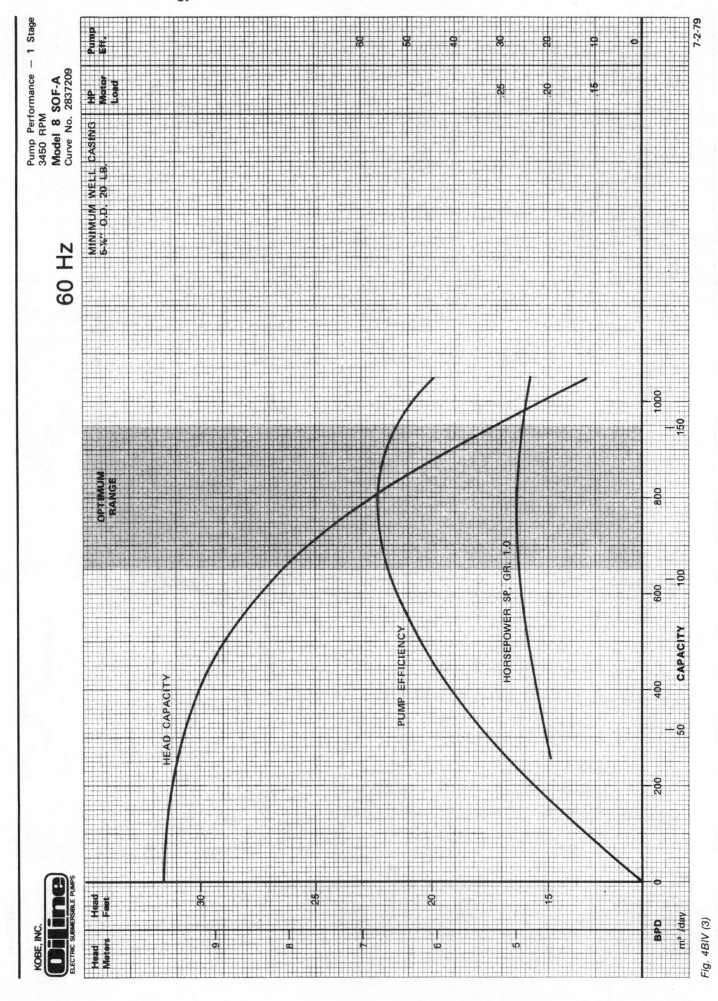

KOBE, INC.
Oiline
ELECTRIC SUBMERSIBLE PUMPS

Pump Performance — 1 Stage
3450 RPM
Model 8 SOF-A
Curve No. 2837209

60 Hz

MINIMUM WELL CASING
5-½" O.D. 20 LB.

Fig. 4BIV (3)

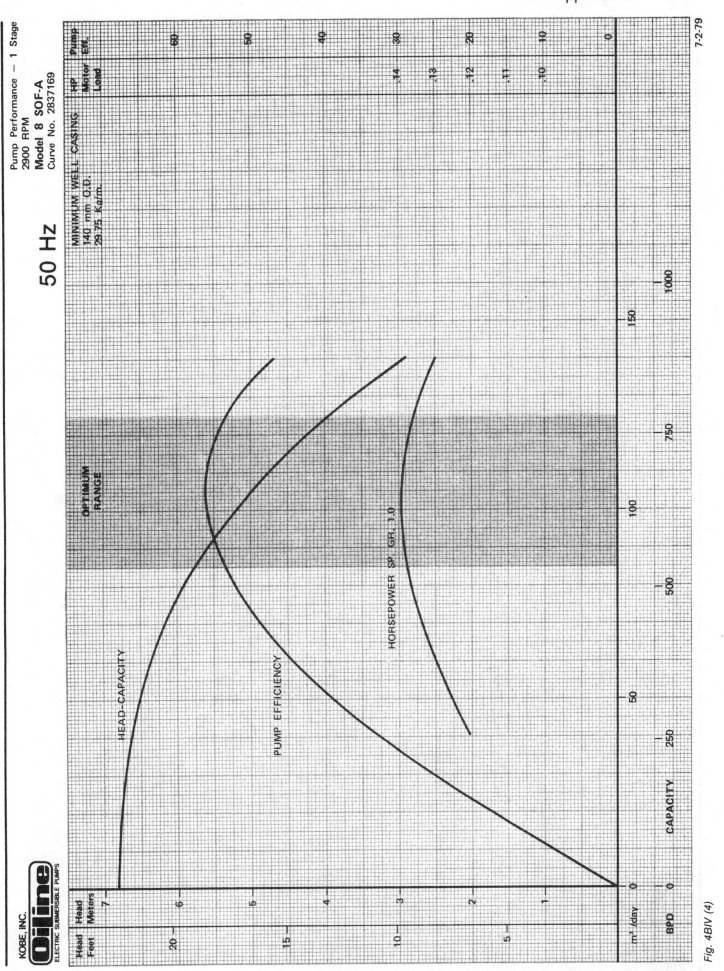

KOBE, INC.

Oil line
ELECTRIC SUBMERSIBLE PUMPS

Pump Performance — 1 Stage
2900 RPM
Model 8 SOF-A
Curve No. 2837169

50 Hz

MINIMUM WELL CASING
140 mm O.D.
29.75 Kg/m.

OPTIMUM RANGE

HEAD-CAPACITY

PUMP EFFICIENCY

HORSEPOWER SP. GR. 1.0

7-2-79

Fig. 4BIV (4)

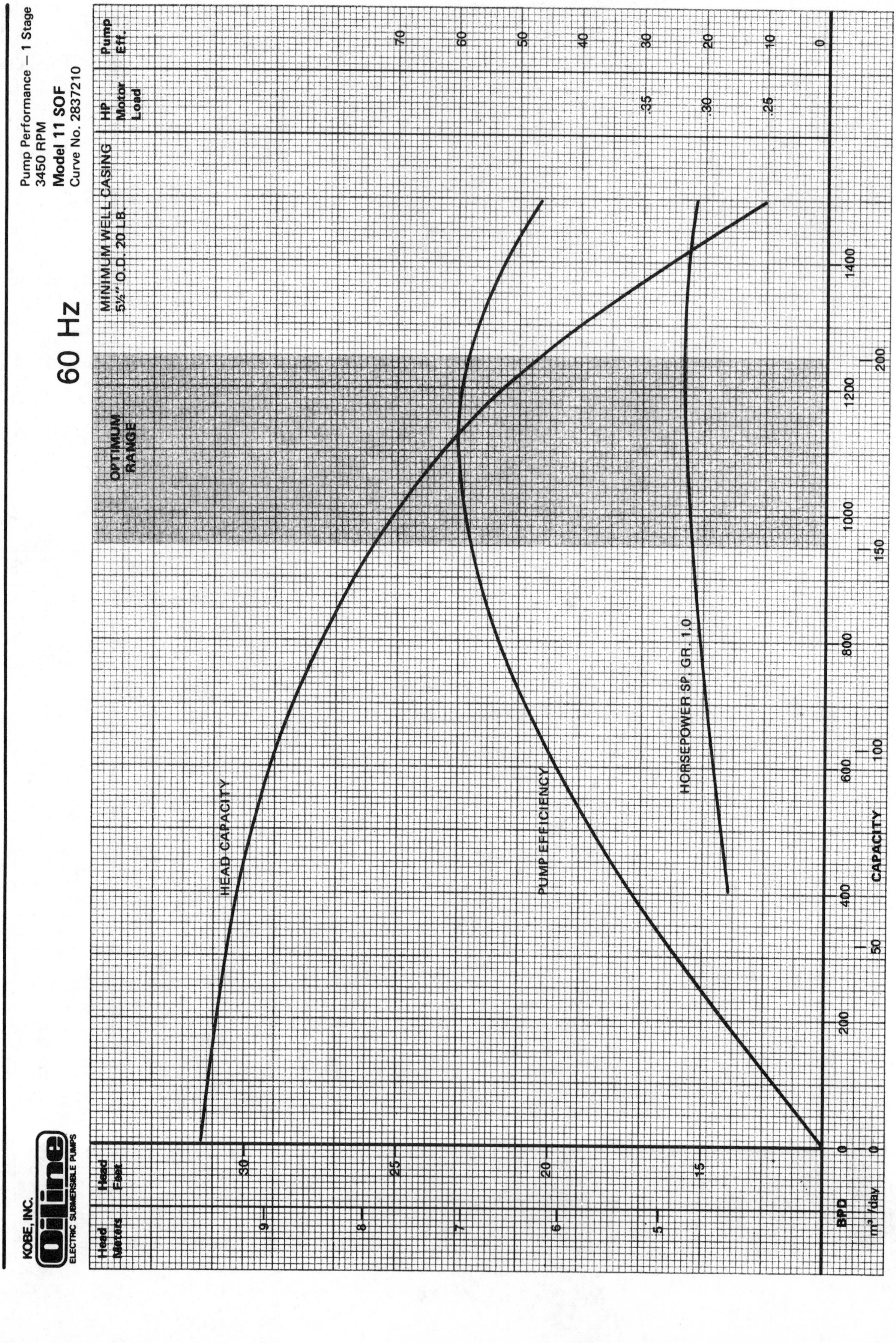

Pump Performance — 1 Stage
3450 RPM
Model 11 SOF
Curve No. 2837210

60 Hz

MINIMUM WELL CASING
5½″ O.D. 20 LB.

OPTIMUM RANGE

HEAD CAPACITY

PUMP EFFICIENCY

HORSEPOWER SP. GR. 1.0

CAPACITY

BPD

m³/day

Head
Meters

Head
Feet

Pump
Eff.

HP
Motor
Load

KOBE, INC.
Oiline
ELECTRIC SUBMERSIBLE PUMPS

7-2-79

Fig. 4BIV (5)

KOBE, INC.

Oilline
ELECTRIC SUBMERSIBLE PUMPS

Pump Performance — 1 Stage
2900 RPM
Model 11 SOF
Curve No. 2837170

50 Hz

MINIMUM WELL CASING
140 mm O.D.
29.75 Kg/m

PUMP EFFICIENCY

HEAD-CAPACITY

HORSEPOWER SP. GR. 1.0

OPTIMUM RANGE

CAPACITY

7-2-79

Fig. 4BIV (6)

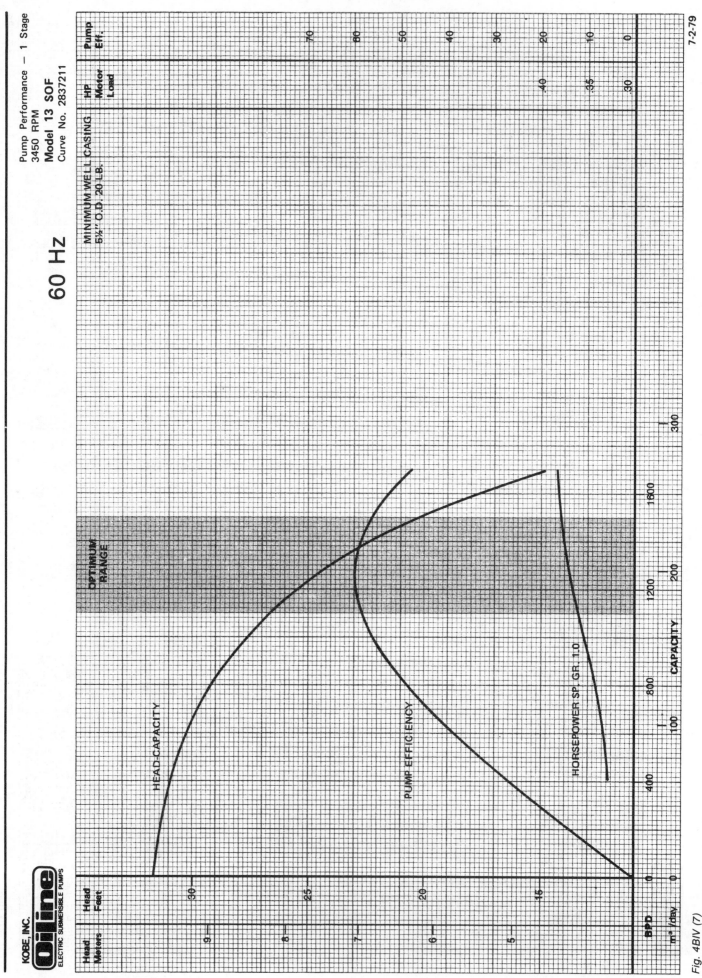

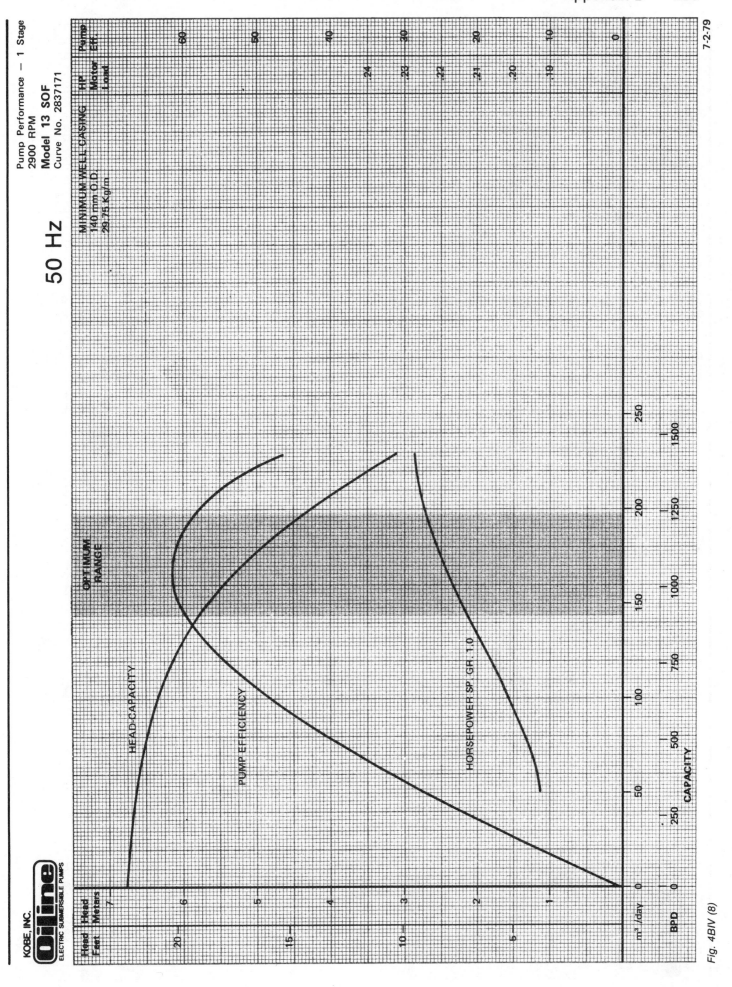

KOBE, INC.
oiline
ELECTRIC SUBMERSIBLE PUMPS

Pump Performance — 1 Stage
2900 RPM
Model 13 SOF
Curve No. 2837171

50 Hz

7-2-79

Fig. 4BIV (8)

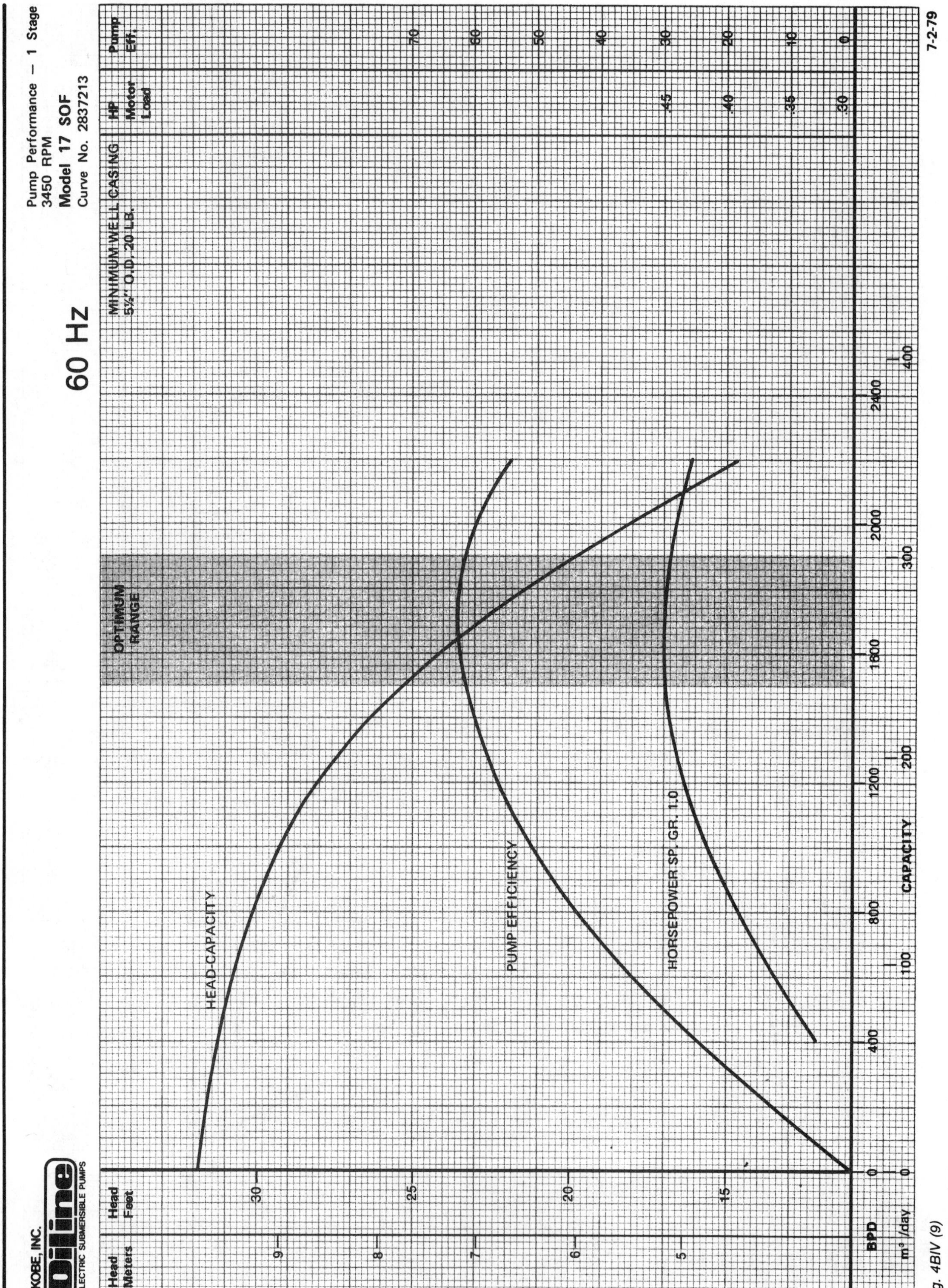

KOBE, INC.
Oilline
ELECTRIC SUBMERSIBLE PUMPS
Pump Performance — 1 Stage
3450 RPM
Model 17 SOF
Curve No. 2837213
60 Hz
MINIMUM WELL CASING
5½" O.D. 20 LB.
OPTIMUM RANGE
HEAD-CAPACITY
PUMP EFFICIENCY
HORSEPOWER SP. GR. 1.0
Head Meters
Head Feet
HP Motor Load
Pump Eff.
BPD
m³/day
CAPACITY
7-2-79

Fig. 4BIV (9)

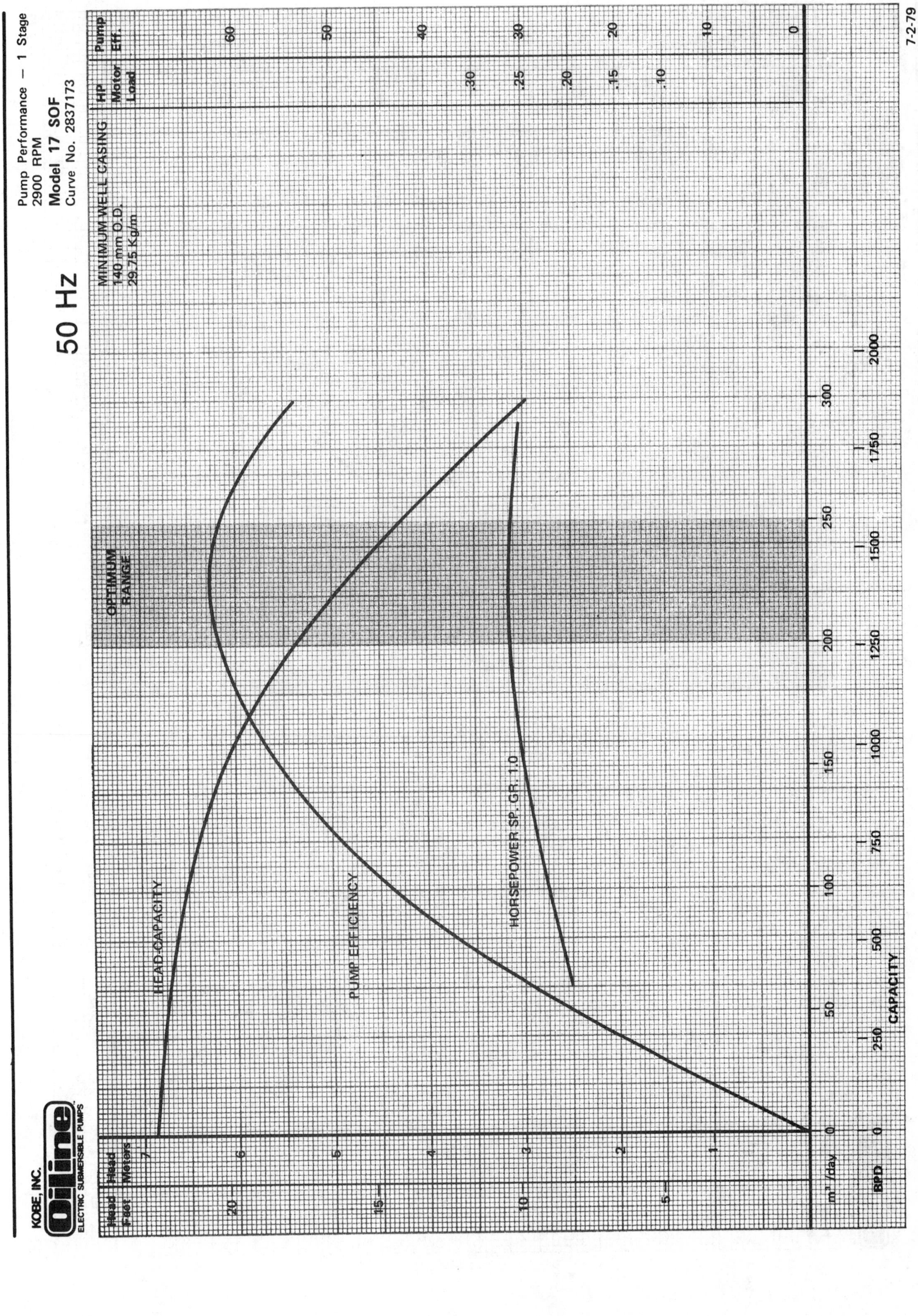

KOBE, INC.
Oilline
ELECTRIC SUBMERSIBLE PUMPS
Pump Performance — 1 Stage
2900 RPM
Model 17 SOF
Curve No. 2837173
50 Hz
MINIMUM WELL CASING
140 mm O.D.
29.75 Kg/m
HP Motor Load
Pump Eff.
OPTIMUM RANGE
HEAD-CAPACITY
PUMP EFFICIENCY
HORSEPOWER SP. GR. 1.0
Head Feet
Head Meters
m³/day
BPD
CAPACITY
7-2-79

Fig. 4BIV (10)

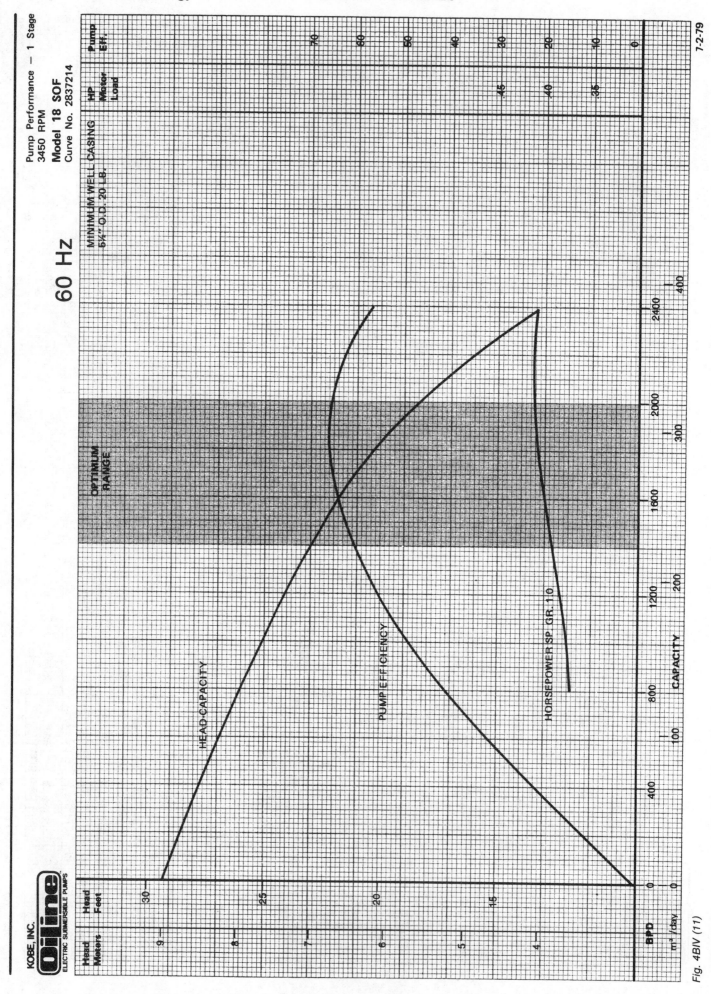

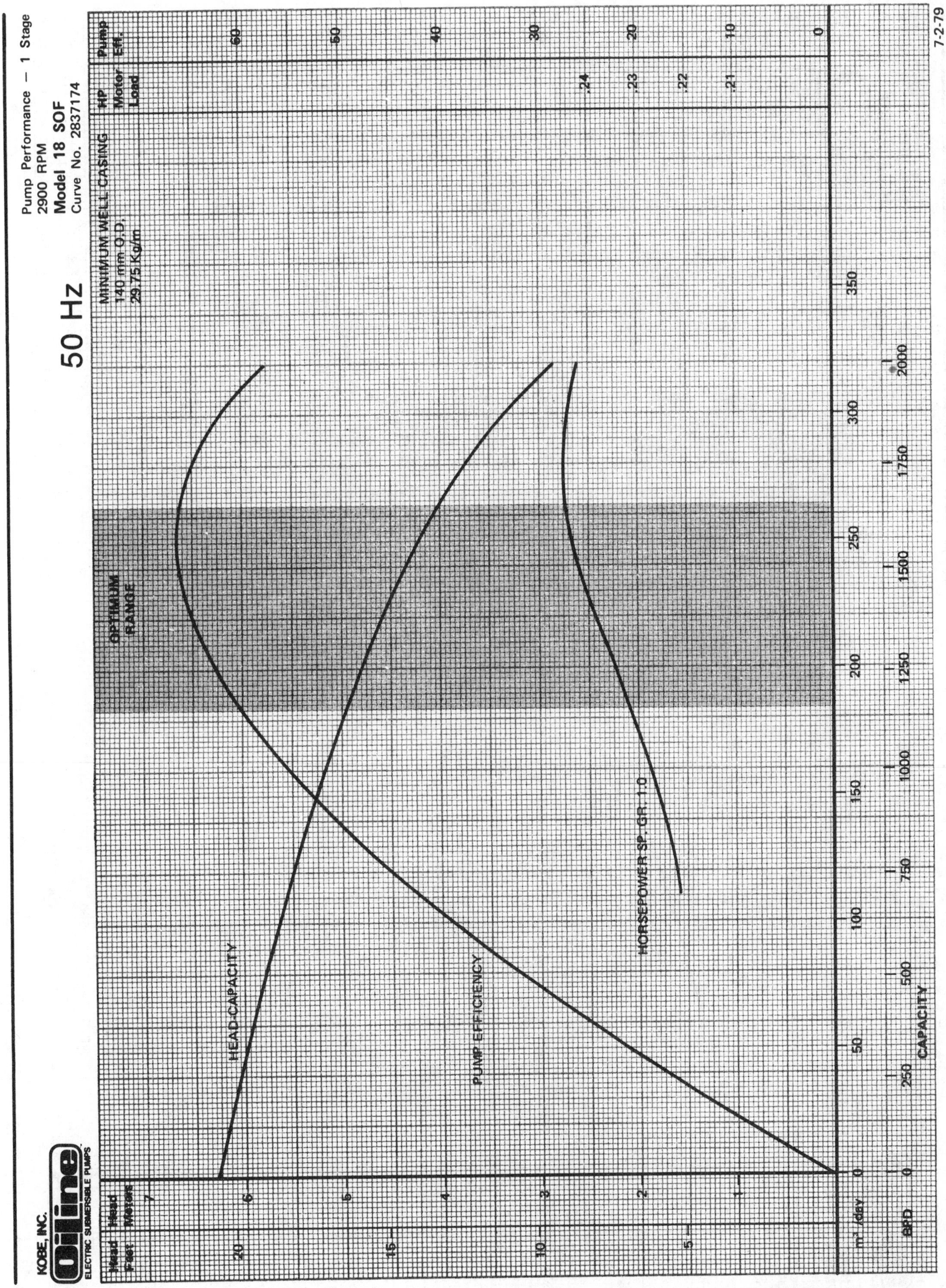

KOBE, INC.
Oilline
ELECTRIC SUBMERSIBLE PUMPS
Pump Performance — 1 Stage
2900 RPM
Model 18 SOF
Curve No. 2837174
50 Hz
MINIMUM WELL CASING
140 mm O.D.
29.75 Kg/m
OPTIMUM RANGE
HEAD-CAPACITY
PUMP EFFICIENCY
HORSEPOWER SP. GR. 1.0
Head Feet
Head Meters
HP Motor Load
Pump Eff.
m³/day
BPD
CAPACITY
7-2-79

Fig. 4BIV (12)

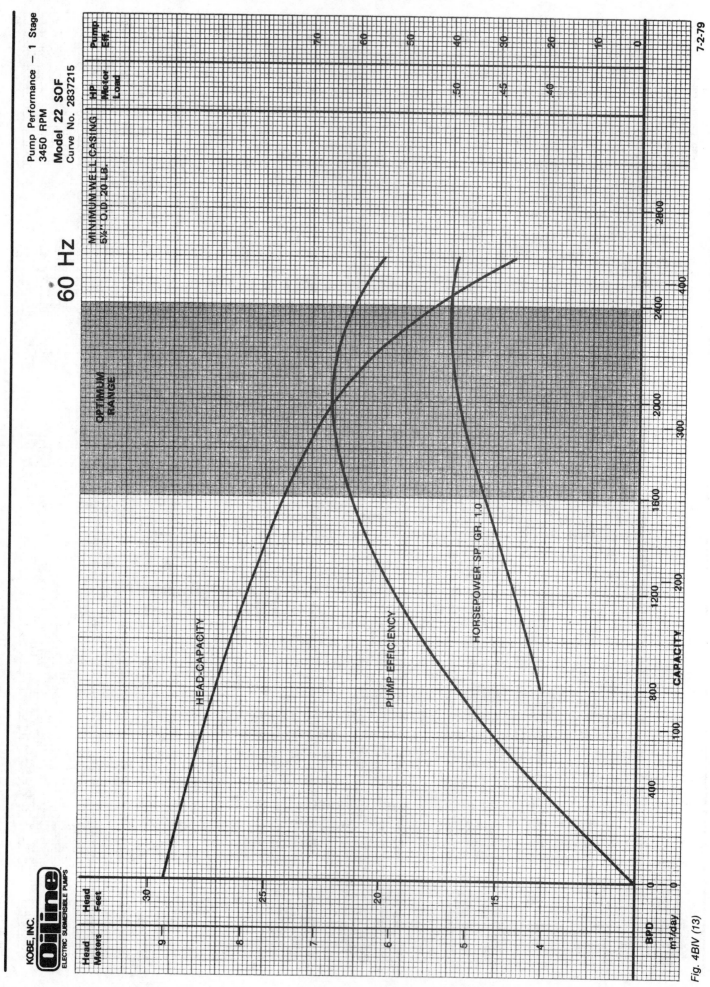

Fig. 4BIV (13)

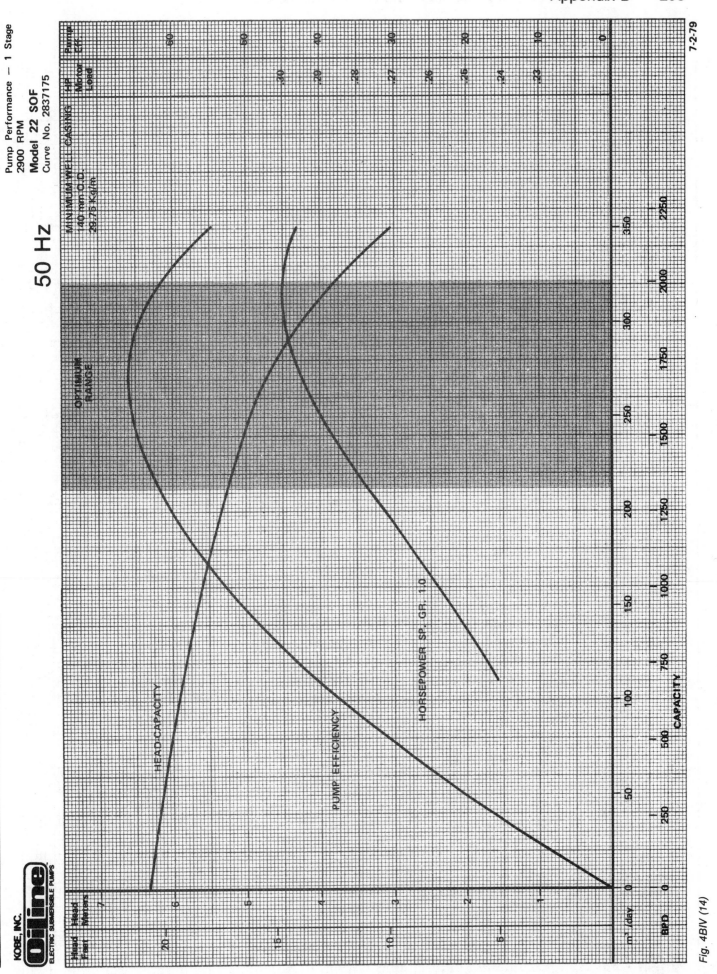

Pump Performance — 1 Stage
2900 RPM
Model 22 SOF
Curve No. 2837175

50 Hz

KOBE, INC.
Oiline
ELECTRIC SUBMERSIBLE PUMPS

MINIMUM WELL CASING
140 mm O.D.
29.75 Kg/m

Fig. 4BIV (14)

7-2-79

Fig. 4BIV (15)

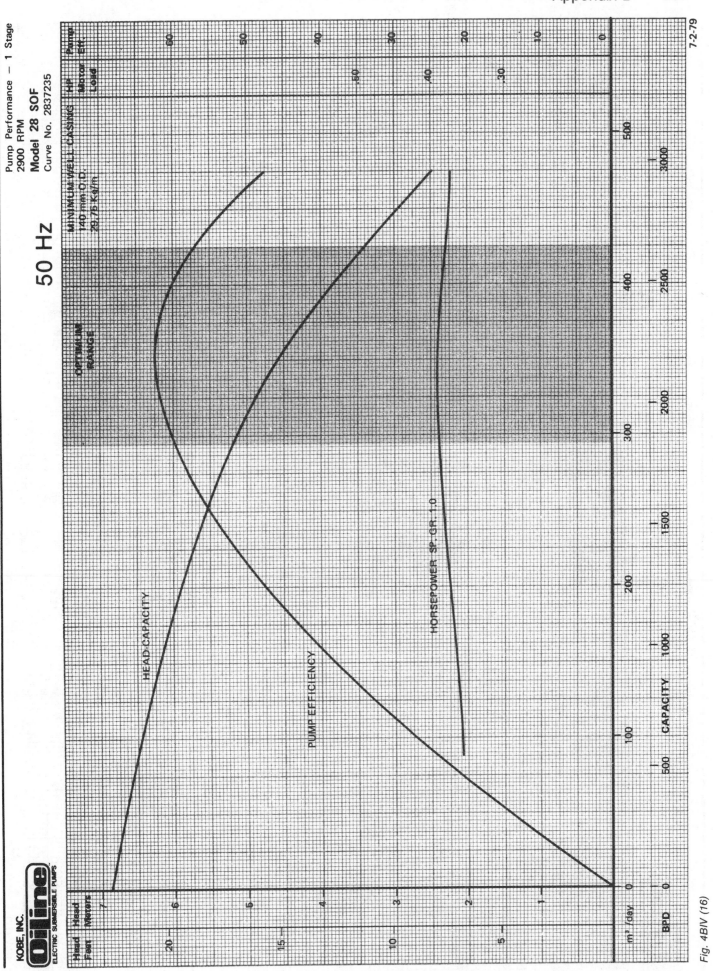

KOBE, INC.

Pump Performance — 1 Stage
2900 RPM
Model 28 SOF
Curve No. 2837235

50 Hz

MINIMUM WELL CASING
140 mm O.D.
29.76 Kg/m

OPTIMUM RANGE

HEAD-CAPACITY

PUMP EFFICIENCY

HORSEPOWER SP. GR. 1.0

7-2-79

Fig. 4BIV (16)

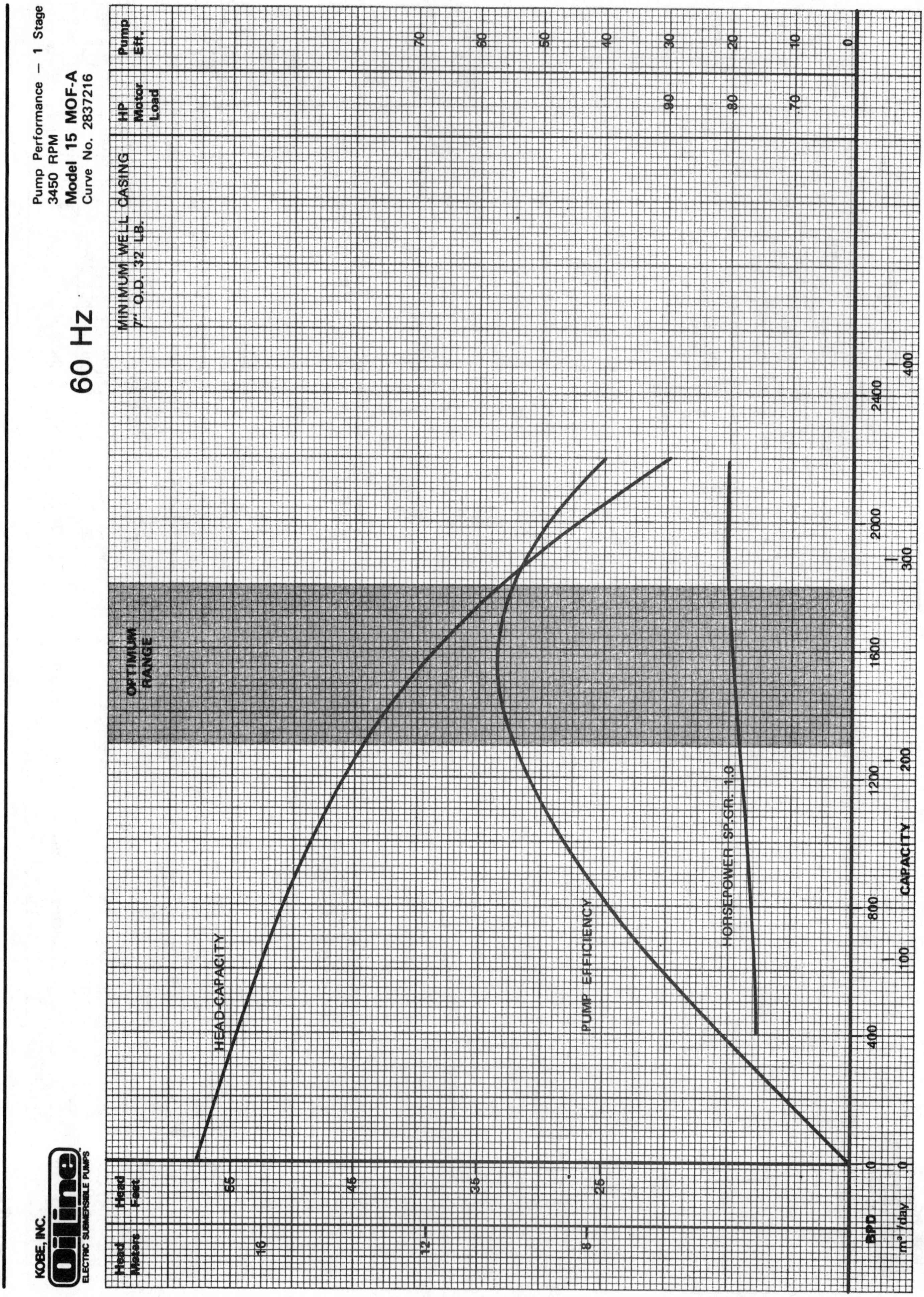
KOBE, INC.
Oilline
ELECTRIC SUBMERSIBLE PUMPS
Pump Performance — 1 Stage
3450 RPM
Model 15 MOF-A
Curve No. 2837216
60 Hz
MINIMUM WELL CASING
7" O.D. 32 LB.
OPTIMUM RANGE
HEAD-CAPACITY
PUMP EFFICIENCY
HORSEPOWER SP.GR. 1.0
Head Meters
16
12
8
Head Feet
55
45
35
25
HP Motor Load
.90
.80
.70
Pump Eff.
70
60
50
40
30
20
10
0
BPD
0
400
800
1200
1600
2000
2400
m³/day
0
100
200
300
400
CAPACITY
7-2-79

Fig. 4BIV (17)

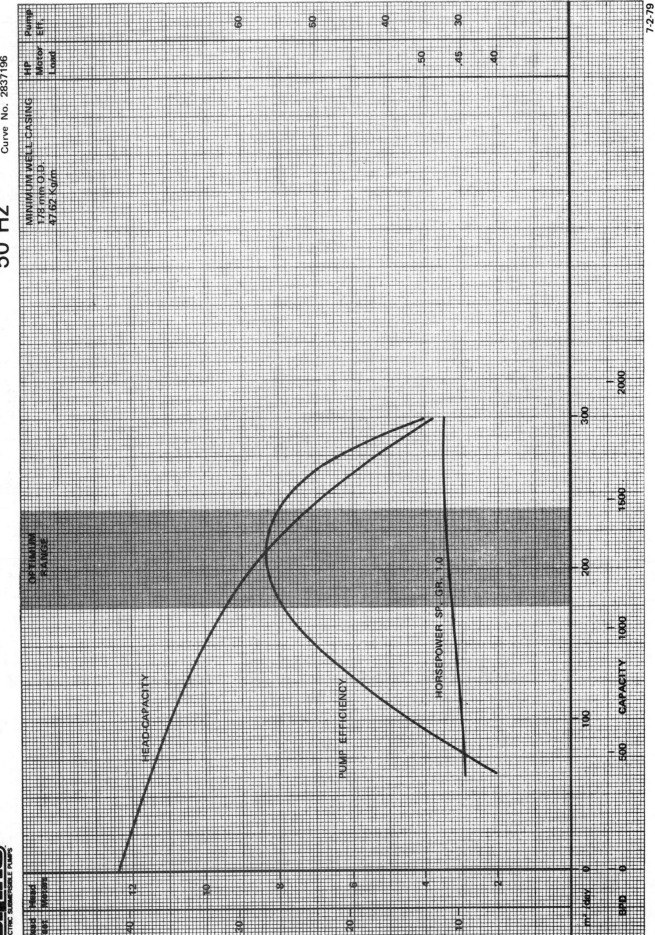

KOBE, INC.

oilline
ELECTRIC SUBMERSIBLE PUMPS

Pump Performance — 1 Stage
2900 RPM
Model 15 MOF-A
Curve No. 2837196

50 Hz

MINIMUM WELL CASING
178 mm O.D.
47.62 Kg/m

HEAD-CAPACITY

PUMP EFFICIENCY

HORSEPOWER SP. GR. 1.0

OPTIMUM RANGE

Fig. 4BIV (18)

7-2-79

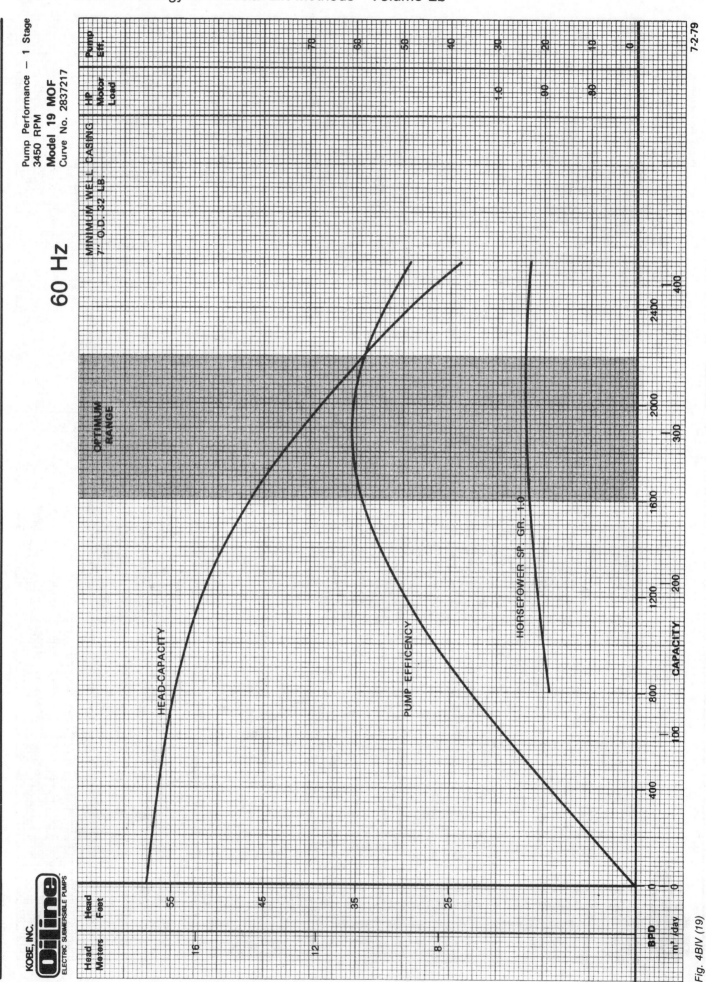

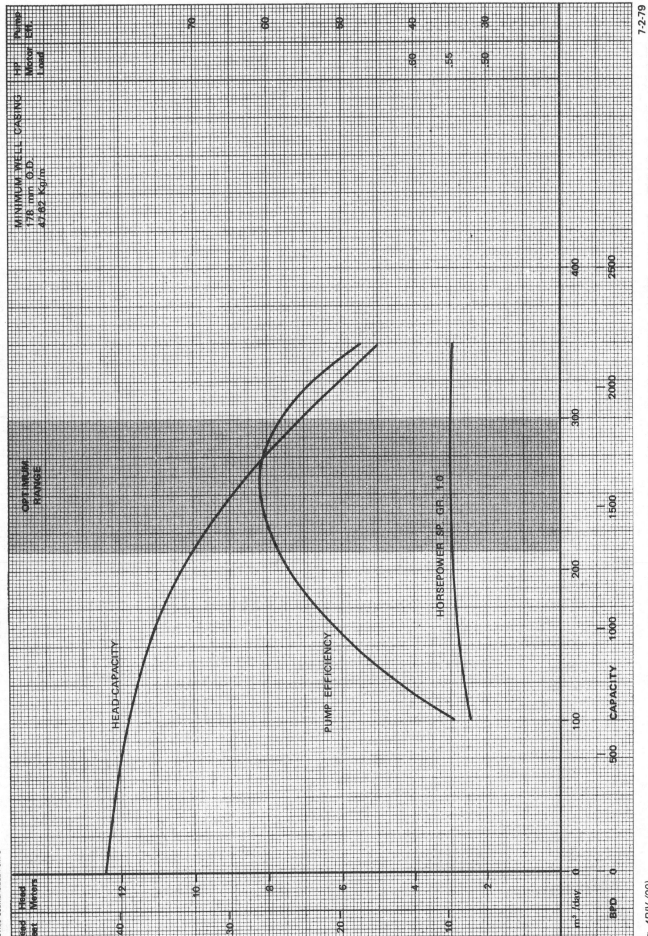

KOBE, INC.

Oil ine
ELECTRIC SUBMERSIBLE PUMPS

Pump Performance — 1 Stage
2900 RPM
Model 19 MOF
Curve No. 2837197

50 Hz

MINIMUM WELL CASING
178 mm O.D.
47.62 Kg/m

Fig. 4BIV (20)

7-2-79

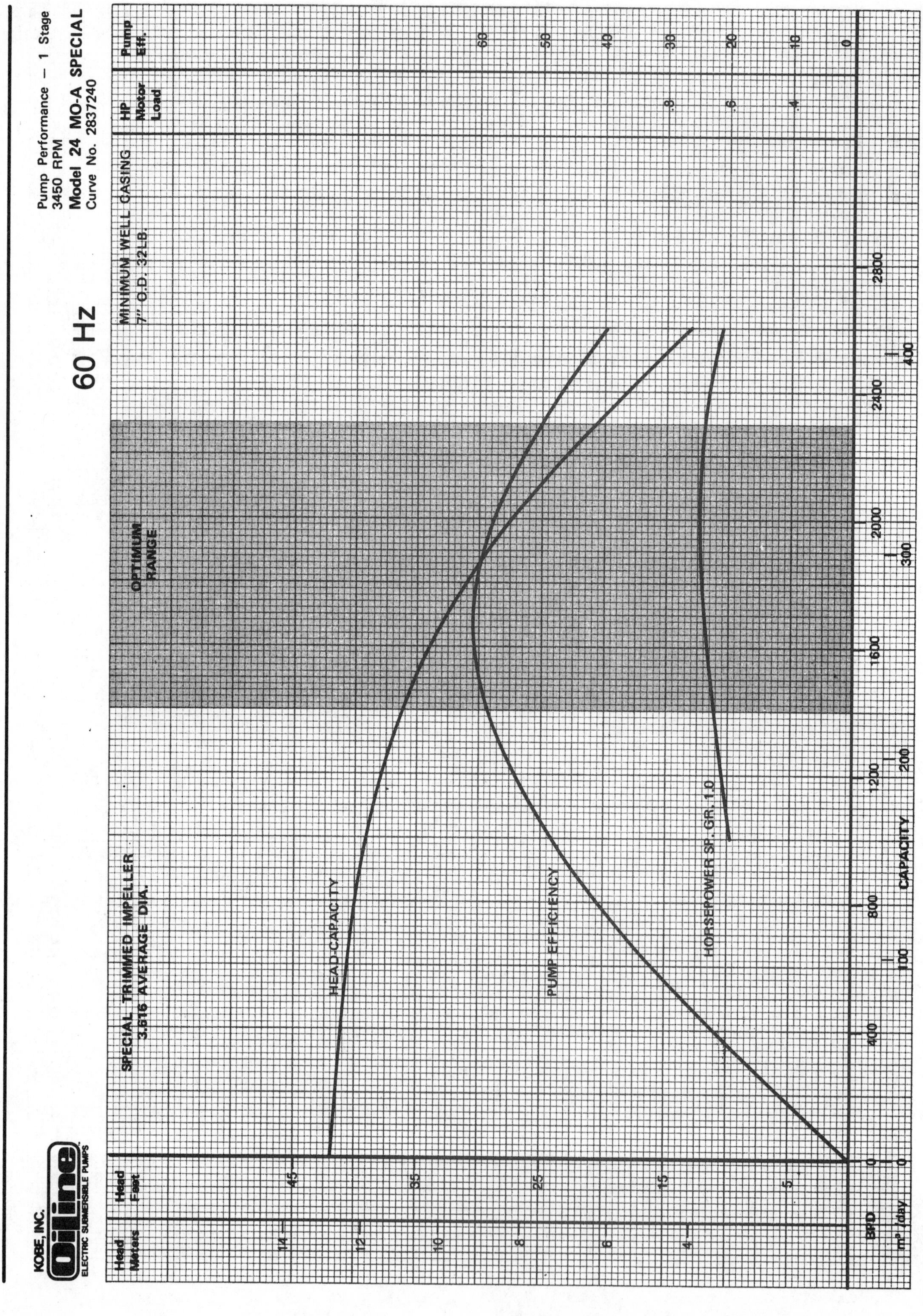
KOBE, INC.
Oiline
ELECTRIC SUBMERSIBLE PUMPS
60 Hz
Pump Performance – 1 Stage
3450 RPM
Model 24 MO-A SPECIAL
Curve No. 2837240
SPECIAL TRIMMED IMPELLER
3.616 AVERAGE DIA.
OPTIMUM RANGE
MINIMUM WELL CASING
7" O.D. 32LB.
HEAD-CAPACITY
PUMP EFFICIENCY
HORSEPOWER SP. GR. 1.0
Head Meters
Head Feet
HP Motor Load
Pump Eff.
BPD
m³/day
CAPACITY
7-2-79

Fig. 4BIV (21)

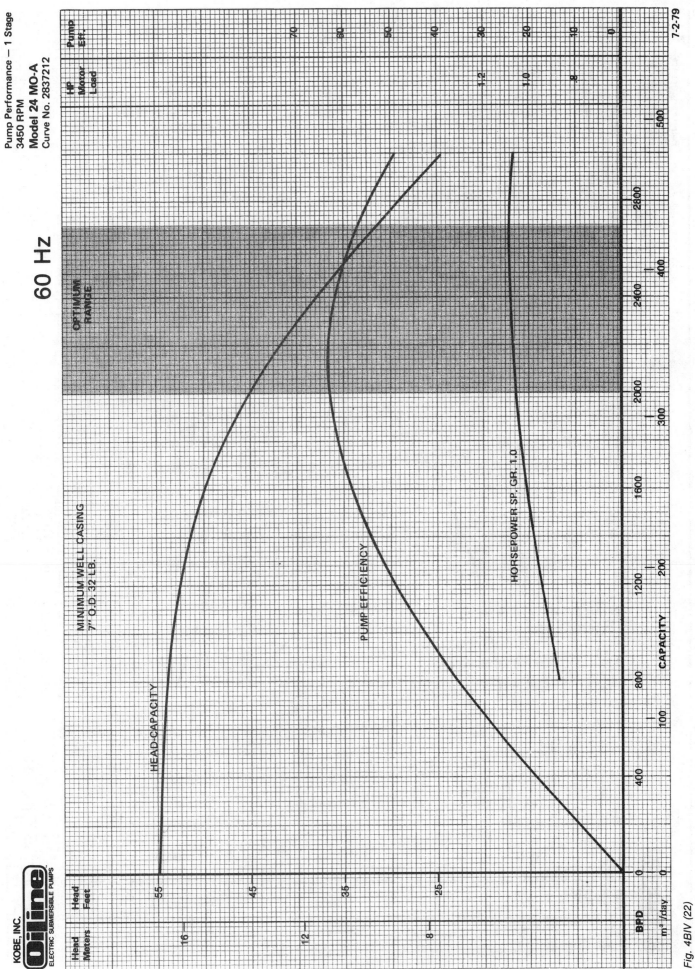

KOBE, INC.

Oiline
ELECTRIC SUBMERSIBLE PUMPS

Pump Performance — 1 Stage
3450 RPM
Model 24 MO-A
Curve No. 2837212

60 Hz

HEAD-CAPACITY

PUMP EFFICIENCY

HORSEPOWER SP. GR. 1.0

MINIMUM WELL CASING
7″ O.D. 32 LB.

OPTIMUM RANGE

Head Feet	Head Meters		Pump Eff.	HP Motor Load

Fig. 4BIV (22)

7-2-79

BPD   CAPACITY
m³/day

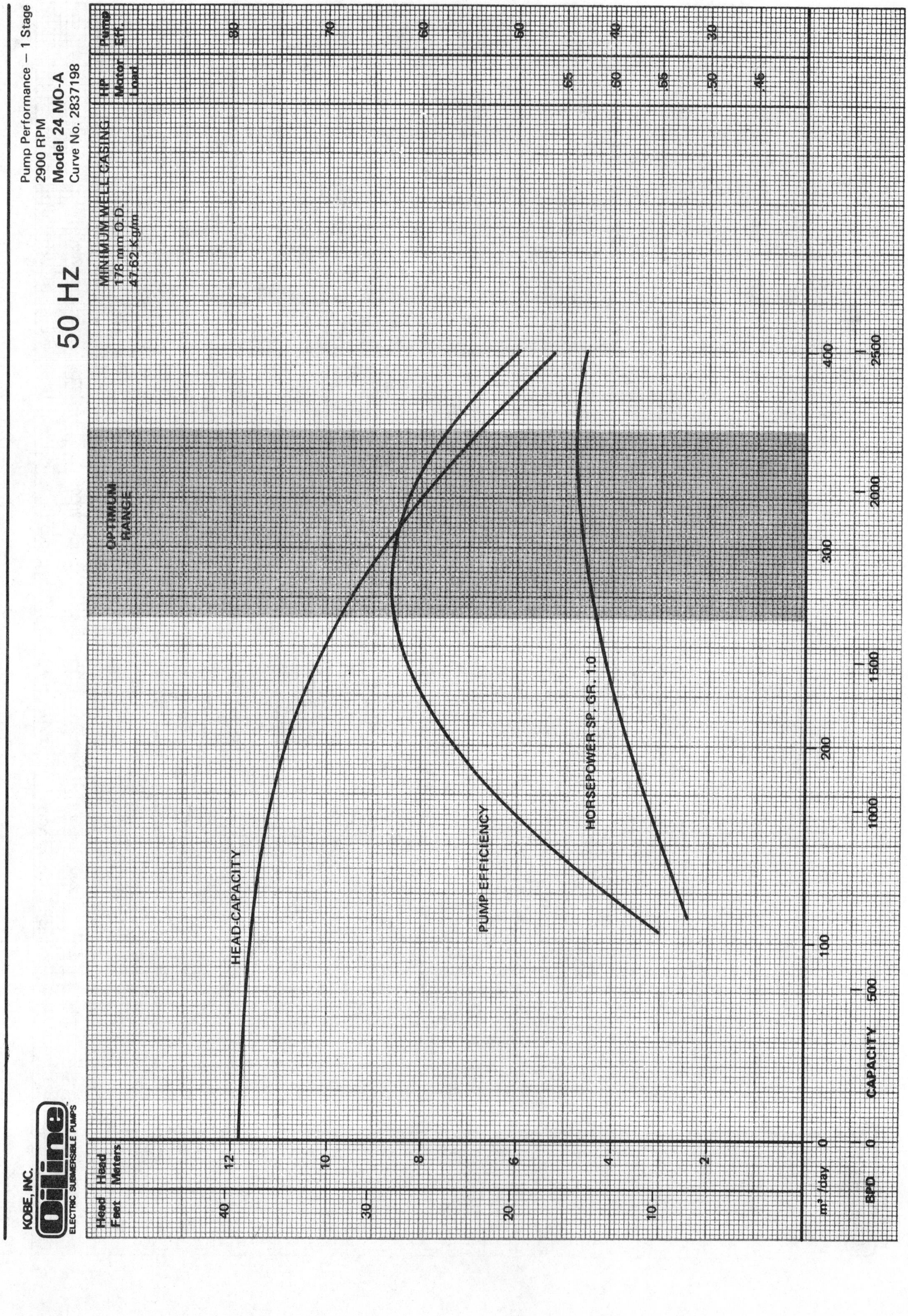

KOBE, INC.
Oiline
ELECTRIC SUBMERSIBLE PUMPS

Pump Performance — 1 Stage
2900 RPM
Model 24 MO-A
Curve No. 2837198

50 Hz

MINIMUM WELL CASING
178 mm O.D.
47.62 Kg/m

Fig. 4BIV (23)

7-2-79

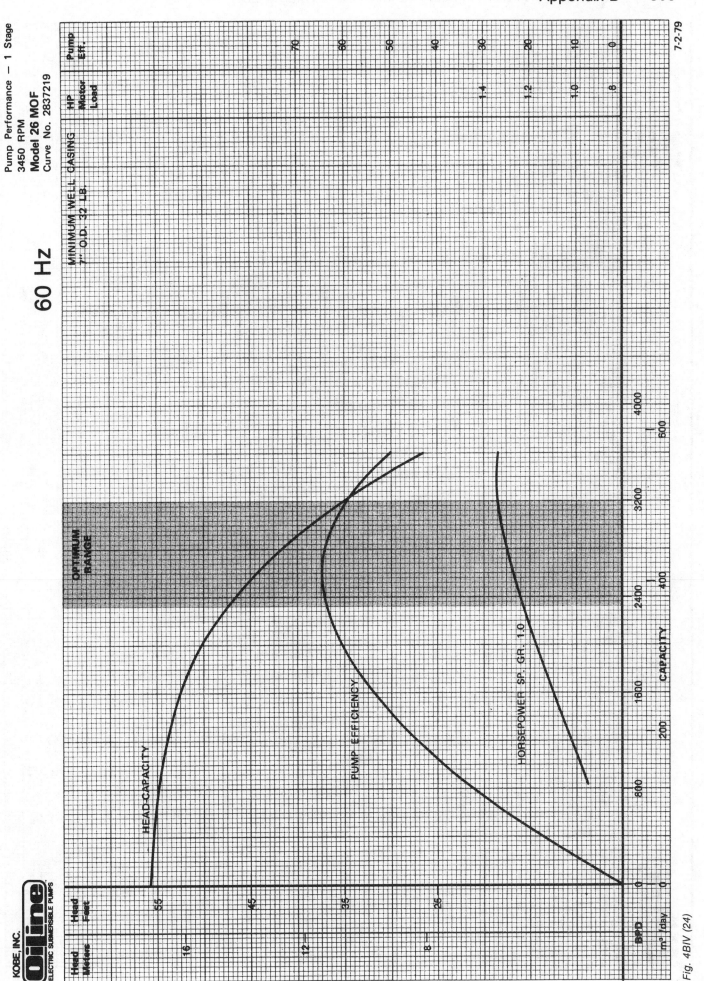

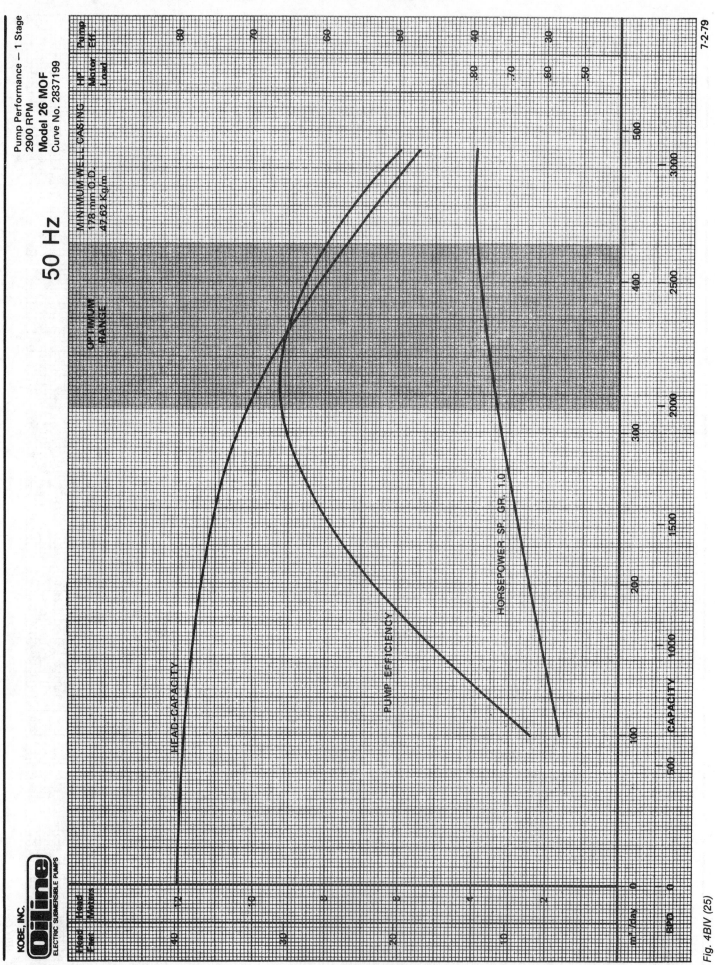

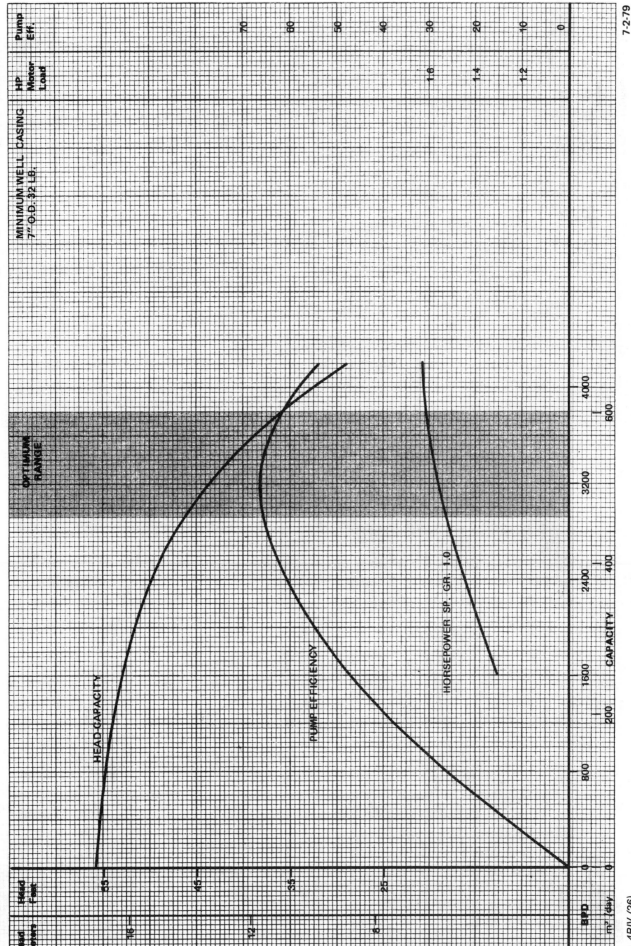

7-2-79

Fig. 4BIV (26)

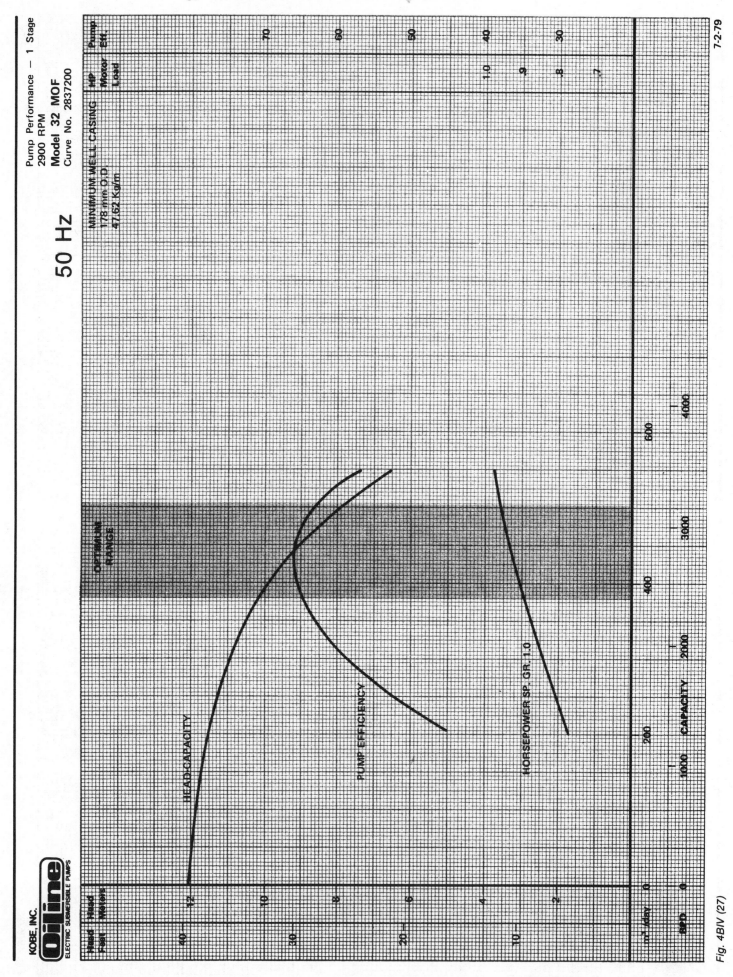

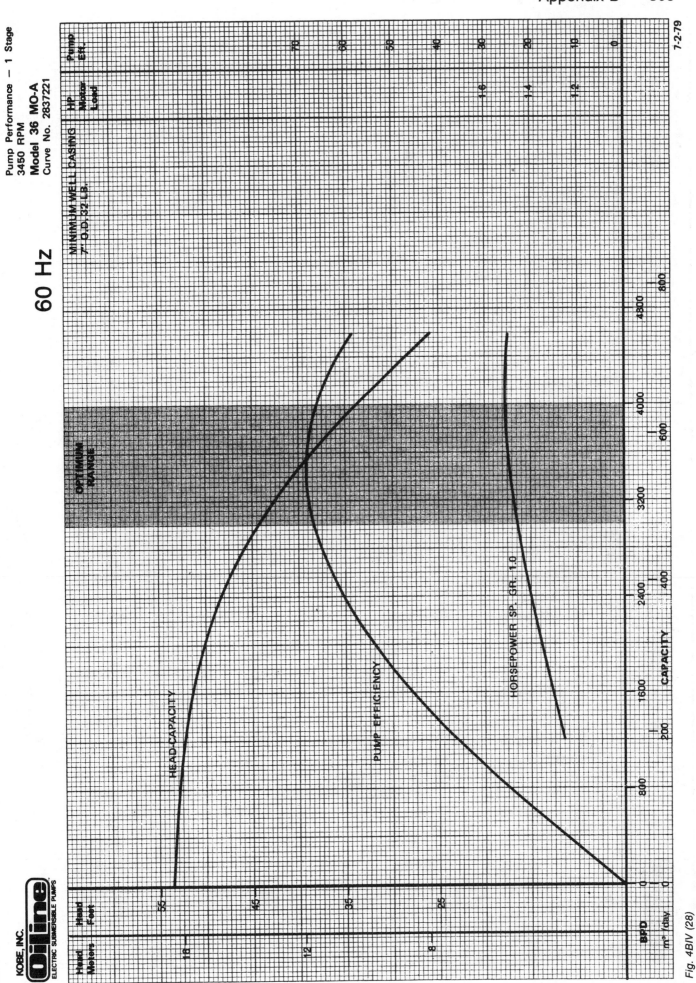

KOBE, INC.

Pump Performance — 1 Stage
3450 RPM
Model 36 MO-A
Curve No. 2837221

60 Hz

MINIMUM WELL CASING
7" O.D. 32 LB.

Fig. 4BIV (28)

7-2-79

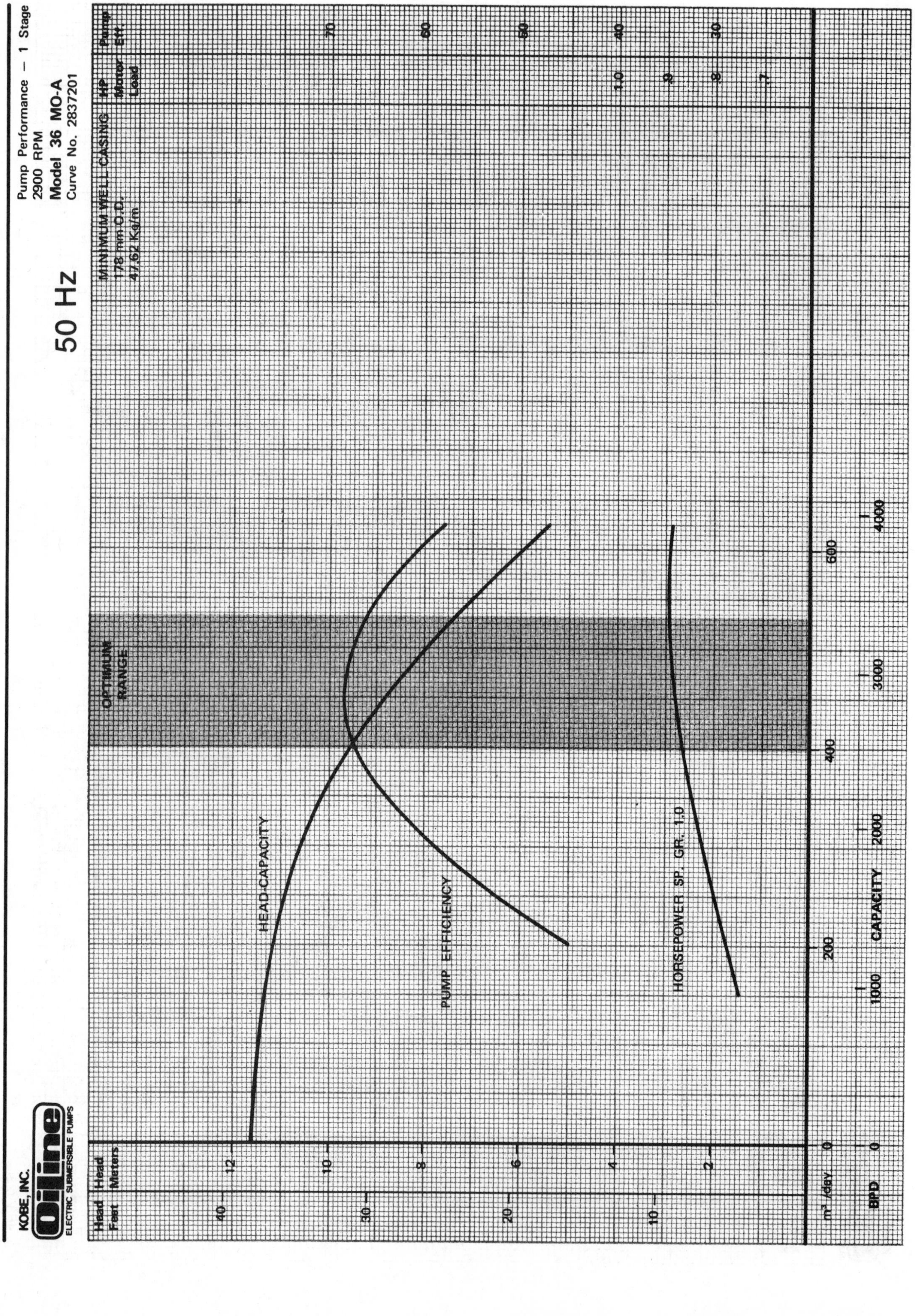
KOBE, INC.
Oiline
ELECTRIC SUBMERSIBLE PUMPS
Pump Performance — 1 Stage
2900 RPM
Model 36 MO-A
Curve No. 2837201
50 Hz
MINIMUM WELL CASING
178 mm O.D.
47,62 Kg/m
OPTIMUM RANGE
HEAD-CAPACITY
PUMP EFFICIENCY
HORSEPOWER SP. GR. 1.0
Head Feet
Head Meters
HP Motor Load
Pump Eff.
m³ /day
BPD
CAPACITY
7-2-79

Fig. 4B/V (29)

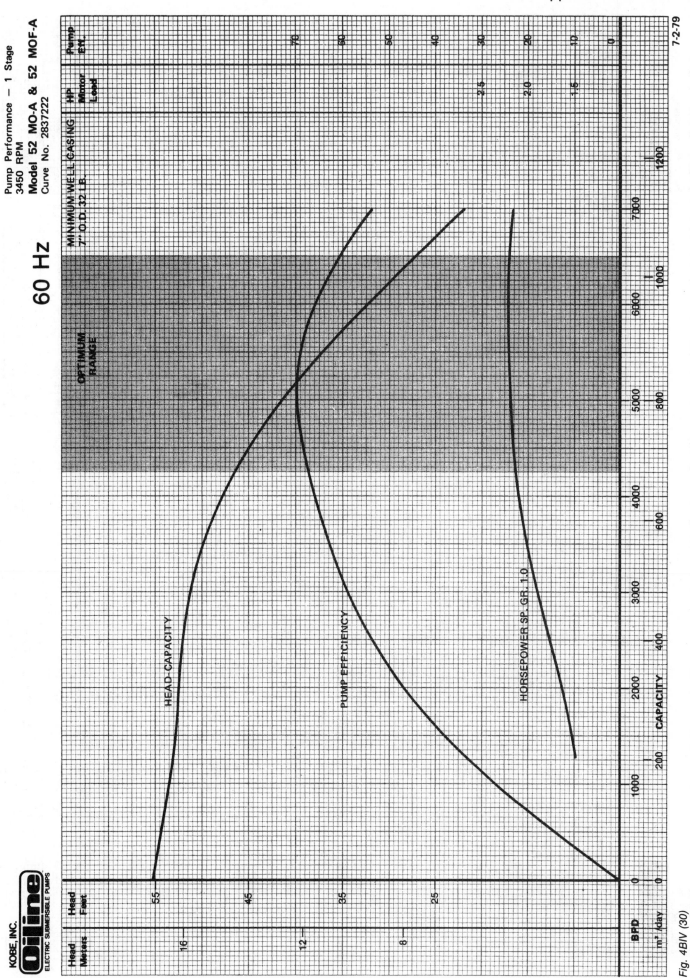

KOBE, INC.

Oilline
ELECTRIC SUBMERSIBLE PUMPS

Pump Performance — 1 Stage
3450 RPM
Model 52 MO-A & 52 MOF-A
Curve No. 2837222

60 Hz

MINIMUM WELL CASING
7" O.D. 32 LB.

OPTIMUM RANGE

HEAD-CAPACITY

PUMP-EFFICIENCY

HORSEPOWER SP. GR. 1.0

Head Meters

Head Feet

BPD

m³ /day

CAPACITY

Pump Eff.

HP Motor Load

7-2-79

Fig. 4BIV (30)

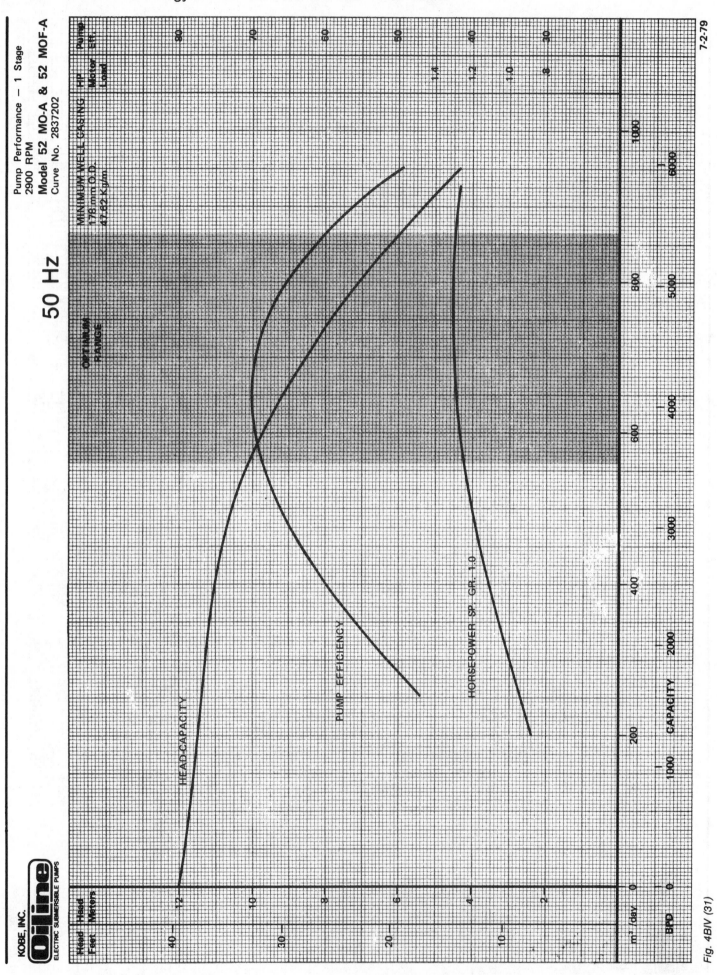

KOBE, INC.

Pump Performance — 1 Stage
2900 RPM
Model 52 MO-A & 52 MOF-A
Curve No. 2837202

50 Hz

HEAD-CAPACITY

PUMP EFFICIENCY

HORSEPOWER SP. GR. 1.0

OPTIMUM RANGE

MINIMUM WELL CASING
178 mm O.D.
47-62 Kg/m.

Fig. 4BIV (31)

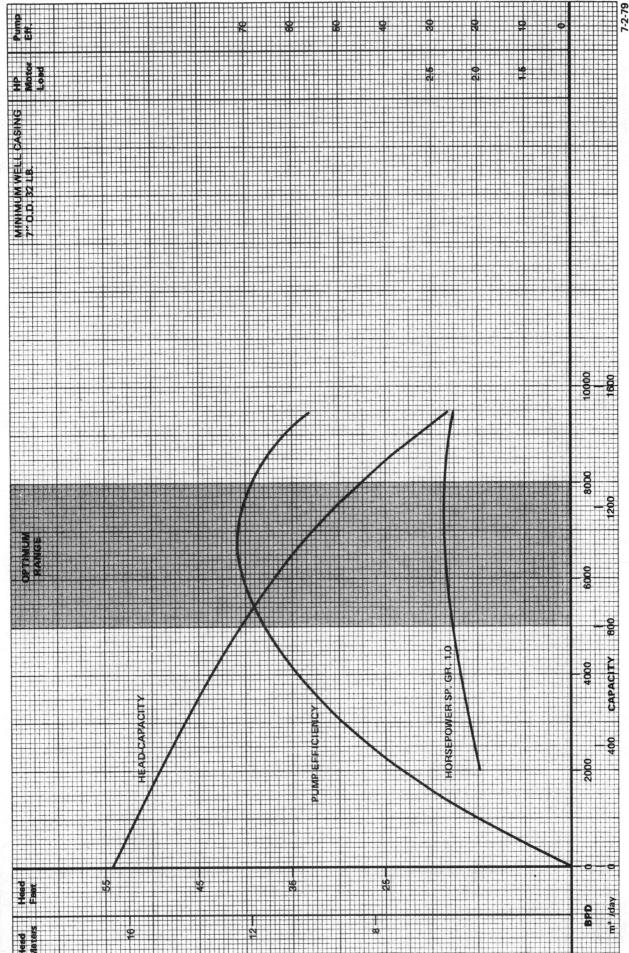

KOBE, INC.
Oilline
ELECTRIC SUBMERSIBLE PUMPS

Pump Performance — 1 Stage
3450 RPM
Model 68 MO-A & 68 MOF-A
Curve No. 2837223

MINIMUM WELL CASING
7" O.D. 32 LB.

60 Hz

OPTIMUM RANGE

HEAD-CAPACITY

PUMP EFFICIENCY

HORSEPOWER SP. GR. 1.0

	Head Feet	Head Meters
	55	16
	45	12
	35	8
	25	
BPD	0	0
m³/day	0	

CAPACITY

2000   400   4000   800   6000   1200   8000   1600   10000

Pump Eff.	HP Motor Load
70	
60	
50	
40	
30	2.5
20	2.0
10	1.5
0	

7-2-79

Fig. 4BIV (32)

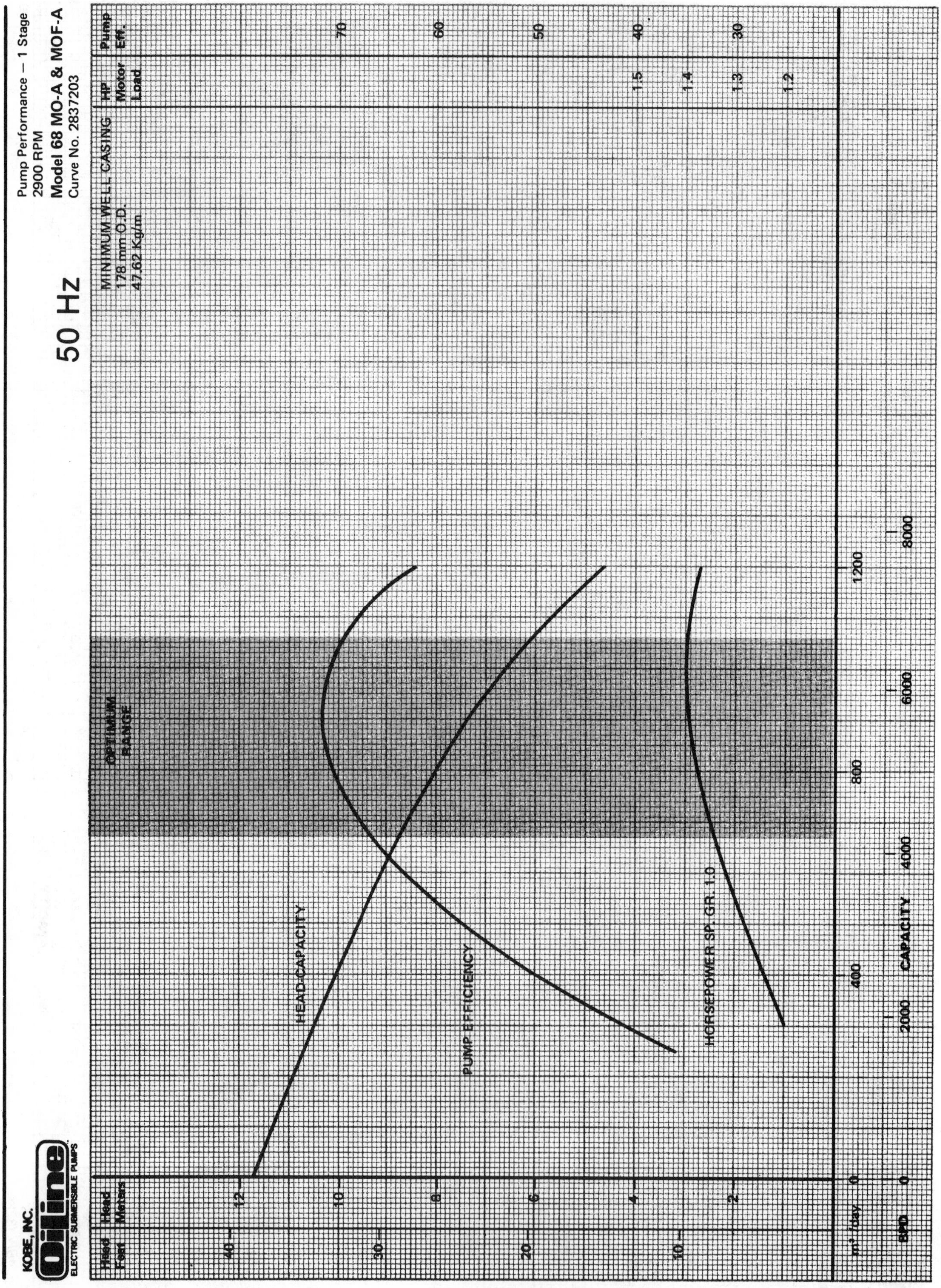

KOBE, INC.
oil line
ELECTRIC SUBMERSIBLE PUMPS

Pump Performance — 1 Stage
2900 RPM
Model 68 MO-A & MOF-A
Curve No. 2837203

50 Hz

MINIMUM WELL CASING
178 mm O.D.
47.62 Kg/m

Pump Eff.	HP Motor Load
70	
60	
50	
40	1.5
	1.4
30	1.3
	1.2

OPTIMUM RANGE

HEAD-CAPACITY

PUMP EFFICIENCY

HORSEPOWER SP. GR. 1.0

Head Feet / Meters

40 — 12
30 — 10
         8
20 — 6
10 — 4
         2
0 — 0

m³/day   0
BPD   0   2000   4000   6000   8000
                    CAPACITY
                    400   800   1200

7-2-79

Fig. 4BIV (33)

FRICTION LOSS BPD

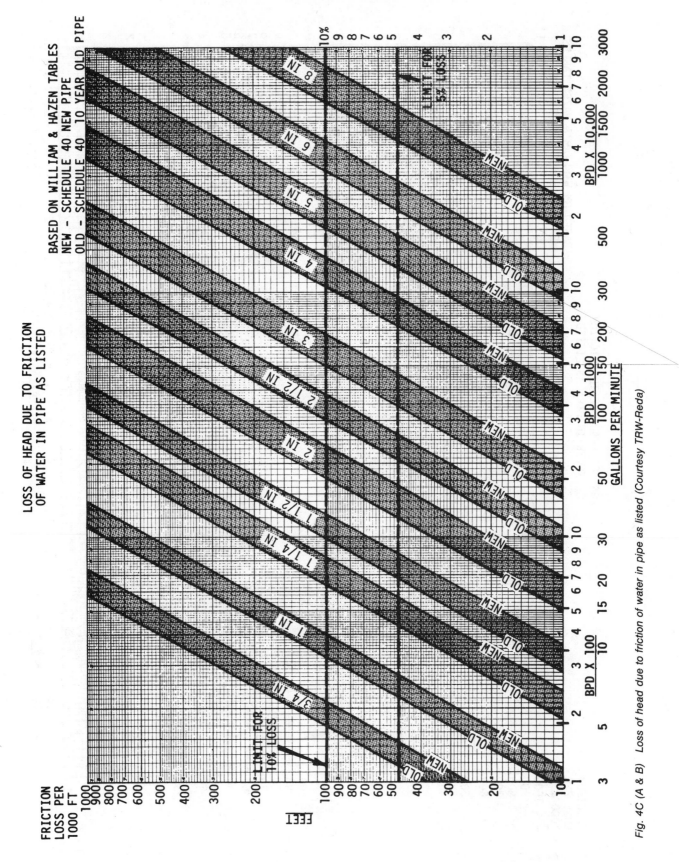

Fig. 4C (A & B)   Loss of head due to friction of water in pipe as listed (Courtesy TRW-Reda)

FRICTION LOSS m³/day

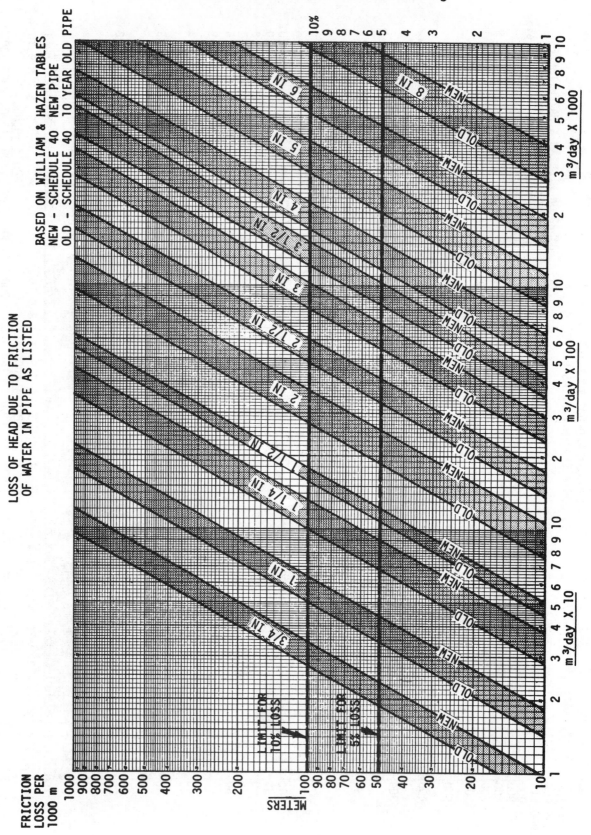

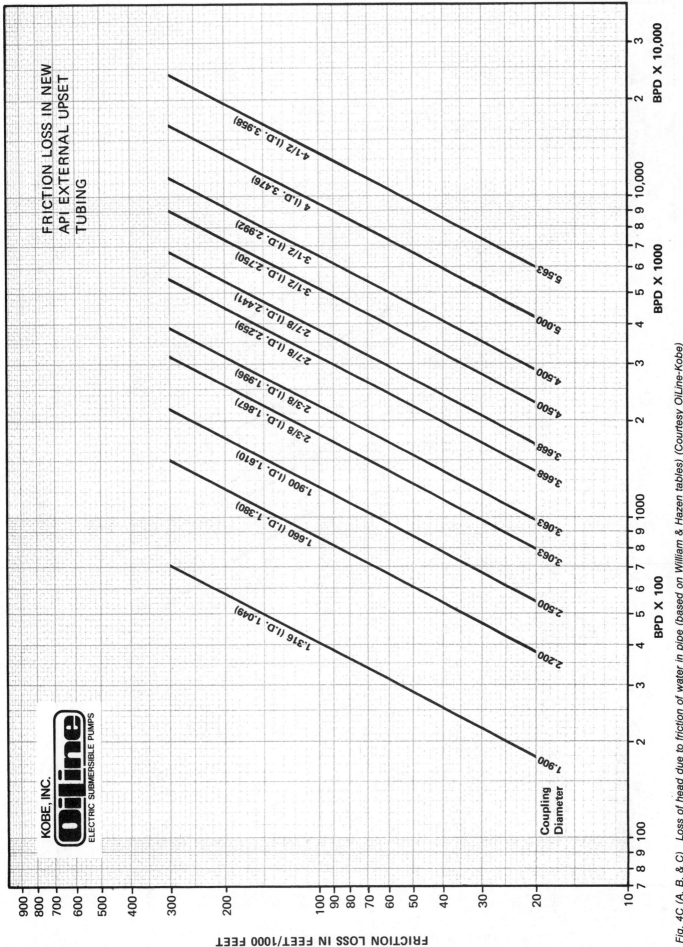

KOBE, INC.

OilLine
ELECTRIC SUBMERSIBLE PUMPS

FRICTION LOSS IN NEW
API EXTERNAL UPSET
TUBING

Coupling
Diameter

1.316 (I.D. 1.049)

1.660 (I.D. 1.380)

1.900 (I.D. 1.610)

2-3/8 (I.D. 1.867)

2-3/8 (I.D. 1.996)

2-7/8 (I.D. 2.259)

2-7/8 (I.D. 2.441)

3-1/2 (I.D. 2.750)

3-1/2 (I.D. 2.992)

4 (I.D. 3.476)

4-1/2 (I.D. 3.958)

1.900

2.200

2.500

3.063

3.063

3.668

3.668

4.500

4.500

5.000

5.563

FRICTION LOSS IN FEET/1000 FEET

BPD X 100

BPD X 1000

BPD X 10,000

Fig. 4C (A, B, & C)   Loss of head due to friction of water in pipe (based on William & Hazen tables) (Courtesy OilLine-Kobe)

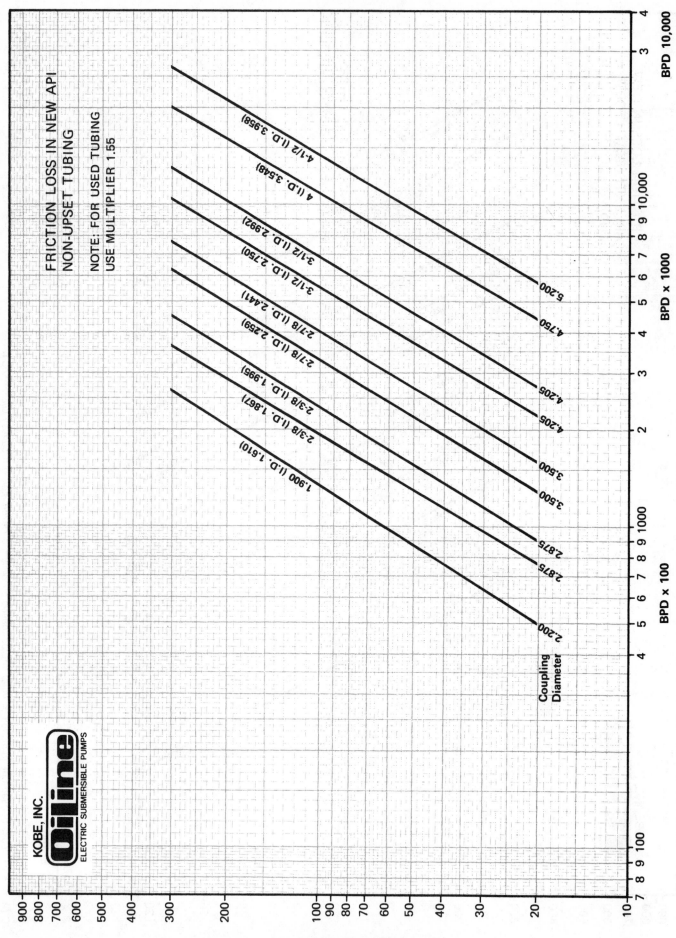

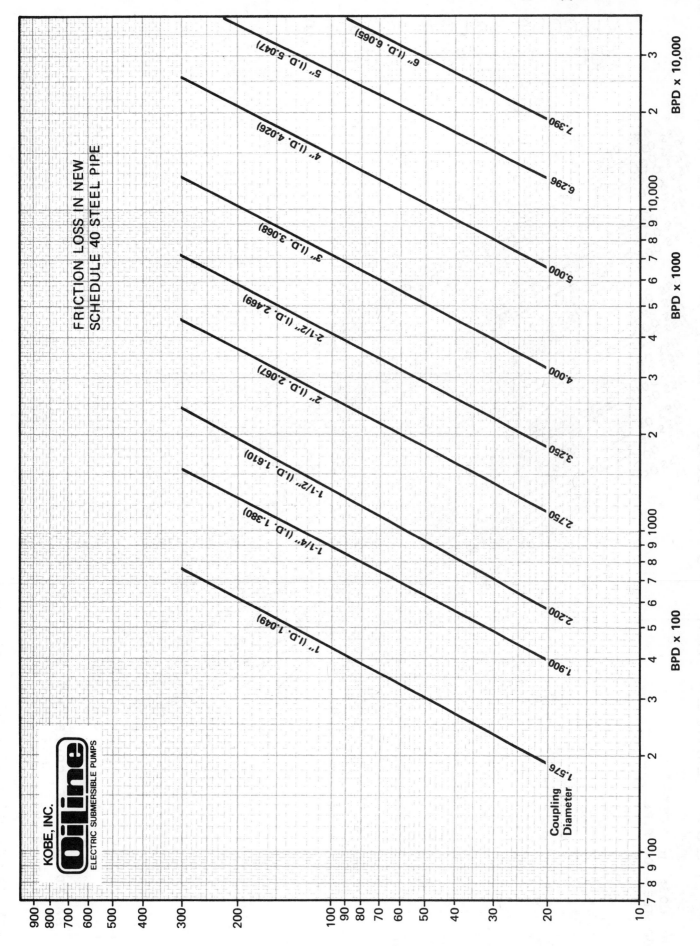

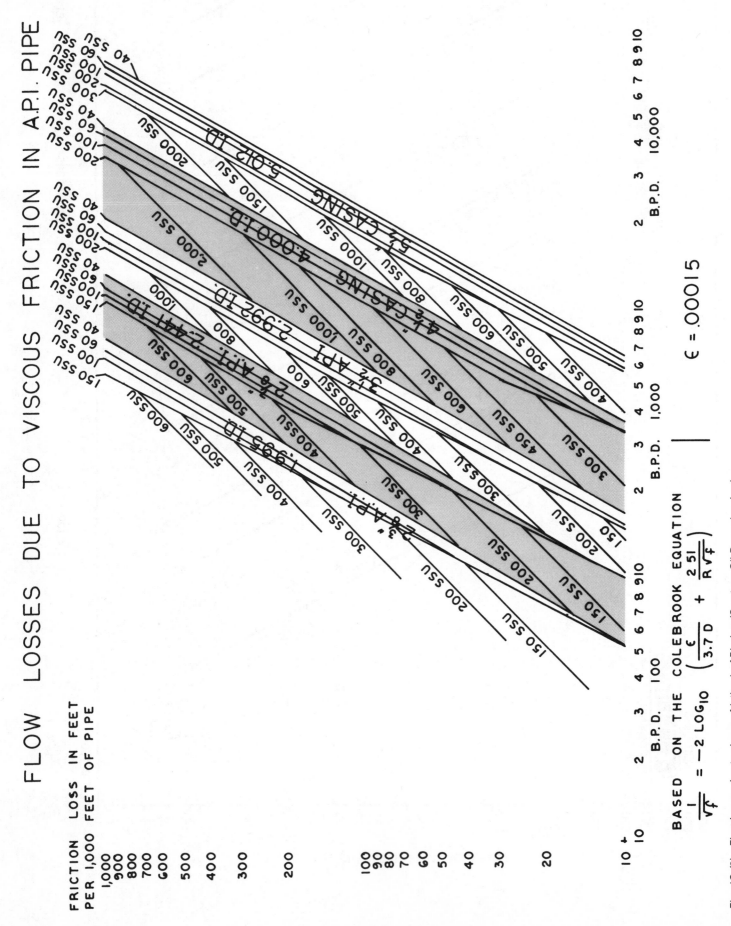

Fig. 4C (3) Flow losses due to viscous friction in API pipe (Courtesy Oil Dynamics, Inc.)

Appendix 4D.1

TRW REDA   **SWITCHBOARDS**

GENERAL DATA

Class	Type	Size	Maximum volts	HP	Maximum full load amps
DFH-2	72	2	600	25	50
		3	600	50	90
		4	600	100	135
		5	600	200	270
45 MFH	76B	2	1000	70	45
120 MFH	76A	3	1000	160	120
100 MDFH	76A	3	1500	160	100
150 MDFH	76A	4	1500	250	150
RP-1	76A		1500	350	360
1512	76C		2400	700	360
RP-2			2400	600	360

STANDARD FEATURES

1. Enclosures are NEME - 3R suitable for outdoor application.
2. Separate compartments are provided for the high and low voltage equipment in all switchboards except the DFH-2 model.
3. "Hand-off-auto" selection switch "start" push-button and disconnect switches.
4. Full protection is provided by Electro-Mechanical or Solid State protection systems.
   A. Electro-Mechanical Systems
      1. Overcurrent or overload protection is provided by three magnetic inverse time delay relays with hand reset contacts.
      2. Type SC undercurrent relays provides pump-off, gas-lock and closed discharge protection.
   B. Solid State Production Control Center
      This solid state device provides overcurrent under-load and single phase protection with pressure switch operation and automatic restart features.
5. Recording ammeter with a combination 24 hour, 7-day clock.
6. Automatic restart feature provides for restarting after expiration of a preset time interval after shutdown due to underload.
7. Switchboards are suitable for use with external control devices at 120 V.
8. Three current limiting fuses for heavy fault current protection on switchboards rated at 1000 volts or higher.
9. Air break contactor.
10. Visable blade fuseable disconnect switch on all except RP type models.
11. Lightning arrestors are provided.

Appendix 4D.2

TRW REDA   **SWITCHBOARDS**

TYPE DHF-2 600 VOLTS MAXIMUM
DFH-2 SIZE 2 25 HORSEPOWER 600 VOLT MAXIMUM 60 HZ

Dimension			Amp load	Part number electro-mechanical	Part number solid-state
Height	36.5 in.	0.93 m	10.5-12	30806-4	30807-2
Width	22.0 in.	0.56 m	12.1-16	30808-0	30809-8
Depth	8.3 in.	0.21 m	16.1-20	30810-6	30811-4
Weight	130 lbs.	59.0 kg	20.1-24	30812-2	30813-0
			24.1-28	30814-8	30815-5
Recommended:			28.1-32	30816-3	30817-1
3-Phase 900 volt maximum service			32.1-40	30818-9	30819-7
Surge Capacitor			40.1-48	30820-5	30821-3
PART NUMBER 51084-2			48.1-50	30822-1	30823-9

DFH-2 SIZE 3 50 HORSEPOWER 600 VOLT MAXIMUM 60 HZ

Dimension			Amp load	Part number electro-mechanical	Part number solid-state
Height	46.5 in.	1.18 m	24.1-28	30824-7	30825-4
Width	23.0 in.	0.58 m	28.1-32	30892-4	30893-2
Depth	8.3 in.	0.21 m	32.1-40	31393-2	31394-0
Weight	180 lbs	81.6 kg	40.1-48	31395-7	31396-5
			48.1-56	31397-3	31398-1
			56.1-64	31399-9	31400-5
Recommended:			64.1-72	31401-3	31402-1
3-phase 900 v maximum service			72.1-80	31403-9	31404-7
Surge capacitor			80.1-88	31405-4	31406-2
PART NUMBER: 51084-2			88.1-100	31407-0	31408-8

Appendix 4D.3

TRW REDA   **SWITCHBOARDS**
TYPE DHF-2 600 VOLTS MAXIMUM
DFH-2 SIZE 4 100 HORSEPOWER 600 VOLT MAXIMUM 60 HZ

Dimension			Amp load	Part number electro-mechanical	Part number solid-state
Height	56.5 in.	1.44 m	24.1-28	31409-6	31410-4
Width	25.0 in.	.64 m	28.1-32	31411-2	31412-0
Depth	8.9 in.	.23 m	32.1-40	31413-8	31414-6
Weight	262 lbs	118.8 kg	40.1-48	31415-3	31416-1
			48.1-56	31417-9	31418-7
			56.1-64	31419-5	31420-3
Recommended:			64.1-72	31421-1	31422-9
3-Phase 900 volt maximum service			72.1-80	31428-7	31424-5
Surge Capacitor			80.1-88	31425-2	31426-0
PART NUMBER: 51084-2			88.1-100	31427-8	31428-6
			100.1-110	31429-4	31430-2
			110.1-120	31431-0	31432-8
			120.1-135	31433-6	31434-4
			135.1-150	31435-1	31436-9

DFH-2 SIZE 5 200 HORSEPOWER 600 VOLT MAXIMUM 60 HZ

Dimension			Amp load	Part number electro-mechanical	Part number solid-state
Height	71.5 in.	1.82 m	88.1-100	31437-7	31438-5
Width	30.0 in.	.76 m	101.1-110	31439-3	31440-1
Depth	13.0 in.	.33 m	110.1-120	31441-9	31442-7
Weight	600 lbs	272.2 kg	121.0-140	31443-5	31444-3
			141.0-160	31445-0	31446-8
			161.0-180	31447-6	31448-4
Recommended:			181.0-200	31449-2	31450-0
3-Phase 900 volt maximum service			201.0-220	31451-8	31425-6
Surge Capacitor			221.0-240	31453-4	31454-2
PART NUMBER: 51084-2			241.0-270	31455-9	31456-7

Appendix 4D.4

TRW REDA   **SWITCHBOARDS**
1000 VOLTS MAXIMUM
MFH SIZE 2 70 HORSEPOWER 1000 VOLT MAXIMUM 60 HZ

Dimension			Amp load	Part number electro-mechanical	Part number solid-state
Height	37.0 in.	.94 m	17	31078-9	31079-7
Width	34.0 in.	.86 m	18-23	31080-5	31081-3
Depth	12.0 in	.30 m	24-26	31534-1	31535-8
Weight	75 lbs	34.0 kg	27-40	31082-1	31083-9
Recommended:			41-45	31536-6	31537-4
3-Phase 2400 volt maximum service					
Surge Capacitor					
PART NUMBER: 51083-4					

MFH SIZE 3 70 HORSEPOWER 1000 VOLT MAXIMUM 60 HZ

Dimension			Amp load		
Height	68.0 in.	1.73 m	17	30862-7	30863-5
Width	26.3 in.	.67 m	18-23	30864-3	30865-0
Depth	20.3 in.	.52 m	27-40	30866-8	30867-6
Weight	530 lbs	240.4 kg			
Recommended:					
3-Phase 2400 volt maximum service					
Surge Capacitor					
PART NUMBER: 51083-4					

Appendix 4D.5

TRW REDA   **SWITCHBOARDS**
1000-1500 VOLTS MAXIMUM
MFH SIZE 3 160 HORSEPOWER 1000 VOLT MAXIMUM 60 HZ

Dimension			Amp load	Part number electro-mechanical	Part number solid-state
Height	68.0 in.	1.73 m	41-57	30868-4	30869-2
Width	26.3 in.	.67 m	58-77	30870-0	30871-8
Depth	20.3 in.	.52 m	78-120	30872-6	30873-4
Weight	530 lbs.	240.4 kg			
Recommended:					
3-Phase 2400 volt maximum service					
Surge Capacitor					
PART NUMBER: 51083-4					

MDFH SIZE 3 150 HORSEPOWER 1500 VOLT MAXIMUM 60 HZ

Dimension			Amp load	Part number electro-mechanical	Part number solid-state
Height	68.0 in.	1.73 m	20	31457-5	31458-3
Width	26.3 in.	.67 m	21-26	31459-1	31460-9
Depth	20.3 in.	.52 m	27-40	31461-7	31462-5
Weight	530 lbs	240.4 kg	41-57	31463-3	31464-1
Recommended:			58-77	31465-8	31466-6
3-Phase 2400 volt maximum service			78-100	31467-4	31468-2
Surge Capacitor					
PART NUMBER: 51083-4					

Appendix 4D.6

TRW REDA   **SWITCHBOARDS**

1500 VOLTS MAXIMUM

MDFH SIZE 4 250 HORSEPOWER 1500 VOLT MAXIMUM 60 HZ

	Dimension		Amp load	Part number electro-mechanical	Part number solid-state
Height	68.0 in.	1.73 m	41-57	31469-0	31470-8
Width	26.3 in.	.67 m	58-77	31471-6	31472-4
Depth	20.3 in.	.52 m	78-100	31473-2	31474-0
Weight	550 lbs	249.5 kg	101-130	31475-7	31476-5
Recommended:			131-150	31477-3	31478-1
3-Phase 2400 volt maximum service					
Surge Capacitor					
PART NUMBER: 51083-4					

RP 1 350 HORSEPOWER 1500 VOLT MAXIMUM 60 HZ

	Dimension		Amp load	Part number electro-mechanical	Part number solid-state
Height	56.5 in.	1.44 m	23-40	30879-1	30880-9
Width	38.5 in.	.98 m	41-60	30881-7	30882-5
Depth	34.5 in.	.88 m	61-80	30883-3	30884-1
Weight	1100 lbs	499.0 kg	81-100	30885-8	30886-6
Recommended:			101-120	30887-4	30888-2
3-Phase 2400 volt maximum service		121-160	30889-0	30890-8
Surge Capacitor					
PART NUMBER: 51083-4					

Appendix 4D.7

TRW REDA   **SWITCHBOARDS**

2400 VOLTS MAXIMUM

RP 2 (NEMA 3R/12) 600 HORSEPOWER 2400 VOLT MAXIMUM 60 HZ

	Dimension		Amp load	Part number electro-mechanical	Part number solid-state
Height	56.5 in.	1.44 m	35	30894-0	30895-7
Width	38.5 in.	.98 m	36-50	30896-5	30897-3
Depth	34.5 in.	.88 m	51-70	30898-1	30899-9
Weight	1100 lbs	499.0 kg	71-100	30900-5	30910-3
Recommended:			101-115	30902-1	30903-9
3-Phase 2400 volt maximum service		116-135	30904-7	30905-4
Surge Capacitor			136-140	30906-2	30907-0
PART NUMBER: 51083-4					

1512 700 HORSEPOWER 2400 VOLT MAXIMUM 60 HZ

	Dimension		Amp load	Part number electro-mechanical	Part number solid-state
Height	68.6 in.	1.74 m	35	31366-8	31367-6
Width	39.0 in.	.99 m	36-50	31368-4	31369-2
Depth	38.6 in.	.98 m	51-70	31370-0	31371-8
Weight	1095 lbs	496.7 kg	71-100	31372-6	31373-4
Recommended:			101-115	31374-2	31375-9
3-Phase 2400 volt maximum service		116-135	31376-7	31377-5
Surge Capacitor			136-140	31378-3	31379-1
PART NUMBER: 51083-4					

Appendix 4D.8

**CENTRILIFT®
SUBMERSIBLE
PUMPS
TYPE I SWITCHBOARD**   **SWITCHBOARD PRICES**

			RATING			DIMENSIONS, in.			
Type	Model	Price*	Max. Volts	Max. Amps	Max. H.P.	Height	Width	Depth	Weight, Lbs.
2C-1	D-35431	$1,260.00	480	50	25	53	30	11	385
3C-1	D-35432	1,470.00	480	100	50	53	30	11	390
4C-1	D-35433	1,770.00	480	150	100	53	30	11	395
5C-1	D-35434	2,450.00	480	300	200	53	30	11	400
2B-1	D-35429	1,470.00	880	50	40	53	30	11	390
3B-1	D-35430	1,625.00	880	100	75	53	30	11	395
3AC-1	B-36025	2,625.00	1500	100	150	64	27	19	625
4AC-1	B-36027	3,150.00	1500	150	250	64	27	19	630
6W-1	D-35936	5,355.00	2500	200	400	62	29	31	1050

*Price includes as standard control equipment: undercurrent shutdown; automatic restart, and recording ammeter, and over current shutdown.

CENTRIGARD® SWITCHBOARD

			RATING			DIMENSIONS, in.			
Type	Model	Price**	Volts	Amps	H.P.	Height	Width	Depth	Weight, Lbs.
2C-CG	C-39137	$1,470.00	480	50	25	53	30	11	375
3C-CG	C-38995	1,680.00	480	100	50	53	30	11	380
4C-CG	C-39077	1,880.00	480	150	100	53	30	11	385
5C-CG	C-39078	2,660.00	480	300	200	53	30	11	390
2B-CG	A-39419	1,680.00	880	50	40	53	30	11	380
3B-CG	A-39420	1,835.00	880	100	75	53	30	11	385
3AC-CG	B-39001	2,825.00	1500	100	150	64	27	19	615
4AC-CG	B-39076	3,360.00	1500	150	250	64	27	19	620
6W-CG	B-39079	5,565.00	2500	200	400	62	29	31	1050

**Price includes, as an integral part of the Centrigard® motor controller, undercurrent and overcurrent shutdown on each phase and direct three-phase current readout, together with an optional recording ammeter, and automatic restart.
Purchase of Centrigard® motor controller (including mounting bracket, pigtail and terminal block) only - $980.00
No control circuit wiring harness available for switchboards except those manufactured by Centrilift, Inc.
Service for installation of Centrigard® Controller charged at published rates on Page E—4.
Repair and exchange of Centrigard® Motor Controller - $450.00

Appendix 4D.9
SWITCHBOARDS & ACCESSORIES
440 VOLT, 880 VOLT, 1500 VOLT, 2500 VOLT
DESCRIPTION & PRICES

ODI SWITCHBOARDS

ODI Switchboards are designed for use with submersible motors. They are weatherproof (NEMA 3), enclosed, and are readily adaptable to the use of external controls, such as float or pressure shut down. Standard Equipment, with all types of Switchboards includes:

1. 3 Pole, fused disconnect switch with interlock to inhibit opening to power components while energized. Except Size 5—Circuit Breaker.
2. 3 Pole Magnetic Contactor.
3. 110 Volt Control Circuit
4. Hand-Off-Automatic Selector and Start Push-button switches.
5. Inverse Time Characteristic Overload relays.
6. Undercurrent Shutdown.
7. Automatic Restart (after undercurrent shutdown).
8. Recording Ammeter.
9. Lightning Arrestors.

SPECIAL FEATURES
TYPES S and M (440 & 880 Volt)

1. Disconnect is snap acting type. (See Note 8 for Size 5.)
2. Same compartment for power and control circuits.
3. Overloads relays are magnetic dash-pot type.
4. Undercurrent relay is magnetic type.
5. Automatic restart is separate timer type.
6. Lightning arrestor is 30 externally mounted.
7. Optional solid state RELIATROL, highest reliability, low maintenance control with run, overload and underload indicating lights, and including single phase protection at additional cost. (See OPTIONS—Page 5.52).
8. For Size 5-440 Volt, Circuit Breaker and RELIATROL are standard.
9. Pole mounted (foot mounting optional).

TYPE H (1500 Volt)

1. Disconnect is isolated from enclosure and is snap acting.
2. Separate compartments for the power and control circuits are included. An inner door is electrically and mechanically interlocked to inhibit entry into the power compartment while running.
3. The solid state RELIATROL is standard with 1500 volt switchboards. The overload, underload, restart timing, auxiliary control contacts, back spin protection as well as single phase protection are all included in RELIATROLLED 1500 volt switchboard.
4. Indicating lights to show RUN, OVERLOAD and UNDERLOAD conditions are included.
5. The 1500 volt switchboard has been designed to be as maintenance free as possible. Problems such as dusty or corroded control circuit contacts are virtually non-existent.
6. Foot Mounted.

SPECIAL FEATURES
TYPE HH (2500 Volt)

1. Isolating switch blades operate to disconnect power source for servicing.
2. Separate compartments for the power and control circuits are included. An inner door is electrically and mechanically interlocked to inhibit entry into the power compartment while running.
3. The solid state RELIATROL is standard with 2500 volt switchboards. The overload, underload, restart timing, auxiliary control contacts, back spin protection as well as single phase protection are all included in RELIATROLLED 2500 volt switchboard.
4. Indicating lights to show RUN, OVERLOAD and UNDERLOAD conditions are included.
5. The 2500 volt switchboard has been designed to be as maintenance free as possible. Problems such as dusty or corroded control circuit contacts are virtually non-existent.
6. Current limiting fuses are provided.
7. Foot Mounted.

RATINGS, SPECIFICATIONS AND PRICES

Size	Model	Volts	Max. Amps	Max. HP	Height	Width	Depth	(Pounds)	PRICE
S2	1099A	440	50	25	37	24½	9½	215	$1205.00
S3	1099A	440	100	50	54	28	10½	290	1360.00
S4	1099A	440	150	100	54	28	10½	310	1690.00
S5	1099R	440	300	200	54	28	10½	350	2575.00
M2	1099A	880	50	40	54	28	10½	290	1480.00
M3	1099A	880	100	75	54	28	10½	295	1710.00
H3	1550	1500	100	150	68	28	20	680	2750.00
H4	1550	1500	150	225	68	28	20	710	3260.00
HH3	1102D	2500	200	700	68	36	34	930	5550.00

See Page 5.52 for optional switchboard equipment and accessories

Appendix 4D.10 Kobe Control Panels
TYPE AO
480 VOLT CONTROL PANELS

Dimensions:	Height 69½"	Width 26"	Depth 21"	Weight 300#
Size	3	4	5	
Maximum Amps	100	150	300	

TYPE DO
1500 VOLT CONTROL PANELS

Dimensions:	Height 69½"	Width 30"	Depth 25"	Weight 625#
Size	3	4		
Maximum Amps	100	150		

TYPE EO
3000 VOLT CONTROL PANELS

Dimensions:	Height 92"	Width 38"	Depth 34"	Weight 2200#
Size	1			
Maximum Amps	200			

CONTROL PANEL DESCRIPTION
TYPE AO (480 VOLT)
BASIC PANEL:

NEMA Type 3 (Outdoor Type) Pad Mounted Enclosure. 2" Conduit Hub. 3 Pole fused disconnect switch for 3 phase short circuit protection. 3 Pole magnetic contactor. Current transformers, each phase. 110 volt control transformer for 110 volt control circuit. Overload relay, each phase. Adjustable underload shutdown. Automatic restart. Recording ammeter w/Bristol circular chart and 7 day wind up drive. Hand-off automatic selector switch and start push button. Float or pressure switch terminal. Motor Protection Package.

TYPE DO (1500 VOLT)
BASIC PANEL:

NEMA Type 3 (Outdoor Type) 2 door Enclosure. Pad Mounted. 3 Pole interlocked disconnect switch. 2-3 Pole magnetic contactors with 110 Volt coil. Current transformers, each phase. 1500/110 Volt secondary control transformer. Overload relay, each phase. Recording ammeter w/Bristol circular chart and 7 day wind up drive. Adjustable underload shutdown. Automatic restart interval timer. Hand-off automatic selector switch and start push buttom. Float or pressure switch terminal. Motor Protection Package.

CONTROL PANEL INFORMATION
TYPE EO (3000 VOLT)
BASIC PANEL:

NEMA Type 3 (Outdoor Type) 2 Door Enclosure. Pad Mounted. 3 Pole, 50,000 KVA interrupting capacity, draw out type contactor having inherent line disconnect, actuated by overcurrent relay in event of short circuit. Current transformers, each phase. Fused 110 volt control transformers. Overload relay, each phase. Recording ammeter w/Bristol circular chart and 7 day wind up drive. Hand-off automatic selector switch and start push putton. Adjustable underload shutdown. Automatic restart interval timer. Float or pressure switch terminal. Motor Protection Package.

KRATOS PROTECTION/CONTROL CENTER

KRATOS Protection/Control Centers are designed to control and protect downhole motor-pump systems. These completely solid-state devices are used to replace all electro-mechanical control, undercurrent, and overcurrent relays; automatic restart and backspin timers; and all time delay devices normally found in motor-pump controller or starter panels. By replacing electro-mechanical devices with highly reliable, solid-state electronic components, down-hole protection is improved, well downtime is reduced, and equipment maintenance and repair costs are minimized. In addition, improved monitoring concepts provide extra insurance against pump failures.

Protection of downhole motor-pump systems during underload conditions caused by well pump-off, is accomplished by automatic shutdown for an adjustable, preset time period. At the end of this time delay, automatic restart occurs. This sequence of shutdown and time-delayed restart will continue for as long as the underload condition persists. Maximum underload protection is provided by sensing power rather than current since power more accurately reflects underload conditions. A standard, five second delay is incorporated to prevent unnecessary shutdown when gas bubbles or other short duration underload conditions occur. Other delay times are available.

Sensing of overload or single-phase conditions is accomplished by monitoring all three phase currents and comparing the highest current with an adjustable trip point. When any phase current exceeds the trip point, shutdown occurs, and restart is prevented until the unit is manually reset. Because sensing is accomplished by means of potential and current transformers, KRATOS Protection/Control Centers may be operated with motor-pumps of any size merely by selecting proper transformer ratios.

Provisions for remote indicating lights (Run, Overload, Underload) and remote control contacts (pressure, float, interlock) are standard.

KRATOS Protection/Control Centers are completely solid-state, and provide long life with excellent reliability in environmental conditions from polar to desert regions including extremely humid atmospheres, and require no periodic maintenance.

Appendix 4E Transformers

Appendix 4E.1

TRW REDA   **TRANSFORMERS**

SINGLE PHASE OISC 60 HERTZ
55 DEGREES CENTIGRADE RISE

| Size | Height | | Width | | Depth | | Weight | | Primary | Secondary | Part |
KVA	in	m	in	m	in	m	lbs	kg	volts	volts	number
25	40.0	1.02	22.5	.57	24.8	.63	460	209	12,500	480/960	66270-0
25	40.0	1.02	22.5	.57	24.8	.63	460	209	14,400/24,900	480/960	69445-5
50	50.0	1.27	31.3	.80	27.3	.69	815	370	12,500	600/1200	69133-7
50	50.0	1.27	31.3	.80	27.3	.69	815	370	12,500	700/1400*	69373-9
50	50.0	1.27	31.3	.80	27.3	.69	815	370	12,500	1200/2400*	69268-1
50	50.0	1.27	31.3	.80	27.3	.69	815	370	14,400/24,900	600/1200	69925-6
50	50.0	1.27	31.3	.80	27.3	.69	815	370	14,400/24,900	700/1400*	69828-2
50	50.0	1.27	31.3	.80	27.3	.69	815	370	14,400/24,900	1200/2400*	78995-8
75	53.0	1.35	32.9	.84	28.6	.73	1135	515	12,500	1200/2400*	69313-5
75	53.0	1.35	32.9	.84	28.6	.73	1135	515	14,400/24,900	1200/2400*	69532-0
100	55.0	1.40	3.40	.86	29.0	.74	1350	612	12,500	1200/2400*	75198-2
100	55.0	1.40	3.40	.86	29.0	.74	1350	612	14,400/24,900	1200/2400*	70940-2
150	58.0	1.47	36.0	.91	33.5	.85	1850	839	12,500	1200/2400*	70941-0
150	58.0	1.47	36.0	.91	33.5	.85	1850	839	14,400/24,900	1200/2400*	70944-4
200	66.0	1.68	42.0	1.07	40.0	1.02	2400	1089	12,500	1200/2400*	80261-1
200	66.0	1.68	42.0	1.07	40.0	1.02	2400	1089	14,400/24,900	1200/2400*	70025-2

*Supplied with high voltage secondary bushings

Oil-immersed, self-cooled (OISC), single-phase, 60 cycle, 55 degrees Centigrade rise with low voltage ratings of 480/960, 600/1200 volts, 700/1400, 1200/2400 Standard tap arrangement provides two 5% taps below normal and two 5% taps above normal.

Note: 1. Special transformers for voltages and KVA sizes other than those shown are available.

2. A 480-volt reduced capacity metering tap can be provided by special order.
3. High voltage secondary bushings are furnished standard for 2400 volt secondaries. High voltage secondary bushings are required for all wye connected secondaries and are available on special order.

Appendix 4E.2

TRW REDA   **TRANSFORMERS**

THREE PHASE AUTO TRANSFORMER
OISC 60 HERTZ

| Size | Height | | Width | | Depth | | Weight | | Primary | Secondary | Part |
KVA	in	m	in	m	in	m	lbs	kg	volts	volts	number
30	50.0	1.27	31.0	.79	20.8	.53	800	363	440/480	700/900	70946-9
50	49.0	1.24	34.0	.86	20.8	.53	870	395	440/480	800/1000	69481-0
75	54.0	1.37	39.0	.99	23.6	.60	1195	542	440/480	800/1000	69950-4
100	55.0	1.40	38.0	.97	27.6	.70	1450	658	440/480	800/1000	69136-0
125	55.0	1.40	38.0	.97	27.6	.70	1925	873	440/480	800/1000	69336-6
150	59.0	1.50	44.0	1.12	29.8	.76	1995	905	440/480	800/1000	70254-8
200	49.5	1.26	66.0	1.68	32.0	.81	2450	1111	440/480	800/1000	78958-6
225	49.5	1.26	66.0	1.68	32.0	.81	2550	1157	440/480	800/1000	77921-5
250	49.5	1.26	66.0	1.68	32.0	.81	2775	1259	440/480	800/1000	80266-0

Oil-immersed, self-cooled (OISC), three-phase, 60-cycle autotransformer for use with 440/480-volt power supply. The low voltage 440/480 is adjustable by means of an internal terminal board and the high voltage 800/850/900/940/1000 is obtained by adjustment of an accessible tap changer. Special autotransformers with secondary (high voltage) voltages and/or frequencies other than shown are available.

Appendix 4E.3

CENTRILIFT®
SUBMERSIBLE
PUMPS
CENTRILIFT TRANSFORMERS

TRANSFORMER PRICES

Single-Phase, Oil-Filled, Outdoor (OFO) Type Transformers With Special Centrilift Winding
(Bank of three single-phase transformers required for three-phase operation.)

Size KVA	Price Each	Drawing Number	Voltage Primary	Voltage Delta Secondary	Line Amps Series Secondary	Dimensions Height	Dimensions Diameter	Weight, Pounds
28	$ 667.	D-32334	2,400/4,160	15 possible voltages 912-1310	37	42″	16″	480
28	677.	D-31437	7,200/12,470	1250-1750	28			
28	708.	D-32787	14,400/24,900	1900-2500	20	42″	16″	480
	810.	same			50			
37½	850.	as above	same as above	same as above	37	42″	16″	625
	887.				27			
	940.	same			68			
50	980.	as above	same as above	same as above	50	42″	20″	810
	1030.				35			
	1300.	same			100			
75	1320.	as above	same as above	same as above	75	46″	21″	1,025
	1340.				54			
					132			
100	1695.	D-33920	7,200/12,470	same as above	100	48″	23″	1,050
	1745.	D-33870	14,400/24,940		72			
					178			
135	2240.	D-35884	7,200/12,470	same as above	135	52″	32″	1,350
	2260.	D-35883	14,400/24,940		96			
					220			
167	2320.	D-35882	7,200/12,470	same as above	167	52″	32″	1,500
	2340.	D-35881	14,400/24,940		120			

Three-Phase, Oil-Filled, Outdoor (OFO) Type Auto Transformers With Special Centrilift Winding

KVA	Each	Number	Voltage Primary	Voltage Secondary	Line Amps Secondary	Dimensions Height	Dimensions Width	Dimensions Depth	Pounds
50	$1030.	D-34013	440/460/480	750-1050	27	32″	26″	16″	650
	1125.	D-35975		800-1250	23				
75	1280.	D-34013	440/460/480	750-1050	41	32″	28″	16″	900
	1330.	D-35975		800-1250	35				
	1435.	D-34014		850-1450	30				
100	1530.	D-34013	440/460/480	750-1050	55	36″	30″	18″	1,120
	1610.	D-35975		800-1250	46				
	1660.	D-34014		850-1450	40				
125	1785.	D-34013	440/460/480	750-1050	69	36″	32″	18″	1,260
	1830.	D-35975		800-1250	58				
	1875.	D-34014		850-1450	50				
150	2195.	D-34013	440/460/480	750-1050	82	36″	38″	22″	1,400
	2240.	D-35975		800-1250	69				
	2280.	D-34014		850-1450	60				
200	3225.	D-35975	440/460/480	800-1250	92	36″	38″	22″	1,880
	3320.	D-34014		850-1450	80				
250	3580.	D-35975	440/460/480	800-1250	115	40″	38″	22″	1,920
	3760.	D-34014		850-1450	110				

For isolated wound auto transformers, multiply above prices by 1.54.

Appendix 4E.4

CENTRILIFT®
SUBMERSIBLE
PUMPS
CENTRILIFT TRANSFORMERS

TRANSFORMER PRICES

Single-Phase, Oil-Immersed-Self-Cooled (OISC)Type Transformers With Centrilift Windings
(Bank of three single-phase transformers required for three-phase operation.)
NOTE: ALL TRANSFORMERS HAVE 2400 VOLT SECONDARY BUSHINGS

Size KVA	Price Each	Drawing Number	Voltage Primary	Delta Secondary	Maximum Sec. 3φ Line Amps	Maximum Horsepower 450	544	Weight, Pounds
		D-38551	2400/4160Y					
	$ 705.	D-38552	7200/12,470Y					
		D-38553	7620/13,200Y					
30					35	60	75	390
		D-38554	13,200/22,860Y					
	780.	D-38555	14,400/24.940Y					
		D	2400/4160Y					
	790.	D-38557	7200/12,470Y					
		D	7620/13,200Y					
37.5					43	85	85	620
		D-39335	13,200/22,860Y					
	870.	D	14,400/24,940Y					
				400-1500				
		D	2400/4160Y					
	940.	D-38563	7200/12,470Y					
		D	7620/13,200Y					
50					58	100	100	675
		D-39336	13,200/22,860Y					
	1030.	D	14,400/24,940Y					
		D-38558	2400/4160Y					
	1095.	D-38559	7200/12,470Y					
		D-38560	7620/13,200Y					
60					70	120	150	682
		D-38561	13,200/22,860Y					
	1200.	D-38562	14,400/24,940Y					
		D	2400/4160Y					
	$1200.	D-38564	7200/12,470Y					
		D	7620/13,200Y					
60					43	120	125	682
		D	13,200/22,860Y					
	1320.	D	14,400/24,940Y					
		D-38565	2400/4160Y					
	1515.	D-38566	7200/12,470Y					
		D-38567	7620/13,200Y					
100				666-2500	70	200	225	1230
		D-38568	13,200/22,860Y					
	1700.	D-38569	14,400/24,940Y					
		D-38572	2400/4160Y					
	2080.	D-38573	7200/12,470Y					
		D-38574	7620/13,200Y					
150					105	200	400	1560
		D-38575	13,200/22,860Y					
	2270.	D-38576	14,400/24,940Y					

Appendix 4E.5

TRANSFORMERS
SINGLE-PHASE, OIL-IMMERSED SELF COOLED TYPE TRANSFORMERS

Transformers are available in the primary voltages of 2,400/4,160 Y; 7,200/12,470 Y; 13,200; 14,400/24,940 Y.

Additional primary voltages and frequencies are available. Contact ODI for price and delivery.

NOTE: As shown in the tabulation below, ODI Power Transformers have a wide range of secondary voltages to accommodate a wide range of motor sizes and voltages in the event it is desired to change pumping equipment size, move to another location, etc.

KVA Size	Secondary Voltage	Height	Width	Depth	Weight
25	456/1310	36″	21″	24″	400#
37,5	625/1750	39″	23″	26″	550#
50	625/1750	45″	23″	26″	670#
75	625/1750	49″	28″	29″	980#
100	625/1750	49″	33″	29″	1,125#

NOTE: Secondary voltages above are those obtained with Delta connected secondary. Taps are provided for numerous intermediate voltages. Additional secondary voltages can be obtained by WYE connecting the secondary.

THREE PHASE AUTO TRANSFORMERS, OIL IMMERSED SELF COOLED

KVA	Primary Voltage	Secondary Voltage	Secondary Amps From H.V. Tap Setting	To L.V. Tap Setting	Dimensions Height	Width	Depth	Weight
			From	To				
50	440/460/	750-1050	27	38	33″	27″	28″	650#
50	480	800-1250	23	36				
75	″	800-1250	35	55	33″	27″	32″	900#
75		850-1450	30	51				
100	″	850-1450	40	68	37″	35″	32″	1120#
125		850-1450	50	85	37″	35″	34″	1250#
150	″	850-1450	60	102	42″	33″	34″	1400#
200		850-1450	80	136	43″	40″	37″	1880#
250		850-1450	100	170	43″	40″	37″	1920#

NOTES:
1. Pole hangers not availabe for 200 & 250 KVA.
2. Voltage steps are 50 V. on 750-1050, 75 V. on 800-1250 and 100V. on 850-1450 secondary voltages. However, intermediate voltages are available by taking advantage of the three primary taps.

Appendix 4E.6
KOBE Transformer Specifications

Single phase transformer

These transformers are designed and built expressly for motor starting service and for continuous operation at full load. The secondary voltages are chosen to match the motors with which they will be used. The transformers are pole mounted, oil immersed, self cooled (OISC).

The following sizes and voltages are standard. Other voltages are available on special order. Primary Voltage 2400/4160Y or 7200/12470Y or 14400/24940Y

KVA	Secondary Voltage	Weight Pounds
25	1100 V.C.T.	390
37½	1300 V.C.T.	600
50	1350 V.C.T.	685
75	1350 V.C.T.	875
75	2400 V.C.T.	900
100	1350 V.C.T.	1200
100	2400 V.C.T.	1225

Three phase auto and isolation transformer

These transformers are designed and built expressly for motor starting service and for continuous operation at full load. The transformers are slab mounted, oil immersed, self cooled (OISC).

The following sizes and voltages are standard. Other voltages are available on special order. Primary Voltage 460/480 volts.

KVA	Secondary Voltage	Weight Pounds
50	1100 V.C.T.	1200
75	1100 V.C.T.	1600
100	1300 V.C.T.	1850
112½	1300 V.C.T.	1875
125	1300 V.C.T.	1900
150	1350 V.C.T.	2050
200	1350 V.C.T.	2950
225	1350 V.C.T.	2975
250	1350 V.C.T.	3000
300	1350 V.C.T.	3350

Appendix 4F Motors

Appendix 4F.1

TRW REDA

MOTORS 60 Hz

375 SERIES (3.75" OD)

HP	Volts	Amp
7.5	415	13.5
10.5	400	20
	690	12
15	330	34
	415	27
19.5	415	35
	650	22.5
22.5	440	38.5
	750	22.5
25.5	650	29.5
	780	24.5

TANDEM MOTORS

HP	Volts	Amp
30	630	35.5
39	575	51
	774	38
45	660	51.5
51	740	51
	1000	37
	1250	31
58.5	860	51
67.5	990	51.5
76.5	1110	51
90	1320	51.5
102	1480	51
112.5	1650	51.5
127.5	1850	51

738 SERIES (7.38" OD)

HP	Volts	Amp
200	2300	53
220	1350	97
	2300	57
240	2300	64
260	2300	70

TANDEM MOTORS

HP	Volts	Amp
400	2300	106
440	2000	131
480	2200	134
520	2300	140
600	3450	106
680	3200	140
720	3300	134

456 SERIES (4.56" OD)

HP	Volts	Amp
10	440	15
15	440	23
	750	14
20	460	28
	760	17
25	420	38
	700	22
30	440	43
	765	25
35	400	55
	690	32
	800	27.5
40	450	57
	675	38
	790	32.5
	900	28.5
50	700	45.5
	840	38
	980	32.5
60	670	57
	775	50
	840	45
	1000	38
70	785	57
	980	45
	1170	38
80	900	57
	1120	45
	1350	38
90	1000	57
	1260	45
	1500	38
	2000	29
100	970	66
	1120	57
	1400	45
	2250	29
110	1080	65
	1240	57
120	1000	77
	1170	66
	1350	57
	2300	34

TANDEM MOTORS

HP	Volts	Amp
140	1080	82.5
	2270	39
160	1270	80
	2160	47.5
180	2270	50
200	2140	59
220	2380	60
240	2250	70

540 SERIES (5.43" OD)

HP	Volts	Amp
20	445	29
	762	17
30	445	44
	720	27.5
40	445	59
	670	39
	740	36
	890	30
50	430	75
	740	44
	920	33
60	445	87
	665	58
	755	52
	890	44
	990	39
70	775	58
	880	51
	1035	44
80	685	76
	770	68
	890	58
	1185	44
100	740	85
	855	74
	960	66
	1100	58
	2200	29
120	770	98
	890	85
	1330	57
	2200	32
130	835	98
	965	84
150	965	97
	1150	75
	2150	43
160	1015	99
	2230	45
180	1000	113
	2000	57
200	1160	105
	2200	53
225	1200	120
	2300	62.5

TANDEM MOTORS

HP	Volts	Amp
240	2060	73
260	2250	67
300	2150	87
320	2230	88.5
360	1890	120
400	2200	115
480	2475	122
600	3300	115

Appendix 4F.2
MOTORS 50 Hz

375 SERIES (3.75" OD)				456 SERIES (4.56" OD)				540 SERIES (5.43" OD)		
HP	Volts	Amp		HP	Volts	Amp		HP	Volts	Amp
6.25	345	13.5		8.5	367	15		16.5	371	29.0
8.8	333	20		12.5	367	23			635	17
	575	12			625	14		25	371	44
12.5	275	34		16.5	383	28			600	27.5
	346	27			633	17		33.5	371	59
16.3	346	35		21	350	38			558	39
	541	22.5			583	22			616	36
18.8	367	38.5		25	367	43			741	30
	625	22.5			637	25		41.5	358	75
21.8	541	29.5		29.5	333	55			616	44
	650	24.5			575	32			766	33
					666	27.5		50	371	87
TANDEM MOTORS				33.5	375	57			554	58
					562	38			629	52
25	525	35.5			658	32.5			741	44
32.5	479	51			750	28.5			825	39
	645	38		41.5	583	45.5		58.5	646	58
37.5	550	51.5			700	38			733	51
42.5	616	51			816	32.5			862	44
	833	37		50	558	57		66.5	571	76
	1041	31			646	50			641	68
48.8	716	51			700	45			741	58
56.3	825	51.5			833	38			987	44
63.8	925	51		58.5	654	57		83.5	616	85
75	1100	51.5			816	45			712	74
85	1233	51			975	38			800	66
93.8	1374	51.5		66.5	750	57			916	58
106.3	1541	51			933	45			1833	29
					1125	38		100	641	98
738 SERIES (7.38" OD)				75	833	57			741	85
					1050	45			1108	57
HP	Volts	Amp			1250	38			1833	32
					1666	29		108	696	98
167	1916	53		83.5	808	66			804	84
183	1916	57			933	57		125	804	97
	1125	97			1166	45			960	75
200	1916	64			1874	29			1791	43
216	1916	70		91.5	900	65		133	845	99
					1033	57			1858	45
TANDEM MOTORS				100	833	77		150	833	113
					975	66			1666	57
334	1916	106			1125	57		167	966	105
336	1666	131			1916	34			1833	53
400	1833	134						187	1000	120
432	1916	140		**TANDEM MOTORS**					1916	62.5
500	2875	106								
566	2666	140		117	900	82.5		**TANDEM MOTORS**		
600	2750	134			1891	39				
				133	1800	47.5		200	1716	73
					1058	80		216	1874	67
				150	1891	50		250	1791	87
				167	1783	59		266	1858	88.5
				183	1983	60		300	1514	120
				200	1874	70		334	1833	115
								400	2062	122
								500	2750	115

Appendix 4F.3
CENTRILIFT® SUBMERSIBLE PUMPS
450 SERIES MODEL FMP MOTOR PRICES
450 SERIES—FOR 5½″ O.D. 20# OR LARGER CASING SIZES—MODEL FMP

Size	Volts/amps	Length (shipping)	Weight (lbs.)
15 HP	440/22	7′2″	269
20 HP	420/31	8′5″	332
25 HP	430/38 750/22	9′9″	396
30 HP	430/46 740/27	11′	459
35 HP	430/53 960/24	12′3″	523
40 HP	420/62 965/27	13′6″	586
50 HP	1200/27	16′	713
60 HP	1270/30	18′7″	840
75 HP	1130/43	22′4″	1031
85 HP	1290/43	24′10″	1158
100 HP	1150/56 2080/31	28′7″	1305
120 HP	1200/64 2080/37	33′7″	1605
Motor models below are two-piece construction			
* 135 HP	1525/56	28′7″, 11′6″	1823
* 150 HP	1690/56	28′7″, 15′3″	2025
* 175 HP	1965/56	28′7″, 21′7″	2363
* 200 HP	2240/56	28′7″, 27′10″	2700

* Requires 1 Motor to Motor Flat Cable.

SEAL SECTION FOR 450 SERIES MOTOR

Series	Model	Part no.	Length	Weight (lbs.)
400	FSD	35268	5′	120

Appendix 4F.4
CENTRILIFT® SUBMERSIBLE PUMPS
544 SERIES MODEL GMP MOTOR PRICES
544 SERIES—FOR 6⅝″ O.D. 26# OR LARGER CASING SIZES—MODEL GMP

Size	Volts/amps	Length (shipping)	Weight (lbs.)
20 HP	440/28 760/17	4′11″	277
30 HP	440/40 760/23	6′4″	376
40 HP	440/60 740/35	9′6″	519
50 HP	430/72 745/42	10′11″	624
60 HP	1010/34	12′3″	718
75 HP	1350/35	15′	917
85 HP	1270/42	16′5″	962
100 HP	1300/46	17′9″	1116
125 HP	1125/68 2270/35	20′7″	1315
150 HP	1300/70 1950/46	24′8″	1613
180 HP	1180/92 2040/53	28′10″	1912
200 HP	1130/105 2270/53	31′7″	2076
225 HP	1210/110 2100/63	34′	2240
Motor models below are two-piece construction			
* 250 HP	1405/105	31′7″, 11′	2595
* 275 HP	1565/105	31′7″, 13′9″	2928
* 300 HP	1675/105 2265/82	31′7″, 16′6″	3226
* 350 HP	1950/105	31′7″, 23′5″	3724
* 375 HP	2105/105	31′7″, 27′7″	3893
* 400 HP	2225/105	31′7″, 30′4″	4222

*Requires 1 Motor to Motor Flat Cable.

SEAL SECTION FOR 544 SERIES MOTOR

Series	Model	Part no.	Length	Weight (lbs.)
513	GSC	38227	6′9″	265
513*	GSCP	39144	6′9″	265
513-400 Adaptor**	GSC	—	4″	10
675	HSCP	37712	7′9″	480

*For use with 150 horsepower motor and larger.
**Required when 400 Series Pump is installed with 544 Series Motor and 513 Series Seal Section.

Appendix 4F.5
MOTORS (60 CYCLE, 3500 RPM)
RATING, SPECIFICATIONS AND PRICES
55 SERIES FOR 5½" O.D AND LARGER CASING

HP	Volts	Amps	Length	Weight, lbs.
15	430	22	5.8'	263
22½	430	33	7.5'	350
	700	21		
30	430	44	9.0'	425
	720	27		
37½	450	53	10.8'	512
	720	33		
	890	28		
45	750	38	12.5'	595
	1080	27		
52½	750	44	14.3'	687
	1130	30		
60	720	53	15.8'	763
	1150	33		
75	1250	38	19.2'	935
90	1290	45	22.5'	1095
	2150	27		
105	1260	53	25.9'	1265
	2270	30		
120	1150	66	29.3'	1430
	2280	34		

70 SERIES MOTORS FOR 7" O.D. AND LARGER CASING

HP	Volts	Amps	Length	Weight, lbs.
30	430	45	5.1'	350
	725	27		
40	445	60	6.0'	415
	720	37		
	935	28		
50	750	41	7.7'	535
	975	32		
60	765	49	8.2'	590
	1110	33		
75	730	63	9.3'	685
	1160	40		
90	1165	47	11.1'	825
100	1200	51	12.9'	960
	2250	27		
125	1270	60	14.7'	1095
	2180	35		
150	1300	70	18.0'	1370
	2250	40		
180	1260	85	21.5'	1640
	2180	49		
200	2100	57	23.0'	1780
225	2320	58	25.0'	1915
245	2150	69	26.7'	2055
275	2300	72	30.1'	2325

Oil Dynamics, Inc., Tulsa, Oklahoma 74145

Appendix 4F.6
KOBE MOTOR SPECIFICATIONS (60 HZ)
SO MOTORS (4.56 INCHES O.D.)

Size horsepower	Voltage	Amperes	Length	Weight pounds
30	425	49	15'-4"	490
	735	29		
	960	22		
40	440	62	17'-4"	620
	715	38		
	980	28		
50	710	47	19'-4"	710
	1225	27		
60	1070	37	21'-4"	810
	1180	33		
70	1085	43	23'-4"	920
	1250	37		
80	1060	50	25'-4"	1030
	1240	43		
90	1190	49	27'-4"	1080
100	1100	59	29'-10"	1190
	1300	50		
120	1070	72	33'-10"	1420
	1300	59		
	2200	37		

TANDEM MOTORS

Size horsepower	Voltage	Amperes	Length	Weight pounds
140	2170	43	40'-4"	1840
	2500	37		
160	2120	50	44'-4"	2060
	2480	43		
180	1980	59	48'-4"	2180
	2380	49		
200	2200	59	52'-4"	2400

Appendix 4F.7
MO MOTORS (5.5 INCHES O.D.)

Size horsepower	Voltage	Amperes	Length	Weight pounds
30	425	45	11'-9"	450
45	425	67	13'-3"	590
	700	41		
60	940	41	14'-9"	730
	1140	34		
75	1060	45	16'-3"	840
	1180	40		
90	1140	50	17'-9"	960
	1250	44		
105	1160	57	19'-9"	1060
	1325	50		
120	1140	65	21'-3"	1180
	1325	57		
	2200	37		
135	1060	80	22'-9"	1320
	1280	67		
	2300	39		
150	940	100	24'-3"	1490
	1180	80		
	2300	44		
165	1220	85	25'-9"	1510
180	1065	106	27'-3"	1680
	1300	85		
	2300	50		
210	1270	106	30'-3"	2020
	2300	58		
240	1250	117	31'-9"	2360
	2300	63		

TANDEM MOTORS

Size horsepower	Voltage	Amperes	Length	Weight pounds
300	2600	77	44'-3"	2940
360	2700	90	50'-3"	3320
420	2600	105	56'-3"	4000

Appendix 4G

Appendix 4G.1
CASING INTERNAL DIAMETERS

Size	Wt/Ft	I.D.	Drift Diameter		Size	Wt/Ft	I.D.	Drift Diameter
4½″	9.50	4.090	3.965		8⅝″	44.00	7.625	7.500
4½″	10.50	4.052	3.927		8⅝″	49.00	7.511	7.386
4½″	11.60	4.000	3.875		9⅝″	29.30	9.063	8.907
4½″	13.50	3.920	3.795		9⅝″	32.30	9.001	8.845
4½″	15.10	3.826	3.701		9⅝″	36.00	8.921	8.765
5″	11.50	4.560	4.435		9⅝″	40.00	8.835	8.679
5″	13.00	4.494	4.369		9⅝″	43.50	8.755	8.599
5″	15.00	4.408	4.283		9⅝″	47.00	8.681	8.525
5″	18.00	4.276	4.151		9⅝″	53.50	8.535	8.379
5″	20.80	4.156	4.031		9⅝″	58.40	8.435	8.279
5″	24.20	4.000	3.875		9⅝″	61.10	8.375	8.219
5½″	14.00	5.012	4.887		9⅝″	71.80	8.125	7.969
5½″	15.50	4.950	4.825		10¾″	32.75	10.192	10.036
5½″	17.00	4.892	4.767		10¾″	40.50	10.050	9.894
5½″	20.00	4.778	4.653		10¾″	45.50	9.950	9.794
5½″	23.00	4.670	4.545		10¾″	51.00	9.850	9.694
6⅝″	20	6.094	5.924		10¾″	55.50	9.760	9.604
6⅝″	24	5.921	5.796		10¾″	60.70	9.660	9.504
6⅝″	28	5.791	5.666		10¾″	65.70	9.560	9.404
6⅝″	32	5.675	5.550		10¾″	71.10	9.450	9.294
7″	17	6.538	6.413		11¾″	42.00	11.084	10.928
7″	20	6.456	6.331		11¾″	47.00	11.000	10.844
7″	23	6.366	6.241		11¾″	54.00	10.880	10.724
7″	26	6.276	6.151		11¾″	60.00	10.772	10.616
7″	29	6.184	6.059		13⅜″	48.00	12.715	12.559
7″	32	6.094	5.969		13⅜″	54.50	12.615	12.459
7″	35	6.004	5.897		13⅜″	61.00	12.515	12.359
7″	38	5.920	5.795		13⅜″	68.00	12.415	12.259
7⅝″	20	7.125	7.000		13⅜″	72.00	12.347	12.191
7⅝″	24	7.025	6.900		13⅜″	77.00	12.275	12.119
7⅝″	26.40	6.969	6.844		13⅜″	85.00	12.159	12.003
7⅝″	29.70	6.875	6.750		13⅜″	98.00	11.937	11.781
7⅝″	33.70	6.765	6.640		16″	65.00	15.250	15.062
7⅝″	39.00	6.625	6.500		16″	75.00	15.124	14.936
7⅝″	45.30	6.435	6.310		16″	84.00	15.020	14.822
8⅝″	24.00	8.097	7.972		16″	109.00	14.688	14.500
8⅝″	28.00	8.017	7.892		20″	94.00	19.124	18.936
8⅝″	32.00	7.921	7.796		20″	106.50	19.000	18.812
8⅝″	36.00	7.825	7.700		20″	133.50	18.730	18.542
8⅝″	40.00	7.725	7.600					

Appendix 4G.2
CASING & TUBING SIZES AND CAPACITIES

Nominal Size	A.P.I. Size (O.D.)	Weight Per Lineal Foot	Inside Diameter	Barrels Per 1000 Ft.	Gals. Per 1000 Ft.
2	2.375	4.65	1.995	3.869	162.5
2½	2.875	6.54	2.441	5.793	243.3
3	3.500	9.25	2.992	8.704	365.5
4	4.500	12.68	3.958	15.23	639.6
5½		20.00	4.773	22.2	931.3
6⅝		20.00	6.049	35.5	1493
7		23.00	6.366	39.4	1653
7⅝		24.00	7.025	48.0	2013
8⅝		32.00	7.921	60.9	2560
9⅝		36.00	8.921	77.3	3247
10¾		40.50	10.050	98.1	4121
11¾		47.00	11.000	117.5	4936
13⅜		61.00	12.515	152.2	6390
16		65.00	15.250	225.9	9489

Appendix 4G.3
STRETCH OF SUSPENDED CASING, TUBING OR DRILL PIPE
Stretch in Air, Fresh Water, Salt Water

Length of Pipe (Feet)	Air	8.34	9	9.625
1000	.68	.56	.55	.54
2000	2.72	2.22	2.18	2.14
3000	6.12	5.00	4.91	4.82
4000	10.88	8.88	8.72	8.57
5000	17.00	13.88	13.63	13.40
6000	24.48	19.98	19.62	19.29
7000	33.32	27.20	26.71	26.26
8000	43.51	35.53	34.88	34.30
9000	55.07	44.96	44.15	43.41
10000	67.99	55.51	54.50	53.59
11000	82.27	67.17	65.95	64.85
12000	92.91	79.94	78.49	77.17
13000	114.91	93.82	92.11	90.57
14000	133.26	108.80	106.83	105.04
15000	152.98	124.90	122.63	120.58
16000	174.06	142.11	139.53	137.20
17000	196.50	160.43	157.52	154.88
18000	220.29	179.86	176.59	173.64
19000	245.45	200.40	196.76	193.47
20000	271.97	222.05	218.02	214.37
Pounds Per Cubic Foot		62.4	67.3	72.0
Specific Gravity		1.00	1.08	1.154

Appendix 4H.1
CONVERSIONS FROM °API TO SPECIFIC WEIGHTS

API Gravity	Specific Gravity	Specific Weight (Density)					API Gravity	Specific Gravity	Specific Weight (Density)				
		lbs/cu ft	lbs/gal	lbs/bbl	gm/cc Ton/m³	kg/m³			lbs/cu ft	lbs/gal	lbs/bbl	gm/cc Ton/m³	kg/m³
100	0.6112	38.14	5.10	213.9	0.6112	611.2	54	0.7628	47.60	6.36	267.0	0.7629	762.9
99	0.6139	38.31	5.12	214.9	0.6139	613.9	53	0.7669	47.85	6.40	268.4	0.7669	766.9
98	0.6166	38.48	5.14	215.8	0.6166	616.6	52	0.7711	48.12	6.43	269.9	0.7711	771.1
97	0.6193	38.64	5.16	216.8	0.6193	619.3	51	0.7753	48.38	6.47	271.4	0.7753	775.3
96	0.6220	38.81	5.19	217.7	0.6220	622.0	50	0.7796	48.65	6.50	272.9	0.7796	779.6
95	0.6247	38.98	5.21	218.6	0.6247	624.7	49	0.7839	48.92	6.54	274.4	0.7839	783.9
94	0.6275	39.16	5.23	219.6	0.6275	627.5	48	0.7883	49.19	6.57	275.9	0.7883	788.3
93	0.6303	39.33	5.26	220.6	0.6303	630.3	47	0.7927	49.46	6.61	277.4	0.7927	792.7
92	0.6331	39.51	5.28	221.6	0.6331	633.1	46	0.7972	49.75	6.65	279.0	0.7972	797.2
91	0.6360	39.69	5.30	222.6	0.6360	636.0	45	0.8017	50.03	6.69	280.6	0.8017	801.7
90	0.6388	39.86	5.33	233.6	0.6388	638.8	44	0.8063	50.31	6.72	282.2	0.8063	806.3
89	0.6417	40.04	5.35	224.6	0.6417	641.7	43	0.8109	50.60	6.76	283.8	0.8109	810.9
88	0.6446	40.22	5.38	225.6	0.6446	644.6	42	0.8155	50.89	6.80	285.4	0.8155	815.5
87	0.6476	40.41	6.40	226.7	0.6476	647.6	41	0.8203	51.19	6.84	287.1	0.8203	820.3
86	0.6506	40.60	5.43	227.7	0.6506	650.6	40	0.8251	51.49	6.88	288.8	0.8251	825.1
85	0.6536	40.78	5.45	228.8	0.6536	653.6	39	0.8299	51.79	6.92	290.5	0.8299	829.9
84	0.6566	40.97	5.48	229.8	0.6566	656.6	38	0.8348	52.09	6.96	292.2	0.8348	834.8
83	0.6597	41.17	5.50	230.9	0.6597	659.7	37	0.8398	52.40	7.00	293.9	0.8398	839.8
82	0.6628	41.36	5.53	232.0	0.6628	662.8	36	0.8448	52.72	7.05	295.7	0.8448	844.8
81	0.6659	41.55	5.55	233.1	0.6659	665.9	35	0.8498	53.03	7.09	297.4	0.8498	849.8
80	0.6690	41.75	5.58	234.2	0.6690	669.0	34	0.8550	53.35	7.13	299.3	0.8550	855.0
79	0.6722	41.95	5.61	235.3	0.6722	672.2	33	0.8602	53.68	7.17	301.1	0.8602	860.2
78	0.6754	42.14	5.63	236.4	0.6754	675.4	32	0.8654	54.00	7.22	302.9	0.8654	865.4
77	0.6787	42.35	5.66	237.5	0.6787	678.7	31	0.8708	54.34	7.26	304.8	0.8708	870.8
76	0.6819	42.55	5.69	238.7	0.6819	681.9	30	0.8762	54.67	7.31	306.7	0.8762	876.2
75	0.6852	42.76	5.71	239.8	0.6852	685.2	29	0.8816	55.01	7.35	308.6	0.8816	881.6
74	0.6886	42.97	5.74	241.0	0.6886	688.6	28	0.8871	55.36	7.40	310.5	0.8871	887.1
73	0.6919	43.17	5.77	242.2	0.6919	691.9	27	0.8927	55.70	7.45	312.4	0.8927	892.7
72	0.6952	43.38	5.80	243.3	0.6952	695.2	26	0.8984	56.06	7.49	314.4	0.8984	898.4
71	0.6988	43.61	5.83	244.6	0.6988	698.8	25	0.9042	56.42	7.54	316.5	0.9042	904.2
70	0.7022	43.82	5.86	245.8	0.7022	702.2	24	0.9100	56.78	7.59	318.5	0.9100	910.0
69	0.7057	44.04	5.89	247.0	0.7057	705.7	23	0.9159	57.15	7.64	320.6	0.9159	915.9
68	0.7093	44.26	5.92	248.3	0.7093	709.3	22	0.9218	57.52	7.69	322.6	0.9218	921.8
67	0.7128	44.48	5.94	249.5	0.7128	712.8	21	0.9279	57.90	7.74	324.8	0.9279	927.9
66	0.7165	44.71	5.98	250.8	0.7165	716.5	20	0.9340	58.28	7.79	326.9	0.9340	934.0
65	0.7201	44.93	6.01	252.0	0.7201	720.1	19	0.9402	58.67	7.84	329.1	0.9402	940.2
64	0.7238	45.17	6.04	253.3	0.7238	723.8	18	0.9465	59.06	7.89	331.3	0.9465	946.5
63	0.7275	45.40	6.07	254.6	0.7275	727.5	17	0.9529	59.46	7.95	333.5	0.9529	952.9
62	0.7313	45.63	6.10	256.0	0.7313	731.3	16	0.9593	59.86	8.00	335.8	0.9593	959.3
61	0.7351	45.87	6.13	257.3	0.7351	735.1	15	0.9659	60.27	8.06	338.1	0.9659	965.9
60	0.7389	46.11	6.16	258.6	0.7389	738.9	14	0.9725	60.68	8.11	340.4	0.9725	972.5
59	0.7428	46.35	6.19	260.0	0.7428	742.8	13	0.9792	61.10	8.17	342.7	0.9792	979.2
58	0.7467	46.59	6.23	261.3	0.7467	746.7	12	0.9861	61.53	8.22	345.1	0.9861	986.1
57	0.7507	46.84	6.26	262.7	0.7507	750.7	11	0.9930	61.96	8.28	347.6	0.9930	993.0
56	0.7547	47.09	6.29	264.1	0.7547	754.7	10	1.0000	62.40	8.34	350.0	1.0000	1000.0
55	0.7587	47.34	6.33	265.5	0.7587	758.7							

Appendix 4I.1
EQUIVALENTS OF ABSOLUTE VISCOSITY

Absolute or dynamic viscosity		Centipoise (μ)	Poise Gram = Cm Sec Dyne Sec Cm² $(100\,\mu)$	Slugs Ft Sec *Pound$_f$ Sec Ft² (μ'_e)	+ Pound$_m$ Ft Sec Poundal Sec Ft² (μ_e)
Centipoise	(μ)	1	0.01	$2.09(10^{-5})$	$6.72(10^{-4})$
Poise $\dfrac{\text{Gram}}{\text{Cm Sec}}$ $\dfrac{\text{Dyne Sec}}{\text{Cm}^2}$	(100μ)	100	1	$2.09(10^{-3})$	0.0672
$\dfrac{\text{Slugs}}{\text{Ft Sec}}$ $\dfrac{\text{*Pound}_f\text{ Sec}}{\text{Ft}^2}$	(μ'_e)	47900	479	1	g or 32.2
$\dfrac{\text{+ Pound}_m}{\text{Ft Sec}}$ $\dfrac{\text{Poundal Sec}}{\text{Ft}^2}$	(μ)	1487	14.87	$\dfrac{1}{g}$ or .0311	1

*Pound$_f$ = Pound of Force   + Pound$_m$ = Pound of Mass

Appendix 4I.2
EQUIVALENTS OF KINEMATIC VISCOSITY

Kinematic viscosity		Centistokes (ϑ)	Stokes Cm² Sec $(100\,\vartheta)$	Ft² Sec (ϑ^1)
Centistokes	(ϑ)	1	0.01	$1.076(10^{-5})$
Stokes $\dfrac{\text{Cm}^2}{\text{Sec}}$	$(100\,\vartheta)$	100	1	$1.076(10^{-3})$
$\dfrac{\text{Ft}^2}{\text{Sec}}$	(ϑ^1)	92900	929	1

Appendix 4I.3
EQUIVALENTS OF KINEMATIC AND SAYBOLT UNIVERSAL VISCOSITY

Kinematic viscosity (Centistokes ϑ)	Equivalent Saybolt Universal viscosity, sec — At 100 F basic values	At 210 F	Kinematic viscosity (Centistokes ϑ)	Equivalent Saybolt Universal viscosity, sec — At 100 F basic values	At 210 F
2.0	32.6	32.9	29	136.9	137.9
2.5	34.4	34.7	30	141.3	142.3
3.0	36.0	36.3	31	145.7	146.8
3.5	37.6	37.9	32	150.2	151.2
4.0	39.1	39.4	33	154.7	155.8
4.5	40.8	41.0	34	159.2	160.3
5	42.4	42.7	35	163.7	164.9
6	45.6	45.9	36	168.2	169.4
7	48.8	49.1	37	172.7	173.9
8	52.1	52.5	38	177.3	178.5
9	55.5	55.9	39	181.8	183.0
10	58.9	59.3	40	186.3	187.6
11	62.4	62.9	41	190.8	192.1
12	66.0	66.5	42	195.3	196.7
13	69.8	70.3	43	199.8	201.2
14	73.6	74.1	44	204.4	205.9
15	77.4	77.9	45	209.1	210.5
16	81.3	81.9	46	213.7	215.2
17	85.3	85.9	47	218.3	219.8
18	89.4	90.1	48	222.9	224.5
19	93.6	94.2	49	227.5	229.1
20	97.8	98.5	50	232.1	233.8
21	102.0	102.8	55	255.2	257.0
22	106.4	107.1	60	278.3	280.2
23	110.7	111.4	65	301.4	303.5
24	115.0	115.8	70	324.4	326.7
25	119.3	120.1		Saybolt	
26	123.7	124.5	Over 70	Seconds equal	
27	128.1	129.0		Centistokes	
28	132.5	133.4		× 4.635 × 4.667	

Appendix 4I.4
EQUIVALENTS OF KINEMATIC AND SAYBOLT
FUROL VISCOSITY AT 122° F

Kinematic viscosity Centistokes ϑ	Equivalent Saybolt furol viscosity sec	Kinematic viscosity Centistokes ϑ	Equivalent Saybolt furol viscosity sec
48	25.3	150	71.7
50	26.1	155	74.0
52	27.0	160	76.3
54	27.9	165	78.7
56	28.8	170	81.0
58	29.7	175	83.3
60	30.6	180	85.6
62	31.5	185	88.0
64	32.4	190	90.3
66	33.3	195	92.6
68	34.2	200	95.0
70	35.1	210	99.7
72	36.0	220	104.3
74	36.9	230	109.0
76	37.8	240	113.7
78	38.7	250	118.4
80	39.6	260	123.0
82	40.5	270	127.7
84	41.4	280	132.4
86	42.3	290	137.1
88	43.2	300	141.8
90	44.1	310	146.5
92	45.0	320	151.2
94	45.9	330	155.9
96	46.8	340	160.6
98	47.7	350	165.3
100	48.6	360	170.0
105	50.9	370	174.7
110	53.2	380	179.4
115	55.5	390	184.1
120	57.8	400	188.8
125	60.1		Saybolt
130	62.4		Furol
135	64.7	Over 400	Seconds =
140	67.0		Centistokes
145	69.4		× 0.4717

Equivalents of Kinematic, Saybolt Universal, Saybolt Furol, and Absolute Viscosity

$$\mu = \nu S$$

The empirical relation between Saybolt Universal Viscosity and Saybolt Furol Viscosity at 100 F and 122 F, respectively, and Kinematic Viscosity is taken from A.S.T.M. D2161-63T. At other temperatures, the Saybolt Viscosities vary only slightly.

Saybolt Viscosities above those shown are given by the relationships:

Saybolt Universal Seconds = Centistokes x 4.6347
Saybolt Furol Seconds = Centistokes x 0.4717

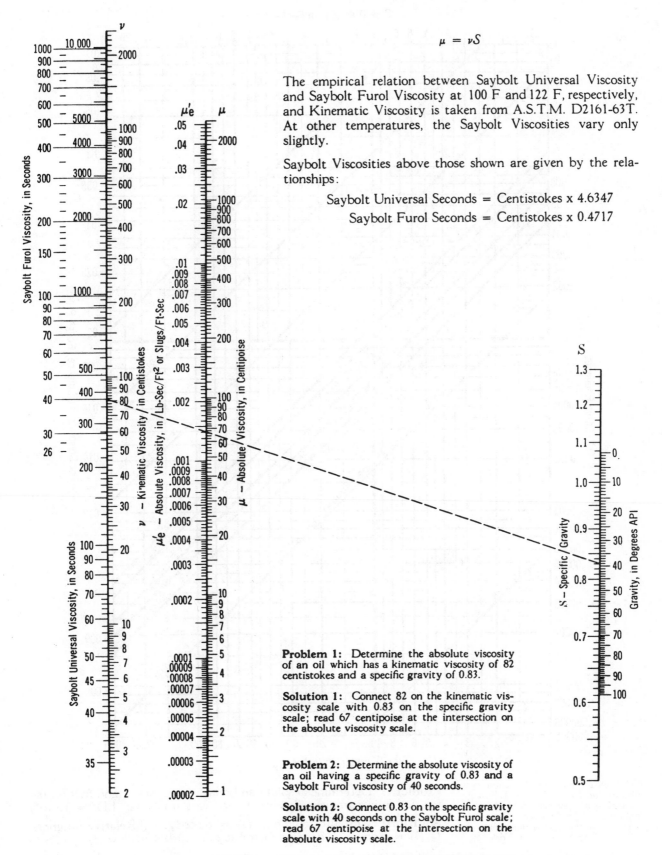

Problem 1: Determine the absolute viscosity of an oil which has a kinematic viscosity of 82 centistokes and a specific gravity of 0.83.

Solution 1: Connect 82 on the kinematic viscosity scale with 0.83 on the specific gravity scale; read 67 centipoise at the intersection on the absolute viscosity scale.

Problem 2: Determine the absolute viscosity of an oil having a specific gravity of 0.83 and a Saybolt Furol viscosity of 40 seconds.

Solution 2: Connect 0.83 on the specific gravity scale with 40 seconds on the Saybolt Furol scale; read 67 centipoise at the intersection on the absolute viscosity scale.

Fig. 4I (5)   Equivalents of kinematic, saybolt universal, saybolt furol, and absolute viscosity

Relative Roughness of Pipe Materials and Friction Factors
For Complete Turbulence[18]

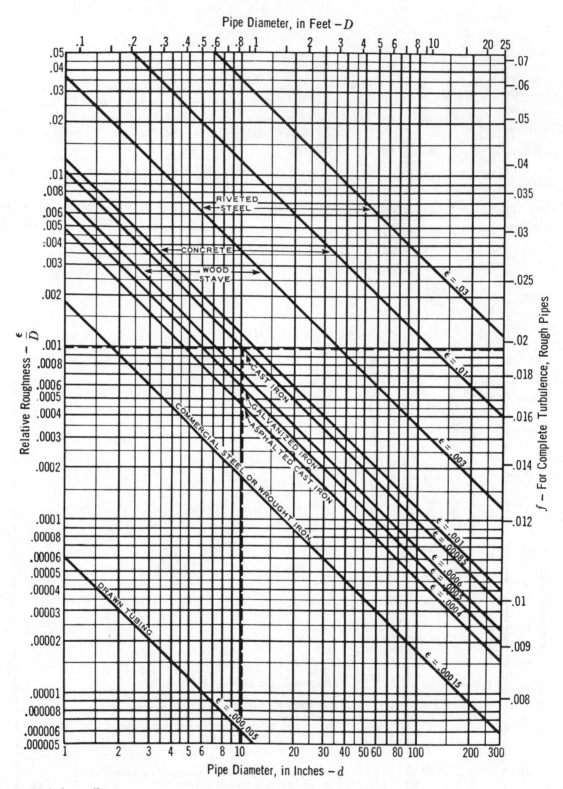

Data extracted from *Friction Factors for Pipe Flow* by L.F. Moody, with permission of the publisher, The American Society of Mechanical Engineers. 29 West 39th Street, New York.

Problem: Determine absolute and relative roughness, and friction factor, for fully turbulent flow in 10-inch cast iron pipe (I.D. = 10.16″).

Solution: Absolute roughness (ϵ) = 0.00085 . . . . . Relative roughness (ϵ/D) = 0.001 . . . . . Friction factor at fully turbulent flow (f) = 0.0196.

Fig. 4J (1)   Relative roughness of pipe materials and friction factors for complete turbulence

Friction Factors for Any Type of Commercial Pipe[18]

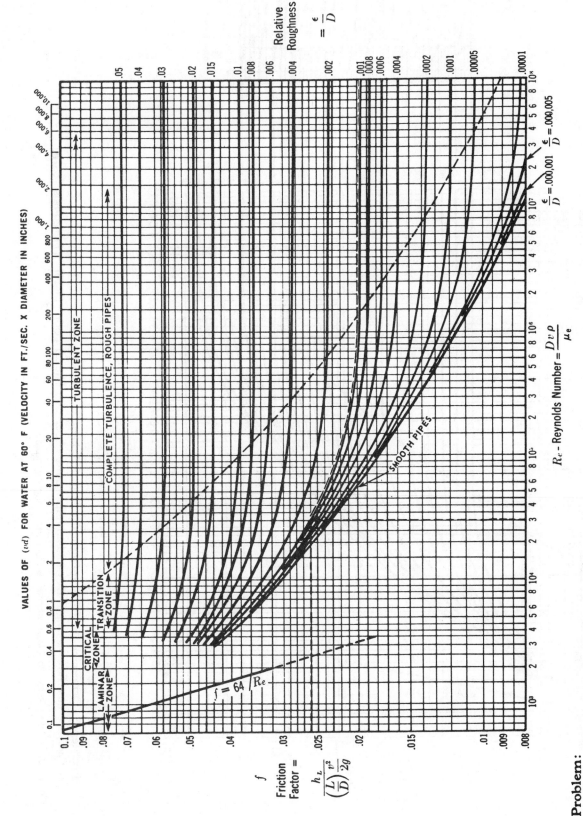

Problem:
Determine the friction factor for 10-inch cast iron pipe (10.16" I.D.) at a Reynolds number flow of 30,000.

Solution: The relative roughness (see page A-23) is 0.001. Then, the friction factor (f) equals 0.026.

For other forms of the R_e equation, see page 3-2.

Data extracted from *Friction Factors for Pipe Flow* by L. F. Moody, with permission of the publisher, The American Society of Mechanical Engineers, 29 West 39th Street, New York 18, N. Y.

Fig. 4J (2) Friction factors for any type of commercial pipe

Friction Factors for Clean Commercial Steel and Wrought Iron Pipe[18]

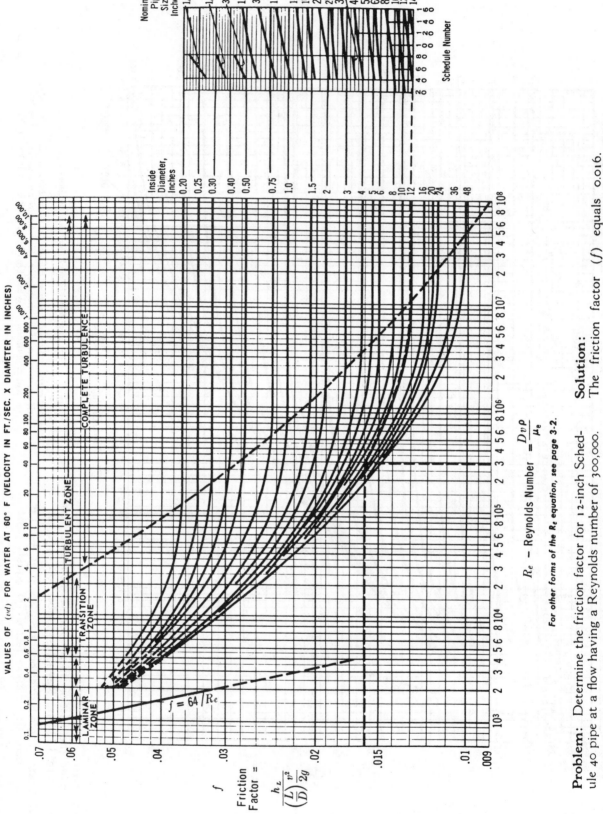

Problem: Determine the friction factor for 12-inch Schedule 40 pipe at a flow having a Reynolds number of 300,000.

Solution: The friction factor (f) equals 0.016.

Fig. 4J (3)   Friction factors for clean commercial steel and wrought iron pipe

Appendix 4K
METHOD FOR ESTIMATING THE PERCENT OF FREE GAS THAT WILL GO THRU A PUMP
(by Don Rhoads)

General:

An important factor in the selection of a pump and motor for installation in a gassy well is the amount of free gas that the pump must handle. The amount of free gas, i.e., gas that is not in solution, determines the density of the liquid and gas mixture that is pumped. If there is enough of it, the free gas will even cause the pump to gas lock and stop pumping altogether. Therefore, it is necessary when selecting pumping equipment for gassy wells to make as good an estimate as possible of how much free gas will go thru the pump and how much will go up the annulus.

Purpose:

This section will formulate a method for estimating how much of the free gas that approaches the pump will also go through the pump.

This section is intended to provide a method that can be refined and improved as additional field and laboratory experience is accumulated. In no way is it intended to be a final treatise on the subject.

Factors considered:

Three factors will be considered:
1. The velocity of the flow as it approaches the pump.
2. The size of the free gas bubbles.
3. The viscosity of the liquid.

EFFECT OF THE VELOCITY OF FLOW APPROACHING THE PUMP

Assume:

1. A well with:

Bubble point pressure (P_{bp})	= 2,400 psig
Pump intake pressure (Pi)	= 2,280 psig
Viscosity (dead oil at intake temp.)	= 2.0 cp
Water cut	= 40%

2. That, for the above well, the following relationships exist between the velocity approaching the pump (Va), the % of free gas that will go thru the pump, and whether there is or is not a gas separator.
   a. No gas separator - Va = 0 - 0% free gas thru pump.
   b. No gas separator - Va = 16 ft/sec. - 90% free gas thru pump.
   c. With gas separator - Va = 0 - 0% free gas thru pump.
   d. With gas separator - Va = 16 ft/sec. - 45% free gas thru pump.
   e. The % of free gas that will go thru the pump will decrease from the maximum at 16 ft/sec. in proportion to the velocity squared. This is because the upward (or buoyant) force on the gas bubble stays the same for all liquid velocities *but* the pressure gradient along the path of flow into the pump decreases as the velocity squared *and* therefore, the ratio of upward force to the force along the path into the pump changes at a greater *rate* as the velocity reduces. This is probably modified some due to the liquid being thrown outward more than the gas as it makes the bend to turn into the pump. These relationships in formula form are:

TABLE 1
VELOCITY OF APPROACH

Intake Flow (B/D)	APPROACH VELOCITY (Va) (Ft./Sec.)										
	338 Pump 338 Seal Sect.		400 Pump 400 Seal Sect.			513 Pump 513 Seal Sect.				675 Pump 675 Seal Sect.	513 Seal Sect.
	4½ Casing	5 Casing	5½ Casing	6⅝ Casing	7 Casing	6⅝ Casing	7 Casing	7⅝ Casing	8⅝ Casing	8⅝ Casing	8⅝ Casing
500	1.3	0.7	0.7	0.3	0.3						
1000	2.6	1.5	1.5	0.6	0.5	1.4	0.9	0.6	0.4		
2000	5.2	3.0	3.0	1.3	1.0	2.7	1.8	1.1	0.7		
3000			4.5	1.9	1.5	4.1	2.7	1.7	1.1	2.5	1.1
4000			6.0	2.5	2.0	5.4	3.6	2.3	1.4	3.4	1.4
5000						6.8	4.6	2.8	1.8	4.2	1.8
6000						8.2	5.5	3.4	2.1	5.1	2.1
7000						9.5	6.4	4.0	2.5	5.9	2.5
8000						10.9	7.3	4.6	2.9	6.8	2.9
9000						12.3	8.2	5.1	3.2	7.6	3.2
10000						13.6	9.1	5.7	3.6	8.4	3.6
12000						16.3	10.9	6.8	4.3	10.1	4.3
14000										11.8	5.0
16000										13.5	5.7
18000										15.2	6.4
20000										16.9	7.1
22000										18.6	7.9
24000										20.3	8.6
26000										21.9	9.3
28000										23.6	10.0
30000										25.3	10.7

NOTE: Table 1 is based on the area between the O.D. of the Seal Section and the I.D. of casings having the following weights:

4½ in. casing - 11.60 lbs/ft
5 in. casing - 15.00 lbs/ft
5½ in. casing - 17.00 lbs/ft
6⅝ in. casing - 24.00 lbs/ft
7 in. casing - 26.00 lbs/ft
7⅝ in. casing - 29.70 lbs/ft
8⅝ in. casing - 40.00 lbs/ft

% of free gas thru pump $= 90 - \left(\dfrac{16 - Va}{16}\right)^2 90$ for no gas separator, % of free gas thru pump $= 45 - \left(\dfrac{16 - Va}{16}\right)^2 45$ for a gas separator.

3. The relationships in assumption 2 can be adjusted to eliminate the effects of bubble size and viscosity in accordance with the methods developed later in this paper. For $P_{bp} = 2,400$ psig and

$$\frac{Pi}{P_{bp}} = \frac{2,280}{2,400} = 0.95$$

From Table 3, effect of bubble size $= -1.0\%$
For viscosity = 2.0 cp and
water cut = 40%
From Table 4, effect of viscosity $= +0.5\%$
Net effect of bubble size and
viscosity $= -0.5\%$
Therefore, in formula form the relationships of 2. adjusted to eliminate the effects of bubble size and viscosity are:

% of free gas thru pump $= 90.5 - \left(\dfrac{16 - Va}{16}\right) 90$ for no gas separator, and % of free gas thru pump $= 45.5 - \left(\dfrac{16 - Va}{16}\right) 45$ for a gas separator used.

% of free gas thru pump =

$90.5 - \left(\dfrac{16 - Va}{16}\right)^2 90$ (No gas separator)

$40.5 - \left(\dfrac{16 - Va}{16}\right)^2 45$ (Gas separator)

EFFECT OF BUBBLE SIZE

Assume:

1. The tendency for the bubble to be thrown outward, i.e. to be separated, as it turns into the pump is directly proportional to its diameter cubed. This is because for $F = Ma = \dfrac{W}{g}\dfrac{V^2}{r}$, W is directly proportional to its diameter cubed (Vol. of sphere = $\frac{1}{6}\pi D^3$).
2. The resistance to movement of the bubble outward is directly proportional to its surface area and, therefore, to its diameter squared. (Surface area of a sphere = πD^2.)
3. ∴ The net tendency for the bubble to move outward is proportional to $D^3 - D^2 = D$, which is proportional to the cube root of the volume.
4. Bubble size can account for as much as 5% variation in "% of free gas thru the pump."
5. The effect of bubble size on % of gas thru the pump can be determined by first calculating the relative bubble size (volume) at different pressures by assuming bubble volume is directly proportional to pressure (Table 2). The relative effect of the bubble size, then, is the cube root of the values in table 2, and, in accordance with 4. above, is proportioned upward until the maximum actual effect is 5% (Table 3).

Assume:
1. The tendency for the bubble to be thrown outward (i.e. to be separated) as it turns into the pump is inversely proportional to the viscosity of the liquid. This is because resistance to flow (f) (in the laminar flow region) decreases almost directly as Reynold's number increases and Reynold's number, $R = \left(\dfrac{DV_p}{\mu}\right)$, increases directly as viscosity (μ) decreases.
2. Viscosity can account for as much as 10% variation in "% thru pump" over the range from 0.7 CP to 16.0 CP (approximately 30 SSU to 100 SSU).

3. The viscosity of oil and water in combination is the average of the oil viscosity and the water viscosity weighted according to their volumes.

TABLE 2
RELATIVE BUBBLE SIZE

1.0	0	0	0	0	0	0	0	0	0	0	0
0.9	0.50	0.45	0.40	0.35	0.30	0.25	0.20	0.15	0.10	0.05	0
0.8	1.00	0.90	0.80	0.70	0.60	0.50	0.40	0.30	0.20	0.10	0
0.7	1.50	1.35	1.20	1.05	0.90	0.75	0.60	0.45	0.30	0.15	0
0.6	2.00	1.80	1.60	1.40	1.20	1.00	0.80	0.60	0.40	0.20	0
0.5	2.50	2.25	2.00	1.75	1.50	1.25	1.00	0.75	0.50	0.25	0
0.4	3.00	2.70	2.40	2.10	1.80	1.50	1.20	0.90	0.60	0.30	0
0.3	3.50	3.15	2.80	2.45	2.10	1.75	1.40	1.05	0.70	0.35	0
0.2	4.00	3.60	3.20	2.80	2.40	2.00	1.60	1.20	0.80	0.40	0
0.1	4.50	4.05	3.60	3.15	2.70	2.25	1.80	1.35	0.90	0.45	0
0.0	5.00	4.50	4.00	3.50	3.00	2.50	2.00	1.50	1.00	0.50	0

NOTE: Table 2 is based on the assumption that at any depth there will be bubbles of various sizes ranging from infinitely small (those that have just come out of solution) to those that came out of solution near P_{bp} and grew larger as the pressure reduced at the shallower depth. Table 2 shows the average bubble size at the depth indicated by P_{bp} and P_i. A scale of 0 to 5 has been arbitrarily selected for the range of average bubble size.

TABLE 3
EFFECT OF BUBBLE SIZE

1.0	0	0	0	0	0	0	0	0	0	0	0
0.9	2.3	2.2	2.2	2.1	2.0	1.8	1.7	1.6	1.4	1.1	0
0.8	2.9	2.8	2.7	2.6	2.5	2.3	2.2	2.0	1.7	1.4	0
0.7	3.3	3.2	3.1	3.0	2.8	2.7	2.5	2.2	2.0	1.6	0
0.6	3.7	3.6	3.4	3.3	3.1	2.9	2.7	2.5	2.2	1.7	0
0.5	4.0	3.8	3.7	3.5	3.3	3.2	2.9	2.7	2.3	1.8	0
0.4	4.2	4.1	3.9	3.7	3.6	3.3	3.1	2.8	2.5	2.0	0
0.3	4.4	4.3	4.1	3.9	3.7	3.5	3.3	3.0	2.6	2.1	0
0.2	4.6	4.5	4.3	4.1	3.9	3.7	3.4	3.1	2.7	2.2	0
0.1	4.8	4.7	4.5	4.3	4.1	3.8	3.6	3.2	2.8	2.2	0
0.0	5.0	4.8	4.6	4.4	4.2	4.0	3.7	3.3	2.9	2.3	0

NOTE: The values in Table 3 are the percentages of the free gas that will *not* go thru the pump. They are found by taking the cube root of the valves in Table 2, then multiplying by the factor that will make the maximum effect equal 5.0%.

TABLE 4
EFFECT OF VISCOSITY

Viscosity (CP)	All Oil 0	10	20	30	40	50	60	70	80	90	100
0.7	0	0	0	0	0	0	0	0	0	0	0
1	0.2	0.2	0.2	0.1	0.1	0.1	0.1	0.1	0	0	0
2	0.8	0.8	0.7	0.6	0.5	0.4	0.3	0.3	0.2	0.1	0
3	1.5	1.4	1.2	1.1	0.9	0.8	0.6	0.5	0.3	0.1	0
4	2.2	1.9	1.7	1.5	1.3	1.1	0.9	0.6	0.4	0.2	0
5	2.8	2.5	2.2	2.0	1.7	1.4	1.1	0.8	0.6	0.3	0
6	3.5	3.1	2.8	2.4	2.1	1.7	1.4	1.0	0.7	0.3	0
7	4.1	3.7	3.3	2.9	2.5	2.1	1.6	1.2	0.8	0.4	0
8	4.8	4.3	3.8	3.3	2.9	2.4	1.9	1.4	1.0	0.5	0
9	5.4	4.9	4.3	3.8	3.3	2.7	2.2	1.6	1.1	0.5	0
10	6.1	5.5	4.9	4.3	3.6	3.0	2.4	1.8	1.2	0.6	0
11	6.7	6.1	5.4	4.7	4.0	3.4	2.7	2.0	1.3	0.7	0
12	7.4	6.6	5.9	5.2	4.4	3.7	3.0	2.2	1.5	0.7	0
13	8.0	7.2	6.4	5.6	4.8	4.0	3.2	2.4	1.6	0.8	0
14	8.7	7.8	7.0	6.1	5.2	4.3	3.5	2.6	1.7	0.9	0
15	9.3	8.4	7.5	6.5	5.6	4.7	3.7	2.8	1.9	0.9	0
16	10.0	9.0	8.0	7.0	6.0	5.0	4.0	3.0	2.0	1.0	0

NOTE: The valves in Table 4 are the percentages of the free gas that will go thru the pump.

EXAMPLE

Given:
Bubble Point (P_{bp}) = 2800 PSIG
Pump Intake Pressure (P_i) = 2660 PSIG
Oil Viscosity = 2.0 CP
Water Cut = 40%
Pump Series = 513
Seal Section Series = 513
Casing O.D. = 7 inch
Intake Flow = 6000 B/D
Gas Separator   Yes

From Table 1 for: 513 series pump V_a = 5.5 ft/sec.
                    513 Series seal section
                    7 in. casing O.D.
                    6000 b/d intake flow

From Figure 1 for: With gas separator and V_a = 5.5 ft/sec.
% of free gas thru pump = 45.5 −
$$\left(\frac{16 - 5.5}{16}\right)^2 45 = +26.1\%$$

From Table 3 for: P_{bp} = 2800 PSIG
                  P_i = 2660 PSIG = 0.95
                  P_{bp} 2800 PSIG
% of free gas separated due to bubble size   = − 1.0%

From Table 4 for: 2.0 CP oil viscosity
                  40 % water cut
% of free gas thru pump due to viscosity   = + 0.5%
% of free gas thru pump (net)   = + 25.6%

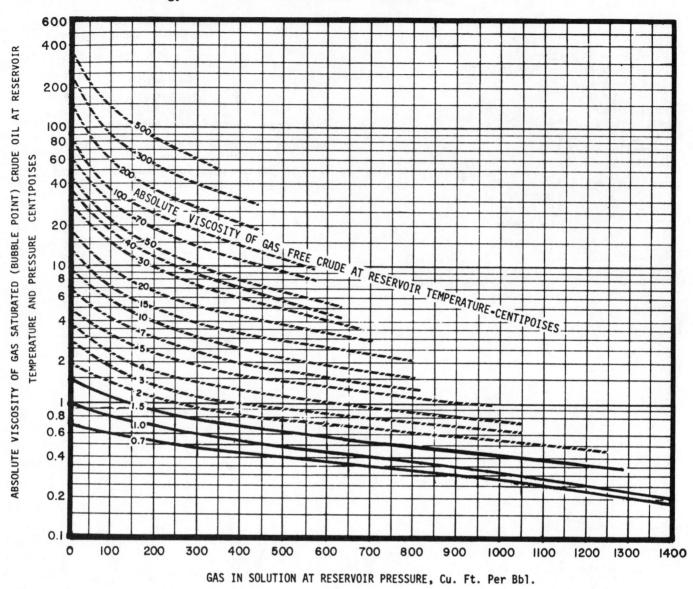

Fig. 4L (1)   Viscosity of gas saturated crude oil at reservoir temperature & pressure

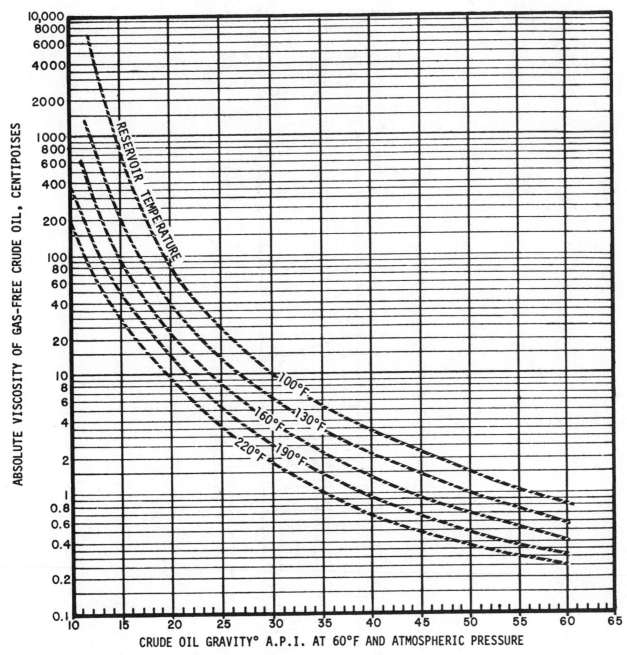

Fig. 4L (2)   The viscosity of gas-free oil at oil-field temperatures

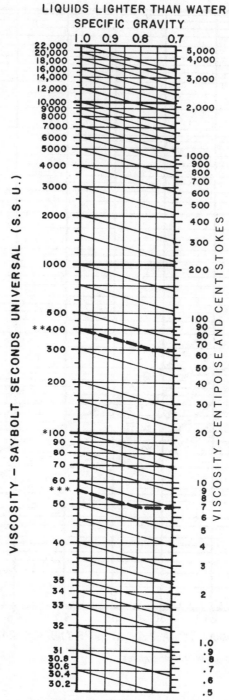

Fig. 4L (3)   Specific gravity of liquids
* Horizontal line converts SSU to cp. (Example 100 SSU = 20 cp)
** Diagonal line plus horizontal line converts SSU to centistokes (Example 400 SSU ($\gamma g = .76$) = 65 cs
*** 55 SSU ($\gamma g = 0.8$) = 7 cs

Appendix 4M.1
USEFUL PUMP DATA

EFFECT OF SMALL CHANGES OF PUMP SPEED

1. The capacity varies directly as the speed.
2. The head varies as the square of the speed.
3. The brake horsepower varies as the cube of the speed.

EFFECT OF SMALL CHANGES OF IMPELLER DIAMETER (BY TRIMMING)

1. The capacity varies directly as the diameter.
2. The head varies as the square of the diameter.
3. The brake horsepower varies as the cube of the diameter.

EFFECT OF SPECIFIC GRAVITY

Brake horsepower varies directly with specific gravity. If the liquid has a specific gravity other than water (1.0) multiply the brake horsepower for the water by the specific gravity of the liquid to be handled.

A centrifugal pump always develops the same head in feet no matter what the specific gravity of the liquid pumped. However, the pressure (in pounds per square inch) will be increased or decreased in direct proportion to the specific gravity.

EFFECT OF VISCOSITY

Viscous liquids tend to reduce pump capacity head and efficiency and to increase pump brake horsepower and increase line pipe line friction. Consult factory for recommendation when pumping viscous liquids.

HORSEPOWER

Liquid HP Equals - HP required to lift liquid at a definite rate to a given distance assuming 100% efficiency.

$$= \frac{GPM \times total\ head\ (in\ ft) \times SPGR}{3960}$$

Brake HP equals - HP delivered by motor
- HP required by pump
- HP input × motor efficiency

$$= \frac{Liquid\ HP}{Pump\ efficiency}$$

HP Input equals - HP input to motor
= 1.341 × kilowatts input to motor

EFFI EFFICIENCY

$$Efficiency\ equals - \frac{power\ output}{power\ input}$$

$$Motor\ efficiency\ equals - \frac{HP\ Output}{1.341 \times KW\ Input}$$

$$Pump\ efficiency\ equals = \frac{GPM \times total\ head\ (ft) \times SPGR}{3960 \times BHP}$$

CONVERSION OF 60 CYCLE PERFORMANCE CURVES TO 50 CYCLE PERFORMANCE

Head capcity curve - (a) Shut-off head will be 69.444
(b) Wide open capacity will be 83.33%
(c) Taking any point on the 60 cycle curve, multiply capacity by 0.8333 and head by 0.6944 to determine 50 cycle performance.

Horsepower - Take any point on 60 cycle curve (related to capacity), multiply by 0.578 and locate on 50 cycle curve at 83.33% of the 60 cycle capacity.

Efficiency - 50 cycle efficiency will be the same magnitude as 60 cycle efficiency, but locate points at 83.33% of the 60 cycle capacity.

Appendix 4M.2
FORMULAS FOR DETERMINING KILOWATTS, KVA, HORSEPOWER AND AMPERES

Desired data	Alternating Current	
	Single phase	Three phase
Kilowatts	$\dfrac{Volts \times AMP \times PF}{1000}$	$\dfrac{1.73 \times Volts \times AMP \times PF}{1000}$
Kilovolt-ampere	$\dfrac{Volts \times AMP}{1000}$	$\dfrac{1.73 \times Volts \times AMP}{1000}$
Horsepower (Output)	$\dfrac{Volts \times AMP \times EFF \times PF}{746 \times 100}$	$\dfrac{1.73 \times Volts \times AMP \times EFF \times PF}{746 \times 100}$
Amperes (When horsepower is known)	$\dfrac{HP \times 746 \times 100}{Volts \times EFF \times HP}$	$\dfrac{HP \times 746 \times 100}{1.73 \times Volts \times EFF \times PF}$
Amperes (When kilowatts are known)	$\dfrac{Kilowatts \times 1000}{Volts \times PF}$	$\dfrac{Kilowatts \times 1000}{1.73 \times Volts \times PF}$

(EFF) Efficiency in the above formulas is expressed in percent—such as 95%.
(PF) Power Factor in the above formulas is expressed as a decimal—such as 0.85.

Appendix 4M.3
USEFUL FORMULAS

To Find	Formula
Brake hp	$= \dfrac{\text{G.P.M.} \times \text{head feet} \times \text{sp. gr.}}{3960 \times \text{pump \% eff.}}$
Brake hp	$= \dfrac{\text{B.P.D.} \times \text{head feet} \times \text{sp. gr.}}{136{,}000 \times \text{pump \% eff.}}$
hp input (three-phase)	$= \dfrac{I \times v \times 1.73 \times \text{p.f.}}{746}$
hp input (three-phase)	$= \text{kw} \times 1.34$

Where hp = horsepower; G.P.M. = gallons per minute; sp. gr. = specific gravity; pump % eff. = pump percent efficiency; B.P.D. = barrels per day; I = amperes; v = surface voltage p.f. = power factor; kw = kilowatts.

CONVERSION FACTORS

General:

$$\text{B.P.D.} = \text{G.P.M.} \times 34.3$$

$$\text{Head Feet} = \frac{2.309 \times \text{P.S.I.}}{\text{Sp. Gr.}}$$

$$\text{kva} = \frac{v \times I \times 1.73}{1000}$$

Converting 60 Cycle
to 50 Cycle:

50-Cycle Head = 69.44% × 60-Cycle Head
50-Cycle Capacity = 83.33% × 60-Cycle Capacity
50-Cycle Horsepower = 57.80% × 60-Cycle Horsepower
50-Cycle Efficiency = Same as 60-Cycle Efficiency

ELECTRICAL TERMS AND DEFINITIONS

AMPERES—unit of current or rate of flow of electricity.
VOLT—unit of electromotive force.
OHM—unit of resistance.

OHM'S LAW:

$$\text{Current} = \frac{\text{Electromotive force}}{\text{Resistance}}$$

$$\text{Ohms} = \frac{\text{Volts}}{\text{Amperes}}$$

VOLT AMPERES—unit of apparent power.
MEGOHM—1,000,000 Ohms.
WATT—unit of true power = volt amperes × power factor.
POWER FACTOR—ratio of true to apparent power.
WATTHOUR—unit of electrical work; indicates expenditure of electrical power of one watt for one hour.
HORSEPOWER—a measure of time-rate of doing work; equivalent to raising 33,000-lbs. one ft. in one min. 1 hp = 746 watts.

Appendix 4M.4
USEFUL FORMULAS

$$\text{G.P.M.} = \frac{\text{lbs/hr}}{500 \times \text{sp. gr.}}$$

$$\text{Feet of head} = \frac{2.31 \text{ P.S.I.}}{\text{sp. gr. of fluid}}$$

$$\text{Spec. grav.} = \frac{141.5}{131.5 + \,^{\circ}\text{A.P.I.}}$$

$$\text{B.H.P.} = \frac{\text{G.P.M.} \times \text{T.D.H.} \times \text{sp. gr.}}{3960 \times \text{EFF.}}$$

$$\text{B.H.P.} = \frac{\text{G.P.M.} \times \text{PSI}}{1715 \times \text{EFF.}}$$

$$= \frac{1840 \, K \sqrt{H}}{\text{rpm}}$$

$$\text{Ns} = \frac{\text{rpm} \sqrt{Q}}{H^{\,3/4}}$$

$$\text{Torque in Ft. Lbs.} = \frac{\text{BHP} \times 5250}{\text{rpm}}$$

$$\text{EFF} = \frac{\text{G.P.M.} \times H \times \text{Sp. Gr.}}{3960 \times \text{BHP}}$$

Ns = Specific Speed
K = Constant Varying Between 1.0 to 1.25
H. = Feet Head @ B.E.P.

Appendix 4M.5
USEFUL FORMULAS

$$\text{HP at meter disc constant} = \frac{R \times K \times M}{0.2072 \times t}$$

$$V = \frac{\text{G.P.M.}}{2.45 D^2} = \frac{0.321 \times \text{G.P.M.}}{A}$$

$$\text{G.P.M.} = \frac{A.V.}{0.321} = 2.45 \, VD^2$$

$$D = \sqrt{\frac{\text{G.P.M.}}{2.45 \, V}} \quad \text{or} \quad D^2 = \frac{\text{G.P.M.}}{2.45 \, V}$$

$$\text{G.P.D.} = 3530 \times V \times D^2$$

A = area, sq in.
D = dia. pipe, in.
V = Vel., ft/sec
R = revolutions of meter disc
K = meter disc constant
M = multiplier—product of PT & CT ratio
t = time for number of revolutions

USEFUL FORMULAS

At constant diameter, head is proportional to GPM.
EXAMPLE: 1000 GPM produces 12 - ft friction in a pipe, how much will 2000 GPM produce?

$$\text{ANSWER: } \left(\frac{2000}{1000}\right)^2 \times 12 = 48 \text{ ft}$$

At constant diameter, capacity is proportional to $\sqrt{h}$.
EXAMPLE: Dia. 10 inches h – 16 feet per 100 feet length. What is capacity? From friction tables, the nearest figure is 10.0 feet for 3500 GPM. The capacity will be greater, hence.

$$\text{CAPACITY: } \frac{3500 \times \sqrt{16}}{\sqrt{10}} = 4430$$

At constant head, the capacity of a pipe is proportional to $d^{2.5}$
EXAMPLE: A 6″ pipe discharges 400 GPM. How much will a 3″ pipe deliver under the same conditions?

$$(6)^{2.5} = 88.20 \qquad (3)^{2.5} = 15.57 \qquad d^{2.5} = d^2 \sqrt{d}$$

$$\text{ANSWER: } \frac{400 \times 15.57}{88.20} = 70.60 \text{ GPM}$$

$$V = \frac{GPM}{2.45D^2} = \frac{.321 \times GPM}{A}$$

$$GPM = \frac{AV}{.321} = 2.45VD^2$$

$$D = \sqrt{\frac{GPM}{2.45V}} \quad D^2 \frac{GPM}{2.45V}$$

Gals per 24 hours = $3530 \times V \times D^2$

A = Area in sq in.
D = Diam. pipe, in.
GPM = gallons per minute
V = Vel. ft/sec
g = 32.16 ft/sec
h = Velocity head.

$$h = \frac{V^2}{2g}$$

Appendix 4M.6
HYDRAULIC EQUATIONS

Pounds per square inch = $0.434 \times$ head of water, ft
Head in feet = $2.31 \times$ lbs/sq in

$$\text{Approximate loss of head due to friction in clean iron pipes} = \frac{0.02 \times L \times V_2}{64.4 \, D} \, \text{ft}$$

Where L = length of pipe in feet. D = diameter in feet. V = velocity of flow in fps. In calculating the total head to be pumped against, it is common to consider it equal to the sum of the friction head and the actual head.

$$\text{Horsepower of waterfall} = \frac{62 \times A \times V \times H}{33,000}$$

Where A = cross section of water in square feet. V = velocity of flow in fpm. H = head of fall in feet.

HEAT AND ENERGY EQUIVALENTS

1 kilowatt-hour = 1.341 horsepower-hours
2,655,217 foot-pounds
3413 British thermal units

1 horsepower-hour = 0.7457 kilowatt-hours (745.7 watt-hours)
1,980,000 foot-pounds (33,000 × 60)
2545 British thermal units

1 British thermal unit = 777.97 foot-pounds
1054.8 Joules or watt-seconds
0.000293 kilowatt-hours = 0.293 watt-hours
0.000393 horsepower-hours

1 kilowatt = 1.341 horsepower
44.254 foot-pounds per minute
56.883 Btu per minute

1 horsepower = 0.7457 kilowatt = 745.7 watts
33,000 foot-pounds per minute
42,418 Btu per minute
1.0139 metric horsepower

HORSEPOWER OF PUMP SHAFTS

$$B.H.P. = \frac{Stress \times R.P.M.}{321000} \times d^3 = \frac{(Dia.\ at\ coupling)^3}{64} \times rpm$$

$$d = \sqrt[3]{\frac{64 B.H.P.}{R.P.M.}}$$

B.H.P. = Brake Horsepower
rpm = Revolutions per minute

$$\text{Stress in shaft} = \frac{B.H.P. \times 321000}{rpm \times d^3}$$

$$B.H.P. = \frac{(torque\ in\ in.\text{-}lbs) \times rpm}{63000}$$

$$T = \text{torque, in.-lbs} = 0.196 \times S \times d^3$$

$$T_1 = \text{torque, ft/lbs} = \frac{0.196 \times S \times d^3}{5250}$$

$$\text{Then } B.H.P. = \frac{(torque\ in\ ft\text{-}lbs.) \times rpm}{5250}$$

S = Stress, lbs/sq in. = 5020
d = Diameter of shaft, in.
T = Torque, in.-lbs
T_1 = Torque, ft-lbs

Appendix 4M.7
POWER TO HOIST A LOAD

$$H.P. = \frac{\text{weight (lbs)} \times \text{ft/min} \times \sin\phi}{33,000}$$

ϕ = Angle of hoist, with horizontal

POWER TO DRIVE PUMPS

$$H.P. = \frac{G.P.M. \times \text{total head (including friction)}}{3,960 \times \text{Eff. of pump}} \times Sp.\ Gr.$$

Approximate efficiency as follows:
500 to 1000 G.P.M. = 70 to 75%
1000 to 1500 G.P.M. = 75 to 80%
More than 1500 G.P.M. = 80 to 85%

NOTE: H.P. varies as the cube of the speed (rpm)

Appendix 4N.1
TEMPERATURE RISE IN PUMPS

t = Temperature rise of liquid in pump, degrees F.

$$t = \frac{5.1 \times BHP\ loss}{Sp.\ Gr. \times Sp.\ Heat \times GPM}$$

$$= \frac{BHP\ (100 - Pump\ Eff.) \times .707 \times 60}{GPM \times 8.33 \times Sp.\ Gr. \times Sp.\ Heat}$$

BHP loss is found by subtracting eff. at rated capacity from 100 and multiplying by BHP at that capacity.
Specific heat for water is 1.0 (for all practical purposes)
Limit rise in temperature to 35°F., maximum
$$1\ hp = 42.41\ BTU/min$$

$$Also\ t = \frac{H\left[\left(\frac{100}{E} - 1\right)\right]}{780 \times sp.\ heat}$$

where:

H = head, ft
E = Pump eff. expressed by whole number

Appendix 4N.2
TEMPERATURE RISE IN PUMP (APPROXIMATION)
EXAMPLE

Heat rise in model W-18 pump at peak efficiency and assuming a 3000 ft head.

$$Pump\ eff. = 60\%$$
$$Capacity = 600\ b/d = 17.5\ GPM$$

Assume: specific heat of 0.5

$$hp = \frac{17.5\ GPM \times 3000\ ft}{3960} = 13.5\ hp$$

Brake hp = hp (output) 13.5 ÷ 60% = 22 hp

Assume: 83% efficiency the actual horsepower input =

$$\frac{22\ hp}{0.83\%} = 37\ hp$$

All but 13.5 hp of this is lost in the form of heat
Power lost = 37 − 13.5 = 23.5 hp

$$23.5\ hp \times \frac{42.44\ BTU/min}{hp} = 1000\ Btu/min$$

The heat will be carried away as the fluid is pumped by the motor and through the pump.

$$17.5\ GPM \times 8.1\ lb/gal = 142\ lb/min$$

$$heat\ added\ per\ lb = \frac{1000\ Btu/min}{142\ lb/min} = 7.05\ Btu/lb$$

$$Temp.\ rise = \Delta T = \frac{7.05}{0.5} = 14.0°\ F.$$

ELECTRICAL SYMBOLS

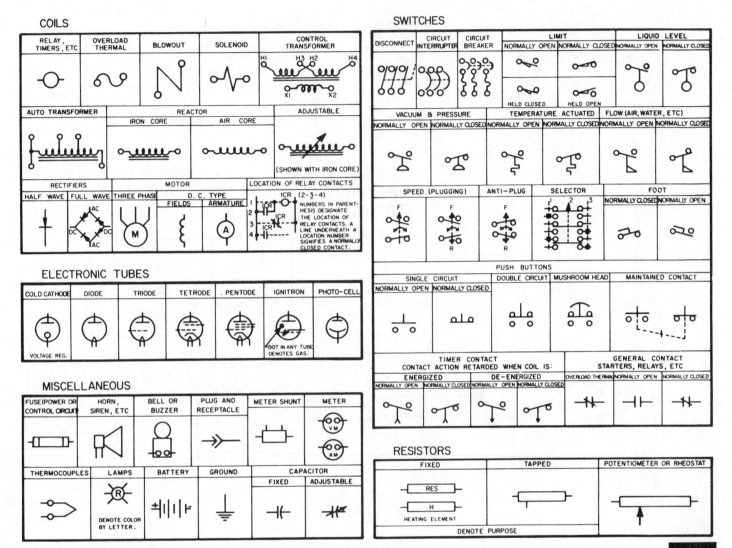

Fig. 4O (1)   Electrical symbols

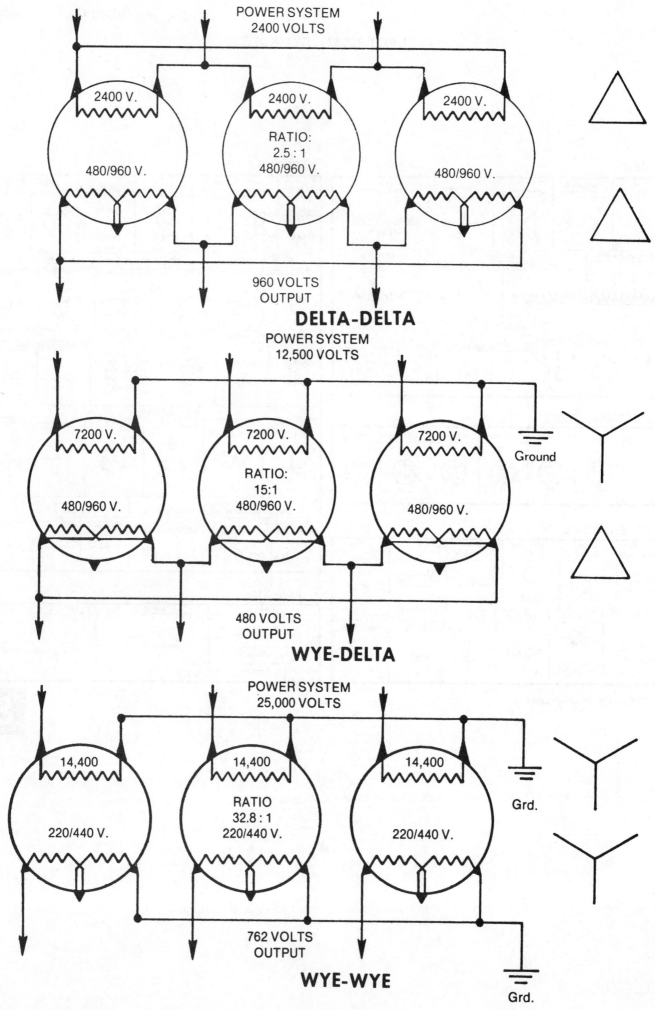

POWER SYSTEM
2400 VOLTS

2400 V.

2400 V.

RATIO:
2.5 : 1
480/960 V.

2400 V.

480/960 V.

480/960 V.

960 VOLTS
OUTPUT

DELTA-DELTA

POWER SYSTEM
12,500 VOLTS

7200 V.

7200 V.

RATIO:
15:1
480/960 V.

7200 V.

480/960 V.

480/960 V.

Ground

480 VOLTS
OUTPUT

WYE-DELTA

POWER SYSTEM
25,000 VOLTS

14,400

14,400

RATIO
32.8 : 1
220/440 V.

14,400

220/440 V.

220/440 V.

Grd.

762 VOLTS
OUTPUT

WYE-WYE

Grd.

Fig. 4P (1)   Transformer connections

Chapter 5

Hydraulic pumping—piston type

by Phil Wilson

5.1 INTRODUCTION

A subsurface hydraulic piston pump is a closely coupled reciprocating engine and pump. The unit is installed below the working fluid level in a well, as shown in Fig. 5.1. High pressure power fluid is directed to the engine through one conduit and spent power fluid and well production are directed to the surface through another conduit. The high pressure power fluid causes the engine to reciprocate much like a steam engine except the power fluid is oil or water instead of steam. The pump, driven by the engine, pumps the fluid from the well bore. Originally, the complete engine and pump was designated "Production Unit," but in prac-

tice it was always called a "pump." Current usage refers to the engine as "the engine end of the pump" and to the pump as "the pump end of the pump."

The piston-type hydraulic pumping represents the deepest method of lift today (18,000 ft in south Louisiana). Also, the pump can be circulated out for repairs, thereby eliminating pulling unit operations. Some of its other advantages are:

(1) has good flexibility on rates
(2) can handle deviated wells
(3) easily adapted to automation
(4) easy to add inhibitor
(5) is suitable for pumping heavy crudes
(6) one well or multiple well units are available
(7) simple well heads accommodate closely spaced wells, covered or cellered well heads and wells in visually sensitive areas.

There are 14 models of pumps shown in the Appendix [Figs. 5.B(1) through 5.B(14)] and each is unique in its design of engine end and/or pump end. Only one, the Kobe A pump (engine end, Fig. 5.2) will be detailed. High pressure power fluid is directed to the top of the engine piston while exhausted power fluid from the lower side of the piston is directed to the relieved area of the engine valve where it is discharged.

When the piston reaches the end of the downstroke, the reduced diameter at the top of the valve rod allows high pressure fluid to enter under the engine valve, as shown in Fig. 5.3. Because the valve has a larger area at its bottom than at its top, it will move upwards.

With the engine valve in the up position, as shown in Fig. 5.4, the flow paths to the piston are reversed. The pump, therefore, begins its upstroke.

When the piston reaches the end of the upstroke, as shown in Fig. 5.5, the reduced diameter near the lower end of the valve rod connects the area under the valve to the discharge, or low pressure, side of the engine. With high pressure on the top of the valve and only exhaust pressure at the bottom, the valve will move to its down position and the cycle will be repeated.

The pump end of the pump is shown making a down-

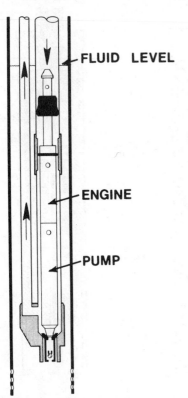

Fig. 5.1 Subsurface hydraulic pump - piston type

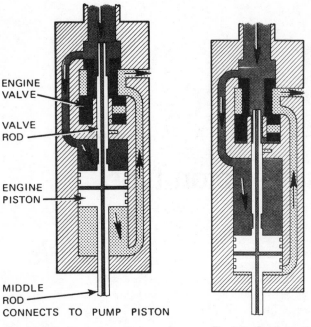

ENGINE VALVE

VALVE ROD

ENGINE PISTON

MIDDLE ROD

CONNECTS TO PUMP PISTON

Fig. 5.2   Engine end

Fig. 5.3   Engine end of downstroke

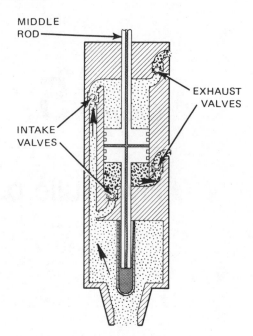

MIDDLE ROD

EXHAUST VALVES

INTAKE VALVES

Fig. 5.6   Pump end in downstroke

stroke in Fig. 5.6. This pump is double-acting, i.e., it pumps on the upstroke and on the downstroke. The arrows show that well fluid is entering on the left and filling the upper part of the cylinder while the well fluid below the piston is being discharged through the ball check valve at the lower right.

The complete pump is shown in Fig. 5.7. On the upstroke well fluid enters the lower part of the cylinder while being discharged from the upper part of the cylinder. The purpose of the hollow lower rod is to balance the areas (forces) on the upstroke and the downstroke.

Most hydraulic pumps are installed as Free Pumps, i.e., they are free to be circulated in and out of the well, as the sequence in Fig. 5.8 illustrates.

A complete hydraulic pumping system is shown in Fig. 5.9. Power fluid systems, represented by the tank at (A), are covered in the next section. Subsequent

sections cover surface pumps (B), control manifolds (C), well heads (D), and tubing arrangements (E).

5.2   POWER FLUID SYSTEMS

Power fluid quality, especially solids content, is an important factor contributing to pump life and repair costs. Power fluid leakage through a pump's fits and clearances is a function of wear caused by abrasive

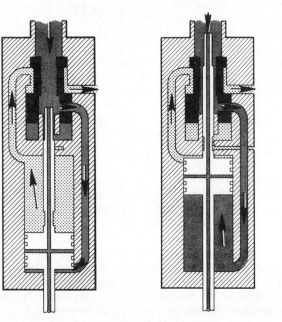

Fig. 5.4   Engine - upstroke

Fig. 5.5   Engine - end of upstroke

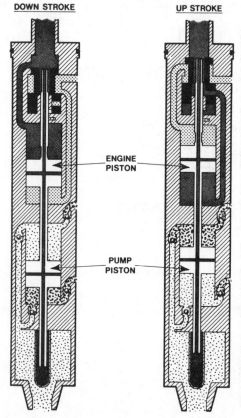

DOWN STROKE   UP STROKE

ENGINE PISTON

PUMP PISTON

Fig. 5.7   Complete pump

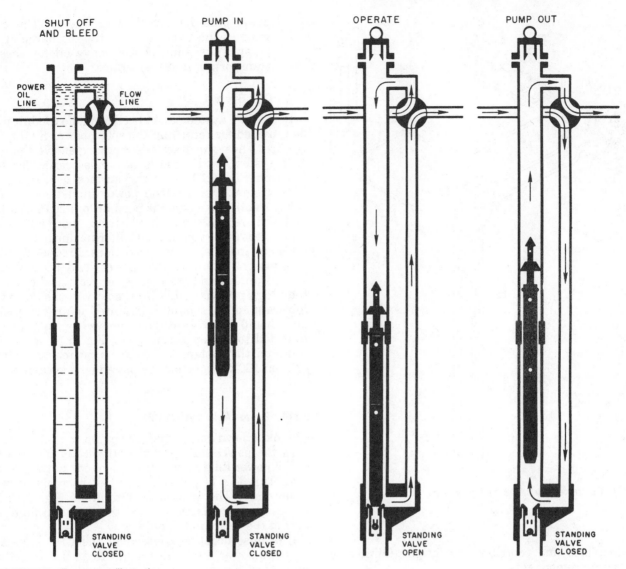

SHUT OFF AND BLEED | PUMP IN | OPERATE | PUMP OUT

POWER OIL LINE   FLOW LINE

STANDING VALVE CLOSED | STANDING VALVE CLOSED | STANDING VALVE OPEN | STANDING VALVE CLOSED

Fig. 5.8   Free pump illustration

solids and of the viscosity of the power fluid. The permissible solids content varies somewhat depending upon the definition of "acceptable pump life" and also on the viscosity but 10-15 ppm is usually acceptable for 30°-40° API gravity oils. For heavier oils, more wear, and consequently more solids, may be tolerated, while for water, less wear and fewer solids are usually the rule. The maximum particle size should not exceed 15 microns. The maximum salt content should not exceed 12 lb/1000 bbl in power oil.

There are two basic types of power fluid systems:

(1) The closed power fluid (CPF) system where the surface and subsurface power fluid stays in a closed circuit and does not mix with the produced fluid.

(2) The open power fluid (OPF) system where the power fluid mixes with the production down hole and returns to the surface as commingled power fluid and production.

The choice of oil or water for power fluid can be based on a number of factors. Following is a list of most of the factors involved in this choice:

(1) Water is preferred for safety and environmental reasons.

(2) For CPF installations the addition of chemicals to power water for lubrication and corrosion is not a large cost factor. (Fresh water is frequently used in CPF installations.)

(3) For OPF installations the addition of chemicals to power water can be a significant cost factor because the power water is commingled with the production. This requires continual injection of chemicals which will add to the operating costs.

(4) Treating power oil is seldom a large cost factor mainly because it seldom needs chemical additives for lubricity. One exception is when high gravity oils are used at very high bottom hole temperatures. When these two factors produce a viscosity below 1.0 centistoke (Fig. 5.46) a lubricant may be necessary for long pump life.

(5) Maintenance on surface pumps is less when using oil because metal to metal plungers and liners are usually used instead of packing. Also, valves last longer and are usually the ball and seat type rather than the disc or poppet type normally used for water. Additionally, the low bulk modulus of water causes much larger pressure pulses than oil, and these pulses are detrimental to pipe con-

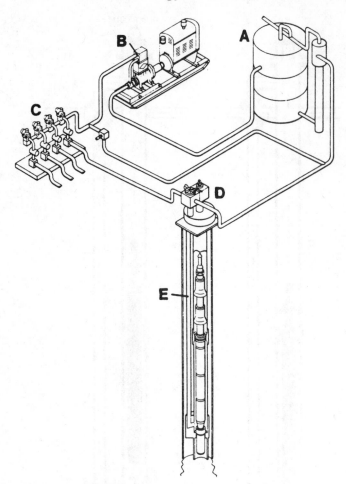

Fig. 5.9   Complete hydraulic pumping system

nections and contribute to fatigue failures of pump components.

(6) Subsurface pumps are sensitive to viscosity and lubricating qualities of the power fluid. Because water has practically no lubricating ability at bottom hole temperatures, it can, if not adequately treated, contribute to shorter pump life. Leakage of power fluid past the various sliding fits in the pump is a function of the viscosity and is greater with water than with most crude oils.

(7) Testing a well for oil production is subject to an added source of error when oil is used for power fluid. (This statement is not true when using the Well Site Power Plant described in the next section.) The power oil must be metered in (metered in and out on a CPF system) and small errors in metering can be significant when the ratio of power oil to produced oil is large, as when the well is producing a large percentage of water. For instance, if the ratio of power oil to produced oil is 10:1, an error of 2% in the power oil meter translates into a 20% error in produced oil.

(8) Usually, the surface pressure required will be less when using power water as compared to using power oil.

(9) Although hydraulic pumps handle viscous crudes (7-20° API) very well, it has sometimes served other purposes to use a higher gravity oil for

power fluid and use the OPF system. This commingles the two crudes at the discharge of the pump, thus diluting the heavy oil for ease in transporting it at the surface.

5.21   CPF System

In the CPF system, an extra down hole conduit must be provided for returning the spent power fluid to the surface. Thus, the system is more expensive than the OPF system, so its use is not widespread. The bold lines on Fig. 5.10 show the surface facilities for a CPF system. Because the power fluid tank is relatively small, this system is popular for urban locations and offshore platforms where surface space is at a premium. It is quite popular in California due to the many townsite and offshore well locations. Frequently, CPF systems use water for the power fluid because it is less hazardous and presents fewer ecological problems than high pressure oil. Water, however, should have a lubricant added, should be inhibited against corrosion, and should have all oxygen removed—considerations that add to the operating costs. Fig. 5.10 shows two wells on the system but there is no reason that 30 or even 100 wells cannot be placed on a system of this type.

5.211   Power fluid tank (CPF)

In most down hole pumps, the pump end is lubricated with the power fluid and part (typically 2% to 10%) of the power fluid is purposely "leaked" to the production. This loss of power fluid must be replaced with clean fluid. The power fluid tank in Fig. 5.10 removes abrasive particles from the make-up fluid and part of the recirculated fluid.

One misconception concerning the CPF system is that the power fluid will remain clean because it has no source of contamination. In actual practice, three factors are constantly working together to corrupt this theory:

(1) The power fluid tank does not completely remove all of the solid particles from the make-up fluid —cleanliness is relative, not absolute.

(2) The power fluid is not completely non-corrosive. Again, this factor is relative, not absolute, and the products of corrosion are generally abrasive solids.

(3) When fluid containing solids, even a very small percentage of solids, is leaked through a long closely fitted clearance space, as in a downhole pump, the solids tend to be held back. This means that the fluid emerging from the fit is cleaner than the fluid trying to enter the fit. The tendency, then, is for the power fluid circuit to lose clean fluid and to retain the solid particles.

Over a period of time these three factors allow the power fluid in the closed circuit to become "dirtier" than the fluid emerging or the fluid entering the closed circuit, unless a part (10% is reasonable) of the recirculated power fluid is continually cleaned by the power fluid settling tank (Fig. 5.10). This "continuous cleaning of part of the recirculated power fluid" is an important feature in the design of the CPF system.

When water is used for the power fluid, filters may be used instead of settling tanks for the clean-

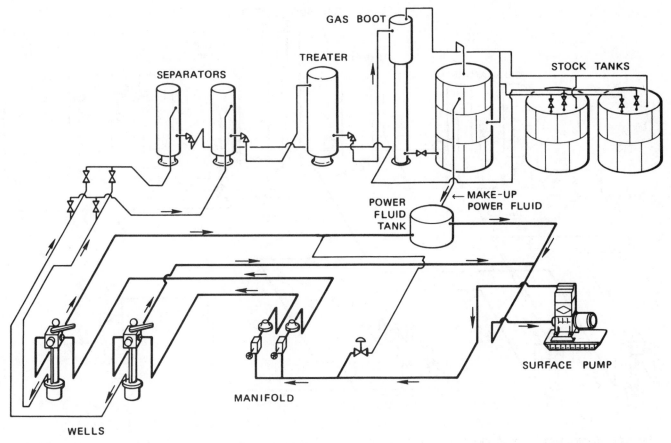

GAS BOOT

TREATER

SEPARATORS

STOCK TANKS

← MAKE-UP
POWER FLUID

POWER
FLUID
TANK

SURFACE PUMP

MANIFOLD

WELLS

Fig. 5.10   Surface facilities for a closed power fluid (CPF) system

ing process. These filters should remove particles down to 10 microns. When oil is used, experience has shown that the settling tank should be large enough to keep the upward velocity of the oil below 1 ft/hr for oils below 30° API gravity and below 2 ft/hr for oils above 30° API gravity.

5.22   OPF system

In the OPF system only two down hole paths are needed: one for conducting power fluid to the engine and one for conducting spent power fluid plus production to the surface. These conduits can be two strings of tubing or one tubing string and the tubing/casing annulus. Simplicity and economy are the important features of the OPF system. When water is used for power fluid in the OPF system, the chemicals added (for lubrication, corrosion inhibition and oxygen scavenging) are generally lost when mixed with the production and must be added continuously. The bold lines on Fig. 5.11 show the surface facilities for an OPF system with 2 wells. Central plants of this type can be used for any number of wells.

Usually, the triplex pump and control manifold are located at the central tank battery, but control manifolds can be located at satellite locations. Even triplex pumps can be located at satellite locations if a small pump is used at the battery to get the fluid to the suction of the triplex pump. Satellite manifolds reduce the footage of surface power fluid lines and satellite pumps reduce the footage of high pressure surface power fluid lines.

5.221   Power fluid tank (OPF)

The OPF power fluid tank shown in Figs. 5.12 and 5.13 has been proven over the years to be an excellent design. This design and one with a slight modification are almost universally used.

Oil generally enters the gas boot in surges and contains gas not removed in the treater—gas that was in solution at the 30 psi treater pressure. The gas boot removes the last remnants of gas which would otherwise keep the tank stirred up. The top section of the boot should be 36 in. to be effective and, even with this diameter, surges frequently occur that cause the oil to be carried over the top through the gas line. To prevent this oil carry-over from going into the top of the tank and by-passing the settling process, a loop with riser ties the boot gas line to the tank gas line.

Dead oil (gas-free) then enters the bottom of the tank which should have a level spreader. The oil entering here is power oil plus production. At the vertical midpoint, production is drawn off through the outside riser that keeps the tank full. From the mid-point up, the power oil settling process takes place. The light solids settled out are carried with the production to stock, while the heavier particles are settled to bottom and must be removed periodically.

One modification allows the production to be taken from the boot or between the boot and the tank—still with a riser to keep the tank full. This modification allows full use of the tank for power oil only, but imbalances in the column weight of the riser and of the tank can occasionally cause seemingly strange fluid

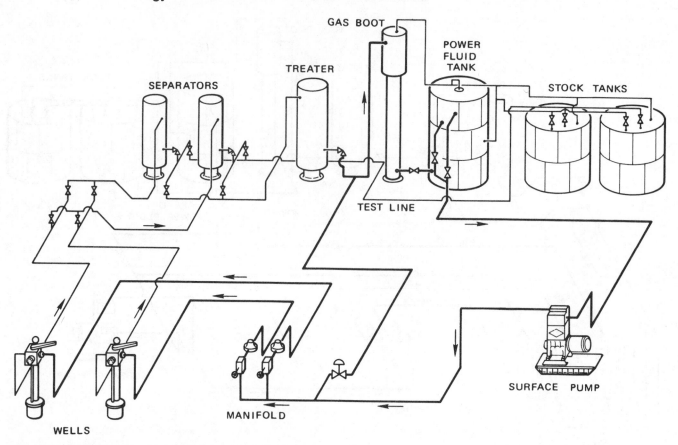

Fig. 5.11 Surface facilities for an open power fluid (OPF) system

level variations in the tank. With this modification, the solids should be drained from the bottom more frequently. The boot can be placed inside the tank, but experience has shown cases of pitting and corrosion at the fluid level. Therefore, the outside location is preferred.

To ensure adequate particle settling, the power oil tank should be sized to allow the upward velocity in the top half to be less than 2 ft/hr. This velocity is 1,500 b/d in a 750-bbl, 24-ft high tank. The velocity should be lower for oils heavier than 30° API gravity and for operations in extremely cold climates. Some operators use an auxiliary pump/cyclone package to continually circulate and remove solids from the power oil settling tank.

5.3 INDIVIDUAL WELL SITE POWER PLANT

The use of individual well site power plants is becoming more and more popular, and these units are competitive with other types of artificial lift. Typical well site units are noted in Figs. 5.14 and 5.15.

A well-site power plant is a package of components, installed at or near a well site, that accomplishes the functions normally performed by a central plant. The basic components consist of a liquid-gas separator, centrifugal separators for removing solids from the power fluid (oil or water), and a surface pump. These units are portable; they require a minimum of installation labor and eliminate the need for advance long-range planning of a central power plant. They are always used with an OPF tubing arrangement, but they have one feature similar to a CPF system: the net production from the well goes into the flow line while the power fluid is recirculated at the well-site. This fea-

ture simplifies well testing and does not increase the load on the treating system at the tank battery. Simple, flexible, compact, and portable are features of the well-site power plant that are of great interest to the design engineer, the production foreman, and to the lease operator.

Generally, there will be a choice of either a central system or an individual well-site system. Some choices are obvious, such as a central system for an offshore platform, or any cluster of wells, such as in a downtown area or islands constructed for that purpose.

For those wells that are isolated or on wide spacing, the individual well-site system will probably be preferred.

Flow schematics in Figs. 5.16, 5.17 and 5.18 show models of Kobe (Solo Unit), fluid packed (Unidraulic) and Johnson-Fagg (Econodraulic) well-site power plants. These units must provide gas-free fluid to the surface pump, provide means to choose oil or water for the power fluid, remove the solids from the power fluid, and provide surge and reserve capacity for circulating a subsurface pump to the surface after a pump failure (Vessel sizes range from 40 in. × 10 ft to 60 in. × 20 ft).

If we assume that a well produces 600 b/d of water, 200 b/d of oil and uses 2000 b/d of power water, these flow rates are shown on the simplified schematics of Figs. 5.19, 5.20 and 5.21.

Removing solids from the power fluid is usually accomplished by cyclone centrifugal separators, Fig. 5.22. These cyclones require 30-60 psi pressure drop from inlet (feed) to top outlet (overflow). The ratio of overflow to underflow out the apex of the cone is controlled by the relationship of overflow pressure to underflow pressure. Usually the overflow must be 5-10 psi

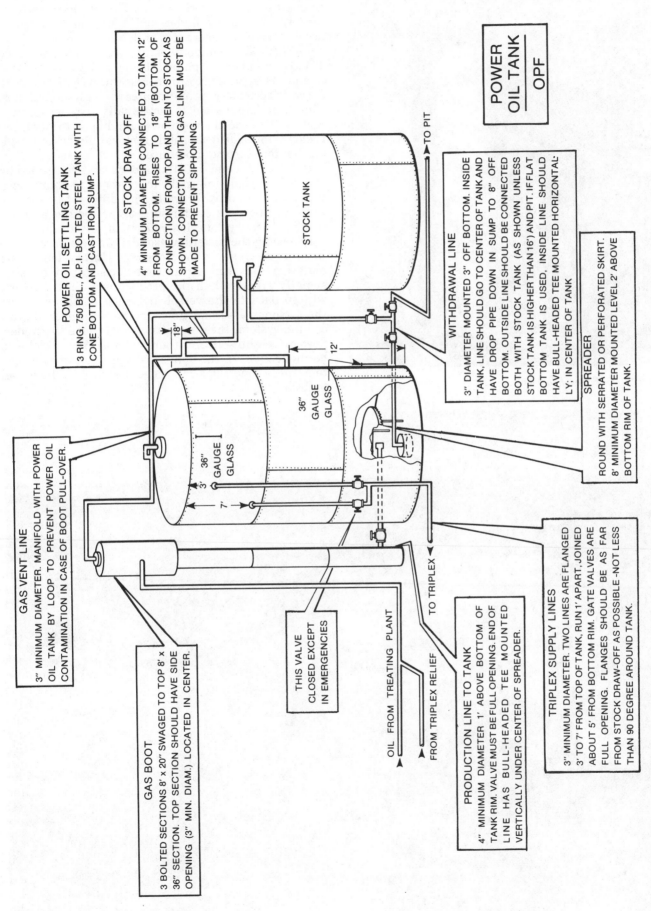

POWER
OIL TANK

OPF

STOCK DRAW OFF

4" MINIMUM DIAMETER CONNECTED TO TANK 12' FROM BOTTOM. RISES TO 18" (BOTTOM OF CONNECTION) FROM TOP AND THEN TO STOCK AS SHOWN. CONNECTION WITH GAS LINE MUST BE MADE TO PREVENT SIPHONING.

POWER OIL SETTLING TANK

3 RING, 750 BBL., A.P.I. BOLTED STEEL TANK WITH CONE BOTTOM AND CAST IRON SUMP.

STOCK TANK

WITHDRAWAL LINE

3" DIAMETER MOUNTED 3" OFF BOTTOM. INSIDE TANK, LINE SHOULD GO TO CENTER OF TANK AND HAVE DROP PIPE DOWN IN SUMP TO 8" OFF BOTTOM. OUTSIDE LINE SHOULD BE CONNECTED BOTH WITH STOCK TANK (AS SHOWN UNLESS STOCK TANK IS HIGHER THAN 16') AND PIT. IF FLAT BOTTOM TANK IS USED, INSIDE LINE SHOULD HAVE BULL-HEADED TEE MOUNTED HORIZONTAL-LY; IN CENTER OF TANK

TO PIT

GAS VENT LINE

3" MINIMUM DIAMETER. MANIFOLD WITH POWER OIL TANK BY LOOP TO PREVENT POWER OIL CONTAMINATION IN CASE OF BOOT PULL-OVER.

36"
GAUGE
GLASS

36"
GAUGE
GLASS

3'

7'

SPREADER

ROUND WITH SERRATED OR PERFORATED SKIRT. 8' MINIMUM DIAMETER MOUNTED LEVEL 2' ABOVE BOTTOM RIM OF TANK.

GAS BOOT

3 BOLTED SECTIONS 8' x 20' SWAGED TO TOP 8' x 36" SECTION. TOP SECTION SHOULD HAVE SIDE OPENING (3" MIN. DIAM.) LOCATED IN CENTER.

THIS VALVE
CLOSED EXCEPT
IN EMERGENCIES

OIL FROM TREATING PLANT

FROM TRIPLEX RELIEF

TO TRIPLEX

PRODUCTION LINE TO TANK

4" MINIMUM DIAMETER 1' ABOVE BOTTOM OF TANK RIM. VALVE MUST BE FULL OPENING. END OF LINE HAS BULL-HEADED TEE MOUNTED VERTICALLY UNDER CENTER OF SPREADER.

TRIPLEX SUPPLY LINES

3" MINIMUM DIAMETER. TWO LINES ARE FLANGED 3' TO 7' FROM TOP OF TANK, RUN 1' APART, JOINED ABOUT 5' FROM BOTTOM RIM. GATE VALVES ARE FULL OPENING. FLANGES SHOULD BE AS FAR FROM STOCK DRAW-OFF AS POSSIBLE -NOT LESS THAN 90 DEGREE AROUND TANK.

18"

12'

Fig. 5.12   Schematic of OPF power fluid tank

Fig. 5.13   OPF power fluid tank

greater than the underflow to insure a positive rather than a negative underflow rate. The cyclone internals, feed nozzle, vortex finder and apex can be sized to accommodate various rates of flow.

In Fig. 5.19 the underflow rate is controlled by Valve 1. Valve number 2, a pump internal relief valve, is used to adjust the pressure drop across the cyclones to 40-50 psi which produces optimum rates through the cyclones. This unit is designed to continually recycle approximately ⅔ of the cleaned power fluid back through the cyclones for additional solids removal. Refer to Appendix 5.F for start up procedures.

In Figs. 5.20 and 5.21 the underflow is controlled by Valve 1. Cyclone internals are changed to obtain optimum pressure drop for wells with different flow rates. Refer to Appendix 5.F for start up procedures and Appendix 5.E for cyclone flow rates with various internals.

Disposing the underflow (solids) can be a problem on low volume wells. In the Kobe Solo Unit system, if the underflow is set at a rate that is constantly or occasionally greater than the production rate, the excess will go back to the vessel and must be separated by the cyclones a second time. In the fluid packed Unidraulic system, these conditions can cause the underflow to be shut-off and thereby cause the cyclone to be ineffective. In the Johnson-Fagg Econodraulic system, these conditions can cause the level in the vessel to drop.

Fig. 5.14   Individual well site power plant

Fig. 5.15   *Individual well site power plant*

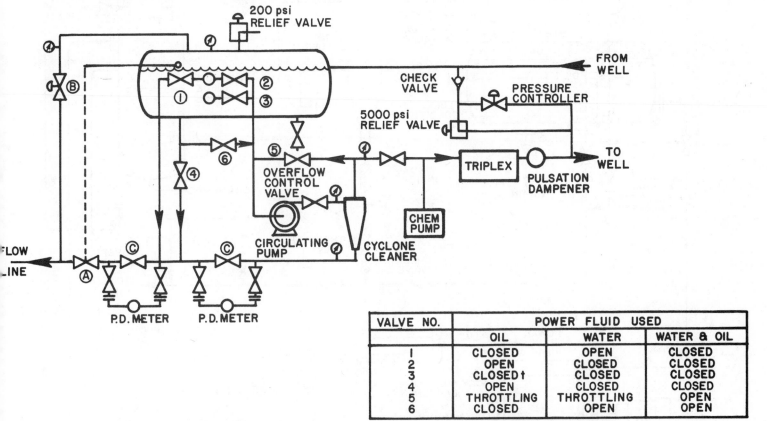

VALVE NO.	POWER FLUID USED		
	OIL	WATER	WATER & OIL
1	CLOSED	OPEN	CLOSED
2	OPEN	CLOSED	CLOSED
3	CLOSED†	CLOSED	CLOSED
4	OPEN	CLOSED	CLOSED
5	THROTTLING	THROTTLING	OPEN
6	CLOSED	OPEN	OPEN

† OPEN FOR RESERVE POWER OIL SUPPLY

OTHER VALVES SHOWN (A) FLOAT-OPERATED DUMP VALVE
(B) VESSEL BACK-PRESSURE VALVE
(C) METER BY-PASS VALVE

Fig. 5.16   *Kobe solo unit*

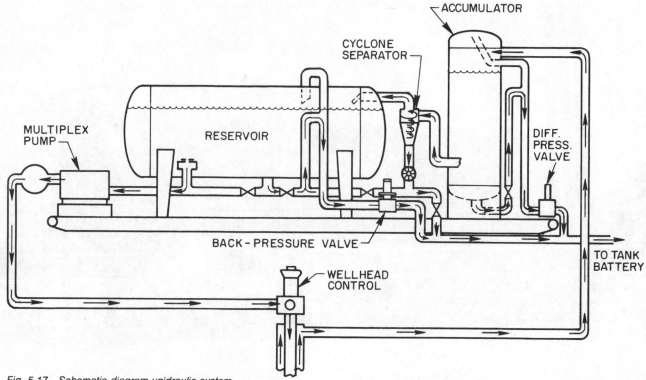

Fig. 5.17   Schematic diagram unidraulic system

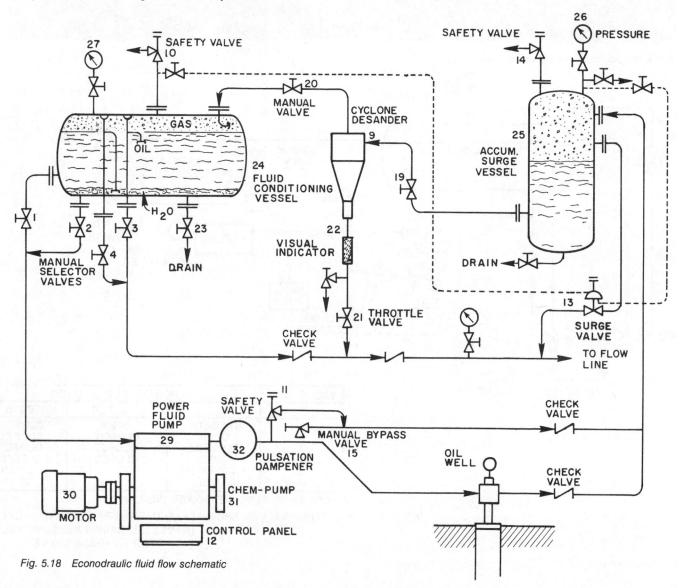

Fig. 5.18   Econodraulic fluid flow schematic

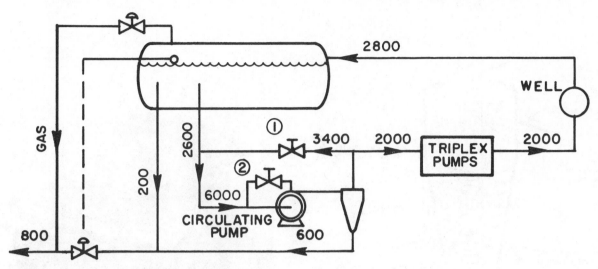

Fig. 5.19   Kobe "solo unit" flow diagram

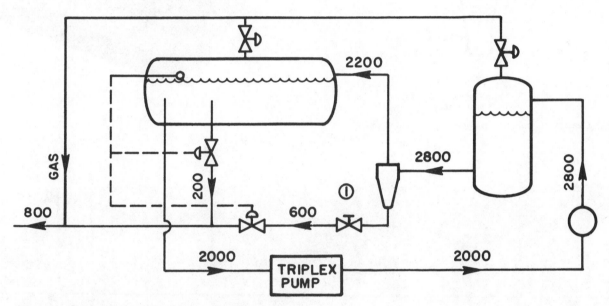

Fig. 5.20   Fluid packed "unidraulic" flow diagram

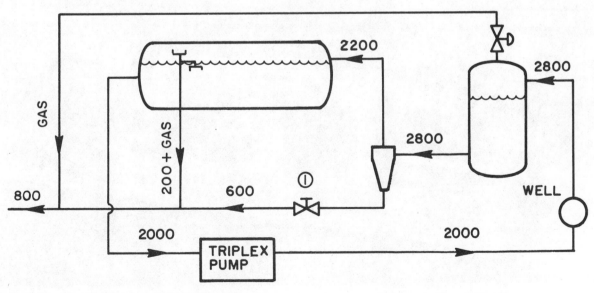

Fig. 5.21   Johnson-Fagg "econodraulic" flow diagram

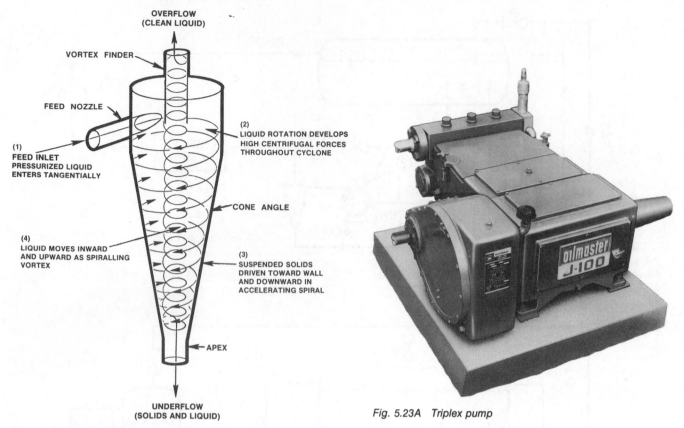

OVERFLOW
(CLEAN LIQUID)

VORTEX FINDER

FEED NOZZLE

(1)
FEED INLET
PRESSURIZED LIQUID
ENTERS TANGENTIALLY

(2)
LIQUID ROTATION DEVELOPS
HIGH CENTRIFUGAL FORCES
THROUGHOUT CYCLONE

CONE ANGLE

(4)
LIQUID MOVES INWARD
AND UPWARD AS SPIRALLING
VORTEX

(3)
SUSPENDED SOLIDS
DRIVEN TOWARD WALL
AND DOWNWARD IN
ACCELERATING SPIRAL

APEX

UNDERFLOW
(SOLIDS AND LIQUID)

Fig. 5.22   Cyclone centrifugal separator

Fig. 5.23A   Triplex pump

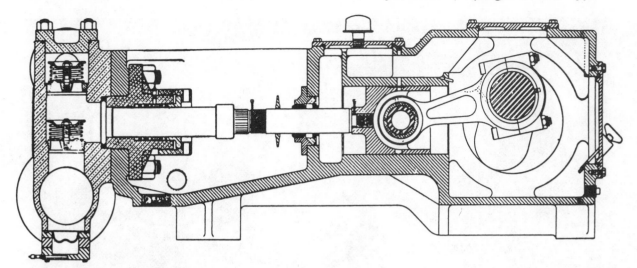

Fig. 5.24   Cross section of surface pump

5.4 SURFACE PUMPS

The surface pumps commonly used are designed specifically for power fluid service and are supplied by the down hole hydraulic pump manufacturers. For high pressure clean oil service these pumps usually use metal to metal plungers and liners, and ball type valves—components which require little or no maintenance. For water service, plungers and liners with packing are usually used. Auxillary items required include a relief valve, pressure gauges and safety switches. Most surface pumps are skid mounted with electric motors or gas engines.

The discharge lines from the relief valve and back pressure control valve should not be connected directly to the suction line of the pump but should be connected to a separate line going back to the tank. The reason for this is that when oil, even dead oil, is suddenly taken from high pressure to low pressure, some gas will flash out of solution. This gas will cause a loss of volumetric efficiency if allowed to enter the pump.

A pulsation dampener may be required in some cases. The pulsations will be more pronounced with water.

For long suction lines an accumulation chamber may be required to prevent separating the liquid into slugs.

Typical triplex pumps are noted in Figs. 5.23A and B with a cross-section of a pump in Fig. 5.24.

5.5 CONTROL MANIFOLDS

5.51 Manifolds

Power fluid distribution manifolds, used at central plants, are supplied by the down hole pump manufacturers and are made in modular header sections that can be added to or subtracted from the manifold easily (Figs. 5.25A and B). These manifolds usually contain pilot-operated control valves that keep the volume of power fluid going to each well constant, regardless of pressure changes in the system. A pressure controller (back pressure regulator) is also used to maintain a constant pressure on the surface pump. Additionally, high pressure meters and pressure gauges are included for each well. Down hole pump speed control, pressure checks, engine end efficiency checks, and trouble shooting can be accomplished at the control manifold.

The following summarize the purpose of these manifolds:
(1) Distribute the flow of power fluid to the individual wells
(2) Regulate the flow rate to the individual wells
(3) Provide a means of metering the flow to each individual well
(4) Provide a means of measuring pressure to each individual well
(5) Provide a means for running soluble plugs in surface lines.
(6) Provide a manual or automatic valve to control manifold pressure by by-passing excess power fluid

Generally, 100 to 300 psi more fluid pressure is brought to the manifold than goes to the wells. For example, if the maximum pressure required by the wells is 3,000 psi, 3,200 psi can be used at the manifold. Dual manifolds may also be used for wells of different zones. For example, one group of wells may require 3,500 psi whereas the other group requires 2,500 psi.

5.52 Constant flow control valves

Constant flow control valves used in these manifolds work on the principle of a constant pressure drop across the main control valve (characterized valve) as shown

Fig. 5.25A   Control manifold

in Fig. 5.26. A spring/diaphragm/pilot valve combination maintains the constant pressure differential across the main valve regardless of changes in the upstream and downstream pressures.

These flow control valves offer excellent control and have found uses in other places where a constant liquid rate is needed.

5.6 WELL HEADS

The well head for a free pump should provide the following functions:
  (1) Direct power fluid down the tubing for "pump in and operate"
  (2) Direct power fluid down the proper conduit for "pump out"
  (3) Shut power fluid line and provide a means to bleed pressure from the tubing
  (4) Catch and hold the pump
  (5) Be a safety device to prevent high pressure from accidentally being applied to the casing

The 4-way valve shown in Fig. 5.27 provides all five of these functions.

The well head for a fixed type pump is relatively simple and straightforward as illustrated in Fig. 5.28.

5.7 TUBING ARRANGEMENTS

When the pump is screwed onto the power tubing and lowered into the well by that tubing, it is called a fixed type pump. When the pump fits inside of the power tubing and is free to be circulated to bottom and back out again, it is called a Free Pump. Either type can be a CPF or OPF system. Two pumps connected together in tandem (in parallel hydraulically) can be installed in a Free Pump system to double capacity. Hydraulic pumps are particularly suitable for deep wells, directionally drilled wells, multiple completed wells, and offshore platform wells.

5.71 Fixed insert

Fixed insert is the name applied to the tubing arrangement shown in Fig. 5.29. In this arrangement gas is vented through the casing.

5.72 Fixed casing

Fixed casing is the name applied to the arrangement shown in Fig. 5.30, where the casing is used for one of the flow paths. In this arrangement, the gas must be handled by the pump. Installations of this type generally use large ($3^{13}/_{16}$ in. OD) pumps. Sometimes a separate tubing string is used to vent the gas from beneath the packer, as shown in Fig. 5.31. Venting is necessary for wells producing below the bubble point with high gas-liquid ratios.

5.73 Parallel free

Parallel free installations are shown in Figs. 5.32, 5.33 and 5.34. The pump in Fig. 5.34 is unseated through the power return tubing; hence it is referred to as power return unseat (PRU). The pumps in Figs. 5.32 and 5.33 are the conventional production unseat type (raw production enters and contaminates the clean power fluid circuit when the production unseat is unseated). Gas is vented through the casing in these arrangements.

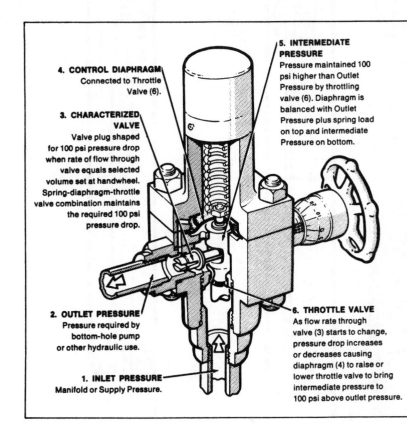

5. INTERMEDIATE PRESSURE
Pressure maintained 100 psi higher than Outlet Pressure by throttling valve (6). Diaphragm is balanced with Outlet Pressure plus spring load on top and intermediate Pressure on bottom.

4. CONTROL DIAPHRAGM Connected to Throttle Valve (6).

3. CHARACTERIZED VALVE
Valve plug shaped for 100 psi pressure drop when rate of flow through valve equals selected volume set at handwheel. Spring-diaphragm-throttle valve combination maintains the required 100 psi pressure drop.

2. OUTLET PRESSURE
Pressure required by bottom-hole pump or other hydraulic use.

1. INLET PRESSURE Manifold or Supply Pressure.

6. THROTTLE VALVE
As flow rate through valve (3) starts to change, pressure drop increases or decreases causing diaphragm (4) to raise or lower throttle valve to bring intermediate pressure to 100 psi above outlet pressure.

HOW IT WORKS

As illustrated at left, there are three separate pressures involved in the operation of the Kobe Constant Flow Controller—inlet, intermediate and outlet. The spring acts on the diaphragm with a force equivalent to 100 psi pressure, therefore the diaphragm is in equilibrium when the intermediate pressure is 100 psi greater than the outlet pressure. The outlet valve thus has a 100 psi pressure drop maintained across it at all times which insures a constant flow rate. The characterized outlet valve is shaped to allow, at this 100 psi pressure drop, the rate of flow selected by the handwheel.

Fig. 5.26 Constant flow control valve

Fig. 5.27   Four-way valve and well head for free pump

Fig. 5.28   Well head for a fixed pump type

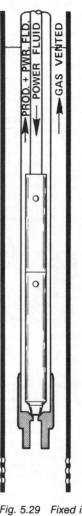

Fig. 5.29   Fixed insert tubing
arrangement (OPF)

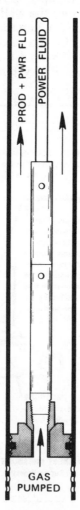

Fig. 5.30   Fixed casing tubing
arrangement

5.74 Casing free

Casing free installations are shown in Figs. 5.35, 5.36, 5.37 and 5.38. Fig. 5.36 is sometimes used instead of Fig. 5.32 to reduce friction, since the return column is the large annular flow area. All of the gas must be handled by the pumps in Figs. 5.35, 5.37 and 5.38. More hydraulic pumps are installed, as shown in Fig. 5.35, than in any other type of installation because it is the lowest-cost type. When venting of gas is necessary, Figs. 5.32 and 5.33 are the most popular installations. For large production rates in small casing, Fig. 5.30 is the most popular.

5.75 Other tubing arrangements

5.751 Introduction

Hydraulic pumps can be adapted to almost any tubing arrangement that can be conceived. Generally speaking, if the tubing will fit in the casing, a pump or pumps can be adapted to it. Pumps are available to fit inside of 1¼ in. I.D. to 4½ in. O.D. tubing.

5.752 Reverse circulation

Fig. 5.39 shows a variation of Fig. 5.32, where power fluid is directed down the small string and production up the large string. This system allows the largest flow rate, power oil plus production, to use the largest tubing string to reduce the overall fluid friction in the system. The pump requires an automatic latching device to hold it down during the pumping operation and requires a releasing tool to be dropped before the pump can be pumped to the surface.

5.753 Dual wells

There are many variations possible for dual wells. Two separate power fluid tubes are almost always used because the separate zones will undoubtedly require different surface operating pressures. If only one tube were to feed two pumps, speed control would be hopeless. One possible tubing arrangement is shown in Fig. 5.40. With 7 in. casing and 2 strings of 2⅜ in. tubing, some of the possible arrangements are 2 pumps of Fig. 5.29, 2 pumps of Fig. 5.32 or 1 of Fig. 5.32 and 1 of Fig. 5.35.

5.754 Tandem pumps

Fig. 5.41 is a method of installing 2 pumps in a single zone well to double the capability of the equipment.

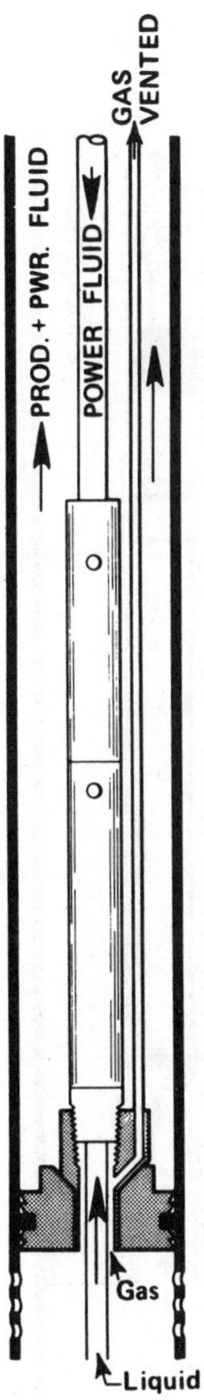

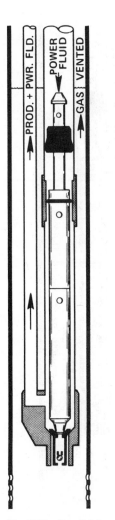

Fig. 5.31 Fixed casing with gas vent (OPF)

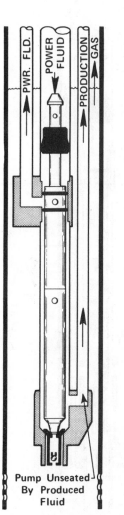

Fig. 5.32 Parallel free tubing arrangements

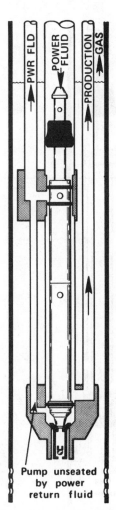

Fig. 5.33 Parallel free tubing arrangements

Fig. 5.34 Parallel free tubing arrangements

5.755 Safety valves

Offshore wells and urban townsite wells usually require subsurface safety valves in the tubing. These valves require an auxiliary pressure source to keep them open. If disaster strikes and the well head is broken off or damaged these valves, set some distance down the tubing, close and keep the well under control. Fig. 5.42 shows such a valve set between the packer and a hydraulic pump. The actuating pressure is obtained from the high pressure power fluid. If disaster strikes at the surface, the power fluid pressure will be released and the safety valve will shut in the tubing and, in Fig. 5.42, the casing as well.

5.8 DESIGN CONSIDERATIONS AND CALCULATIONS

When designing a hydraulic pumping installation the following decisions must be made:

(1) Decide on an OPF or a CPF system.
(2) Decide whether to vent the gas or to pump the gas.
(3) Choose a down hole tubing arrangement.
(4) Choose a pump to fit the tubing and the well requirements.
(5) Choose a central or a well site power plant.
(6) Choose a surface pump.
(7) Design the power fluid cleaning system.

5.81 OPF or CPF

If surface space at the battery is limited, as in a town-site location or on an offshore platform, or if ecological or cosmetic factors are important, choose a closed system. Using water will minimize the hazard of leaks causing ecological and fire problems but will cause the surface pump to be more expensive and will require considerable operating expense for additives (lubricant and oxygen scavenger) to the power water.

If none of these factors are compelling, then choose an open power fluid system. Oil should generally be chosen because the chemical additives for water are lost in the open system and require continuous injection.

5.82 Vent gas or pump gas

5.821 Introduction

The lowest-cost installations are those that do not vent the gas (Figs. 5.30, 5.35, 5.37 and 5.38) but these

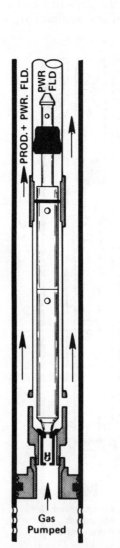

Fig. 5.35 Casing free tubing arrangement

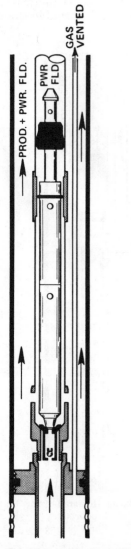

Fig. 5.36 Casing free tubing arrangements

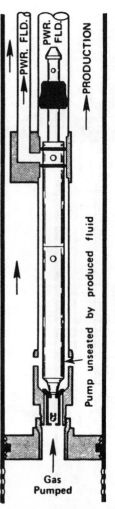

Fig. 5.37 Casing free tubing arrangements

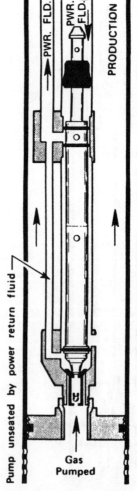

Fig. 5.38 Casing free tubing arrangements

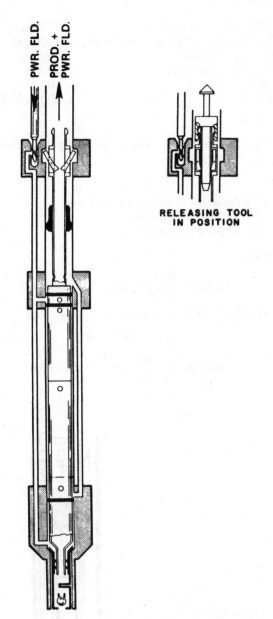

RELEASING TOOL
IN POSITION

Fig. 5.39 Reverse circulation tubing arrangement

TABLE 5.1
COMPARATIVE RESULTS OF VENTED & NON-VENTED SYSTEMS

No. of wells	Production with vent string closed—b/d			Production with vent string open—b/d			Production increase—b/d
	Oil	Water	Total	Oil	Water	Total	(total fluid)
11	125	227	352	265	228	493	141

No change in lift equipment—3 two zone rod pumps, 7 dual hydraulics and 1 dual rod pump.

7 of 11 wells had significant production increases.

2 high water cut wells showed little change, but already had high efficiencies of 75% - 80%.

3 wells showed no production as soon as the vent string was closed, indicating an immediate gas lock.

No. of wells	Production without a vent string—b/d			Production with 1″ vent string—b/d			Production increase—b/d
	Oil	Water	Total	Oil	Water	Total	(total fluid)
1	240	220	460	532	440	972	512
1	185	900	1085	220	1000	1220	135

Addition of 1 in. vent string to hydraulic pumping installation

No. of wells	Production with 1″ vent string—b/d			Production with annulus venting—b/d			Production increase—b/d
	Oil	Water	Total	Oil	Water	Total	(total fluid)
16	168	172	340	231	189	420	80

Production was initially from lower zone (vented) of dual well. Well then commingled permitting the annulus to be used for venting.

Note: 1¼ in. tubing will vent approximately twice the volume of gas as 1 in. tubing with the same amount of pressure loss.

installations are undesirable in wells that have both low producing bottom hole pressures and high gas-oil ratios. Usually an installation that vents the gas (Figs. 5.29, 5.31, 5.32, 5.33, 5.34 and 5.36) is a necessity when the gas-liquid ratio is over 500 SCF/b and the pumping bottom hole pressure is lower than 400 psi.

Based on efficiency, do not design for efficiencies less than 50% if possible and use 30% as a minimum value. Vent the gas to obtain efficiencies greater than these values. A hydraulic pump is well suited to pump the gas without gas locking problems, but the efficiency is much better if the gas can be vented. The pumps have a built-in governor to prevent destruction if attempting to pump gas.

5.822 Field tests on venting gas

Table 5.1 shows field results of tests that were run to compare venting and not venting the gas.

Refer to Appendix 5.A for gas pressure losses in 1 and 1¼ in. vent lines.

Note from Table 5.1 that the production increase is 141 b/d, of which 54% is oil for the first 11 wells. This represents another 76 b/d of oil from 11 wells. In the second group the first well showed an increase of 512 b/d of fluid, of which 280 b/d was oil. Of the six wells in case three, an increase of 80 b/d was noted, of which 44 b/d was oil. Case 2 shows an obvious economic return with cases 1 and 3 also proving to be economical. Larger pumps may have produced similar results.

5.823 Venting gas

Installations that vent the gas are shown in Figs. 5.29, 5.31, 5.32, 5.33, 5.34, 5.36 and 5.39. If friction in the return tubing were too great for the simple parallel free (Fig. 5.32), then the casing free with gas vent (Fig. 5.36), the reverse circulation (Fig. 5.39), or the fixed casing with gas vent (Fig. 5.31) could be used. Usually more gas will be vented through the casing than through a tubing vent, but fluid friction considerations usually favor the tubing vent type of installation. The order in which these are listed (5.32, 5.36, 5.39 and 5.31) is the order of increasing cost.

5.824 Pumping gas

For installations that require the pump to compress free gas, Fig. 5.43 gives the theoretical liquid pump and displacements at different gas oil ratios and bottom hole pressures. If the indicated displacement is low (30-50%), the gas should be vented instead of pumped. At this point the IPR curve for the well should be consulted to determine if a higher bottom hole pressure can be allowed. (If the well is being produced below the bubble point and we refer to Vogel's reference curve for solution gas drive wells, we see that 60% drawdown allows 80% of maximum production and 80% drawdown allows over 90% of maximum production.)

EXAMPLE PROBLEM #1 TO DETERMINE PUMP CAPACITY FOR WELL PRODUCING GAS THROUGH PUMP

Given:

Reservoir pressure, $\overline{P}_R$	= 2000 psi
Flowing bottom hole pressure, P_{wf}	= 1000 psi
Flowing rate, oil	= 100 b/d
Flowing rate, water	= 50 b/d
Oil API gravity	= 40°
GOR	= 500 SCF/b

Assume:

Solution gas drive well with bubble point above 2000 psi

Find:

Pump capacity required to produce:
(A) 150 b/d oil and water
(B) 180 b/d oil and water

Solution:

$$P_{wf}/P_R = \frac{1000}{2000} = 0.5 \qquad \text{(at 150 b/d)}$$

From Vogel's (5.1) reference curve (Fig. 5.44)

$$q_o/(q_o)\,\text{max} = 0.7$$

$$(q_o)\,\text{max} = \frac{q_o}{0.7} = \frac{150}{0.7} = 214 \text{ b/d}$$

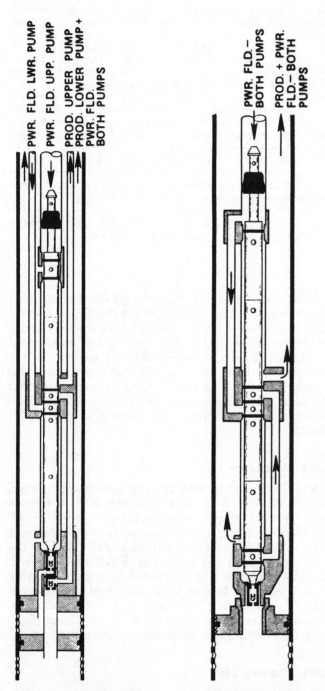

Fig. 5.40   Dual well tubing arrangement

Fig. 5.41   Tandem pumps

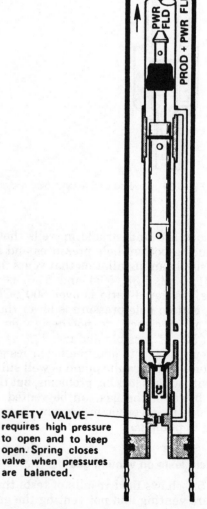

SAFETY VALVE—
requires high pressure to open and to keep open. Spring closes valve when pressures are balanced.

Fig. 5.42   Safety valve arrangement

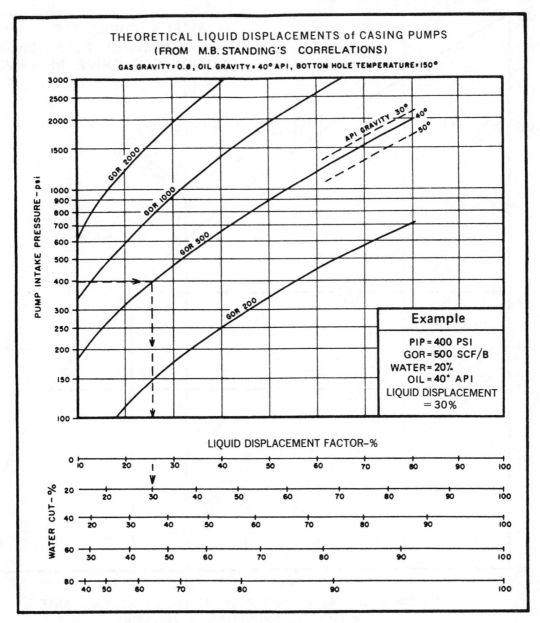

THEORETICAL LIQUID DISPLACEMENTS of CASING PUMPS
(FROM M.B. STANDING'S CORRELATIONS)
GAS GRAVITY = 0.8, OIL GRAVITY = 40° API, BOTTOM HOLE TEMPERATURE = 150°

Example

PIP = 400 PSI
GOR = 500 SCF/B
WATER = 20%
OIL = 40° API
LIQUID DISPLACEMENT
= 30%

Fig. 5.43   Theoretical volumetric efficiencies of casing pumps

We were given $P_{wf} = 1000$ psi for 150 b/d
For 180 b/d:

$$q_o/(q_o)\max = \frac{180}{214} = 0.84$$

From Vogel's reference curve (Fig. 5.44)

$$P_{wf}/P_R = 0.33 \quad \text{(at 180 b/d)}$$
$$P_{wf} = 0.33 \times 2{,}000 = 660 \text{ psi}$$

From Fig. 5.43 for 1000 psi, 500 GOR and 33% water, the liquid displacement is approximately 65%. Therefore:

(A) Required pump capacity $= \dfrac{150}{0.65} = 231$ b/d

For 660 psi, 500 GOR and 33% water, Fig. 5.43 gives a liquid displacement of approximately 52%. Therefore:

(B) Required pump capacity $= \dfrac{180}{0.52} = 346$ b/d

This 346 b/d is 180 b/d of stock tank oil and water and 166 b/d of gas (gas at 600 psi).

CLASS PROBLEM #1-A: How to determine the required pump capacity for a casing type installation.

Given:

Solution gas drive reservoir

$\overline{P}_R = 2000$ psi
$P_{wf} = 1500$ psi
$q_o = 100$ b/d (All Oil)
Oil = 40° API Gravity

Assume:

Solution gas drive well with bubble point above 2000 psi.

Find:

Pump capacity for:
(a) 100 b/d when GOR = 600 SCF/b
(b) 150 b/d when GOR = 1000 SCF/b

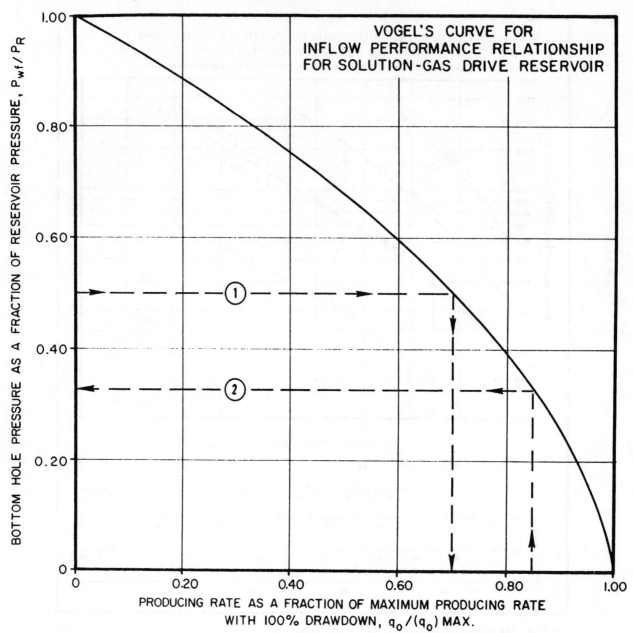

Fig. 5.44 *Vogel's reference curve*

5.83 Pumps

5.831 Introduction

Schematic drawings and specification tables for the pumps presently available are shown in the Appendix in Figs. 5.B(1) through 5.B(14). The schematic drawings show each pump making an upstroke in an OPF casing free installation. Some of the pumps have two engine pistons and some have two pump pistons. Engine reversing valves are located at the top of some pumps, in the middle of some pumps, and in the engine piston of other pumps. Most tubing sizes are available and length may range from 10-30 ft.

5.832 Pump selection

In many cases the proper pump for a given well can be chosen directly from the specification tables. The first column lists the pump size, which also identifies the tubing size that it will run in. The second column lists

values for a quantity called P/E which will be explained in detail in Section 5.835. These values are related to the surface pressure required for a given lift. To limit surface pressure to the generally acceptable maximum of 5000 psi, use the following rule-of-thumb equation:

$$\text{Maximum P/E} = \frac{10,000}{\text{Net Lift, ft}} \qquad (5.1)$$

The third column of the specification tables lists the maximum pump displacement. It is good practice to design for 85% or less of the pump's maximum rated capacity.

Usually when two or more pump sizes can be used, the one with the greatest maximum fluid lift capability (lowest P/E value) will be chosen. This is because it will require less surface power fluid pressure to operate. This will be easier on the surface pump and will have less high pressure power fluid slippage in the bottom hole pump itself.

The power fluid rate required to produce a given

amount of production will be covered in the next section; it depends on the values in columns four and five of the pump specification tables.

EXAMPLE PROBLEM #2: TO CHOOSE THE PROPER PUMP SIZE

Given:

Desired pump capacity = 400 b/d
Fluid lift = 7000 ft

Assume:

A pump to run in 2⅞" OD tubing

Find:

Pump size from Appendix 5.B(1)
Pump size from Appendix 5.B(2)
Pump size from Appendix 5.B(3)
Pump size from Appendix 5.B(8)

Solution:

Maximum allowed P/E = 10,000/7,000 = 1.43

From Fig. 5.B(1) examine the 2½ in. nominal size pumps (2½ in. nominal fits in 2⅞ in. OD tubing) and find that the first two pumps, 2½ in. × 1¼ in. and 2½ in. × 1½ in., cannot produce 400 b/d. (The 1¼ in. and 1½ in. refer to pump piston, or plunger diameter.) The fourth pump, a 2½ in. × 2 in., has a P/E value above 1.43 so the choice must be the 2½ in. × 1¾ in.

From Fig. 5.B(2) only the first 2½ in. pump, a 2½ in. × 2 in. × 1¼ in., cannot meet the 400 b/d requirement while all of them can meet the P/E requirement. This pump lists nominal diameter × engine piston diameter × pump piston diameter.

From Fig. 5.B(3) (VFR25 identifies pumps to fit 2⅞ in. tubing) all of the 2½ in. pumps meet the requirements. A VFR 252017 description identifies the nominal pump size as 2.5 in. the engine piston as 2.0 in., and the pump piston as 1.7 in.

From Fig. 5.B(8) the first two 2½ in. pumps, 2½ in. × 1¼ in. − 1 in. and 2½ in. × 1¼ in. − 1⅛ in., do not meet the 400 b/d requirement. All the rest do, but the 2½ in. × 1¼ in. − 1⁷/₁₆ in. must operate right at its maximum lift capabilities because its P/E value is the maximum allowed.

CLASS PROBLEM #2-A

Select all the pumps for example problem no. 2 from Appendix Figs. 5.B(4), (5), (6), (7), (9), (10), (11), (12), (13), and (14).

CLASS PROBLEM #2-B

q = 650 b/d − Lift 8,000 ft 2⅞" O.D. tubing
Select any 2 pumps that will operate at 80% capacity

5.833 Power fluid rate

Power fluid rate is a function of pump end efficiency, engine end efficiency, and the displacements per SPM from the specification tables.
The following symbols will be used:

q_1 = engine end displacement per SPM, b/d per SPM
Q'_1 = theoretical power fluid rate, b/d (q_1 × SPM)
Q_1 = actual power fluid rate, b/d

q_4 = pump end displacement per SPM, b/d per SPM
Q'_4 = theoretical production rate, b/d (q_4 × SPM)
Q_4 = actual production rate, b/d ($Q_4 = Q_5 + Q_6$)
Q_5 = oil production rate, b/d
Q_6 = water production rate, b/d
Q'_1/Q_1 = engine end efficiency
Q_4/Q'_4 = pump end efficiency

The values for q_1 and q_4 are obtained from columns four and five of the pump specification tables. A new pump has an engine end efficiency of around 95% and a pump end efficiency above 90%. Good design practice is to use 90% engine end and 85% pump end efficiencies and to select a pump that will operate below 85% of its rated speed. (Some operators use 70%.)

If the pump is pumping from beneath a packer and consequently handling gas, the pump end efficiency should be obtained from Fig. 5.43. The above definitions can be written:

$$Q_4 = Q'_4 \,(Q_4/Q'_4) = (q_4 \times SPM)\,(Q_4/Q'_4) \qquad (5.2)$$

$$Q_1 = \frac{Q'_1}{Q'_1/Q_1} = \frac{q_1 \times SPM}{Q'_1/Q_1} \qquad (5.3)$$

Overall volumetric efficiency, $N\nu$, is pump end efficiency multiplied by engine end efficiency. Therefore:

$$N\eta = Q_4/Q'_4 \times Q'_1/Q_1 = Q_4/Q_1 \times Q'_1/Q'_4 = Q_4/Q_1 \times q_1/q_1 \qquad (5.4)$$

(See also Appendix 5.G(1))

EXAMPLE PROBLEM #3: CALCULATE POWER FLUID RATE

Given:

Q_4 = 500 b/d
q_1 = 16.5 b/d per SPM
q_4 = 13.4 b/d per SPM

Assume:

Q'_1/Q_1 = 90%
Q_4/Q'_4 = 85%

Find:

Power fluid rate, Q_1

Solution #1:

SPM from equation 5.2:

$$SPM = \frac{Q_4}{q_4\,(Q_4/Q'_4)} = \frac{500 \text{ b/d}}{13.4 \text{ b/d per SPM} \times 0.85} = 43.9$$

Q_1 from equation 5.3:

$$Q_1 = \frac{q_1 \times SPM}{Q'_1/Q_1} = \frac{(16.5 \text{ b/d per SPM}) \times 43.9}{0.9} = 805 \text{ b/d}$$

Solution #2:

Q_1 from Equation 5.4:

$$Q_1 = Q_4/N\nu \times q_1/q_4 = \frac{Q_4}{Q_4/Q'_4 \times Q'_1/Q_1} \times \frac{q_1}{q_4}$$

$$Q_1 = \frac{500 \text{ b/d}}{0.85 \times 0.9} \times \frac{16.5}{13.4} = 805 \text{ b/d}$$

CLASS PROBLEM #3-A: TO CALCULATE POWER FLUID RATE

Given:

Q_4 = 500 b/d
q_1 = 30.80 b/d per SPM
q_4 = 23.60 b/d per SPM

Assume:

$Q'_1/Q_1 = 85\%$
$Q_4/Q'_4 = 75\%$

Find:

Q_1

5.834 Pump friction

The pressure required to operate a hydraulic pump under "no load" conditions is shown in Fig. 5.45. This chart represents the mechanical and hydraulic friction in the pump. From the curves in Figs. 5.46 and 5.47, the power fluid viscosity at the bottom hole tempera-

PRESSURE INCREASE DUE TO MECHANICAL AND HYDRAULIC FRICTION IN PUMP AND BOTTOM HOLE ASSEMBLY
vs
PERCENT OF RATED SPEED

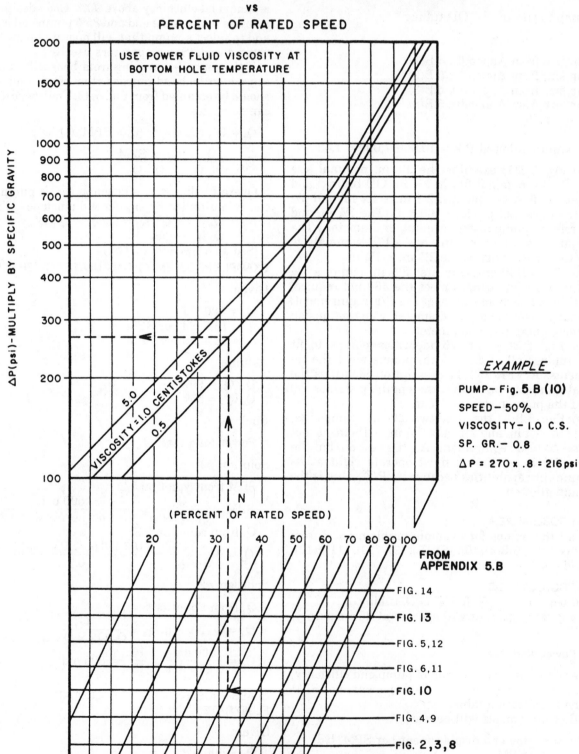

EXAMPLE

PUMP- Fig. 5.B (10)
SPEED- 50%
VISCOSITY- 1.0 C.S.
SP. GR.- 0.8
$\Delta P = 270 \times .8 = 216\,psi$

Fig. 5.45   Pressure required to operate a hydraulic pump under no load conditions

ture can be obtained to use with the pump friction chart. Conversions from API gravity to specific gravity can be made from Table 5.2. The values obtained from Fig. 5.45 show maximum values based on the largest pump piston operating at 100% pump end efficiency. When the fluid rate through the pump end is reduced by smaller pistons or by gas, the total friction will be somewhat lower than the chart predicts. This is because approximately 25% of the total friction is fluid friction in the pump end of the pump. This value is not well documented for all pumps but can be used to estimate the reduction in pump friction due to actual pump end liquid rate. In equation form the ΔP from Fig. 5.45 is:

$$\Delta P = F_{EE} + F_{PE}$$

where:

F_{EE} = Engine end friction = 0.75 ΔP
F_{PE} = Pump end friction   = 0.25 ΔP

In the example shown in Fig. 5.45 the ΔP is 216 psi. Therefore:

F_{EE} = 0.75 × 216 = 162 psi
F_{PE} = 0.25 × 216 = 54 psi

Suppose this is to be a 2½" pump from Fig. 5.B(10). If it is a 2½ in. × 1¾ in. − 1¾ in. pump (the largest 2½ in. pump) and is operating at 100% pump end efficiency, the 54 psi is correct. But if it is a 2½ in. × 1¾ in. — 1½ in. pump operating at 80% pump end efficiency, Q_4 will be less than that used to construct the chart. Because the correction to F_{PE} is a small quan-

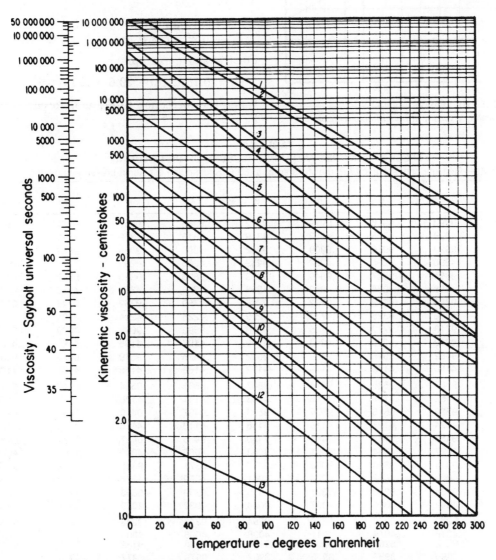

Fig. 5.46 *Power fluid viscosity at bottom hole temperature*

Line	Gravity °API	Field	Line	Gravity °API	Field
1	9.5	Boscan, Venezuela	8	30.6	Ventura, Calif.
2	10.7	Boscan, Venezuela	9	31.1	Kettleman Hills, Calif.
3	14.0	Maricopa, Calif.	10	36.4	Oklahoma City, Okla.
4	15.0	Wilmington, Calif.	11	34.6	Kettleman Hills, Calif.
5	19.8	Sansinena, Calif.	12	44.0	Denton, New Mexico
6	25.6	Scholem Alechem, Okla.	13	50.7	Kettleman Hills, Calif.
7	26.8	Seal Beach, Calif.			

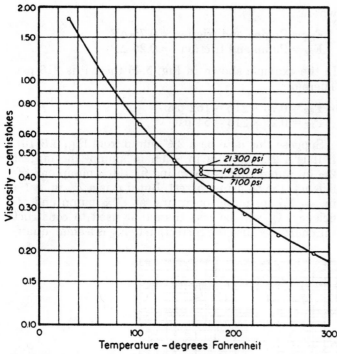

Fig. 5.47 Power fluid (water) viscosity at bottom hole temperature

tity, it is customary to ignore it. The error thus introduced is on the safe side.

If F_{PE} is to be corrected it should be a direct proportion to fluid rates through the pump end:

$$F_{PE} = 0.25 \frac{\Delta P \times q_4 \text{ of piston used} \times (Q_4/Q'_4)}{q_4 \text{ of max. size piston}} \quad (5.5)$$

If the pump in the example in Fig. 5.45 were a 2½ in. × 1¾ in. − 1½ in. [from Fig. 5.B(10)] and the pump end efficiency were 80%:

$$F_{PE} = (0.25)(216) \left(\frac{7.44}{10.86} \right) (0.8) = 29.6 \text{ psi}$$

Actual total pump friction, F_p, thus becomes:

$$F_p = F_{EE} + F_{PE} \quad (5.6)$$

where:

$F_{EE} = 0.75 \, \Delta P$ from Fig. 5.45
$F_{PE} = 0.25 \, \Delta P \times q_4/q_{4 \text{ max.}} \times Q_4/Q'_4$ (from Eq. 5.5)

For the above example:

$$F_p = (0.75)(216) + 29.6 = 191.6 \text{ psi}$$

When the correction to F_{PE} is ignored, as it usually is, $F_p = \Delta P$.

TABLE 5.2
SPECIFIC GRAVITIES AND UNIT PRESSURE OF OIL COLUMNS
Note—First line opposite each API gravity is sp gr at 60°F. Second line is column pressure in psi/ft

Degrees A.P.I.	0	.1	.2	.3	.4	.5	.6	.7	.8	.9
10	1.0000	.9993	.9986	.9979	.9972	.9965	.9958	.9951	.9944	.9937
	.4331	.4328	.4325	.4322	.4319	.4316	.4313	.4310	.4307	.4304
11	.9930	.9923	.9916	.9909	.9902	.9895	.9888	.9881	.9874	.9868
	.4301	.4298	.4295	.4292	.4289	.4286	.4282	.4279	.4276	.4274
12	.9861	.9854	.9847	.9840	.9833	.9826	.9820	.9813	.9806	.9799
	.4271	.4268	.4265	.4262	.4259	.4256	.4253	.4250	.4247	.4244
13	.9792	.9786	.9779	.9772	.9765	.9759	.9752	.9745	.9738	.9732
	.4241	.4238	.4325	.4232	.4229	.4226	.4224	.4221	.4218	.4215
14	.9725	.9718	.9712	.9705	.9698	.9692	.9685	.9679	.9672	.9665
	.4212	.4209	.4206	.4203	.4200	.4198	.4195	.4192	.4189	.4186
15	.9659	.9652	.9646	.9639	.9632	.9626	.9619	.9613	.9606	.9600
	.4183	.4180	.4178	.4175	.4172	.4169	.4166	.4163	.4160	.4158
16	.9593	.9587	.9580	.9574	.9567	.9561	.9554	.9548	.9541	.9535
	.4155	.4152	.4149	.4146	.4143	.4141	.4138	.4135	.4132	.4130
17	.9529	.9522	.9516	.9509	.9503	.9497	.9490	.9484	.9478	.9471
	.4127	.4124	.4121	.4118	.4116	.4113	.4110	.4108	.4105	.4102
18	.9465	.9459	.9452	.9446	.9440	.9433	.9427	.9421	.9415	.9408
	.4099	.4097	.4094	.4091	.4088	.4085	.4083	.4080	.4078	.4075
19	.9402	.9396	.9390	.9383	.9377	.9371	.9365	.9358	.9352	.9346
	.4072	.4069	.4067	.4064	.4061	.4059	.4056	.4053	.4050	.4048
20	.9340	.9334	.9328	.9321	.9315	.9309	.9303	.9297	.9291	.9285
	.4045	.4043	.4040	.4037	.4034	.4032	.4029	.4027	.4024	.4021
21	.9279	.9273	.9267	.9260	.9254	.9248	.9242	.9236	.9230	.9224
	.4019	.4016	.4014	.4011	.4008	.4005	.4003	.4000	.3998	.3995
22	.9218	.9212	.9206	.9200	.9194	.9188	.9182	.9176	.9170	.9165
	.3992	.3990	.3987	.3985	.3982	.3979	.3977	.3974	.3972	.3969
23	.9159	.9513	.9147	.9141	.9135	.9129	.9123	.9117	.9111	.9106
	.3967	.3964	.3962	.3959	.3956	.3954	.3951	.3949	.3946	.3944
24	.9100	.9094	.9088	.9082	.9076	.9071	.9065	.9059	.9053	.9047
	.3941	.3939	.3936	.3933	.3931	.3929	.3926	.3923	.3921	.3918
23	.9042	.9036	.9030	.9024	.9018	.9013	.9007	.9001	.8996	.8990
	.3916	.3913	.3911	.3908	.3906	.3904	.3901	.3898	.3896	.3894
26	.8984	.8978	.8973	.8967	.8961	.8956	.8950	.8944	.8939	.8933
	.3891	.3888	.3886	.3884	.3881	.3879	.3876	.3874	.3871	.3869
27	.8927	.8922	.8916	.8911	.8905	.8899	.8894	.8888	.8883	.8877
	.3866	.3864	.3862	.3859	.3857	.3854	.3852	.3849	.3847	.3845
28	.8871	.8866	.8860	.8855	.8849	.8844	.8838	.8833	.8827	.8822
	.3842	.3840	.3837	.3835	.3833	.3830	.3828	.3826	.3823	.3821

TABLE 5.2 (CONTINUED)
SPECIFIC GRAVITIES AND UNIT PRESSURE OF OIL COLUMNS
Note—First line opposite each API gravity is sp gr at 60°F. Second line is column pressure in psi/ft

Degrees A.P.I.	0	.1	.2	.3	.4	.5	.6	.7	.8	.9
29	.8816	.8811	.8805	.8800	.8794	.8789	.8783	.8778	.8772	.8767
	.3818	.3816	.3813	.3811	.3809	.3807	.3804	.3802	.3799	.3797
30	.8762	.8756	.8751	.8745	.8740	.8735	.8729	.8724	.8718	.8713
	.3795	.3792	.3790	.3787	.3785	.3783	.3781	.3778	.3776	.3774
31	.8708	.8702	.8697	.8692	.8686	.8681	.8676	.8670	.8665	.8660
	.3771	.3769	.3767	.3765	.3762	.3760	.3758	.3755	.3753	.3751
32	.8654	.8649	.8644	.8639	.8633	.8628	.8623	.8618	.8612	.8607
	.3748	.3746	.3744	.3742	.3739	.3737	.3735	.3732	.3730	.3728
33	.8602	.8597	.8591	.8586	.8581	.8576	.8571	.8565	.8560	.8555
	.3726	.3723	.3721	.3719	.3716	.3714	.3712	.3710	.3707	.3705
34	.8550	.8545	.8540	.8534	.3529	.8524	.8519	.8514	.8509	.8504
	.3703	.3701	.3699	.3696	.3694	.3692	.3690	.3687	.3685	.3683
35	.8498	.8493	.8488	.8483	.8478	.8473	.8468	.8463	.8458	.8453
	.3680	.3678	.3676	.3674	.3672	.3670	.3667	.3665	.3663	.3661
36	.8448	.8443	.8438	.8433	.8428	.8423	.8418	.8413	.8408	.8403
	.3659	.3657	.3654	.3652	.3650	.3648	.3646	.3644	.3642	.3639
37	.8398	.8393	.8388	.8383	.8378	.8373	.8368	.8363	.8358	.8353
	.3637	.3635	.3633	.3631	.3629	.3626	.3624	.3622	.3620	.3618
38	.3848	.8343	.8338	.8333	.8328	.8324	.8319	.8314	.8309	.8304
	.3616	.3613	.3611	.3609	.3607	.3605	.3603	.3601	.3599	.3596
39	.8299	.8294	.8289	.8285	.8280	.8275	.8270	.8265	.8260	.8256
	.3594	.3592	.3590	.3588	.3586	.3584	.3582	.3580	.3577	.3576
40	.8251	.8248	.8241	.8236	.8232	.8227	.8222	.8217	.8212	.8208
	.3574	.3571	.3569	.3567	.3565	.3563	.3561	.3559	.3557	.3555
41	.8203	.8198	.8193	.8189	.8184	.8179	.8174	.8170	.8165	.8160
	.3553	.3551	.3548	.3547	.3544	.3542	.3540	.3538	.3536	.3534
42	.8155	.8151	.8146	.8142	.8137	.8132	.8128	.8123	.8118	.8114
	.3532	.3530	.3528	.3526	.3524	.3522	.3520	.3518	.3516	.3514
43	.8109	.8104	.8100	.8095	.8090	.8086	.8081	.8076	.8072	.8067
	.3512	.3510	.3508	.3506	.3504	.3502	.3500	.3498	.3496	.3494
44	.8063	.8058	.8054	.8049	.8044	.8040	.8035	.8031	.8026	.8022
	.3492	.3490	.3488	.3486	.3484	.3482	.3480	.3478	.3476	.3474
45	.8017	.8012	.8008	.8003	.7999	.7994	.7990	.7985	.7981	.7976
	.3472	.3470	.3468	.3466	.3464	.3462	.3460	.3458	.3457	.3554
46	.7972	.7967	.7963	.7958	.7954	.7949	.7945	.7941	.7936	.7932
	.3453	.3451	.3449	.3447	.3445	.3443	.3441	.3439	.3437	.3435
47	.7927	.7923	.7918	.7914	.7909	.7905	.7901	.7896	.7892	.7887
	.3433	.3431	.3429	.3428	.3425	.3424	.3422	.3420	.3918	.3416
48	.7883	.7879	.7874	.7870	.7865	.7861	.7857	.7852	.7848	.7844
	.3414	.3412	.3410	.3408	.3406	.3405	.3403	.3401	.3399	.3397
49	.7839	.7835	.7831	.7826	.7822	.7818	.7813	.7809	.7805	.7800
	.3395	.3393	.3392	.3389	.3388	.3386	.3384	.3382	.3380	.3378
50	.7796	.7792	.7788	.7783	.7779	.7775	.7770	.7766	.7762	.7758
	.3376	.3375	.3373	.3371	.3369	.3367	.3365	.3363	.3362	.3360
51	.7753	.7749	.7745	.7741	.7736	.7732	.7728	.7724	.7720	.7715
	.3358	.3356	.3354	.3353	.3350	.3349	.3347	.3345	.3344	.3341
52	.7711	.7707	.7703	.7699	.7694	.7690	.7686	.7682	.7678	.7674
	.3340	.3338	.3336	.3334	.3332	.3331	.3329	.3327	.3325	.3324
53	.7669	.7665	.7661	.7657	.7653	.7649	.7645	.7640	.7636	.7632
	.3321	.3320	.3318	.3316	.3315	.3313	.3311	.3309	.3307	.3305
54	.7628	.7624	.7620	.7616	.7612	.7608	.7603	.7599	.7595	.7591
	.3304	.3302	.3300	.3298	.3297	.3295	.3293	.3291	.3289	.3288
55	.7587	.7583	.7579	.7575	.7571	.7567	.7563	.7559	.7555	.7551
	.3286	.3284	.3282	.3281	.3279	.3277	.3276	.3274	.3272	.3270
56	.7547	.7543	.7539	.7535	.7531	.7527	.7523	.7519	.7515	.7511
	.3269	.3267	.3265	.3263	.3262	.3260	.3258	.3256	.3255	.3253
57	.7507	.7503	.7499	.7495	.7491	.7487	.7483	.7479	.7475	.7471
	.3251	.3250	.3248	.3246	.3244	.3243	.3241	.3239	.3237	.3236
58	.7467	.7463	.7459	.7455	.7451	.7447	.7443	.7440	.7436	.7432
	.3234	.3232	.3230	.3229	.3227	.3225	.3224	.3222	.3221	.3219
59	.7428	.7424	.7420	.7416	.7412	.7408	.7405	.7401	.7397	.7393
	.3217	.3215	.3214	.3212	.3210	.3208	.3207	.3205	.3204	.3202
60	.7389	.7385	.7381	.7377	.7374	.7370	.7366	.7362	.7358	.7354
	.3200	.3198	.3197	.3195	.3194	.3192	.3190	.3188	.3187	.3185

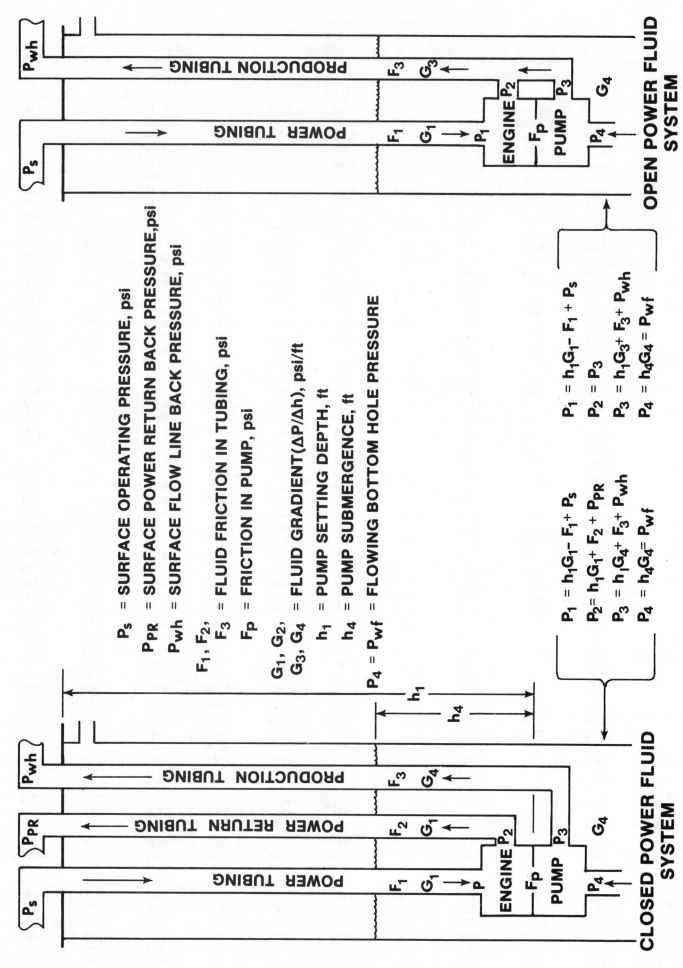

P_s = SURFACE OPERATING PRESSURE, psi

P_{PR} = SURFACE POWER RETURN BACK PRESSURE, psi

P_{wh} = SURFACE FLOW LINE BACK PRESSURE, psi

F_1, F_2, F_3 = FLUID FRICTION IN TUBING, psi

F_p = FRICTION IN PUMP, psi

G_1, G_2, G_3, G_4 = FLUID GRADIENT($\Delta P/\Delta h$), psi/ft

h_1 = PUMP SETTING DEPTH, ft

h_4 = PUMP SUBMERGENCE, ft

$P_4 = P_{wf}$ = FLOWING BOTTOM HOLE PRESSURE

CLOSED POWER FLUID SYSTEM

$P_1 = h_1G_1 - F_1 + P_s$

$P_2 = h_1G_1 + F_2 + P_{PR}$

$P_3 = h_1G_4 + F_3 + P_{wh}$

$P_4 = h_4G_4 = P_{wf}$

OPEN POWER FLUID SYSTEM

$P_1 = h_1G_1 - F_1 + P_s$

$P_2 = P_3$

$P_3 = h_1G_3 + F_3 + P_{wh}$

$P_4 = h_4G_4 = P_{wf}$

Fig. 5.48   *Pressure & Friction losses affecting hydraulic pumps*

5.835 Pressure calculations

The various pressures, friction losses, and fluid densities involved in CPF and OPF systems are shown in Fig. 5.48. The total pressure available to drive the engine is P_1, while the total pressure the engine must discharge against is P_2. The pump end must discharge against P_3 while being filled by P_4.

Fig. 5.49 illustrates those cross section areas of the Kobe A pump [Fig. 5.B(9)] which are involved with the various pressures. (Other pumps have different configurations.) If we now add the forces on the upstroke and assign the plus sign to upward acting forces we obtain:

$$-P_1 A_R - P_2(A_E - A_R) + P_1(A_E - A_R) - P_3(A_p - A_R)$$
$$+ P_4(A_p - A_R) + P_1 A_R = 0$$

$$(P_1 - P_2)(A_E - A_R) - (P_3 - P_4)(A_p - A_R) = 0$$

$$P_1 - P_2 - (P_3 - P_4)\frac{A_P - A_R}{A_E - A_R} = 0$$

Pump friction, F_P, is not shown in Fig. 5.49 because it does not operate against an area. It is a function of pump speed, fluid passageways in the pump, and mechanical friction in the pump. Since it opposes motion, it will have a negative sign and our equation becomes:

$$P_1 - P_2 - (P_3 - P_4)\frac{A_p - A_R}{A_E - A_R} - F_p = 0$$

The quantity $\dfrac{A_p - A_R}{A_E - A_R}$ is the ratio of net pump area to net engine area and, for this pump, it is the same for the downstroke and the upstroke. Since this is a double acting pump, this ratio is also the volume ratio—pump end displacement to engine end displacement—and is referred to as the "pump to engine ratio" or the "P over E ratio" (P/E). Single acting pumps require about 20% extra power fluid to make the unproductive downstroke (to fill the pump end), so their volume ratio is different from their P/E. For this reason single acting pumps refer to P/E as the "pressure ratio." The specification tables list the numerical values of P/E for each pump size. The algebraic equation relating pump areas to P/E is different for different pumps, so we must substitute P/E in the above equation to arrive at the following general CPF equation for all hydraulic pumps:

$$P_1 - P_2 - (P_3 - P_4)\,P/E - F_p = 0 \qquad (5.7)$$

Substituting the values for P_1, P_2, P_3 and P_4 from Fig. 5.48 gives:

$$(h_1 G_1 - F_1 + P_s) - (h_1 G_1 + F_2 + P_{PR}) -$$
$$(h_1 G_4 + F_3 + P_{wh} - h_4 G_4)\,P/E - F_p = 0$$

In words (for an upstroke):

$$\begin{pmatrix}\text{Total pressure}\\ \text{on bottom of}\\ \text{engine piston}\end{pmatrix}\begin{pmatrix}\text{Effective}\\ \text{area of}\\ \text{engine} = 1\end{pmatrix} - \begin{pmatrix}\text{Total pressure}\\ \text{on top of}\\ \text{engine piston}\end{pmatrix}$$

$$\begin{pmatrix}\text{Effective}\\ \text{area of}\\ \text{engine} = 1\end{pmatrix} - \begin{pmatrix}\text{Net pressure across}\\ \text{pump piston} =\\ \text{discharge-intake}\end{pmatrix}\begin{pmatrix}\text{effective}\\ \text{area of}\\ \text{pump} = P/E\end{pmatrix}$$

$$- \begin{pmatrix}\text{Pump}\\ \text{friction}\end{pmatrix} = 0$$

Solving for P_s gives:

$$P_s = F_1 + F_2 + P_{PR} + F_p + [(h_1 - h_4)G_4 + F_3 + P_{wh}]P/E \qquad (5.8)$$

Some pumps have an additional term in their CPF equations that is a function of $(P_2 - P_3)$, but the term is always ignored because it is very small. For Fig. 5.B(11) the term is:

$$(P_2 - P_3)\left(\frac{A_{R2} - A_{R1}}{A_E - A_{R2}}\right) = (P_2 - P_3)(0.07)$$

The value of $P_2 - P_3$ is usually less than 500 psi, so this term is usually less than 35 psi. (For the OPF system, $P_2 = P_3$, so this term drops out).

Some pumps have different values for the upstroke P/E and the downstroke P/E. When the difference is small, an average value is given in the specification table; when it is significant, the larger value is given and the pump has a power fluid restriction on either the upstroke or the downstroke in order to provide essentially the same operating pressure for both directions.

For the OPF system ($P_2 = P_3$, as shown in Fig. 5.48) the general equation for all hydraulic pumps becomes:

$$P_1 - P_3 - (P_3 - P_4)\,P/E - F_p = 0 \qquad (5.9)$$

Rearranging:

$$P_1 = P_3 + (P_3 - P_4)\,P/E + F_p$$

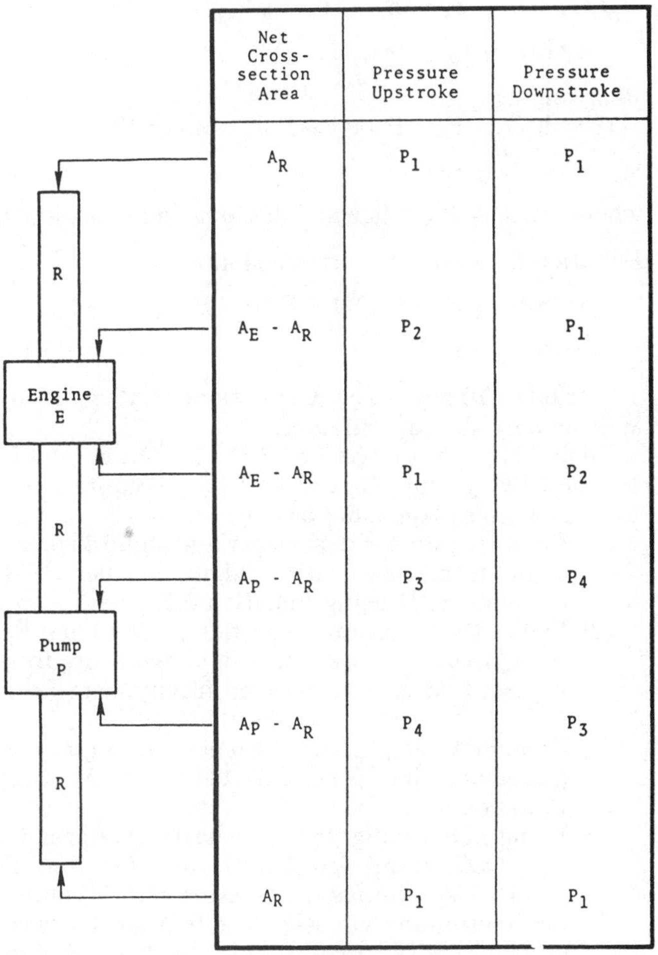

Net Cross-section Area	Pressure Upstroke	Pressure Downstroke
A_R	P_1	P_1
$A_E - A_R$	P_2	P_1
$A_E - A_R$	P_1	P_2
$A_p - A_R$	P_3	P_4
$A_p - A_R$	P_4	P_3
A_R	P_1	P_1

Fig. 5.49 *Pressures acting on a Kobe type A pump*

In words:

$$\begin{pmatrix} \text{Pressure at bot-} \\ \text{tom of power} \\ \text{fluid column} \end{pmatrix} \begin{pmatrix} \text{Effective} \\ \text{area of} \\ \text{engine-1} \end{pmatrix} = \begin{pmatrix} \text{Pressure} \\ \text{at bottom of} \\ \text{return column} \end{pmatrix}$$

$$\begin{pmatrix} \text{Effective} \\ \text{area of} \\ \text{engine} = 1 \end{pmatrix} + \begin{pmatrix} \text{Net pressure} \\ \text{across pump} \\ \text{piston} \end{pmatrix} \begin{pmatrix} \text{Effective} \\ \text{area of} \\ \text{pump} = P/E \end{pmatrix} +$$

$$\begin{pmatrix} \text{Pump} \\ \text{friction} \end{pmatrix}$$

Rearranging again:

$$P_1 = P_3(1 + P/E) - P_4(P/E) + F_p$$

In words:

$$\begin{pmatrix} \text{Pressure at bot-} \\ \text{tom of power} \\ \text{fluid column} \end{pmatrix} \begin{pmatrix} \text{Effective} \\ \text{area of} \\ \text{engine} = 1 \end{pmatrix} = \begin{pmatrix} \text{Pressure at} \\ \text{bottom of} \\ \text{return column} \end{pmatrix}$$

$$\begin{pmatrix} \text{Effective area of:} \\ \text{engine} (= 1) \text{ plus} \\ \text{pump} (= P/E) \end{pmatrix} - \begin{pmatrix} \text{Flowing} \\ \text{bottom hole} \\ \text{pressure} \end{pmatrix} \begin{pmatrix} \text{Effective} \\ \text{area of} \\ \text{pump} = P/E \end{pmatrix}$$

$$+ \begin{pmatrix} \text{Pump} \\ \text{friction} \end{pmatrix}$$

This form of the equation illustrates that P_3 is acting against both the engine piston and the pump piston. Because of this, an increase in surface flow line back pressure, P_{wh}, causes a $(1 + P/E)$ increase in surface operating pressure, P_s.

Substituting for P_1, P_3 and P_4 from Fig. 5.48:

$$h_1G_1 - F_1 + P_s = (h_1G_3 + F_3 + P_{wh})(1 + P/E) -$$
$$h_4G_4(P/E)F_p$$

Rearranging:

$$P_s = (h_1G_3 + F_3 + P_{wh})(1 + P/E) - h_4G_4(P/E) +$$
$$F_p + F_1 - h_1G_1 \qquad (5.10)$$

where $h_4G_4 = P_{wf} = $ flowing bottom hole pressure.

Equation 5.10 can be written as follows:

$$P_s = (h_1G_3 + F_3 + P_{wh})(1 + P/E) - P_{wf}(P/E) + F_p$$
$$- (h_1G_1 - F_1) \qquad (5.11)$$

Equations 5.10 and 5.11 give the same results and are written only slightly different.

To find P_s we must first find SPM, F_p, Q_1, F_1 (and F_2 for the CPF system), G_3 and F_3. The procedure in detail is: [See also Appendix 5.G(2)]

(1) From Q_4, pump end efficiency and pump displacement (from specification tables, b/d per SPM) calculate SPM using Equation 5.2.

(2) Follow the procedure in Section 5.834 to find F_p. (Use viscosity at bottom hole temperature from Figures 5.46 or 5.47. Specific gravity from Table 5.2)

(3) From SPM, engine end efficiency and engine displacement (b/d per SPM) calculate Q_1 using Equation 5.3.

(4) Using the tubing friction charts of Appendix Figs. 5.C(1) through 5.C(27) and Q_1, find F_1 and F_2. Use average temperature of fluid column for determining viscosity. The friction charts require viscosity in centistokes for the turbulent flow regions and Soybolt Universal Seconds for

the laminar flow regions. Fig. 5.46 includes a conversion scale for these units.

(5) Calculate G_3 using:

$$G_3 = \frac{Q_1G_1 + Q_5G_5 + Q_6G_6}{Q_1 + Q_4} \qquad (5.12)$$

Where:

$Q_4 = Q_5$ (oil production) $+ Q_6$ (water production)

(6) Using the tubing friction charts of Figs. 5.C(1) through 5.C(27) and Q_4 (CPF) or Q_3 (OPF) find F_3, where the specific gravity of Q_3 is obtained by dividing G_3 by 0.433. The viscosity is obtained by:

$$v_3 = \frac{Q_1v_1 + Q_5v_5 + Q_6v_6}{Q_1 + Q_4} \qquad (5.13)$$

(7) Substitute in Equation 5.7 or 5.9 and solve for P_s.

Steps 5 and 6 can be eliminated by using multiphase flow correlations for determining P_3, but in this chapter we will use steps 5 and 6. This procedure will not account for gas in the production column but it is a conservative procedure that will provide a safe design.

EXAMPLE PROBLEM #4: TO CALCULATE SURFACE OPERATING PRESSURE

Given:

$Q_5 = 200$ b/d of 40° API gravity oil   Power fluid = 40°
$Q_6 = 100$ b/d of 1.03 sp. gr. water   API gravity oil
$P_4 = 600$ psi   Bottom hole tem-
$h_1 = 10,000$ ft   perature =
$P_{FL} = 75$ psi   180°F.

Assume:

252016 pump from Fig. 5.B(7); OPF parallel free installation using 2⅞ in. OD power tubing and 2⅜ in. OD production tubing, with gas to be vented to the surface through the casing annulus so GOR is not required information;

$Q'_1/Q_1 = 90\%$
$Q_4/Q'_4 = 85\%$

Find:

P_s

From Fig. 5.B(7)

$q_1 = 16.5$ b/d per SPM
$q_4 = 10.6$ b/d per SPM
$P/E = 0.64$

Solution:

(1) $SPM = \dfrac{Q_4}{q_4 \times Q_4/Q'_4}$ (from Equation 5.2)

$$= \frac{300}{10.6 \times 0.85} = 33.3$$

(2) From Fig. 5.46, viscosity for 40° API gravity oil at 180°F is between lines 11 and 12 at approximately 1.5 centistokes. From Table 5.2, specific gravity of power oil is 0.8251. From Fig. 5.B(7), rated pump speed is 51 SPM.

Percent of rated speed $= \dfrac{33.3}{51} \times 100 = 65$

From Fig. 5.45, ΔP is approximately 410 psi multiplied by specific gravity. We will ignore the error in F_{PE} due to reduced volume through the pump end, so $\Delta P = F_p$.

$$F_p = \Delta P = 410 \times 0.8251 = 338 \text{ psi}$$

(3) $Q_1 = \dfrac{q_1 \times \text{SPM}}{Q'_1/Q_1}$ (from Equation 5.3)

$$= \dfrac{16.5 \times 33.3}{0.9} = 610 \text{ b/d}$$

(4) In Step 2 the viscosity of the power fluid at bottom hole temperature (where the pump is located) was found, but to be precise for tubing friction calculations, estimate the average temperature of the fluid from the bottom to the surface. Assume the fluid reaches the surface at 100°F. which will give us an average temperature of 140°F.

From Fig. 5.46 the average oil viscosity, v_1, is approximately 2.1 centistokes. From Fig. 5.C(8) at 610 b/d and 2.1 centistokes, the pressure drop is approximately 1.6 psi per 1,000 ft multiplied by specific gravity.

$$F_1 = 1.6 \times 10 \times 0.8251 = 13 \text{ psi}$$

(5) From Table 5.2, $G_1 = G_5 = 0.3574$ psi/ft

$$G_6 = 1.03 \times 0.433 = 0.446 \text{ psi/ft}$$

$$G_3 = \dfrac{(610)(0.3574) + (200)(0.3574) + (100)(0.446)}{610 + 300}$$

$$= \dfrac{334}{910} = 0.367 \text{ psi/ft}$$

(6) Specific gravity of $Q_3 = \dfrac{0.367}{0.433} = 0.848$

The viscosity of water, v_6, at 140°F. (from Fig. 5.47) is 0.46 centistokes.

$$v_3 = \dfrac{(610)(2.1) + (200)(2.1) + (100)(0.46)}{910}$$

$$= \dfrac{1281 + 420 + 46}{910} = \dfrac{1747}{910} = 1.92 \text{ centistokes}$$

From Fig. 5.C(7) at 910 b/d and 1.92 centistokes, the tubing friction in the return string of 2⅜ in. tubing (power oil plus production) is 11 psi per 1,000 ft multiplied by specific gravity.

$$F_3 = 11 \times 10 \times 0.848 = 93 \text{ psi}$$

(7) From Steps 4 and 5, $F_1 = 13$ and $G_1 = 0.3574$. Therefore:

$$P_1 = h_1G_1 - F_1 + P_s$$
$$= (10,000 \times 0.3574) - 13 + P_s$$
$$= 3561 \text{ psi} + P_s$$

From Steps 5 and 6, $G_3 = 0.367$ and $F_3 = 93$. Therefore:

$$P_3 = h_1G_3 + F_3 + P_{FL}$$
$$= (10,000 \times 0.367) + 93 + 75 = 3838 \text{ psi}$$

From Step 2, $F_p = 338$ psi. We were given $P_4 = 600$ psi and we found $P/E = 0.64$ from the specification tables. Substituting in Equation 5.9:

$$P_1 - P_3 - (P_3 - P_4)P/E - F_p = 0$$
$$(P_s + 3561) - 3838 - (3838 - 600)\,0.64$$
$$\quad - 338 = 0$$
$$P_s - 615 - (3238)\,0.64 = 0$$
$$P_s = 2072 + 615 = 2687 \text{ psi}$$

CLASS PROBLEM #4A: TO CALCULATE SURFACE OPERATING PRESSURE

Given:

Same problem as example problem #4 but assume the casing size will not allow the 2⅞ in. and 2⅜ in. tubing to be used, so we must use 2⅞ in. power tubing and 1¼ in. production tubing.

Find:

P_s

CLASS PROBLEM #4B: TO CALCULATE SURFACE OPERATING PRESSURE

Given:

Same problem as example problem #4 but assume the return tubing string size must be 1¼ in. and the pump to be used is number 252020 from Fig. 5.B(7).

Find:

P_s

EXAMPLE PROBLEM #5: TO CALCULATE SURFACE OPERATING PRESSURE

Given:

Same conditions as example problem #4.
The following additional data is known: GOR = 200 SCF/B and $P_{PR} = 30$ psi.

Assume:

A CPF casing free installation with 2⅞ in. power fluid tubing, 1¼ in. power return tubing and 7 in. casing anulus as the production string (Fig. 5.37). Choose the 2½ in. × 1⁷⁄₁₆ in. – 1¼ in. pump shown in Fig. 5.B(8)—Appendix 5.

Find:

P_s

From Fig. 5.B(8).

$q_1 = 7.13$ b/d per SPM
$q_4 = 4.92$ b/d per SPM
$P/E = 0.700$

Because this casing-free installation will require the pump to pump the gas as well as the oil and water, first determine the required pump displacement. From Fig. 5.43, 700 psi pump intake pressure, a GOR of 200 SCF/b, and extrapolating the 33% water cut between the 20% and 40% scales, the theoretical pump end displacement is 80%. Assume a pump end efficiency due to slippage of 85%, as in the previous example problem, then total pump end efficiency becomes:

$$Q_4/Q'_4 = 0.80 \times 0.85 = 0.68$$

(1) $SPM = \dfrac{Q_4}{q_4 \times Q_4/Q'_4} = \dfrac{300}{4.92 \times 0.68} = 89.7$

(2)   From Fig. 5.B(8)—Appendix, the maximum rated speed is 100 SPM.

Percent of rated speed $= \dfrac{89.7}{100} \times 100 = 89.7\%$

From Fig. 5.45, using the previously determined values of 1.5 centistokes and 0.8251 specific gravity, ΔP is approximately 460 psi. Again ignore the error in F_{PE} due to reduced volume through the pump end.

$$F_p = \Delta P = 460 \times 0.8251 = 380 \text{ psi}$$

(3) $Q_1 = \dfrac{q_1 \times SPM}{Q'_1/Q_1} = \dfrac{7.13 \times 89.7}{0.9} = 711 \text{ b/d}$

(4)   From Fig. 5.C(8)—Appendix at 711 b/d

$$F_1 = 2 \times 10 \times 0.8251 = 17 \text{ psi}$$

From Fig. 5.C(4)—Appendix at 711 b/d

$$F_2 = 32 \times 10 \times 0.8251 = 264 \text{ psi}$$

(5) $G_4 = \dfrac{Q_5 G_5 + Q_6 G_6}{Q_4}$

$\quad = \dfrac{(200)(0.3574) + (100)(0.446)}{300}$

$\quad = \dfrac{71.5 + 44.6}{300} = 0.387 \text{ psi/ft}$

(6)   Specific gravity of $Q_4 = \dfrac{0.387}{0.433} = 0.894$

$v_4 = \dfrac{(200)(2.1) + (100)(0.46)}{300} = \dfrac{466}{300}$

$\quad = 1.55 \text{ centistokes}$

The friction charts of Figures 5.C(1) through 5.C(27) of Appendix 5 do not list a casing anulus with two tubing strings but Fig. 5.C(24) (7 in. × 3½ in.) is close enough to use and it indicates that the 300 b/d of production will have insignificant friction losses up the casing anulus.

$$F_3 = 0$$

(7) $P_1 = h_1 G_1 - F_1 + P_s$
$\quad = (10{,}000 \times 0.3574) - 17 + P_s = 3557 + P_s$
$P_2 = h_1 G_1 + F_2 + P_{PR}$
$\quad = (10{,}000 \times 0.3574) + 264 + 30 = 3868 \text{ psi}$
$P_3 = h_1 G_4 + F_3 + P_{wh}$
$\quad = (10{,}000 \times 0.387) + 0 + 75 = 3945 \text{ psi}$

$P_1 - P_2 - (P_3 - P_4)P/E - F_p = 0$
$(P_s + 3557) - 3868 - (3945 - 600)0.7 - 380 = 0$
$P_s - 691 - (3345)0.7 = 0$
$P_s = 691 + 2342 = 3033 \text{ psi}$

CLASS PROBLEM #5-A: TO CALCULATE SURFACE OPERATING PRESSURE

Given:

Same problem as above Problem #5 except use the 2½ in. × 2 in. × 1½ in. pump of Fig. 5.B(2) Appendix

Find:

P_s

Equations 5.7 and 5.9 are frequently used to calculate the flowing bottom hole pressure (P_4) in existing hydraulic pumping installations. This calculation is similar to finding P_s, except steps 1, 2 and 3 are eliminated because SPM and Q_1 are given. Often the last-stroke method is used. The procedure for this method is to close the valve on the power fluid supply line to the well and record the pressure when the pump stops stroking.

This last-stroke pressure is the operating pressure at zero pump speed and zero fluid flow. It is the pressure where F_p, F_1, F_2 and F_3 are zero. It takes less than a minute for the pump to stop stroking so P_4 and $h_1 G_3$ do not change appreciably from their producing values. If P_{LS} is the last-stroke pressure, Equations 5.7 and 5.9 become:

$(P_{LS} + h_1 G_1) - (h_1 G_1 + P_{PR}) - (h_1 G_4 + P_{wh} - P_4)P/E$
$\quad = 0$
$P_{LS} - P_{PR} - (h_1 G_4 + P_{wh} - P_4)P/E = 0 \qquad (5.14)$
$(P_{LS} + h_1 G_1) - (h_1 G_3 + P_{wh}) - (h_1 G_3 + P_{FL} - P_4)P/E$
$\quad = 0$
$P_{LS} + h_1 G_1 - (h_1 G_3 + P_{wh})(1 + P/E) - (P_4)P/E = 0$
$\qquad\qquad (5.15)$

where $P_4 = P_{wf} =$ flowing bottom hole pressure.

Using these equations eliminates the need for calculating the various friction losses in the system. This last-stroke method must not be used with a severely worn pump, however, because fluid slippage in the pump end of the pump can cause the pump to continue stroking past its balance pressure.

If precise bottom hole pressure data is required, a pressure bomb can be attached to most hydraulic pumps to record actual producing bottom hole pressures while the pump is operating. This method is preferred to calculations based on Equations 5.7, 5.9, 5.14 or 5.15. The principal sources of error in calculations using the equations are:

(1)   Pump friction using Equations 5.7 and 5.9. The values for F_p using Fig. 5.45 are sometimes 200 - 400 psi high. This is because pump friction varies with clearance, concentricity, and surface finish of the sliding parts, and these factors vary due to manufacturing tolerances and pump wear. The values shown on the chart are purposely on the high side of the range to allow a safety factor for design purposes. Using Equations 5.14 and 5.15 solves this problem but G_3 and G_4 become sources of error as explained below.

(2)   When gas is present in the production column, multiphase flow correlations must be used with Equations 5.7 and 5.9 because even a 200 scf/B gas-liquid ratio will reduce P_3 by 600-1200 psi at depths below 5,000 ft. This error provides a safety factor when designing an installation but is intolerable for bottom hole pressure calculations. Multiphase flow correlations should not be used with Equations 5.14 and 5.15, however, because G_3 and G_4, as applied to these equations, are not static, or no-flow, gradients. They are full flow gradients (with friction subtracted out) because the last stroke pressure is recorded before the column density has had time to change significantly. But the column density term (elevation term) of multiphase flow correlations is a

function of fluid velocity and consequently is different at different flow rates. An alternate procedure for h_1G_3 and h_1G_4 in Equations 5.14 and 5.15 is to subtract the friction loss as obtained from Figs. 5.C(1) through 5.C(27) from the full flow multiphase flow correlations.

Another approach, contributed by S. G. Gibbs and K. B. Nolen (Nabla Corporation, Midland, Texas), is to first subtract Equation 5.14 from Equation 5.7 (and 5.15 from 5.9) giving:[3]

$$P_s - P_{LS} - F_1 - F_2 - (F_3)P/E - F_p = 0 \quad (5.16)$$
$$P_s - P_{LS} - F_1 - F_3(1 + P/E) - F_p = 0 \quad (5.17)$$

Using these equations, calculate F_p for use in Equations 5.7 and 5.9.

5.836   Horsepower calculations

A useful oil field hydraulic horsepower equation is:

$$\text{Horsepower} = \Delta P \times Q \times 1.7 \times 10^{-5} \quad (5.18)$$

where:

$$\Delta P = \text{Change in pressure, psi}$$
$$Q = \text{Liquid rate, b/d}$$

This equation can be used for surface horsepower and for work done by the pump end of the down hole pump. The surface horsepowers required for problems #4 and #5 in the previous section are:

Problem #4 surface horsepower

$$2687 \text{ psi} \times 610 \text{ b/d} \times 1.7 \times 10^{-5} = 27.9 \text{ hp}$$

Problem #5 surface horsepower

$$3{,}033 \text{ psi} \times 711 \text{ b/d} \times 1.7 \times 10^{-5} = 36.6 \text{ hp}$$

The work done by the pumps is that to raise 300 b/d from $P_{wf} = 600$ psi to P_3 (3838 psi for Problem #4 and 3945 psi for Problem #5).

Problem #4 pump end horsepower

$$(3838 - 600) \text{ psi} \times 300 \text{ b/d} \times 1.7 \times 10^{-5} = 16.5 \text{ hp}$$

Problem #5 pump end horsepower

$$(3945 - 600) \text{ psi} \times 300 \text{ b/d} \times 1.7 \times 10^{-5} = 17.0 \text{ hp}$$

Without friction and gas, the 2 pumps would have required the same surface horsepower. This would not be true, however, if single and double-acting pumps were compared because single-acting pumps do no work when making a downstroke. A measure of the unproductive work can be obtained by multiplying $P/E \times q_1/q_4$. If this ratio is 1.0, no power is being wasted. A value of 1.2 means that 20% of the supplied horsepower is being wasted. We can call this the "power ratio," and for the first pump in Figs. 5.B(2) and 5.B(8) of the Appendix it is:

Fig. 5.B(2)   Power ratio $= 0.52\left(\dfrac{15.08}{6.45}\right) = 1.21$

Fig. 5.B(8)   Power ratio $= 0.545\left(\dfrac{2.15}{1.15}\right) = 1.02$

5.9   DESIGN OF COMPLETE SYSTEM

5.91   Introduction

Previous sections have covered individual components making up the complete pumping system. Also,

a decision must be made on the type of power fluid to use (water or oil). There may be some question as whether to use a central or one-well system.

5.92   Procedure for the design of equipment for one well

This procedure serves as a guide to select a pump and determine the surface pressure needed for one well. A central or individual surface power unit can be used. (Refer also to Appendix 5.G)
Procedure:

(1) Determine the required flowing pressure for the desired rate.
(2) Decide upon the type of installation and whether or not to vent the gas.
(3) Find the pump displacement to produce the desired rate. We will find the fluid displacement factor and use a pump efficiency of 80%.
(4) Select a tentative pump to handle the required displacement. Generally more than one pump will handle the rate desired. Normally we try to select a pump such that the desired displacement rate is no greater than 85% of the maximum pump capacity.
(5) Check the required pumping speed.
(6) Determine the power oil requirements assuming an engine volumetric efficiency of 80%.
(7) Determine the total volume of return fluid and the pressure exerted by the return fluid column.
(8) Determine the friction loss of power fluid going downwards.
(9) Determine pressure loss due to friction for return fluid.
(10) Find total return fluid pressure.
(11) Find effective pressure of column of power oil.
(12) Determine pump friction.
(13) Determine surface operating pressure of power oil.
(14) Select an appropriate triplex pump.

EXAMPLE PROBLEM #6

Given:

7000 ft
2⅜ in. tubing 30° API
5½ in. casing 25% water, ($\gamma w = 1.07$)
GOR = 350 SCF/B
$\overline{P}_R = 1530$ psi
$P_{wh} = 100$ psi
$q_L = 430$ b/d
PI = 1.0 (assume linear)

Solution Procedure:

(1) Determine flowing pressure

$$\Delta P = \frac{q_L}{PI} = \frac{430}{1.0} = 430 \text{ psi}$$

$$P_{wf} = \overline{P}_R - \Delta P = 1530 - 430 = 1100 \text{ psi}$$

(2) Decide upon type of installation and whether or not to vent the gas.
   Where possible use the most economical and the most common installation—try casing free, OPF system. See Fig. 5.43 for theoretical pump displacement $\cong 80\%$—pump at bottom therefore we will pump gas (no vent required).

(3) Find the actual pump displacement to produce fluids. Assume pump eff. = 80%. (80% provides some safety).

Pump displacement =
$$\frac{q_L}{(\text{pump eff.})(\text{fluid displacement factor})}$$
$$= \frac{430}{(0.80)(0.80)} = 672 \text{ b/d}$$

(4) Select a tentative unit to handle this well, see Figures 5.B(1) through 5.B(14) (Appendix). From Fig. 5.B(3) try pump VFR 201616.

P/E = 1.32
Eng. displ. = 4.24 b/d/SPM
Pump displ. = 4.49 b/d/SPM
Max. rated speed = 150 SPM
Pump displ. = (150)(4.49) = 673.5 b/d.

Because this pump will be required to operate at 100% of rated speed we should attempt to select another pump such as one from Fig. 5.B(13) or 5.B(14). However, we will select this pump. (The others are more complex)

(5) Check required pumping speed = $\frac{672}{4.49}$ = 149.7
$\cong$ 150 SPM.

(6) Power oil requirements, assume engine vol. eff. = 80%.

(A) Volume power oil = $\frac{(\text{Eng. displ.})(\text{SPM})}{\text{efficiency}}$ =
$\frac{(4.24)(150)}{0.80}$ = 795 b/d

(B) Find pressure exerted by power oil at bottom
30° API = 0.3795 psi/ft (Pg. 383)
(0.3795)(7000) = 2656 psi.

(7) Return fluid
(A) Volume = prod. + power oil = 430 + 795 = 1225 b/d
(B) Water percentage in return fluid =
$\frac{(430)(.25)}{1225}$ = 9%

Gradient of water + oil = 0.387 $\frac{\text{psi}}{\text{ft}}$
(C) Weight of return fluid column.
Neglecting gas = (0.387)(7000) = 2709 psi.

Step 7 has determined the pressure at the bottom of the return fluid column by neglecting any gas in the column. If the gas is included, find a gas-liquid ratio of:

$$G/L = \frac{(430)(0.75)(350)}{1225} = 92 \text{ SCF/b}$$

By including the gas and calculating the pressure from the Hagedorn and Brown correlation we find 2350 psi. This compares to 2709 psi calculated by the commonly recommended procedure of the pump companies. This 359 psi difference provides some safety factor. However, for fairly high gas-oil ratio wells the pressure should be determined from a multiphase flow correlation. Use the 2709 psi value.

(8) Friction losses
(A) Power fluid going down 2⅜ in. tubing (see Fig. 5.C(7) Appendix). Need viscosity in SSU. From Fig. 5.46, for 30° API @ 100°F SSU $\cong$ 50. 30° API $\cong$ 10 cs @ 100°F. From Fig. 5.C(7) and for a power oil rate = 795 b/d, friction loss = 9 psi/1000 ft. Total loss = 9 × 7 = 63 psi. Multiply by γ_o = 0.876. (0.876)(63) = 55 psi friction loss.

(9) Return fluid coming up 2⅜ in. × 5½ in. annulus, 1225 b/d. Refer to Fig. 5.C(19), Appendix. Assume SSU = 50

$$\Delta P \cong 1.0 \text{ psi}/1000 \text{ ft} = (1)(7) = 7 \text{ psi}.$$

(10) Find total return fluid lift = (static head) + friction + P_{wh} = 2709 + 7 + 100 = 2816 psi.
(11) Find effective pressure of column of power oil = (static head) − (friction) = 2656 − 55 = 2601 psi.
(12) Determine operating pressure at well-head of power oil. Use Equation 5.11.

$$P_s = (2709 + 7 + 100)(1 + 1.32) - (1100)(1.32)$$
$$- (-55 + 2650) + \text{pump friction} = 2480 + F_p$$

To find pump friction see Fig. 5.45, pump of Fig. 5.B(3).

F_p = 650 psi × (.876) = 570 psi
P_s = 2480 + 570 = 3050 psi.

Approximation to find surface power oil pressure

$$P_s = (\text{net lift})(G_3)(P/E) + F_p + P_{wh}(1 + P/E)$$

May simplify further

$$P_s = (\text{N.L.})(0.40)(P/E) + 570$$

For example: N.L. = $7000 - \frac{(1100 - 100)}{.38}$ = 4368

P_s = (4368)(0.40)(1.32) + 570 = 2876 psi compared to 3050 psi.

(13) Select triplex, refer to Fig. 5.D(1) and 5.D(2), Appendix.

q_L = 795 b/d power oil
P_s = 3050 psi

From Fig. 5.D(2) in Appendix, Select J-60-H, 1⅜ in. plunger (3200 psi at 400 rpm) where q = 795 b/d.
(14) Determine hydraulic hp and prime mover hp

3050 × 795 × 0.000017 = 41.2 hydraulic hp output

prime mover input = $\frac{41.2}{0.9}$ = 46 hp

CLASS PROBLEM #6-A

Given:

8000 ft
PI = 1.5 (assume constant)
q_L = 800 b/d
30% water
oil = 40 °API (30 cp at 100°F.)
2⅜-in. tubing
7-in. casing

$\overline{P}_R = 1600$ psi
GOR = 400 SCF/bbl
$P_{wh} = 120$ psi
Design a complete system:
  set pump on bottom; select 2 different pumps.

CLASS PROBLEM #6-B

Given:

  7000 ft
  2⅜ in. tubing
  40% water ($\gamma_w = 1.07$)
  Rate = 300 b/d
  $\overline{P}_R = 1600$ psi
  GOR = 300 SCF/b
  $P_{wh} = 110$ psi
  35° API
  5½ in. casing
  PI = 1.0 (assume constant)
Design a complete system:
  pump on bottom; select two different pumps

5.93 Design of a four-well system

We will now proceed to design a complete 4-well installation. Most installations consist of wells that are fairly similar but assume dissimilar well conditions. This will demonstrate the various parameters and their effects on the design. Each well will be designed individually, but remember that the four wells will be operated from a central system.

EXAMPLE PROBLEM #7: TO DESIGN A COMPLETE 4-WELL HYDRAULIC PUMPING SYSTEM

Given:
  h = 10,000 ft
  $P_{wh} = 75$ psi
  Oil = 40° API gravity
  Water = 1.05 specific gravity

Information on each well:

	Well #1	Well #2	Well #3	Well #4
Casing size	5½	5½	7	7
Tubing size	2⅜	2⅜	2⅞	2⅜
Q_5, bo/d	80	400	100	600
Q_6, bw/d	120	600	500	0
GOR, SCF/b	500	500	500	1000
P_{wf}, psi	500	1000	500	500

Assume:

  $Q'_1/Q_1 = 0.9$
  OPF Power Fluid System

Find:

  Pump sizes, types of down-hole equipment and surface equipment.

First, decide if the gas must be vented or pumped. Assume a pump end efficiency due to slippage of 0.85 and arrive at the efficiency due to gas from Fig. 5.43. Multiplying these two values gives Q_4/Q'_4, as tabulated in Table 5.3.

TABLE 5.3
SUMMARY TABULATION TO DETERMINE WHETHER OR NOT TO VENT GAS

Well #	Q_4	Efficiency due to slippage	Eff. due to gas	Q_4/Q'_4	Q'_4
1	200	0.85	0.55	0.46	435
2	1000	0.85	0.75	0.64	1563
3	600	0.85	0.73	0.64	937
4	600	0.85	0.17	0.14	4286 pumping gas
4	600	0.85		0.85	706 venting gas

The pump end efficiencies appear acceptable for wells 1, 2 and 3, but the low efficiency for well 4 tells us we must vent the gas.

Using the rule-of-thumb Equation 5.1 shown in Section 5.832, estimate the maximum P/E values. For wells 1, 3, and 4 the P_{wf} value of 500 psi equals a column height of 1,100 ft to 1,400 ft. Use a value of 1,200 ft, giving a net lift of 8,800 ft. For well 2, use a net lift of 7600 ft.

$$\text{P/E max.} = \frac{10,000}{\text{Net Lift, ft}}$$

For wells 1, 3 and 4

$$\text{P/E max.} = \frac{10,000}{8,800} = 1.14$$

For well 2

$$\text{P/E max.} = \frac{10,000}{7,600} = 1.32$$

With Q'_4 (required pump end displacement) and P/E maximum, go to the specification tables of Appendix 5.B and choose the pump sizes.

First, restrict the problem by considering only Kobe pumps [(Figs. 5.B(8) through 5.B(14)]. These pumps are listed in order of ascending price, so start with Fig. 5.B(8) and go no further if a suitable pump is found.

For well 1 with 2⅜ in. tubing, the first 2 in. pump that meets the requirement is the 2 in. × 1⅜ in. − 1⅜ in. pump in Fig. 5.B(10). For well 2 with 2⅜ in. tubing, there is no 2 in. pump with the required 1563 b/d displacement, so consider the following alternatives:

(1) Change the tubing to 2⅞ in. or 3½ in.
(2) Vent the gas
(3) Use a pump larger than the I.D. of the tubing and run it as a fixed type pump

Changing the tubing is more expensive than going to a fixed-type pump. Venting the gas is also expensive and the largest side string of tubing that can be run in 5½ in. casing is 1 in., which could be too small for the amount of returned fluid or for the amount of gas to be vented. Based on these assumptions, run a fixed casing type installation (Fig. 5.25). The first pumps to have over 1563 b/d displacement and under 1.32 for the P/E ratio are the 4 in. pumps in Fig. 5.B(8). Select the 4 in. × 2⅜ in. − 2⅜ in. pump.

For well 3 (2⅞ in. tubing) the first 2½ in. pump to have over 937 b/d displacement and under 1.14 P/E ratio is the 2½ in. × 1¾ in. − 1¾ in. pump from Fig. 5.B(10). For well 4 (2⅜ in. tubing) the first 2 in. pumps to have 706 b/d displacement with a P/E ratio of under

1.14 are the pumps from Fig. 5.B(13). Use the 2 in. $\times$ $1^3/_{16}$ in. $\times$ 1⅜ in. − 1⅜ in. $\times$ 1⅜ in. pump.

TABLE 5.4
PUMP SPECIFICATIONS

Well #	Pump size	P/E	q_1	q_4	Max. SPM
1	2″ × 1⅜″ − 1⅜″	1.0	4.54	4.50	121
2	4″ × 2⅜″ − 2⅜″	1.0	32.94	32.50	77
3	2½″ × 1¾″ − 1¾″	1.0	10.96	10.86	100
4	2″ × $1^3/_{16}$″ × 1⅜″ − 1⅜″ × 1⅜″	0.976	7.79	7.55	121

Only one more decision must be made before starting calculations. For well 4 the gas must be vented, but method and tubing size are undetermined. Since the well has 2⅜ in. tubing inside of 7 in. casing, there is enough room to run a second string of 2⅜″ tubing. Choose the parallel free installation shown in Fig. 5.32 with two strings of 2⅜ in. tubing. This will minimize return fluid friction and will allow the gas to vent up the casing annulus.

Now, follow steps 1 through 7, presented in section 5.835, to calculate the required power oil rate and power oil surface pressure for each well.

(1)

Well 1 $SPM = \dfrac{Q'_4}{q_4} = \dfrac{435}{4.50} = 97$

Well 2 $SPM = \dfrac{1563}{32.50} = 48$

Well 3 $SPM = \dfrac{937}{10.86} = 86$

Well 4 $SPM = \dfrac{706}{7.55} = 94$

(2)

From Fig. 5.46 the viscosity of 40° API gravity oil at 180°F is 1.5 CS. From Table 5.2 the specific gravity of the oil is 0.8251. Divide the SPM calculated above by the maximum SPM listed in the specification tables for percent of rated speed. With this value and viscosity, obtain pump friction from Fig. 5.45.

TABLE 5.5
PUMP FRICTION

	SPM	Max. SPM	$\dfrac{SPM}{Max.\ SPM}$	Pump fig. #	*ΔP
Well #1	97	121	0.80	5.B(10)	450 psi
Well #2	48	77	0.62	5.B(8)	280 psi
Well #3	86	100	0.86	5.B(10)	540 psi
Well #4	94	121	0.78	5.B(13)	650 psi

*Values must be multiplied by specific gravity as per Fig. 5.45

Ignore the error in F_{PE} due to reduced volume through the pump end so $\Delta P = F_p$

Well 1 $F_p = 450 \times 0.8251 = 371$ psi
Well 2 $F_p = 280 \times 0.8251 = 231$ psi
Well 3 $F_p = 540 \times 0.8251 = 446$ psi
Well 4 $F_p = 650 \times 0.8251 = 536$ psi

(3)

$$Q_1 = \dfrac{q_q \times SPM}{Q_1/Q_1} \qquad \text{(Equation 5.3)}$$

Well 1 $Q_1 = \dfrac{4.54 \times 97}{0.9} = 489$ b/d

Well 2 $Q_1 = \dfrac{32.94 \times 48}{0.9} = 1757$ b/d

Well 3 $Q_1 = \dfrac{10.96 \times 86}{0.9} = 1047$ b/d

Well 4 $Q_1 = \dfrac{7.79 \times 94}{0.9} = 814$ b/d

(4)

From Fig. 5.46 the average viscosity from surface to bottom (average temperature = 140°F) is 2.1 cs. From the tubing friction charts, multiply the pressure drop per 1,000 ft by the specific gravity of 0.8251 as listed below:

TABLE 5.6.
FRICTION IN POWER TUBING (F_1)

	Friction chart	Q_1	Press. drop × 10	F_1
Well #1	Fig. 5.C(7)	489	24 psi	20 psi
Well #2	Fig. 5.C(7)	1757	250 psi	206 psi
Well #3	Fig. 5.C(8)	1047	45 psi	37 psi
Well #4	Fig. 5.C(7)	814	65 psi	54 psi

(5)

From Table 5.2, $G_1 = G_5 = 0.3574$ psi/ft

$G_6 = 1.05 \times 0.433 = 0.455$ psi/ft

Well 1 $G_3 = \dfrac{(0.3574)(489 + 80) + (0.455 \times 120)}{489 + 200}$

$= \dfrac{203 + 54.6}{689} = 0.374$ psi/ft

Well 2 $G_3 = \dfrac{(0.3574)(1757 + 400) + (0.455 \times 600)}{1757 + 100}$

$= \dfrac{771 + 273}{2757} = 0.379$ psi/ft

Well 3 $G_3 = \dfrac{(0.3574)(1047 + 100) + (0.455 \times 500)}{1047 + 600}$

$= \dfrac{410 + 228}{1647} = 0.387$ psi/ft

Well 4 $G_3 = \dfrac{(0.3574)(814 + 600) + (0.455 \times 0)}{814 + 600}$

$= \dfrac{505}{1414} = 0.3574$ psi/ft

(6)

From Fig. 5.47 the viscosity of water at 140°F. is 0.46 CS.

Well 1 $\nu_3 = \dfrac{(2.1)(489 + 80) + (0.46)(120)}{489 + 200}$

$= \dfrac{1195 + 55}{689} = 1.8$ cs

Well 2 $\nu_3 = \dfrac{(2.1)(1757 + 400) + (0.46)(600)}{1757 + 1000}$

$= \dfrac{4530 + 276}{2757} = 1.74$ cs

Well 3 $v_3 = \dfrac{(2.1)(1047 - 100) + (0.46)(500)}{1047 + 600}$

$= \dfrac{2409 + 230}{1647} = 1.6 \text{ cs}$

Well 4 $v_3 = 2.1 \text{ cs}$

Next divide G_3 by 0.433 to obtain the specific gravity of Q_3. With this value, v_3 and the value of Q_3, obtain F_3 from the friction charts.

TABLE 5.7.
FRICTION IN PRODUCTION TUBING (F_3)

	Friction chart	Q_3	v_3	Press. drop $\times 10$	Sp. gr.	F_3
Well 1	Fig. 5.C(19)	689	1.8	2.5	0.866	2 psi
Well 2	Fig. 5.C(19)	2757	1.73	30	0.878	26 psi
Well 3	Fig. 5.C(23)	1647	1.6	2.5	0.896	2 psi
Well 4	Fig. 5.C(7)	1414	2.1	280	0.825	231 psi

(7)

$$P_1 = h_1 G_1 - F_1 + P_s$$

Well 1 $P_1 = 3574 - 20 + P_s = 3554 + P_s$
Well 2 $P_1 = 3374 - 206 + P_s = 3368 + P_s$
Well 3 $P_1 = 3574 - 37 + P_s = 3537 + P_s$
Well 4 $P_1 = 3574 - 54 + P_s = 3628 + P_s$

$$P_3 = h_1 G_3 + F_3 + P_{FL}$$

Well 1 $P_3 = 3750 + 2 + 75 = 3827$
Well 2 $P_3 = 3790 + 26 + 75 = 3891$
Well 3 $P_3 = 3870 + 2 + 75 = 3947$
Well 4 $P_3 = 3574 + 231 + 75 = 3880$

$P_1 - P_3 - (P_3 - P_4)P/E - F_p = 0$ (Equation 5.9)
$P_s + (h_1 G_1 - F_1) - P_3(1 + P/E) + P_4(P/E) - F_p = 0$
$P_s = -(h_1 G_1 - F_1) + P_3(1 + P/E) - P_4(P/E) - F_p$

Well 1 $P_s = -3554 + 3817(2) - 500 + 371$
$= -3554 + 7634 - 129 = 3951 \text{ psi}$

Well 2 $P_s = -3368 + 3891(2) - 1000 + 231$
$= -4368 + 7782 + 231 = 3645 \text{ psi}$

Well 3 $P_s = -3537 + 3957(2) - 500 + 446$
$= -4037 + 7914 + 446 = 4323 \text{ psi}$

Well 4 $P_s = -3628 + 3880(1.976) - 500(0.976) + 536$
$= -4116 + 7667 + 536 = 4087 \text{ psi}$

Use Equation 5.18 for the individual well power requirements and for the total horsepower. For total horsepower requirements using a central plant to operate all four wells, add the power oil rates for each well. For the pressure use the highest pressure required by any one well.

TABLE 5.8
THEORETICAL HORSEPOWER REQUIRED

	Q_1	P_s	Theoretical horsepower
Well 1	489	3951	33
Well 2	1757	3645	109
Well 3	1047	4323	77
Well 4	814	4087	57
Total	4107	4323	302

In actual practice the surface pumps would be designed for about 200 psi over the pressure required by the down hole pumps. This would provide the pressure drop across the surface flow control valves for satisfactory control.

TABLE 5.9
ACTUAL HORSEPOWER REQUIREMENTS

	Q_1	$P_s + 200$	Actual horsepower
Well 1	489	4151	34.5
Well 2	1757	3845	115
Well 3	1047	4523	80.5
Well 4	814	4287	59.3
Total	4107	4523	316

Suppose surface pumps are available with 30, 90, and 160 hydraulic hp outputs. If we use "well site power plants," accept the 30 for well 1 (assuming calculations are on the high side), the 160 for well 2, and the 90 for wells 3 and 4. If these wells will be put on pump when they quit flowing, considerable time may elapse between the first pump going on and the last one. This would probably convince us to choose the "well site power plant" approach and to purchase the equipment only when needed.

If all four wells need pumps at about the same time, install a central power plant because it will be more economical. In this case use two 160 hp surface pumps. Tanks 24 ft high are available with 300, 750, 1,500 and 3,000 b capacity. These are adequate settling tanks for 600, 1,500, 3,000 and 6,000 b/d power oil rates, respectively. Since power oil rate is 4107 b/d, use one 3,000 b tank or two 1,500 b tanks. Two tanks, if used, must be connected in parallel. If they are connected in series, the velocity through each one will be 4107 b/d and particles not settled out of the first one will not settle out in the second one. Not only must two tanks be connected in parallel, they must not have their outlets connected together because there will then be no insurance that the rates through them will be equal. The only way to control the power oil rates through the two tanks in parallel is to have one surface pump for each tank, which rules out the possibility of our using a 1500 and a 750 tank for the 4107 b/d power oil rate.

CLASS PROBLEM #7-A: TO CALCULATE SURFACE HORSEPOWER

Given:

Well 3 of example problem.
Use the 2½ in. × 1⁷/₁₆ × 1¾ in. – 1¾ in. pump from Fig. 5.B(14)

Assume:

Add 200 psi to P_s to allow for proper operation of surface control valves.

Find:

Surface horsepower for well 3 and total surface horsepower for the four wells assuming a central power plant.

5.10 A FIELD CASE OF ARTIFICIAL LIFT WITH HYDRAULIC PUMPS AT GREAT DEPTHS LAKE BARRE FIELD, TERREBONNE PARISH, LOUISIANA, BY A.E. GAROFALO, JR. TEXACO, INC.

5.101 Introduction

Increased producing rates and ultimate recoveries are being achieved with hydraulic pumps where other methods of artificial lift, such as gas lift, are inefficient or unsuccessful. Seven wells in the Lake Barre field are equipped with hydraulic pump installations at depths ranging from 12,700 to 17,640 ft. Total production from these wells is 1605 barrels per day of oil with individual well rates ranging from 100 to 420 bo/d.

Both surface and subsurface equipment design and operation are discussed with operating controls for power fluid and gas production. To minimize costs per completion, all wells except one are part of three and four-well systems operated from a central control plant with a 4,000 psig maximum recommended operating pressure.

5.102 Developing pumping installations

Table 5.10 represents the seven wells completed with subsurface hydraulic pumps in the Lake Barre field. The first downhole hydraulic pump was installed in Unit 39-19 in December 1971. This well was originally completed as a dual installation and produced water-free by natural flow for six years. The long string production had declined to 130 bo/d, with a 50 psig tubing pressure before it ceased to flow. The short string production had declined to 70 bo/d water-free with a 50 psig tubing pressure. Total oil produced from the two zones was 310,000 and 457,000 bo from the long and short strings, respectively. Remedial action was undertaken to increase recovery by singly completing the subject well to produce both zones simultaneously by means of a subsurface hydraulic pump. This mode of artificial lift was deemed most efficient based on the low reservoir pressure, the depth, the gas-oil ratio, and the volume of production desired. The initial production from pumping averaged 400 bo/d and four years hence is presently averaging 160 bo/d.

In October 1972, three additional wells (Units 34-7, 34-10 and 34-15) were equipped with downhole hydraulic pumps to form a three-well system with centrally-located surface equipment. These three wells are producing from the same Upper M-4 Sand reservoir. Table 5.11 shows reservoir and fluid data of interest. Table 5.12 manifests daily oil production before and immediately after the hydraulic pumps were installed, as well as cumulative oil production before and after pump installation. Unit 34-7 was producing 200 bo/d on gas lift before going off production, unable to gas lift due to low reservoir pressure. Unit 34-10 was producing 160 bo/d, no water, with a GOR of 4800 SCF/bbl, with 200 psig tubing pressure before ceasing to flow. Unit 34-15 was producing 227 bo/d, water-free, with a GOR of 2154 SCF/bbl, with 150 psig tubing pressure before ceasing to flow. In each of the aforementioned wells, installation of a downhole hydraulic pump not only returned the well to a producing status, but resulted in an immediate increase over the previous daily producing rate. A total of 477,000 b of oil has been recovered from these three wells with the aid of hydraulic pumps.

TABLE 5.10
HYDRAULIC PUMP WELLS
LAKE BARRE FIELD

Well	Pump Depth	Started Pumping	Current Production	Cumul. Prod. From Pumping
39-19	15,100	Dec. 1971	160 BOPD	200,000 Bbl.
34-7	12,700	Oct. 1972	325 BOPD	181,000 Bbl.
34-10	13,050	Oct. 1972	420 BOPD	206,000 Bbl.
34-15	12,880	Oct. 1972	100 BOPD	90,000 Bbl.
39-22	14,625	Oct. 1972	240 BOPD	326,000 Bbl.
38-24	14,700	Aug. 1974	230 BOPD	37,000 Bbl.
43-4	17,640	Mar. 1974	130 BOPD	14,000 Bbl.
	14,385 (Avg.)		1,605 BOPD	1,054,000 Bbl.

TABLE 5.11
UPPER M-4 SAND
RESERVOIR DATA

Mean Reservoir Depth	13,000 Ft
Reservoir Temp.	243°F
Orig. Reservoir Press	6130 psia
Curr. Reservoir Press.	1500 psia
Porosity	26 Percent
Permeability	100-200 md
Gas-Oil Ratio	2000-4000 Scf/Bbl
Gravity	33.6° API

TABLE 5.12
HYDRAULIC PUMP—THREE-WELL SYSTEM
LAKE BARRE FIELD

Well	Production (BOPD)		Total Production (Bbl. Oil)	
	Before Pump	Initial Pumping	Before Pump	Since Pumping
34-7	200	500	469,000	181,000
34-10	160	500	559,000	206,000
34-15	227	289	511,000	90,000
	587	1,289	1,539,000	477,000

Unit 38-24, a newly drilled well, (Table 5.10) is producing 230 bo/d and no water with the pump set at 14,700 ft. This well is part of a four-well system which includes Unit 39-19 (mentioned previously), Unit 39-22, and a well to be included at a future date. Unit 39-22 is presently producing 240 bo/d with the pump set at 14,625 ft.

The deepest pump installation at Lake Barre is Unit 43-4. The pump is set at 17,640 ft, and the top of the producing interval is 18,231 ft. The producing zone is a very fine-grained, slightly shaly sandstone with 20% porosity, 65 md permeability and 3,400 psi reservoir pressure. This well initially produced this zone by gas lift with the bottom intermittent valve located at 9322 ft. It produced 27,800 b of oil and was producing 43 bo/d by gas lift before going off production. The well was recompleted with a hydraulic pump and a single well surface unit. The well produced 192 bo/d initially and is now producing 130 bo/d.

5.103 Subsurface equipment

A typical Lake Barre hydraulic pump installation is the same as shown in Figure 5.36. Each well is a casing free, gas-vent installation (except Units 38-24 and 39-19, which do not have vent strings to release free gas) with 2⅞ in. OD tubing for the power oil string and a parallel string of 2¹/₁₆ in. or 1¼ in. OD tubing for the gas vent string. The pump used in each case is a 2½ in. tandem engine Armco VFR pump. Current operating conditions range from 66 SPM with a wellhead pressure of 2000 psi, up to 94 SPM with a wellhead pressure of 3100 psi.

One consideration in designing these installations is the producing gas-oil ratio. In a simple casing-free installation, all of the gas that is produced must be displaced through the pump. In this case, the pump efficiency will drop rapidly as the volume of free gas increases or as the pumping bottom hole pressure decreases. Knowing the anticipated GOR and estimated pumping bottom hole pressure, theoretical volumetric efficiencies of casing pumps were determined from charts found in the literature.

Based on these efficiencies and the maximum displacement rating of the hydraulic unit selected, the simple casing-free installation design is economically impractical from a production yield standpoint in five of the seven installations analyzed. The casing-free gas-vent installation was, therefore, selected as the simplest and most efficient type of downhole system for those wells exhibiting gas-oil ratios in excess of 2000 scf/b.

In the gas-vent system, the gas is gathered from under the packer and vented up a parallel string, while the combined exhaust power oil and production is produced up the casing annulus. This installation was selected as the best system because of the high fluid volumes desired, since the pressure drop on the combined streams of production and exhaust power oil is relatively low in the large casing annulus.

5.104 Surface facilities

Power oil fluid used to operate the subsurface pump is supplied by one of two different surface equipment systems. One type of system consists of centrally located pumping and treating facilities (similar to Fig. 5.11) which service several wells at some distance from this location. The other system consists of a single-well portable unit (similar to Fig. 5.15), located near the wellhead, which contains all of the necessary equipment to pump and condition the power fluid.

5.105 Steps taken to maintain and improve equipment operation

During the early stages of operation, several of the pump wells produced various amounts of sand. Unit 34-10 produced for six months by pumping and then experienced sand production. The pump had to be changed out several times when it stopped stroking, due to sand in the engine end of the pump. Also, the pump became stuck and could not be reversed out due to sand on top of it. The well had to be washed out with 1 in. tubing and the pump fished out of the tubing.

At the same time, Unit 34-7 experienced the same problem with a stuck pump due to fine sand settling out of the power fluids inside the tubing string. Both wells are in the same system and the power oil directed to both wells is out of the same power oil tank.

A workover was performed on Unit 34-10, the source of the sand production, in April 1973. The producing interval was plasticized to control the sand production, and the well was recompleted with a hydraulic pump. Since then, there have been no further sand problems with this well or with Unit 34-7. Four centrifugal de-sanders were added to the four-well, centrally-located system in February 1975 to provide additional removal of solids from the power oil. Another action taken to enhance operation was to service pumps on a regular basis, at least once every two months. Previously, pumps were changed only when a drop in production occurred. Also, whenever the system is to be down for any length of time, the pump is circulated to the surface to prevent any sand from settling on the pump.

5.106 Conclusions

Hydraulic pumps are effective in producing large volumes from deep low-pressured reservoirs. To achieve the highest efficiency possible, a method of venting free gas is necessary.

Maintaining a clean power fluid is the key to trouble-free pump operation. This, along with servicing and/or changing pumps out at least once every two months, will prevent any serious problems from developing which may lead to a costly workover.

5.11 OPERATING CONSIDERATIONS

There are several inherent features of hydraulic pumps unique to this type of pumping system, and an understanding of these features will allow operating personnel to get the maximum results from the pumps.

5.111 Well testing OPF

In the OPF system, where power oil is returned mixed with the production, the power oil must be measured and subtracted from the returned oil to determine net production. This sounds simple, but under some circumstances it is a source of frustration. For instance, suppose a well makes 95 b/d water and 5 b/d oil while using 200 b/d power oil. If the power oil meter reading is 2½% low, the test will show 10 b/d oil production—a 100% error.

Another problem can result when water is determined by centrifuging the return fluid. The water cut thus determined is a percent of the total fluid returned —produced oil, produced water, and power oil. In the example used above, the "shake out" would show 31.7% water (95/300 × 100); if the power oil were neglected this would indicate only about 2½ b of water. This sounds like a simple problem, but to some field personnel it can be mystifying.

These problems do not exist, however, when a well site power plant is used.

5.112 Well testing CPF

In the CPF system, power fluid is not mixed with production, so testing would seem to be as simple as with a flowing well. To be precise, though, the power fluid loss must be accounted for. This loss is to production and is usually less than 5% of the power fluid rate. To measure the loss may require power fluid meters for power fluid to the well and for power fluid return. With two meters the chances of inaccuracies are increased.

5.113 Power fluid quality control

Keeping solids out of the power fluid is probably the most important function in controlling repair costs for hydraulic pumps. For sale to the pipe line, oil that is 0.1% (0.001) B.S. and W. is considered good clean oil. This is about as low as most centrifuges can measure accurately and is 1,000.0 parts per million. By contrast, power fluid frequently has about 10.0 ppm solids—impossible to measure with the common centrifuge. A rule of thumb for good power oil quality is: solids—10 to 15 ppm; salt—10 to 15 lbs/1000 b. With crude oil of 30° API gravity or less, more solids can be tolerated because the viscosity of the oil retards power fluid leakage through the pump fits.

To insure full benefits from a settling tank, it should have an efficient and level spreader and gas must be removed before the oil enters. Also, the bottom foot or two of oil should be removed periodically.

To remove salt from the oil, fresh water can be injected continuously into the power fluid stream. An alternate but less desirable method is to keep a foot or two of fresh water in the bottom of the power oil tank. This second method has two hazards: it may not be changed often enough, and it allows an emulsion interface to build up between the oil and water. This interface traps solids that might otherwise settle to the bottom.

For well site power units, cleaning the power fluid is usually done by cyclone centrifuges. These cyclones are most effective with 30 to 50 psi pressure differential between feed (inlet) pressure and overflow (clean fluid outlet) pressure. Solids are discharged out the "underflow" or apex of the cyclone. The quantity and size of particles carried over with the overflow are reduced by increasing the pressure drop from feed to overflow —accomplished by increasing the feed rate—and increasing the underflow rate.

5.114 Speed control

Speed, or production, control should be accomplished by power fluid rate and not by power fluid pressure. Operating pressure can change due to changes in bottom hole pressure, pump intake restrictions, paraffin build up in any of the flow paths and changes in the oil-water-gas ratios. Pump speed in SPM or power fluid rate in b/d are proper production rate control factors.

5.115 Analysis and troubleshooting

5.1151 Introduction

Most of the failures (perhaps 75%) occur in the engine end of the pump, and most of these are due to dirty power fluid. When using water as the power fluid, the packings in the liners of surface pumps give more trouble.

When looking for solutions to short pump runs, low production or high repair costs, one should first determine whether the problem is engine end or pump end. The logical starting place then is to calculate the engine end and pump end efficiencies of the pump in question. Low engine end efficiency points to poor power fluid quality, while low pump end efficiency can mean excessive speed, restricted pump intake passages, gas interference or corrosion damage.

In most cases the life of the pump end of the unit should compare to a rod pump. One of the best methods of analysis is to observe the surface pump pressure at the well. One can observe any changes in pressures taking place. Note that the pressure does not tell you how fast the pump is pumping (SPM). Also, on start-up the pressure may be lower than later due to more initial help from the formation (P_{wf} is higher).

Sometimes it is difficult to detect SPM. We may use an oscilloscope or possibly a strain gauge. Also, the last stroke analysis may be used to detect a pump leak. If the pressure continues to decline rapidly after shut-down it indicates a leaking pump.

5.1152 Subsurface troubleshooting guide

The following listing will serve as a guide for analyzing and trouble-shooting the sub-surface pumping unit.

INDICATION	CAUSE	REMEDY
1. Sudden increase in operating pressure—pump stroking.	(a) Lowered fluid level which causes more net lift.	(a) If necessary, slow pump down.
	(b) Paraffin build-up or obstruction in power oil line, flow line or valve.	(b) Run soluble plug, hot oil or remove obstruction.
	(c) Pumping heavy material, such as salt water or mud.	(c) Keep pump stroking—do not shut down.
	(d) Pump beginning to fail.	(d) Retrieve pump and repair.
2. Gradual increase in operating pressure—pump stroking.	(a) Gradually lowering fluid level. Standing valve or formation plugging up.	(a) Surface pump and check. Retrieve standing valve.
	(b) Slow build-up of paraffin.	(b) Run soluble plug or hot oil.
	(c) Increasing water production.	(c) Raise pump SPM and watch pressure.
3. Sudden increase in operating pressure—pump not stroking.	(a) Pump stuck or stalled.	(a) Alternately increase and decrease pressure. If necessary, unseat and reseat pump. If this fails to start pump, surface and repair.

INDICATION	CAUSE	REMEDY
	(b) Sudden change in well conditions requiring operating pressure in excess of triplex relief valve setting.	(b) Raise setting on relief valve.
	(c) Sudden change in power oil-emulsion, etc.	(c) Check power oil supply.
	(d) Closed valve or obstruction in production line.	(d) Locate and correct.
4. Sudden decrease in operating pressure—pump stroking. (Speed could be increased or reduced).	(a) Rising fluid level—pump efficiency up.	
	(b) Failure of pump so that part of power oil is bypassed.	(b) Surface pump and repair.
	(c) Gas passing through pump.	
	(d) Tubular failure—downhole or in surface power oil line. Speed reduced.	(d) Check tubulars.
	(e) Broken plunger rod. Increased speed.	(e) Surface pump and repair.
	(f) Seal sleeve in bottom hole assembly washed or failed. Speed reduced.	(f) Pull tubing and repair bottom hole assembly.
5. Sudden decrease in operating pressure—pump not stroking.	(a) Pump not on seat.	(a) Circulate pump back on seat.
	(b) Failure of production unit or external seal.	(b) Surface pump and repair.
	(c) Bad leak in power oil tubing string.	(c) Check tubing and pull and repair if leaking.
	(d) Bad leak in surface power oil line.	(d) Locate and repair.
	(e) Not enough power oil supply at manifold.	(e) Check volume of fluid discharged from triplex. Valve failure, plugged supply line, low power oil supply, excess bypassing, etc., all of which could reduce available volume.
6. Drop in production—pump speed constant.	(a) Failure of pump end of production unit.	(a) Surface pump and repair.
	(b) Leak in gas vent tubing string.	(b) Check gas vent system.
	(c) Well pumped off—pump speeded up.	(c) Decrease pump speed.
	(d) Leak in production return line.	(d) Locate and repair.
	(e) Change in well conditions.	
	(f) Pump or standing valve plugging.	(f) Surface pump and check. Retrieve standing valve.
	(g) Pump handling free gas.	(g) Test to determine best operating speed.
7. Gradual or sudden increase in power oil required to maintain pump speed. Low engine efficiency.	(a) Engine wear.	(a) Surface pump and repair.
	(b) Leak in tubulars—power oil tubing, bottom hole assembly, seals or power oil line.	(b) Locate and repair.
8. Erratic stroking at widely varying pressures.	(a) Caused by failure or plugging of engine.	(a) Surface pump and repair.
9. Stroke "down-kicking" instead of "up-kicking."	(a) Well pumped off—pump speeded up.	(a) Decrease pump speed. Consider changing to smaller pump end.
	(b) Pump intake or downhole equipment plugged.	(b) Surface pump and clean up. If in downhole equipment, pull standing valve and back flush well.
	(c) Pump failure (balls and seats).	(c) Surface pump and repair.
	(d) Pump handling free gas.	
10. Apparent loss of, or unable to account for, system fluid.	(a) System not full of oil when pump was started due to water in annulus U-tubing after circulating, well flowing or standing valve leaking.	(a) Continue pumping to fill up system. Pull standing valve if pump surfacing is slow and cups look good.
	(b) Inaccurate meters or measurement.	(b) Recheck meters. Repair if necessary.

INDICATION	CAUSE	REMEDY
	(c) Leaking valve, power oil or production line or packer.	(c) Locate and repair.
	(d) Affect of gas on production metering.	(d) Improve gas separation.
	(e) Pump not deep enough.	(e) Lower pump.
11. Well not producing— (a) Pressure increase, stroking. (b) Pressure loss, stroking.	(a) Engine plugging, flow line plugged, broken engine rod, suction plugged. (b) Standing valve leaking. Tubular leak.	(a) Surface unit and repair. Locate restriction in flow line. Pull standing valve. (b) Pull standing valve. Check tubulars.

5.1153 Power oil plunger pumps troubleshooting guide

I. KNOCKING OR POUNDING IN FLUID END AND PIPING

Possible Cause	Correction
Suction line restricted by:	
a. Trash, scale build-up, etc.	Locate and remove.
b. Partially closed valve in suction line.	Locate and correct.
c. Meters, filters, check valves, non-full opening, cut-off valves or other restrictions.	Rework suction line to eliminate.
d. Sharp 90° bends or 90° blind tees.	Rework suction line to eliminate.
Air entering suction line through valve stem packing.	Tighten or repack valve stem packing.
Air entering suction line through loose connection or faulty pipe.	Locate and correct.
Air or vapor trapped in suction.	Locate rise or trap and correct by straightening line, providing enough slope to permit escape and prevent build-up.
Low fluid level.	Increase supply and install automatic low level shutdown switch.
Suction dampener not operating.	Inspect and repair as required.
Worn pump valves or broken spring.	Inspect and repair as required.
Entrained gas or air in fluid.	Provide gas boot or scrubber for fluid.
Inadequate sized suction line.	Replace with individual suction line or next size larger than inlet of pump.
Leaking pressure relief valve that has been piped back into pump suction.	Repair valve and rework piping to return to supply tank—not suction line.
Bypass piped back to suction.	Rework to return bypassed fluid back to supply tank—not supply line.
Broken plunger.	Inspect when rotating pump by hand and replace as required.
Worn crosshead pin or connecting rod.	Locate and replace as required.

II. KNOCK IN POWER END

Possible Cause	Correction
Worn crosshead pin or connecting rod.	Locate and replace as required. Check oil quality and level.
Worn main bearings.	Replace as required. Check oil quality and level.
Loose plunger—intermediate rod—crosshead connection.	Inspect for damage—replace as required and tighten.

III. RAPID VALVE WEAR OR FAILURE

Possible Cause	Correction
Cavitation.	Predominate cause of short valve life and is always a result of poor suction conditions. This situation can be corrected by following appropriate recommendations as listed under No. I.
Corrosion.	Treat fluid as required.
Abrasives in fluid.	Treat to remove harmful solids.

IV. FLUID SEAL PLUNGER WEAR, LEAKAGE OR FAILURE

Possible Cause	Correction
Solids in power oil.	This is likely to cause greatest amount of wear. Power oil should be analyzed for amount and type of solids content. Proper treating to remove solids should be instigated.
Improper installation.	Follow written instructions and use proper tools. Remember, plunger and liner are matched sets. Assure proper lubrication at start up. (Be sure air is bled out of fluid end before starting up.)

V. REDUCED VOLUME OR PRESSURE

Possible Cause	Correction
Bypassing fluid.	Locate and correct.
Air in fluid end of triplex.	Bleed off.
Inaccurate meter or pressure gauge.	Check and correct.
Pump suction cavitation due to improper hook-up, suction restriction or entrained gas.	Locate and correct.
Valves worn or broken.	Replace.
Plungers and liners worn.	Replace.
Reduced prime mover speed due to increased load, fuel or other conditions.	Determine cause and correct. (May be increased pressure due to paraffin, temperature change, etc.)

5.116 Installation practices

To prevent possible problems, the following installation practices should be adhered to. First, power oil tubing should be free from dirt, scale, or other insolubles; have clean threads, and should pressure test used tubing to check for leaks. Second, for fixed installations, check accessories for operation, limit packer setting weight, and circulate the system to clean up. Next, free pump installations should check clearance for parallel strings, circulate brush type cleaning tool, drift tubing with dummy pump to insure pump clearance, and use the minimum packer setting weight. Finally, surface connections should keep surface lines free of sharp bends, have short straight suction lines, use full opening valves on suction, and prevent obstructions or turns close to triplex discharge.

5.12 SUMMARY AND CONCLUSIONS

Hydraulic piston type pumps have the following advantages:
(1) The free pump can be circulated into and out of the well, thus eliminating well servicing units.
(2) They are unaffected by very crooked wells.
(3) They have virtually no depth limitation.
(4) Pump speed can be easily changed.
(5) Pump size can be easily changed due to the free pump.
(6) Bottom hole pressure data can be obtained at different producing rates.
(7) If there is very little free gas or, if the gas is vented, hydraulic pumps usually have the highest efficiency (lowest input horsepower) of any artificial lift system.
(8) Well heads are compact and ideal for platform operations.
(9) Chemicals can be added to the power fluid to protect tubing and casing from corrosion and to help break emulsions.

REFERENCES

1. Vogel, J. V., "Inflow Performance Relationship for Solution Gas Drive Wells," *Journal Petroleum Tech.* January 1968, pp. 83-93.
2. Garofalo, A. E. Jr., Texaco, Inc., personal communication.
3. Gibbs, S. G. and K. B. Nolen, Nabla Corp., personal communication.

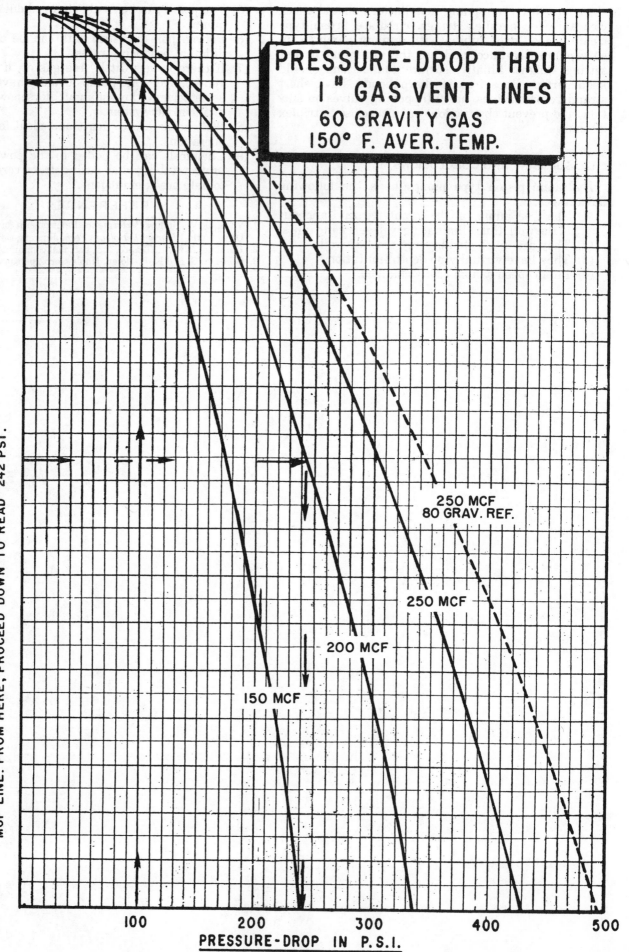

EXAMPLE: TO FIND MINIMUM BOTTOM HOLE PRESSURE AT WHICH 200 MCF / DAY CAN BE VENTED WITH 100 PSI BACK-PRESSURE ON EXHAUST AT WELL HEAD. ASSUME PACKER SET AT 8000 FT.
ENTER CHART AT BOTTOM AT 100 PSI AND PROCEED VERTICALLY TO 200 MCF LINE. FROM INTERSECTION, PROJECT TO THE BORDER. MEASURE DOWN 1/2" PER 1000 FT. (4") AND FOLLOW HORIZONTAL TO NEW POINT ON 200 MCF LINE. FROM HERE, PROCEED DOWN TO READ 242 PSI.

PRESSURE-DROP THRU 1" GAS VENT LINES
60 GRAVITY GAS
150° F. AVER. TEMP.

250 MCF
80 GRAV. REF.

250 MCF

200 MCF

150 MCF

PRESSURE-DROP IN P.S.I.

Fig. 5A (1)   Pressure drop through 1" gas vent line (Courtesy Fluid
Packed Pump Co.)

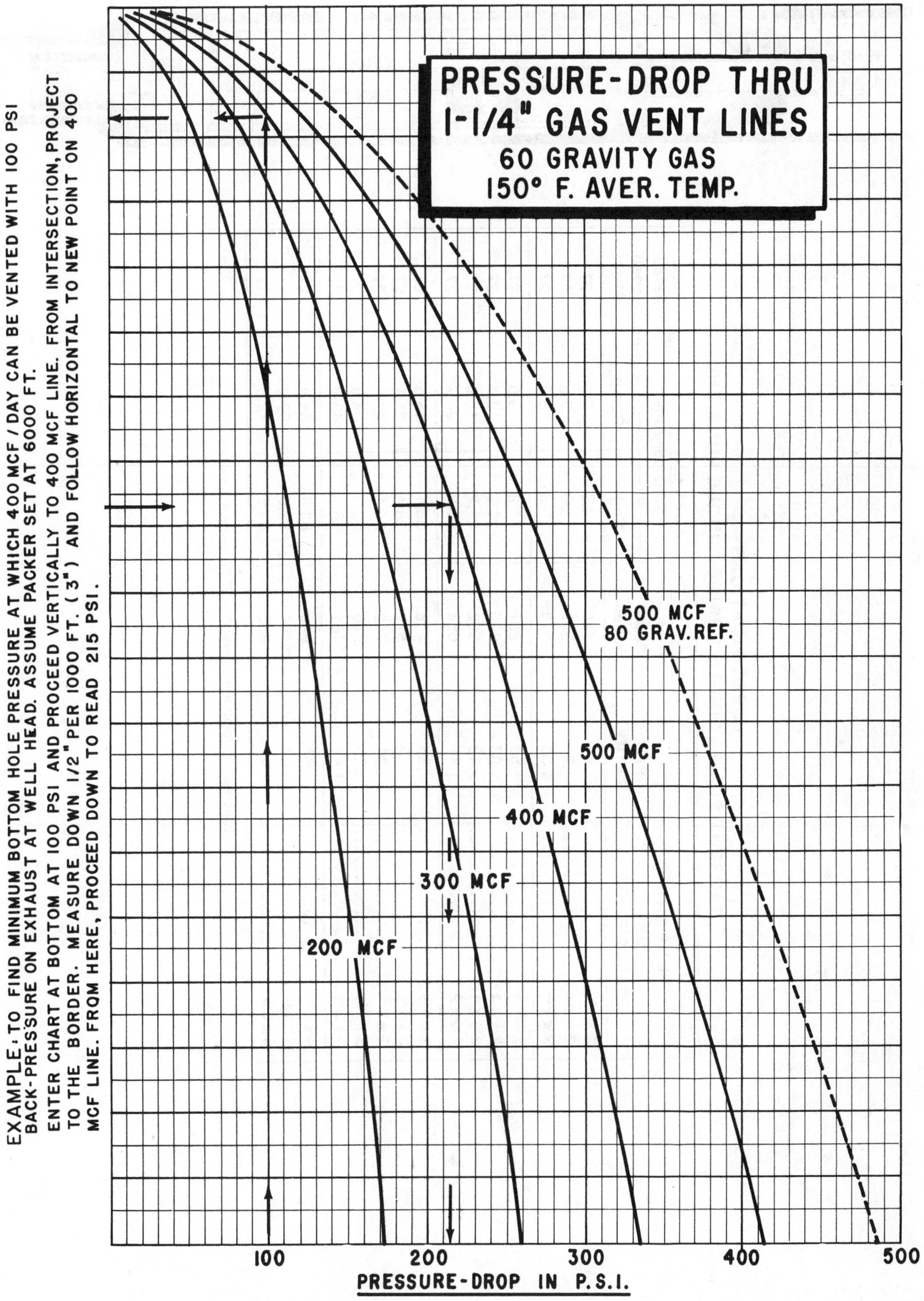

Appendix A   401

EXAMPLE: TO FIND MINIMUM BOTTOM HOLE PRESSURE AT WHICH 400 MCF / DAY CAN BE VENTED WITH 100 PSI BACK-PRESSURE ON EXHAUST AT WELL HEAD. ASSUME PACKER SET AT 6000 FT. ENTER CHART AT BOTTOM AT 100 PSI AND PROCEED VERTICALLY TO 400 MCF LINE. FROM INTERSECTION, PROJECT TO THE BORDER. MEASURE DOWN 1/2" PER 1000 FT. (3") AND FOLLOW HORIZONTAL TO NEW POINT ON 400 MCF LINE. FROM HERE, PROCEED DOWN TO READ 215 PSI.

PRESSURE-DROP THRU 1-1/4" GAS VENT LINES
60 GRAVITY GAS
150° F. AVER. TEMP.

500 MCF 80 GRAV. REF.

500 MCF

400 MCF

300 MCF

200 MCF

PRESSURE-DROP IN P.S.I.

Fig. 5A (2)   Pressure drop through 1¼" gas vent lines (Courtesy fluid Packed Pump Co.)

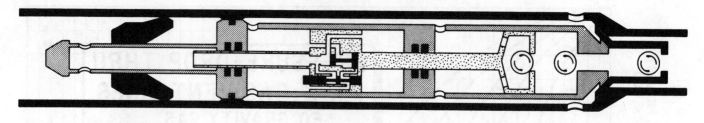

SPECIFICATIONS—
SARGENT PUMP—SINGLE ENGINE, SINGLE PUMP END

PUMP SIZE OR DESCRIPTION	P/E RATIO	DISPLACEMENT				Max. Rated Speed (SPM)
		At Rated Speed	BPD per SPM			
			Eng.	Pump		
2 x 1¹⁄₁₆	.67	186	13.3	6.9		27
2 x 1¼	.93	259	13.3	9.6		27
2 x 1½	1.33	373	13.3	13.8		27
2½ x 1¼	.58	257	21.2	9.5		27
2½ x 1½	.83	370	21.2	13.7		27
2½ x 1¾	1.13	502	21.2	18.6		27
2½ x 2	1.47	653	21.2	24.2		27
3 x 1½	.53	418	36.1	15.5		27
3 x 1¾	.72	570	36.1	21.1		27
3 x 2	.94	742	36.1	27.5		27
3 x 2¼	1.20	940	36.1	34.8		27
3 x 2½	1.47	1161	36.1	43.0		27
4 x 2¼	.68	940	63.5	34.8		27
4 x 2¾	1.01	1404	63.5	52.0		27
4 x 3¼	1.41	1960	63.5	72.6		27

(Courtesy Sargent Industries, Oilwell Equipment Division)

Fig. 5B (1)

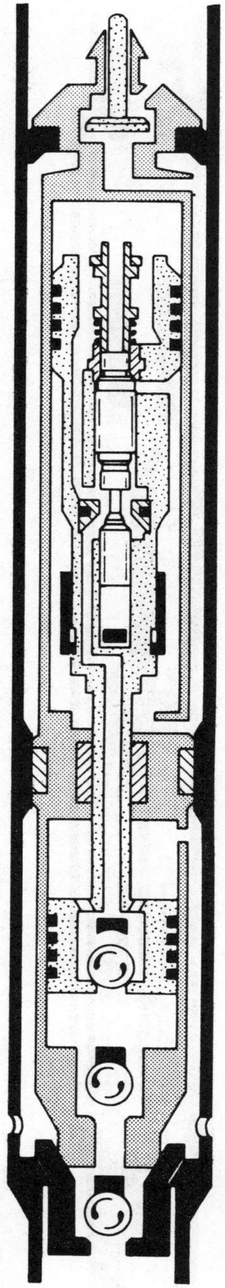

SPECIFICATIONS—
JOHNSON—FAGG PUMP—SINGLE ENGINE, SINGLE PUMP END

PUMP SIZE OR DESCRIPTION	P/E RATIO	DISPLACEMENT At Rated Speed	BPD per SPM Eng.	BPD per SPM Pump	Max. Rated Speed (SPM)
2 x 1⅝ x 1¹⁄₁₆	.52	225	15.08	6.45	35
2 x 1⅝ x 1¼	.72	312	15.08	8.92	35
2 x 1⅝ x 1½	1.03	450	15.08	12.85	35
2 x 1⅝ x 1⅝	1.21	528	15.08	15.08	35
2½ x 2 x 1¼	.44	264	30.80	12.02	22
2½ x 2 x 1½	.68	467	30.80	17.30	27
2½ x 2 x 1⅝	.80	547	30.80	20.30	27
2½ x 2 x 1¾	.93	637	30.80	23.60	27
2½ x 2 x 2	1.21	831	30.80	30.80	27
3 x 2½ x 1¾	.59	643	43.71	21.42	30
3 x 2½ x 2	.77	840	43.71	27.98	30
3 x 2½ x 2¼	.98	1062	43.71	35.41	30
3 x 2½ x 2½	1.21	1311	43.71	43.71	30
4 x 2 ¹⁵⁄₁₆ x 2	.57	840	60.35	27.98	30
4 x 2 ¹⁵⁄₁₆ x 2¼	.72	1062	60.35	35.41	30
4 x 2 ¹⁵⁄₁₆ x 2½	.89	1311	60.35	43.71	30
4 x 2 ¹⁵⁄₁₆ x 2¾	1.08	1587	60.35	52.90	30
4 x 2 ¹⁵⁄₁₆ x 2 ¹⁵⁄₁₆	1.22	1810	60.35	60.35	30

(Courtesy Oil Field Products Division—Dresser Industries, Inc.)

Fig. 5B (2)

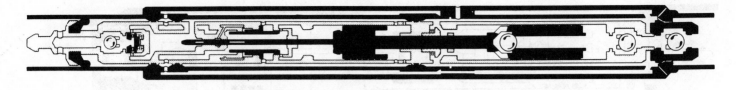

SPECIFICATIONS—
FLUID PACKED VFR PUMP—SINGLE ENGINE, SINGLE PUMP END

PUMP SIZE OR DESCRIPTION	P/E RATIO	DISPLACEMENT			Max. Rated Speed (SPM)
		At Rated Speed	BPD per SPM		
			Eng.	Pump	
VFR 201611	.62	318	4.24	2.12	150
VFR 201613	.87	444	4.24	2.96	150
VFR 201616	1.32	673	4.24	4.49	150
VFR 252015	.74	630	8.89	5.25	120
VFR 252017	1.00	858	8.89	7.15	120
VFR 252020	1.32	1119	8.89	9.33	120
VFR 302424	1.28	1612	12.99	13.44	120

(Courtesy Armco-Fluid Packed Pump)

Fig. 5B (3)

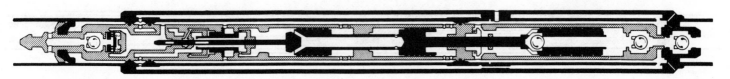

SPECIFICATIONS—
FLUID PACKED VFR PUMP—TANDEM ENGINE, SINGLE PUMP END

PUMP SIZE OR DESCRIPTION	P/E RATIO	DISPLACEMENT BPD per SPM			Max. Rated Speed (SPM)
		At Rated Speed	Eng.	Pump	
VFR 20161613	.54	444	6.86	2.96	150
VFR 20161616	.81	673	6.86	4.49	150
VFR 25202015	.41	630	15.16	5.25	120
VFR 25202017	.56	858	15.16	7.15	120
VFR 25202020	.73	1119	15.16	9.33	120

(Courtesy Armco-Fluid Packed Pump)

Fig. 5B (4)

SPECIFICATIONS—
FLUID PACKED V-11 PUMP—SINGLE ENGINE, SINGLE PUMP END

PUMP SIZE OR DESCRIPTION	P/E RATIO	DISPLACEMENT			Max. Rated Speed (SPM)
		At Rated Speed	BPD per SPM		
			Eng.	Pump	
V-25-11-118	1.18	1419	5.33	6.31	225
V-25-11-095	.95	1299	6.66	6.31	206

(Courtesy Armco-Fluid Packed Pump)

Fig. 5B (5)

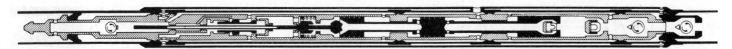

SPECIFICATIONS—
FLUID PACKED V-21 PUMP—TANDEM ENGINE, SINGLE PUMP END

| PUMP SIZE OR DESCRIPTION | P/E RATIO | DISPLACEMENT | | | Max. Rated Speed (SPM) |
| | | At Rated Speed | BPD per SPM | | |
			Eng.	Pump	
V25-21-075	.75	1173	8.38	6.31	186
V25-21-063	.63	1072	10.00	6.31	170

(Courtesy Armco-Fluid Packed Pump)

Fig. 5B (6)

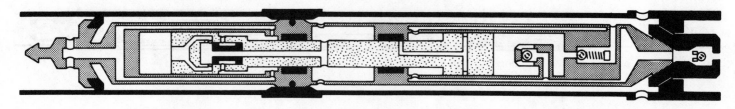

SPECIFICATIONS—
FLUID PACKED F, FE & FEB PUMPS

PUMP SIZE OR DESCRIPTION	P/E RATIO	DISPLACEMENT			Max. Rated Speed (SPM)
		At Rated Speed	BPD per SPM		
			Eng.	Pump	
201311	.71	204	4.2	3.0	68
201313	1.00	285	4.2	4.2	68
201611	.47	204	6.4	3.0	68
201613	.66	285	6.4	4.2	68
201616	1.00	517	9.4	9.4	55
251611	.47	214	7.0	3.3	65
251613	.66	299	7.0	4.6	65
251616	1.00	455	7.0	7.0	65
252016	.64	540	16.5	10.6	51
252018	.81	683	16.5	13.4	51
252020	1.00	841	16.5	16.5	51
402422	.88	1269	32.1	28.2	45

(Courtesy Armco-Fluid Packed Pump)

Fig. 5B (7)

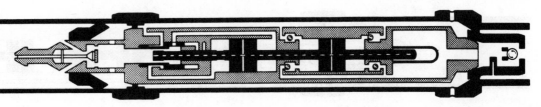

SPECIFICATIONS—
KOBE TYPE A PUMP—SINGLE ENGINE, SINGLE PUMP END

PUMP SIZE OR DESCRIPTION	P/E RATIO	DISPLACEMENT			Max. Rated Speed (SPM)
		At Rated Speed	BPD per SPM		
			Eng.	Pump	
2 x 1—¹³/₁₆	.545	139	2.15	1.15	121
2 x 1—1	1.000	254	2.15	2.10	121
2 x 1—1³/₁₆	1.546	393	2.15	3.25	121
2 x 1³/₁₆—1	.647	254	3.30	2.10	121
2 x 1³/₁₆—1³/₁₆	1.000	393	3.30	3.25	121
2½ x 1¼—1	.520	256	5.02	2.56	100
2½ x 1¼—1⅛	.746	367	5.02	3.67	100
2½ x 1¼—1¼	1.000	492	5.02	4.92	100
2½ x 1¼—1⁷/₁₆	1.431	703	5.02	7.03	100
2½ x 1⁷/₁₆—1¼	.700	492	7.13	4.92	100
2½ x 1⁷/₁₆—1⁷/₁₆	1.000	703	7.13	7.03	100
3 x 1½—1¼	.592	486	9.61	5.59	87
3 x 1½—1⅜	.787	646	9.61	7.43	87
3 x 1½—1½	1.000	821	9.61	9.44	87
3 x 1½—1¾	1.480	1218	9.61	14.00	87
3 x 1¾—1½	.676	821	14.17	9.44	87
3 x 1¾—1¾	1.000	1218	14.17	14.00	87
4 x 2—1¾	.687	1108	21.44	14.40	77
4 x 2—2	1.000	1617	21.44	21.00	77
4 x 2—2⅜	1.541	2502	21.44	32.50	77
4 x 2⅜—2	.649	1617	32.94	21.00	77
4 x 2⅜—2⅜	1.000	2502	32.94	32.50	77

(Courtesy Kobe, Inc.—Subsidiary of Baker International Corp.)

Fig. 5B (8)

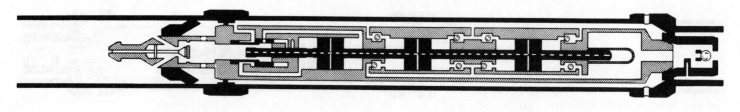

SPECIFICATIONS—
KOBE TYPE A PUMP—SINGLE ENGINE, DOUBLE PUMP END

PUMP SIZE OR DESCRIPTION	P/E RATIO	DISPLACEMENT At Rated Speed	BPD per SPM Eng.	BPD per SPM Pump	Max. Rated Speed (SPM)
2 x 1³/₁₆ —1 x 1	1.290	508	3.30	4.20	121
2 x 1³/₁₆ —1³/₁₆ x 1	1.647	647	3.30	5.35	121
2 x 1³/₁₆ x 1³/₁₆	2.000	786	3.30	6.50	121
2½ x 1⁷/₁₆ —1¼ x 1¼	1.400	984	7.13	9.84	100
2½ x 1⁷/₁₆ —1⁷/₁₆ x 1¼	1.701	1195	7.13	11.95	100
2½ x 1⁷/₁₆ —1⁷/₁₆ x 1⁷/₁₆	2.000	1406	7.13	14.06	100
3 x 1³/₄ —1¼ x 1¼	.800	972	14.17	11.18	87
3 x 1³/₄ —1½ x 1½	1.351	1642	14.17	18.88	87
3 x 1³/₄ —1³/₄ x 1½	1.675	2039	14.17	23.44	87
3 x 1³/₄ —1³/₄ x 1³/₄	2.000	2436	14.17	28.00	87
4 x 2³/₈ —2 x 1³/₄	1.094	2725	32.94	35.40	77
4 x 2³/₈ —2 x 2	1.299	3234	32.94	42.00	77
4 x 2³/₈ —2³/₈ x 2	1.650	4119	32.94	53.50	77
4 x 2³/₈ —2³/₈ x 2³/₈	2.000	5005	32.94	65.00	77

(Courtesy Kobe, Inc.—Subsidiary of Baker International Corp.)

Fig. 5B (9)

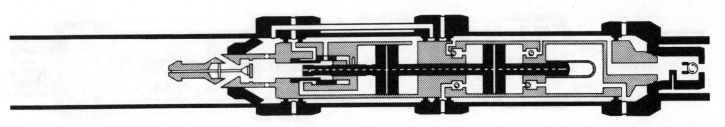

SPECIFICATIONS—
KOBE TYPE B PUMP—SINGLE ENGINE, SINGLE PUMP END

PUMP SIZE OR DESCRIPTION	P/E RATIO	DISPLACEMENT			Max. Rated Speed (SPM)
		At Rated Speed	BPD per SPM		
			Eng.	Pump	
2 x 1³/₈ — 1³/₁₆	.700	381	4.54	3.15	121
2 x 1³/₈ — 1³/₈	1.000	544	4.54	4.50	121
2¹/₂ x 1³/₄ — 1¹/₂	.685	744	10.96	7.44	100
2¹/₂ x 1³/₄ — 1³/₄	1.000	1086	10.96	10.86	100
3 x 2¹/₈ — 1⁷/₈	.740	1388	21.75	15.96	87
3 x 2¹/₈ — 2¹/₈	1.000	1874	21.75	21.55	87

(Courtesy Kobe, Inc.—Subsidiary of Baker International Corp.)

Fig. 5B (10)

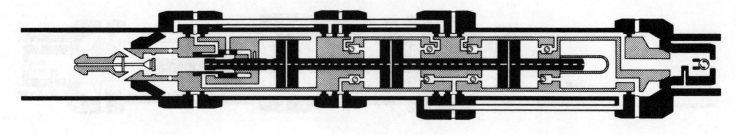

SPECIFICATIONS—
KOBE TYPE B PUMP—SINGLE ENGINE, DOUBLE PUMP END

PUMP SIZE OR DESCRIPTION	P/E RATIO	DISPLACEMENT				Max. Rated Speed (SPM)
		At Rated Speed	BPD per SPM			
			Eng.	Pump		
2 x 1⅜—1³/₁₆ x 1³/₁₆	1.380	751	4.54	6.21		121
2 x 1⅜—1⅜ x 1³/₁₆	1.680	913	4.54	7.55		121
2 x 1⅜—1⅜ x 1⅜	1.980	1076	4.54	8.90		121
2½ x 1¾—1½ x 1½	1.336	1452	10.96	14.52		100
2½ x 1¾—1¾ x 1½	1.652	1794	10.96	17.94		100
2½ x 1¾—1¾ x 1¾	1.957	2136	10.96	21.36		100
3 x 2⅛—1⅞ x 1⅞	1.454	2726	21.75	31.34		87
3 x 2⅛—2⅛ x 1⅞	1.714	3213	21.75	36.94		87
3 x 2⅛—2⅛ x 2⅛	1.974	3700	21.75	42.53		87

(Courtesy Kobe, Inc.—Subsidiary of Baker International Corp.)

Fig. 5B (11)

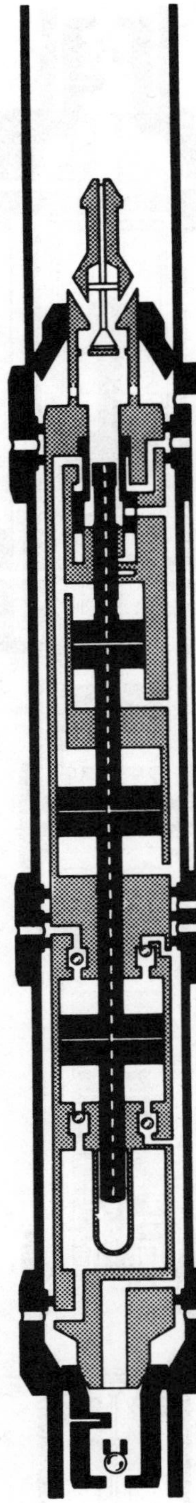

SPECIFICATIONS—
KOBE TYPE D PUMP—DOUBLE ENGINE, SINGLE PUMP END

PUMP SIZE OR DESCRIPTION	P/E RATIO	DISPLACEMENT At Rated Speed	DISPLACEMENT BPD per SPM Eng.	DISPLACEMENT BPD per SPM Pump	Max. Rated Speed (SPM)
2 x 1 3/16 x 1 3/8—1 3/16	.407	381	7.79	3.15	121
2 x 1 3/16 x 1 3/8—1 3/8	.581	544	7.79	4.50	121
2 1/2 x 1 7/16 x 1 3/4—1 1/2	.411	744	17.99	7.44	100
2 1/2 x 1 7/16 x 1 3/4—1 3/4	.608	1086	17.99	10.86	100
3 x 1 3/4 x 2 1/8—1 7/8	.449	1357	35.74	15.96	87
3 x 1 3/4 x 2 1/8—2 1/8	.606	1874	35.74	21.55	87

(Courtesy Kobe, Inc.—Subsidiary of Baker International Corp.)

Fig. 5B (12)

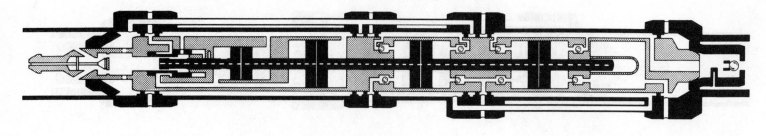

SPECIFICATIONS—
KOBE TYPE D PUMP—DOUBLE ENGINE, DOUBLE PUMP END

PUMP SIZE OR DESCRIPTION	P/E RATIO	DISPLACEMENT			Max. Rated Speed (SPM)
		At Rated Speed	BPD per SPM		
			Eng.	Pump	
2 x 1³/₁₆ x 1³/₈—1³/₁₆ x 1³/₁₆ . .	.802	751	7.79	6.21	121
2 x 1³/₁₆ x 1³/₈—1³/₈ x 1³/₁₆ . .	.976	913	7.79	7.55	121
2 x 1³/₁₆ x 1³/₈—1³/₈ x 1³/₈ ..	1.150	1076	7.79	8.90	121
2½ x 1⁷/₁₆ x 1³/₄—1½ x 1½	.813	1452	17.99	14.52	100
2½ x 1⁷/₁₆ x 1³/₄—1³/₄ x 1½	.976	1794	17.99	17.94	100
2½ x 1⁷/₁₆ x 1³/₄—1³/₄ x 1³/₄	1.196	2136	17.99	21.36	100
3 x 1³/₄ x 2⅛—1⁷/₈ x 1⁷/₈ ...	.882	2726	35.74	31.34	87
3 x 1³/₄ x 2⅛—2⅛ x 1⁷/₈ ...	1.039	3213	35.74	36.94	87
3 x 1³/₄ x 2⅛—2⅛ x 2⅛ ...	1.197	3700	35.74	42.53	87

(Courtesy Kobe, Inc.—Subsidiary of Baker International Corp.)

Fig. 5B (13)

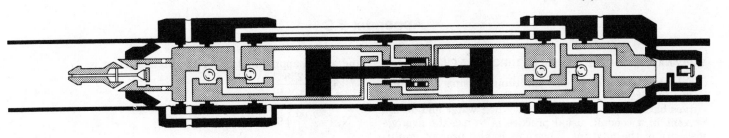

SPECIFICATIONS—
KOBE TYPE E PUMP

PUMP SIZE OR DESCRIPTION	P/E RATIO	DISPLACEMENT			Max. Rated Speed (SPM)
		At Rated Speed	BPD per SPM		
			Eng.	Pump	
2 x 1⅜..............	1.152	1311	18.35	21.15	62
2½ x 1¾..............	1.146	2397	37.35	42.81	56
3 x 2⅛..............	1.142	4015	66.32	75.76	53

(Courtesy Kobe, Inc.—Subsidiary of Baker International Corp.)

Fig. 5B (14)

Appendix 5.C

EXPLANATIONS OF FIGURES 5.C (1) THROUGH 5.C (27)

Friction curves for tubing and tubing/casing annuli are shown in Figs. 5.C(1) through 5.C(27). These are theoretical curves based on single phase flow. When water and oil are present in a conduit, usual practice is to use the average viscosity of the mixture. When gas is also present, multiphase flow correlations should be used.

For Figs. 5.C(1) through 5.C(27) the following equations were used:

$$\Delta P = 0.433\ f\frac{L}{D}\ \frac{v^2}{2g}$$

$$f = \frac{64}{N_{R_e}}\ (\text{When } N_{R_e} \text{ is less than 1200})$$

In normal oil field practice, N_{R_e} does not exceed 50,000

and a good approximation for the friction factor is:

$$f = \frac{0.236}{(N_{R_e})^{0.21}}$$

$$N_{R_e} = \left(\frac{vd}{\nu}\right) 9.290 \times 10^4$$

where:

ΔP = pressure drop, psi
f = friction factor, dimensionless
L = length of pipe, ft
v = velocity, ft/sec
d = diameter of pipe, ft
N_{R_e} = Reynolds number, dimensionless
ν = kinematic viscosity, cs

PRESSURE DROP IN PIPE
½" STANDARD PIPE (0.62" I.D.)

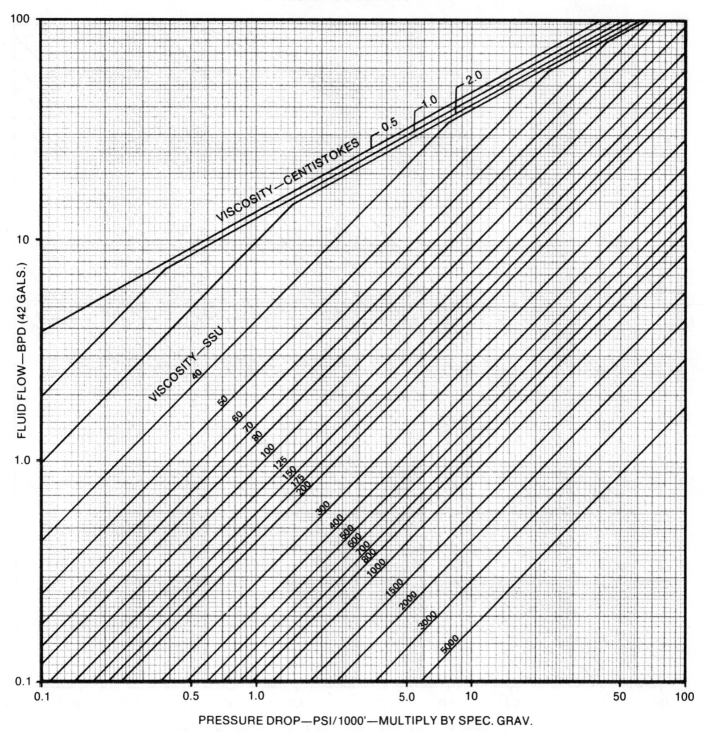

PRESSURE DROP—PSI/1000'—MULTIPLY BY SPEC. GRAV.

Fig. 5C (1)

PRESSURE DROP IN PIPE
¾" STANDARD PIPE (0.82" I.D.)

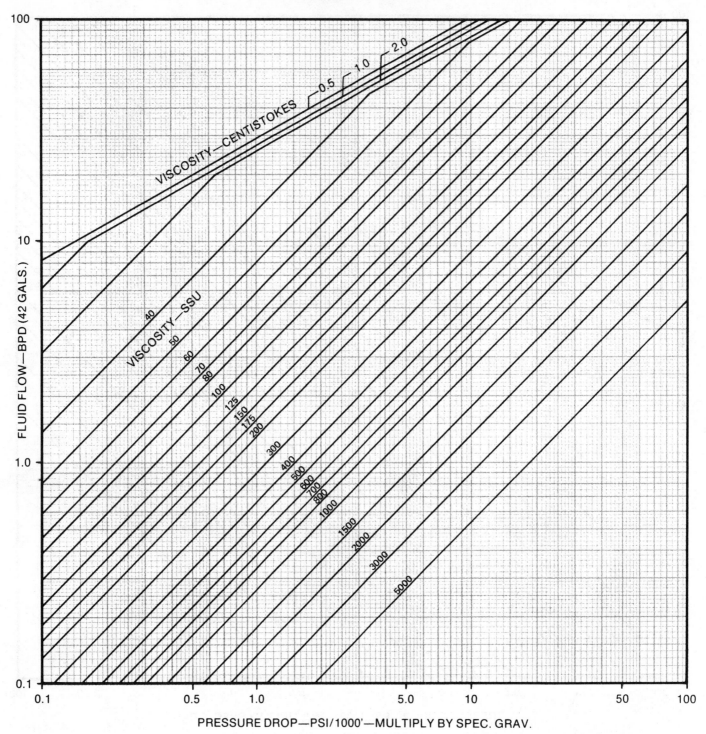

Fig. 5C (2)

PRESSURE DROP IN PIPE
1" STANDARD PIPE (1.05" I.D.)

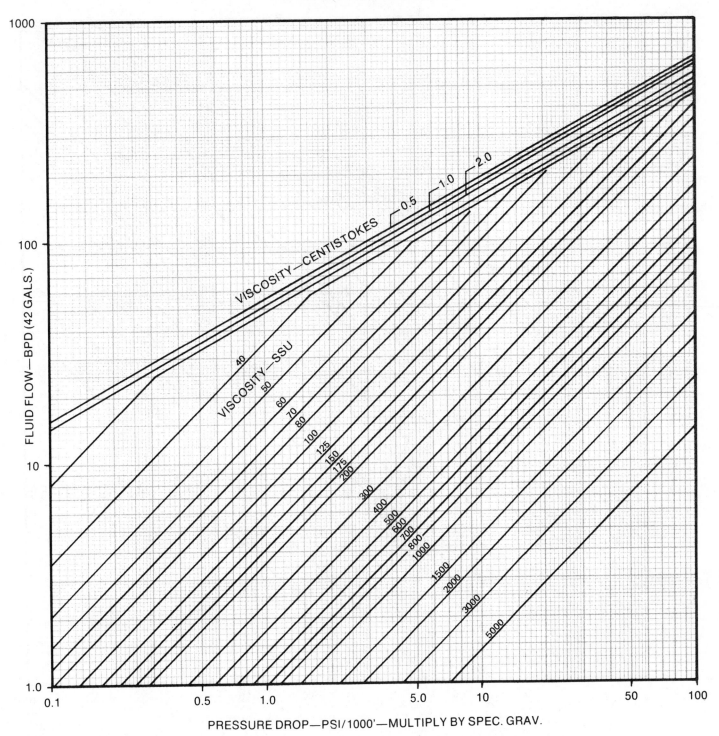

PRESSURE DROP—PSI/1000'—MULTIPLY BY SPEC. GRAV.

Fig. 5C (3)

PRESSURE DROP IN TUBING
1-¼" EU API TUBING

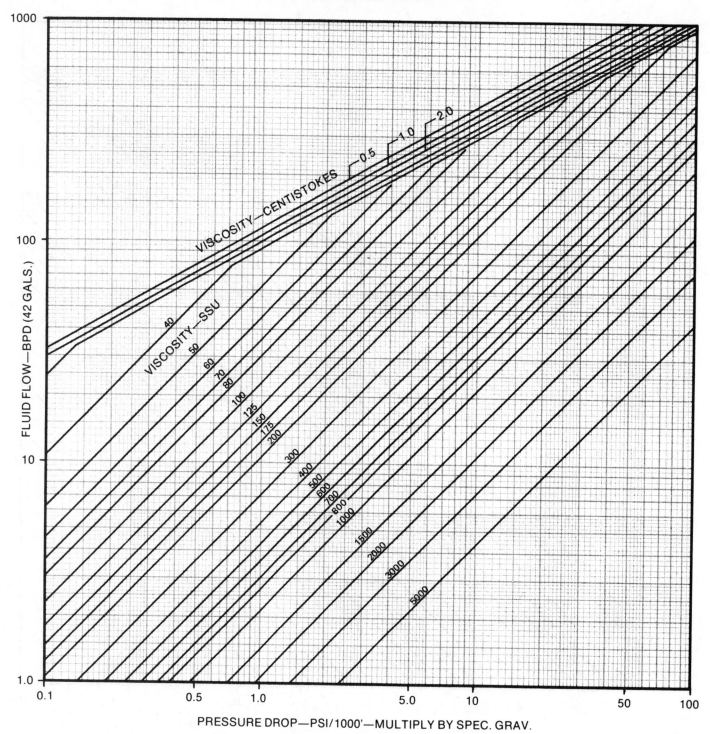

Fig. 5C (4)

PRESSURE DROP IN TUBING
1-½" EU API TUBING

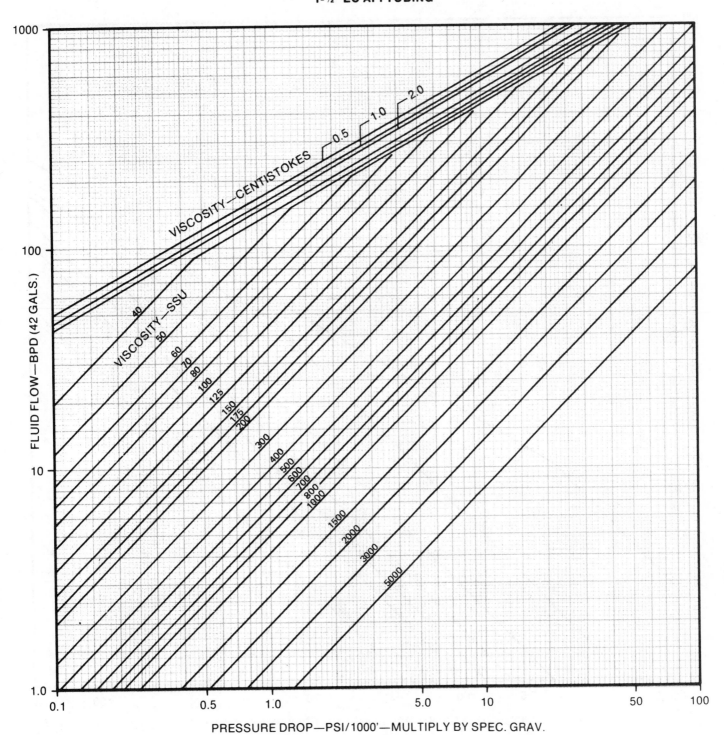

Fig. 5C (5)

PRESSURE DROP IN TUBING
2-1/16" API TUBING

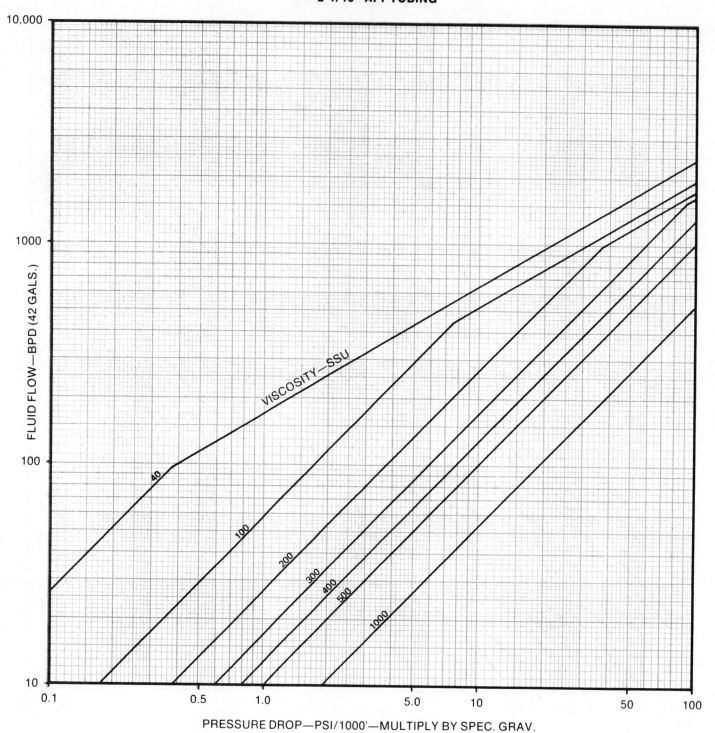

PRESSURE DROP—PSI/1000'—MULTIPLY BY SPEC. GRAV.

Fig. 5C (6)

PRESSURE DROP IN TUBING
2-⅜" EU API TUBING

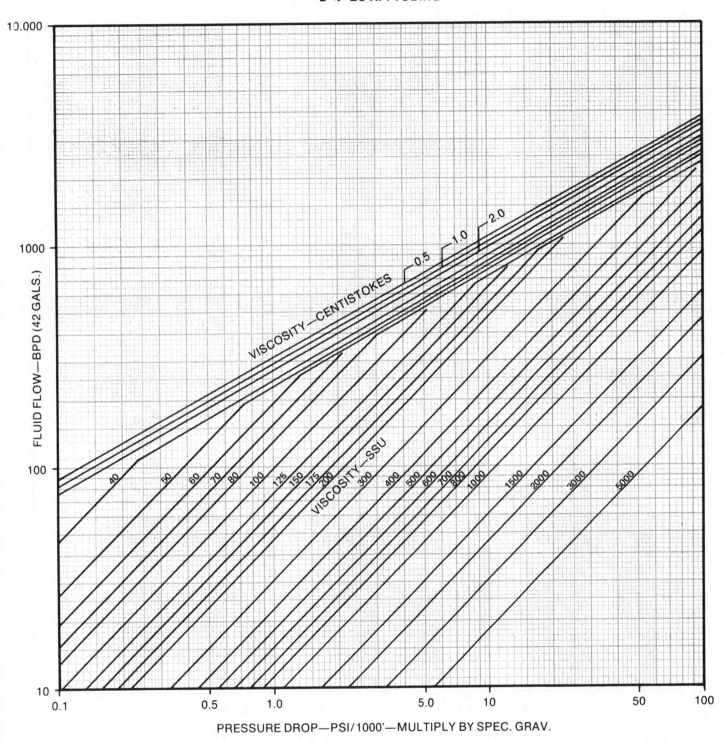

FLUID FLOW—BPD (42 GALS.)

PRESSURE DROP—PSI/1000'—MULTIPLY BY SPEC. GRAV.

Fig. 5C (7)

PRESSURE DROP IN TUBING
2-⅞" EU API TUBING

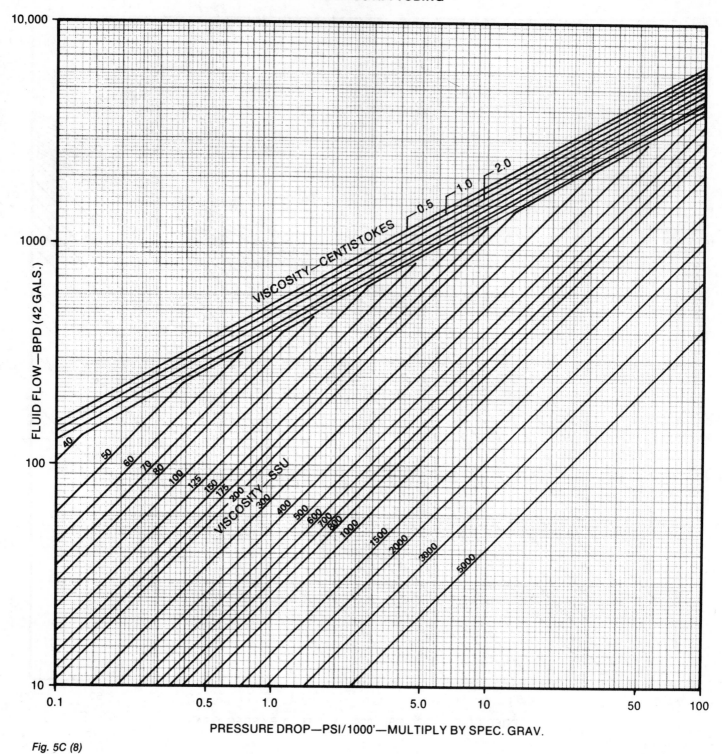

PRESSURE DROP—PSI/1000'—MULTIPLY BY SPEC. GRAV.

Fig. 5C (8)

PRESSURE DROP IN TUBING
3-½" EU API TUBING

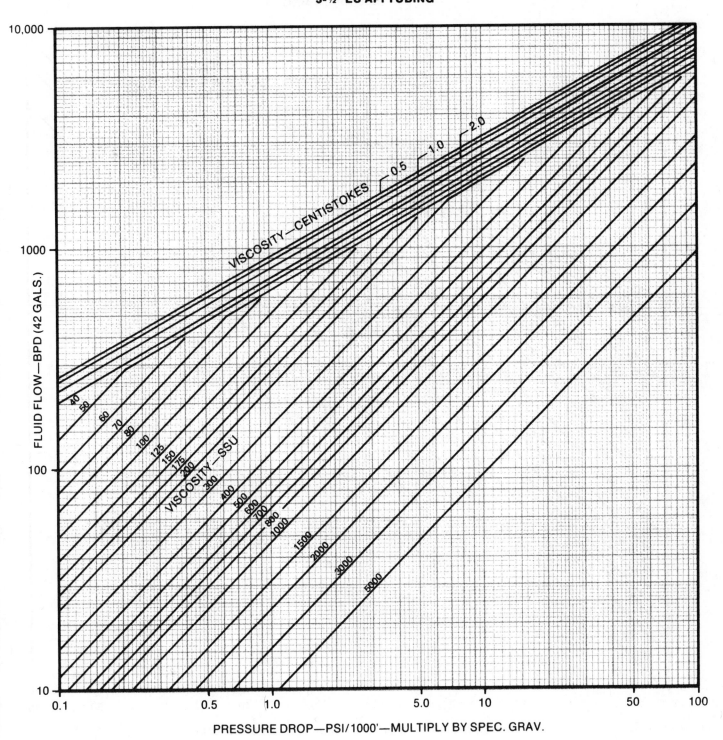

FLUID FLOW—BPD (42 GALS.)

PRESSURE DROP—PSI/1000'—MULTIPLY BY SPEC. GRAV.

VISCOSITY—CENTISTOKES

VISCOSITY—SSU

Fig. 5C (9)

PRESSURE DROP IN TUBING
4-½" EU API TUBING

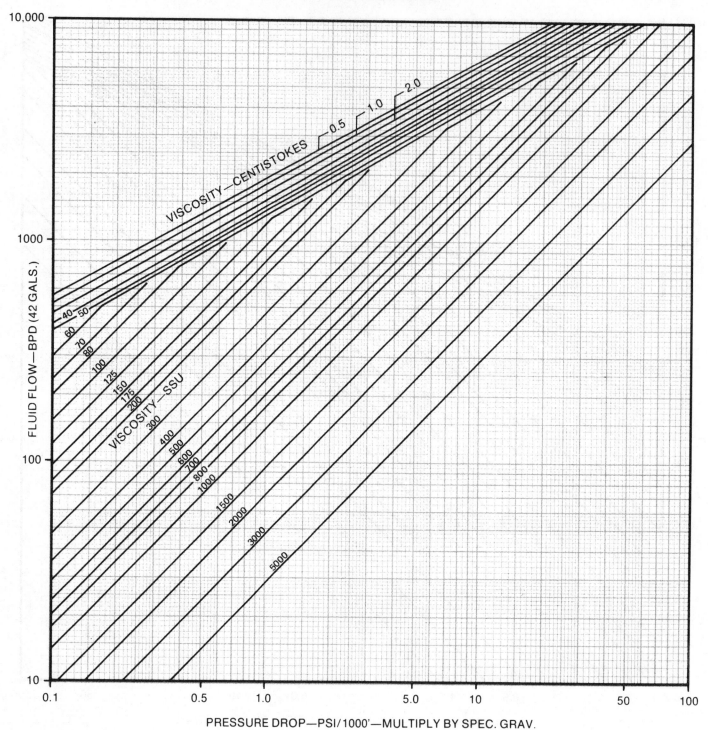

PRESSURE DROP—PSI/1000'—MULTIPLY BY SPEC. GRAV.

Fig. 5C (10)

PRESSURE DROP IN TUBING
FLOW BETWEEN 2-⅜" EU TUBING & ¾" (1.050" O.D.) PIPE

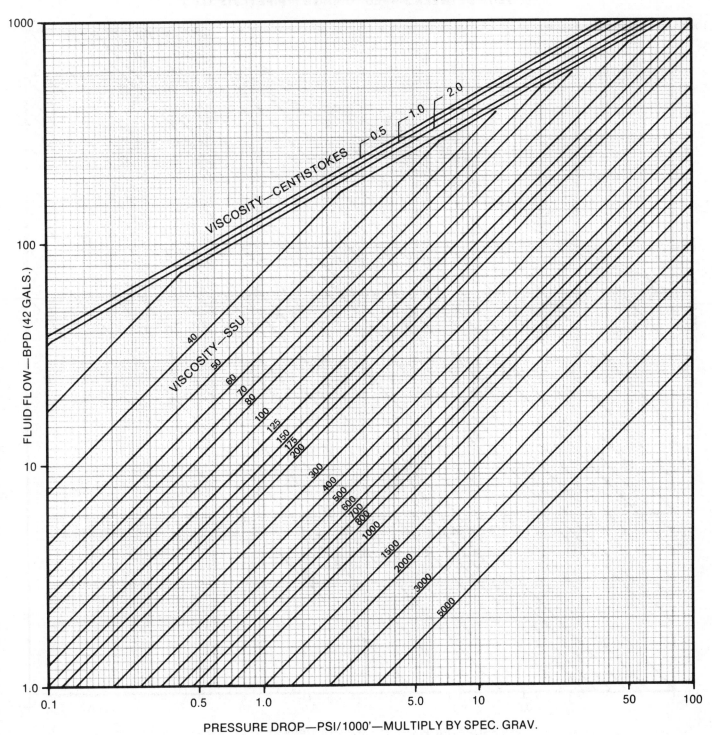

FLUID FLOW—BPD (42 GALS.)

PRESSURE DROP—PSI/1000'—MULTIPLY BY SPEC. GRAV.

Fig. 5C (11)

PRESSURE DROP IN TUBING
FLOW BETWEEN 2-⅞" EU TUBING & 1" PIPE (1.315" O.D.)

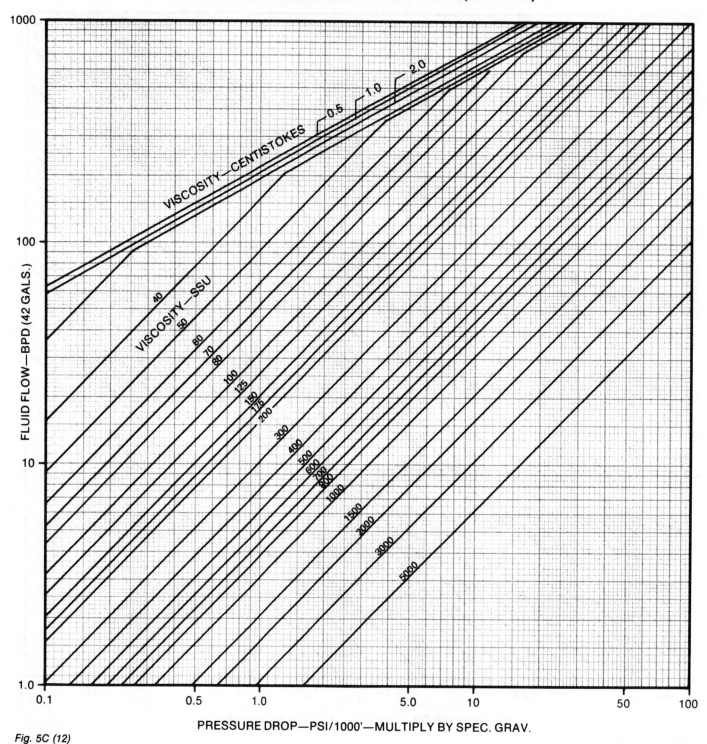

PRESSURE DROP—PSI/1000'—MULTIPLY BY SPEC. GRAV.

Fig. 5C (12)

PRESSURE DROP IN TUBING
FLOW BETWEEN 2-⅞" & 1-¼" EU API TUBING

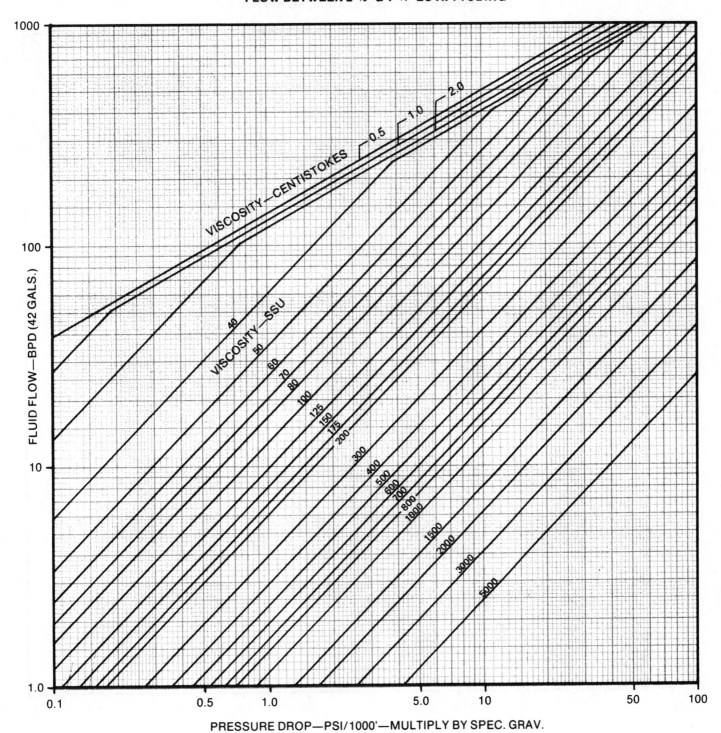

Fig. 5C (13)

PRESSURE DROP IN TUBING
FLOW BETWEEN 3-½" & 1-¼" EU API TUBING

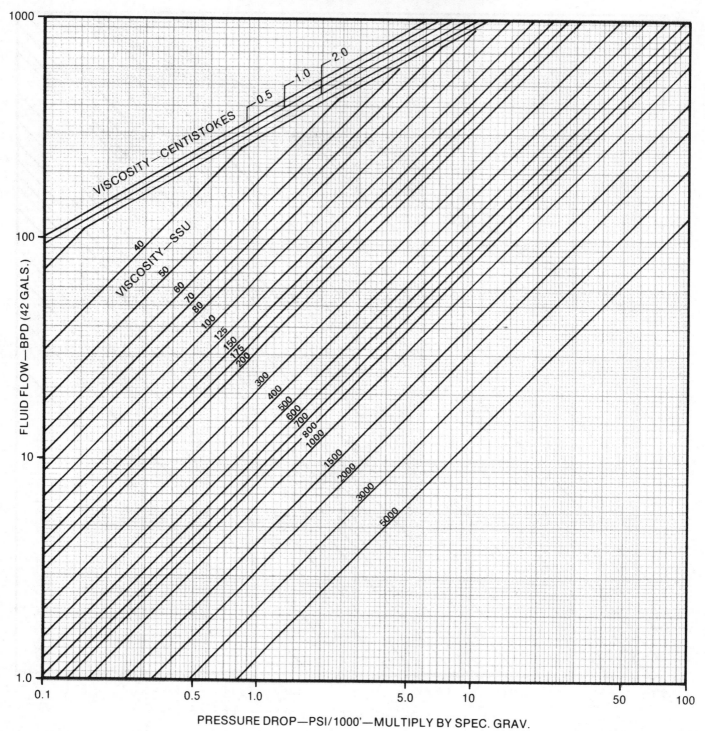

PRESSURE DROP—PSI/1000'—MULTIPLY BY SPEC. GRAV.

Fig. 5C (14)

PRESSURE DROP IN TUBING
FLOW BETWEEN 3-½" & 1-½" EU API TUBING

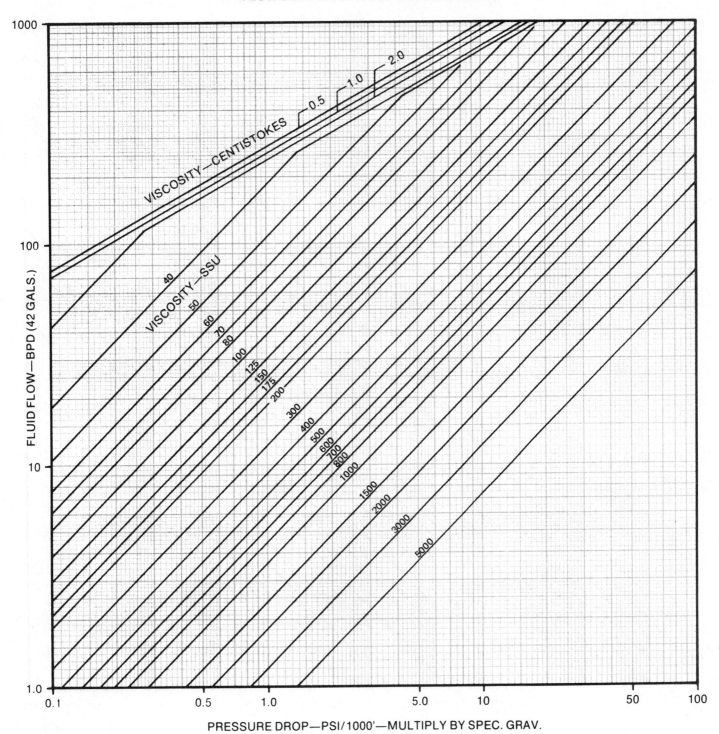

PRESSURE DROP—PSI/1000'—MULTIPLY BY SPEC. GRAV.

Fig. 5C (15)

PRESSURE DROP IN TUBING
FLOW BETWEEN 3-½" & 2-⅜" EU API TUBING

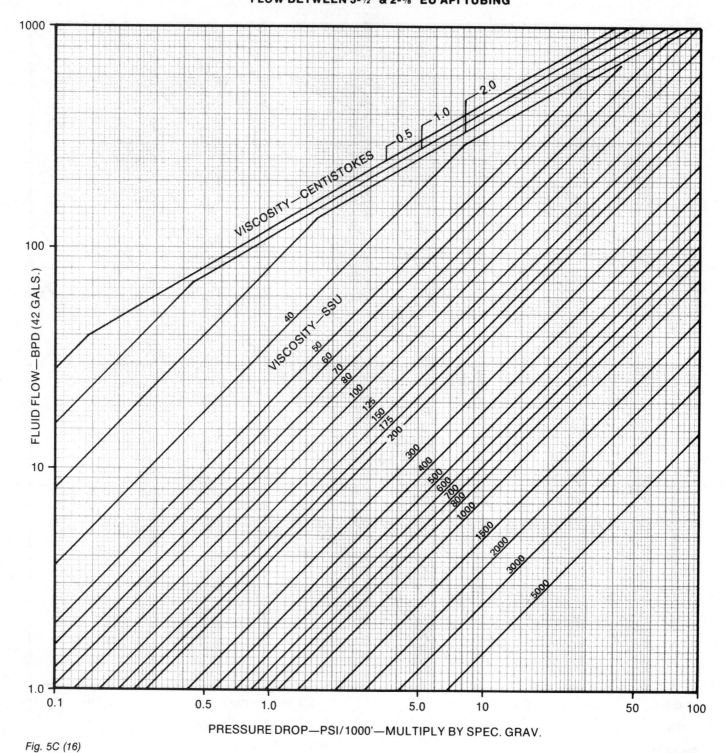

Fig. 5C (16)

PRESSURE DROP IN TUBING
FLOW BETWEEN 4-½" & 2-⅜" EU API TUBING

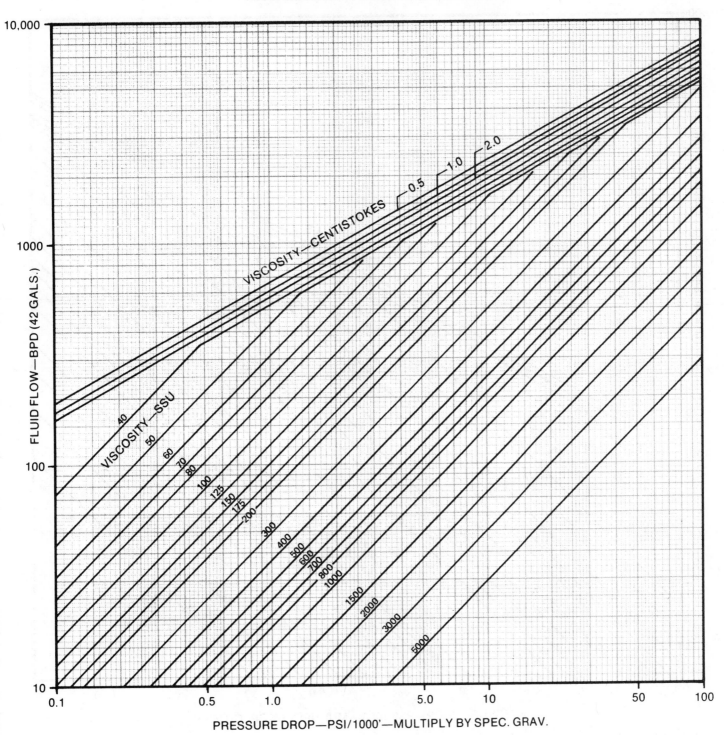

Fig. 5C (17)

PRESSURE DROP IN TUBING
FLOW BETWEEN 4-½" & 2-⅞" EU API TUBING

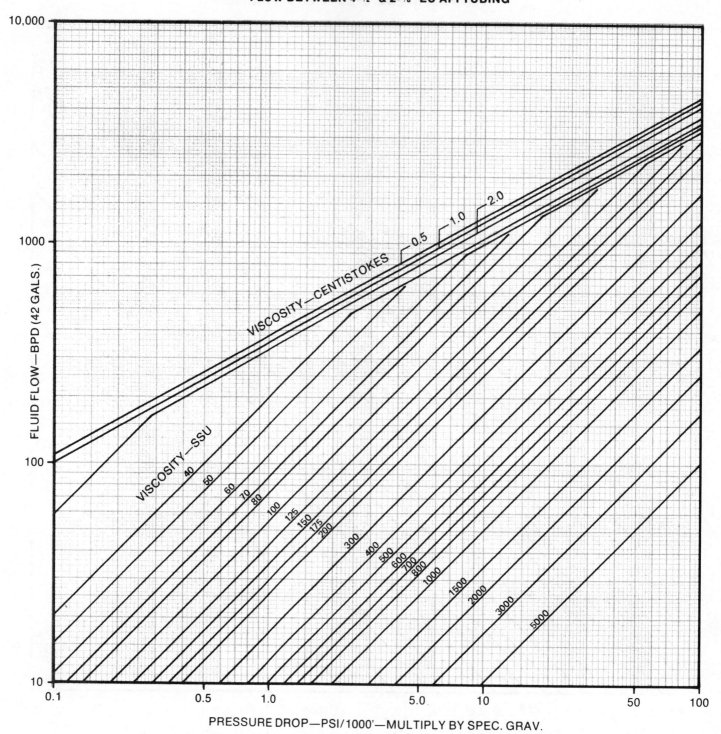

PRESSURE DROP—PSI/1000'—MULTIPLY BY SPEC. GRAV.

Fig. 5C (18)

PRESSURE DROP IN TUBING
FLOW BETWEEN 5-½", 20# CSG. & 2-⅜" O.D. EU TUBING

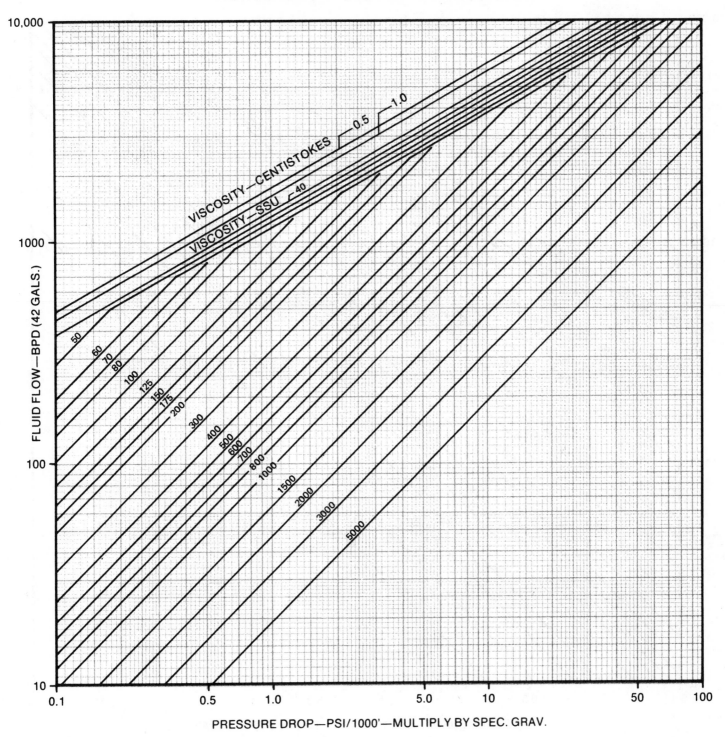

Fig. 5C (19)

PRESSURE DROP IN TUBING
FLOW BETWEEN 5-½", 14# CSG. & 2-⅞" O.D. EU TUBING

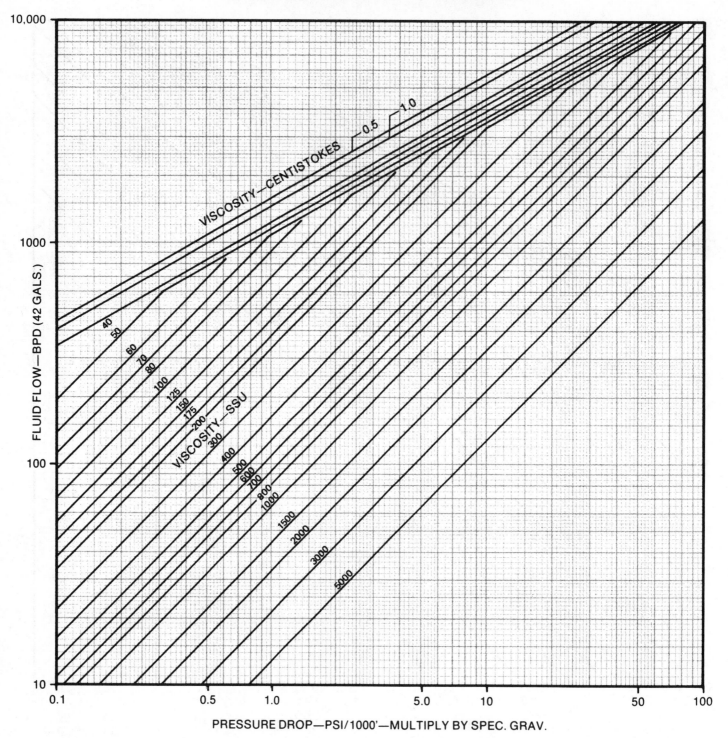

PRESSURE DROP—PSI/1000'—MULTIPLY BY SPEC. GRAV.

Fig. 5C (20)

PRESSURE DROP IN TUBING
FLOW BETWEEN 6-⅝", 28# CSG. & 3-½" EU TUBING

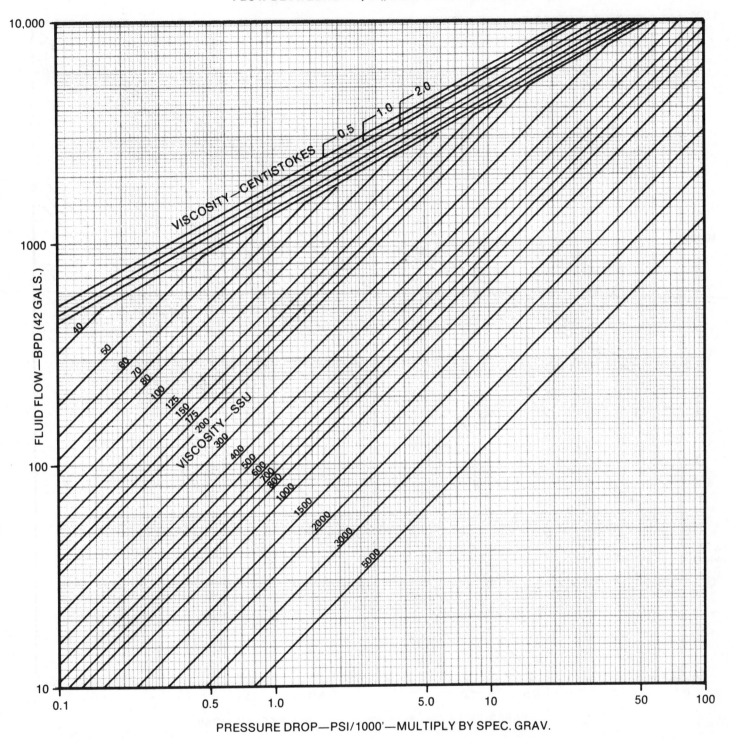

PRESSURE DROP—PSI/1000'—MULTIPLY BY SPEC. GRAV.

Fig. 5C (21)

PRESSURE DROP IN TUBING
FLOW BETWEEN 6-⅝", 28# CSG. & 4-½" EU TUBING

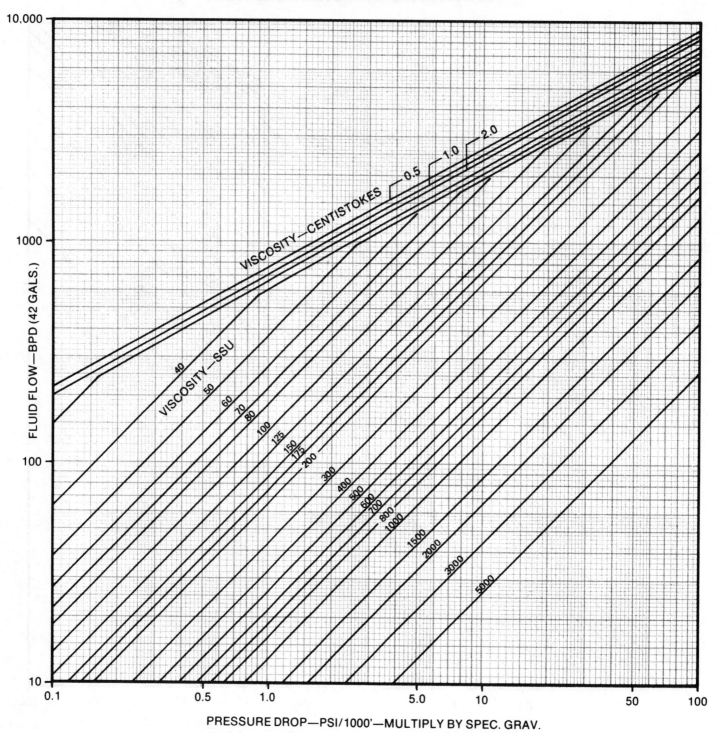

PRESSURE DROP—PSI/1000'—MULTIPLY BY SPEC. GRAV.

Fig. 5C (22)

PRESSURE DROP IN TUBING
FLOW BETWEEN 7", 26# CSG. & 2-⅞" O.D. EU TUBING

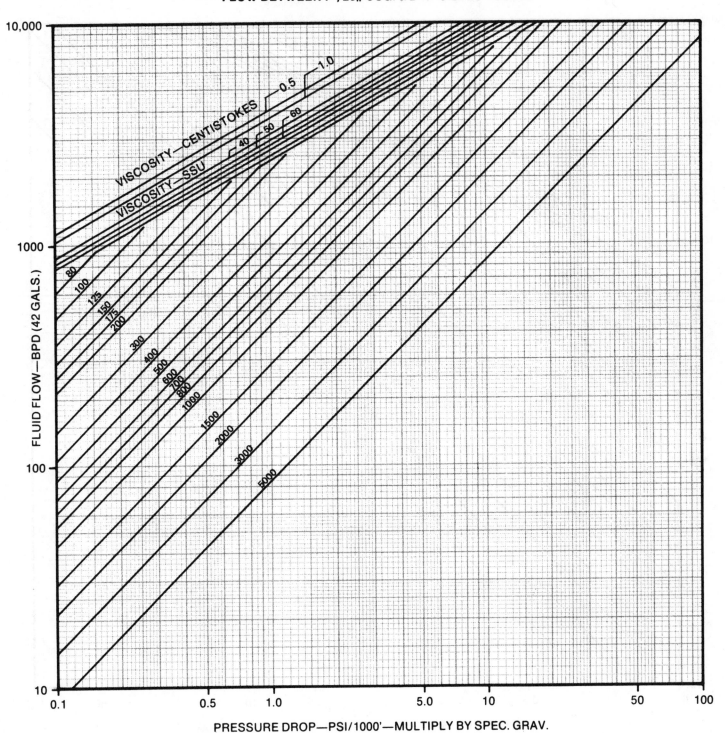

Fig. 5C (23)

PRESSURE DROP IN TUBING
FLOW BETWEEN 7", 26# CSG. & 3-½" O.D. EU TUBING

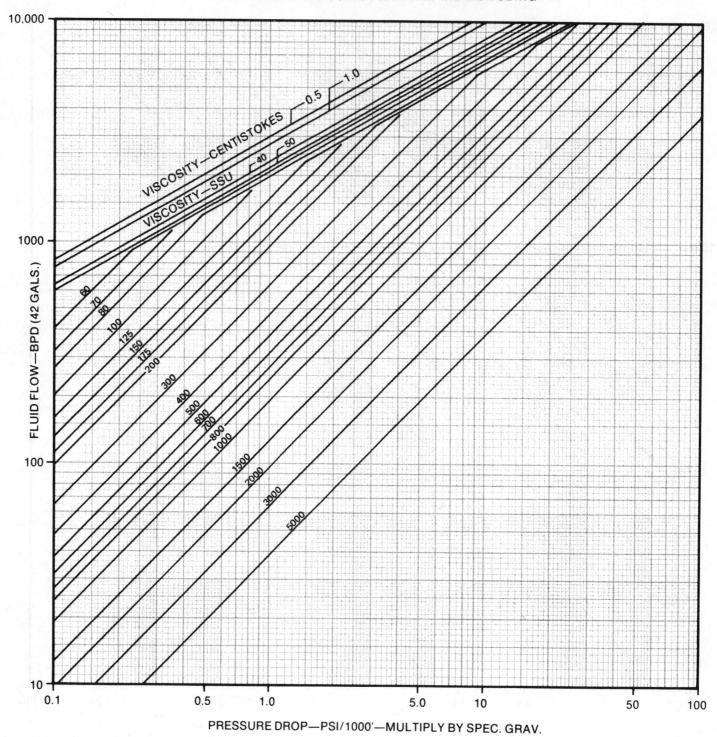

PRESSURE DROP—PSI/1000'—MULTIPLY BY SPEC. GRAV.

Fig. 5C (24)

PRESSURE DROP IN TUBING
FLOW BETWEEN 7", 30# CSG. & 4-½" EU TUBING

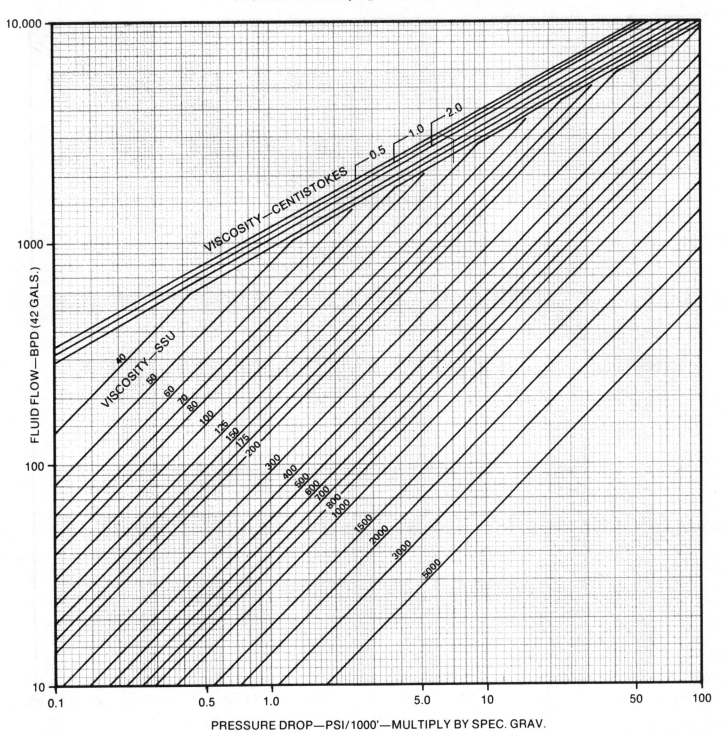

PRESSURE DROP—PSI/1000'—MULTIPLY BY SPEC. GRAV.

Fig. 5C (25)

PRESSURE DROP IN TUBING
FLOW BETWEEN 9-⅝", 43-½# CSG. & 3-½" O.D. EU TUBING

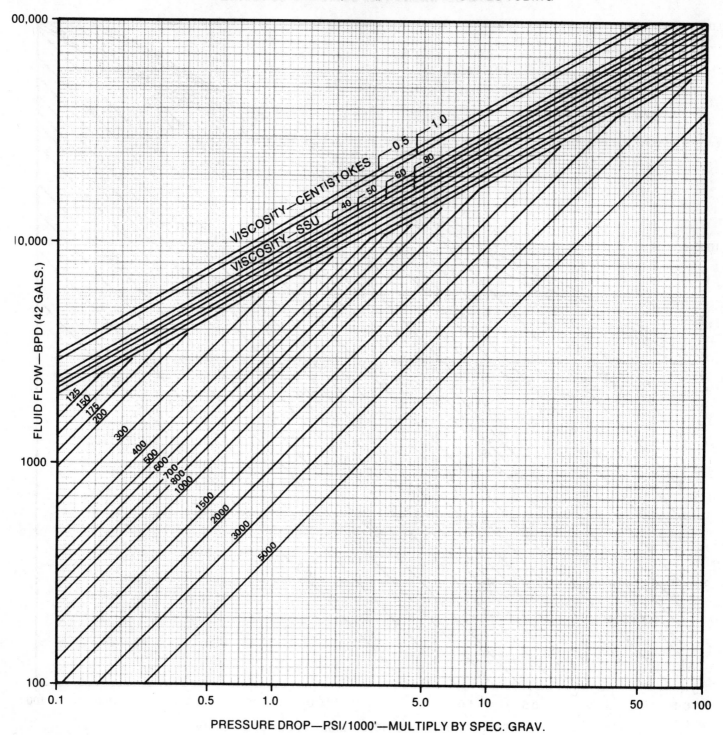

PRESSURE DROP—PSI/1000'—MULTIPLY BY SPEC. GRAV.

Fig. 5C (26)

PRESSURE DROP IN TUBING
FLOW BETWEEN 9-⅝", 43-½# CSG. & 4-½" O.D. EU TUBING

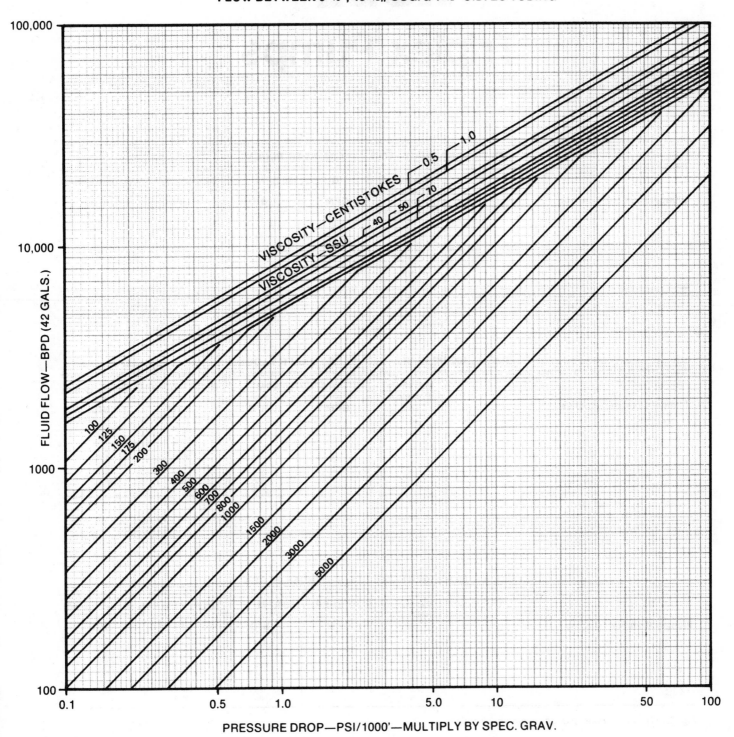

PRESSURE DROP—PSI/1000'—MULTIPLY BY SPEC. GRAV.

Fig. 5C (27)

KOBE
Triplex Pumps

SIZE 2—23 HP INPUT—5000 PSI CYLINDER BLOCK

Plunger Diameter—Inches	Maximum Pressure psi	DISPL./100 RPM		DISPL./450 RPM (Rated Speed)	
		GPM	B/D	GPM	B/D
5/8	5000	1.20	41.0	5.40	184
11/16	4710	1.45	49.6	6.53	223
3/4	3960	1.72	59.0	7.74	265
13/16	3380	2.02	69.2	9.09	311
7/8	2910	2.34	80.3	10.53	361
15/16	2540	2.69	92.2	12.10	415
1	2230	3.06	105.	13.77	472
1-1/8	1760	3.87	133.	17.42	597
1-1/4	1430	4.78	164.	21.51	737

SIZE 3—100 HP INPUT—5000 PSI CYLINDER BLOCK

Plunger Diameter—Inches	Maximum Pressure psi	DISPL./100 RPM		DISPL./500 RPM (Rated Speed)	
		GPM	B/D	GPM	B/D
3/4	5000	2.30	79	11.5	395
7/8	5000	3.12	107	15.6	535
1	5000	4.08	140	20.4	700
1-1/8	5000	5.16	177	25.8	885
1-1/4	4560	6.38	219	32.0	1100
1-3/8	3880	7.72	265	37.7	1330
1-1/2	3180	9.18	315	46.0	1580
1-5/8	2720	10.8	369	53.8	1850
1-3/4	2340	12.5	428		

*Above maximum rated capacity of 1850 BPD.

SIZE 4—180 HP INPUT—5000 PSI CYLINDER BLOCK

Plunger Diameter—Inches	Maximum Pressure psi	DISPL./100 RPM		DISPL./450 RPM (Rated Speed)	
		GPM	B/D	GPM	B/D
1-1/2	5000	11.5	393	51.8	1768
1-5/8	4630	13.5	462	60.8	2079
1-3/4	4000	15.6	535	70.2	2408
1-7/8	3480	17.9	615	80.6	2768
2	3060	20.4	700	91.8	3150
2-1/8	2720	23.0	790	104.0	3555
2-1/4	2420	25.8	885	116.0	3983
2-3/8	2170	28.8	986	130.0	4437

Fig. 5D (1)   Kobe triplex plunger pump specifications

NATIONAL J-30
Triplex Plunger Pump

Continuous duty performance

PUMP SIZE	PLNGR DIA. IN.	MAX PRESS. PSI	GPM per RPM	BPD per RPM	200 RPM* 12 HP		250 RPM 15 HP		300 RPM 18 HP		350 RPM 21 HP		400 RPM 24 HP		450 RPM 27 HP		500 RPM 30 HP	
					GPM	B/D	GPM	B/D	GPM	B/D	GPM	B/D	GPM	B/D	GPM	B/D	GPM	B/D
J-30-L LOW PRESSURE	2¼	900	.1033	3.540	20.7	710	25.8	885	31.0	1065	36.2	1240	41.3	1415	46.5	1595	51.6	1770
	2	1150	.0816	2.798	16.3	560	20.4	700	24.5	840	28.6	980	32.6	1120	36.7	1260	40.8	1400
	1¾	1500	.0625	2.142	12.5	430	15.6	535	18.8	645	21.9	750	25.0	855	28.1	965	31.2	1070
	1½	2000	.0459	1.573	9.2	315	11.5	395	13.8	475	16.1	550	18.4	630	20.7	710	22.9	78?
J-30-H HIGH PRESSURE	1½	2000	.0459	1.573	9.2	315	11.5	395	13.8	475	16.1	550	18.4	630	20.7	710	22.9	78?
	1⅜†	2500	.0385	1.322	7.8	265	9.7	330	11.6	395	13.5	465	15.4	530	17.4	595	19.3	660
	1¼	3000	.0319	1.093	6.4	220	8.0	275	9.6	330	11.2	385	12.8	440	14.4	490	15.9	54?
	1⅛†	3750	.0258	.885	5.2	175	6.5	220	7.8	265	9.1	310	10.3	355	11.6	400	12.9	44?
	1	5000	.0204	.699	4.1	140	5.1	175	6.1	210	7.2	245	8.2	280	9.2	315	10.2	35?
	15⁄16†	5000	.0179	.614	3.6	125	4.5	155	5.4	185	6.3	215	7.2	245	8.1	275	9.0	31?

†These sizes on application.
*Below 200 rpm, optional auxiliary lubrication system required.
Input HP—Based on 90% Mechanical Efficiency.
Volume—Based on 100% Volumetric Efficiency.

NATIONAL J-60
Triplex Plunger Pump

Continuous duty performance

PUMP SIZE	PLNGR DIA. IN.	MAX PRESS. PSI	GPM per RPM	BPD per RPM	200 RPM* 24 HP		250 RPM 30 HP		300 RPM 36 HP		350 RPM 42 HP		400 RPM 48 HP		450 RPM 54 HP		500 RPM 60 HP	
					GPM	B/D	GPM	B/D	GPM	B/D	GPM	B/D	GPM	B/D	GPM	B/D	GPM	B/D
J-60-L LOW PRESSURE	3	725	.2754	9.442	55	1890	69	2360	83	2835	96	3304	110	3775	124	4250	138	4720
	2¾†	850	.2314	7.934	46	1585	58	1985	69	2380	81	2775	93	3175	104	3570	116	3965
	2½	1000	.1912	6.557	38	1310	48	1640	57	1965	67	2295	76	2625	86	2950	96	3280
	2¼†	1275	.1549	5.311	31	1060	39	1330	46	1595	54	1860	62	2125	70	2390	77	2655
	2	1600	.1224	4.197	24	840	31	1050	37	1260	43	1470	49	1680	55	1890	61	2100
J-60-M MEDIUM PRESSURE	2	1600	.1224	4.197	24	840	31	1050	37	1260	43	1470	49	1680	55	1890	61	2100
	1¾	2050	.0937	3.213	19	645	23	805	28	965	33	1125	37	1285	42	1445	47	1605
	1⅝†	2450	.0808	2.770	16	555	20	690	24	830	28	970	32	1110	36	1245	40	1385
	1½	2900	.0688	2.360	14	470	17	590	21	710	24	825	28	945	31	1060	34	1180
	1⅜	3200	.0579	1.983	12	395	14	495	17	595	20	695	23	795	26	890	29	990
J-60-H HIGH PRESSURE	1⅜	3200	.0579	1.983	12	395	14	495	17	595	20	695	23	795	26	890	29	990
	1¼	3900	.0478	1.639	10	330	12	410	14	490	17	575	19	655	22	735	24	820
	1⅛	4800	.0387	1.328	8	265	9.7	345	12	400	14	485	15	530	17	620	19	665
	1	5000	.0306	1.049	6	210	7.7	260	9	315	11	370	12	420	14	475	15	525

†These sizes on application.
*Below 200 rpm, optional auxiliary lubrication system required.
Input HP—Based on 90% Mechanical Efficiency.
Volume—Based on 100% Volumetric Efficiency.

Fig. 5D (2)   National triplex plunger pump specifications

NATIONAL J-100
Triplex Plunger Pump

Continuous duty performance

PUMP SIZE	PLNGR DIA. IN.	MAX PRESS. PSI	GPM per RPM	BPD per RPM	VOLUME—INPUT HORSEPOWER DATA 150 RPM* 34 HP		200 RPM 44 HP		250 RPM 56 HP		300 RPM 67 HP		350 RPM 78 HP		400 RPM 89 HP		450 RPM 100 HP	
					GPM	B/D	GPM	B/D	GPM	B/D	GPM	B/D	GPM	B/D	GPM	B/D	GPM	B/D
J-100-L LOW PRESSURE	3⅜	750	.4647	15.934	70	2390	93	3185	116	3985	139	4780	163	5575	186	6375	209	7170
	3⅛†	850	.3984	13.660	60	2050	80	2730	100	3415	120	4100	139	4780	159	5465	179	6145
	2⅞	1000	.3372	11.562	51	1735	67	2310	84	2890	101	3470	118	4045	135	4625	152	5205
	2⅝†	1200	.2811	9.639	42	1445	56	1930	70	2410	84	2890	98	3375	112	3855	127	4335
	2⅜	1500	.2301	7.890	35	1185	46	1580	58	1970	69	2365	81	2760	92	3155	104	3550
J-100-M MEDIUM PRESSURE	2⅛	2000	.1842	6.316	28	950	37	1260	46	1575	55	1895	65	2210	74	2525	83	2840
	2†	2200	.1632	5.595	24	840	33	1120	41	1400	49	1675	57	1960	65	2240	73	2515
	1⅞	2500	.1434	4.918	22	740	29	980	36	1225	43	1475	50	1720	57	1965	65	2210
	1¾†	2850	.1250	4.284	19	645	25	855	31	1070	37	1280	44	1500	50	1710	56	1925
	1⅝	3200	.1077	3.694	16	555	22	735	27	920	32	1105	38	1295	43	1480	48	1660
J-100-H HIGH PRESSURE	1¾†	2850	.1250	4.284	19	645	25	855	31	1070	37	1280	44	1500	50	1710	56	1925
	1⅝	3200	.1077	3.694	16	555	22	735	27	920	32	1105	38	1295	43	1480	48	1660
	1½	3750	.0918	3.147	13.8	470	18.3	625	22.9	785	27.5	945	32.1	1100	36.7	1260	41.3	1415
	1⅜	4450	.0771	2.644	11.6	395	15.4	530	19.2	660	23.1	790	26.9	920	30.8	1055	34.7	1190
	1¼	5000	.0637	2.186	9.6	330	12.7	435	15.9	545	19.1	655	22.3	765	25.5	875	28.6	980

†These sizes on application.
*Below 200 rpm, optional auxiliary lubrication system required.
Input HP—Based on 90% Mechanical Efficiency.
Volume—Based on 100% Volumetric Efficiency.

NATIONAL J-150 Triplex Plunger Pump

Continuous duty performance

PUMP SIZE	PLNGR DIA. IN.	MAX PRESS. PSI	GPM per RPM	BDP per RPM	100 RPM* 37.5 HP		150 RPM* 56 HP		200 RPM 75 HP		250 RPM 94 HP		300 RPM 112.5 HP		350 RPM 131 HP		400 RPM 150 HP	
					GPM	B/D	GPM	B/D	GPM	B/D	GPM	B/D	GPM	B/D	GPM	B/D	GPM	B/D
J-150-L LOW PRESSURE	4†	700	.816	27.976	82	2800	122	4195	163	5595	204	6995	245	8395	286	9790	—	—
	3¾	800	.717	24.590	72	2460	108	3690	143	4915	179	6145	215	7380	251	8605	287	9835
	3½	950	.625	21.420	62	2145	94	3215	125	4280	156	5355	187	6425	219	7495	250	8570
	3¼	1100	.539	18.469	54	1850	81	2770	108	3695	135	4620	162	5540	189	6465	215	7390
	3†	1250	.459	15.737	46	1575	69	2360	92	3150	115	3935	138	4720	161	5510	184	6295
	2¾	1500	.386	13.223	39	1325	58	1985	77	2645	96	3305	116	3965	135	4630	154	5290
J-150-M MEDIUM PRESSURE	2¾	1500	.386	13.223	39	1325	58	1985	77	2645	96	3305	116	3965	135	4630	154	5290
	2⅝†	1650	.351	12.024	35	1205	53	1805	70	2410	88	3010	105	3615	123	4215	141	4820
	2½	1800	.319	10.928	32	1095	48	1640	64	2185	80	2735	96	3280	112	3825	127	4370
	2⅜†	2000	.288	9.863	29	985	43	1480	58	1975	72	2465	86	2960	101	3450	115	3945
	2¼	2250	.258	8.852	26	885	39	1325	52	1770	65	2215	77	2655	90	3100	103	3540
	2⅛†	2500	.230	7.896	23	790	35	1185	46	1580	58	1975	69	2370	81	2765	92	3160
	2	2850	.204	6.994	20	700	31	1050	41	1400	51	1750	61	2100	71	2450	82	2800
J-150-H HIGH PRESSURE	2	2850	.204	6.994	20	700	31	1050	41	1400	51	1750	61	2100	71	2450	82	2800
	1⅞	3250	.179	6.147	18	615	27	920	36	1230	45	1535	54	1845	63	2150	72	2460
	1¾	3700	.156	5.355	16	535	23	800	31	1070	39	1340	47	1610	55	1875	62	2140
	1⅝	4300	.1347	4.617	14	460	20	695	27	925	34	1155	40	1385	47	1615	54	1845
	1½	5000	.115	3.934	11	395	17	590	23	785	29	985	34	1180	40	1375	46	1575

†These sizes on application.
*Below 200 rpm, optional auxiliary lubrication system required.
Input HP—Based on 90% Mechanical Efficiency.
Volume—Based on 100% Volumetric Efficiency.

3" PIONEER CYCLONE

Effect of Feed Nozzle Size and Vortex Finder on Capacity:

Feed Nozzle I. D.	Vortex Type & I. D.	Apex Size	Pressure Drop	Feed Volume G. P. M.	B. P. D.
.500"	Standard .75" ID	.625"	30 PSI	18 GPM	612
			40 PSI	21 GPM	714
			50 PSI	23 GPM	782
.500"	Spiral .75" ID	.625"	30 PSI	20 GPM	680
			40 PSI	24 GPM	816
			50 PSI	26 GPM	884
.500"	Spiral 1.00" ID	.625"	30 PSI	24 GPM	816
			40 PSI	27 GPM	918
			50 PSI	30 GPM	936
.500"	Spiral 1.25" ID	.625"	30 PSI	25 GPM	850
			40 PSI	29 GPM	838
			50 PSI	33 GPM	1,122
.600"	Standard .75" ID	.625"	30 PSI	23 GPM	782
			40 PSI	26 GPM	884
			50 PSI	28 GPM	952
.600"	Spiral .75" ID	.625"	30 PSI	25 GPM	850
			40 PSI	28 GPM	952
			50 PSI	32 GPM	1,088
.600"	Spiral 1.00" ID	.625"	30 PSI	30 GPM	1,020
			40 PSI	35 GPM	1,190
			50 PSI	38 GPM	1,292
.600"	Spiral 1.25" ID	.625"	30 PSI	34 GPM	1,156
			40 PSI	40 GPM	1,360
			50 PSI	44 GPM	1,496

Note: .75" Vortex should be used if possible.

Fig. 5E (1)   Specifications for cyclones

4" PIONEER CYCLONE

Effect of Feed Nozzle Size and Vortex Finder on Capacity:

Feed Nozzle I. D.	Vortex I. D.	Apex Size	Pressure Drop	Feed Volume G. P. M.	B. P. D
.500"	1-1/2" Standard	.688"	20 PSI	26 GPM	857
			30 PSI	32 GPM	1,074
			40 PSI	36 GPM	1,226
.600"	1-1/2" Standard	.688"	20 PSI	33 GPM	1,131
			30 PSI	39 GPM	1,334
			40 PSI	45 GPM	1,532
.700"	1-1/2" Standard	.688"	20 PSI	36 GPM	1,227
			30 PSI	43 GPM	1,467
			40 PSI	49 GPM	1,651
.800"	1-1/2" Standard	.688"	20 PSI	37 GPM	1,255
			30 PSI	44 GPM	1,499
			40 PSI	50 GPM	1,717
.500"	1-1/2" Spiral	.688"	20 PSI	27 GPM	924
			30 PSI	36 GPM	1,234
			40 PSI	41 GPM	1,389
.600"	1-1/2" Spiral	.688"	20 PSI	37 GPM	1,260
			30 PSI	47 GPM	1,608
			40 PSI	52 GPM	1,776
.700"	1-1/2" Spiral	.688"	20 PSI	42 GPM	1,430
			30 PSI	53 GPM	1,783
			40 PSI	58 GPM	1,975
.800"	1-1/2" Spiral	.688"	20 PSI	44 GPM	1,474
			30 PSI	54 GPM	1,819
			40 PSI	60 GPM	2,039

Fig. 5E (2)   Specifications for cyclones

Appendix 5F(1) Kobe Solo Unit

GENERAL DESCRIPTION

The Kobe Solo Unit is basically a large three-phase separator. It has sufficient fluid capacity to operate a hydraulic subsurface pump and perform the necessary procedures of circulating the pump in and out of the well, trouble-shooting, etc. The external valves and plumbing are arranged so either water or oil can be used for power fluid. This is accomplished by selecting the appropriate discharge level for expelling the produced fluid (oil and water) which will maintain the vessel nearly full of the fluid used for power fluid.

When operating on oil power fluid, production fluid is discharged from the bottom of the vessel through valve 4, thereby removing all the heavier produced water with produced oil, maintaining the vessel full of oil. Power oil is withdrawn from a point just below the float controlled fluid level through valve 2.

When operating on water power fluid, production fluid is withdrawn from a point just below the fluid level through valve 1, thereby maintaining the vessel nearly full of the (heavier) water. Power water is withdrawn from the bottom of the vessel through valve 6 (with valve 4 closed).

The power fluid is not delivered directly into the suction of the triplex. It is first picked up by a circulating pump and passed through cyclone centrifugal separators which remove the solids and abrasives. These solids are directed into the flowline by controlling a small amount of the well's production out the cyclone(s) underflow. A portion of the cleaned fluid is introduced into the triplex suction and the rest is recirculated back through the cyclones. The ratio of cleaned fluid recirculated to that supplied to the triplex suction is designed to be between two and three to one. This "multiple circulation" of the cyclones increases their effectiveness in removing offensive material from the power fluid.

KOBE SOLO UNIT START UP AND OPERATING PROCEDURES

1. Fill the entire system (vessel, tubing(s) and casing) with power fluid.
2. Set all valves for the power fluid used, as shown in the table.
3. Open triplex and circulating pump intake and discharge and wellhead valves.
4. Start triplex and set booster pump relief valve to provide 45-50 psi discharge pressure.
5. Throttle triplex discharge valve and set relief valve to 10% above the anticipated operating pressure. Set pressure controller to the operating pressure.
6. Well tubing strings may now be circulated and subsurface pump put in operation.
7. Throttle overflow control valve 5 to provide at least 200 b/d through the cones' underflow, as measured at the underflow meter.
8. Set the back pressure valve B on the vessel gas outlet line to about 15 psi higher than the flow line pressure, being sure to not exceed the maximum working pressure shown on the vessel nameplate.
9. Set the relief valve on top of the separator. Back out the adjusting screw until the valve opens, then screw it down about ¾ of a turn.
10. When gas appears in the top of the vessel, adjust the dump valve A control rod length to provide a fluid level 3-5 in. from the top of the upper sight glass.
11. After flow conditions have stabilized, recheck the underflow rate meter, adjusting the overflow control valve 5 if necessary. Open the meter bypass valve C and close the two meter valves.
    This check should also be made about once a week, during operation.
12. The only time valve C is closed is when the underflow meter is in the flow circuit. It should never be throttled.

APPENDIX 5.G(1)
SELECTION GUIDELINES

PUMP SELECTION:

1. Determine Maximum Acceptable P/E (Eq. 5.1)

$$\text{Max. Accp. P/E} = \frac{10,000}{\text{Net Fluid Lift}} \times .80 \quad \text{(Assume 20\% System Loss)}$$

$$\frac{10,000 \times .8}{\underline{\qquad} \text{(NFL)}} = \underline{\qquad} \text{P/E}$$

2. Determine Volumetric Correction Factor (VCF) from Theoretical Volumetric Efficiency Chart (Fig. 5.43)

VCF = _____ %   (If gas is vented use 95% for VCF)

3. Calculate Required Pump Displacement (Q'_4)

$$Q'_4 = \frac{Q_4, \text{Required Fluid Production (Oil \& Water), B/D}}{.85 \times \text{VCF}}$$

$$\frac{\underline{\qquad} \text{B/D}}{.85 \times \underline{\qquad}} = \underline{\qquad} \text{B/D}$$

4. Pump Selected (Appendix B)

Type _____

Size _____ × _____ × _____ – _____ × _____

Pump Displacement Factor (q_4) _____ BD/SPM

Engine Displacement Factor (q_1) _____ BD/SPM

P/E _____

Maximum Rated Speed _____

5. Calculate Pump Speed

$$\text{SPM} = \frac{(Q'_4)}{(q_4)} = \underline{\qquad}$$

% of Rated Speed _____

6. Calculate Power Fluid Required (Q'_1)

$$Q'_1 = \frac{(\text{SPM})x \qquad (q_1)}{.90} = \underline{\qquad} \text{B/D}$$

APPENDIX 5.G(2)

OPERATING PRESSURE CALCULATIONS:

1. a. Return Fluid Column Pressure $(h_1 G_3) \times (1 + P/E)$

    _____ = _____ PSI

   b. Friction Loss, Return Column $(F_3) \times (1 + P/E)$

    _____ = _____ PSI

   c. Surface Back Pressure $(P_{wh}) \times (1 + P/E)$

    _____ = _____ PSI

   d. Friction Loss, Power Tubing (F_1)

    _____ = _____ PSI

   e. Pump Friction (F_p) @ Operating Speed (%)

    _____ = _____ PSI

   f. Surface Pressure Loss (control valves, etc.) (assume 200 psi)

    _____ = _____ PSI

   Total 1 (a + b + c + d + e + f) = _____ PSI

2. a. Pump Intake Pressure $(P_{wf}) \times (P/E)$

    _____ = _____ PSI

   b. Power Column Pressure $(h_1 G_1)$

    _____ = _____ PSI

   Total 2 (a + b) = _____ PSI

3. Operating Pressure (P_s) = Total 1—Total 2

   (1) – (2) = _____ PSI

APPENDIX 5.G(3)

TRIPLEX SELECTION:

1. Determine Triplex Displacement Required (TDR)

$$TDR = \frac{PFR}{.90}$$

_____ (PFR) / .90 = _____ B/D (TDR)

2. Hydraulic Horsepower Required (HHP)

$$HHP = TDR \times OP \times 1.7 \times 10^{-5}$$

_____ (TDR) × _____ (OP) × 1.7×10^{-5} _____ HHP

3. Triplex Selected

    Triplex Size   ☐ 2   ☐ 3   ☐ 4

                        ☐ Vertical ☐ Horizontal

                        ☐ Solo   ☐ Non-Solo

    Plunger and Liner: Size _____

                        ☐ Lap Fit   ☐ Packed

    Gear Ratio _____

    Prime Mover Speed _____ RPM

    Triplex Speed _____ RPM

4. Minimum Prime Mover Horsepower Required (MPMH)

    $MPMH = HHP \times 1.10$

    _____ (HHP) × 1.10 = _____ HP

Chapter 6

Jet pumping

by Hal Petrie

6.1 INTRODUCTION

Subsurface jet pumps are a special class of subsurface hydraulic pumps. All conventional pumps operate by means of a positive displacement reciprocating pump piston driven by a coupled engine piston. The jet pump, however, employs no moving parts and achieves its pumping action by means of momentum transfer between the power fluid and produced fluid.

A typical example of a subsurface jet pump is shown in Fig. 6.1. The power fluid enters the top of the pump from the pump tubing and passes through the nozzle where virtually all of the total pressure of the power fluid is converted to a velocity head. The jet from the nozzle discharges into the production inlet chamber which is connected to the formation. The production fluid is entrained by the power fluid and the combined fluids enter the throat of the pump.

In the confines of the throat, which is always of larger diameter than the nozzle, complete mixing of the power fluid and production fluid takes place. During this process, the power fluid loses momentum and energy and the production fluid gains momentum and energy. The resultant mixed fluid exiting the throat has sufficient total head to flow against the production return column gradient. Much of this total head, however, is still in the form of a velocity head. The final working section of the jet pump is, therefore, a carefully shaped diffuser section of expanding area which converts the velocity head to a static pressure head greater than the static column head, allowing flow to the surface.

The potential advantages of such a pumping system are numerous. Principal among these is the absence of closely fitted reciprocating parts, which allows the jet pump to tolerate power and production fluids of much poorer quality than those normally required for reasonable life in a subsurface hydraulic pump. Another advantage of the jet pump results from the compactness of the working section—the nozzle, inlet chamber, throat, and diffuser. This allows the jet pump to be adapted to almost any bottom hole assembly, including free pump subsea completions with a five foot radius loop at the wellhead. Frequently, much higher liquid (and gas) flow rates can be obtained with the jet pump compared to a conventional subsurface hydraulic pump in the same tubing size.

Because of these advantages, subsurface jet pumps are finding increased application, particularly in high

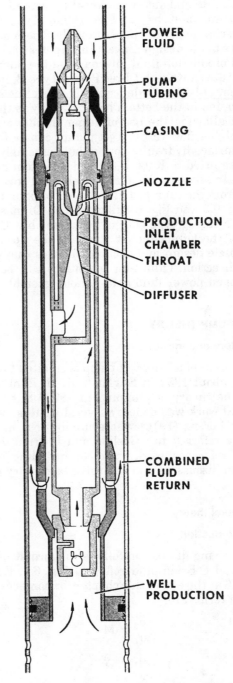

Fig. 6.1 Type A jet free pump, casing type

453

volume, gassy, or dirty wells. They are not, however, applicable to all wells. The two primary characteristics of the subsurface jet pump limiting its employment are its need for a relatively high suction pressure to avoid cavitation and its lower mechanical efficiency, typically requiring higher input horsepower than a conventional hydraulic pump. These characteristics will be discussed in detail in the section on cavitation and sizing of the jet pump to a well.

6.2   TYPES OF JET PUMPS

Subsurface jet pumps for oilfield application are currently offered by Kobe and National Production. The basic geometries of the working sections of the pumps from both manufacturers are very similar, the principal differences being in the manner in which the fluid are routed into and out of the working sections. Figure 6.1 illustrates a Kobe Jet Free Pump, Type A, in a casing discharge bottom hole assembly. The A designation refers to the design concept that routes both power fluid and production fluid internally in the pump. Fig. 6.2 illustrates a Kobe Jet Free Pump, Type B, in a casing discharge bottom hole assembly. Here the suction fluid is routed in the bottom hole assembly to the inlet chamber, allowing the use of larger throats and nozzles for higher flow rates. A jet pumps can be fitted to any subsurface hydraulic bottom hole assembly. B jet pumps are fitted to Kobe B, D, and E bottom hole assemblies. In all cases the nozzle receives its power fliud directly from the tubing string above and points downward. Fig. 6.3 and Fig. 6.4 illustrate an Oilmaster jet pump in a casing type bottom hole assembly. Characteristic of the Oilmaster jet pumps is the upward directed nozzle design and suction passages that do not reverse direction. Oilmaster jet pumps can be fitted to most open power fluid bottom hole assemblies.

6.3   JET PUMP THEORY

6.31   History and introduction

The first use of a water jet pump is credited to James Thomson about 1852 in England.[1] J. M. Rankine developed the theory of pumping in 1870.[2] Subsequent theoretical work was done by several writers, notably including Lorenz (1910) whose mixing loss model is commonly referred to.[3] Gosline and O'Brien did the standard reference work in 1933, including both a theoretical discussion and extensive laboratory tests.[4]

6.32   General theory

6.321   Introduction

The following discussion follows the format used by Gosline and O'Brien, and refers to Fig. 6.5. Refer to Section 6.6 at the end of this chapter for nomenclature.

Defined terms are:

$$\frac{q_3}{q_1} = M \qquad (6.4)$$

$$q_1 = \frac{q_3}{M} \qquad (6.4a)$$

$$\frac{A_j}{A_t} = R \qquad (6.5)$$

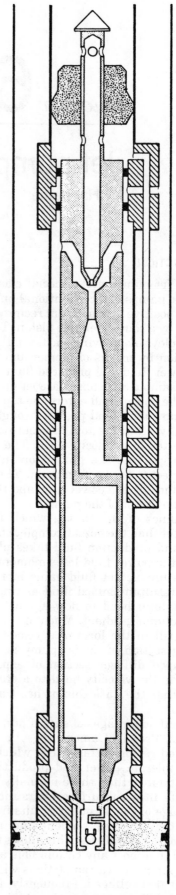

Fig. 6.2   Type B jet free pump, casing type

The continuity relations are:

$$q_1 = A_j V_j \tag{6.6}$$

$$q_3 = A_s V_s \tag{6.7}$$

$$q_3 + q_1 = A_t V_t = q_2 \tag{6.8}$$

$$A_s + A_j = A_t \tag{6.9}$$

Equations 6.8 and 6.9 lead to:

$$V_t = \frac{q_3 + q_1}{A_s + A_j} = \frac{q_3 + q_1}{A_t} \tag{6.10}$$

Equations 6.5 and 6.9 lead to:

$$\frac{A_s}{A_j} = \frac{A_t - A_j}{A_j} = \frac{\frac{1}{A_t}(A_t - A_j)}{A_j/A_t}$$

$$\frac{A_s}{A_j} = \frac{1 - \frac{A_j}{A_t}}{A_j/A_t} = \frac{1 - R}{R} \tag{6.11}$$

The Lorenz mixing loss model does not look at the details of the flows, but, on a macroscopic scale, states that the head losses associated with mixing will be proportional to the square of the difference in velocities between the mixing flows. As an energy loss per unit time in the mixing zone of the throat, this is expressed as:

$$L = q_1\rho \frac{(V_j - V_t)^2}{2g} + q_3\rho \frac{(V_s - V_t)^2}{2g} \tag{6.12}$$

The nozzle energy supplied per unit time is:

$$E_j = q_1 \rho (H_1 - H_2) \tag{6.13}$$

The energy per unit time added to the production fluid is:

$$E_s = q_3 \rho (H_2 - H_3) \tag{6.14}$$

The loss of energy per unit time due to frictional resistance at the boundaries of the mixing chamber is approximately:

$$F_t = \rho K_t (q_1 + q_3) \frac{V_t^2}{2g} \tag{6.15}$$

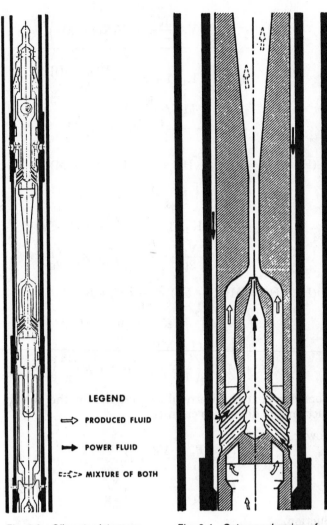

Fig. 6.3 Oilmaster jet pump

LEGEND

⇨ PRODUCED FLUID

➡ POWER FLUID

⟿ MIXTURE OF BOTH

Fig. 6.4 Cutaway drawing of Oilmaster jet pump

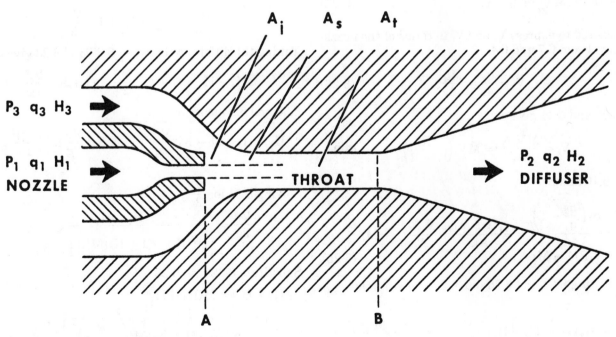

Fig. 6.5 Jet pump nomenclature

where V_t is the average velocity from 6.10 and K_t is a resistance factor computed as for flow in pipes.
Similarly, in the diffuser, suction circuit, and nozzle:

$$F_d = \rho \, K_d \, (q_1 + q_3) \frac{V_t^2}{2g} \tag{6.16}$$

$$F_s = \rho \, K_s \, (q_3) \frac{V_s^2}{2g} \tag{6.17}$$

$$F_j = \rho \, K_j \, (q_1) \frac{V_j^2}{2g} \tag{6.18}$$

The total friction energy loss per unit time is, therefore:

$$F_f = \rho \, K_t \, (q_1 + q_3) \frac{V_t^2}{2g} + \rho \, K_d \, (q_1 + q_3) \frac{V_t^2}{2g}$$
$$+ \rho \, K_s \, (q_3) \frac{V_s^2}{2g} + \rho \, K_j \, (q_1) \frac{V_j^2}{2g} \tag{6.19}$$

Simplifying 6.19 gives:

$$F_f = \rho \, (K_t + K_d)(q_1 + q_3) \frac{V_t^2}{2g} + \rho \, K_s (q_3) \frac{V_s^2}{2g}$$
$$+ \rho \, K_j (q_1) \frac{V_j^2}{2g} \tag{6.20}$$

Utilizing 6.12, 6.13, 6.14 and 6.20, the power supplied is equated with the work done per unit time plus the mixing losses plus the friction losses.

$$q_1 \rho \, (H_1 - H_2) = q_3 \rho \, (H_2 - H_3) + q_1 \rho \, \frac{(V_j - V_t)^2}{2g}$$
$$+ q_3 \rho \, \frac{(V_s - V_t)^2}{2g} + \rho \, (K_t + K_d)(q_1 + q_3) \frac{V_t^2}{2g}$$
$$+ \rho \, K_s (q_3) \frac{V_s^2}{2g} + \rho \, K_j (q_1) \frac{V_j^2}{2g} \tag{6.21}$$

Substituting 6.4 and simplifying leads to:

$$(\rho)(H_1 - H_2) = M(\rho)(H_2 - H_3) + (\rho) \frac{(V_j - V_t)^2}{2g}$$
$$+ M(\rho) \frac{(V_s - V_t)^2}{2g} + (\rho)(K_t + K_d)(1 + M) \frac{V_t^2}{2g}$$
$$+ (\rho) \, K_s M \frac{V_s^2}{2g} + (\rho) \, K_1 \frac{V_j^2}{2g} \tag{6.22}$$

It is desired to express V_s and V_t in terms of the nozzle velocity. From 6.7 and 6.4

$$V_s = \frac{q_3}{A_s} = \frac{M q_1}{A_s} \tag{6.23}$$

6.6, 6.23 and 6.11 give

$$V_s = \frac{M V_j A_j}{A_s} = M V_j \left(\frac{R}{1 - R} \right) \tag{6.24}$$

From 6.10 and 6.4

$$V_t = \frac{q_1 \left(\dfrac{q_3}{q_1} + 1 \right)}{A_t} = \frac{q_1 (1 + M)}{A_t} = \frac{V_j A_j}{A_t}(1 + M)$$
$$V_t = V_j R(1 + M) \tag{6.25}$$

Substituting 6.24 and 6.25 into 6.22 and eliminating ρ leads to:

$$(H_1 - H_2) = M(H_2 - H_3) + \frac{[V_j - V_j R(1 + M)]^2}{2g}$$

$$+ M \frac{\left(M V_j \dfrac{R}{1 - R} - V_j R(1 + M) \right)^2}{2g}$$
$$+ (K_d + K_t)(1 + M) \frac{[V_j R(1 + M)]^2}{2g}$$
$$+ K_s M \frac{\left(M V_j \dfrac{R}{1 - R} \right)^2}{2g} + K_j \frac{V_j^2}{2g}$$

$$(H_1 - H_2) = M(H_2 - H_3) + \frac{V_j^2}{2g}[1 - R(1 + M)]^2$$
$$+ \frac{V_j^2}{2g} M \left[M \frac{R}{1 - R} - R(1 + M) \right]^2$$
$$+ (K_t + K_d)(1 + M) \frac{V_j^2}{2g}[R(1 + M)]^2$$
$$+ K_s M \frac{V_j^2}{2g} \left(\frac{MR}{1 - R} \right)^2 + K_j \frac{V_j^2}{2g}$$

$$(H_1 - H_2) = M(H_2 - H_3) + \frac{V_j^2}{2g}\bigg\{ K_j + K_s M^3$$
$$\left(\frac{R}{1 - R} \right)^2 + (K_t + K_d)(1 + M)^3 R^2$$
$$+ [1 - R(1 + M)]^2$$
$$+ M \left[M \left(\frac{R}{1 - R} \right) - R(1 + M) \right]^2 \bigg\} \tag{6.26}$$

Consider next the Bernoulli equations for the power, suction and discharge circuits:

Power,

$$H_1 = \frac{P_a}{\rho} + \frac{V_j^2}{2g} + K_j \frac{V_j^2}{2g}$$
$$H_1 = \frac{P_a}{\rho} + (1 + K_j) \frac{V_j^2}{2g} \tag{6.27}$$

Suction,

$$H_3 = \frac{P_a}{\rho} + \frac{V_s^2}{2g} + K_s \frac{V_s^2}{2g}$$
$$H_3 = \frac{P_a}{\rho} + (1 + K_s) \frac{V_s^2}{2g} \tag{6.28}$$

Discharge,

$$\frac{P_b}{\rho} + \frac{V_t^2}{2g} = H_2 + K_d \frac{V_t^2}{2g} \tag{6.29}$$

Taking the difference between 6.27 and 6.28 gives:

$$H_1 - H_3 = (1 + K_j) \frac{V_j^2}{2g} - (1 + K_s) \frac{V_s^2}{2g} \tag{6.30}$$

6.30 and 6.24 lead to:

$$H_1 - H_3 = (1 + K_j) \frac{V_j^2}{2g} - (1 + K_s) \frac{V_j^2}{2g} M^2 \left(\frac{R}{1 - R} \right)^2 \tag{6.31}$$

Solving 6.31 for $\dfrac{V_j^2}{2g}$,

$$\frac{V_j^2}{2g} = \frac{(H_1 - H_3)}{(1 + K_j) - (1 + K_s) M^2 \left(\dfrac{R}{1 - R} \right)^2} \tag{6.32}$$

Substitute 6.32 into 6.26:

$$(H_1 - H_2) = M(H_2 - H_3)$$
$$+ \frac{(H_1 - H_3)}{(1 + K_j) - (1 + K_s) M^2 \left(\dfrac{R}{1 - R} \right)^2} \bigg\{ K_j$$

$$+ K_s M^3 \left(\frac{R}{1-R}\right)^2 + (K_t + K_d)(1 + M)^3 R^2$$
$$+ [1 - R(1 + M)]^2 + M\left[M\left(\frac{R}{1-R}\right)\right.$$
$$\left.- R(1 + M)\right]^2\Bigg\}$$

Simplifying,

$$(H_1 - H_2) = M(H_2 - H_3)$$
$$+ \frac{(H_1 - H_3)}{(1 + K_j) - (1 + K_s)M^2\left(\frac{R}{1-R}\right)^2}\Bigg\{ K_j$$
$$+ K_s M^3 \left(\frac{R}{1-R}\right)^2 + (K_t + K_d)(1 + M^3)R^2$$
$$+ 1 - 2R(1 + M) + R^2(1 + M)^2$$
$$+ M^3 \left(\frac{R}{1-R}\right)^2 - 2M^2(1 + M)\frac{R^2}{1-R}$$
$$+ M(1 + M)^2 R^2\Bigg\}$$

$$(H_1 - H_2) = M(H_2 - H_3)$$
$$+ \frac{(H_1 - H_3)}{(1 + K_j) - (1 + K_s)M^2\left(\frac{R}{1-R}\right)^2}\Bigg\{ (1 + K_j)$$
$$+ (1 + K_s)M^3 \left(\frac{R}{1-R}\right)^2 + (K_t + K_d)(1 + M)^3 R^2$$
$$- 2R - 2RM + R^2 + 2R^2M + R^2M^2 - 2\frac{R^2}{1-R}M^2$$
$$- 2\frac{R^2}{1-R}M^3 + R^2M + 2R^2M^2 + R^2M^3\Bigg\}$$

$$(H_1 - H_2) = M(H_2 - H_3)$$
$$+ \frac{(H_1 - H_3)}{(1 + K_j) - (1 + K_s)M^2\left(\frac{R}{1-R}\right)^2}\Bigg\{ (1 + K_j)$$
$$+ (1 + K_s)M^3 \left(\frac{R}{1-R}\right)^2 + (K_t + K_d)(1 + M)^3 R^2$$
$$+ (R^2 + 3R^2M + 3R^2M^2 + R^2M^3) - 2R - 2RM$$
$$- 2\frac{R^2}{1-R}M^2 - 2\frac{R^2}{1-R}M^3\Bigg\}$$

$$(H_1 - H_2) = M(H_2 - H_3)$$
$$+ \frac{(H_1 - H_3)}{(1 + K_j) - (1 + K_s)M^2\left(\frac{R}{1-R}\right)^2}\Bigg\{ (1 + K_j)$$
$$+ (1 + K_s)M^3 \left(\frac{R}{1-R}\right)^2 + (K_t + K_d)(1 + M)^3 R^2$$
$$+ R^2(1 + M)^3 - 2R(1 + M) - 2\frac{R^2}{1-R}M^2(1 + M)\Bigg\}$$

$$(H_1 - H_2) = M(H_2 - H_3)$$
$$+ \frac{(H_1 - H_3)}{(1 + K_j) + (1 + K_s)M^2\left(\frac{R}{1-R}\right)^2}\Bigg\{ (1 + K_j)$$
$$+ (1 + K_s)M^3 \left(\frac{R}{1-R}\right)^2 + (1 + K_t + K_d)(1 + M)^3 R^2$$
$$- 2R(1 + M) - 2\frac{R^2}{1-R}M^2(1 + M)\Bigg\} \qquad (6.33)$$

Equation 6.33 is now in the form of:

$$(H_1 - H_2) = M(H_2 - H_3) + N(H_1 - H_3) \qquad (6.34)$$

Where

$$N = \left[(1 + K_t) + (1 + K_s)M^3\left(\frac{R}{1-R}\right)^2\right.$$
$$+ (1 + K_t + K_d)(1 + M)^3 R^2 - 2R(1 + M)$$

$$\left.- 2\frac{R^2}{(1-R)}M^2(1 + M)\right] \Bigg/ \left[(1 + K_j)\right.$$
$$\left.- (1 + K_s)M^2\left(\frac{R}{1-R}\right)^2\right]$$
$$(6.35)$$

Rearranging Equation 6.34 gives:

$$1 = \frac{H_2 - H_3}{H_1 - H_2}M + \frac{N(H_1 - H_3)}{H_1 - H_2} \qquad (6.36)$$

Note that

$$\frac{H_2 - H_3}{H_1 - H_2} + 1 = \frac{H_1 - H_3}{H_1 - H_2} \qquad (6.37)$$

Substituting 6.37 in 6.36 gives:

$$1 = \frac{H_2 - H_3}{H_1 - H_2}M + \left(\frac{H_2 - H_3}{H_1 - H_2} + 1\right)N \qquad (6.38)$$

Define

$$H = \frac{H_2 - H_3}{H_1 - H_2}$$

Then

$$1 = HM + (H + 1)N = HM + HN + N$$
$$H = \frac{1 - N}{M + N} = \frac{H_2 - H_3}{H_1 - H_2} \qquad (6.39)$$

Where M is defined by Equation 6.4 and N is defined by Equation 6.35.

An examination of 6.35 shows that the parameters K_j, K_s, K_t, K_d, and R are all geometric characteristics of the pump, whereas the remaining parameter, M, is a function of the flows in the pump since

$$M = \frac{q_3}{q_1} \text{ from 6.4}$$

Equation 6.39, therefore, is a function only of M for a given pump. Further, with regard to Equation 6.39, the total head is very closely approximated by the static pressure in hydraulic pump application, so

$$H = \frac{P_2 - P_3}{P_1 - P_2} = \frac{1 - N}{N + M} = f(M) \qquad (6.40)$$

Hereafter, the total head H_i will be assumed equal to the static pressure P_i. The physical significance of the parameter H can be seen as the ratio of head or pressure rise experienced by the production fluid in the pump to the head or pressure loss suffered by the power fluid in the pump. It should be apparent that for high discharge heads, as in a deep well, the geometry of the pump (described by R) and the flow ratio M should be chosen to give a high value of H.

6.322 Efficiency

The efficiency of a jet pump is defined as the ratio of the power added to the produced fluid to the power lost by the power fluid.

The power added to the produced well fluid is:

$$(HP)_{q_3} \propto q_3(P_2 - P_3) \qquad (6.41)$$

And the power lost by the power fluid is:

$$(HP)_{q_1} \propto q_1(P_1 - P_2) \qquad (6.42)$$

Equations 6.41 and 6.42 lead to the efficiency:

$$E = \frac{(HP)_{q_3}}{(HP)_{q_1}} = \frac{q_3(P_2 - P_3)}{q_1(P_1 - P_2)} \qquad (6.43)$$

Note that the right hand side of 6.43 is:

$$M \times H = \frac{q_3}{q_1} \times \frac{(P_2 - P_3)}{(P_1 - P_2)}$$

Therefore:

$$\text{Efficiency} = E = MH = \frac{q_3(P_2 - P_3)}{q_1(P_1 - P_2)} \qquad (6.44)$$

6.323 Dimensionless performance curves

The performance of geometrically similar jet pumps operating at the same Reynolds Number is described by Equations 6.35, 6.40 and 6.44. A plot of these equations showing H versus M for several values of R is contained in Fig. 6.6. The respective efficiencies are also plotted as a function of M. These curves were plotted using the following loss coefficients found to be typical by Gosline and O'Brien:

$$K_j = .15, \quad K_s = 0, \quad K_t = .28, \quad K_d = .10$$

The area ratios selected cover the range from a relatively high head, low flow rate pump (A ratio R = 0.410) to a relatively low head, high flow rate pump (E ratio R = 0.168) Refer to Table 6.1 for the various nozzle and throat diameters and areas.

TABLE 6.1
NOZZLE AND THROAT DIAMETERS AND AREAS

No.	Nozzles Area	Dia.	No.	Throats Area	Dia.
1	0.00371	0.06869	1	0.00905	0.10733
2	0.00463	0.07680	2	0.01131	0.12000
3	0.00579	0.08587	3	0.01414	0.13416
4	0.00724	0.09600	4	0.01767	0.15000
5	0.00905	0.10733	5	0.02209	0.16771
6	0.01131	0.12000	6	0.02761	0.18750
7	0.01414	0.13416	7	0.03451	0.20963
8	0.01767	0.15000	8	0.04314	0.23438
9	0.02209	0.16771	9	0.05393	0.26204
10	0.02761	0.18750	10	0.06741	0.29297
11	0.03451	0.20963	11	0.08426	0.32755
12	0.04314	0.23438	12	0.10533	0.36621
13	0.05393	0.26204	13	0.13166	0.40944
14	0.06741	0.29297	14	0.16458	0.45776
15	0.08426	0.32755	15	0.20572	0.51180
16	0.10533	0.36621	16	0.25715	0.57220
17	0.13166	0.40944	17	0.32144	0.64974
18	0.16458	0.45776	18	0.40180	0.71526
19	0.20572	0.51180	19	0.50225	0.79968
20	0.25715	0.57220	20	0.62782	0.89407
			21	0.78477	0.99960
			22	0.98096	1.11759
			23	1.22620	1.24950
			24	1.53275	1.39698

The high head pump would typically be employed in a deep well with a high lift. Note that the maximum

efficiency point for the A ratio occurs at M = 0.5. This means that for every barrel of production (q_3), two barrels of power fluid (q_1) must be supplied. The high flow rate pump, on the other hand would typically be applied in a shallow well with a low lift and would require only about 0.7 barrel of power fluid for each barrel of production at the maximum efficiency point of M = 1.45.

It must be emphasized that the curves presented in Fig. 6.6 delineate the only permissable operating points for a non-cavitating jet pump with the particular loss coefficients and area ratios listed.

EXAMPLE PROBLEM #1

Given:

$P_1 = 6000$ psi
$P_2 = 3000$ psi
$P_3 = 1000$ psi

Find M and the efficiency for A, B, C, D and E ratio pumps.

Solution:

From Fig. 6.6, an A ratio pump with pressure of $P_1 = 6000$ psi, $P_2 = 3000$ psi, and $P_3 = 1000$ psi will have an H value of:

$$H = \frac{P_2 - P_3}{P_1 - P_2} = \frac{3000 - 1000}{6000 - 3000} = 0.667$$

and operates only at an M = q_3/q_1 value of .285 and an efficiency of 19%.

With a B ratio pump at the same pressures, H is still 0.667, but the operating M point is 0.16 at an efficiency of 10.7%.

The C, D, and E ratio pumps do not have sufficient head regain characteristics to pump at H = .667 and would actually show back flow out the suction port. In a hydraulic pump installation this back flow could not in practice take place because the standing valve would close.

CLASS PROBLEMS

Given:

#1A
  $P_1 = 5000$ psi
  $P_2 = 2500$ psi
  $P_3 = 750$ psi

Find M and the efficiency for A, B, C, D and E ratio pumps.

Given:

#1B
  $P_1 = 6000$ psi
  $P_2 = 4000$ psi
  $P_3 = 1000$ psi

Find M and the efficiency for A, B, C, D and E ratio pumps.

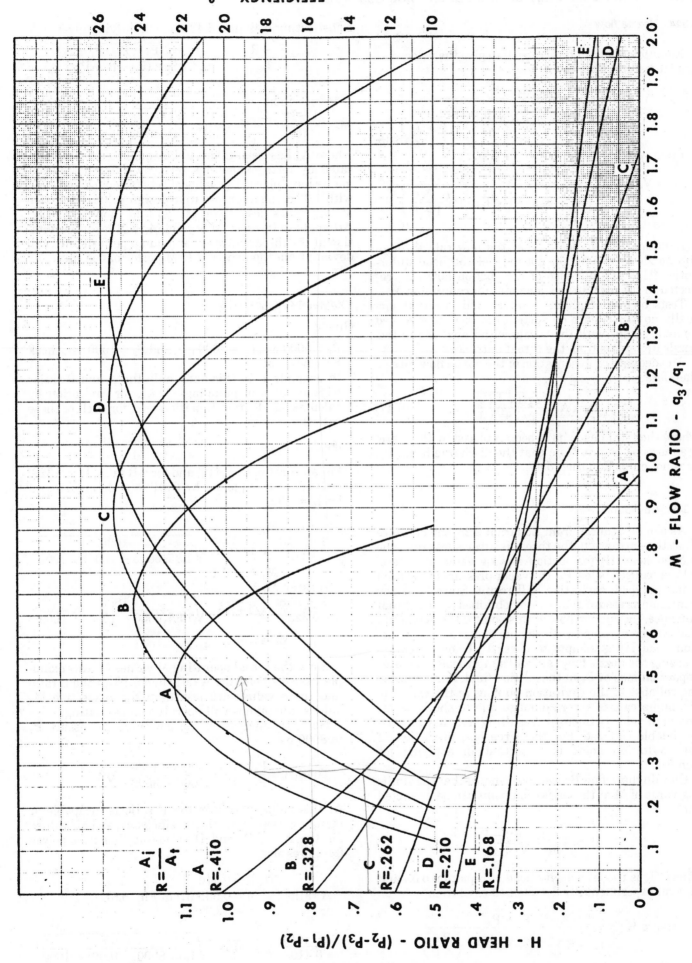

EFFICIENCY - %

26 24 22 20 18 16 14 12 10

M - FLOW RATIO - q_3/q_1

$\dfrac{A_i}{A_t}$

R=

A
R=.410

B.
R=.328

C
R=.262

D
R=.210

E
R=.168

H - HEAD RATIO - $(P_2-P_3)/(P_1-P_2)$

1.1 1.0 .9 .8 .7 .6 .5 .4 .3 .2 .1

Fig. 6.6   M-flow ratio - q_3/q_1

6.324 Nozzle flow

Equation 6.32 can be arranged to give the velocity of the jet issuing from the nozzle

$$V_j = \sqrt{\frac{2g(H_1 - H_3)}{(1 + K_j) - (1 + K_s)M^2\left(\dfrac{R}{1 - R}\right)^2}}$$

(6.45)

From which:

$$q_1 = V_j A_j = A_j \sqrt{\frac{2g(H_1 - H_3)}{(1 + K_j) - (1 + K_s)M^2\left(\dfrac{R}{1 - R}\right)^2}}$$

(6.46)

Equation 6.46 indicates that the flow rate through the nozzle is a function not only of the head differential $(H_1 - H_3)$ but also of the suction flow rate reflected in the term in the denominator containing M.

Tests by Cunningham, however, indicate that virtually no dependence on suction flow rate is observed for real pumps with the nozzle withdrawn one to two nozzle diameters from the throat entrance.[5]

An adequate representation of the nozzle flow rate is, therefore:

$$q_1 = A_j \sqrt{\frac{2g(P_1 - P_3)}{\rho(1 + K_j)}}$$

(6.47)

where the total heads have been replaced by the static pressures as before. Note that the discharge head, P_2, does not enter Equation 6.47.

6.325 Cavitation

As can be deduced from Equation 6.28, the pressure P_a at the entrance to the throat is always less than the suction head H_3 for suction flows greater than zero. If P_a is reduced below P_v, the vapor pressure of the fluid being pumped, cavitation will result. Since P_v is the minimum pressure that can be obtained at the throat entrance, the suction flow at this point is the maximum that can be obtained with the particular value of suction head H_3. Attempts to lower P_a below P_v by increasing the nozzle flow rate will simply lead to greater vapor volumes at P_v in the suction fluid. Furthermore, the collapse of the cavitation bubbles in the throat of the pump causes severe damage due to shock waves and high velocity microjets resulting from asymmetrical bubble collapse. For these reasons, prediction of the cavitation point is important in applying jet pumps.

Cunningham and Brown have shown that the limiting value of M at the cavitation point can be predicted by [6]

$$M_c = \frac{1 - R}{R} \sqrt{\frac{P_3 - P_v}{I_c H_v}}$$

where H_v is the jet velocity head of Equation 6.32 and I_c is an experimentally determined cavitation index.

Hence $H_v = \dfrac{P_1 - P_3}{(1 + K_j) - (1 + K_s)M^2\left(\dfrac{R}{1 - R}\right)^2}$

from Equation 6.32

(6.49)

Substituting 6.49 into 6.48 and simplifying gives

$$M_c = \frac{1 - R}{R}\sqrt{1 + K_j}\sqrt{\frac{(P_3 - P_v)/(P_1 - P_3)}{I_c + (P_3 - P_v)/(P_1 - P_3)}}$$

(6.50)

If $P_v = 0$, then

$$M_c = \frac{1 - R}{R}\sqrt{1 + K_j}\sqrt{\frac{P_3}{I_c(P_1 - P_3) + P_3}}$$

(6.51)

where K_s has been taken to be equal to zero, as indicated by tests. Numerous tests by different investigators have placed the value of I_c between 0.8 and 1.67, with 1.35 being a conservative design value. Operation at M values less than M_c will be non-cavitating. Attempts to increase M beyond M_c will lead to cavitation at the throat entrance and pump performance will deviate from the expected H-M performance curve.

EXAMPLE PROBLEM #2

Given:

$P_1 = 6000$ psi, $P_2 = 3000$ psi, and $P_3 = 1000$ psi, check for cavitation.

In example 1 for this data, only the A and B ratios were capable of pumping. Check for cavitation by calculating M_c from Equation 6.51 for each of these ratios:

Take $P_v = 0$. psi.

For an A ratio, R = .410, and for a B ratio, R = 0.328. Use $k_j = 0.15$ as found by Gosline and O'Brien and let $I_c = 1.35$.

A ratio: $M_c =$
$$\frac{1 - .410}{0.410}\sqrt{1 + 0.15}\sqrt{\frac{1000}{1.35(6000 - 1000) + 1000}}$$
$$M_c = 0.554$$
B ratio: $M_c =$
$$\frac{1 - .328}{0.328}\sqrt{1 + .15}\sqrt{\frac{1000}{1.35(6000 - 1000) + 1000}}$$

$$M_c = .789$$

Since the actual operating M values for this example were 0.285 for the A ratio and 0.16 for the B ratio and these values are less than the respective M_c values, the pumps will operate non-cavitating.

Next, try increasing the power fluid pressure P_1 to 8000 psi.

Then $H = \dfrac{P_2 - P_3}{P_1 - P_2} = \dfrac{3000 - 1000}{8000 - 3000} = 0.4$

At this value of H only the E ratio is incapable of pumping. Determine the M, efficiency, and M_c values for the A, B, C, and D ratios at this new power fluid pressure.

R = 0.410
A ratio: M = 0.555 (from Fig. 6.6)
E = 22.2%
$M_c =$
$$\frac{1 - .410}{0.410}\sqrt{1 + 0.15}\sqrt{\frac{1000}{1.35(8000 - 1000) + 1000}}$$
$$M_c = 0.477$$

Since the expected M is greater than M_c, the pump will cavitate.

$$R = 0.328$$

B ratio: $M = 0.605$

$$E = 24.2\%$$

$$M_c =$$

$$\frac{1 - .328}{.328} \sqrt{1+0.15} \sqrt{\frac{1000}{1.35(8000 - 1000) + 1000}}$$

$$M_c = 0.680$$

Since M is less than M_c, the pump will not cavitate and the efficiency is higher with this ratio.

$$R = 0.262$$

C ratio: $M = 0.53$

$$E = 21.2\%$$

$$M_c = 0.934$$

Again M is less than M_c and the pump will not cavitate, but the efficiency is lower than with the B ratio.

$$R = 0.210$$

D ratio: $M = 0.245$

$$E = 9.8\%$$

$$M_c = 1.248$$

This ratio is the farthest from cavitation, but the efficiency has dropped off significantly.

The previous examples have been included to illustrate several important facts:

(1) Increasing the power fluid pressure drives a given ratio jet pump closer to cavitation.

(2) For a given value of H, there is at least one ratio that gives maximum efficiency. This will be the ratio that gives the maximum value of M.

(3) At given values of P_1, P_2, and P_3, the smaller ratios will give better cavitation protection. For example, if H = 0.47, then both A and B ratios will operate at the same M and efficiency, but the B ratio will have a higher M_c value because of the $\frac{1 - R}{R}$ term in Equation 6.50. Physically, this occurs because the smaller ratios have bigger throat areas, hence lower suction port velocities.

CLASS PROBLEMS

Class Problem 2A: Assume $P_1 = 8000$ psi and $P_2 = 3000$ psi as in the example above, but decrease P_2 from 3000 psi to 2400 psi. Which ratio gives the most efficient non-cavitating performance? What is the effect ing suction pressure on the cavitation tendency of a jet pump?

Class Problem 2B: Assume $P_1 = 8000$ psi and $P_3 = 1000$ psi as in the example above, but decrease $P_2 =$ from 3000 psi to 2400 psi. Which Ratio gives the most efficient non-cavitating performance. What is the effect of reduced discharge pressure on the cavitation tendency of a jet pump?

6.326   Effect of discharge back pressure

Note in Fig. 6.6 that the dimensionless performance curves for the different ratios cross over each other. For example, at the M value of 0.7 and an H value of 0.265, the performances of the A and E ratios are identical. It has previously been shown however that in such a case the cavitation characteristics of the two pumps would not be the same. Similarly, the responses to a change in the discharge pressure will be different for the two different ratio pumps.

Consider A ratio and E ratio pumps operating at the following pressures:

$$P_1 = 6000 \text{ psi}, P_2 = 3000 \text{ psi, and } P_3 = 2205 \text{ psi}$$

From Equation 6.40

$$H = \frac{3000 - 2205}{6000 - 3000} = 0.265, \text{ hence M} = .7$$

for both ratio pumps.

Increasing P_2 by 5% leads to

$$H = \frac{3150 - 2205}{6000 - 3150} = 0.332$$

At H = 0.332 the A ratio will operate at an M value of 0.64 while the E ratio will operate at an M value of only about 0.16. In the case of the A ratio, a 5% increase in discharge pressure has led to a 9% decrease in M and hence the production flow rate, q_3. With the E ratio, however, the decrease in production flow was 77% for the same 5% increase in discharge pressure.

In practice, however, an operator would increase P_1 in an effort to regain the lost production. For this reason, a more appropriate way to look at the sensitivities to back pressure of the various ratios is to ask how much extra surface power fluid pressure will have to be supplied to regain production after some incremental increase in back pressure. Mathematically, this is the rate of change of P_1 with respect to P_2 with P_3 and q_3 held constant. That is:

$$\text{Sensitivity to back pressure} = X = \left(\frac{\delta P_1}{\delta P_2}\right)_{P_3, q_3 = \text{constant}}$$

An examination of Fig. 6.6 indicates that the H-M curves can be approximated by straight lines of the form:

$$H = I - m \times M \qquad (6.52)$$

where I is the vertical axis intercept as illustrated in Fig. 6.6a and m is the slope of the line.

Then from Equation 6.52

$$M = \frac{H - I}{- m} = \frac{I - H}{m} \qquad (6.53)$$

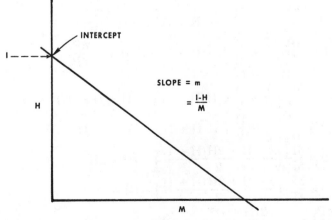

Fig. 6.6A   Approximation of H-M curves

Equations 6.4 and 6.53 lead to:

$$\frac{q_3}{q_1} = \frac{I - H}{m} \qquad (6.54)$$

For a given pump, Equation 6.47 can be written as

$$q_1 = K\sqrt{P_1 - P_3} \qquad (6.55)$$

and when combined with 6.54 gives:

$$q_3 = K\sqrt{P_1 - P_3}\left(\frac{I - H}{m}\right) \qquad (6.56)$$

Substituting the definition of H (Eqn. 6.40) into 6.56 leads to:

$$q_3 = \frac{K\sqrt{P_1 - P_3}}{m}\left(I - \frac{P_2 - P_3}{P_1 - P_2}\right)$$

which can be rearranged to:

$$q_3\left(\frac{m}{K}\right) = \sqrt{P_1 - P_3}\left[I - \left(\frac{P_2 - P_3}{P_1 - P_2}\right)\right] \qquad (6.57)$$

Equation 6.57 now has q_3 in terms of the two pressures of interest, P_1 and P_2. m, K, I and P_3 are constants. Taking the partial derivative of each side of 6.57 with respect to P_2 and holding q_3 constant gives

$$0 = \frac{1}{2\sqrt{P_1 - P_3}}\frac{\delta P_1}{\delta P_2}(I) - \frac{1}{2}\frac{\dfrac{\delta P_1}{\delta P_2}}{\sqrt{P_1 - P_3}}H$$
$$+ \sqrt{P_1 - P_3}\frac{\delta}{\delta P_2}\left(I - \frac{P_2 - P_3}{P_1 - P_2}\right)$$

$$0 = \frac{\delta P_1}{\delta P_2}\frac{I}{2} - \frac{H}{2}\frac{\delta P_1}{\delta P_2} + (P_1 - P_3)\frac{\delta}{\delta P_2}\left(I - \frac{P_2 - P_3}{P_1 - P_2}\right)$$

$$0 = \frac{\delta P_1}{\delta P_2}\left(\frac{I}{2} - \frac{H}{2}\right) + (P_1 - P_3)\left\{0 - \right.$$

$$\left.\left[\frac{(P_1 - P_2) - (P_2 - P_3)\left(\dfrac{\delta P_1}{\delta P_2} - 1\right)}{(P_1 - P_2)^2}\right]\right\}$$

$$0 = \frac{\delta P_1}{\delta P_2}\left(\frac{I}{2} - \frac{H}{2}\right) - (P_1 - P_3)\left[\frac{1 - H\left(\dfrac{\delta P_1}{\delta P_2} - 1\right)}{P_1 - P_2}\right]$$

$$0 = \frac{\delta P_1}{\delta P_2}\left(\frac{I}{2} - \frac{H}{2}\right) - \frac{P_1 - P_3}{P_1 - P_2}\left[1 - H\left(\frac{\delta P_1}{\delta P_2} - 1\right)\right]$$

But $\dfrac{P_1 - P_3}{P_1 - P_2} = H + 1$ from Equation 6.37, so

$$0 = \frac{\delta P_1}{\delta P_2}\left(\frac{I}{2} - \frac{H}{2}\right) - (H + 1)\left[1 - H\left(\frac{\delta P_1}{\delta P_2} - 1\right)\right]$$

$$0 = \frac{\delta P_1}{\delta P_2}\left(\frac{I}{2} - \frac{H}{2}\right) - (H + 1) + H(H + 1)\frac{\delta P_1}{\delta P_2}$$
$$- H(H + 1)$$

$$0 = \frac{\delta P_1}{\delta P_2}\left[\frac{I}{2} - \frac{H}{2} + H(H + 1)\right] - (H + 1) - H(H + 1)$$

$$0 = \frac{\delta P_1}{\delta P_2}\left[\frac{I}{2} - \frac{H}{2} + H(H + 1)\right] - H - 1 - H^2 - H$$

$$0 = \frac{\delta P_1}{\delta P_2}\left(\frac{I - H + 2H(H + 1)}{2}\right) - H^2 - 2H - 1$$

$$0 = \frac{\delta P_1}{\delta P_2}\left(\frac{I - H + 2H(H + 1)}{2}\right) - (H + 1)^2$$

$$\frac{\delta P_1}{\delta P_2} = \frac{2(H + 1)^2}{I - H + 2H(H + 1)} = X \qquad (6.58)$$

Equation 6.58 has several interesting features. First, for a given value of H, smaller values of I give greater increases in P_1 with respect to increases in P_2. Thus, the E ratio with $I \cong 0.35$ will have $\dfrac{\delta P_1}{\delta P_2} = 4.24$ for $H = 0.265$ while the A ratio with $I \cong 1$ has $\dfrac{\delta P_1}{\delta P_2} = 2.28$.

Therefore, the E ratio would require an increase in triplex pressure of $150 \times 4.24 = 636$ psi for the 150 psi P_2 increase used in the earlier example. The A ratio, on the other hand, would require an increase of $150(2.28) = 342$ psi. to maintain the same production.

Another prediction of Equation 6.58 is that the sensitivity, X, is a function of where on its particular H-M curve the pump is operating. Taking the E ratio, if $H = 0.35$, then

$$X = 3.86$$

But if $H = 0$

$$X = 5.71$$

Table 6.2 gives the X values for the various ratios at their maximum efficiency points. "I" was found by noting the intercept of a straight line tangent to the H-M curve at its maximum efficiency point.

TABLE 6.2
SENSITIVITY VALUES (X) AT MAXIMUM EFFICIENCY POINTS

Ratio	I	H	X
A	0.94	0.47	2.33
B	0.73	0.37	2.73
C	0.57	0.28	3.26
D	0.45	0.22	3.88
E	0.35	0.18	4.68

Table 6.2 illustrates the importance of minimizing the pump discharge pressure to achieve low operating horsepowers. Further, it should be noted that any errors in well data that affect pump discharge pressure, such as fluid gradients, flow line back pressure, and gas-oil ratio, will lead to greater errors in predicted performance with the small ratios such as E than with the larger ratios such as the A.

6.4   DESIGN OF JET PUMP INSTALLATIONS

6.41   Nozzles and throats for field application

In sizing a jet pump for a specific well, the optimum nozzle size and ratio must be determined. Since desired production rates may vary, depending on the well, from less than 100 B/D to wells capable of more than 15,000 b/d, a range of nozzle sizes must be available.

Such a set of nozzles is presented in Table 6.1. The nozzle flow areas, A_j increase from nozzle 1 to nozzle 20 in 25% increments. That is, nozzle 2 has a flow area 25% greater than 1 and nozzle 3 has a flow area 25% greater than 2. The range of nozzle sizes is such that the smallest will flow 200 to 300 b/d in a typical well and the largest will flow 16,000 to 18,000 b/d in a typical well. The actual flow rate of each nozzle is, of course, a function of the pressures P_1 and P_3, its flow area, and the specific gravity of the power fluid.

Equation 6.47 can be reformulated to give answers in common oilfield units as follows:

$$q_1 = 1214.5 A_j \sqrt{\frac{P_1 - P_3}{\gamma_1}} \qquad (6.59)$$

where:

q_1, b/d
A_j, in.^{2}
P_1 and P_3, PSI
γ is specific gravity
K_j has been assumed = 0.15

$$\text{Alternatively } A_j = \frac{q_1}{1214.5\sqrt{\frac{P_1 - P_3}{\gamma_1}}} \qquad (6.60)$$

The throats listed in Table 6.1 are also arranged in order of increasing size and are dimensioned in such a manner as to yield the following relationship:

Let Y be a given nozzle.

Then:

Nozzle No. Y and Throat No. Y form the A ratio Jet Pump (R = 0.410)

Nozzle No. Y and Throat No. (Y + 1) form the B ratio (R = 0.328)

Nozzle No. Y and Throat No. (Y + 2) form the C ratio (R = 0.262)

Nozzle No. Y and Throat No. (Y + 3) form the D ratio (R = 0.210)

Nozzle No. Y and Throat No. (Y + 4) form the E ratio (R = 0.168)

This relationship will hold for any value of Y from 1 to 20.

TABLE 6.3
NOZZLE SIZES AVAILABLE

Nominal tubing size	Kobe		National production systems		
	Nozzles	Throats	Nozzles	Throats	
2"	1- 9	1-12 (Type A)	4- 9	4-10	Three seal series
	3-11	3-14 (Type B)	4- 7	4- 8	Single seal series
			4- 9	4-10	Standard series
			4-12	4-13	High volume series
2½"	3-11	1-12 (Type A)	4-12	4-13	Three seal series
	5-13	5-17 (Type B)	4- 9	4-10	Single seal series
			4-12	4-13	Standard series
			4-16	4-17	High volume series
3"	5-13	5-16 (Type A)	4-18	4-19	High volume series
	7-15	7-19 (Type B)			

Note 1: The nozzles, throats and ratios listed in Table 6.1 may not exactly agree in size with the parts offered by Kobe or National Production listed above. Consult the respective manufacturers for precise information.

Note 2: Nozzles larger than 15 are offered by Kobe for special applications.

Note 3: Three seal series—interchangeable with 2 in. and 2½ in. units with top engine discharge.

Single seal series—interchangeable with all 2 in. and 2½ in. National Production and FE series and Kobe A.

Standard and high volume series—special bottom hole assemblies

Not all of the nozzles and throats are practical in a given tubing string since they are commonly employed in free pumps whose overall outside diameters are limited by the constraint of having to pass through the tubing string. Table 6.3 lists the sizes of nozzles currently offered by fluid packed pumps and Kobe.

EXAMPLE PROBLEM #3

If P_1 = 5500 psi, P_2 = 2500 psi and P_3 = 1250 psi, how much fluid can be produced by a 7 Nozzle, A ratio jet pump? Assume a specific gravity of .8.

(1) Calculate q_1 from Equation 6.59

$$q_1 = 1214.5 \, (0.01414)\sqrt{\frac{5500 - 1250}{0.8}}$$
$$= 1252 \text{ b/d}$$

(2) Calculate H from Equation 6.40

$$H = \left(\frac{2500 - 1250}{5500 - 2500}\right) = 0.417$$

(3) Obtain M from Fig. 6.6

$$M = 0.54$$

(4) Calculate q_3, the production rate, using the relations of Equation 6.4.

$$M = \frac{q_3}{q_1}; q_3 = q_1 \ (M)$$
$$q_3 = 1252 \, (0.54) = 676 \text{ b/d}$$

CLASS PROBLEM 3A:

Using the pressures of problem #3, which throat number should be used with the No. 7 nozzle to obtain the maximum production? What is the maximum production? Assume a specific gravity of 0.8.

EXAMPLE PROBLEM #4

It is desired to produce 1000 b/d with a pump intake pressure (P_3) of 650 psi and a discharge pressure (P_2) of 2000 psi. What is the most efficient non-cavitating nozzle and throat combination such that P_1 is less than 7000 psi? What is the power fluid pressure (P_1)? What is the power fluid rate q_1? Assume a specific gravity of 0.8, I_c = 1.35, K_j = 0.15, and P_v = 0.

(1) Determine the H for maximum efficiency for each ratio. From Fig. 6.6, the following Table 6.4 can be constructed.

TABLE 6.4
VALUES OF M, H, AND EFFICIENCY FOR EXAMPLE PROBLEM #4

Ratio	M @ maximum efficiency	H @ maximum efficiency	Efficiency
A	0.475	0.475	22.6%
B	0.675	0.360	24.6
C	0.900	0.282	25.4
D	1.15	0.225	25.6
E	1.425	0.180	25.6

(2) Solve Equation 6.40 for P_1.

$$H = \frac{P_2 - P_3}{P_1 - P_2} \qquad P_1 = \frac{1}{H}[P_2(1 + H) - P_3] \qquad (6.61)$$

(3) Using the expression for P_1 above, calculate the P_1 values for each H determined in 1).

A Ratio: $P_1 = \left(\dfrac{1}{0.475}\right)[2000(1 + 0.475) - 650] = 4842$ psi

B Ratio: $P_1 = \left(\dfrac{1}{0.360}\right)[2000(1 + 0.360) - 650] = 5750$ psi

C Ratio: $P_1 = \left(\dfrac{1}{0.282}\right)[2000(1 + 0.282) - 650] = 6787$ psi

D Ratio: $P_1 = \left(\dfrac{1}{0.225}\right)[2000(1 + 0.225) - 650] = 8000$ psi

E Ratio: $P_1 = \left(\dfrac{1}{0.180}\right)[2000(1 + 0.180) - 650] = 9500$ psi

Eliminate D and E since their P_1's exceed 7000 psi.

(4) Using Equation 6.51, calculate the cavitation limiting value M_c and compare it with the values of M obtained in step (1) for A, B, and C ratios.

A Ratio: $M_c = \left(\dfrac{1 - .410}{0.410}\right)\sqrt{1 + 0.15 \times \dfrac{650}{1.35(4842 - 650) + 650}}$

$M_c = .495$

B Ratio: $M_c = \left(\dfrac{1 - .328}{0.328}\right)\sqrt{1 + 0.15 \times \dfrac{650}{1.35(5750 - 650) + 650}}$

$M_c = .645$

C Ratio: $M_c = \left(\dfrac{1 - .262}{0.262}\right)\sqrt{1 + 0.15 \times \dfrac{650}{1.35(6787 - 650) + 650}}$

$M_c = 0.815$

By comparing the M_c values obtained with the M values at maximum efficiency in (1) above, note that the B and C ratios are cavitating at their maximum efficiency points but the A ratio is not. However, the efficiency of the B ratio @ $M = M_c$ is 24.5% and the C ratio @ $M = M_c$ is 25.4% while the maximum obtainable efficiency with an A ratio is 22.6%.

(5) Calculate P_1 for a C ratio at $M = 0.815$ since it is the most efficient.

From Fig. 6.6, H = .31

$P_1 = \dfrac{1}{.310}[2000(1 + .310) - 650] = 6355$ psi

(6) Calculate q_1 from $M = \dfrac{q_3}{q_1}$ for a C ratio

$q_1 = \dfrac{q_3}{M} = \dfrac{1000}{.815} = 1227$ b/d

(7) Select a nozzle using Equation 6.60

$A_j = \dfrac{q_1}{1214.5\sqrt{\dfrac{P_1 - P_3}{\gamma_1}}}$

$A_j = 0.01196$ in^2

Referring to Table 6.1, note that the closest nozzle is a No. 6.

This application would best be pumped, therefore, with a 6-C jet pump using about 1227 b/d power fluid at about 6355 psi. However, this pump is on the verge of cavitation and no greater production flow could be obtained if desired.

CLASS PROBLEM #4A:

It is desired to pump 2640 b/d production using 1760 b/d power fluid. The discharge pressure P_2 is 2000 psi and the suction pressure P_3 is 1200 psi. What is the most efficient ratio? What size nozzle should be used and what will the operating pressure be? Assume a specific gravity of 0.8, $I_c = 1.35$, $K_j = 0.15$, and $P_v = 0$.

CLASS PROBLEM 4B:

Can the above pumping requirements be met if $P_v = 600$ psi?

6.42   Sizing a jet pump for a well

6.421   Introduction

The preceding sections have dealt only with the jet pump itself and have not considered the effects of such well variables as pressures caused by fluid column gradients, temperatures, gas-oil ratios, or power fluid supply pressures or flow limitations. The example problems and class problems have, however, used pressures and flow rates typical of the values one would arrive at in considering well conditions.

Fig. 6.7 shows the pressures and friction losses affecting a jet pump in a well installation. Fig. 6.7 is basically a combination of Fig. 6.5 and Fig. 5.48 of Chapter 5. A jet pump installation is, of course, always an open power fluid system. The installation illustrated is an OPF parallel system, chosen for clarity of the nomenclature, although use of the casing annulus for the return column is more common (Fig. 5.30).

The procedure of sizing a jet pump for a well can be approached in a number of ways. One could, for example, try all of the combinations of nozzles and throats listed in Table 6.1 at different power fluid pressures and see which combination gives the optimum operating parameters for the desired production. With five ratios for each nozzle size, however, the calculations would be extremely lengthy and tedious. Clearly, a procedure that eliminates those nozzles and throats that are not practical is desirable. Then from the remaining practical combinations, detailed calculations could be made to select the optimum combination.

Previous discussions and problems have pointed out that operating pressures and cavitation limitations often severely restrict the number of practical nozzle and throat combinations that could be used in a given well. The selection procedure should, therefore, emphasize these considerations.

6.422   Cavitation and percent submergence in a well

A convenient procedure for estimating the cavitation limitation of a jet pump in a well involves the concept of *percent submergence*. Referring to Fig. 6.7, note that the total lift is h_1 and the submergence is h_3. Define the percent submergence (f_{h_3}) as

$$f_{h_3} = \dfrac{h_3}{h_1} \tag{6.62}$$

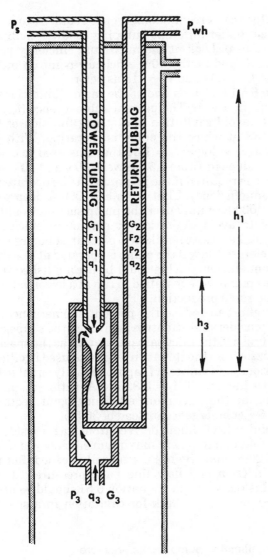

P_s = SURFACE OPERATING PRESSURE, PSI

P_{wh} = SURFACE FLOW LINE BACK PRESSURE, PSI

F_1 = FRICTION IN POWER TUBING, PSI

F_2 = FRICTION IN RETURN COLUMN, PSI

G_1 = FLUID GRADIENT IN POWER TUBING, PSI/FT

G_2 = FLUID GRADIENT IN RETURN COLUMN, PSI/FT

G_3 = FLUID GRADIENT OF FORMATION FLUID, PSI/FT

h_1 = PUMP SETTING DEPTH, FT

h_3 = PUMP SUBMERGENCE, FT

$P_1 = h_1 G_1 - F_1 + P_s$

$P_2 = h_1 G_2 + F_2 + P_{wh}$

$P_3 = h_3 G_3$

Fig. 6.7   Pressures and friction losses affecting a jet pump

Further note that if $G_2 = G_3$,

$$\frac{h_3}{h_1} = \frac{P_3}{P_2} = f_{h_3} \qquad (6.63)$$

providing return column friction, F_2, and flow line back pressure, P_{wh}, are neglected.

Equation 6.48 gives

$$M_c = \frac{1-R}{R}\sqrt{\frac{P_3 - P_v}{I_c H_v}} = \frac{1-R}{R}\frac{1}{\sqrt{I_c}}\sqrt{\frac{P_3 - P_v}{H_v}} \qquad (6.64)$$

and 6.49 is

$$H_v = \frac{P_1 - P_3}{(1 + K_j) - (1 + K_s)\dfrac{M_c^2 R^2}{(1-R)^2}} \qquad (6.49)$$

Substituting 6.49 into 6.64 gives

$$M_c = \frac{1-R}{R}\frac{1}{\sqrt{I_c}}\sqrt{\frac{P_3 - P_v}{(P_1 - P_3)\Big/\left(1 + K_j - (1+K_s)\dfrac{M_c^2 R^2}{(1-R)^2}\right)}} \qquad (6.64a)$$

Equation 6.63 gives:

$$P_3 = f_{h_3} P_2 \qquad (6.65)$$

Substituting 6.58 into 6.40, the expression for H, gives

$$H = \frac{P_2 - P_3}{P_1 - P_2} = \frac{P_2 - f_{h_3} P_2}{P_1 - P_2} \qquad (6.66)$$

Solving 6.66 for P_1 gives

$$HP_1 - HP_2 = P_2 - f_{h_3} P_2$$

$$P_1 = \frac{1}{H}(P_2 - f_{h_3} P_2 + HP_2)$$

$$P_1 = \frac{1}{H}[P_2(1 - f_{h_3} + H)] \qquad (6.67)$$

Substituting 6.65 and 6.67 into 6.64 gives M_c in terms of f_{h_3} and P_2

$$M_c =$$

$$\frac{1-R}{R}\frac{1}{\sqrt{I_c}}\sqrt{\frac{(f_{h_3}P_2 - P_v)\Big(1 + K_j - (1+K_s)\dfrac{M_c^2 R^2}{(1-R)^2}\Big)}{\dfrac{1}{H}[P_2(1 - f_{h_3} + H)] - f_{h_3}P_2}}$$

and, if $P_v = 0$ and $K_s = 0$ as before,

$$M_c = \left(\frac{1-R}{R}\right) \times \frac{1}{\sqrt{I_c}} \times$$

$$\sqrt{\frac{f_{h_3} P_2}{P_2\left(\frac{1}{H}(1 - f_{h_3} + H) - f_{h_3}\right)}} \left[1 + K_j - \left(\frac{M_c R}{1-R}\right)^2\right]$$

$$M_c = \frac{1-R}{R}\frac{1}{\sqrt{I_c}}\sqrt{\frac{f_{h_3}}{\frac{1}{H} - \frac{f_{h_3}}{H} + 1 - f_{h_3}}} \times$$

$$\sqrt{1 + K_j - \left(\frac{M_c R}{1-R}\right)^2} \qquad (6.68)$$

Solving Equation 6.61 for f_{h_3}:

$$\left(\frac{M_c \sqrt{I_c} R}{1-R}\right)^2 = \left(\frac{f_{h_3}}{\frac{1}{H} - f_{h_3}\left(\frac{1}{H} + 1\right) + 1}\right) \times$$

$$\left(1 + K_j - \left(\frac{M_c R}{1-R}\right)^2\right)$$

$$\frac{f_{h_3}}{\left(\frac{1}{H} + 1\right) - f_{h_3}\left(\frac{1}{H} + 1\right)} = \frac{f_{h_3}\left(\frac{M_c R}{1-R}\right)^2}{1 + K_j - \left(\frac{M_c R}{1-R}\right)^2}$$

$$= \frac{f_{h_3}}{\left(\frac{1}{H} + 1\right)(1 - f_{h_3})}$$

Let $\beta_R = \dfrac{I_c\left(\dfrac{M_c R}{1-R}\right)^2}{1 + K_j - \left(\dfrac{M_c R}{1-R}\right)^2}$ \qquad (6.69)

Then $\dfrac{f_{h_3}}{\left(\dfrac{1}{H} + 1\right)(1 - f_{h_3})} = \beta_R$

and $f_{h_3} = \left(\dfrac{1}{H} + 1\right)\beta_R - \left(\dfrac{1}{H} + 1\right)f_{h_3}\beta_R$

and $f_{h_3}\left[1 + \left(\dfrac{1}{H} + 1\right)\beta_R\right] = \left(\dfrac{1}{H} + 1\right)\beta_R$

$$f_{h_3} = \frac{\left(\dfrac{1}{H} + 1\right)\beta_R}{1 + \left(\dfrac{1}{H} + 1\right)\beta_R} \text{ where } \beta_R \text{ is defined by Eq. } 6.69$$

$$(6.70)$$

We now have an expression for f_{h_3}, the percent submergence, in terms of R, which is a geometric characteristic of the pump; K_j and I_c, which are experimentally determined coefficients; and M_c which is the M value at which cavitation inception occurs.

A table can be constructed from Equation 6.70 which illustrates the percent submergence required to avoid cavitation under different pumping conditions (different M values). Look at the maximum efficiency point for each R, and at the 20% efficiency point on either side. Let $I_c = 1.35$ and $K_j = .15$ as before.

Table 6.5 illustrates one of the major limitations of jet pumps in a well. The percent submergence (f_{h_3}) required to avoid cavitation is a strong function of the value of M at which the pump is operating. With the conservative value of $I_c = 1.35$ that was used to calculate the values in Table 6.5, between 30% and 40% submergence is required if the jet pump is to be operated at maximum efficiency. That is, in a 10,000 ft deep well, 3000 to 4000 ft of fluid over the pump suction would be required to avoid cavitation at maximum efficiency. Lower intake pressures (lower percent submergence) can be tolerated only if the pump is operated at less than maximum efficiency. This could be accomplished if an oversized pump were installed to run at a lower value of M for the same production.

In an effort to reduce the percent submergence required, considerable attention is given to the shape and surface finish of the inlet passages as well as the spacing of the nozzle from the throat on jet pumps used in oilfield applications, and values of I_c approximately equal to 0.8 can be achieved. If $I_c = 0.8$, an A ratio jet pump operating at the maximum efficiency point requires only 20.6% submergence instead of 30.5%.

A useful rule of thumb derived from such considerations is that to pump near maximum efficiency, a minimum of 20% submergence is generally needed. Return column friction and flow line back pressure can increase this requirement. Equation 6.51 should be used, therefore, as a final check for cavitation in a specific well.

6.4221 Estimating pump intake pressure

Particularly because of the damage problems associated with cavitation, it is often desirable to have a means of checking the pump intake pressure while the pump is operating. Other types of artificial lift equipment are constructed so that an estimate of the pump discharge pressure must be made before a pump intake pressure can be calculated. Since the pump discharge pressure often depends on vertical multiphase flow pressure gradients, this can be a cumbersome calculation. The design of the jet pump, however, avoids this problem.

As noted in section 6.324, the flow rate through the nozzle does *not* depend on the pump discharge pressure. If Eq. 6.59 is solved for P_3, the pump intake pressure, the following expression results:

$$P_3 = P_1 - \left(\frac{q_1}{1214.5 A_j}\right)^2 \gamma_1 \qquad (6.70a)$$

TABLE 6.5
PERCENT SUBMERGENCE (f_{h_3}) TO AVOID CAVITATION

E	A R = .410			B R = .328			C R = .262			D R = .210			E R = .168		
	M	H	f_{h_3}(%)	M	H	f_{h_3}(%)	M	H	f_{h_3}(%)	M	H	f_{h_3}(%)	M	H	f_{h_3}(%)
0.20	0.320	0.628	13.6	0.375	0.530	10.5	0.475	0.419	10.4	0.605	0.329	11.1	0.78	0.256	12.7
Max.	0.475	0.475	30.5	0.675	0.365	34.5	0.900	0.282	37.4	1.150	0.223	39.6	1.425	0.180	40.7
0.20	0.655	0.308	55.7	0.965	0.207	65.3	1.295	0.155	69.4	1.66	0.120	72.0	2.06	0.097	73.0

As indicated in Fig. 6.7, P_1 is equal to the surface operating pressure (P_s) plus the hydrostatic head $(h_1 G_1)$ minus the fluid friction pressure loss in the power fluid conduit (F_1).

Therefore:

$$P_3 = P_s + h_1 G_1 - F_1 - \left(\frac{q_1}{1214.5\,A_j}\right)^2 \gamma_1 \quad (6.71)$$

With this expression, the pump intake pressure can be calculated if the power fluid rate and pressure can be measured. The power fluid friction loss must also be calculated, but the curves of Appendix 5.C can be used to simplify this part of the procedure.

6.423 Sizing procedure for a well producing no gas

6.4231 Estimating fluid column gradients

Since the jet pump is inherently an open power fluid device, the fluid gradient in the return column will be determined by the mixture of power fluid and produced fluid. If the power fluid is the produced oil and there is no water cut, the power fluid column and production column gradients are identical. If any water is present, however, either as a produced water cut or as power fluid, the return column gradient will depend on the value of M at which the pump is operating. Specifically:

$$G_2 = .4331\,\text{psi ft} \left[\frac{q_1(\text{S.G. Power Fluid}) +}{q_1 + q_3}\right.$$
$$\left. + \frac{q_3 f_w(\text{S.G. Prod. Water}) + q_3(1 - f_w)(\text{S.G. oil})}{q_1 + q_3}\right]$$

where f_w is the water cut.

But since $q_1 = \dfrac{q_3}{M}$ from 6.4

$$G_2 = .4331$$
$$\left[\frac{\dfrac{\text{S.G. Power Fluid}}{M} + f_w(\text{S.G. Prod. Water})}{\dfrac{1}{M} + 1}\right.$$
$$\left. + \frac{(1 - f_w)(\text{S.G. oil})}{\dfrac{1}{M} + 1}\right]$$
$$(6.72)$$

For the initial estimate, when M is unknown, let M = 1. Then

$$G_2 = .2166[(\text{S.G. Power Fluid}) + f_w(\text{S.G. Prod. Water}) + (1 - f_w)(\text{S.G. oil})] \quad (6.73)$$

6.4232 Ranges of different ratios

Equation 6.40 can be expanded to include the power fluid column and surface operating pressure effects (Fig. 6.7).

$$P_1 = P_s + h_1 G_1 - F_1 \quad (6.74)$$
$$P_2 = h_1 G_2 + F_2 + P_{wh} \quad (6.75)$$

Equations 6.40, 6.74 and 6.75 give

$$H = \frac{(h_1 G_2 + F_2 + P_{wh}) - P_3}{(P_s + h_1 G_1 - F_1) - (h_1 G_2 + F_2 + P_{wh})} \quad (6.76)$$

For initial estimates, the tubing friction loss terms F_1 and F_2 are often neglected.

$$H \cong \frac{(h_1 G_2 + P_{wh}) - P_3}{(P_s + h_1 G_1) - (h_1 G_2 + P_{wh})} \quad (6.76a)$$

As previously discussed, P_3 must generally be equal to 20% of P_2. Thus,

$$H = \frac{.8(h_1 G_2 + F_2 + P_{wh})}{P_s + h_1(G_1 - G_2) - F_1 - F_2 - P_{wh}} \quad (6.77)$$

Neglecting friction,

$$H = \frac{.8(h_1 G_2 + P_{wh})}{P_s + h_1(G_1 - G_2) - P_{wh}} \quad (6.77a)$$

Normally the surface operating pressure, P_s, is between 1000 psi and 4000 psi. Inserting these values into Equation 6.77a with the h_1, G_1, G_2 and P_{wh} values for the well will give the range of H values at which it is possible to operate.

Figure 6.6 shows which ratios are capable of pumping within this range of H's. The following table is included as an example and to illustrate the application areas of the various ratios. Flow line back pressure has been taken to be 80 psi and G_1 has been assumed equal to G_2, with a value of 0.355 psi/ft. Equation 6.77a then reduces to

$$H = \frac{0.8[h_1(.355) + 80]}{P_s - 80} = \frac{0.3465\,h_1 + 64}{P_s - 80} \quad (6.78)$$

Table 6.6 is constructed from Equation 6.78. Note that the lift is 80% of the setting depth. The most efficient ratio is underlined in each case.

TABLE 6.6
RANGE OF OPERATING "H" VALUES

Setting depth (ft)	1000 ft		2000 ft		5000 ft		8000 ft	
Operating pressure (psi)	H	Ratio	H	Ratio	H	Ratio	H	Ratio
1000	.45	ABC	.82	A	1.95	-	3.08	-
2500	.17	ABCDE	.31	ABCDE	.74	AB	1.17	-
4000	.10	ABCDE	.19	ABCDE	.46	ABC	.72	AB

6.4233 Operating pressure

At a setting depth of 2,000 ft, any of the ratios could be used, depending on the operating pressure chosen (see Table 6.6). In such a case, the decision as to which ratio to employ will depend on the nature of the particular installation. Often the operator will prefer to use less power fluid and run at higher pressures (E ratio) to minimize the power fluid rate q_1, thereby reducing tubing friction and the volume of fluid to be handled and treated on the surface. Other operators may prefer to handle larger volumes of fluid on the surface in exchange for the decreased surface equipment maintenance associated with lower operating pressures (A ratio).

The friction losses in the fluid conduits will be less with small volumes of high pressure fluid, and the surface treatment and separation of the mixed power and produced fluid will be easier. For examples in this text, therefore, the larger throats (smaller R values) will be regarded as more desirable. The successful application of these "sensitive" ratios, however, depends on accurate well performance data.

With the above considerations in mind, it is suggested that the installation design be based on the highest pressure considered acceptable for the surface power unit. Examples in this text will use 4000 psi. As is evident from Table 6.6 this choice will determine the optimum ratio.

6.4234   Selecting a ratio and nozzle from the value of H

With $P_s = 4000$ psi and an assumed value of $M = 1$, calculate H from Equation 6.76a (friction neglected).

With this value of H, obtain the most efficient ratio from Figure 6.6 and the associated M value.

This value of M can now be used to correct Equation 6.72 for G_2 and to determine the values for F_1 and F_2.

The G_2, F_1 and F_2 corrections can then be included in Equation 6.76 to more accurately predict H and M. From Equation 6.4, obtain

$$q_1 = \frac{q_3}{M} \qquad (6.4a)$$

And, from equation 6.60, the nozzle flow area is obtained

$$A_j = \frac{q_1}{1214.5\sqrt{\dfrac{P_1 - P_3}{\gamma_1}}} \qquad (6.60)$$

From Table 6.1, select the closest nozzle to that calculated from Equation 6.60. Selecting the nearest smaller nozzle will lead to an operating pressure greater than the assumed value of 4000 psi. The next larger nozzle will lead to an operating pressure lower than the initially assumed P_s.

6.4235   Correcting for the difference between the calculated nozzle and available nozzle sizes

The available equations are:

$$q_1 = 1214.5\, A_j \sqrt{\frac{P_1 - P_3}{\gamma_1}} \qquad (6.59)$$

$$M = \frac{q_3}{q_1} \qquad (6.4)$$

$$P_1 = \frac{1}{H}[P_2(1 + H) - P_3] \qquad (6.61)$$

Subtracting P_3 from both sides of 6.61 leads to

$$P_1 - P_3 = \frac{P_2}{H} + P_2 - \frac{P_3}{H} - P_3 = P_2\left(\frac{1}{H} + 1\right) - P_3\left(\frac{1}{H} + 1\right)$$

$$P_1 - P_3 = \left(\frac{1}{H} + 1\right)(P_2 - P_3) \qquad (6.79)$$

Substituting 6.79 into 6.59 gives

$$q_1 = 1214.5\, A_j \sqrt{\frac{\left(\dfrac{1}{H} + 1\right)(P_2 - P_3)}{\gamma_1}} \qquad (6.80)$$

Equation 6.4 gives

$$q_3 = Mq_1 \qquad (6.81)$$

Equations 6.72 and 6.73 lead to

$$q_3 = M(1214.5)\, A_j \sqrt{\frac{\left(\dfrac{1}{H} + 1\right)(P_2 - P_3)}{\gamma_1}} \qquad (6.82)$$

Rearranging all of the known terms for this problem to the left hand side gives

$$\frac{q_3}{1214.5\, A_j\sqrt{\dfrac{P_2 - P_3}{\gamma_1}}} = M\sqrt{\left(\frac{1}{H} + 1\right)} \qquad (6.83)$$

Define:

$$\frac{q_3}{(1214.5)\, A_j\sqrt{\dfrac{P_2 - P_3}{\gamma_1}}} = \Theta_R \qquad (6.84)$$

The problem has now been reduced to finding an M and H that both satisfy Equation 6.83 and fall on the H-M curve for the particular ratio. Θ_R is plotted against M for the various ratios in Figures 6.8, 6.9, 6.10, 6.11 and 6.12. H is also plotted for easy reference.

Calculate Θ_R for the standard nozzle size and ratio selected. This determines M and H.

P_1 is determined from Equation 6.61 and P_s, the surface operating pressure, is determined from a rearrangement of Equation 6.74.

$$P_s = P_1 - h_1G_1 + F_1 \qquad (6.85)$$

The power fluid rate, q_1, is obtained from Equation 6.4a.

The corrections obtained from this procedure will generally not be sufficiently large as to necessitate recalculating the column densities or frictions.

6.4236   Example problem #5.

Tubing O.D.	= 2⅜ in.
Casing	= 7 in.
Setting depth	= 7600 ft
Separator pressure	= 80 PSI
Flow line pressure	= 80 PSI (short flow line)
Static bottom hole pressure	= 1500 PSI
Productivity index	= 0.2 b/d/PSI (Assume Constant)
API gravity of crude	= 41°
Well head temperature	= 110°F.
Bottom hole temperature	= 167°F.
Water cut	= 0.0%
Solution gas-oil ratio	= 300 scf/B
Desired production	= 200 b/d

Select an appropriate jet pump finding nozzle size, surface operating pressure, power fluid rate and hydraulic HP.

First consider this application assuming that the gas-oil ratio is zero. For the jet pump assume $K_j = 0.15$, $I_c = 1.35$. Further assume that an OPF casing installation will be used.

Fig. 5.46 of Chapter 5 indicates an expected viscosity for the 41° API crude of approximately 2 cs at an average temperature of 138°F. in the power tubing. To pump this well, use no more than 500 to 600 b/d of power fluid. Fig. 5.C(7) (Appendix 5.C of Chapter 5) shows that the pressure drop in 2⅜ tubing at 500 B/D with a viscosity of 2 cs is only 4.4 PSI/1000 ft × sp.

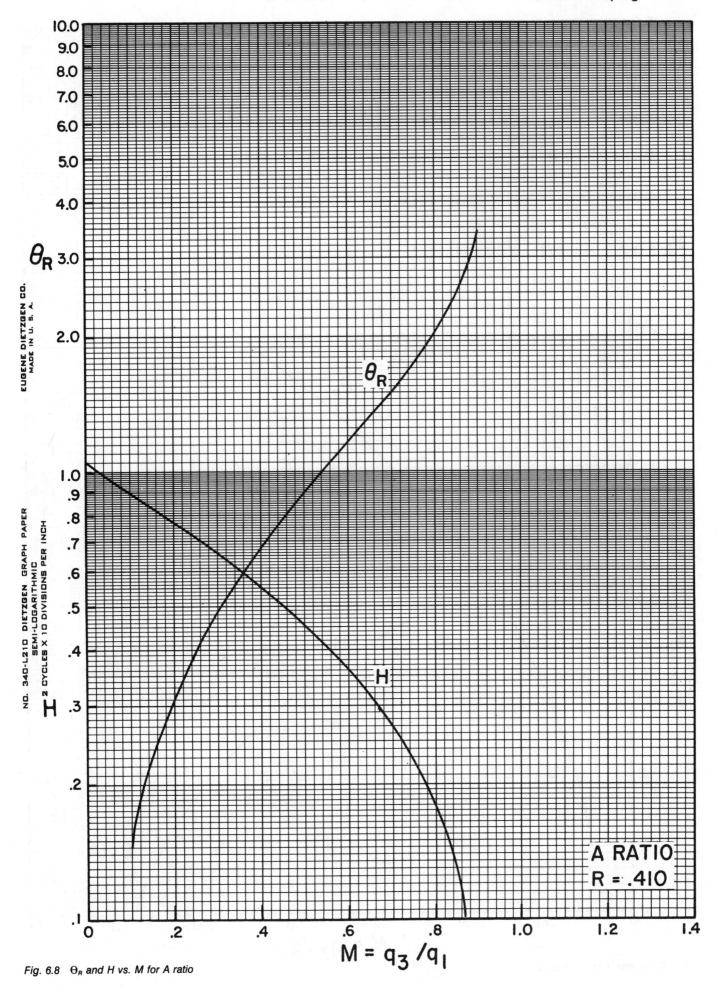

Fig. 6.8   θ_R and H vs. M for A ratio

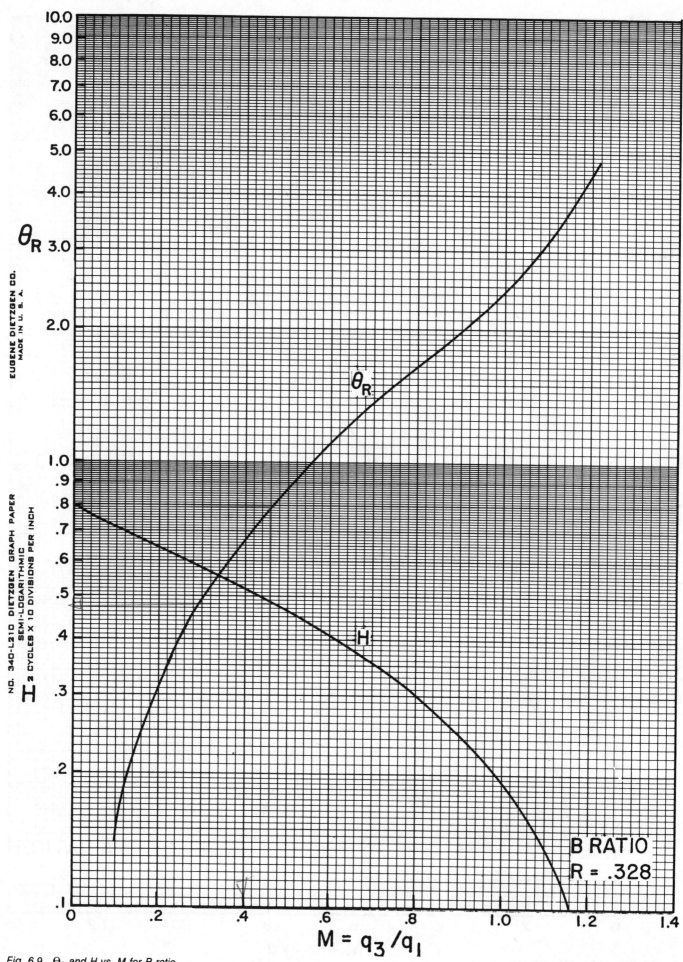

Fig. 6.9 Θ_R and H vs. M for B ratio

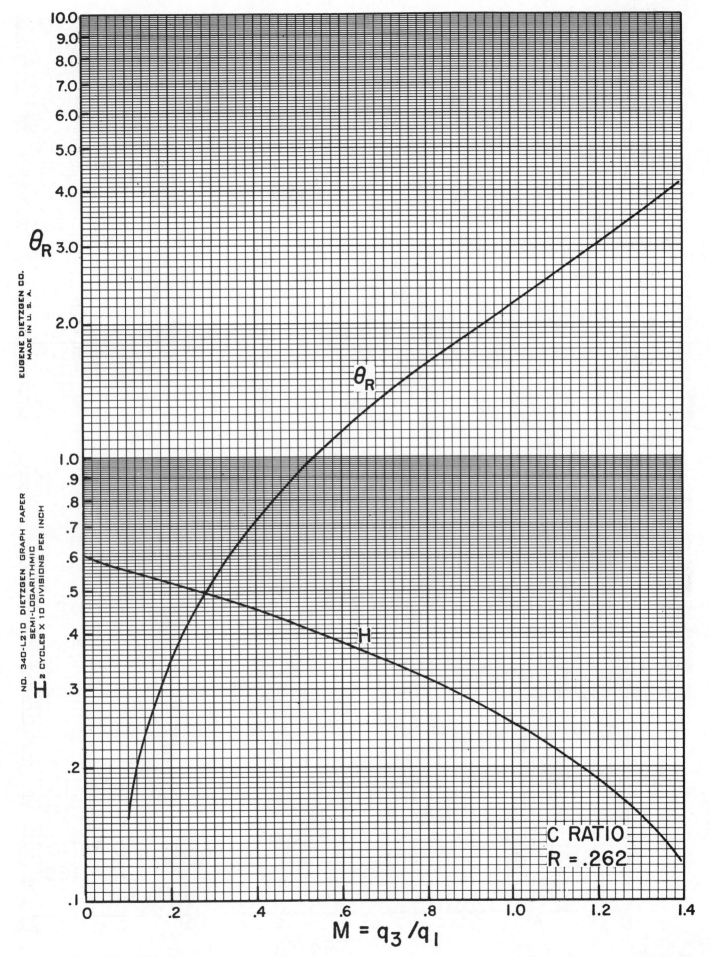

Fig. 6.10 θ_R and H vs. M for C ratio

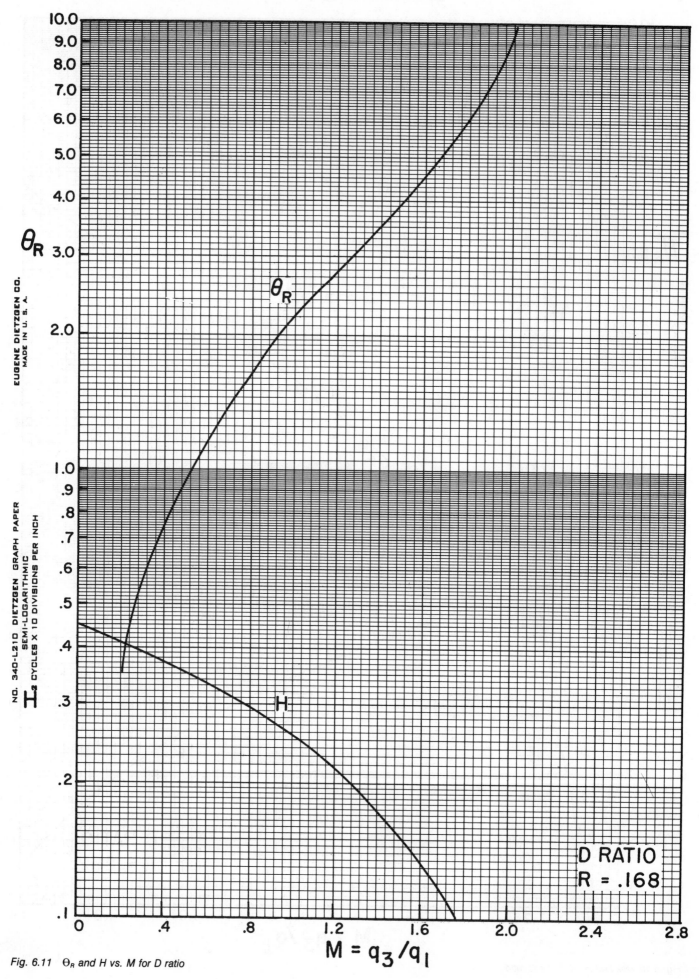

Fig. 6.11 θ_R and H vs. M for D ratio

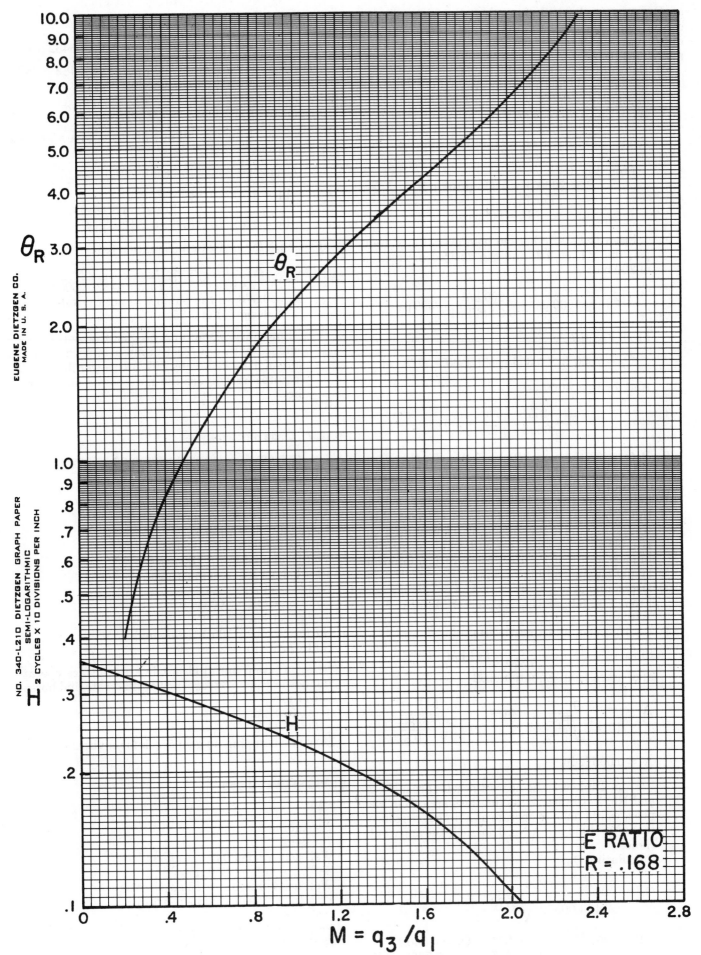

Fig. 6.12 θ_R and H vs. M for E ratio

gr. For our initial calculations, then, the friction terms can be neglected in the power string. The casing annulus return will have even less friction loss and can also be neglected.

(1) Determine H. Assume $P_s = 4000$ psi.

41°API crude has a specific gravity of 0.8203 and a gradient of 0.355 psi/ft from Table 5.2, of Chapter 5. Since there is no water present,

$$G_1 = G_2 = G_3 = 0.355$$

From Fig. 6.7
$$P_1 = 7600 \text{ ft} \times 0.355 \text{ psi/ft} + P_s$$
$$\text{For } P_s = 4000 \text{ psi}$$
$$P_1 = 6698 \text{ psi}$$

From the static bottom hole pressure of 1500 psi, the P.I. of 0.2, and the desired production of 200 b/d,

$$P_3 = 1500 \text{ psi} - \frac{200 \text{ b/d}}{0.2 \text{ b/d/PSI}} = 500 \text{ psi}$$

From Fig. 6.7

$$P_2 = 7600 \text{ ft} \times 0.355 \text{ psi/ft} + 80 = 2778 \text{ psi}$$

And for $P_s = 4000$ psi

$$H = \frac{2778 - 500}{6698 - 2778} = .581$$

(2) Determine the most efficient ratios for the value of H calculated in 1.

From Fig. 6.6, only the A and B ratios can pump effectively at H = 0.581 or greater, and in this range the A ratio is the most efficient. Furthermore, since the efficiency is declining at H = 0.581 or greater, we will want to use the highest value of P_s (4000 psi) which minimizes H and maximizes efficiency.

(3) Determine M.

From Fig. 6.6, at H = .581 for an A ratio, M = .370.

(4) Determine q_1 and a nozzle size.

Since $q_1 = \frac{q_3}{M}$, $q_1 = \frac{200}{0.370} = 541$ b/d

and, since from Equation 6.60

$$A_j = \frac{q_1}{1214.5\sqrt{\frac{P_1 - P_3}{\gamma_1}}}, A_j = \frac{541}{1214.5\sqrt{\frac{6698 - 500}{0.8203}}}$$

$$A_j = 0.00512 \text{ in.}^2$$

From Table 6.1, the nearest nozzles are a No. 2 and No. 3. the No. 2, having less area than we need, would have to operate at a higher pressure than $P_s = 4000$ psi. A No. 3 nozzle is the apparent choice.

Calculating Θ_R:

$$\Theta_R = \frac{200}{1214.5(.00579)\sqrt{\frac{2778 - 500}{.8203}}} = .540$$

From Fig. 6.8   M = 0.335 at Θ_R = 0.540
and H = 0.615 at M = 0.335

From Equation 6.61

$$P_1 = \frac{(1)}{(0.615)}[2778(1.615) - 500] = 6482 \text{ psi}$$
$$P_s = P_1 - 7600 \text{ ft} \times .355 \text{ psi/ft} = 6482 - 2698$$
$$= 3784 \text{ psi (neglecting friction)}.$$

And, since $q_1 = \frac{q_3}{M}$; $q_1 = \frac{200}{0.335} = 597$ b/d

(Note: Using Equation 6.59 for q_1 gives 600 b/d as a check)

the surface horsepower is

$$hp = 1.7 \times 10^{-5} q_1 \times P_s$$
$$\text{(Equation 5.18 of Chapter 5)}$$
$$hp = 1.7 \times 10^{-5} (597)(3784) = 38.4$$

(6) Check for cavitation (Equation 6.51)

$$M_c = \frac{(1 - .410)}{0.410}\sqrt{1.15}\sqrt{\frac{500}{1.35(6482 - 500) + 500}}$$
$$M_c = 0.373, \text{ so the pump will not cavitate at the design M of .335.}$$

Summary: Assuming no gas, the well can be pumped with a 3-A jet pump at a surface pressure of 3784 PSI using 597 B/D power fluid. This requires 38.4 hydraulic horsepower at the surface.

6.4237 Class problems

Class Problem #5A (Assume pumping no gas)

Given:

2⅜″ tubing
5½″ casing
Depth of well = 8000 ft
Setting depth = 7950 ft
Wellhead pressure = 100 psi
PI = 0.3 b/d psi (Assume constant)
Crude = 35°API
Solution GOR = 400 scf/b
Temp. bottom = 170°F.
Surface temp. = 105°F. (Flowing)
Static pressure = 2000 psi

Select a pump and find nozzle size, surface operating power fluid pressure, power fluid rate and horsepower for:
(a) 300 b/d
(b) 450 b/d
(Assume pumping no gas) check for cavitation

CLASS PROBLEM #5B

Given:

2⅞″ tubing
7″ casing
Well depth = 9000 ft
Setting depth = 8850 ft
Well head pressure = 120 psi
PI = 0.5 b/d/psi (Assume constant)
Crude = 35°API
Producing 50% water ($\gamma_w = 1.07$)
GOR = 400 scf/b
GLR = 200 scf/b

Temp. bottom = 200°F.
Flowing temp. surface = 115°F.
Static pressure = 2200 psi
(Assume pumping no gas)

Select a pump and find nozzle size, surface power pressure, power fluid rate and HP for:
   (a) 300 b/d
   (b) 600 b/d
   (c) 750 b/d
check all for cavitation.
Approximate the maximum rate possible from this well with a reasonable efficiency.

6.424 Pumping the well if the gas-oil ratio is greater than zero

6.4241 Introduction

Introducing solution gas alters the problem of applying the jet pump to the well in four significant respects.

The first consideration is the type of bottom hole assembly to use. With no gas, the casing type bottom hole assembly was chosen for its simplicity and low cost. With this type of installation, however, all of the solution and free gas must pass through the pump. An alternative is to run a parallel return string and allow the free gas to vent up the casing annulus. In this arrangement the pump would, in principle, only have to handle the solution gas remaining at the pumping bottom hole pressure.

In practice, however, the efficiency of separation of the free gas from the fluid is hard to predict. Furthermore, the power fluid rate to the jet pump can be increased to help pump free gas, a procedure often unacceptable with a reciprocating pump because of the resulting higher stroke rates which tend to shorten pump life markedly. As a consequence, most jet pump installations employ the casing type bottom hole assembly, and such an installation will be assumed for this well.

A second consideration is the effect of the gas on the return column gradient. The use of multiphase flow correlations is necessary, but in practice this is complicated by the fact that the ratio of production to power fluid (M) is not constant with the jet pump. This in turn means that the gas-liquid ratio depends on M. A similar problem manifests itself when water is present, either as formation water or as power fluid. In such a case, the return column water cut is a function of M, again complicating friction, gradient, or multiphase flow calculations.

The conception of cavitation becomes very difficult to deal with if gas is present. In laboratory tests with water, the onset of cavitation is fairly abrupt and predictable. Similarly, in laboratory tests oil is fairly stable at pressures down to the point of cavitation. Crude oil with solution gas, however, will liberate gas continuously as the pressure is lowered below the bubble point. In terms of pump performance, this creates a gradually increasing choking effect on the pump as the pressure decreases, analogous to cavitation choking. True cavitation may not be taking place, however, and even if it is there is evidence to indicate that the presence of free gas greatly reduces the resultant damage.

A final question arises with the presence of gas; and it relates to how the presence of two phases affects the mixing and pressure recovery of the jet pump. A given amount of free gas associated with the liquid phase will occupy some portion of the throat, thereby increasing the fluid velocities. This can have significant effects on the mixing loss terms, the length of throat required for complete mixing, and on the friction terms. Additionally, the performance of diffusers is difficult to predict with two phase flow, particularly with the associated uncertainty of the rates at which free gas goes back into solution. All of these effects are shape sensitive and may vary considerably among pumps by different manufacturers.

Approximate solutions can be obtained, however, which illustrate the nature of the analytical prediction techniques and which give a reasonable estimate on the feasibility of using a jet pump and what its power requirements would be.

An assumption that leads to reasonable predictions of jet pump performance with gas present is that the pump will produce gas and liquid equally well on a volume basis. This assumption is only an approximation, but it holds reasonably well up to about 10 parts of gas to one part liquid at bottom hole conditions. Figure 5.43 of Chapter 5 can be used for jet pumps as well as piston pumps for this calculation.

The procedure, then, is to assume a value for P_s, calculate H, and to make a correction to the M value based on the expected volumetric efficiency obtained from Fig. 5.43 of Chapter 5. The gas-lift effect in the return column, however, can markedly change the value of the pump discharge pressure (P_2) thus changing H. The first step, therefore, must be to calculate P_2 using an appropriate multiphase flow correlation or use gradient curves such as found in Appendix 3B of Chapter 3. A trial value of M = 0.5 is suggested when gas is present since the volumetric efficiency of the pump is diminished.

The gas-liquid ratio at the return column fluid is a function of M given by:

$$GLR = \frac{Total\ Gas}{Power\ Fluid + Production\ Fluid}$$

$$GLR = \frac{(G.O.R.)\,(Production\ Oil)}{q_1 + q_3}$$

$$GLR = \frac{(G.O.R.)\,(1 - f_w)\,q_3}{q_1 + q_3}$$

$$GLR = \frac{(G.O.R.)\,(1 - f_w)\,q_3}{\dfrac{q_3}{M} + q_3} = \frac{G.O.R.\,(1 - f_w)}{\dfrac{1}{M} + 1}$$

$$GLR = \frac{M(G.O.R.)(1 - f_w)}{1 + M} \tag{6.86}$$

The water cut in the return column is given by:

$$f_{w_2} = \frac{Total\ Water}{Total\ Fluid}$$

For oil power fluid:

$$f_{w_2} = \frac{(f_w)\,q_3}{q_1 + q_3}$$

$$f_{w_2} = \frac{(f_w)\,q_3}{\dfrac{q_3}{M} + q_3} = \frac{M(f_w)}{1 + M} \tag{6.87}$$

For water power fluid:

$$f_{w_2} = \frac{q_1 + f_w(q_3)}{q_1 + q_3} = \frac{\dfrac{q_3}{M} + f_w(q_3)}{\dfrac{q_3}{M} + q_3}$$

$$f_{w_2} = \frac{1 + M(f_w)}{1 + M} \qquad (6.88)$$

With the values obtained from Equations 6.86, 6.87 and 6.88, the value of H can be calculated. As in the no-gas case, Fig. 6.6 will show which ratio provides the best efficiency at the calculated value of H. However, the value of M must be multiplied by the efficiency value obtained from Fig. 5.43 of Chapter 5 to give the actual *liquid* M value at which the pump will operate.

The value of M thus obtained must then be used to recalculate Equations 6.86, 6.87, and 6.89, along with vertical multiphase flow pressure gradients, leading to an improved estimate of P_2, H, and M. This iterative process is repeated until the desired degree of accuracy is obtained. Usually, agreement between successive values of M to within 5% is sufficient.

The nozzle size is then selected based on the "liquid" M value by means of Equations 6.4a and 6.60 as in the no-gas case.

There are many potential sources of error in the calculation sequence described above, including inaccuracies in calculating the pump discharge pressure and the approximation involved in using Figure 5.43 of Chapter 5. In addition, inaccurate field data, particularly the gas-oil ratio, can greatly reduced the significance of the calculations.

For reasons such as these, the above sizing calculations should be regarded as leading to a reasonable first estimate and the refinement of using the Θ_R curves is not necessary. Subsequently in field testing, size and ratio changes can be made to determine the optimum combination. Such individual well testing is more practical with jet pumps than with any other type of pump because of the ease of surfacing a free pump and the fact that nozzles and throats can be changed at the well location.

6.4242 Example problem #6.

Size the nozzle and throat with the same data as Example #5 but with a gas-oil ratio = 300 scf/B

(1) Assuming M = .5, from Eq. 6.86
$$GLR = \frac{0.5(300)(1 - 0)}{1.5} = 100 \text{ SCF/bbl}$$

(2) If M = .5, q_1 = 400 b/d and F_1 = 2 psi/1,000 ft from Figure 5.C(7) of Appendix 5.C (Chapter 5). Therefore, P_1 = 7600 ft × (.355 psi/ft) − 2 psi/1000 ft × 7.6 + 4000
P_1 = 6683 psi
Note that the friction, F_1, is only 15 psi and the variation of F_1 with respect to M can be neglected in this case.

(3) $q_1 + q_3$ = 400 + 200 = 600 b/d
From the appropriate multiphase flow correlation;

P_2 = 2760 psi

(4) Assuming a linear P.I. as before
$$P_3 = 1500 \text{ psi} - \frac{200 \text{ b/d}}{0.2 \text{ b/d/psi}} = 500 \text{ psi}$$

(5) $$H = \frac{P_2 - P_3}{P_1 - P_2} = \frac{2760 - 500}{6683 - 2760} = 0.570$$

Note that this is lower than the value of 0.581 calculated for the no-gas case.

(6) From Figure 6.6, the gas plus liquid M for H = 0.500 is M = 0.38 (A ratio).

(7) From Figure 5.43 of Chapter 5 using the reservoir solution gas-oil ratio, the volumetric efficiency is 52%.
Therefore, the "liquid" M value is

$$M = 0.52 \times 0.38 = 0.198$$

Note that despite the lower value of H resulting from the gas lift effect in the return column, the value of M is less than in the no-gas case.

(8) Recalculate GLR at M = .198

$$GLR = \frac{0.198(300)(1 - 0)}{1.198} = 50 \text{ SCF/BBL}$$

(9) Recalculate P_2:
At M = .198, q_1 = 1010 b/d and $q_1 + q_3$ = 1210 b/d

From the appropriate multiphase flow correlation:

P_2 = 2800 psi
$$H = \frac{2800 - 500}{6683 - 2800} = 0.592$$

(10) M = 0.36 from Figure 6.6
M = 0.52(0.36) = 0.187

This is sufficiently close to the previous value of M.

Use

$$M = 0.19, \quad q_1 = \frac{200}{0.19} = 1053$$

(11) From Equation 6.60

$$A_j = \frac{q_1}{1214.5\sqrt{\dfrac{P_1 - P_3}{\gamma_1}}}$$

$$A_j = \frac{1053}{1214.5\sqrt{\dfrac{6683 - 500}{0.8203}}} = 0.0100 \text{ in}^2$$

This is between a 5 and 6 nozzle. In such a case the larger nozzle should be selected. This leads to a 6-A jet pump. The power fluid rate will be approximately

$$q_1 = 1214.5 \, A_j\sqrt{\dfrac{P_1 - P_3}{\gamma_1}}$$

$$q_1 = 1214.5(.01131)\sqrt{\dfrac{6683 - 500}{0.8203}}$$

q_1 = 1193 BPD

(12) Check for cavitation

$$M_c = \frac{(1 - .410)}{0.410}\sqrt{1.15}\sqrt{\dfrac{500}{1.35(6683 - 500) + 500}}$$

M_c = 0.367 which is greater than the calculated value of M = 0.36 so the pump will not cavitate.

In reality, wells producing gas have less tendency to cavitate than wells producing a non-gaseous liquid such as water. Evidently, there is a form of cushioning effect that occurs as the free gas passes through the pump. Additional experimental work is necessary in this area. Cavitation should always be checked for wells producing water.

(13) And from equation 5.18 of Chapter 5.

$HP = 1.7 \times 10^{-5}(1193 \text{ b/d})(4000 \text{ psi})$

$HP = 81 \text{ hp}$

Note that this is more than twice the horsepower required with no gas present. The relatively low producing pump intake pressure leads to a significant volume of free gas which decreases the liquid pumping efficiency of the pump. At the same time, there is little return column lightening from the produced gas to reduce the pump discharge pressure. In some well installations, the relative magnitudes of the choking and gas lifting effects may be reversed, allowing the use of smaller values of R and smaller nozzles. The following example problem illustrates this.

6.4243   Example problem #7

Given:

Tubing O.D.	= 2⅞ in.
Casing size	= 7 in.
Water cut	= 50%
Gas oil ratio	= 300 scf/B
Static bottom hole pressure	= 1920 psi
Productivity index	= 4 b/d/psi
Wellhead pressure	= 120 psi
Setting depth	= 8000 ft
Wellhead temperature	= 110°F.
Bottom hole temperature	= 170°F.
Desired production	= 800 b/d (oil plus water)
API gravity of crude	= 41°

Since water is present already, the use of water as power fluid is a reasonable option. From the discussions of Section 6.326 on the effects of discharge pressure, however, one can deduce that, in general, the use of water power fluid will lead to higher operating pressures. Although the denser power fluid column gives a higher value of P_1 for a given surface operating pressure, the denser return column more than offsets this by increasing the operating pressure by two to five times the increase in discharge pressure. Therefore, in the absence of high tubing friction losses, safety reasons, or other considerations, produced oil is the suggested power fluid.

(1) Assuming M = 0.5, from Eq. 6.86 GLR: =
$$\frac{0.5(300)(1 - .5)}{1.5} = 50 \text{ scf/bbl}$$

(2) If M = 0.5, q_1 = 1600 B/D and F_1 = 7.8 psi/1000 ft from Figure 5.C(8) of Appendix 5.C (Chapter 5).

Therefore:

$P_1 = 8000 \text{ ft } (0.355 \text{ psi/ft}) - 7.8 \text{ psi} \times$
$$\frac{8000}{1000} + 4000$$

P_1 = 6778 psi for an assumed surface operating pressure of 4000 psi.

(3) $q_1 + q_3$ = 2400 b/d
$$f_{w_2} = \frac{0.5(0.5)}{1.5} = 0.167 \text{ from Eq. 6.87}$$

Multiphase Flow calculations show P_2 = 2740 psi

(4) Assuming a linear P.I. as before,

$$P_3 = 1920 \text{ psi} - \frac{800 \text{ b/d}}{4 \text{ b/d/psi}} = 1720 \text{ psi}$$

(5) $H = \dfrac{P_2 - P_3}{P_1 - P_2} = \dfrac{2740 - 1720}{6778 - 2740} = 0.253$

(6) From Figure 6.6, the gas plus liquid M for H = 0.253 is M = 1.0 (C or D ratio)

(7) From Figure 5.43 using the reservoir solution gas-oil ratio and water cut, the volumetric efficiency is 100%. This means that at P_3 = 1,720 psi, the pump will operate above the bubble point and no gas choking effects in the pump will occur.

(8) Recalculate P_1 at q_1 = 800 B/D (M = 1.0)

F_1 = 2.22 psi/1,000 ft
$P_1 = 8,000 \times 0.355 - (2.22 \times 8) + 4,000$
$P_1 = 6,822$

(9) Recalculate GLR at M = 1.0 from Eq. 6.86

$$GLR = \frac{1.0 (300)(1 - 0.5)}{1 + 1.0} = 75 \text{ scf/bbl}$$

(10) Recalculate P_2:

At M = 1.0, q_1 = 800 B/D and $q_1 + q_3$ = 1,600 B/D

And from Eq. 6.87

$$f_{w_2} = \frac{1(.5)}{1 + 1} = 0.25$$

Multiphase flow calculations show P_2 = 2,669 psi for water cut = 25%.

(11) Recalculate H

$$H = \frac{2669 - 1720}{6822 - 2669} = 0.229$$

(12) From Figure 6.6, the gas plus liquid M for H = .229 is M = 1.1 (D ratio)

At this point in the iteration, it is apparent that the changes in M are sufficiently small that no additional vertical multiphase flow calculations are necessary. Therefore, assume M = 1.1

$$q_1 = \frac{800}{1.1} = 727 \text{ B/D}$$

(13) From Equation 6.60

$$A_j = \frac{727}{1214.5\sqrt{\dfrac{6822 - 1720}{0.8203}}} = 0.00759 \text{ in}^2$$

This is in between a #4 and a #5 nozzle. Selecting the larger nozzle, the power fluid rate will be approximately

$$q_1 = 1214.5(0.00905)\sqrt{\frac{6822 - 1720}{0.8203}}$$

q_1 = 867 b/d

and the horsepower will be

HP = 1.7 × 10⁻⁵ (867 b/d)(4000 psi)
HP = 60 horsepower for a 5-D Jet Pump.

Since water is present, check for cavitation, using Equation 6.51.

$$M_c = \frac{1 - R}{R} \sqrt{1 + K_j} \sqrt{\frac{P_3}{I_c(P_1 - P_3) + P_3}}$$

$$M_c = \frac{1 - .210}{.210} \sqrt{1.15} \sqrt{\frac{1720}{1.35(6822 - 1720) + 1720}}$$

$$M_c = 1.80$$

$M_c = 1.80$, which is greater than the operating value of M = 1.1, so the flow is non-cavitating.

The methods illustrated previously rely heavily on the basic equations governing jet pump performance and, consequently, are rather tedious. The manufacturers of jet pumps have developed various techniques involving selector charts, nomographs, and computer programs for predicting the performance of their pumps under different well conditions. Such techniques involve the exact loss coefficients, nozzle and throat sizes, cavitation parameters, and multiphase performance corrections of their products. Often, their pumps, especially when Reynolds Number corrections are included, will show higher efficiencies than those discussed in the first part of this chapter.

In general, the procedures employed in these sections lead to conservative predictions of jet pump performance in a given well and should be sufficiently accurate to compare this mode of hydraulic pumping with other artificial lift methods.

Section 6.425 shows a more rigorous computer solution to two example problems.

6.425 Computer solutions

The jet pump design problem is more easily handled by computer. In particular various variables can be checked very quickly. To illustrate this procedure two example problems are shown.

EXAMPLE PROBLEM #8

Given data:

2⅜" O.D. tubing,
7" casing
GOR = 400 scf/B
Depth = 7600 ft
P_{wh} = 80 psi
$\overline{P}_R$ = 1500 psi
P_b = 1500 psi
Well test shows 150 b/d at P_{wf} = 750 psi
Crude = 35°API
Temp. bottom = 170°F.
Temp. surface = 109°F.

The inflow performance is assumed to follow that of Vogel's solution which in equation form is as follows:

$$q_o/q_{o(max)} = 1 - 0.2 \frac{P_{wf}}{\overline{P}_R} - 0.8\left(\frac{P_{wf}}{\overline{P}_R}\right)^2$$

Various computer runs were made for all the pump sizes shown in Table 6.7. Production rates for surface

TABLE 6.7
RESULTS OF EXAMPLE PROBLEM #8

Size	Production @ 4000 psi or cavitation	HP	Production @ 5000 psi or cavitation	HP
3A	116	29	116	29
3B	130	29	134	32
3C	130	29	154	39
3D	87	29	140	39
5B	162	49	163	49
5C	153	49	170	60
7B	178	73	178	73
7C	156	82	178	93
9B	175	114	175	114
9C	170	135	182	150

pressures of 4000 psi and 5000 psi are shown along with the hp.

The optimum size was selected as 7B and the computer plot for this size is shown in Fig. 6.13. Several things are noted on Fig. 6.13. The surface production rate (abscissa) is plotted vs. the power fluid rate (ordinate) with power fluid pressure and pump intake pressures being included as parameters. The inflow performance curve is superimposed on Fig. 6.13, and the maximum rate of approximately 180 b/d just before cavitation can be noted on Fig. 6.13.

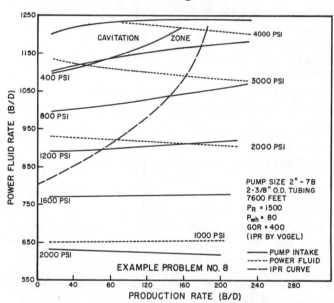

Fig. 6.13 Computer solution for example problem no. 8 (Courtesy Kobe, Inc.)

6.426 Class problems

CLASS PROBLEM #8A

Given:

2⅜" tubing
5½" casing
GOR = 300 scf/B
Producing 100% oil (35° API)
8000 ft
$\overline{P}_R$ = 1600 psi
P_{wh} = 100
PI = 0.4 (Assume Constant)
Temperature bottom = 180°F.
Surface temp. = 115°F.

Design a pump to handle 400 b/d (assume pumping all gas). Select a size, surface pressure, and HP and check for cavitation.

CLASS PROBLEM #8B

Work example 8A assuming Vogel solution where a test shows 200 b/d at a flowing pressure of 1100 psi. Also assume that 50% of the gas can be vented. Design for 300 b/d and for the maximum rate.

EXAMPLE PROBLEM #9

A higher flow rate well is given in this example.

Given:

Tubing size = 2⅞″ O.D.
Casing size = 7 in.
GOR = 400 scf/b
Producing 50% water (γ_w = 1.07)
GLR = 200 scf/b
$\overline{P}_R$ = 1920 psi
PI = 5 to bubble point pressure of 1500 psi after which Vogel's solution is followed
Crude = 35°API
P_{wh} = 120 psi
Depth = 8000 ft
Temp. bottom = 170°F.

Again, as in Problem #4, several computer runs were made for different pump sizes and the results are shown in Table 6.8. The optimum size is selected as pump 13B at a production rate of 2760 B/D for 345 hp and 4000 psi surface operating pressure. The computer plot for the size 13B is noted in Fig. 6.14 with the same type of information as noted in Fig. 6.13.

TABLE 6.8
RESULTS OF EXAMPLE PROBLEM #9

Size	Production @ 4000 psi or cavitation	HP	Production @ 5000 psi or cavitation	HP
11A	1650	204	1,650	204
11B	2050	209		
11C	2220	211		
12B	2400	267	2,650	364
12C	2450	269		
13B	2760	345		
13C	2575	343		

CLASS PROBLEM #9A

Given:

2⅞″ tubing
Casing = 7″
GOR = 600 scf/B
Producing 50% water (γ_w = 1.07)
Depth = 9,000 ft
$\overline{P}_R$ = 2200 psi
PI = 6 (constant)
Crude 35°API
P_{wh} = 140 psi
Temp. bottom = 210°F.
Temp. surface = 130°F.

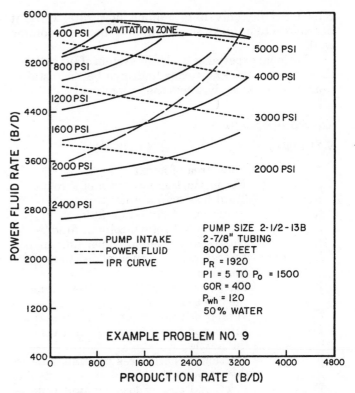

Fig. 6.14 Computer solution for example problem no. 9 (Courtesy Kobe, Inc.)

Design a jet pumping system selecting size, surface pressure, HP, etc. for 2000 B/D, 3000 B/D and maximum rate.

CLASS PROBLEM #9B

Work Example #9A for:
Producing 100% oil (pump all gas) for 2000 b/d, 3000 B/D and maximum rate.

CLASS PROBLEM #9C

Work problem #9A if 50% of the gas is vented and for
(1) P_{wh} = 140 psi
(2) P_{wh} = 80 psi

6.5 SUMMARY AND CONCLUSIONS

The jet pump is presently becoming more popular in the area of artificial lift and, according to one manufacturer, it is presently being placed on about 50% of the wells that use hydraulic lift of some type.

Its simplicity in incorporating no moving parts increases its popularity. Along with gas lift it is one of the two lift methods that can be retrieved by pump down techniques (Chapter 8). This makes it a method to consider for ocean floor completions.

Although this chapter presents sufficient charts and procedures to design a jet pump system it is recommended that computer solutions be obtained from the manufacturer if not available otherwise. Wells should be carefully selected and, if such things as sufficient submergence cannot be obtained, some other method of lift should be used. However, its range of applicability appears to be quite good. Although it was originally considered for only high volume flow rate wells, it has shown to be quite successful in some lower volume wells.

Additional experience with this system of lift will further increase the knowledge to design better installations and to correctly select the proper well.

6.6   NOMENCLATURE

VARIABLE	DEFINITION
A_j	Flow area of nozzle
A_s	Net production flow area of throat
A_t	Total flow area of throat
f_w	Water cut of formation fluid, %
f_{w_2}	Water cut of return column fluid, %
E_j	Nozzle energy supplied per unit time
E_s	Energy per unit time added to production fluid
F_d	Frictional energy loss per unit time in the diffuser
F_f	Total frictional energy loss per unit time
F_j	Frictional energy loss per unit time in the nozzle
F_s	Frictional energy loss per unit time in the suction circuit
F_t	Frictional energy loss per unit time in the throat
F_1	Pressure loss in power fluid tubing, psi
F_2	Pressure loss in return circuit, psi
GLR	Gas-liquid ratio, scf/bbl
GOR	Gas-oil ratio, scf/bbl
G_1	Fluid gradient in power fluid circuit, psi/ft
G_2	Fluid gradient in return column circuit, psi/ft
G_3	Fluid gradient of formation fluid, psi/ft
g	Acceleration of gravity
H	Dimensionless head recovery ratio
HP	Horsepower
HP_{q_1}	Horsepower supplied in power fluid
HP_{q_3}	Horsepower added to produced fluid
H_1	Total head, power fluid
H_2	Total head, discharge fluid
H_3	Total head, suction fluid
h_1	Pump setting depth, ft
h_3	Fluid level over pump suction, ft
I	Vertical axis intercept of straight line approximation of H-M curve.
K	Constant
K_j	Nozzle loss coefficient
K_d	Diffuser loss coefficient
K_s	Suction loss coefficient
K_t	Throat loss coefficient
L	Mixing energy loss per unit time
M	Dimensionless flow ratio, q_3/q_1
M_c	Cavitation limited flow ratio
N	Intermediate algebraic variable used in defining H
P_a	Pressure at the entrance of the throat, psi
P_b	Pressure at the exit of the throat, psi
P_{wh}	Flow line back pressure, psi
P_v	Vapor pressure, psi
q_1	Power fluid flow rate
q_2	Production flow rate, power fluid & suction flows
q_3	Suction flow rate
R	Area ratio, A_j/A_t
m	Slope of straight line approximation of H-M curve
γ	Specific gravity
V_j	Velocity of fluid in nozzle
V_s	Velocity of fluid in suction area around tip of nozzle
V_t	Velocity of fluid in throat
Y	Nozzle number
H_v	Velocity head of power fluid in nozzle
$f_{h_3} = h_3/h_1$	Percent submergence
β_R	Intermediate algebraic variable to calculate f_{h_3}
ρ	Unit weight of fluid (specific weight)
E	Efficiency
Θ_R	Dimensionless variable for determining M and H
X	Sensitivity to back pressure
I_c	Cavitation index

REFERENCES

1. Thomson, James, 1852 Rep. Brit. Ass'n., 1853 Rep. Brit. Ass'n.
2. Rankine, J.M., 1870 Proc. Roy. Soc. No. 123
3. Lorenz, Hans., *Technische Hydromechanik,* 1910.
4. Gosline, James E. and Morrough P. O'Brien, Morrough P., "The Water Jet Pump," Univ. of Calif. Pub. in Eng. (1933).
5. Cunningham, R.G. "Jet Pump Theory and Performance with Fluids of High Viscosity," 1956 Proc. of ASME, Paper No. 56-A-58.
6. Cunningham, R.G. and F.B. Brown, "Oil Jet Pump Cavitation," 1970 ASME Cavitation Forum.

Appendix 6

Appendix 6A (1)

DATA FOR FIGURE 6.6

H - M Curves $1 + K_j = 1.15$
$1 + K_D + K_t = 1.38$
$1 + K_s = 1.0$

	R = .4096		R = .32768		R = .26214		R = .20972		R = .16777	
M	H	N	H	N	H	N	H	N	H	N
0	1.045	0	.789	0	.596	0	.453	0	.348	0
.2	.763	.153	.640	.128	.517	.103	.410	.082	.323	.065
.4	.547	.219	.515	.206	.445	.178	.369	.148	.300	.120
.6	.359	.215	.403	.242	.378	.227	.330	.198	.276	.166
.8	.174	.139	.296	.236	.314	.251	.291	.233	.253	.202
1.0	− .029	− .029	.188	.188	.250	.250	.252	.252	.230	.230
1.2			.075	.090	.186	.223	.214	.256	.206	.247
1.4			− .052	− .073	.119	.166	.174	.244	.182	.255
1.6					.047	.075	.133	.213	.157	.252
1.8					− .032	− .058	.090	.162	.132	.237
2.0							.044	.088	.105	.210
2.2							− .005	− .011	.077	.170
2.4									.048	.116
2.6									.017	.044
2.8									− .016	− .045
3.0										

Appendix 6A (2)

DATA FOR FIGURES 6.8 THROUGH 6.12

	A		B		C		D		E	
M	H	Θ_R	H	Θ_R	H	Θ_R	H	Θ_R	H	Θ_R
.0	1.045	0	.789	0	.596	0	.453	0	.348	0
.2	.763	.304	.640	.320	.517	.343	.410	.371	.323	.405
.4	.547	.673	.515	.686	.445	.721	.369	.770	.300	.833
.6	.359	1.167	.403	1.120	.378	1.146	.330	1.205	.276	1.290
.8	.174	2.078	.296	1.674	.314	1.637	.291	1.685	.253	1.780
1.0	− .029	---	.188	2.514	.250	2.236	.252	2.229	.230	2.313
1.2			.075	4.543	.186	3.030	.214	2.858	.206	2.903
1.4			− .052	---	.119	4.293	.174	3.637	.182	3.568
1.6					.047	7.552	.133	4.670	.157	4.343
1.8					− .032	---	.090	6.264	.132	5.271
2.0							.044	9.742	.105	6.488
2.2							− .005	---	.077	8.228
2.4									.048	11.214
2.6									− .017	20.11
2.8									− .016	---

Chapter 7

Plunger lift

by Bolling Abercrombie

7.1 INTRODUCTION

The plunger method of lift incorporates a piston which normally travels the entire length of the tubing string in a cyclic manner. Generally, plunger lift is classified as a separate and distinct method of artificial lift although in many instances it serves as only a temporary means of keeping a well flowing prior to the installation of another method of lift. Some of the most common applications are as follows:

(1) Used in a high gas-oil ratio oil well to maintain production by cycling.
(2) Used in a gas well to unload accumulated liquids.
(3) Used in conjunction with intermittent gas lift to reduce liquid fall-back.
(4) Used in an oil or gas well to keep the tubing clean of paraffin, scale, etc.

Plunger lift when used with gas lift is a separate form of the intermitting production method that introduces the plunger to the lift cycle to provide a solid and sealing interface between the lifting gas and the produced liquid.

The interface provided by the plunger in gas lift changes the flow pattern of the gas during a lifting cycle from the familiar ballistic shape of gas penetration of the liquid slug to a pattern whereby gas flow is possible only in the annular space between the tubing walls and the plunger outside diameter.

Since the lift gas pressure under the plunger must be greater than the pressure created by the liquid load above the plunger, the small quantity of gas that bypasses the plunger flows up through the annular space and acts as a "sweep," thus minimizing any tendency for liquid "fall-back." The elimination of possible gas penetration through the center of the liquid slug and the minimization of any liquid "fall-back" makes plunger application the most efficient form of intermitting production available.

7.11 GOR control applications

Most attempts to reduce the GOR on a well with a plunger will not be successful unless the high ratio is being created by the method of production. Wells being produced in a manner that will not allow sufficient gas velocity to produce the well's liquids to the surface may be helped—not due to decreased total gas production, but rather increased liquid production without an associated change in total gas production.

7.12 Highly emulsified wells

Gas lifting a well with a bad emulsion problem, either by constant flow or intermittent lift, is often an inefficient operation. The emulsion prevents establishment of the required lifting velocity in the tubing and this, together with the friction caused by the high viscosity of the emulsified liquid, allows gas to channel through the liquid column, losing lift efficiency.

By introducing a plunger as a solid interface that directs any gas passing the plunger through the annulus between the tubing wall and the plunger O.D., the channeling action and liquid fall-back is minimized and lift efficiency is greatly improved.

7.2 WELL INSTALLATION & MECHANICAL EQUIPMENT REQUIRED

Fig. 7.1 shows a typical example of the most widely used plunger installation. Depending on the requirements of the well, a large number of variations on both surface and downhole equipment are possible. Some of the surface variations are shown in Figs. 7.2 through 7.7 and downhole variations in Figs. 7.8 through 7.11.

Regardless of the equipment chosen for a well, the first items to be considered for a plunger installation are the type of master valve and the condition of the tubing.

7.21 Suitable master valve

The master valve on the well must have a full bore equal to the tubing size to allow plunger passage. The valve must not be oversize, since this would allow excessive gas passage around the plunger and possibly prevent the plunger from being lifted into the lubricator. The plunger must reach the lubricator to allow removal for service and, where installed, to activate a plunger arrival system.

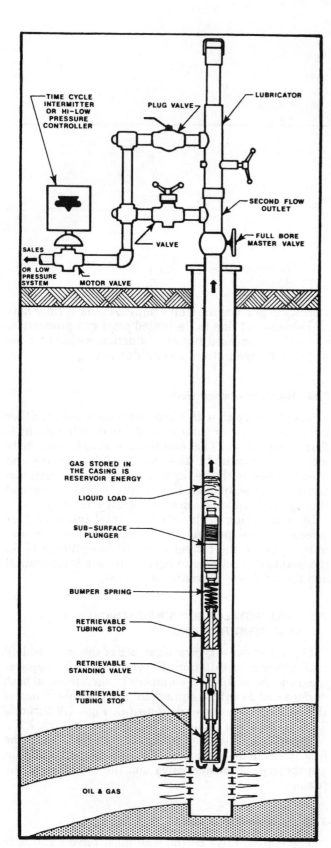

Fig. 7.1   Typical plunger installation (Courtesy McMurry Oil Tools, Inc.)

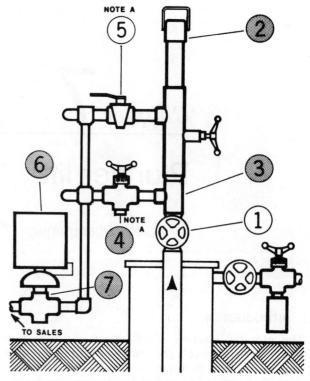

Fig. 7.2   Plunger surface equipment - for well with excess gas from formation - all flow through tubing (Courtesy McMurry Oil Tools, Inc.)

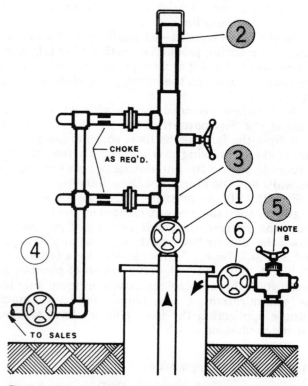

Fig. 7.3   Plunger surface equipment for well with insufficient gas from formation - well being gas lifted on packer - all flow through tubing (Courtesy McMurry Oil Tools, Inc.)

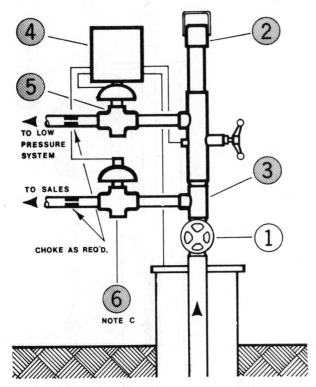

Fig. 7.4   Plunger surface equipment for well with excess gas from formation - liquid cannot be lifted against sales line pressure (Courtesy McMurry Oil Tools, Inc.)

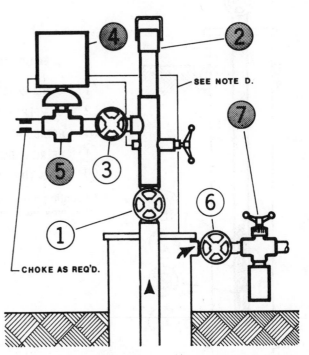

Fig. 7.6   Plunger surface equipment for well with excess gas from formation - gas flow from casing to sales - liquid flow from tubing to low pressure system or sales (Courtesy McMurry Oil Tools, Inc.)

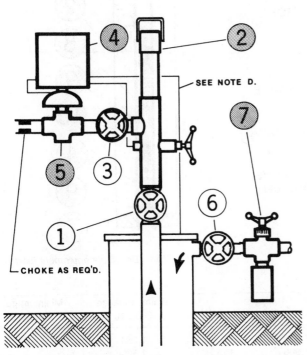

Fig. 7.5   Plunger surface equipment with proper gas from formation or gas supplement through volume control valve.

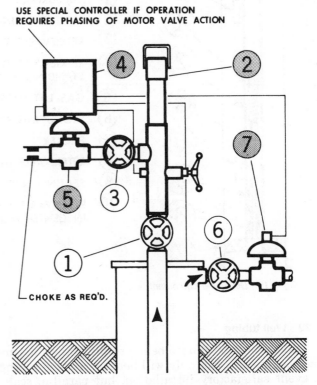

Fig. 7.7   Plunger surface equipment for well with excess gas from formation - gas flow from casing to sales - liquid flow from tubing to low pressure system (Courtesy McMurry Oil Tools, Inc.)

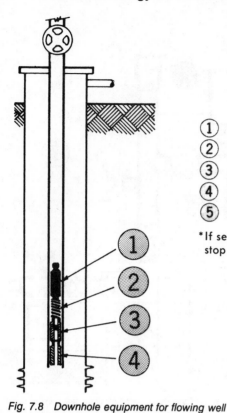

Fig. 7.8   Downhole equipment for flowing well

EQUIPMENT REQUIRED

(1)   SUBSURFACE PLUNGER

(2)   BUMPER SPRING

(3)   RETRIEVABLE STANDING VALVE

(4)   RETRIEVABLE TUBING STOP*

(5)   GAS LIFT VALVES

*If seating nipple is installed in well, tubing stop may be eliminated.

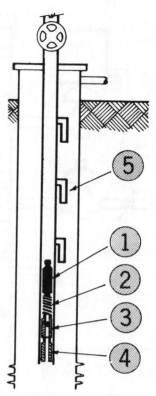

Fig. 7.9   Downhole equipment with gas lift

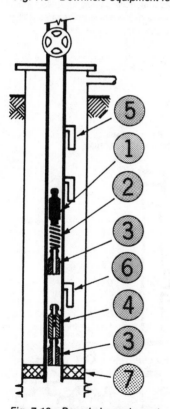

Fig. 7.10   Downhole equipment

EQUIPMENT REQUIRED

(1)   SUBSURFACE PLUNGER

(2)   BUMPER SPRING

(3)   RETRIEVABLE TUBING STOP

(4)   RETRIEVABLE DUPLEX STANDING VALVE

(5)   GAS LIFT VALVES

(6)   PRODUCING GAS LIFT VALVE

(7)   PACKER

(8)   RETRIEVABLE STANDING VALVE

(9)   SEATING NIPPLE

(10) RETRIEVABLE GAS LIFT VALVE IN CENTER MOUNT MANDREL

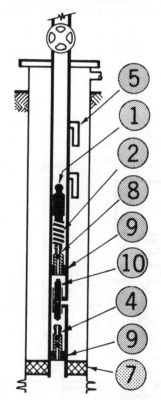

Fig. 7.11   Downhole equipment for gas lift

7.22   Well tubing

Tubing must be gauged before running any sub-surface equipment. Not only will bent or crushed tubing prevent satisfactory installation, but paraffin, scale, etc., can prevent initial operation.

The gauges recommended for the various tubing sizes are as follows:

| Tubing size | | Minimum gauge | Minimum gauge |
O.D., in.	nominal	O.D., in.	length, ft
1.660	1¼	1.250	2
1.900	1½	1.500	2
2.063	2¹/₁₆	1.630	2
2.375	2⅜	1.900	2
2.875	2⅞	2.312	2

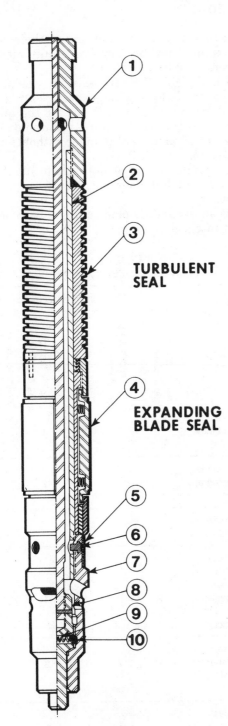

TURBULENT
SEAL

EXPANDING
BLADE SEAL

TYPICAL PLUNGER PARTS

1	MANDREL HEADPIECE
2	MANDREL TUBE
3	SPIRAL PACKING ELEMENT
4	EXPANDING PACKING ELEMENT
5	LOCK PIN KEEPER RING
6	LOCK PIN
7	VALVE BODY
8	VALVE STEM
9	SPRING
10	¼ DIAMETER BALL, CHROME STEEL

TYPICAL PLUNGER

Fig. 7.12 *Typical plunger with integral valve rod showing both turbulent seal and expanding blade seal*

There may be some variations in tubing gauge requirements from the preceding table. The operator should check with the equipment manufacturer to determine the correct gauge size.

7.23 Bottom of tubing

Note that the typical installation, Fig. 7.1, does not have a packer. Plunger lift with a packer is very seldom used (see Paragraph 7.33).

If the tubing is set on a packer and the operator does not desire to pull the tubing and packer, either unseat the packer and work around it, or shoot holes in the tubing to allow circulation between casing-tubing annulus and the tubing.

If the annulus is loaded with a liquid and possible formation damage may result from allowing liquid contact with the formation, method 2 should be used and a tubing plug set before the holes are shot. The annulus liquid can then be swabbed out without contacting the formation.

7.24 Plunger mechanical design

Three operating characteristics are necessary for any plunger, regardless of type of operation: high degree of repeatability of plunger valve operation, high shock and wear resistance, and resistance to sticking in the tubing. Two other characteristics desirable for plunger operation are ability to fall rapidly through gas and

Fig. 7.13   Brush type plunger without integral valve rod

classified either 1, 2, or 3 according to their relative effectiveness in fulfilling the five requirements for plunger operations listed above.

To show how Table 7.1 can assist in selection of a plunger type, the following two examples are given:

(1) A well produces excessive gas for the quantity of liquid produced but requires very rapid cycling:
   Use a turbulent seal plunger without integral valve rod.

(2) A well produces just enough gas to lift the produced liquid but does not require rapid cycle operation:
   Use an expanding blade plunger with or without integral valve rod.

TABLE 7.1
PLUNGER CLASSIFICATIONS

Type of plunger	Operating characteristics				
	High degree of repeatability of valve operation	High shock & wear resistance	Resistance to sticking in tubing	Ability to fall rapidly through gas & liquid	Ability to provide good seal against tubing during upward travel
(1) Expanding blade seal without integral valve rod	2	2	2	1	1
(2) Expanding blade seal with integral valve rod	1	2	2	2	1
(3) Expanding blade seal without valve	-	1	2	3	1
(4) Turbulent seal, wobble-washer, etc. without integral valve rod (valve actuating rod is part of lubricator)	2	2	1	1	2
(5) Turbulent seal, wobble-washer, etc. with integral valve rod	1	2	1	2	2
(6) Turbulent seal, wobble-washer, etc. without valve	-	1	1	3	2

liquid, and ability to provide a good seal against the tubing during upward travel.

There are essentially six different types of plungers available. The mechanical variations are based on the operating requirements. If gas is limited, an expanding blade seal provides a closer fit to the tubing wall, thus minimizing gas bypass around the annulus as compared to the turbulent seal, wobble washer, etc. If rapid cycling is required, the plunger without an integral valve rod provides less resistance to falling as compared to the plunger with an integral valve rod or the plunger without a valve.

Figs. 7.12, 7.13, and 7.14 show three different plunger types. In Table 7.1 six plunger types are listed and

7.25   Lubricator

The lubricator is an essential part of any plunger installation. The various parts of typical lubricators are shown in Figs. 7.15 and 7.16.

The cap (1) contains a spring for resisting the force of the rising plunger (2). The striker pad (3) is the initial contact of the plunger with the lubricator. With an integral rod plunger, the valve rod is activated and the plunger valve is opened by contact with the striker pad. Where plunger without valve rod is used, the striker pad contains a rod for activation of the plunger valve.

In the lubricator shown, the cap, bumper spring, and striker pad are removed as a unit for easy access to the plunger for examination and repair. The flow body (4) is self-explanatory. The catcher assembly (5) is used to hold the plunger in the lubricator for easy removal.

Where the operating characteristics of a plunger installation require rapid dissipation of tailgas or flow to a lower pressure system to lift the plunger and accumulated liquids, a separate outlet is installed below the lubricator (Figs. 7.2, 7.3 and 7.4).

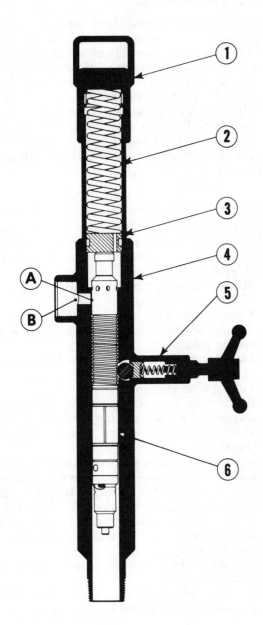

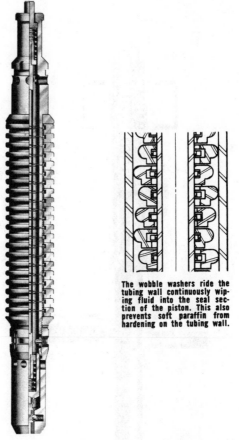

The wobble washers ride the tubing wall continuously wiping fluid into the seal section of the piston. This also prevents soft paraffin from hardening on the tubing wall.

Fig. 7.14   Wobblewasher type plunger with integral valve rod

TYPICAL LUBRICATOR PARTS

1	CAP
2	BUMPER SPRING
3	STRIKER PAD
4	FLOW BODY
5	CATCHER ASSEMBLY
6	PLUNGER
** 7	SECOND FLOW OUTLET

** NOT ALWAYS REQUIRED IN PLUNGER INSTALLATION.

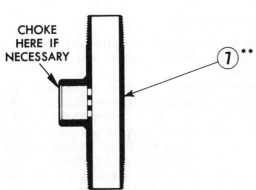

CHOKE HERE IF NECESSARY

Fig. 7.15   Typical lubricator

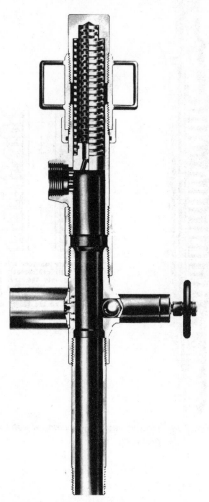

Fig. 7.16   Typical lubricator

Fig. 7.17   Typical bumper spring   Fig. 7.18   Typical tybing stop

7.26   Bumper spring

A bumper spring, as shown in Fig. 7.17, is an essential part of a plunger installation that prevents excessive shock load on the plunger when falling to bottom, particularly if the well does not have liquid above the tubing stop.

7.27   Tubing or collar stops

Where the well tubing string is not equipped with a seating nipple, a tubing or collar stop can be used for positioning the bumper spring and standing valve. Fig. 7.18 shows a typical tubing stop. Note should be made that if the plunger can fall to bottom dry, an individual stop should be used to set the standing valve independently of the bumper spring. Experience has shown that a plunger falling dry on a bumper spring, standing valve, and stop set together will set up a vibration that rapidly causes a failure of the standing valve ball and seat.

7.28   Plunger running procedure

Below is a complete list of sequential operations involved in running a plunger installation, assuming that the well is set on a packer, will not be pulled, and has fluid behind the packer.

(1) Master valve checked for proper sizing.
(2) Well tubing is gauged.
(3) Tubing plug is set just above packer.
(4) Tubing is perforated a known distance above tubing plug.
(5) Liquid from the annulus is swabbed out of well, or unloaded by gas.
(6) Tubing plug is pulled.
(7) Retrievable stop and standing valve are set immediately above tubing perforations. (Note: This standing valve and tubing stop are optional.)
(8) Retrievable stop is set above standing valve or tubing perforations (Note: Due to the difficulty in obtaining proper jarring action through the bumper spring to set the stop, the stop and bumper spring should be run independently).
(9) Retrievable bumper spring is run and latched to the previously set stop.
(10) Plunger is run to bottom on wireline to insure no tight spots.
(11) Wireline lubricator is removed from wellhead and operation commences.

7.29   Surface controls

Two basic types of surface controls are used for plunger installations: time or pressure control, or a combination of both.

Pressure control is designed to maximize plunger cycles per day, maximize liquid production, and minimize gas production. It is, therefore, best suited for (1) oil wells which do not have excess gas available for plunger lift operations and (2) oil wells with high productivity indices even if they have excess gas available. In cases where there is excess gas, it should be bled off continuously from the casing.

Time control should generally be used on gas wells

and oil wells with very high gas-liquid ratios. Pressure control (designed to minimize gas production) of such wells results in unnecessarily long shut-in times and thus prevents maximum gas production.

Most plunger manufacturers provide a system for closing the control valve when the plunger arrives in the lubricator. This feature is highly desirable when the well does not have excess gas, since gas flow is shut off after plunger arrival and all gas can be used to lift the liquid.

There are a number of variations on the plunger ar-

rival systems but the primary ones use differential pressure signals, magnetic signals, or mechanical signals to shut the control valve.

Although time and pressure are the basic controls necessary for plunger operation, a large number of variations are aviable in possible surface installations to fulfill the requirements of specific wells and surface systems. Refer to Figs. 7.2 through 7.7 for six possibilities.

To help coordinate the choice of equipment with well characteristics, refer to Table 7.2.

TABLE 7.2 CHOICE OF EQUIPMENT DEPENDING ON WELL CHARACTERISTICS

	Conditions							Recommendations		
Gas quantity	Gas flow		Liquid flow (all tubing flow) To	Type of wellhead installation	Type of downhole installation	Type of plunger (note 4)	'Plunger arrival' signal required	Type of surface controller	Procedure or comments
	From	To							
More gas volume than required to lift liquid	Tubing	Sales (note 1)	Sales (note 1)	Fig. 7.2 or Fig. 7.3	Fig. 7.8	4 5 6	No	Time cycle intermitter or pressure controller	Plunger remains in lubricator normally
	Tubing	Sales	LP system (note 2)	Fig. 7.4	Fig. 7.8	4 5 6	Yes	Time cycle (dual control) (motor valve on sales line may be necessary to allow proper plunger cycling)	
	Casing	Sales	LP system	Fig. 7.6 or Fig. 7.7	Fig. 7.8	4 5 6	Yes	Time cycle w/low casing pressure shutoff (optional) or pressure controller	
	Casing	Sales	Sales	Fig. 7.6	Fig. 7.8	4 5 6	Yes	Time cycle intermitter or pressure controller	
Proper gas volume to lift liquid	Tubing	Sales	Sales	Fig. 7.5	Fig. 7.8	1 2 3	Yes	Pressure controller	
	Tubing	Sales	Sales	Fig. 7.5	Fig. 7.9	1 2 3	Yes	Pressure controller	
	Tubing	Sales	LP system	Fig. 7.7	Fig. 7.9	1 2 3	Yes	Time cycle controller with twin wheel (note 3)	
	Tubing	Sales	Sales	Fig. 7.5	Fig. 7.8	1 2 3	Yes	Pressure controller	
	Tubing	Sales	Sales	Fig. 7.5	Fig. 7.9	1 2 3	Yes	Pressure controller with volume control valve on casing inlet	
Insufficient gas volume to lift liquid	Tubing	Sales	Sales	Fig. 7.5	Fig. 7.10	1 2 3	No	Pressure controller with volume control valve or time cycle intermitter on casing inlet	
	Tubing	Sales	Sales	Fig. 7.5	Fig. 7.11	1 2 3	No	Pressure controller with volume control valve on time cycle intermitter on casing inlet	
	Tubing	Sales	Sales	Fig. 7.3	Fig. 7.10 or Fig. 7.11	1 2 3	No	Pressure controller with volume control valve or time cycle intermitter on casing inlet	

Note 1: Where gas and liquid are produced to different pressure systems, the gas outlet will be referred to as sales and the liquid outlet as a lp or low pressure system.
Note 2: Where gas and liquid are produced to the same system, this system will be referred to as sales.
Note 3: Twin wheel time controller allows motor valve on sales line to be shut in ahead of time. Casing can then build up to sufficient pressure to run plunger when valve on lp line opens. Upon plunger arrival, mv on lp line closes, and motor valve on sales line reopens.
Note 4: Refers to number, from Table 7.1. First column—rapid cycling; second column—intermediate cycling; third column—slow cycling.

7.3   TYPES OF PLUNGER INSTALLATIONS

7.31   Introduction

Wells where plunger installations are being considered can be categorized into three types. These types do not differentiate between a gas or oil well but only whether the quantity of gas produced by the formation is (1) excessive, (2) the required amount, or (3) insufficient to lift the produced liquid to the surface.

In the first and second categories only, the formation gas energy is utilized to lift the produced liquid. In the third catagory, additional gas volume must be supplied to the casing to lift the produced liquid. (Installations utilizing a plunger in conjunction with gas lift valves are in this third catagory).

Plunger installations themselves can be classified as three different types:

(1) Intermittent gas lift with a plunger
(2) Plunger lift with a packer
(3) Conventional plunger lift without a packer. This catagory is by far the most important and widely applied and presents the greatest difficulty in determining applicability. This section is primarily devoted to installations of this type.

7.32   Intermittent gas lift with a plunger

This type of plunger application is normally used on a well with a bottomhole pressure so low that the height of liquid fill-in from the formation is not sufficient to prevent gas breakthrough of the liquid column during an intermitting lift cycle.

All gas is provided by a supplementary source which involves an outside source of power. The plunger allows much greater utilization of this energy and less fallback, thus a corresponding decrease in bottomhole pressure and an increase in fluid inflow.

The choice of equipment depends on the rapidity of cycles required to achieve maximum production. Typical examples are covered under equipment selection tables.

7.33   Plunger lift with a packer

An installation of this nature is very seldom used. All gas must come directly from the formation during the lift cycle; thus, the GLR requirements are greatly in excess of those required for conventional plunger lift. The volume of gas from the formation, although possibly sufficient for conventional plunger lift, may not be instantaneously great enough to give the necessary volume to lift the plunger and produced liquids from the well.

Normally, plunger lift with a packer is applicable to gas wells only. Operation would be for the plunger to remain in the lubricator with flow to sales. Time cycle control is used. The flow line is closed, the plunger drops to bottom, the flow line is reopened, the plunger and liquid load are surfaced, and the cycle recommences.

7.34   Conventional plunger lift without a packer

Conventional plunger lift without a packer is used normally only where the well supplies all of the energy, although more systems using additional gas for a supplement are presently being installed. There are a large number of variations possible in the methods of flow, i.e., casing flow of gas to sales, liquid flow to a lower pressure system; tubing flow, gas and liquid, to sales; tubing flow, gas to sales, and liquid to a lower pressure system. The type of flow pattern established depends on the well's characteristics.

Determining whether a well is suitable for conventional plunger lift and what flow system should be used is one of the major problems of plunger lift application. Tables 7A.1 through 7A.12 in the Appendix are provided to help solve these problems. These tables, based on the Foss and Gaul paper, are presented in a form to allow easy evaluation of a well's plunger potential.[1]

The information required to utilize Tables 7A.1 through 7A.12 of Appendix 7 are listed as follows:

(1) Depth of well
(2) Wellhead back pressure
(3) GLR—Required gas: liquid ratio in MCF per barrel
(4) MCF—Required gas in MCF per cycle
(5) P_c—Required surface casing pressure (average)
(6) BPD—Maximum production attainable in b/d
(7) ~—Maximum cycles per day

To obtain the above, the parameters used and the methods of determination are considered in detail in the following paragraphs.

(1) *Depth of Well*—Self-explanatory.
(2) *Wellhead Back Pressure*—The lowest pressure reading that will be seen at the wellhead during continued operation—normally separator pressure.
(3) *GLR*—From past or present production records of oil, water and gas production.
(4) *MCF Required Per Cycle*—Gas required per cycle is a function of depth and liquid load size for a given tubing diameter. The fundamental volume required per cycle is the sum of the following:
  (a) Volume contained in the tubing just before the flow line control valve opens. This volume is at a surface pressure equal to the maximum tubing pressure buildup which is equal to or slightly less than the maximum casing pressure buildup.
  (b) Volume of gas that slips past the plunger during upward travel.
  (c) Volume of gas that flows to the surface after plunger arrival but before the flowline control valve closes (This volume can be minimized by use of a plunger arrival shutoff system).

In practice a good approximation to the volume required per cycle is equal to 1.15 times the gas contained in the tubing with a pressure equal to the maximum surface casing pressure buildup (assumed to be 1.05 times the average casing pressure).

(5) *Average Casing Pressure*—To evaluate the average casing pressure, we will quote directly from the Foss and Gaul paper:

DETERMINATION OF AVERAGE CASING PRESSURE

A general pressure balance equation for plunger lift operation when the plunger is at any point in the tubing string and is ascending with its liquid load can be expressed as follows:

Casing pressure + pressure due to weight gas column − gas friction pressure loss (casing-tubing annulus) = gas friction pressure loss in tubing underneath the plunger + pressure due to weight of gas column underneath the plunger + plunger friction pressure loss + pressure required to lift weight of (7.1) plunger + pressure required to lift weight of liquid + liquid friction pressure loss + gas friction pressure loss above plunger + pressure due to weight of gas column above the plunger + surface tubing back pressure + pressure to account for entry of produced liquid underneath plunger.

During the period of time when the plunger travels upward, most of the above factors are in a state of change interacting with one another to satisfy the equation. Significant among these are the following:

(1) Tubing pressure decreases drastically from a pressure equal to (or nearly equal to) maximum casing buildup pressure to a minimum pressure which is controlled by separator pressure, the length of flow line and a relatively low gas flow rate.
(2) Gas flow rate decreases from a relatively high value (during tubing bleed-down) to an ever-decreasing rate as the plunger nears the surface.
(3) Casing pressure ordinarily decreases from a maximum to a minimum pressure (with 2½ in. tubing, 7 in. casing, the change averages about 10%).
(4) Plunger and liquid velocity changes from zero to approximately 1000 ft/min average velocity.

With a given set of well conditions it is possible to operate with a wide range of average casing pressures, but it is desirable to operate at the lowest possible casing pressure in order to achieve the greatest well drawdown since the bottom hole operating pressure is a direct function of casing pressure. Further, as seen in the performance curves, gas requirements per cycle decrease with decreasing casing pressures. In order to establish the value of the minimum possible average casing pressure it is necessary to know at which point in the plunger's upward travel a stallout is most likely to occur. If this point is known, then Equation 7.1 needs to be solved only once in order to determine the minimum casing pressure required.

It seems likely that the most critical point in the plunger's upward travel would be either when the plunger is nearing the surface with the liquid load or when the fluid is surfacing and is passing through the wellhead. At this time the casing pressure is at its lowest value and the gas column pressure benefit in the casing-tubing annulus is nearly cancelled by the gas column pressure effect in the tubing under the plunger. This theory is borne out by the fact that in field operations gas flow rates are consistently observed to be decreasing as the plunger nears the surface indicating the everdecreasing energy available for lift. Furthermore, at this time the greatest pressure effect from liquid production (in the tubing under the plunger) could be expected.

Assuming that the critically low casing pressure occurs when the plunger is nearing the surface with its liquid load, Equation 7.1 can be restated and simplified as follows:

Minimum casing pressure = gas friction loss in entire length of tubing + pressure required to lift weight of plunger + pressure (7.2) required to lift weight of liquid + liquid friction pressure loss + surface tubing back-pressure.

This equation ignores the pressure effects of plunger friction, slight gas column pressure differences between tubing and casing-tubing annulus, the pressure effect of liquid entry beneath the plunger, and casing-tubing annulus gas friction pressure loss. These factors, though present, are considered to be of small effect when the casing-tubing annulus cross-sectional area is relatively large compared to the tubing cross-sectional area and gas flow rates are high enough to sustain annular-ring or mist flow.

Restating Equation 7.2 in terms of average casing pressure and substituting an approximation for gas frictional pressure losses:

$$P_c(avg.) = \left[\frac{1 + P_c(max) - P_c(min)}{2} \right] \left[P_p + P_t(min) + (P_{Lh} + P_{Lf})L \right] \left[1 + \frac{D}{K} \right] \quad (7.3)$$

where:

P_c = casing pressure (psig)
P_p = pressure required to lift weight of plunger
P_{LH} = pressure required to lift weight of liquid (per b)
P_{Lf} = liquid frictional pressure loss (per b)
L = load size (b)
P_t = flow line pressure (psig)
D = depth of tubing (ft)
K = constant

Assuming a constant temperature and liquid velocity, the term $(P_{Lh} + P_{Lf})$ becomes a constant for a given tubing size and liquid type. Substituting average or typical values for the variables, Equation 7.3 can be restated as follows:

$$P_c(avg.) = 1.05 \left[5 + P_t(min) + (P_{Lh} + P_{Lf})L \right] \left[1 + \frac{D}{K} \right] \quad (7.4)$$

Where: $(P_{Lh} + P_{Lf})$ have values as follows:

	2⅜″ OD	2⅞″ OD	3½″ OD
$(P_{Lh} + P_{Lf})$	165	102	63
K	33,500	45,000	57,600

and other assumptions are as follows:

Liquid: 30° API crude with 15% water cut, pressure gradient 0.39 psi/ft, kinematic viscosity 11 cs at 60°F., 3.0 cs at 200°F.
Plunger and Liquid Velocities: 1000 ft/min
Temperature: 150°F.
Pressure Required To Lift Weight of Plunger: 5 PSI
Casing Pressure Range: 10% (corresponds to 2⅞″ tubing in 7″ casing)
Flow Line: Free from restrictions and ID equal to or greater than tubing ID with length 2000 feet or less

Equation 7.4 evolved from more rigorous calculations which more correctly ascertain gas and oil friction pressure losses and more accurately correct for temperature effects at various depths. Ordinarily, Equation 7.4 was found to compare within two percent of these more complex calculations.

In order to calculate the liquid and gas frictional pressure losses, a plunger ascending velocity of 1000 feet per minute was used. Average ascending plunger velocities of 1070 feet per minute were measured in 24 wells. The velocities ranged between 700 and 1400 feet per minute with most of the wells ranging between 900 and 1200 feet per minute. These velocities were determined by measuring the total time lapse between flow line control valve opening and arrival of the plunger at the surface and are probably higher than actual near-surface plunger velocities. Correspondingly, the use of the 1000 feet per minute velocity is conservative in that it predicts higher gas and oil friction pressure losses than would be predicted with lower velocities. These higher than actual pressure losses should compensate to a degree for the pressure losses that are not accounted for in Equations 7.2, 7.3 and 7.4.

(6) *Surface Operating Casing Pressure*—Two types of wells are normally under consideration for sub-surface plunger installations: pumping wells and wells that have been flowing but have died. The above include not only oil wells but also gas wells, whether tubing or casing flowing the gas.

Past or present production information on the wells can allow an accurate determination of the gas-liquid ratio and minimum fluid production, but it may not allow evaluation of what surface operating casing pressure can be expected during plunger lift operation.

FOR A PUMPING WELL

To determine the probable surface operating casing pressure of a pumping well, three possible methods can be used, each with certain limitations. All methods of pressure determination need accurate measurements of oil, water and gas volumes.

(a) With the well producing normally, obtain the working fluid level in the casing with a well sounder, then by use of flowing gradient curves or a multiphase flow correlation for the recorded gas-liquid ratio, project from the working fluid level to the perforations to determine the flowing bottomhole pressure. With gas column weight charts, the probable surface operating casing pressure can be obtained. This method is subject to considerable error due to foaming, dependence on flowing gradient correlations and unknown composition of the annulus fluid, i.e., oil or water, paraffin, etc.

(b) A second method of flowing bottomhole pressure determination is to run a bottomhole pressure gauge below the pump to record pressures while the well is being produced. The bottomhole pressure is then projected to the surface by gas weight charts. This is an accurate method, but is not widely used because it is time-consuming and expensive, involving two pump pulling jobs.

(c) A third method of determining the probable surface operating casing pressure is to produce the pumping well in a normal manner, record oil, water and gas production. With a two-pen recorder on the casing and tubing, shut in the casing and leave the pump running. As casing pressure builds up, the working liquid level in the annulus is depressed. When the gas column in the annulus depresses the liquid to the pump inlet, gas

will break through the pump and can be recognized by increases in tubing pressure and corresponding decreases in casing pressure on the surface two-pen recorder. The highest casing pressure reading on the two-pen chart occurs immediately before the gas breakthrough; this is the probable surface operating casing pressure while flowing. If this method does not establish gas break around the end of the tubing, then test data gathered can be used as outlined in (a) above to determine probable surface operating casing pressure.

An alternative would be to use a well sounder to determine the working fluid level in the annulus. The sounding should be done only after the casing pressure of the well has stabilized. By projecting the surface casing pressure to the working fluid level with use of gas weight charts, then to bottom with the liquid gradient, a flowing bottomhole pressure can be estimated. Liquid production rates should be determined after the well has stabilized with the casing shut in. The estimated flowing bottomhole pressure, together with the static bottomhole pressure, allows the operator to determine the productivity characteristics of the well and from this, the maximum depth from which a subsurface plunger can operate.

FOR A FLOWING WELL

To determine the probable surface operating casing pressure of a well that has been flowing and has died, three methods are available to the operator:

(a) The first method is to swab the well in again, then run a bottomhole pressure survey while the well is flowing. The flowing bottomhole pressure of the well is then projected to the surface by use of gas weight charts and is the probable surface operating casing pressure. The use of this method is dependent on being able to reestablish flowing conditions in the well.

(b) The second method of determining the probable surface operating casing pressure is dependent on accurate well test information. From this information, oil, water and gas volumes and flowing wellhead back pressure, the operator can estimate the probable flowing bottomhole pressure of the well by proper application of available flowing gradient charts or multiphase flow computer programs. Use of the gas weight charts will allow projection of the flowing bottomhole pressure to the surface to determine the probable operating surface casing pressure.

(c) The third method is to test the well while swabbing. If (a) above was tried and flowing conditions were not reestablished, then this third method is applicable. Place a two-pen recorder on the casing and tubing and while swabbing, record the depth of swab run, liquid produced per swab, water percent, gas produced, and casing and tubing pressures.

Also record static fluid level and fluid level after each swab run.

Swabbing should be done with the casing shut in and the tubing wing valve pinched so as to prevent the swab from being blown up the hole should gas break around the end of the tubing. This breakaround is recognized by a rapid decrease in casing pressure with a corresponding rapid increase in tubing pressure. The highest casing pressure reading occurs immediately before the gas break-around, and this is the probable surface operating casing pressure.

If swabbing does not establish this gas break around the end of the tubing, then the test data gathered may be used as outlined in (b) above to determine the probable surface operating pressure.

(7) *Maximum Cycles Per Day*—A complete plunger cycle consists of the time required for the plunger to rise to the surface, time for the surface control mechanism to shut in, and time for the plunger to fall through the gas and liquid to reach bottom.

Each of the above are variables to a certain extent. The rising velocity depends on the differential created across the plunger and fluid load when the surface control opens; surface closing time depends on the type of control used; falling time depends on type of plunger used and how much liquid is present in the tubing.

To determine the maximum cycle frequency, the following values are used:

Rising velocity = 1000 ft/min
Surface control time = Negligible
Falling through gas = 1000 ft/min
(Foss & Gaul used a 2000 ft/min falling velocity to determine the maximum cycle rate and thus the maximum production. Field experience has shown that 1000 to 1250 ft/min is more realistic)
Fall through liquid = 172 ft/min

Using the above values, an equation for maximum cycle frequency may be written:

$$\text{Maximum cycles/day} = \cfrac{1440}{\cfrac{2(\text{depth})}{1000} + \cfrac{\text{Length of 1 bbl load} \times \text{load size (bbl)}}{172}} \tag{7.5}$$

The times involved are predicated on the use of an expanding blade plunger without an integral valve rod.

7.4 IDENTIFYING SUBSURFACE PLUNGER APPLICATIONS AND EXAMPLE PROBLEMS

7.41 Minimum GLR

For any given well candidate for sub-surface plunger applications, the first thing to identify is whether the GLR is sufficient to meet the minimum requirements. This minimum GLR depends only on tubing size and flow line pressure, and it is the quantity of gas that enters the well bore that is capable of just lifting the simultaneously entering liquid (no extra weight) to the surface. If gas required to lift the plunger is not present, plunger lift is impossible. This minimum GLR decreases as the flow line pressure decreases and as the tubing size increases.

Tables 7A.1 through 7A.12 (Appendix 7A) can be used to determine this minimum GLR. The formula is shown below:

$$\text{Minimum GLR} = \cfrac{\text{MCF/cycle @ 2 b/cycle} - \text{MCF/cycle @ 1 b/cycle}}{2 \text{ b/cycle} - 1 \text{ b/cycle}} \tag{7.6}$$

As an example, a 10,000 ft well with 2⅜ in. OD tubing having a 60 PSI wellhead pressure would have a minimum of (10.0 MCF/cycle – 5.6 MCF/cycle) or 4.4 MCF/bbl. Refer to Table 7A.3, and read at 10,000 ft. the MCF at a liquid surface delivery of 2 bbl and 1 bbl. These are 10 and 5.6 MCF, respectively.

7.42 Optimum GLR

Where it has been determined that the well being evaluated has a GLR greater than the required minimum, an optimum GLR can be determined if IPR data is available on the well. This optimum GLR is important in determining the method of producing the well, whether gas should be injected into or removed from the casing-tubing annulus during operations and is dependent on the mechanical characteristics of the plunger equipment and the IPR of the well. The optimum GLR will produce the well at the maximum possible liquid rate. At either a higher or lower GLR, a lower production rate will result.

The producing tendency of plunger lift is directly opposed to that of the well. Plunger lift requires an increase in casing pressure for increased production, whereas the well itself requires a decrease in casing pressure for increased production. The compromise that yields the greatest production is always found when cycling the plunger at the maximum frequency possible without killing the well. With an optimum GLR the plunger operates at maximum cycle frequency (leaves bottom on arrival); therefore, maximum production will be obtained.

To produce a well at the optimum GLR, whether adding or removing gas in the annulus, is extremely difficult because of inaccuracies in measuring the gas and liquid production. Since the average casing pressure is in direct correlation with the GLR of a well, it is possible to utilize the average casing pressure to make a field determination that optimum operation is being achieved. The average casing pressure can also be used to determine the optimum GLR.

EXAMPLE PROBLEM

Given:

Depth = 10,000 ft
Tubing = 2⅜" OD
P_{wh} = 60 PSI
IPR = Curve as in Fig. 7.19
Gas gravity = 0.65

First, plot the average surface operating pressure corresponding to the IPR curve for producing bottom-

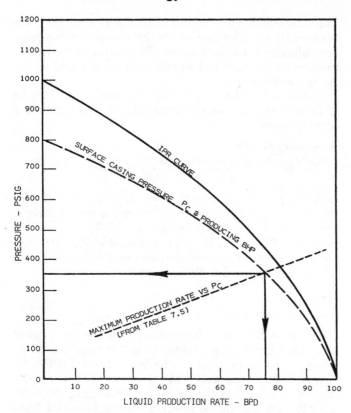

Fig. 7.19 Optimum GLR determination

hole pressure. Assuming no liquid is in the annulus, use the existing gas weight charts for the gas gravity in the annulus. (Refer to Fig. 3A.1) Note Figure 7.19.

Next, plot the maximum production in b/d vs. P_c, required surface casing pressure (average) from Table 7A.3. The intersection of the maximum production curve and the average surface casing pressure curve indicate that, at optimum operation, the well will produce 76 b/d with an average surface casing pressure of 350 PSI. The plunger would cycle 66 times per day.

To determine the optimum GLR, extrapolate between the values from table 7A.3 for 1.00 bbl and 1.25 b liquid surface delivery.

$$5.4 \text{ MCF/bbl} + (5.6 \text{ MCF/bbl} -$$
$$5.4 \text{ MCF/bbl}) \left(\frac{76 \text{ b/d} - 67 \text{ b/d}}{82 \text{ b/d} - 67 \text{ b/d}} \right) = 5.5 \text{ MCF/bbl}$$

7.43 Less than optimum GLR

With the GLR less than optimum, the casing pressure present when the plunger reaches bottom is not adequate to lift the liquid load. The plunger must wait on bottom until the casing pressure is high enough although the in-flow to the well is causing an increase in liquid load.

Assume the well used in the previous examples produces only 4.9 MCF/bbl. We know that the minimum GLR is 4.4 MCF/bbl (see paragraph 7.41) and the optimum GLR is 5.5 MCF/bbl (see paragraph 7.42).

To determine the expected production rate for the well without use of supplementary gas, we again utilize the IPR curve. Entering Table 7A.3 at the 10,000 ft depth on the GLR line, move to the right to the 4.9 MCF/bbl (Note: always use the heaviest known surface delivery per cycle if the GLR is repeated). Reading

down we determine the required P_c for this GLR to be 652 PSI. Plotting this on Fig. 7.20, see that the well should produce 31 b/d with a 4.9 MCF/bbl ratio. (An additional injected gas volume of 0.6 MCF/bbl would allow production of 76 b/d). The liquid surface delivery would be 2.50 b/cycle; therefore, the plunger would cycle:

$$\frac{32 \text{ b/d}}{2.5 \text{ b/cycle}} \text{ or } 12.8 \text{ times per day}$$

7.44 Greater than optimum GLR

A well that produces at a GLR greater than the optimum can limit liquid production. This reduction is caused by excessive gas pressure in the annulus; if the reduction is significant, the excess can be produced from the annulus.

Casing and tubing annulus volume are important factors in the pressure variations since with a large annulus volume, the pressure effect of excess gas will cause considerably less buildup than a well with a small annulus. The best determination of production decrease is the actual deviation from the optimum production as shown in the previous examples.

7.45 Foss & Gaul example

A Foss & Gaul example is presented below using the same data as previously given to show the variations in results from the prepared tables and the original solution of Foss & Gaul:

Depth = 10,000 ft
Tubing = 2⅜" OD
P_{wh} = 60 PSI

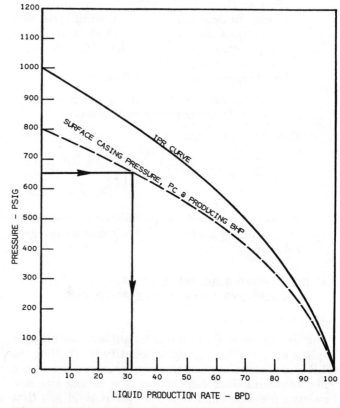

Fig. 7.20 Liquid production rate at less than optimum GLR

IPR = Curve as in Fig. 7.23
Gas gravity = 0.65

Fig. 7.21 is the Foss & Gaul curve necessary to determine the operating characteristics of the well. Note that Figure 7.21 is for 2⅜ in. tubing and 60 PSI flow line pressure.

7.451 Minimum GLR

From the right-hand curves of Figure 7.21, read across at the 10,000-ft level the gas required per cycle for various loads. This data is tabulated in Table 7.3 below and plotted on Fig. 7.22.

TABLE 7.3
LOAD VS. GAS REQUIRED

Load size b/cycle	Gas required MCF/bbl
0.00	1.6
0.25	2.5
0.50	3.5
0.75	4.5
1.00	5.6
1.25	6.7
1.50	7.8
1.75	8.9
2.00	10.0

The minimum GLR is equal to the slope of the straight line portion of the curve shown in Figure 7.22.

$$\text{GLR min.} = \frac{10.0 - 1.2 \text{ (MCF/cycle)}}{2.00 - 0 \text{ (b/cycle)}} = 4.40 \text{ MCF/bbl}$$

7.452 Optimum GLR

Use the left-hand curves of Fig. 7.21 which plot production at maximum cycle frequency versus average surface casing pressure. Enter on the 10,000-ft depth line and tabulate as shown in Table 7.4

TABLE 7.4
RATE VS. CASING PRESSURE

Producing Rate B/D	Average Surface Casing Pressure PSI
25	145
50	210
75	280
100	353
150	504

Plot the data from Table 7.41 on Fig. 7.23, the IPR curve of the well with surface pressure shown corresponding to the producing BHP. The example well would produce 81 b/d total liquid at an average surface casing pressure of 300 PSI, 0.93 b/cycle and cycle $\frac{81}{0.93}$ or 87 cycles per day.

Reentering Figure 7.21 on the right hand side, we determine that 5.2 MCF/cycle is required to lift a 0.93

bbl load. The optimum GLR is therefore $\frac{5.2 \text{ MCF/cycle}}{0.93 \text{ b/cycle}}$ or 5.6 MCF/bbl.

Following are the tabulated results of use of Table 7A.3 and the Foss & Gaul example for the same well:

	Table 7A.3	Foss & Gaul
Optimum GLR	5.5	5.6
Average surface operating pressure	350	300
Maximum production	76	81
Maximum cycles	67	87
Minimum GLR	4.4	4.4

The variation in the results are due to the difference in the assumed falling velocity of the plunger, which means fewer cycles per day, slightly less production and a slightly higher average surface operating pressure. Foss & Gaul stated that falling velocities measure between 900 and 3,000 ft/min depending on plunger configuration. Although the faster falling velocity can be achieved, a 1000-ft/min falling velocity is more probable and gives a more conservative estimate of possible production. The prepared Tables 7A.1 through 7A.12 can be used with confidence.

7.46 Additional example problems

7.461 Gas well loading with liquids—low-pressure system available

An 8000-ft deep gas well being produced up the casing-tubing annulus produces 200,000 SCF/day and 20 b/d liquid to a 205 PSI sales line pressure immediately after the tubing is opened to a low pressure, 30 PSI gas system (blown down). Over a period of time, the gas and liquid production decreases until the well dies. Will a plunger operate and what type of equipment should be used?

Given:

Depth = 8000 ft
Tubing = 2⅜″ OD
q_g = 200 MCF/d
q_L = 20 b/d
Sales line pressure = 205 PSI

Solution:

The sales line pressure of 205 PSI determines the operating surface casing pressure of this installation. The producing gas liquid ratio is

$$\frac{q_g}{q_L} \text{ or } \frac{200 \text{ MCF/d}}{20 \text{ b/d}} \text{ or } 10,000:1 \text{ (SCF/B)}$$

Refer to Table 7A.2 for 2⅜ in. OD tubing and 30 PSI wellhead pressure. From the Table at 8,000 ft and 205 PSI P_c, we can ascertain the following information:

(1) Maximum liquid surface delivery per cycle = 0.75 bbl
(2) Required gas-liquid ratio = 4200 SCF/B
(3) SCF required per cycle = 3100
(4) Maximum production = 63 b/d
(5) Maximum cycles per day = 84

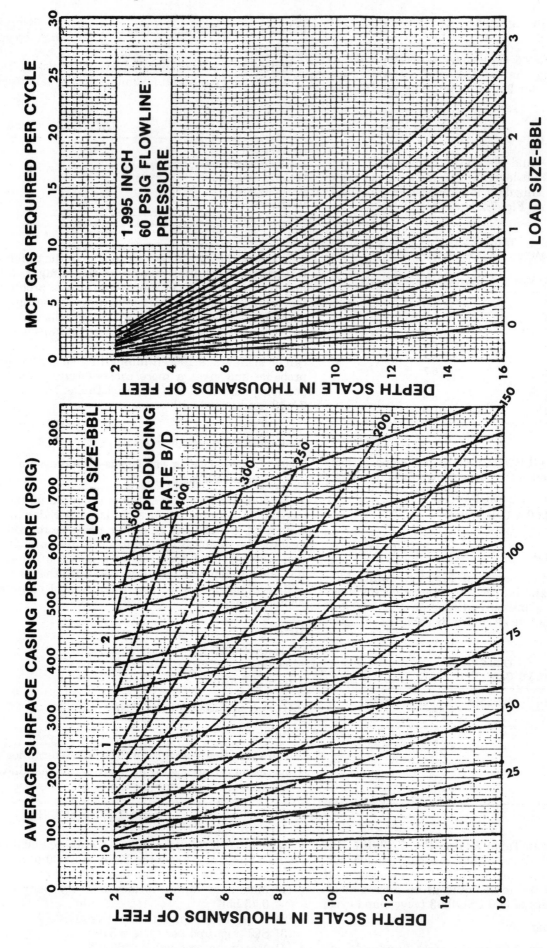

Fig. 7.21 Foss and Gaul curve

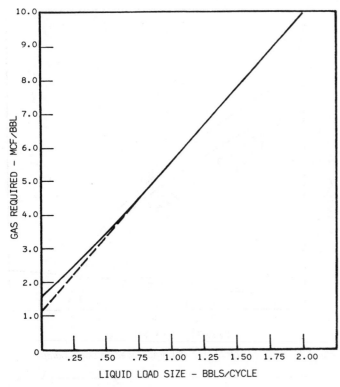

Fig. 7.22   Minimum gas:liquid ratio (Fossand Gaul example)

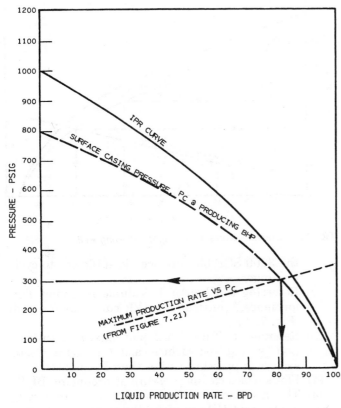

Fig. 7.23   Optimum GLR Determination (Foss and Gaul example)

The following conclusions can be reached:
(1) The well produces at 10,000:1 GLR, the necessary gas is available.
(2) With a 0.75 b/cycle surface delivery $\dfrac{20\ \text{b/d}}{0.75\ \text{b/cycle}}$ or 27 cycles per day are required.

(3) At 27 cycles per day, (27 cycles/day × 3100 SCF/cycle) 83,700 SCF/day will be produced to the low pressure system.
(4) Equipment choice:
  (a) Surface equipment
      Install as shown in Fig. 7.6 without volume control valve from casing. Time cycle control is necessary.
  (b) Downhole equipment
      Install as shown in Fig. 7.8. Since the cycle time is $\dfrac{1440\ \text{min/d}}{27\ \text{cycles/d}}$ or 53 min/cycle, the plunger fall rate is not critical. Use either an expanding blade plunger with integral valve rod or an expanding blade plunger without valve. The expanding blade is used to minimize the amount of gas produced to the low pressure system.

7.462 Gas well loading with liquids—no low pressure system available

Using the same well information as in Section 7.461 with the exception that a low pressure system is not available, all flow must be to the sales line. Static bottomhole pressure is 600 PSI.

Solution:

In this example, the well should be flowed through the tubing. Refer to Table 7A.6 for 2⅜" OD tubing at 200 PSI wellhead pressure.

The governing factor of the well is the gas-liquid ratio. Examining data from the table indicates that for a gas-liquid ratio of 10,000:1 (given and assumed to remain constant at all input volumes), the minimum liquid surface delivery per cycle will be 0.65 (extrapolated) which requires an average surface casing pressure of 405 PSI. Since the 405 PSI is considerably greater than the 205 PSI P_c used in the previous example, the produced gas volume will be reduced together with the produced liquid volume. An IPR curve of the well would allow a reasonable estimate of production.

The equipment choice should be as follows:
(1) *Surface Equipment*
    Install as shown in Fig. 7.2. Pressure control is recommended.
(2) *Downhole Equipment*
    Install as shown in Fig. 7.8. Operation would be to shut well in, allowing plunger to fall to bottom. Motor valve on flow line opens on casing pressure buildup and closes on casing pressure drop sometime after the plunger surfaces. The pressure settings should be adjusted down in small increments until the plunger will just run. This casing pressure is the minimum possible for operation, giving the minimum flowing bottomhole pressure and thus, maximum production.

7.463 Well on beam pump

The following information is given:

Depth = 6,000 ft
Production = 28 b/d of 35° API oil

GOR = 1000 scf/B
Separator pressure = 30 PSI
Tubing = 2⅜″ OD
$\overline{P}_R$ = 800 PSI
Paraffin causes rod job once a month
Outside source of gas available
P_{wh} = 30 PSI

A sounder indicates that the pumping fluid level of the well is 5,000 feet.

Determine probable change in production if a plunger is installed.

Solution:

(1) A quick examination of Table 7A.2 shows that the 1000 SCF/bbl produced is not sufficient to cycle a plunger. Supplementary gas will be required.

(2) The flowing bottomhole pressure of the well (gas flow through the liquid is neglected) is P_c + (Total depth − Fluid level) (Gradient of liquid—PSI/ft) or 30 + (6000 − 5000) (.370) = 400 PSI.

(3) Referring to Chapter 1 of Volume I of this book series, "Inflow Performance," an IPR curve for the well can be constructed (Fig. 7.24). Since the estimated flowing bottomhole pressure is 400 PSI or 50% of reservoir pressure, the well is producing at 70% of maximum, which is $\dfrac{28\ b/d}{0.7}$ or 40 b/d.

(4) Referring to Table 7A.2, plot maximum production vs. required surface casing pressure on Fig. 7.24.

(5) Plot surface casing pressure @ producing BHP.

(6) The intersection of the surface casing pressure @ producing BHP curve and the maximum production vs. P_c curve indicates that the well should produce greater than 38 b/d with a surface casing pressure of 105 PSI.

(7) To determine the required GLR, extrapolate between the values from Table 7A.2 for 0.25 bbl and 0.50 bbl liquid = surface delivery per cycle.
3.2 MCF/bbl + (4.4 MCF/bbl −

$$3.2\ \text{MCF/bbl})\left(\frac{38\ b/d - 29\ b/d}{56\ b/d - 29\ b/d}\right) = 2.6\ \text{MCF/bbl}$$

(2.6 MCF/bbl) (38 b/d) = 98.8 MCF/day must be added to produce the well. Remember that 1.0 MCF/bbl is available from the well.

7.464 Flowing well

The following information is given:

Depth = 7000 ft
Production = 50 bo/d
GOR = 3700 SCF/bbl
$\overline{P}_R$ = 600 PSI
P_{wf} = 400 PSI
Tubing = 2⅜″ OD
P_{wh} = 50 PSI
P_{sep} = 30 PSI

No outside source of gas is available. Determine the probable change in production if a plunger is installed.

Solution:

(1) A quick examination of Table 7A.2 shows that

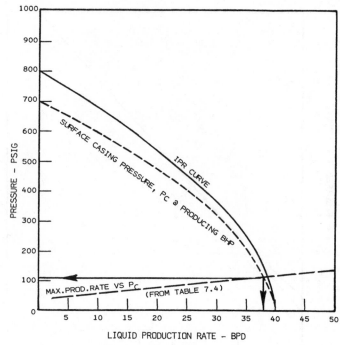

Fig. 7.24   Example problem 7.463 graph (well on beam pump)

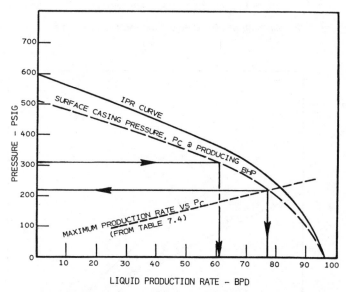

Fig. 7.25   Example problem 7.464 graph (flowing well)

the 3200 SCF/bbl produced is sufficient to cycle a plunger.

(2) Referring to Chapter 1, Volume I, "Inflow Performance," construct an IPR curve for the well (Fig. 7.25).

(3) Referring to Table 7A.2, plot maximum production vs. required surface casing pressure. (Note: separator pressure rather than P_{wh} is used.)

(4) Plot surface casing pressure at producing BHP.

(5) The intersection of the surface casing pressure @ producing BHP curve and the maximum production vs. P_c curve indicates the well should produce 77 b/d with a surface casing pressure of 220 PSI. This would be optimum production.

(6) Referring again to Table 7A.2, note that the 3200 SCF/bbl produced is not sufficient to lift the well at the optimum rate; therefore, determine the expected production rate without supplementary gas.

(7) Entering Table 7A.2 at the 700-ft depth on the GLR line, move to the right to the 3.2 MCF/bbl. Reading down, determine the required P_c for this GLR to be 305 PSI. Plotting this on Fig. 7.25, determine that the well should produce 61 b/d with the 3.2 MCF/bbl ratio.

(8) If supplementary gas were available, the amount required could be calculated as noted in section 7.463.

7.465 Well on intermittent gas lift

Where a well is being gas lifted and a plunger installation is contemplated, the bottom valve should be set so that there is no possibility of opening the next upper valve. Probably, the best way to insure this is to set the bottom valve so the casing opening pressure with minimum load (possibly separator pressure) is the same as the casing closing pressure of the next upper valve.

If the valve set pressures on an existing installation are properly set, use downhole equipment as shown in Fig. 7.10 and surface equipment as shown in Fig. 7.3.

If the well must be pulled, the downhole equipment can be installed as shown in Fig. 7.11 and the surface equipment as shown in Fig. 7.3.

7.466 Operation and analysis of plunger lift

In the same manner as discussed for gas lift, the best way to analyze and adjust the operation of a plunger is to install a two-pen surface recorder. By recording both the tubing and casing pressure, we can determine (1) whether or not the plunger is running, and (2) whether or not we have a good operation.

Fig. 7.26 shows a gas lift well with a plunger installed to improve efficiency. The injection is controlled by time cycle. Note the similarity between this chart and charts showing intermitting lift without a plunger (see Chapter 3).

Fig. 7.27 shows a well on pressure control with a plunger arrival system. This chart indicates how the backup function of a high-low controller works when the plunger does not arrive in the lubricator to shut off the controller. This can be a very important type of back-up system for a plunger. If the well was not adjusted to close on a low casing pressure, it would continue to flow and load up and die. Note from Fig. 7.27 that the well continues to produce and that the plunger does arrive on the next cycle.

Fig. 7.28 shows the same well as in Fig. 7.27, four years later. Note the difference in casing operating pressure due to the decline in bottom-hole pressure. However, the plunger is continuing to travel each cycle and closure occurs due to the plunger arrival shut-in system.

Fig. 7.29 shows a typical gas lift well operating with a plunger. Production is approximately reduced by one-half when the plunger is not running.

Fig. 7.30 shows a plunger being used to unload liquids from a gas well. Note the higher tubing pressures when the plunger is running. This indicates a higher production rate. Many weak gas wells have been greatly improved by the installation of a plunger. A few are "free running" installations, that is, the well does not have to be closed to cause the plunger to fall (Chapter 8).

Fig. 7.31(A) shows a gas well utilizing a plunger to unload a small volume of liquids per day (11 b/d). Without the plunger, this liquid will eventually load up the well and cause the gas production rate to decrease considerably. The well makes 235 MCF/D. The well produces two hours and is closed in one hour. During the sixteen hours of production, the rate is 352 MCF/D, making 252 MCF/D for the full 24 hours. Fig. 7.31(B) shows the corresponding surface flowing temperature chart.

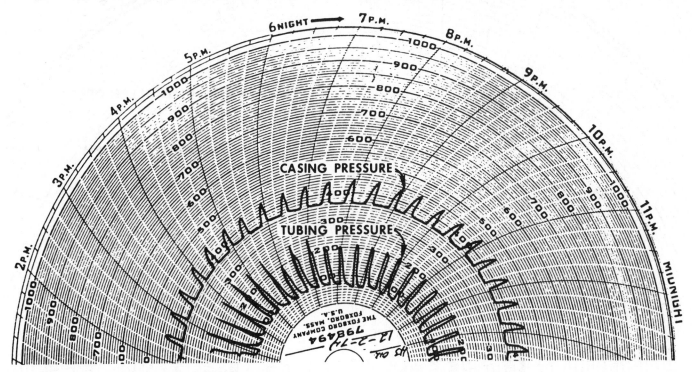

Fig. 7.26 Two pen recorder chart of plunger used with intermittent gas lift

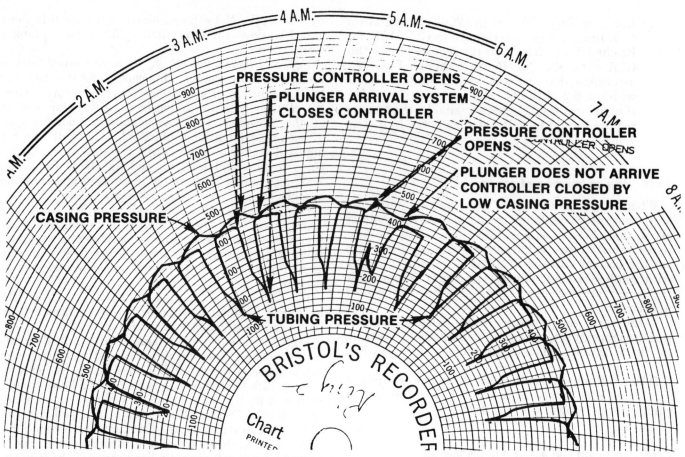

Fig. 7.27   *Two pen recorder chart of plunger in flowing well - combination plunger arrival low casing pressure shut-off shows case of plunger not reaching top and drop in casing pressure must close well in.*

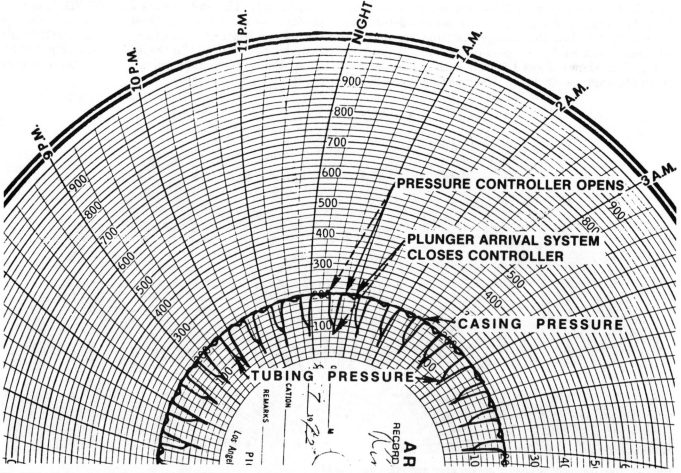

Fig. 7.28   *Two pen recorder chart of plunger in flowing well - High-low pressure system with plunger arrival shut-off (Same well as in Fig. 7.27 but bottom hole pressure has declined)*

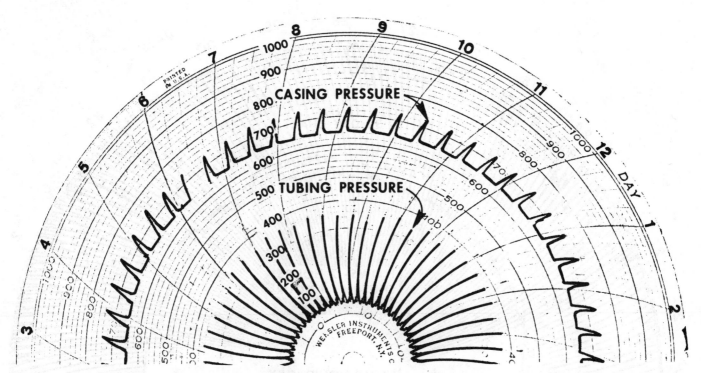

Fig. 7.29   Two pen recorder chart plunger in combination with intermittent gas lift - 8000 ft. depth, P_R = 400 PSI q_L = 165 B/D (35 B/D Oil), production per cycle = 2.75+ bbl., G/L = 2075 SCF

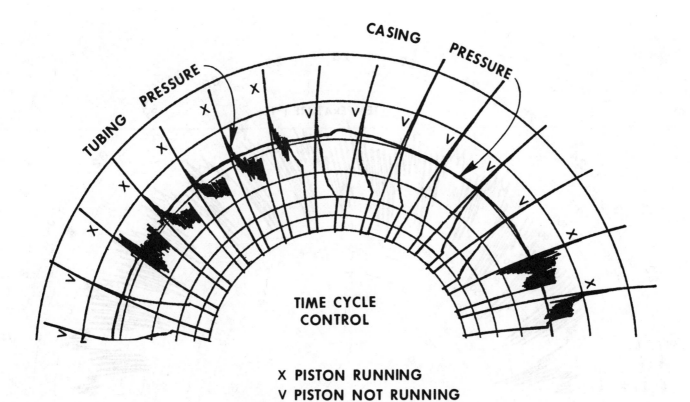

X PISTON RUNNING
V PISTON NOT RUNNING

Fig. 7.30   Piston in a gas well plunger running only part of the time. Note higher tubing pressure and hence greater production when piston is running.

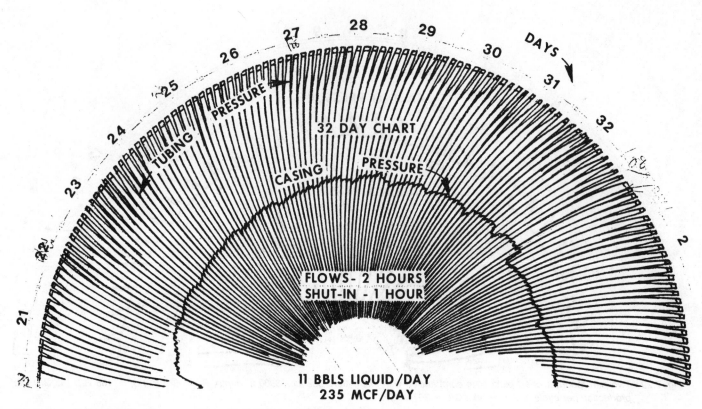

Fig. 7.31A   Gas well with plunger used to unload 11 bbl. of liquid per day

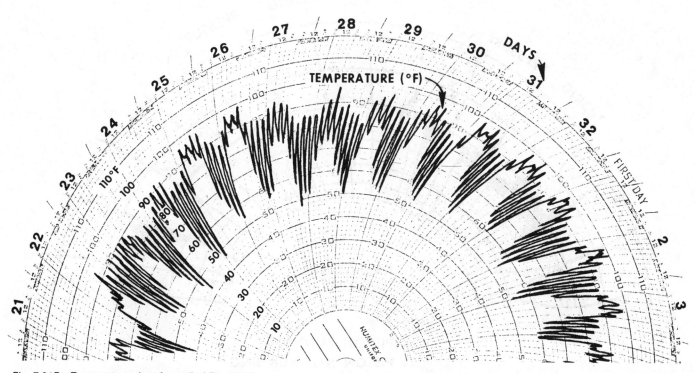

Fig. 7.31B   Temperature chart for well of Fig. 7.31A

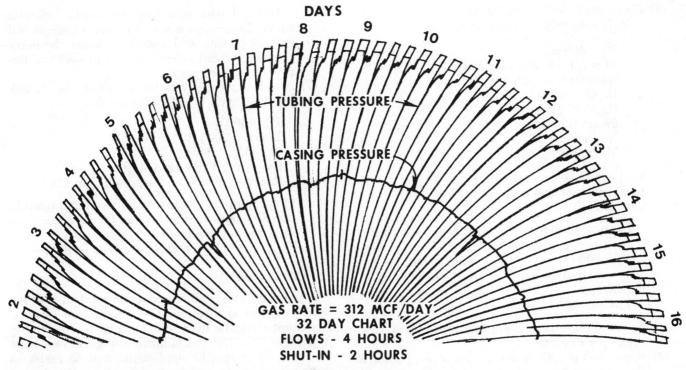

Fig. 7.32A Gas well with plunger unloading 6.5 bbl. water per day and 17 bls. oil per day

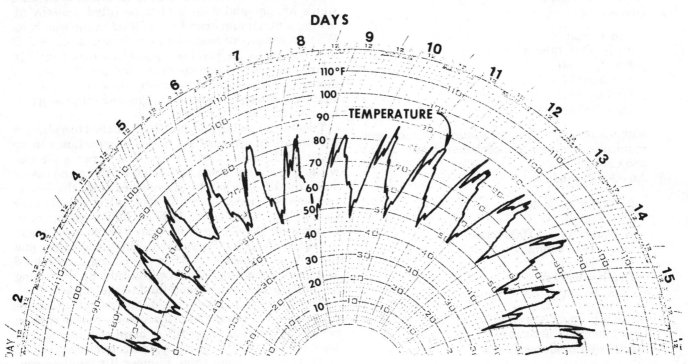

Fig. 7.32B Temperature chart for well of Fig. 7.32A

Fig. 7.32(A) shows another gas well being unloaded of 6.5 bbl water and 17 bo/d with an average gas production rate of 312 MCF/D and a rate of 468 MCF/D during the actual flow time. The well flows for 4 hours then is shut in for 2 hours. Fig. 7.32(B) is the corresponding temperature chart for this well.

7.467 Class problems

(1) An 8000-ft deep gas well with 2⅞ in. tubing is being produced up the casing-tubing annulus against a 150 PSI sales line pressure. The well is loading up. The well produces 200 MCF/day and 25 b/d liquid immediately after being tubing flowed to a 60 PSI system. The well is equipped with 7″ casing.

Determine:
(1) Method of operation and equipment required.
(2) Surface liquid delivery per cycle, and total liquid production.
(3) Gas produced to low pressure system.

(2) Given the following well data:
Well presently on beam pump

$\overline{P}_R$ = 400 psi
PI = 1.0 (assume constant)
Gas-oil ratio = 2500 SCF/b
Depth = 6000 ft
P_{wh} = 40 psi
2⅜ in. tubing
No packer
5½ in. casing

Will a plunger operate? If so, make the equipment selection and predict production rate.

(3) Given:

Depth = 8000 ft
GOR = 4000 scf/b
$\overline{P}_R$ = 600 psi
q_o = 40 B/D for P_{wf} = 300 psi
7 in. casing

IPR behaves according to Vogel's reference curve.
2⅜ in. O.D. tubing
P_{wh} = 50 psi

Select plunger equipment and predict flow rate, cycles per day, etc.

(4) A gas well is 7000-ft deep. Presently a packer is set and no communication exists between the tubing and casing.
Given:

7 in. casing
2⅞ in. O.D. tubing
For P_{wh} = 60
q_g = 300,000 scf/day
Well produces 30 bo/d
$\overline{P}_R$ = 700 psi

Make a plunger recommendation. Should communication be established between the tubing and casing?

(5) An oil well produces

20 b/d for P_{wf} = 200 psi
$\overline{P}_R$ = 600 psi
Gas-oil ratio = 10,000 scf/B
2⅜ in. O.D. tubing
5½ in. casing
Depth = 6000 ft
P_{wh} = 60 psi

Make a plunger selection. Should gas be added to or taken off the casing?

(6) An oil well produces 60 b/d oil

Gas-oil ratio = 1500 scf/B
Depth = 8000 ft
$\overline{P}_R$ = 1000 psi
P_{wf} = 600 psi for 60 b/d
2⅞" O.D. tubing by 7" casing

IPR behaves according to Vogel's reference curve. Determine whether or not a plunger will operate in this well and, if so, make the necessary equipment selections and predict the flow rate and cycles per day.

(7) A well is presently on intermittent gas lift making 3.5 b/cycle for 24 cycles per day.

P_{so} = 600 psi
Depth = 7000 ft
2⅞" O.D. tubing
Operating valve at 7000 ft
Recovery = 50% under present conditions.

Make equipment selection for a plunger installation in conjunction with gas lift. Estimate any increase in production.

$\overline{P}_R$ = 800 psi and P_{wf} average is 400 psi

7.5 CONCLUSIONS

Plunger lift is a viable production method that has wide potential application. The information given in this chapter can be used to determine the feasibility of plunger lift, the mechanical configurations most desirable for a specific application and to assist in economic evaluations of a complete plunger package.

The plunger provides a sealing interface between a well's lift gas and fluid load to be lifted, thereby allowing a maximum drawdown of fluid in the well bore. When a velocity of more than 300 ft/min is reached, it also serves as an excellent paraffin cutting tool. In many cases a plunger operation for paraffin cutting alone is very profitable in savings on mechanical cutting, hot oiling, and increased gas and oil production by maintaining a clean well bore.

The well's capabilities should be matched to a plunger lift installation before a successful operation can be expected. It is practically impossible to run a plunger without communication between the tubing and casing since the lift gas has to be produced from the formation and is generally not sufficient unless stored in the casing. An installation without tubing-casing communication usually must be vented to the atmosphere to create enough differential to lift the plunger and the fluid to the surface. When the casing can be used as a gas volume storage, gas will u-tube up the tubing thus lifting the plunger and fluid to the surface.

One factor, probably the most important for successful plunger lift, has not been discussed—field operating personnel. Since plunger lift is more of a mechanical process than gas lift, operating personnel must spend proportionately more time. Plunger lift requires constant and diligent surveillance so minor mechanical or production problems are discovered and acted on before they develop into major problems. Where the field people are interested, plunger lift is successful. The converse also holds true.

2⅜ O.D. TUBING
$P_{wh} = 0$ PSI

TABLE 7A.1
SUB-SURFACE PLUNGER APPLICATION INFORMATION

DEPTH (FEET)		LIQUID SURFACE DELIVERY PER CYCLE—BARRELS															
		0.25	0.50	0.75	1.00	1.25	1.50	1.75	2.00	2.25	2.50	2.75	3.00	3.25	3.50	3.75	4.00
2000	GLR	.8	.8	.8	.8	.7	.7	.7	.7	.7	.7	.7	.7				
	MCF	.2	.4	.6	.8	.9	1.0	1.2	1.4	1.6	1.7	1.9	2.2				
	Pc	52	96	141	190	235	280	328	374	419	464	510	556				
	BPD	83	154	215	269	316	358	395	430	459	487	514	537				
	~	332	308	287	269	253	239	226	215	204	195	187	179				
3000	GLR	1.2	1.2	1.2	1.2	1.2	1.2	1.2	1.2	1.2	1.2	1.2	1.2				
	MCF	.3	.6	.9	1.2	1.4	1.7	1.9	2.2	2.5	2.8	3.2	3.4				
	Pc	53	100	145	193	241	288	336	383	430	476	524	572				
	BPD	56	107	153	196	233	268	301	330	357	382	407	429				
	~	227	215	205	196	187	179	172	165	159	153	148	143				
4000	GLR	2.0	1.6	1.6	1.6	1.5	1.5	1.5	1.5	1.5	1.5	1.5	1.5				
	MCF	.5	.8	1.2	1.6	1.8	2.2	2.6	3.0	3.4	3.8	4.3	4.7				
	Pc	54	102	150	198	247	295	345	394	441	490	538	587				
	BPD	43	83	119	154	185	214	241	268	292	315	338	357				
	~	172	167	159	154	148	143	138	134	130	126	123	119				
5000	GLR	2.4	2.2	2.0	2.0	2.0	2.0	2.0	2.0	2.0	2.0	2.0	2.0				
	MCF	.6	1.1	1.5	2.0	2.4	2.8	3.4	3.8	4.4	5.0	5.5	6.1				
	Pc	55	104	154	204	254	304	354	405	453	503	552	603				
	BPD	34	67	97	126	153	178	203	226	247	267	288	306				
	~	139	134	130	126	123	119	116	113	110	107	105	102				
6000	GLR	2.8	2.6	2.4	2.4	2.4	2.4	2.4	2.4	2.4	2.4	2.4	2.4				
	MCF	.7	1.3	1.8	2.4	3.0	3.5	4.2	4.7	5.3	6.0	6.6	7.4				
	Pc	56	107	158	210	260	311	363	415	465	506	567	618				
	BPD	29	56	82	107	131	153	175	196	213	232	250	267				
	~	116	113	110	107	105	102	100	98	95	93	91	89				
7000	GLR	3.2	3.2	3.0	2.9	2.9	2.8	2.8	2.8	2.8	2.8	2.8	2.8				
	MCF	.8	1.6	2.2	2.9	3.5	4.2	4.9	5.6	6.4	7.2	8.0	8.7				
	Pc	58	110	162	215	267	320	372	426	476	530	582	634				
	BPD	25	49	71	93	113	133	154	172	189	205	222	237				
	~	100	98	95	93	91	89	88	86	84	82	81	79				
8000	GLR	4.0	3.5	3.4	3.3	3.3	3.3	3.3	3.3	3.3	3.3	3.3	3.3				
	MCF	1.0	1.7	2.5	3.3	4.1	4.8	5.6	6.5	7.4	8.3	9.3	10.2				
	Pc	60	113	166	220	273	326	382	436	488	542	595	650				
	BPD	22	43	63	83	101	118	136	154	168	185	200	213				
	~	88	86	84	83	81	79	78	77	75	74	73	71				
9000	GLR	4.8	4.2	3.9	3.8	3.8	3.8	3.8	3.8	3.8	3.9	3.9	3.9				
	MCF	1.2	2.1	2.9	3.8	4.7	5.6	6.6	7.5	8.5	9.6	10.6	11.7				
	Pc	60	115	170	225	280	335	390	446	500	555	610	665				
	BPD	19	38	56	74	91	106	122	138	153	167	181	195				
	~	78	77	75	74	73	71	70	69	68	67	66	65				
10000	GLR	5.2	4.6	4.6	4.4	4.4	4.2	4.3	4.3	4.3	4.4	4.4	4.4				
	MCF	1.3	2.3	3.4	4.4	5.4	6.3	7.4	8.5	9.6	10.8	12.0	13.2				
	Pc	61	118	173	230	286	342	400	456	512	568	624	680				
	BPD	17	34	51	67	82	97	112	126	139	152	165	177				
	~	70	69	68	67	66	65	64	63	62	61	60	59				
11000	GLR	5.6	5.2	5.0	4.9	4.8	.48	4.8	4.8	4.8	4.9	4.9	4.9				
	MCF	1.4	2.6	3.7	4.9	6.0	7.2	8.3	9.5	10.7	12.1	13.4	14.7				
	Pc	62	120	178	236	292	350	409	468	523	580	639	697				
	BPD	16	31	46	61	75	88	103	116	128	140	154	165				
	~	64	63	62	61	60	59	59	58	57	56	56	55				
12000	GLR	6.4	5.6	5.6	5.5	5.4	5.4	5.4	5.4	5.4	5.4	5.5	5.5				
	MCF	1.6	2.8	4.2	5.5	6.7	8.0	9.3	10.6	12.0	13.5	15.0	16.5				
	Pc	63	123	181	241	300	358	418	479	535	593	653	712				
	BPD	14	29	42	56	70	82	94	106	119	130	140	153				
	~	59	58	57	56	56	55	54	53	53	52	51	51				

LEGEND: GLR: — Required Gas Liquid Ratio MCF/BBL.
  MCF — MCF Required Per Cycle.
  Pc — Required Surface Casing Pressure* (average)
  BPD — Maximum Production BBLS./DAY
  ~ — Maximum Cycles Per Day.

*Can be correlated with static pressure of well to determine maximum draw down.

TABLE 7A.2
SUB-SURFACE PLUNGER APPLICATION INFORMATION

2⅜ O.D. TUBING
P_{wh} = 30 PSI

DEPTH (FEET)		LIQUID SURFACE DELIVERY PER CYCLE—BARRELS															
		0.25	0.50	0.75	1.00	1.25	1.50	1.75	2.00	2.25	2.50	2.75	3.00	3.25	3.50	3.75	4.00
2000	GLR	1.2	0.8	0.8	0.8	0.8	0.8	0.8	0.8	0.8	0.8	0.8	0.8				
	MCF	.3	.4	.6	.8	1.0	1.2	1.4	1.6	1.8	2.0	2.2	2.4				
	Pc	82	131	175	220	268	303	360	406	451	496	542	589				
	BPD	83	154	215	269	316	358	395	430	459	487	514	537				
	~	332	308	287	269	253	239	226	215	204	195	187	179				
3000	GLR	2.0	1.4	1.4	1.3	1.3	1.3	1.3	1.3	1.3	1.2	1.2	1.2				
	MCF	.5	.7	1.0	1.3	1.6	1.8	2.2	2.5	2.8	3.0	3.3	3.7				
	Pc	85	134	180	226	275	312	370	416	463	510	558	603				
	BPD	56	107	153	196	233	268	301	330	357	382	407	429				
	~	227	215	205	196	187	179	172	165	159	153	148	143				
4000	GLR	2.4	2.0	1.8	1.8	1.8	1.8	1.8	1.7	1.7	1.7	1.7	1.7				
	MCF	.6	1.0	1.3	1.8	2.2	2.6	3.0	3.3	3.7	4.1	4.6	5.0				
	Pc	88	138	184	233	282	331	380	428	477	525	573	620				
	BPD	43	83	119	154	185	214	241	268	292	315	338	357				
	~	172	167	159	154	148	143	130	134	130	126	123	119				
5000	GLR	3.2	2.6	2.3	2.2	2.2	2.2	2.2	2.2	2.1	2.1	2.1	2.1				
	MCF	.8	1.3	1.7	2.2	2.7	3.3	3.7	4.3	4.7	5.3	5.7	6.4				
	Pc	90	141	190	240	290	340	390	440	490	540	589	638				
	BPD	34	67	97	126	153	178	203	226	247	267	288	306				
	~	139	134	130	126	123	119	116	113	110	107	105	102				
6000	GLR	4.4	3.2	3.0	2.8	2.8	2.7	2.7	2.6	2.6	2.6	2.6	2.6				
	MCF	1.1	1.6	2.2	2.8	3.4	4.0	4.6	5.2	5.7	6.5	7.2	7.8				
	Pc	92	143	195	245	297	350	400	452	502	552	603	654				
	BPD	29	56	82	107	131	153	175	196	213	232	250	267				
	~	116	113	110	107	105	102	100	98	95	93	91	89				
7000	GLR	5.2	3.8	3.5	3.3	3.2	3.2	3.1	3.1	3.1	3.1	3.1	3.1				
	MCF	1.3	1.9	2.6	3.3	4.0	4.7	5.5	6.2	6.8	7.6	8.4	9.3				
	Pc	95	149	200	251	305	358	410	462	514	566	620	671				
	BPD	25	49	71	93	113	133	154	172	189	205	222	237				
	~	100	98	95	93	91	89	88	86	84	82	81	79				
8000	GLR	6.0	4.4	4.2	3.8	3.8	3.7	3.7	3.6	3.6	3.6	3.6	3.6				
	MCF	1.5	2.2	3.1	3.8	4.7	5.5	6.4	7.2	8.0	8.8	9.7	10.6				
	Pc	97	152	205	259	312	367	420	472	528	581	633	689				
	BPD	22	43	63	83	101	118	136	154	168	185	200	213				
	~	88	86	84	83	81	79	78	77	75	74	73	71				
9000	GLR	6.8	5.2	4.7	4.4	4.4	4.2	4.2	4.2	4.1	4.1	4.1	4.1				
	MCF	1.7	2.6	3.5	4.4	5.4	6.3	7.3	8.3	9.2	10.2	11.2	12.2				
	Pc	100	156	210	264	320	374	430	485	540	594	650	705				
	BPD	19	38	56	74	91	106	122	138	153	167	181	195				
	~	78	77	75	74	73	71	70	69	68	67	66	65				
10000	GLR	8.0	5.8	5.4	5.0	4.9	4.8	4.8	4.7	4.7	4.6	4.6	4.6				
	MCF	2.0	2.9	4.0	5.0	6.1	7.2	8.3	9.3	10.4	11.5	12.6	13.8				
	Pc	102	160	214	271	328	384	441	498	552	609	665	721				
	BPD	17	34	51	67	82	97	112	126	139	152	165	177				
	~	70	69	68	67	66	65	64	63	62	61	60	59				
11000	GLR	8.4	6.6	6.0	5.6	5.5	5.4	5.3	5.3	5.2	5.2	5.2	5.2				
	MCF	2.2	3.3	4.5	5.6	6.8	8.1	9.3	10.5	11.6	12.9	14.2	15.5				
	Pc	105	162	220	278	335	392	451	510	565	622	680	738				
	BPD	16	31	46	61	75	88	103	116	128	140	154	165				
	~	64	63	62	61	60	59	59	58	57	56	56	55				
12000	GLR	10.0	7.4	6.7	6.3	6.1	6.0	6.0	5.9	5.8	5.8	5.7	5.7				
	MCF	2.5	3.7	5.0	6.3	7.6	9.0	10.4	11.7	13.0	14.4	15.7	17.2				
	Pc	109	167	224	283	342	402	461	520	580	639	695	756				
	BPD	14	29	42	56	70	82	94	106	119	130	140	153				
	~	59	58	57	56	56	55	54	53	53	52	51	51				

LEGEND: GLR: — Required Gas Liquid Ratio MCF/BBL.
MCF — MCF Required Per Cycle.
Pc — Required Surface Casing Pressure* (average)
BPD — Maximum Production BBLS./DAY
~ — Maximum Cycles Per Day.

*Can be correlated with static pressure of well to determine maximum draw down.

TABLE 7A.3
SUB-SURFACE PLUNGER APPLICATION INFORMATION

2⅜ O.D. TUBING
$P_{wh} = 60$ PSI

DEPTH (FEET)		LIQUID SURFACE DELIVERY PER CYCLE—BARRELS														
		0.25	0.50	0.75	1.00	1.25	1.50	1.75	2.00	2.25	2.50	2.75	3.00	3.25	3.50	3.75	4.00
2000	GLR	1.6	1.2	1.1	.9	.9	.9	.9	.9	.9	.9	.9	.9				
	MCF	.4	.6	.8	.9	1.1	1.3	1.5	1.7	1.9	2.1	2.3	2.5				
	Pc	115	162	207	256	302	346	394	440	486	532	578	622				
	BPD	83	154	215	269	316	358	395	430	459	487	514	537				
	~	332	308	287	269	253	239	226	215	204	195	187	179				
3000	GLR	2.8	1.6	1.6	1.4	1.4	1.4	1.4	1.3	1.3	1.3	1.3	1.3				
	MCF	.7	.8	1.2	1.4	1.7	2.0	2.3	2.6	2.9	3.2	3.5	3.8				
	Pc	119	166	212	262	311	357	404	452	500	546	593	640				
	BPD	56	107	153	196	233	268	301	330	357	382	407	429				
	~	227	215	205	196	187	179	172	165	159	153	148	143				
4000	GLR	3.2	2.4	2.2	1.9	1.9	1.9	1.9	1.8	1.8	1.8	1.8	1.8				
	MCF	.8	1.2	1.6	1.9	2.3	2.7	3.2	3.5	4.0	4.4	4.7	5.2				
	Pc	122	171	219	270	319	367	416	463	512	561	611	658				
	BPD	43	83	119	154	185	214	241	268	292	315	338	357				
	~	172	167	159	154	148	143	138	134	130	126	123	119				
5000	GLR	4.4	3.0	2.7	2.5	2.4	2.4	2.3	2.3	2.3	2.2	2.2	2.2				
	MCF	1.1	1.5	2.0	2.5	3.0	3.5	4.0	4.5	5.0	5.5	6.0	6.6				
	Pc	125	176	225	276	327	378	427	477	527	577	627	677				
	BPD	34	67	97	126	153	178	203	226	247	267	288	306				
	~	139	134	130	126	123	119	116	113	110	107	105	102				
6000	GLR	5.6	3.8	3.4	3.0	3.0	2.9	2.9	2.8	2.8	2.7	2.7	2.7				
	MCF	1.4	1.9	2.5	3.0	3.7	4.3	5.0	5.6	6.2	6.7	7.4	8.2				
	Pc	129	180	230	283	336	386	438	490	540	591	642	693				
	BPD	29	56	82	107	131	153	175	196	213	232	250	267				
	~	116	113	110	107	105	102	100	98	95	93	91	89				
7000	GLR	6.4	4.4	3.9	3.6	3.6	3.4	3.4	3.3	3.3	3.3	3.2	3.2				
	MCF	1.6	2.2	2.9	3.6	4.4	5.1	5.8	6.6	7.4	8.1	8.7	9.6				
	Pc	132	185	237	291	343	396	450	502	553	606	660	711				
	BPD	25	49	71	93	113	133	154	172	189	205	222	237				
	~	100	98	95	93	91	89	88	86	84	82	81	79				
8000	GLR	7.2	5.2	4.6	4.2	4.1	4.0	3.9	3.9	3.8	3.8	3.7	3.7				
	MCF	1.8	2.6	3.4	4.2	5.1	6.0	6.8	7.7	8.5	9.4	10.2	11.2				
	Pc	136	190	242	300	352	406	461	513	568	622	676	730				
	BPD	22	43	63	83	101	118	136	154	168	185	200	213				
	~	88	86	84	83	81	79	78	77	75	74	73	71				
9000	GLR	8.8	6.0	5.2	4.8	4.7	4.6	4.6	4.4	4.4	4.3	4.3	4.3				
	MCF	2.2	3.0	3.9	4.8	5.8	6.8	7.9	8.8	9.7	10.7	11.6	12.7				
	Pc	140	194	250	306	361	416	471	527	581	638	692	747				
	BPD	19	38	56	74	91	106	122	138	153	167	181	195				
	~	78	77	75	74	73	71	70	69	68	67	66	65				
10000	GLR	10.0	6.8	6.0	5.6	5.4	5.2	5.2	5.0	5.0	4.9	4.8	4.8				
	MCF	2.5	3.4	4.5	5.6	6.7	7.8	9.0	10.0	11.1	12.1	13.1	14.4				
	Pc	142	200	255	312	370	426	482	540	595	652	708	763				
	BPD	17	34	51	67	82	97	112	126	139	152	165	177				
	~	70	69	68	67	66	65	64	63	62	61	60	59				
11000	GLR	11.2	7.8	6.7	6.3	6.0	5.9	5.8	5.7	5.6	5.4	5.4	5.4				
	MCF	2.8	3.9	5.0	6.3	7.5	8.8	10.0	11.3	12.5	13.5	14.7	16.2				
	Pc	146	204	261	320	378	436	492	551	609	668	724	781				
	BPD	16	31	46	61	75	88	103	116	128	140	154	165				
	~	64	63	62	61	60	59	59	58	57	56	56	55				
12000	GLR	12.8	9.0	7.6	7.1	6.8	6.6	6.4	6.3	6.2	6.1	6.0	6.0				
	MCF	3.2	4.5	5.7	7.1	8.5	9.8	11.2	12.6	13.9	15.2	16.5	18.1				
	Pc	150	209	267	328	386	445	504	563	622	681	740	800				
	BPD	14	29	42	56	70	82	94	106	119	130	140	153				
	~	59	58	57	56	56	55	54	53	53	52	51	51				

LEGEND: GLR: — Required Gas Liquid Ratio MCF/BBL.
　　　　MCF — MCF Required Per Cycle.
　　　　Pc — Required Surface Casing Pressure* (average)
　　　BPD — Maximum Production BBLS./DAY
　　　~ — Maximum Cycles Per Day.

*Can be correlated with static pressure of well to determine maximum draw down.

TABLE 7A.4
SUB-SURFACE PLUNGER APPLICATION INFORMATION

2⅜ O.D. TUBING
P_{wh} = 100 PSI

DEPTH (FEET)		LIQUID SURFACE DELIVERY PER CYCLE—BARRELS															
		0.25	0.50	0.75	1.00	1.25	1.50	1.75	2.00	2.25	2.50	2.75	3.00	3.25	3.50	3.75	4.00
2000	GLR	1.6	1.2	1.1	1.0	1.0	1.0	1.0	1.0	1.0	1.0	1.0	1.0				
	MCF	.4	.6	.8	1.0	1.2	1.4	1.6	1.8	2.1	2.3	2.5	2.7				
	Pc	161	207	253	299	345	391	438	483	530	573	621	666				
	BPD	83	154	215	269	316	358	395	430	459	487	514	537				
	~	332	308	287	269	253	239	226	215	204	195	187	179				
3000	GLR	3.2	2.2	1.9	1.7	1.7	1.6	1.6	1.5	1.5	1.5	1.5	1.5				
	MCF	.8	1.1	1.4	1.7	2.1	2.3	2.7	3.0	3.3	3.6	4.0	4.3				
	Pc	166	212	260	307	355	401	450	496	542	590	638	683				
	BPD	56	107	153	196	233	268	301	330	357	382	407	429				
	~	227	215	205	196	187	179	172	165	159	153	148	143				
4000	GLR	4.8	3.2	2.7	2.4	2.3	2.2	2.2	2.1	2.0	2.0	2.0	2.0				
	MCF	1.2	1.6	2.0	2.4	2.8	3.3	3.7	4.1	4.5	4.9	5.4	5.7				
	Pc	170	219	268	315	364	412	462	510	559	606	655	703				
	BPD	43	83	119	154	185	214	241	268	292	315	338	357				
	~	172	167	159	154	148	143	138	134	130	126	123	119				
5000	GLR	6.4	4.2	3.4	3.1	2.9	2.8	2.7	2.7	2.6	2.4	2.4	2.4				
	MCF	1.6	2.1	2.5	3.1	3.6	4.2	4.7	5.2	5.7	6.2	6.8	7.4				
	Pc	175	224	273	323	373	422	475	523	572	622	672	722				
	BPD	34	67	97	126	153	178	203	226	247	267	288	306				
	~	139	134	130	126	123	119	116	113	110	107	105	102				
6000	GLR	8.0	5.2	4.3	3.7	3.6	3.4	3.3	3.2	3.2	3.1	3.1	3.0				
	MCF	2.0	2.6	3.2	3.7	4.4	5.0	5.7	6.3	7.0	7.7	8.3	9.0				
	Pc	180	231	282	332	383	434	488	538	589	639	690	741				
	BPD	29	56	82	107	131	153	175	196	213	232	250	267				
	~	116	113	110	107	105	102	100	98	95	93	91	89				
7000	GLR	9.2	6.2	5.1	4.5	4.2	4.0	3.9	3.8	3.7	3.6	3.6	3.6				
	MCF	2.3	3.1	3.8	4.5	5.2	6.0	6.7	7.5	8.2	9.0	9.9	10.7				
	Pc	184	236	290	341	392	447	500	552	602	655	709	760				
	BPD	25	49	71	93	113	133	154	172	189	205	222	237				
	~	100	98	95	93	91	89	88	86	84	82	81	79				
8000	GLR	10.8	7.2	5.9	5.2	5.0	4.7	4.5	4.4	4.3	4.2	4.2	4.1				
	MCF	2.7	3.6	4.4	5.2	6.2	7.0	7.8	8.7	9.6	10.5	11.4	12.3				
	Pc	189	242	296	350	402	458	512	566	618	671	725	780				
	BPD	22	43	63	83	101	118	136	154	168	185	200	213				
	~	88	86	84	83	81	79	78	77	75	74	73	71				
9000	GLR	12.8	8.2	6.7	6.0	5.6	5.4	5.2	5.0	4.9	4.8	4.8	4.7				
	MCF	3.2	4.1	5.0	6.0	7.0	8.0	9.0	10.0	11.0	12.0	13.0	14.0				
	Pc	193	248	303	359	412	468	524	580	632	688	743	800				
	BPD	19	38	56	74	91	106	122	138	153	167	181	195				
	~	78	77	75	74	73	71	70	69	68	67	66	65				
10000	GLR	14.4	9.4	7.6	6.7	6.3	6.0	5.9	5.6	5.5	5.4	5.4	5.3				
	MCF	3.6	4.7	5.7	6.7	7.8	9.0	10.2	11.2	12.3	13.5	14.7	15.8				
	Pc	198	253	310	368	422	480	538	593	648	703	761	818				
	BPD	17	34	51	67	82	97	112	126	139	152	165	177				
	~	70	69	68	67	66	65	64	63	62	61	60	59				
11000	GLR	16.0	10.4	8.6	7.6	7.2	6.8	6.6	6.3	6.2	6.1	6.0	6.0				
	MCF	4.0	5.2	6.4	7.6	8.9	10.2	11.4	12.5	13.8	15.1	16.5	17.8				
	Pc	202	260	318	376	432	490	600	608	662	720	778	837				
	BPD	16	31	46	61	75	88	103	116	128	140	154	165				
	~	64	63	62	61	60	59	59	58	57	56	56	55				
12000	GLR	18.0	11.6	9.6	8.5	8.0	7.6	7.3	7.0	6.9	6.8	6.7	6.6				
	MCF	4.5	5.8	7.2	8.5	9.9	11.3	12.7	14.0	15.4	16.8	18.4	19.8				
	Pc	206	265	325	385	442	502	612	621	678	738	795	854				
	BPD	14	29	42	56	70	82	94	106	119	130	140	153				
	~	59	58	57	56	56	55	54	53	53	52	51	51				

LEGEND: GLR: — Required Gas Liquid Ratio MCF/BBL.
      MCF — MCF Required Per Cycle.
      Pc — Required Surface Casing Pressure* (average)
      BPD — Maximum Production BBLS./DAY
      ~ — Maximum Cycles Per Day.

*Can be correlated with static pressure of well to determine maximum draw down.

TABLE 7A.5
SUB-SURFACE PLUNGER APPLICATION INFORMATION

2⅜ O.D. TUBING
P_{wh} = 150 PSI

DEPTH (FEET)		0.25	0.50	0.75	1.00	1.25	1.50	1.75	2.00	2.25	2.50	2.75	3.00	3.25	3.50	3.75	4.00
							LIQUID SURFACE DELIVERY PER CYCLE—BARRELS										
2000	GLR	4.0	2.4	1.9	1.6	1.5	1.3	1.3	1.2	1.2	1.2	1.1	1.1				
	MCF	1.0	1.2	1.4	1.6	1.8	2.0	2.2	2.4	2.6	2.8	3.0	3.2				
	Pc	219	262	310	356	402	448	494	541	585	631	677	723				
	BPD	83	154	215	269	316	358	395	430	459	487	514	537				
	~	332	308	287	269	253	239	226	215	204	195	187	179				
3000	GLR	6.4	3.8	3.0	2.5	2.3	2.1	2.0	1.9	1.8	1.8	1.7	1.7				
	MCF	1.6	1.9	2.2	2.5	2.8	3.1	3.4	3.7	4.0	4.3	4.6	5.0				
	Pc	223	271	319	366	412	460	505	556	602	650	696	743				
	BPD	56	107	153	196	233	268	301	330	357	382	407	429				
	~	227	215	205	196	187	179	172	165	159	153	148	143				
4000	GLR	8.8	5.2	4.0	3.4	3.1	2.8	2.7	2.5	2.5	2.4	2.3	2.3				
	MCF	2.2	2.6	3.0	3.4	3.8	4.2	4.6	5.0	5.5	5.9	6.3	6.8				
	Pc	230	278	327	377	423	472	521	571	619	667	715	763				
	BPD	43	83	119	154	185	214	241	268	292	315	338	357				
	~	172	167	159	154	148	143	138	134	130	126	123	119				
5000	GLR	10.8	6.4	5.0	4.2	3.9	3.6	3.4	3.2	3.1	3.0	3.0	2.9				
	MCF	2.7	3.2	3.7	4.2	4.8	5.3	5.8	6.4	6.9	7.4	8.0	8.5				
	Pc	237	285	338	387	436	484	536	587	634	683	734	784				
	BPD	34	67	97	126	153	178	203	226	247	267	288	306				
	~	139	134	130	126	123	119	116	113	110	107	105	102				
6000	GLR	13.6	8.0	6.2	5.2	4.8	4.4	4.1	3.9	3.8	3.7	3.6	3.5				
	MCF	3.4	4.0	4.6	5.2	5.9	6.5	7.1	7.8	8.4	9.1	9.7	10.4				
	Pc	242	291	344	396	446	499	550	601	651	702	753	804				
	BPD	29	56	82	107	131	153	175	196	213	232	250	267				
	~	116	113	110	107	105	102	100	98	95	93	91	89				
7000	GLR	16.0	9.4	7.4	6.2	5.6	5.2	4.9	4.6	4.5	4.3	4.2	4.1				
	MCF	4.0	4.7	5.5	6.2	7.0	7.7	8.5	9.2	10.0	10.7	11.5	12.3				
	Pc	249	301	353	406	458	511	563	616	669	720	772	824				
	BPD	25	49	71	93	113	133	154	172	189	205	222	237				
	~	100	98	95	93	91	89	88	86	84	82	81	79				
8000	GLR	18.4	11.0	8.6	7.3	6.6	6.2	5.8	5.5	5.3	5.2	5.1	5.0				
	MCF	4.6	5.5	6.4	7.3	8.2	9.2	10.1	11.0	11.9	12.9	13.8	14.8				
	Pc	254	308	362	417	470	523	578	631	682	738	791	845				
	BPD	22	43	63	83	101	118	136	154	168	185	200	213				
	~	88	86	84	83	81	79	78	77	75	74	73	71				
9000	GLR	21.2	12.4	9.6	8.2	7.4	6.6	6.4	6.1	5.9	5.7	5.6	5.5				
	MCF	5.3	6.2	7.2	8.2	9.2	10.2	11.2	12.2	13.2	14.2	15.2	16.5				
	Pc	260	315	371	427	481	536	592	647	700	756	810	865				
	BPD	19	38	56	74	91	106	122	138	153	167	181	195				
	~	78	77	75	74	73	71	70	69	68	67	66	65				
10000	GLR	24.0	14.2	10.7	9.2	8.2	7.5	7.1	6.7	6.5	6.4	6.3	6.2				
	MCF	6.0	7.1	8.0	9.2	10.2	11.2	12.3	13.4	14.5	15.9	17.2	18.6				
	Pc	268	322	380	437	492	549	607	662	717	773	830	885				
	BPD	17	34	51	67	82	97	112	126	139	152	165	177				
	~	70	69	68	67	66	65	64	63	62	61	60	59				
11000	GLR	27.2	16.0	12.2	10.3	9.2	8.4	7.9	7.5	7.2	7.2	7.1	7.0				
	MCF	6.8	8.0	9.1	10.3	11.5	12.6	13.8	15.0	16.2	17.8	19.3	20.9				
	Pc	273	330	388	447	503	561	621	678	732	791	848	905				
	BPD	16	31	46	61	75	88	103	116	128	140	154	165				
	~	64	63	62	61	60	59	59	58	57	56	56	55				
12000	GLR	30.4	17.8	13.6	11.5	10.3	9.4	8.9	8.4	8.1	8.0	7.9	7.8				
	MCF	7.6	8.9	10.2	11.5	12.8	14.1	15.5	16.7	18.1	19.8	21.5	23.3				
	Pc	280	340	397	457	514	573	633	692	750	810	868	926				
	BPD	14	29	42	56	70	82	94	106	119	130	140	153				
	~	59	58	57	56	56	55	54	53	53	52	51	51				

LEGEND: GLR: — Required Gas Liquid Ratio MCF/BBL.
MCF — MCF Required Per Cycle.
Pc — Required Surface Casing Pressure* (average)
BPD — Maximum Production BBLS./DAY
~ — Maximum Cycles Per Day.

*Can be correlated with static pressure of well to determine maximum draw down.

TABLE 7A.6
SUB-SURFACE PLUNGER APPLICATION INFORMATION

2⅜ O.D. TUBING
P_{wh} = 200 PSI

DEPTH (FEET)		LIQUID SURFACE DELIVERY PER CYCLE—BARRELS														
		0.25	0.50	0.75	1.00	1.25	1.50	1.75	2.00	2.25	2.50	2.75	3.00	3.25	3.50	3.75	4.00
2000	GLR	3.2	2.2	1.9	1.6	1.6	1.5	1.5	1.4	1.4	1.3	1.3	1.3				
	MCF	.8	1.1	1.4	1.6	1.9	2.2	2.5	2.7	3.0	3.2	3.5	3.7				
	Pc	272	319	363	410	457	502	549	594	640	686	732	779				
	BPD	83	154	215	269	316	358	395	430	459	487	514	537				
	~	332	308	287	269	253	239	226	215	204	195	187	179				
3000	GLR	6.0	3.6	3.5	2.6	2.4	2.3	2.2	2.2	2.1	2.1	2.1	2.0				
	MCF	1.5	1.8	2.2	2.6	3.0	3.4	3.8	4.3	4.6	5.1	5.6	6.0				
	Pc	280	328	373	422	470	516	564	611	658	705	752	801				
	BPD	56	107	153	196	233	268	301	330	357	382	407	429				
	~	227	215	205	196	187	179	172	165	159	153	148	143				
4000	GLR	8.4	5.0	4.0	3.6	3.4	3.2	3.0	2.9	2.8	2.8	2.8	2.8				
	MCF	2.1	2.5	3.0	3.6	4.2	4.7	5.2	5.8	6.3	7.0	7.6	8.3				
	Pc	287	337	384	433	482	530	580	627	676	724	773	823				
	BPD	43	83	119	154	185	214	241	268	292	315	338	357				
	~	172	167	159	154	148	143	138	134	130	126	123	119				
5000	GLR	10.4	6.6	5.2	4.6	4.3	4.0	3.9	3.7	3.6	3.6	3.6	3.5				
	MCF	2.6	3.3	3.9	4.6	5.3	6.0	6.7	7.4	8.1	8.9	9.7	10.5				
	Pc	295	346	394	445	495	544	595	644	694	744	794	845				
	BPD	34	67	97	126	153	178	203	226	247	267	288	306				
	~	139	134	130	126	123	119	116	113	110	107	105	102				
6000	GLR	12.8	8.0	6.6	5.8	5.3	5.0	4.8	4.6	4.5	4.4	4.3	4.3				
	MCF	3.2	4.0	4.9	5.8	6.6	7.5	8.3	9.2	10.1	10.9	11.5	12.8				
	Pc	303	355	405	457	508	558	611	661	712	763	814	866				
	BPD	29	56	82	107	131	153	175	196	213	232	250	267				
	~	116	113	110	107	105	102	100	98	95	93	91	89				
7000	GLR	15.2	9.6	7.6	6.8	6.3	5.8	5.6	5.3	5.2	5.2	5.1	5.1				
	MCF	3.8	4.8	5.7	6.8	7.8	8.7	9.7	10.6	11.7	12.8	14.0	15.1				
	Pc	311	364	415	469	520	572	626	678	730	782	835	888				
	BPD	25	49	71	93	113	133	154	172	189	205	222	237				
	~	100	98	95	93	91	89	88	86	84	82	81	79				
8000	GLR	18.0	11.2	9.1	8.0	7.4	6.8	6.5	6.3	6.1	6.0	5.9	5.9				
	MCF	4.5	5.6	6.8	8.0	9.2	10.2	11.3	12.5	13.6	14.9	16.2	17.5				
	Pc	319	373	426	480	533	586	642	695	748	802	855	910				
	BPD	22	43	63	83	101	118	136	154	168	185	200	213				
	~	88	86	84	83	81	79	78	77	75	74	73	71				
9000	GLR	20.8	13.0	10.3	9.2	8.4	7.8	7.4	7.1	7.0	6.8	6.8	6.7				
	MCF	5.2	6.5	7.7	9.2	10.4	11.6	12.9	14.2	15.6	17.0	18.5	20.0				
	Pc	326	382	436	492	546	600	657	711	766	821	876	932				
	BPD	19	38	56	74	91	106	122	138	153	167	181	195				
	~	78	77	75	74	73	71	70	69	68	67	66	65				
10000	GLR	23.6	14.8	11.9	10.4	9.4	8.8	8.4	8.1	7.8	7.7	7.6	7.5				
	MCF	5.9	7.4	8.9	10.4	11.7	13.1	14.6	16.1	17.5	19.2	20.9	22.5				
	Pc	334	391	447	504	558	615	673	728	784	840	897	954				
	BPD	17	34	51	67	82	97	112	126	139	152	165	177				
	~	70	69	68	67	66	65	64	63	62	61	60	59				
11000	GLR	26.8	16.8	13.4	11.6	10.6	9.9	9.5	9.1	8.7	8.6	8.5	8.4				
	MCF	6.7	8.4	10.0	11.6	13.2	14.8	16.5	18.1	19.6	21.5	23.3	25.2				
	Pc	342	400	457	516	571	629	688	745	802	860	915	976				
	BPD	16	31	46	61	75	88	103	116	128	140	154	165				
	~	64	63	62	61	60	59	59	58	57	56	56	55				
12000	GLR	30.0	18.8	15.0	13.0	11.8	11.0	10.5	10.1	9.8	9.6	9.5	9.4				
	MCF	7.5	9.4	11.2	13.0	14.7	16.5	18.3	20.2	21.9	24.0	26.0	28.0				
	Pc	350	409	468	528	584	643	704	762	820	879	938	998				
	BPD	14	29	42	56	70	82	97	106	119	130	140	153				
	~	59	58	57	56	56	55	54	53	53	52	51	51				

LEGEND: GLR: — Required Gas Liquid Ratio MCF/BBL.
    MCF — MCF Required Per Cycle.
    Pc — Required Surface Casing Pressure* (average)
    BPD — Maximum Production BBLS./DAY
    ~ — Maximum Cycles Per Day.

*Can be correlated with static pressure of well to determine maximum draw down.

TABLE 7A.7
SUB-SURFACE PLUNGER APPLICATION INFORMATION

2⅞ O.D. TUBING
$P_{wh} = 0$ PSI

DEPTH (FEET)		LIQUID SURFACE DELIVERY PER CYCLE—BARRELS															
		0.25	0.50	0.75	1.00	1.25	1.50	1.75	2.00	2.25	2.50	2.75	3.00	3.25	3.50	3.75	4.00
2000	GLR	.8	.7	.7	.7	.7	.7	.7	.7	.7	.7	.7	.7	.7	.7	.7	.7
	MCF	.2	.3	.5	.6	.8	.9	1.1	1.2	1.4	1.5	1.7	1.8	2.0	2.1	2.3	2.5
	Pc	31	60	89	116	144	173	201	230	257	285	313	340	370	398	425	454
	BPD	84	160	227	288	342	393	437	480	515	552	585	618	643	672	697	720
	~	339	320	303	288	274	262	250	240	230	221	213	206	198	192	186	180
3000	GLR	1.2	1.0	1.0	1.0	1.0	1.0	1.0	1.0	1.0	1.0	1.0	1.0	1.0	1.0	1.0	1.0
	MCF	.3	.5	.7	1.0	1.2	1.5	1.7	2.0	2.2	2.5	2.7	3.0	3.2	3.4	3.7	4.0
	Pc	32	61	91	118	147	177	205	235	262	291	320	347	378	406	434	465
	BPD	57	110	159	206	247	288	325	360	391	422	451	480	503	528	551	576
	~	230	221	213	206	198	192	186	180	174	169	164	160	155	151	147	144
4000	GLR	1.6	1.4	1.4	1.4	1.4	1.4	1.4	1.4	1.4	1.4	1.4	1.4	1.4	1.4	1.4	1.4
	MCF	.4	.7	1.0	1.4	1.7	2.1	2.4	2.7	3.1	3.4	3.8	4.1	4.4	4.8	5.1	5.5
	Pc	33	62	93	120	150	180	209	239	267	297	326	354	385	414	443	473
	BPD	43	84	123	160	193	226	251	288	315	342	368	393	416	437	457	480
	~	174	169	164	160	155	151	147	144	140	137	134	131	128	125	122	120
5000	GLR	2.0	1.8	1.8	1.8	1.8	1.8	1.8	1.8	1.8	1.8	1.8	1.8	1.8	1.8	1.8	1.8
	MCF	.5	.9	1.4	1.8	2.2	2.6	3.1	3.5	3.8	4.3	4.8	5.2	5.6	6.1	6.5	6.9
	Pc	34	63	95	123	153	184	213	244	273	303	333	361	393	423	452	483
	BPD	35	68	100	131	160	187	213	240	263	287	310	330	351	371	390	400
	~	140	137	134	131	128	125	122	120	117	115	113	110	108	106	104	100
6000	GLR	2.4	2.2	2.2	2.2	2.2	2.2	2.2	2.2	2.2	2.2	2.2	2.2	2.2	2.2	2.1	2.1
	MCF	.6	1.1	1.7	2.2	2.6	3.2	3.7	4.3	4.7	5.3	5.7	6.3	6.8	7.4	7.8	8.4
	Pc	34	65	96	125	155	188	218	249	278	308	340	369	401	431	460	492
	BPD	29	57	84	110	135	159	182	204	227	247	266	288	305	322	341	360
	~	117	115	113	110	108	106	104	102	101	99	97	96	94	92	91	90
7000	GLR	2.8	2.8	2.7	2.5	2.5	2.5	2.5	2.5	2.5	2.5	2.5	2.5	2.5	2.5	2.5	2.5
	MCF	.7	1.4	2.0	2.5	3.1	3.7	4.5	5.0	5.6	6.2	6.8	7.5	8.1	8.6	9.3	10.0
	Pc	35	66	98	128	158	191	222	254	283	314	346	376	408	439	469	502
	BPD	25	49	72	96	117	138	159	180	198	217	236	252	269	287	303	320
	~	101	99	97	96	94	92	91	90	88	87	86	84	83	82	81	80
8000	GLR	3.6	3.0	3.0	3.0	2.9	2.9	2.9	2.9	2.9	2.9	2.9	2.9	2.9	2.9	2.9	2.9
	MCF	.9	1.5	2.2	3.0	3.6	4.4	5.2	5.8	6.5	7.2	8.0	8.6	9.4	10.0	10.7	11.5
	Pc	36	67	100	130	161	195	226	259	289	320	353	383	416	448	478	511
	BPD	22	43	64	84	103	123	141	160	175	192	209	225	240	255	270	284
	~	88	87	86	84	83	82	81	80	78	77	76	75	74	73	72	71
9000	GLR	4.0	3.4	3.4	3.4	3.3	3.3	3.3	3.3	3.3	3.3	3.3	3.3	3.3	3.3	3.3	3.3
	MCF	1.0	1.7	2.6	3.4	4.1	5.0	5.8	6.6	7.4	8.2	9.0	9.8	10.6	11.5	12.2	13.0
	Pc	37	68	102	133	164	199	230	263	294	326	360	390	424	456	487	521
	BPD	19	38	57	75	92	109	126	144	159	177	189	204	217	234	251	264
	~	78	77	76	75	74	73	72	72	71	71	69	68	67	67	67	66
10000	GLR	4.4	4.0	3.7	3.7	3.7	3.7	3.7	3.7	3.7	3.7	3.7	3.7	3.7	3.7	3.7	3.7
	MCF	1.1	2.0	2.8	3.7	4.6	5.6	6.6	7.5	8.4	9.3	10.2	11.1	12.0	12.9	13.8	14.7
	Pc	38	69	104	135	167	202	234	268	299	332	366	397	431	464	496	531
	BPD	17	35	51	68	83	100	117	132	144	160	173	189	198	213	225	240
	~	71	71	69	68	67	67	67	66	64	64	63	63	61	61	60	60
11000	GLR	5.2	4.4	4.3	4.2	4.2	4.2	4.2	4.2	4.2	4.2	4.2	4.1	4.1	4.1	4.1	4.1
	MCF	1.3	2.2	3.2	4.2	5.1	6.2	7.3	8.3	9.4	10.4	11.4	12.3	13.4	14.4	15.4	16.4
	Pc	39	70	106	137	170	206	239	273	305	338	373	405	439	473	505	540
	BPD	16	32	47	63	76	91	105	120	132	145	159	171	185	196	206	220
	~	64	64	63	63	61	61	60	60	59	58	58	57	57	56	55	55
12000	GLR	5.6	5.0	4.8	4.7	4.6	4.6	4.6	4.6	4.6	4.6	4.6	4.6	4.6	4.6	4.6	4.6
	MCF	1.4	2.5	3.6	4.7	5.7	6.8	8.1	9.2	10.3	11.5	12.5	13.6	14.8	16.0	17.0	18.2
	Pc	40	72	108	140	173	210	243	278	310	344	380	412	447	481	514	550
	BPD	14	29	43	57	71	84	96	110	121	135	145	159	169	182	191	204
	~	59	58	58	57	57	56	55	55	54	54	53	53	52	52	51	51

LEGEND: GLR: — Required Gas Liquid Ratio MCF/BBL.
MCF — MCF Required Per Cycle.
Pc — Required Surface Casing Pressure* (average)
BPD — Maximum production BBLS./DAY
~ — Maximum Cycles Per Day.

*Can be correlated with static pressure of well to determine maximum draw down.

TABLE 7A.8
SUB-SURFACE PLUNGER APPLICATION INFORMATION

2⅞ O.D. TUBING
P_{wh} = 30 PSI

DEPTH (FEET)		LIQUID SURFACE DELIVERY PER CYCLE—BARRELS															
		0.25	0.50	0.75	1.00	1.25	1.50	1.75	2.00	2.25	2.50	2.75	3.00	3.25	3.50	3.75	4.00
2000	GLR	1.6	1.0	.8	.8	.8	.8	.8	.7	.7	.7	.7	.7	.7	.7	.7	.7
	MCF	.4	.5	.6	.8	1.0	1.2	1.3	1.4	1.6	1.8	2.0	2.2	2.3	2.4	2.6	2.8
	Pc	63	92	121	150	176	203	233	262	289	317	343	372	401	430	458	487
	BPD	84	160	227	288	342	393	437	480	515	552	585	618	643	672	697	720
	~	339	320	303	288	274	262	250	240	230	221	213	206	198	192	186	180
3000	GLR	2.4	1.6	1.4	1.2	1.2	1.2	1.2	1.2	1.2	1.1	1.1	1.1	1.1	1.1	1.1	1.1
	MCF	.6	.8	1.0	1.2	1.5	1.7	2.0	2.2	2.5	2.7	2.9	3.2	3.5	3.8	4.0	4.2
	Pc	65	94	123	153	180	207	238	267	295	323	350	380	409	439	468	497
	BPD	57	110	159	206	247	288	325	360	391	422	451	480	503	528	551	576
	~	230	221	213	206	198	192	186	180	174	169	164	160	155	151	147	144
4000	GLR	3.2	2.4	1.9	1.7	1.7	1.6	1.6	1.6	1.6	1.5	1.5	1.5	1.5	1.5	1.5	1.5
	MCF	.8	1.2	1.4	1.7	2.1	2.4	2.7	3.1	3.4	3.7	4.1	4.4	4.7	5.1	5.5	5.9
	Pc	66	96	126	156	183	212	243	273	301	330	358	388	418	448	477	507
	BPD	43	84	123	160	193	226	251	288	315	342	368	393	416	437	457	480
	~	174	169	164	160	155	151	147	144	140	137	134	131	128	125	122	120
5000	GLR	4.0	2.8	2.4	2.2	2.1	2.0	2.0	2.0	2.0	1.9	1.9	1.9	1.9	1.9	1.9	1.9
	MCF	1.0	1.4	1.8	2.2	2.6	3.0	3.5	3.9	4.4	4.7	5.2	5.6	6.0	6.5	7.0	7.5
	Pc	68	98	129	159	187	217	248	279	307	337	365	396	427	457	487	518
	BPD	35	68	100	131	160	187	213	240	263	287	310	330	351	371	390	400
	~	140	137	134	131	128	125	122	120	117	115	113	110	108	106	104	100
6000	GLR	4.8	3.4	3.0	2.7	2.6	2.4	2.4	2.4	2.4	2.3	2.3	2.3	2.3	2.3	2.3	2.3
	MCF	1.2	1.7	2.2	2.7	3.2	3.6	4.3	4.7	5.2	5.7	6.2	6.8	7.3	7.9	8.5	9.0
	Pc	70	100	131	162	191	221	253	284	313	343	373	404	435	467	497	528
	BPD	29	57	84	110	135	159	182	204	227	247	266	288	305	322	341	360
	~	117	115	113	110	108	106	104	102	101	99	97	96	94	92	91	90
7000	GLR	6.0	4.0	3.5	3.2	3.1	3.0	2.9	2.8	2.8	2.8	2.7	2.7	2.7	2.7	2.7	2.7
	MCF	1.5	2.0	2.6	3.2	3.8	4.4	5.0	5.6	6.2	6.8	7.4	8.0	8.6	9.4	10.1	10.7
	Pc	71	102	134	165	194	226	258	290	319	350	380	412	444	476	506	538
	BPD	25	49	72	96	117	138	159	180	198	217	236	252	269	287	303	320
	~	101	99	97	96	94	92	91	90	88	87	86	84	83	82	81	80
8000	GLR	6.8	4.8	4.0	3.6	3.6	3.4	3.4	3.2	3.2	3.2	3.2	3.1	3.1	3.1	3.1	3.1
	MCF	1.7	2.4	3.0	3.6	4.4	5.1	5.8	6.5	7.2	7.8	8.6	9.3	10.0	10.8	11.7	12.4
	Pc	73	104	136	168	198	230	263	295	325	356	388	420	452	485	516	549
	BPD	22	43	64	84	103	123	141	160	175	192	209	225	240	255	270	284
	~	88	87	86	84	83	82	81	80	78	77	76	75	74	73	72	71
9000	GLR	7.6	5.4	4.7	4.2	4.0	3.9	3.9	3.8	3.7	3.6	3.6	3.6	3.6	3.6	3.6	3.6
	MCF	1.9	2.7	3.5	4.2	5.0	5.8	6.7	7.5	8.2	9.0	9.7	10.6	11.5	12.4	13.3	14.1
	Pc	75	106	139	171	202	235	268	301	332	363	395	428	461	494	526	559
	BPD	19	38	57	75	92	109	126	144	159	177	189	204	217	234	251	264
	~	78	77	76	75	74	73	72	72	71	71	69	68	67	67	67	66
10000	GLR	8.8	6.2	5.2	4.7	4.6	4.4	4.3	4.2	4.2	4.1	4.0	4.0	4.0	4.0	4.0	4.0
	MCF	2.2	3.1	3.9	4.7	5.7	6.6	7.5	8.4	9.3	10.1	11.0	11.9	12.9	13.9	15.0	15.9
	Pc	76	108	142	174	205	240	273	307	338	370	403	436	470	503	535	569
	BPD	17	35	51	68	83	100	117	132	144	160	173	189	198	213	225	240
	~	71	71	69	68	67	67	67	66	64	64	63	63	61	61	60	60
11000	GLR	9.6	6.8	5.9	5.3	5.1	4.9	4.8	4.8	4.7	4.6	4.5	4.5	4.5	4.5	4.5	4.5
	MCF	2.4	3.4	4.4	5.3	6.3	7.3	8.4	9.5	10.4	11.3	12.3	13.3	14.4	15.5	16.7	17.7
	Pc	78	110	144	177	209	244	278	312	344	376	410	444	478	513	545	580
	BPD	16	32	47	63	76	91	105	120	132	145	159	171	185	196	206	220
	~	64	64	63	63	61	61	60	60	59	58	58	57	57	56	55	55
12000	GLR	10.8	7.6	6.6	6.0	5.7	5.5	5.4	5.3	5.2	5.1	5.0	5.0	5.0	5.0	5.0	5.0
	MCF	2.7	3.8	4.9	6.0	7.1	8.2	9.3	10.5	11.6	12.6	13.7	14.8	16.0	17.2	18.5	19.7
	Pc	80	112	147	180	213	249	283	318	350	383	418	452	487	522	555	590
	BPD	14	29	43	57	71	84	96	110	121	135	145	159	169	182	191	204
	~	59	58	58	57	57	56	55	55	54	54	53	53	52	52	51	51

LEGEND: GLR: — Required Gas Liquid Ratio MCF/BBL.
MCF — MCF Required Per Cycle.
Pc — Required Surface Casing Pressure* (average)
BPD — Maximum Production BBLS./DAY
~ — Maximum Cycles Per Day.

*Can be correlated with static pressure of well to determine maximum draw down.

TABLE 7A.9
SUB-SURFACE PLUNGER APPLICATION INFORMATION

2⅞ O.D. TUBING
P_{wh} = 60 PSI

DEPTH (FEET)		LIQUID SURFACE DELIVERY PER CYCLE—BARRELS															
		0.25	0.50	0.75	1.00	1.25	1.50	1.75	2.00	2.25	2.50	2.75	3.00	3.25	3.50	3.75	4.00
2000	GLR	2.4	1.6	1.4	1.2	1.1	1.0	9.0	9.0	9.0	9.0	9.0	8.0	8.0	8.0	8.0	8.0
	MCF	.6	.8	1.0	1.2	1.3	1.4	1.5	1.7	1.9	2.0	2.2	2.3	2.5	2.7	2.9	3.0
	Pc	100	127	154	182	210	239	268	295	321	349	378	405	434	462	490	520
	BPD	84	160	227	288	342	393	437	480	515	552	585	618	643	672	697	720
	~	339	320	303	288	274	262	250	240	230	221	213	206	198	192	186	180
3000	GLR	4.0	2.4	1.9	1.7	1.6	1.5	1.4	1.3	1.2	1.2	1.2	1.2	1.2	1.2	1.2	1.2
	MCF	1.0	1.2	1.4	1.7	1.9	2.2	2.4	2.6	2.8	3.1	3.4	3.6	3.9	4.2	4.5	4.7
	Pc	102	129	157	186	214	244	273	301	328	356	386	414	443	472	500	531
	BPD	57	110	159	206	247	288	325	360	391	422	451	480	503	528	551	576
	~	230	221	213	206	198	192	186	180	174	169	164	160	155	151	147	144
4000	GLR	5.2	3.2	2.6	2.2	2.1	2.0	1.9	1.8	1.8	1.7	1.7	1.7	1.7	1.6	1.6	1.6
	MCF	1.3	1.6	1.9	2.2	2.6	2.9	3.2	3.6	3.9	4.2	4.5	5.0	5.3	5.6	6.0	6.4
	Pc	104	132	161	189	218	249	279	307	335	364	394	422	452	481	511	542
	BPD	43	84	123	160	193	226	251	288	315	342	368	393	416	437	457	480
	~	174	169	164	160	155	151	147	144	140	137	134	131	128	125	122	120
5000	GLR	6.4	3.2	3.2	2.8	2.6	2.4	2.4	2.3	2.3	2.2	2.1	2.1	2.1	2.1	2.1	2.1
	MCF	1.6	2.0	2.4	2.8	3.2	3.6	4.1	4.5	5.0	5.4	5.8	6.3	6.7	7.2	7.7	8.2
	Pc	106	134	164	193	223	254	284	314	342	371	402	431	462	491	521	553
	BPD	35	68	100	131	160	187	213	240	263	287	310	330	351	371	390	400
	~	140	137	134	131	128	125	122	120	117	115	113	110	108	106	104	100
6000	GLR	7.6	4.8	3.8	3.4	3.1	3.0	2.8	2.8	2.7	2.6	2.6	2.6	2.6	2.5	2.5	2.5
	MCF	1.9	2.4	2.8	3.4	3.8	4.4	4.9	5.5	6.0	6.5	7.1	7.6	8.2	8.7	9.3	9.9
	Pc	108	137	167	197	227	259	290	320	348	378	410	440	471	501	532	564
	BPD	29	57	84	110	135	159	182	204	227	247	266	288	305	322	341	360
	~	117	115	113	110	108	106	104	102	101	99	97	96	94	92	91	90
7000	GLR	8.8	5.4	4.6	4.0	3.7	3.5	3.4	3.2	3.2	3.1	3.1	3.0	3.0	3.0	3.0	3.0
	MCF	2.2	2.7	3.4	4.0	4.6	5.2	5.8	6.4	7.1	7.6	8.4	9.0	9.6	10.3	11.0	11.7
	Pc	110	139	171	201	231	264	295	326	355	386	418	448	480	511	542	575
	BPD	25	49	72	96	117	138	159	180	198	217	236	252	269	287	303	320
	~	101	99	97	96	94	92	91	90	88	87	86	84	83	82	81	80
8000	GLR	10.0	6.4	5.1	4.6	4.3	4.0	3.9	3.8	3.7	3.6	3.5	3.5	3.5	3.4	3.4	3.4
	MCF	2.5	3.2	3.8	4.6	5.3	6.0	6.7	7.5	8.2	8.9	9.6	10.5	11.1	11.8	12.6	13.5
	Pc	112	142	174	205	236	269	301	333	362	393	426	457	490	520	553	586
	BPD	22	43	64	84	103	123	141	160	175	192	209	225	240	255	270	284
	~	88	87	86	84	83	82	81	80	78	77	76	75	74	73	72	71
9000	GLR	11.2	7.2	5.9	5.2	4.8	4.6	4.4	4.3	4.2	4.1	4.0	4.0	3.9	3.9	3.9	3.9
	MCF	2.8	3.6	4.4	5.2	6.0	6.8	7.6	8.5	9.4	10.2	11.0	11.8	12.7	13.5	14.5	15.4
	Pc	114	144	178	208	240	275	306	339	369	401	435	466	499	530	563	597
	BPD	19	38	57	75	92	109	126	144	159	177	189	204	217	234	251	264
	~	78	77	76	75	74	73	72	72	71	71	69	68	67	67	67	66
10000	GLR	12.8	8.2	6.7	5.8	5.5	5.2	5.0	4.8	4.7	4.6	4.5	4.5	4.4	4.4	4.4	4.3
	MCF	3.2	4.1	5.0	5.8	6.8	7.7	8.6	9.6	10.5	11.4	12.4	13.3	14.3	15.2	16.2	17.2
	Pc	116	147	181	212	244	280	312	345	376	408	443	474	508	540	574	608
	BPD	17	35	51	68	83	100	117	132	144	160	173	189	198	213	225	240
	~	71	71	69	68	67	67	67	66	64	64	63	63	61	61	60	60
11000	GLR	14.4	90.2	7.4	6.6	6.1	5.8	5.5	5.4	5.2	5.1	5.1	5.0	5.0	4.9	4.9	4.8
	MCF	3.6	4.6	5.5	6.6	7.6	8.6	9.6	10.7	11.7	12.7	13.8	15.0	16.0	17.0	18.1	19.2
	Pc	118	149	184	216	249	285	317	352	383	415	451	483	517	550	584	619
	BPD	16	32	47	63	76	91	105	120	132	145	159	171	185	196	206	220
	~	64	64	63	63	61	61	60	60	59	58	58	57	57	56	55	55
12000	GLR	16.0	10.2	8.3	7.3	6.8	6.5	6.2	5.9	5.8	5.7	5.6	5.5	5.4	5.4	5.4	5.4
	MCF	4.0	5.1	6.2	7.3	8.4	9.6	10.7	11.8	13.0	14.2	15.4	16.5	17.6	18.8	20.1	21.4
	Pc	120	152	188	220	253	290	323	358	390	423	459	492	527	560	595	630
	BPD	14	29	43	57	71	84	96	110	121	135	145	159	169	182	191	204
	~	59	58	58	57	57	56	55	55	54	54	53	53	52	52	51	51

LEGEND: GLR: — Required Gas Liquid Ratio MCF/BBL.
   MCF — MCF Required Per Cycle.
   Pc — Required Surface Casing Pressure* (average)
BPD — Maximum Production BBLS./DAY
   ~ — Maximum Cycles Per Day.

*Can be correlated with static pressure of well to determine maximum drawdown.

TABLE 7A.10
SUB-SURFACE PLUNGER APPLICATION INFORMATION

2⅞ O.D. TUBING
P_{wh} = 100 PSI

DEPTH (FEET)		LIQUID SURFACE DELIVERY PER CYCLE—BARRELS															
		0.25	0.50	0.75	1.00	1.25	1.50	1.75	2.00	2.25	2.50	2.75	3.00	3.25	3.50	3.75	4.00
2000	GLR	2.4	1.6	1.4	1.1	1.1	1.0	1.0	.9	.9	.9	.8	.8	.8	.8	.8	.8
	MCF	.6	.8	1.0	1.1	1.3	1.4	1.6	1.7	1.9	2.1	2.2	2.3	2.5	2.7	2.9	3.1
	Pc	143	171	199	225	254	282	311	340	367	393	422	450	479	506	533	562
	BPD	84	160	227	288	342	393	437	480	515	552	585	618	643	672	697	720
	~	339	320	303	288	274	262	250	240	230	221	213	206	198	192	186	180
3000	GLR	4.8	2.8	2.2	1.8	1.7	1.6	1.5	1.4	1.4	1.4	1.3	1.3	1.3	1.3	1.3	1.3
	MCF	1.2	1.4	1.6	1.8	2.1	2.3	2.6	2.8	3.0	3.3	3.6	3.8	4.1	4.3	4.6	5.0
	Pc	146	174	203	230	259	288	318	347	374	401	431	459	489	517	544	574
	BPD	57	110	159	206	247	288	325	360	391	422	451	480	503	528	551	576
	~	230	221	213	206	198	192	186	180	174	169	164	160	155	151	147	144
4000	GLR	6.4	4.0	3.1	2.5	2.4	2.2	2.1	2.0	1.9	1.9	1.9	1.8	1.8	1.8	1.8	1.8
	MCF	1.6	2.0	2.3	2.5	2.9	3.2	3.6	3.9	4.2	4.6	5.0	5.3	5.6	6.1	6.4	6.9
	Pc	148	178	207	234	265	294	324	354	382	410	440	469	499	527	556	586
	BPD	43	84	123	160	193	226	251	288	315	342	368	393	416	437	457	480
	~	174	169	164	160	155	151	147	144	140	137	134	131	128	125	122	120
5000	GLR	8.4	5.0	3.9	3.3	3.0	2.8	2.7	2.5	2.4	2.4	2.4	2.4	2.3	2.2	2.2	2.2
	MCF	2.1	2.5	2.9	3.3	3.7	4.1	4.6	5.0	5.4	5.8	6.4	6.8	7.3	7.6	8.3	8.8
	Pc	151	182	211	239	270	300	331	361	390	418	449	478	509	538	567	598
	BPD	35	68	100	131	160	187	213	240	263	287	310	330	351	371	390	400
	~	140	137	134	131	128	125	122	120	117	115	113	110	108	106	104	100
6000	GLR	10.0	6.0	4.7	4.0	3.6	3.4	3.2	3.1	3.0	2.9	2.9	2.8	2.8	2.7	2.7	2.7
	MCF	2.5	3.0	3.5	4.0	4.5	5.1	5.6	6.2	6.7	7.2	7.8	8.3	8.9	9.4	10.0	10.6
	Pc	154	185	215	244	275	306	338	368	397	427	457	488	519	549	579	610
	BPD	29	57	84	110	135	159	182	204	227	247	266	288	305	322	341	360
	~	117	115	113	110	108	106	104	102	101	99	97	96	94	92	91	90
7000	GLR	12.0	7.2	5.6	4.7	4.4	4.0	3.8	3.7	3.6	3.5	3.4	3.3	3.3	3.3	3.2	3.2
	MCF	3.0	3.6	4.2	4.7	5.4	6.0	6.6	7.3	8.0	8.6	9.2	9.9	10.6	11.2	11.9	12.6
	Pc	157	189	219	248	281	312	344	375	405	435	466	497	529	559	590	622
	BPD	25	49	72	96	117	138	159	180	198	217	236	252	269	287	303	320
	~	101	99	97	96	94	92	91	90	88	87	86	84	83	82	81	80
8000	GLR	14.0	8.4	6.6	5.5	5.1	4.7	4.4	4.3	4.1	4.0	3.9	3.9	3.8	3.8	3.7	3.7
	MCF	3.5	4.2	4.9	5.5	6.3	7.0	7.7	8.5	9.2	9.9	10.7	11.5	12.2	13.0	13.7	14.6
	Pc	160	192	223	253	286	318	351	382	412	444	475	507	539	570	602	634
	BPD	22	43	64	84	103	123	141	160	175	192	209	225	240	255	270	284
	~	88	87	86	84	83	82	81	80	78	77	76	75	74	73	72	71
9000	GLR	16.0	9.6	7.5	6.3	5.8	5.4	5.1	4.9	4.7	4.6	4.5	4.4	4.3	4.3	4.2	4.2
	MCF	4.0	4.8	5.6	6.3	7.2	8.0	8.8	9.7	10.5	11.4	12.2	13.1	14.0	14.8	15.6	16.7
	Pc	162	196	228	258	292	324	358	390	420	452	484	516	549	581	613	646
	BPD	19	38	57	75	92	109	126	144	159	177	189	204	217	234	251	264
	~	78	77	76	75	74	73	72	72	71	71	69	68	67	67	67	66
10000	GLR	18.0	11.0	8.4	7.2	6.5	6.0	5.8	5.5	5.3	5.1	5.0	4.9	4.9	4.8	4.7	4.7
	MCF	4.5	5.5	6.3	7.2	8.1	9.0	10.0	10.9	11.8	12.7	13.7	14.7	15.7	16.6	17.6	18.8
	Pc	165	200	232	262	297	330	364	397	428	461	493	526	560	591	625	658
	BPD	17	35	51	68	83	100	117	132	144	160	173	189	198	213	225	240
	~	71	71	69	68	67	67	67	66	64	64	63	63	61	61	60	60
11000	GLR	20.0	12.2	9.5	8.0	7.3	6.8	6.4	6.2	6.0	5.8	5.6	5.5	5.4	5.4	5.3	5.3
	MCF	5.0	6.1	7.1	8.0	9.1	10.2	11.2	12.2	13.3	14.3	15.4	16.5	17.5	18.6	19.7	21.0
	Pc	168	203	236	267	302	336	371	404	435	469	502	535	570	602	636	670
	BPD	16	32	47	63	76	91	105	120	132	145	159	171	185	196	206	220
	~	64	64	63	63	61	61	60	60	59	58	58	57	57	56	55	55
12000	GLR	22.4	13.4	10.4	8.8	8.0	7.5	7.1	6.8	6.6	6.4	6.3	6.1	6.0	6.0	5.9	5.9
	MCF	5.6	6.7	7.8	8.8	10.0	11.2	12.4	13.5	14.7	15.8	17.1	18.3	19.5	20.7	22.0	23.3
	Pc	171	207	240	272	308	342	378	411	443	478	511	545	580	613	648	682
	BPD	14	29	43	57	71	84	96	110	121	135	145	159	169	182	191	204
	~	59	58	58	57	57	56	55	55	54	54	53	53	52	52	51	51

LEGEND: GLR: — Required Gas Liquid Ratio MCF/BBL.
  MCF — MCF Required Per Cycle.
  Pc — Required Surface Casing Pressure* (average)
  BPD — Maximum Production BBLS./DAY
  ~ — Maximum Cycles Per Day.

*Can be correlated with static pressure of well to determine maximum draw down.

TABLE 7A.11
SUB-SURFACE PLUNGER APPLICATION INFORMATION

2⅞ O.D. TUBING
P_{wh} = 150 PSI

DEPTH (FEET)		LIQUID SURFACE DELIVERY PER CYCLE—BARRELS														
		0.25	0.50	0.75	1.00	1.25	1.50	1.75	2.00	2.25	2.50	2.75	3.00	3.25	3.50	3.75	4.00
2000	GLR	4.0	2.6	2.0	1.6	1.4	1.2	1.2	1.1	1.1	1.0	1.0	.9	.9	.9	.9	.9
	MCF	1.2	1.3	1.5	1.6	1.7	1.8	2.0	2.1	2.3	2.4	2.6	2.7	2.9	3.1	3.3	3.5
	Pc	196	223	252	280	309	338	367	396	420	449	478	505	533	562	590	617
	BPD	84	160	227	288	342	393	437	480	515	552	585	618	643	672	697	720
	~	339	320	303	288	274	262	250	240	230	221	213	206	198	192	186	180
3000	GLR	6.8	3.8	3.0	2.4	2.1	1.9	1.8	1.7	1.6	1.6	1.5	1.5	1.5	1.5	1.5	1.4
	MCF	1.7	1.9	2.2	2.4	2.6	2.8	3.1	3.3	3.6	3.8	4.1	4.4	4.7	5.1	5.3	5.6
	PC	200	228	257	286	315	345	374	404	429	458	488	516	544	574	602	630
	BPD	57	110	159	206	247	288	325	360	391	422	451	480	503	528	551	576
	~	230	221	213	206	198	192	186	180	172	169	164	160	155	151	147	144
4000	GLR	9.2	5.2	3.9	3.2	2.9	2.6	2.5	2.3	2.3	2.3	2.1	2.1	2.0	2.0	2.0	1.9
	MCF	2.3	2.6	2.9	3.2	3.6	3.9	4.3	4.6	5.0	5.3	5.6	6.1	6.5	6.8	7.2	7.6
	Pc	204	232	262	292	322	352	382	412	438	468	498	526	556	586	615	643
	BPD	43	84	123	160	193	226	251	288	315	342	368	393	416	437	457	480
	~	174	169	164	160	155	151	147	144	140	137	134	131	128	125	122	120
5000	GLR	11.2	6.4	5.0	4.1	3.6	3.4	3.2	3.0	2.8	2.8	2.7	2.6	2.6	2.5	2.5	2.5
	MCF	2.8	3.2	3.7	4.1	4.5	5.0	5.5	5.9	6.3	6.8	7.2	7.7	8.3	8.7	9.2	9.7
	Pc	209	237	268	298	328	359	390	421	447	477	508	537	567	598	628	657
	BPD	35	68	100	131	160	187	213	240	263	287	310	330	351	371	390	400
	~	140	137	134	131	128	125	122	120	117	115	113	110	108	106	104	100
6000	GLR	14.0	8.0	5.6	5.0	4.4	4.0	3.8	3.6	3.5	3.4	3.2	3.2	3.1	3.1	3.0	3.0
	MCF	3.5	4.0	4.5	5.0	5.5	6.0	6.6	7.2	7.7	8.3	8.8	9.5	10.1	10.6	11.2	11.8
	Pc	213	242	273	304	335	367	397	429	456	486	528	548	579	610	640	670
	BPD	29	57	84	110	135	159	182	204	227	247	266	288	305	322	341	360
	~	117	115	113	110	108	106	104	102	101	99	97	96	94	92	91	90
7000	GLR	16.4	9.4	7.1	6.0	5.3	4.8	4.6	4.3	4.1	4.0	3.9	3.8	3.7	3.7	3.6	3.5
	MCF	4.1	4.7	5.3	6.0	6.6	7.2	8.0	8.6	9.2	9.8	10.5	11.2	12.0	12.6	13.2	13.9
	Pc	217	247	278	310	341	374	405	437	465	496	528	559	590	622	653	683
	BPD	25	49	72	96	117	138	159	180	198	217	236	252	269	287	303	320
	~	101	99	97	96	94	92	91	90	88	87	86	84	83	82	81	80
8000	GLR	18.8	11.0	8.3	7.0	6.1	5.6	5.3	5.0	4.8	4.6	4.4	4.4	4.3	4.2	4.1	4.1
	MCF	4.7	5.5	6.2	7.0	7.6	8.4	9.2	10.0	10.6	11.5	12.1	13.0	13.8	14.6	15.4	16.2
	Pc	222	252	284	316	348	381	412	446	474	505	538	570	602	634	665	697
	BPD	22	43	64	84	103	123	141	160	175	192	209	225	240	255	270	284
	~	88	87	86	84	83	82	81	80	78	77	76	75	74	73	72	71
9000	GLR	22.0	12.4	9.5	8.0	7.0	6.4	6.0	5.7	5.5	5.2	5.1	5.0	4.8	4.8	4.7	4.7
	MCF	5.5	6.2	7.1	8.0	8.7	9.6	10.5	11.4	12.2	13.0	13.9	14.8	15.6	16.6	17.5	18.5
	Pc	226	256	289	322	354	388	420	454	484	515	549	580	613	646	678	710
	BPD	19	38	57	75	92	109	126	144	159	177	189	204	217	234	251	264
	~	78	77	76	75	74	73	72	72	71	71	69	68	67	67	67	66
10000	GLR	24.8	14.2	10.7	9.0	8.0	7.2	6.8	6.4	6.1	5.9	5.7	5.6	5.5	5.4	5.3	5.2
	MCF	6.2	7.1	8.0	9.0	9.9	10.8	11.8	12.8	13.7	14.7	15.6	16.6	17.6	18.6	19.6	20.6
	Pc	230	261	294	328	361	395	428	462	493	524	559	591	625	658	691	723
	BPD	17	35	51	68	83	100	117	132	144	160	173	189	198	213	225	240
	~	71	71	69	68	67	67	67	66	64	64	63	63	61	61	60	60
11000	GLR	28.0	16.0	12.0	10.0	8.9	8.1	7.6	7.2	6.9	6.6	6.4	6.2	6.1	6.0	5.9	5.8
	MCF	7.0	8.0	9.0	10.0	11.1	12.1	13.2	14.3	15.4	16.5	17.5	18.6	19.7	20.8	22.0	23.1
	Pc	235	266	300	334	367	403	435	471	502	533	569	602	636	670	703	737
	BPD	16	32	47	63	76	91	105	120	132	145	159	171	185	196	206	220
	~	64	64	63	63	61	61	60	60	59	58	58	57	57	56	55	55
12000	GLR	30.8	17.6	13.4	11.2	10.0	9.0	8.4	7.9	7.6	7.3	7.1	6.9	6.7	6.6	6.5	6.4
	MCF	7.7	8.8	10.0	11.2	12.4	13.5	14.7	15.8	17.0	18.2	19.4	20.6	21.8	23.1	24.3	25.6
	Pc	239	271	305	340	374	410	443	479	511	543	579	613	648	682	716	750
	BPD	14	29	43	57	71	84	96	110	121	135	145	159	169	182	191	204
	~	59	58	58	57	57	56	55	55	54	54	53	53	52	52	51	51

LEGEND: GLR: — Required Gas Liquid Ratio MCF/BBL.
    MCF — MCF Required Per Cycle.
    Pc — Required Surface Casing Pressure* (average)
  BPD — Maximum Production BBLS./DAY
    ~ — Maximum Cycles Per Day.

*Can be correlated with static pressure of well to determine maximum draw down.

TABLE 7A.12

SUB-SURFACE PLUNGER APPLICATION INFORMATION

2⅞ O.D. TUBING
P_{wh} = 200 PSI

DEPTH (FEET)		0.25	0.50	0.75	1.00	1.25	1.50	1.75	2.00	2.25	2.50	2.75	3.00	3.25	3.50	3.75	4.00
		LIQUID SURFACE DELIVERY PER CYCLE—BARRELS															
2000	GLR	6.0	3.2	2.4	1.9	1.7	1.5	1.4	1.3	1.2	1.2	1.1	1.1	1.1	1.0	1.0	1.0
	MCF	1.5	1.6	1.8	1.9	2.1	2.2	2.3	2.5	2.7	2.8	3.0	3.2	3.4	3.5	3.7	3.8
	Pc	251	279	307	333	362	391	419	448	475	502	532	560	588	615	643	673
	BPD	84	160	227	288	342	393	437	480	515	552	585	618	643	672	697	720
	~	339	320	303	288	274	262	250	240	230	221	213	206	198	192	186	180
3000	GLR	8.8	5.0	3.6	2.9	2.6	2.3	2.1	2.0	1.9	1.8	1.7	1.7	1.6	1.6	1.6	1.6
	MCF	2.2	2.5	2.7	2.9	3.2	3.4	3.6	4.0	4.2	4.4	4.7	5.0	5.2	5.5	5.8	6.2
	Pc	256	285	313	340	370	399	428	457	485	513	543	572	600	628	657	687
	BPD	57	110	159	206	247	288	325	360	391	422	451	480	503	528	551	576
	~	230	221	213	206	198	192	186	180	174	169	164	160	155	151	147	144
4000	GLR	12.0	6.6	4.8	4.0	3.5	3.1	2.9	2.7	2.6	2.5	2.4	2.3	2.3	2.2	2.2	2.2
	MCF	3.0	3.3	3.6	4.0	4.3	4.6	5.0	5.4	5.7	6.1	6.4	6.8	7.3	7.6	8.0	8.5
	Pc	262	291	320	348	378	408	437	467	495	524	555	584	613	642	671	702
	BPD	43	84	123	160	193	226	251	288	315	342	368	393	416	437	457	480
	~	174	169	164	160	155	151	147	144	140	137	134	131	128	125	122	120
5000	GLR	14.8	8.4	6.2	5.0	4.4	4.0	3.7	3.5	3.3	3.2	3.1	2.9	2.9	2.8	2.8	2.7
	MCF	3.7	4.2	4.6	5.0	5.5	5.9	6.4	6.4	7.4	7.8	8.3	8.7	9.2	9.6	10.2	10.8
	Pc	267	297	326	355	386	416	446	476	505	535	566	596	626	655	685	716
	BPD	35	68	100	131	160	187	213	240	263	287	310	330	351	371	390	400
	~	140	137	134	131	128	125	122	120	117	115	113	110	108	106	104	100
6000	GLR	18.4	10.2	7.5	6.2	5.4	4.8	4.5	4.2	4.0	3.8	3.7	3.6	3.5	3.4	3.3	3.3
	MCF	4.6	5.1	5.6	6.2	6.7	7.2	7.8	8.4	9.0	9.5	10.0	10.6	11.2	11.7	12.4	13.2
	Pc	272	302	333	362	394	424	455	486	516	545	577	608	638	668	699	730
	BPD	29	57	84	110	135	159	182	204	227	247	266	288	305	322	341	360
	~	117	115	113	110	108	106	104	102	101	99	97	96	94	92	91	90
7000	GLR	21.6	12.0	8.8	7.3	6.4	5.8	5.4	5.0	4.8	4.5	4.4	4.2	4.1	4.0	3.9	3.9
	MCF	5.4	6.0	6.6	7.3	8.0	8.6	9.3	9.9	10.6	11.2	12.0	12.6	13.3	14.0	14.6	15.5
	Pc	278	308	339	370	402	433	464	495	526	556	589	620	657	682	713	745
	BPD	25	49	72	96	117	138	159	180	198	217	236	252	269	287	303	320
	~	101	99	97	96	94	92	91	90	88	87	86	84	83	82	81	80
8000	GLR	24.8	14.0	10.2	8.5	7.4	6.6	6.2	5.7	5.5	5.2	5.0	4.9	4.8	4.7	4.6	4.5
	MCF	6.2	7.0	7.6	8.5	9.2	9.9	10.7	11.4	12.3	13.0	13.7	14.5	15.4	16.1	16.9	18.0
	Pc	283	314	346	377	410	441	473	505	536	567	600	632	663	695	727	759
	BPD	22	43	64	84	103	123	141	160	175	192	209	225	240	255	270	284
	~	88	87	86	84	83	82	81	80	78	77	76	75	74	73	72	71
9000	GLR	28.4	16.0	11.6	9.6	8.4	7.6	7.0	6.5	6.3	6.0	5.7	5.6	5.4	5.3	5.2	5.2
	MCF	7.1	8.0	8.7	9.6	10.5	11.3	12.2	13.0	14.0	14.9	15.7	16.6	17.5	18.4	19.3	20.5
	Pc	289	320	352	385	418	450	483	514	546	578	612	644	676	709	741	774
	BPD	19	38	57	75	92	109	126	144	159	177	189	204	217	234	251	264
	~	78	77	76	75	74	73	72	72	71	71	69	68	67	67	67	66
10000	GLR	32.0	18.0	13.2	10.8	9.5	8.5	7.9	7.3	7.0	6.7	6.5	6.2	6.1	5.9	5.8	5.8
	MCF	8.0	9.0	9.9	10.8	11.8	12.7	13.7	14.6	15.7	16.7	17.7	18.6	19.6	20.6	21.6	23.0
	Pc	294	326	359	392	426	458	492	524	556	589	623	656	689	722	755	788
	BPD	17	35	51	68	83	100	117	132	144	160	173	189	198	213	225	240
	~	71	71	69	68	67	67	67	66	64	64	63	63	61	61	60	60
11000	GLR	36.0	20.0	14.7	12.1	10.6	9.6	8.8	8.2	7.8	7.5	7.2	7.0	6.8	6.6	6.5	6.4
	MCF	9.0	10.0	11.0	12.1	13.2	14.3	15.4	16.4	17.5	18.6	19.7	20.8	21.9	23.0	24.1	25.5
	Pc	299	332	365	399	434	466	501	533	567	600	634	668	701	735	769	802
	BPD	16	32	47	63	76	91	105	120	132	145	159	171	185	196	206	220
	~	64	64	63	63	61	61	60	60	59	58	58	57	57	56	55	55
12000	GLR	40.0	22.4	16.4	13.5	11.7	10.6	9.8	9.1	8.7	8.3	8.0	7.7	7.5	7.3	7.1	7.1
	MCF	10.0	11.2	12.3	13.5	14.6	15.8	17.0	18.2	19.5	20.6	21.8	23.0	24.2	25.5	26.7	28.4
	Pc	305	338	372	407	442	475	510	543	577	611	646	680	714	749	783	817
	BPD	14	29	43	57	71	84	96	110	121	135	145	159	169	182	191	204
	~	59	58	58	57	57	56	55	55	54	54	53	53	52	52	51	51

LEGEND: GLR: — Required Gas Liquid Ratio MCF/BBL.
MCF — MCF Required Per Cycle.
Pc — Required Surface Casing Pressure* (average)
BPD — Maximum Production BBLS./DAY
~ — Maximum Cycles Per Day.

*Can be correlated with static pressure of well to determine maximum draw down.

Chapter 8

Other methods of artificial lift

8.1 INTRODUCTION

All would agree that the ultimate type of artificial lift has not yet been devised. This chapter presents some of those methods that have been used, or perhaps are still in the experimental stage, and a few that have been proven but are rather limited in application. However, space and lack of knowledge of the existence of all lifts prohibits that discussion in this text.

Some of these methods, such as the Alpha I unit of Bethlehem, should perhaps be under the sucker rod pumping chapter. Due to its uniqueness, though, it is in this chapter.

Some of the systems discussed in this chapter are the ball pump, the gas pump, the Alpha I unit, T and C unit, hydro-gas lift system, Elfarr pump, free running plunger, Chancellor rotating horsehead, pneumatic unit, and pumpdown techniques.

In reality the pumpdown system is not a method of artificial lift but represents a relatively new system of pumping in and pumping out of artificial lift equipment and well servicing equipment. It is presently limited to gas-lift and jet pumping equipment. Due to its uniqueness of operation, it is included in this Chapter.

The persons writing each section are listed as the author of that section.

8.2 THE BALL PUMP, BY PHIL PATTILLO

The ball pump gas lift system was originally developed during the early 1950's by Stanolind Oil and Gas Company*, the first published work on the ball pump appearing in 1952. Because the ball pump did not gain commercial acceptance, a minimum of information on this gas lift system has appeared in the literature. For this reason, most of this discussion has been compiled from research reports written by Stanolind engineers during the early years of the studies of the ball pump.

The ball pump is a system whereby synthetic rubber balls are injected into the lower end of the production tubing string, thus forming a series of solid interfaces between liquid and gaseous phases in an effort to minimize the holdup effect. In 1953, a package designed to perform the functions of injecting balls into the production tubing and recovering the balls at the wellhead was designed and licensed to Pacific Pumps Inc. and the system was placed in operation in several wells in west Texas and Oklahoma. However, the

*Presently Amoco Production Company.

low cost of the lift gas during that period in history reduced the attractiveness of the ball pump, and the system failed to gain widespread industry acceptance. In the same sense, the current economic value of gas for lifting operations has renewed interest in the ball pump as an economic alternative to conventional gas lift.

The following sections introduce the reader to the basic concepts surrounding ball pump gas lift and describe the basic characteristics of the system operating in the intermittent and continuous modes. Following the descriptions of the ball pump in its various operating modes, an experimental evaluation of the ball pump is presented. The final paragraphs contain a summary of the ball pump gas lift system.

8.21 Basic concepts

Given the characteristics of the fluid phases being produced in a particular wellbore, the engineer is confronted with two phenomena that contribute to the loss of energy as the multiphase mixture traverses the vertical production tubing: friction loss and liquid holdup (since the ball pump is basically intended as an alternative to conventional gas lift, all fluid ratios and changes in potential energy are assumed constant in this discussion). In multiphase flow, both friction loss and liquid holdup are intimately related to the flow pattern developed in the tubing during production (refer to Volume I).

Fig. 8.21 illustrates schematically the steady-state flow pattern developed when solid balls are introduced into a multiphase system. In this figure the gas phase will probably be accompanied by a liquid boundary layer wetting the internal surface of the tubing. However, the gas phase should consist of gas only, rather than the less efficient condition associated with the mist flow regime of conventional vertical multiphase flow.

The liquid phase will, in general, contain dissolved gas, but it will not be described by the theory of bubble flow. It follows that the pressure drop calculations developed for conventional multiphase flow phenomena are not directly applicable to ball pump production. This unfortunate circumstance complicates a quantitative evaluation of the ball pump. In particular, the effect of the injected balls on friction loss, if considered rigorously, is quite complex. Nevertheless, one can conclude that unless this friction loss is extremely large, the multiphase energy savings associated with

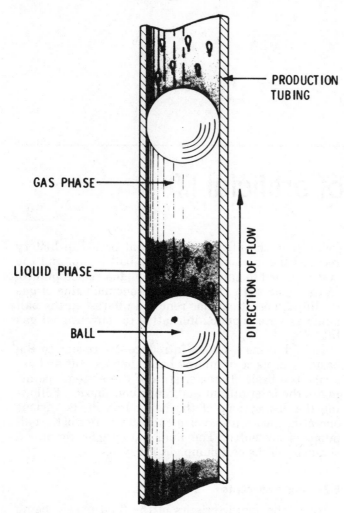

GAS PHASE

LIQUID PHASE

BALL

PRODUCTION TUBING

DIRECTION OF FLOW

Fig. 8.21   Schematic of hypothesized ball pump phenomenon

the elimination of adverse holdup effects render the ball pump a gas lift candidate worthy of further consideration.

8.22   Intermittent lift

Originally, the ball pump was designed to be operated in the intermittent lift mode. In intermittent lift, several balls are circulated in a closed system so one ball is always at the lower end of the production tubing in a position to lift fluids. The waiting time associated with a system such as the plunger lift is eliminated. This cyclic operation also provides an excellent means of circulating various treatment chemicals through the system.

8.221   Mechanical equipment

Fig. 8.22 illustrates schematically the equipment necessary for a typical ball pump installation operating in an intermittent lift mode. The two tubing strings required have equal ID. One tubing string, termed the power tubing or injection string, provides a conductor for the balls and injection gas to the bottom of the well. The other tubing string is the production tubing.

A rubber ball sits on a constriction at the bottom of the injection string. It passes through the constriction when the gas pressure increases sufficiently. This constriction is removable by pumping a ball through the system in reverse, thus carrying the constriction to the surface.

A return bend at the lower end of the assembly serves as a conduit through which the balls and lift gas enter the production tubing from the injection string. The return bend also serves as part of the accumulation chamber for produced liquids.

A wireline removable standing valve is located in the production tubing below the return bend. The standing valve protects the formation from backpressure and loss of lift gas to the casing-tubing annulus during the lifting phase of each cycle.

A ball hopper collects and stores balls leaving the production string prior to their reinjection into the injection string.

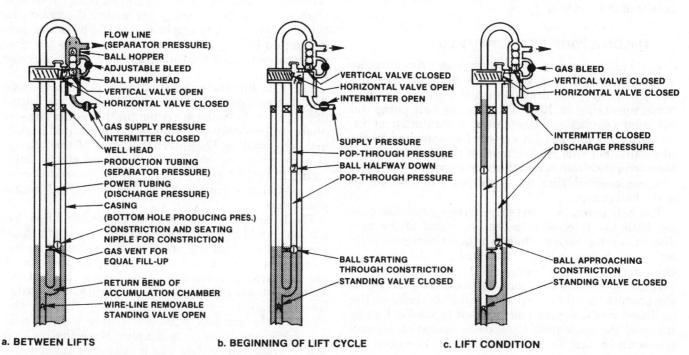

FLOW LINE (SEPARATOR PRESSURE)
BALL HOPPER
ADJUSTABLE BLEED
BALL PUMP HEAD
VERTICAL VALVE OPEN
HORIZONTAL VALVE CLOSED

GAS SUPPLY PRESSURE
INTERMITTER CLOSED
WELL HEAD
PRODUCTION TUBING (SEPARATOR PRESSURE)
POWER TUBING (DISCHARGE PRESSURE)
CASING (BOTTOM HOLE PRODUCING PRES.)
CONSTRICTION AND SEATING NIPPLE FOR CONSTRICTION
GAS VENT FOR EQUAL FILL-UP

RETURN BEND OF ACCUMULATION CHAMBER
WIRE-LINE REMOVABLE STANDING VALVE OPEN

a. BETWEEN LIFTS

VERTICAL VALVE CLOSED
HORIZONTAL VALVE OPEN
INTERMITTER OPEN

SUPPLY PRESSURE
POP-THROUGH PRESSURE
BALL HALFWAY DOWN
POP-THROUGH PRESSURE

BALL STARTING THROUGH CONSTRICTION
STANDING VALVE CLOSED

b. BEGINNING OF LIFT CYCLE

GAS BLEED
VERTICAL VALVE CLOSED
HORIZONTAL VALVE CLOSED

INTERMITTER CLOSED
DISCHARGE PRESSURE

BALL APPROACHING CONSTRICTION
STANDING VALVE CLOSED

c. LIFT CONDITION

Fig. 8.22   Operation of the intermittent system

A pressure-operated ball pump head injects one ball into the injection string ahead of each charge of injected gas. The ball pump head illustrated in Fig. 8.22 is representative of the injection heads used in the original field tests of the ball pump. Since then, simpler heads, such as Fig. 8.23, have been designed. However, the functions performed by the various heads are the same.

Also included are pressure regulators, timing devices, other surface equipment common to intermittent gas lift operations, and a source of lift gas.

8.222 Operation of the intermittent system

Fig. 8.22 illustrates the phases of a typical lift cycle to be described in the paragraphs below.

Between lifts [Fig. 8.22(a)]. Ball 6 has just traversed the production string and has unloaded a column of fluid into the flow line. Gas in the production tubing is essentially at separator pressure. The standing valve is open, allowing fluids to enter the accumulation chamber. Ball 1 is resting on the constriction. The horizontal valve in the injection head is closed, the power tubing being subjected to a pressure sufficient to discharge the next load of fluid but insufficient to force ball 1 through the constriction. Pressure on opposing faces of the vertical valve has been equalized through the adjustable bleed, and gravity has forced the valve to drop open.

Beginning of lift cycle [Fig. 8.22(b)]. At the assigned time, the intermitter opens, forcing the vertical valve upward into its seat. Concurrent with this action, ball 2 is forced into the power tubing along with a slug of lift gas. This motion compresses the gas between ball 2 and ball 1. At some point, the pressure on ball 1 becomes sufficient to force this ball through the constriction.

Lift condition (Fig. 8.22). The intermitter closes, allowing the spring-loaded horizontal valve to seat. Pressure loss through the bleed to the separator will open the vertical valve. The amount of gas in the power tubing will be sufficient to unload fluid produced in the next cycle. At the same time, expansion of the lift gas from pop-through pressure to discharge pressure conveys a load of produced fluids to the surface.

8.23 Continuous lift

Applying the ball pump to continuous gas lift or in a flowing well uses this artificial lift concept in its most basic form. In continuous lift, one has a single objective—to introduce balls into the production tubing to eliminate the effects of holdup. The amount of lift gas and/or the temporal spacing of the balls is no longer of any great consequence. In this regard, the intermitter equipment and the constriction in the injection string can be removed from the system. In fact, one may choose to remove the injection tubing string altogether and simply drop the balls down the tubing-casing annulus to a suitable receptacle at the lower end of the production tubing.

Such a single string installation is illustrated in Fig. 8.24 for the case of a flowing well (the gas lift valve indicated in the figure could be used to unload the well initially). Fig. 8.25 is a schematic of a collector

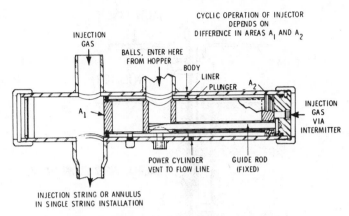

Fig. 8.23   Alternate design of ball injector

tube designed for a single-string installation. Notice the perforated section at the lower end of the collector tube. These perforations provide an avenue of release of formation fluids and prevent the balls from bridging at the mouth of the collector tube.

In a single-string installation, the production tubing lies along the inside wall of the casing, reducing the possibility of the balls bridging in the annulus. This condition can be realized by either decentralizing the tubing mechanically or landing the tubing in compression and allowing helical buckling to result in decentralization.

Operating in the continuous lift mode, the ball pump is indeed a simple and inexpensive installation. However, no real evidence has been presented to justify the use of the ball pump as a method of artificial lift. The paragraphs to follow present experimental evidence of the effectiveness of the system, most notably in comparison to conventional gas lift.

8.24 Evaluating the ball pump

Comparing the ball pump with other artificial lift methods has largely been restricted to a comparison of the ball pump with conventional gas lift, the reasons for which are primarily historical. Inasmuch as the original ball pump did not gain widespread commercial use upon being introduced to the industry, long-term production and economic data are not available. The absence of such information complicates any attempt to compare the ball pump with a dissimilar lift system, such as the rod pump.

Nevertheless, a comparison of the ball pump with conventional gas lift is straightforward—one simply analyzes a production stream both with and without the presence of balls. The introduction of balls into the production string reduces the effects of holdup and results in improved lift efficiency of the supplied gas. The engineer is therefore confronted with a familiar economic decision: will the increased production realized through application of the ball pump offset the cost of installing and maintaining the system?

8.241 Experimental evidence

During the formative years of the ball pump, little analytical work on multiphase flow in vertical conduits had been done. Therefore, the natural avenue for comparing the operating characteristics of the ball pump

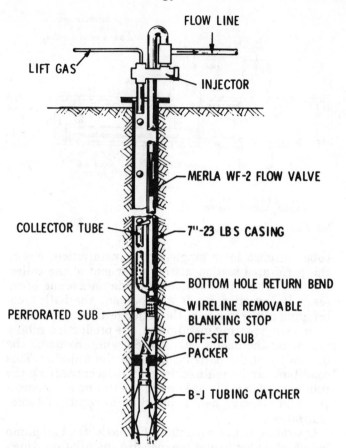

Fig. 8.24   *Schematic diagram of installation of single string pump in a flowing well*

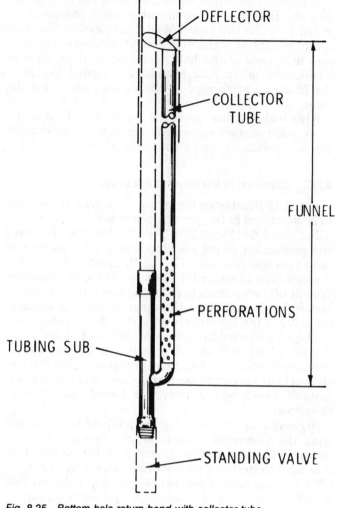

Fig. 8.25   *Bottom hole return bend with collector tube*

with conventional gas lift was through experimentation. To this end, a model of the ball pump was constructed and continuous flow tests of a gas-water system were conducted to relate ball pump performance to conventional gas lift in continuous lift.

8.2411   Experimental apparatus

Fig. 8.26 shows the installation used to test the ball pump in the continuous lift mode. For the test a string of 7 in. oil well casing was hung to a depth of 740 ft inside a string of 10¾ in. casing that had been previously installed in the test well. A single string of 2⅜ in. tubing was installed as the production string. As the test was to be run in the 7 in. casing, a packer was installed at the lower end of the tubing below the bottom-hole return bend to isolate the annulus between the 10¾ in. and 7 in. casing. The water to be produced was maintained in the annulus between the two casing strings. Entrance into the production tubing occurred via a standing valve located below the return bend.

With this arrangement, variations in the depth of the lift and the bottom-hole pressure could be realized by simply varying the height of water in the annulus between the two casing strings. The gas lift valve shown in the figure was installed as a means of initially unloading the well. The fluid system was a closed system; water produced at the surface was immediately reinjected into the 7-10¾ in. casing annulus, thereby maintaining a constant hydrostatic pressure at the return bend. A regulated gas supply pressure of 165 psig was maintained, and the gas injection rates were varied

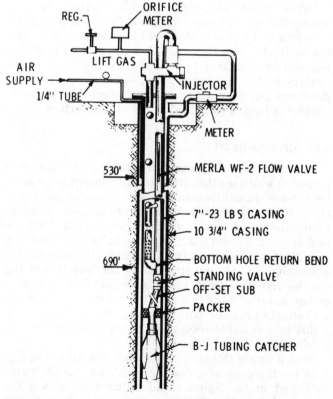

Fig. 8.26   *Schematic diagram of ball pump test installation*

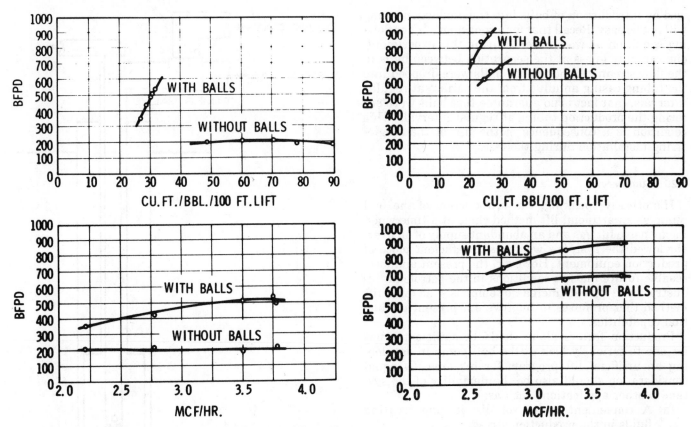

Fig. 8.27   Test results of ball pump, BHP = 70 PSI, depth of lift = 528 ft.

Fig. 8.29   Test results of ball pump, BHP = 110 PSI, depth of lift = 436 ft.

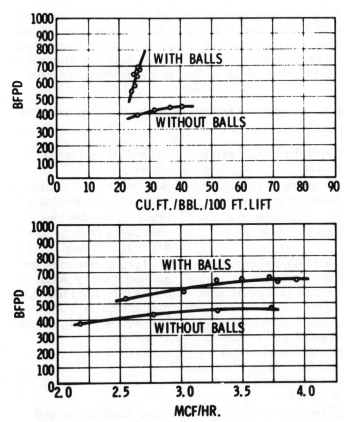

Fig. 8.28   Test results of ball pump, BHP = 90 PSI, depth of lift = 482 ft.

with the use of an adjustable choke valve on the gas input line.

Liquid volumes produced from the well were measured by a positive displacement meter prior to reinjection into the annulus between the casing strings. The volume of gas injected was measured using a standard orifice meter. The bottom-hole pressure, defined for this experiment as the pressure at the bottom-hole return bend, was measured indirectly by measuring the liquid head in the 7-10¾ in. casing annulus. The depth of lift was the distance from the top of this annular liquid to the surface.

8.2412 Testing procedure and results

A series of tests were conducted using differing bottom-hole pressures and, as a consequence of the test apparatus as described above, differing depths of lift. All pertinent variables were measured once the system had reached stabilized conditions. Each set of input conditions was run both with and without balls being injected into the production tubing. Typical results of these tests are presented in Fig. 8.27, 8.28 and 8.29. Notice that in each instance the following two equivalent qualitative observations can be made:

(a) For a given amount of injected gas, the introduction of balls into the flow stream results in increased liquid production.

(b) For a given amount of liquids produced, the introduction of balls into the flow stream results in a decrease in the injected gas requirement.

One additional implication of these test results is that any attempt to predict the degree of improvement realized by introduction of balls into the production string will not be easy. Recall that for a continuous lift or flowing condition as was present during this experiment, one can only guess at the separation between balls in the flow stream. Although the balls were dropped into the tubing-casing annulus from the surface at regular intervals, that fact is no guarantee that balls were entering the production tubing at regular intervals. This condition is a consequence of the use of the single-string installation during the test.

8.25   Summary

Historically speaking, the introduction of the ball pump as an artificial lift method came at a time when inexpensive lift gas and an abundant supply of oil overshadowed the system's worth as an efficient piece of production equipment. However, the present and foreseeable future market conditions in the petroleum industry have placed increased emphasis on any lift method that promises to improve our present production capabilities.

Production factors alone would be sufficient cause for one to seriously look at the ball pump in developing an artificial lift program for his area. Of equal consideration should also be the flexibility of the system in other applications such as:

(a) A convenient means of distributing treating fluids in the production string.

(b) A method of paraffin control in flowing oil wells.

(c) An aid to "heading" wells where some means of artificial lift is occasionally required to reestablish and maintain flow.

(d) A means of utilizing the formation gas more efficiently in lifting wells which are generally referred to as "stop cock" wells.

REFERENCE

1. Vincent, R. P., and L. B. Wilder, "New Gas Lift System," *The Petroleum Engineer* (November 1952), p. B-84.

8.3   THE GAS PUMP, BY C.R. CANALIZO

8.31   Introduction to the gas pump

The gas pump system is a new concept in artificial lift methods. Basically, the pump displaces a barrel of oil with an equal volume of gas at lift depth pressure. This is accomplished in a unique and efficient manner by a simple positive displacement pump that is wireline or tubing retrievable. It is presently in the experimental development stage, but several field installations have been made. Recent modifications show improvements, and commercial use should be forthcoming.

Where gas pressures are too low to lift wells by positive displacement, the gas pump can be allied with gas lift to lift deeper with lower pressures. The combination of the two methods permits affecting a maximum drawdown, as in rod pumping, while simultaneously creating an artificial submergence for gas lift to the extent of the supply gas pressure.

A few gas pump methods have been used in the past. In general, these early methods consisted of alternately injecting gas into a downhole accumulation chamber and bleeding it off to allow it to refill (Fig. 8.30).

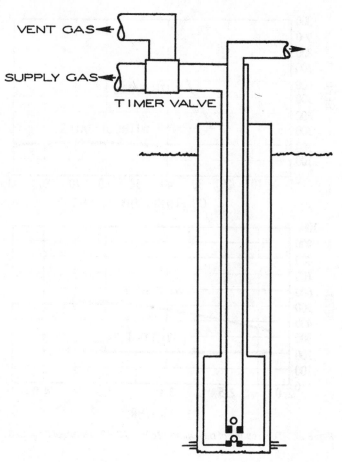

Fig. 8.30   Early gas pump system

Most of these were controlled by a surface intermitting device. Normally, oil field gas pressures are not high enough to permit anything but shallow lifting by this method, and the time lag between filling and venting limited the method to low productivity wells.

8.32   Comparative features of the gas pump vs. rod pump

Gas pump	Rod pump
Centrally located prime mover.	Individual well prime mover.
No tubing or casing wear with pump operation.	Rod, tubing, and casing wear with rod breakage.
Full pump stroke at any depth or cycle rate.	Pump stroke shortens with depth or cycle rate due to rod stretch.
Well inflow rate controls maximum production rate with surface gas control for reducing cycle rate, for complete automatic operation. If well pumps off, pump stops, but will resume operation when formation fluids enter well bore.	Well production must be manually set by pump stroke rate. If set too high, well pumps off. If set too low, well will not produce at desired rate.
Pump may be wirelined retrieved or run on Macaroni tubing or hollow rods.	Pump must be run and pulled with rod unit.

Pump may be combined with gas lift to get maximum formation drawdown at maximum production rate.

Cannot be combined with gas lift.

Minimum (50-100 psi) load differential on plunger, minimizes cup wear.

Full load differential on plunger at all times causes high friction and wear.

Deviated wells may be pumped without any problems in running or pulling the pump with wireline.

Rod, tubing, and casing wear result from beam type pumps.

Offshore platform wells can be easily lifted with the gas pump, as wireline is relatively inexpensive when compared to rig cost to pull rods and tubing.

The large prime movers and expense of rig costs make rod pumping prohibitive for platform operations.

At a later date, it will be possible to set and retrieve a gas pump with pumpdown techniques.

Not possible

8.33 Gas pump assembly

Figs. 8.31(A) and 8.31(B) are schematics of the component parts that make up the pump installation. These are described on Figure 8.31:

(1) Gas injection line
(2) Exhaust line
(3) Unloading valve
(4) Valve assembly
(5) Sleeve valve
(6) Switching mechanism
(7) Plunger rod
(8) Plunger
(9) Working barrel
(10) Standing valve assembly
(11) Wireline lock assembly
(12) Receiver assembly consists of the upper (a) and lower (b) locating nipples connected together by the slotted nipple (c).

Fig. 8.31 (C) shows a more basic schematic illustrating the operation of the gas pump.

8.34 Method of operation

In Fig. 8.31 (A) well fluids enter the pump barrel through the lower check valve, forcing the plunger downward. The sleeve valve is in the exhaust position, permitting the gas below the plunger to be vented out the exhaust line to the casing. The high pressure gas is shut off by the sleeve valve. The upper check valve is holding the fluid in the tubing.

In Fig. 8.31 (B) the sleeve valve has been shifted by the switching mechanism on contact by the plunger. The sleeve valve is now open to the supply gas and has blocked off the exhaust. The high pressure gas is entering the working barrel below the plunger, forcing it upwards. The plunger, in turn, is forcing the well fluids upward, causing the lower check valve to close and the upper check valve to open. The fluids displaced from the working barrel are thus pumped into the

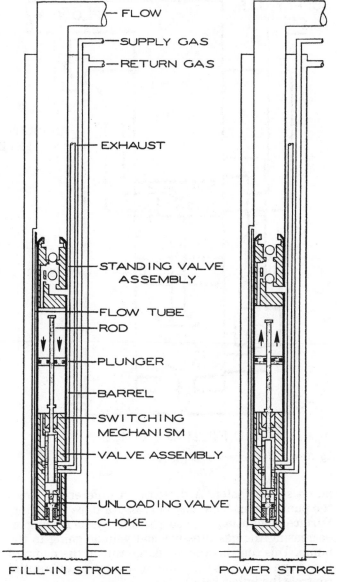

Fig. 8.31A   Fill-in stroke, insert gas pump

Fig. 8.31B   Power stroke, insert gas pump

tubing, which in turn displaces the same amount of fluid from the well to the storage tank.

The extended exhaust line creates a differential across the plunger by the difference of the hydrostatic fluid column in the annulus and the vent gas in the exhaust line.

For example, assume the working fluid level in the annulus is exerting a 50 psi hydrostatic pressure at the pump with the fluid column 150 ft above the pump. The exhaust line is 200 ft above the pump and is venting into the annulus which, in turn, is venting to atmosphere. On the exhaust cycle or fluid fill-in, the lower end of the plunger is exposed to atmospheric pressure through the exhaust line, while the fluid end of the plunger is sensing the 50 psi hydrostatic pressure. This 50 psi pressure differential forces the plunger down while filling the working barrel until the plunger reaches the limit stop, which trips the sleeve valve to the supply gas starting another cycle.

The aforementioned applied forces tend to make the pump self-regulating. A greater differential across the plunger increases the rate of fill in which makes the

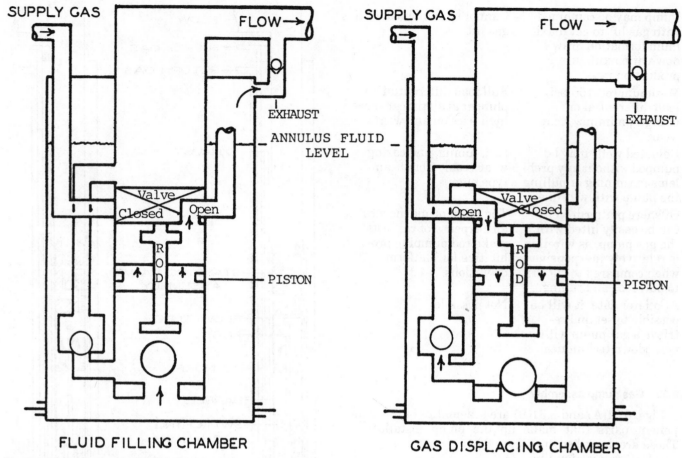

Fig. 8.31C Basic operation of gas pump

pump stroke faster. A decrease in differential slows the pump down. A balance of pressure stops the pump. Surface regulation is also possible by increasing or decreasing the gas pressure and volume going to the pump. This controls the discharge rate of the pump in the same manner that the hydrostatic differential controls the inflow rate.

The unloading valve (Port #3 of Fig. 8.31) serves one specific function: it blows out the fluid that has entered the exhaust line due to having the pump out of the receiver, either for repairs or prior to the initial installation. With the same well fluids in the exhaust line as in the annulus, there cannot be a differential across the plunger—in this condition the pump will not operate. An increase in gas pressure that is greater than the combined forces of the spring and hydrostatic tubing pressure is necessary to open the unloading valve.

With the unloading valve open, gas enters the lower end of the working barrel forcing the plunger up until it reaches the upper rod stop. This actuates the switching mechanism to trip the sleeve valve to the exhaust position. The gas entering through the unloading valve is then directed through the exhaust line, purging it of well fluids. Closing off the surface gas supply permits the unloading valve to close. With the exhaust line clear and the pressure vented out of the lower end of the pump, a working differential again exists across the plunger allowing it to fill the working barrel with well fluids and finally causing the sleeve valve to switch over to the high pressure inlet.

When the surface control is opened to permit the correct gas pressure and volume to operate the pump, it will start. A pressure charged valve, similar to a gas lift valve, may also be used for an unloading valve. It must be set higher than the gas lift valves used for lift, in order for it not to open during normal operation.

8.35 Designing a gas pump installation

8.351 Introduction

The design of a gas pump installation is similar to a gas lift installation. Gas lift will be used in most installations incorporated with the gas pump.

The gas pump can lower the flowing bottom hole pressure to a point equal to a rod pump. It also creates an artificial submergence for gas lift by pumping a fluid column into the eductor tube to within 50 psi of the gas injection pressure. The same well data as required for a gas lift installation is used in a gas pump installation. Designing a gas pump installation graphically is easier and more informative than a mathematical design. Fig. 8.32 shows a graphical design combined with continuous flow gas lift. The advantage in using gas lift with the gas pump is in being able to lift from a greater depth with the available lift gas. This is accomplished by lowering the tubing column gradient by aeration.

8.352 Example problem

The following example problem is a gas pump graphical solution combined with continuous flow gas lift.

Data:

Daily fluid production rate	= 150 b
Normal available gas pressure	= 900 psig
Temporary kick-off pressure	= 960 psig
Gas gravity	= 0.65
Wellhead tubing pressure	= 120 psig
Average formation depth	= 8150 ft
Tubing size	= 2⅜″ O.D.
Gas injection tubing size	= 1¼″ N.U.
Formation temperature	= 168°F.
Surface flowing temperature	= 87°F.
Oil percentage	= 50%
Static fluid gradient	= 0.400
Static fluid level	= 5380 ft
Productivity index (P.I.)	= 0.142
Static bottom hole pressure	= 1230 psig

When plotting a graphical design, the scales used on the abscissa and ordinate should match those of any flowing gradient curves that you are using.

Procedure: (Refer to Fig. 8.32)

(1) Lable coordinates, depth on ordinate, and pressure on abscissa on left side with the temperature scale on far right hand side of abscissa.

(2) Draw a line across the graph paper at the average depth of the perforation. In this example, it would be 8150 feet.

(3) Plot the wellhead tubing pressure (P_{wh}), the operating gas pressure (P_{so}), and the kick-off pressure (P_{ko}) on the pressure scale of the abscissa.

(4) Plot the wellhead flowing temperature on the temperature scale of the abscissa.

(5) Select the descending gas flow chart for the tubing size and pressure required, in this case 1¼ in. tubing, 900 psig operating pressure [Figs. 8.33(A) and 8.33(B)].

Determine the static operating pressure at 8150 ft and plot on the depth line. The static kick-off pressure of 960 at depth may be determined by averaging the 900 and 1000 static pressures and adding the 10 psi difference to the average.

900 at 8150 ft = 1090 psig
1000 at 8150 = 1210 psig

$$\frac{1090 + 1210}{2} = 1150 + 10 = 1160 \text{ psig}$$

Draw broken lines to indicate the static pressure from 900 psig (P_{so}) to 1090 psig at depth and 960 psig (P_{ko}) to 1160 psig at depth.

(6) Calculating the gas requirements of the pump

(a) Convert the volume displacement of the pump in barrels per day to strokes per minute. The problem example shows a desired production rate of 150 b of total fluid (Fig. 8.34). (b/d/strokes/min) to determine the strokes per minute for 150 b/d. The chart shows 5 strokes/min.

(b) The strokes per minute must be converted to cu ft gas/day at one atmosphere (atm). Use Fig. 8.35 and note that 5 strokes/min. is equal to 850 cfd at one atm.

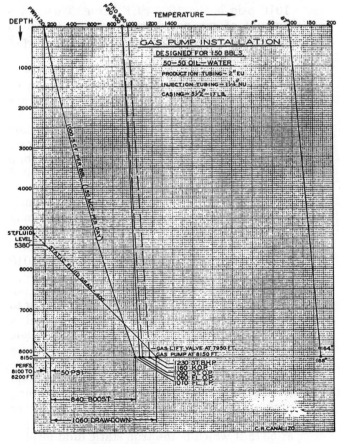

Fig. 8.32   Gas pump installation

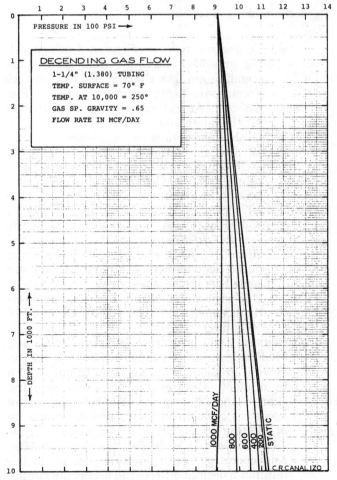

Fig. 8.33A   Descending gas flow

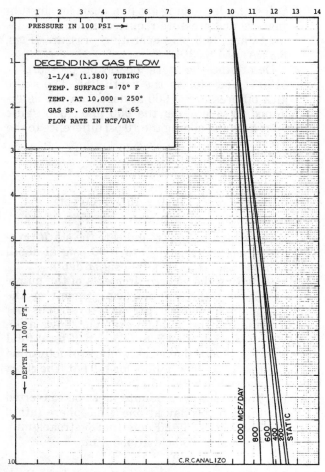

Fig. 8.33B Descending gas flow

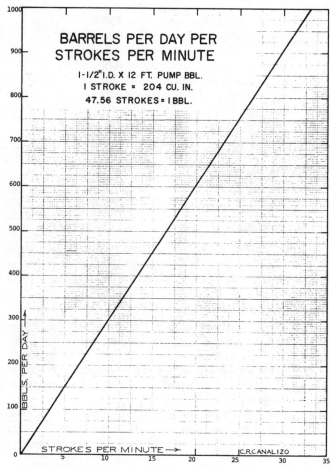

Fig. 8.34 Barrels per day per strokes per minute

(c) To find the volume of gas at depth operating pressure, determine from the 1¼ descending gas flow chart at 900 psig surface pressure the depth operating pressure [Fig. 8.33(A)]. An estimated total requirement for the pump and aeration is assumed that might later require an adjustment if too far off. The estimation in the example is 400 MCF which shows a depth pressure of 1060 psig at 8150 ft. Convert 1060 psig to 1075 psia by adding 1 atm. Fig. 8.36 shows that 1075 psia is 73 atm. Multiply 73 atm by 850 cf at 1 atm equals 62.05 MCF, the amount of gas required by the pump at 100% efficiency. As there will be friction and liquid displacement by formation gas plus some leakage, it is practical to assume 50% efficiency and multiply the calculated gas requirement by two (62.05 × 2 = 124.10 MCF/day).

(7) Plot the 1060 psig depth pressure of the 400 MCF rate per day on the depth line. Select the flowing gradient chart for 2-in. tubing, 150 b fluid, 50-50 oil-water. Lay the graph over the chart so the 120 psig wellhead tubing pressure joints the common tangent curve at a pressure of 120 psig at 1650 ft. At depth, check the gradient curve that is 50 psi less than the 1060 psig operating depth pressure. (50 psi is the minimum desired differential across the pump.) The curve that intersects the depth pressure of

1010 psig is the 1000 SCF curve. This is the amount of gas necessary to aerate the fluid column to produce 150 b of fluid from the depth and pressure available.

Multiplying 150 b by 1000 SCF/b gives 150 MCF/D, the amount of gas needed to aerate. The 150 MCF plus the 124 MCF pump gas equals 274 MCF, the total gas requirement for the well. Checking the 1¼ descending flow chart at 900 psig [Fig. 8.33(A)], the 274 MCF rate shows a 10 psi higher downhole pressure than the estimated 1,060 at the assumed 400 MCF rate.

As the 10 psi is on the safe side and is a small pressure difference, the gas injection line may be drawn in from the 900 psig surface to the 1060 psig depth. Also draw in the 1000 SCF gradient curve from 120 psig surface to 1010 psig at depth.

(8) Draw in the static gradient line. This will join the static bottom hole pressure of 1230 psig to the static fluid level. The static fluid level is determined by dividing the static bottom hole pressure by the static gradient of 0.400. This is 1230/.40 = 3075 ft. The static fluid column is 8150 ft minus 3075 = 5075 ft, the static fluid level at zero atmosphere; however, the tubing pressure is 120 psig and is drawn in to where it intersects the static gradient 300 ft deeper at 5375 ft which is the true static fluid level.

(9) Extend the 120 psig tubing pressure with a

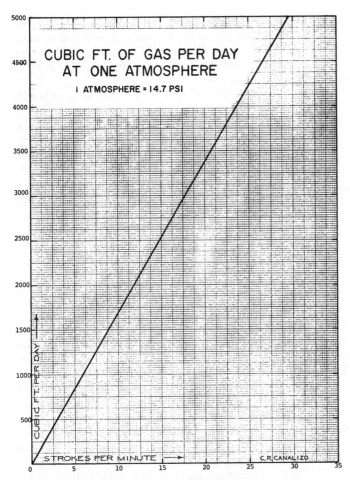

Fig. 8.35 Cubic ft. of gas per day at one atmosphere

orifice size $^8/_{64}$, is the closest and shows a flow rate of 170 MCF per day. Size the choke slightly larger to insure the ability to inject enough gas. It is easier to regulate the surface pressure down a few psi if too much gas is being injected.

(11) A gas lift valve should be installed directly above the exhaust line at 150-200 ft above the pump. In the example it will be 200 ft above the pump. The gas lift valve mandrel is tied directly to the gas injection line. The gas lift valve selected for this example is a tapered stem continuous flow valve. The purpose of the gas lift valve is to be able to inject more gas into the tubing string in order to initially start the well flowing, or in case the well proves to be able to produce more fluid than anticipated.

The tubing back pressure of the gas lift valve is 995 psig at depth. For a surface operating pressure of 900 psig, the base pressure of the valve would be 970 psig with a $^3/_{16}$ in. seat on a 1 in. valve, (R factor 0.0863). When the pump is first started the well is full of fluid to the static fluid level. If the static fluid pressure is greater than the kick-off pressure at depth, it may be necessary to install unloading gas lift valves as in a standard gas lift installation. These may be installed connected to the injection line or preferably the casing annulus. The casing may be pressured up while holding pressure on the 1¼ injection line. This will depress the fluid

broken line to the depth line. Below the depth line, show 50 psi greater than the 120 psig wellhead pressure. This 170 psig indicates the minimum down hole pressure that is possible in order to create the maximum drawdown across the sand face.

Draw a line below the static pressure of 1230 psig to the 170 psig and show this as the drawdown of 1060 psi. From the 170 psig line draw a line to the flowing gradient line of 1010 psig. This represents the pressure difference from the wellbore to the tubing across the pump and is referred to as the boost pressure.

In the example the difference from 170 to 1010 is 840 psi boost pressure. This creates an artificial submergence for gas lift by maintaining, in effect, a fluid column 2100 ft above the pump. When gas is injected into this column at the rate of 1000 SCF/b, the pump will retain the maximum formation drawdown and maintain the tubing column submergence to produce 150 b/d.

(10) To aerate the fluid column, size and install chokes in the pump. As the differential pressure across the pump is from 1060 flowing injection pressure to 1010 flowing tubing pressure, these pressures will be used in determining choke sizes (Appendix 3.C). The daily gas injection rate is 150 MCF. Select the choke chart that will produce 150 MCF with 1060 upstream and 1010 downstream pressures. Figure 3.C8,

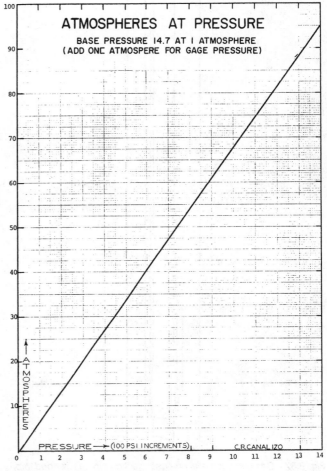

Fig. 8.36 Atmospheres at pressure

out of the casing into the tubing, and the unloading valves will work the well down to the point where the gas from the 1¼ will be able to enter the fluid column and lift the well. At this point the casing pressure may be bled off.

(12) If a pressure-charged unloading valve is used in the pump to purge the liquids out of the exhaust line, set it at a higher pressure (50 psi) than the gas lift valve above the pump. The unloading valve will only be used when the pump is initially started or after the well has been shut in. By setting it higher, it will not be opened under normal operating conditions. In the example a pressure charged unloading valve was used. A valve with a low R factor should be used to minimize the back pressure effect. A valve with an R factor of 0.0216 was used.

The unloading valve is subjected to two pressures. When the main pump valve is in the exhaust position the unloading valve seat area is exposed to the casing pressure, and when the main valve is in the injection position the unloading valve seat area is exposed to the high injection pressure. It is therefore necessary to consider both these pressures when setting the valve to avoid possible interference from the high side which would open the valve every power stroke and might cause the bellows to fatigue and fail.

The gas lift valve was set at a base pressure of 970 psig at 80°F., which under the well conditions would operate at 900 psig surface pressure. To prevent the unloading valve from opening after the well is kicked off, it should be set higher than the gas lift valve. In the example it was set 50 psi higher, or 950 psig surface operating pressure (with the seat area exposed to the lesser casing pressure, or in the exhaust position of the main valve). The minimum pressure that might exist in the casing is the 120 psig tubing pressure, and this is used to set the valve for 950 psig surface opening.

The 950 psig surface pressure at a depth of 8000 ft is 1130 psig. The R factor of 0.0216 multiplied by 120 psig is 3 psi which is added to 1,130 or 1,133 at depth temperature of 168°F. The base pressure at 80°F. is 940 psig, rounding out to the nearest 5 psi. To check for injection pressure interference, multiply the R factor of 0.0216 by the depth flowing operating pressure of 1160 psig, which is 25 psi. Subtract this from 1133 psig at 168°F. which is 1108 psig.

Note the pressure of 930 psig which is well above the operating pressure of the gas lift valve at 900 psig. Therefore, the base setting pressure of the unloading valve is safe at 940 psig at 80°F.

8.36 Methods of installing the gas pump and required equipment

Fig. 8.37 illustrates a gas pump installation with the supply gas tubing strapped to the production tubing and one gas lift valve connected to the gas supply tubing. Gas lift valves for unloading the well are shown on the production tubing. The annulus would be pressured as well as the supply gas tubing to operate the valves. With the well unloaded down to the valve on the supply gas line, the annulus would be bled off as the upper unloading valves would no longer be needed.

In Fig. 8.38 the supply gas tubing is run in separately, after the production tubing, and is stabbed into a dual head where it packs off and locks in place. The dual head is located directly above the top of the exhaust line. The exhaust line and supply line below the dual head are strapped to the production tubing and the gas lift valve is connected to the gas supply line.

Fig. 8.39 is a gas pump installation for a high gas-oil ratio well. It is similar to Figs. 8.37 and 8.38 except that a gas anchor is included.

Fig. 8.310 shows the equipment used on the surface to control and lubricate the gas pump. A variable choke control for regulating the gas volume into the well is recommended, and an upstream regulator may be required where the system gas supply fluctuates abnormally. The lubricator may be a chemical injection pump for injecting oil into the gas in order to lubricate the valve and switching mechanism in the pump. A tie-in is also recommended for checking the pump

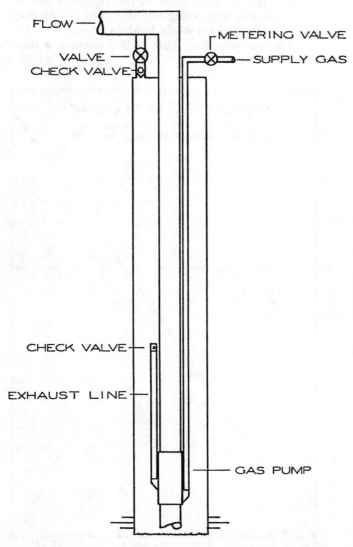

Fig. 8.37 Gas pump installation - gas injection tubing strapped to production tubing

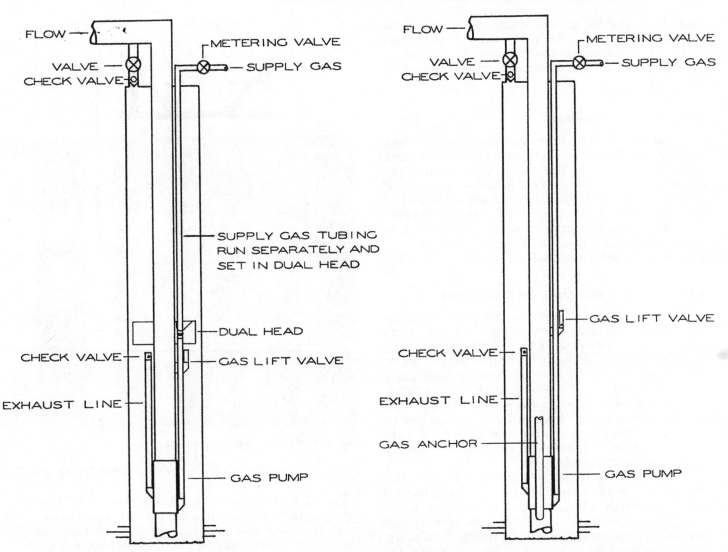

Fig. 8.38 *Gas pump - tubing strings run separately using dual head*

Fig. 8.39 *Gas pump with gas anchor*

operation. This is a meter run, downstream of the control valve.

8.37   Results from field installations

8.371   Shallow well installations

The gas pump has been installed in shallow wells in southeast Texas. The wells in this field were less than 1000 ft deep. The gas supply system was 650 psig, so the wells could be lifted without need of gas lift aeration. The wells had been on beam pump, producing approximately sixty to one hundred barrels per day. The gas pump tripled the production in the two wells

in which they were installed. The gas that was used to lift the wells was recovered from a fire flood.

8.372   Medium depth wells

Medium depth wells in the Hawkins field in eastern Texas, approximately 5000 ft deep, were significantly improved over beam pump and gas lift. In one well that was on beam pump, the production increased from 115 b/d to 380 b/d. This well was later put on gas lift, for comparison purposes, and the production dropped 100 b/d less than when it was on the gas pump. Two other wells were lifted with the gas pump and the production

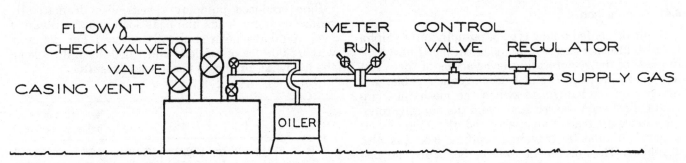

Fig. 8.310 *Surface equipment for gas pump*

more than doubled in both instances. A well in the Los Angeles area was deviated with a vertical depth of 4800 ft. A gas pump was installed and the production increased to 150 b/d from 80 b when it was on gas lift.

8.373   Deep well installation

A deep lift well in southern Louisiana has a gas pump installation. The well is lifting from 9000 feet; the total well depth is 9300 ft. The well is a low productivity, low bottom hole pressure well. It had been on gas lift making 60 to 70 b/d. After the gas pump was installed, the well produced 125 to 135 b/d. The system gas pressure was 950 psig maximum with 900 psig as a dependable operating pressure. The pump was combined with gas lift in this installation.

8.38   Summary—advantages and disadvantages

The gas pump is a means of improving many gas lift wells by reducing the working fluid level while increasing the submergence for gas lift. Wells on deep beam pump can be improved by the gas pump by taking advantage of a full pump stroke. In deep beam pump lift, a good portion of the total stroke is lost to rod stretch. Rod breakage and pump wear due to high loads are an expensive maintenance factor that is greatly reduced by the gas pump. No rods are used and the load differential on the plunger remains less than 100 psi regardless of the depth.

In any field where gas is available, the gas pump can be used to advantage. The problems of individual prime movers, as in beam pumps, are reduced to a centrally located compressor station. This reduces field operational expense as the daily servicing of well equipment is eliminated. When pump failure occurs, the pumps may be easily and economically serviced by wireline methods. All of the parts that are subject to wear and replacement are confined to a single retrievable unit.

The disadvantages of the gas pump are related to additional installation cost over gas lift by the difference of the cost of the pump and gas supply line to the pump. The downhole well installation is comparable to a hydraulic pump installation. The pump parts are subject to wear and periodic replacement. We do have insufficient history on the pump to predict an operational life span. Each well will affect the pump wear in a different manner.

8.4   THE WINCH-TYPE PUMPING UNIT, BY R. H. GAULT

8.41   Introduction

Gault presented a new type of pumping unit, Alpha I, which incorporated several modifications (Fig. 8.41.)[1] in place of the reciprocating lever action of the walking beam unit, the Alpha I uses a winch, cable, and rotary cam mechanism to switch the mechanical advantage between the rod string and the hanging counterbalance. Instead of a polished rod, the Alpha I uses a polished tube. A traveling stuffing box was also developed. The unit has no walking beam and no massive mechanical framework. It weighs about half as

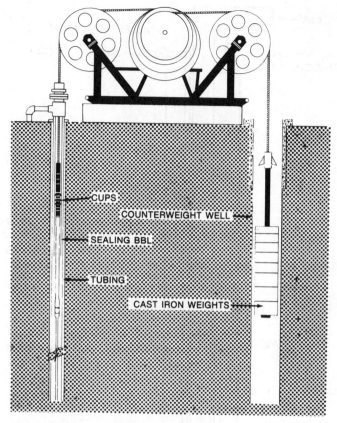

Fig. 8.41   The winch-type pumping unit

much as a conventional unit of the same lifting capacity.

The Alpha I is a conceptually new piece of artificial lift equipment. Its computer-designed geometry makes possible a long, slow pumping stroke with excellent load distribution throughout the pumping cycle.

The use of a long, slow stroke in rod pumped oil wells has been the goal of prudent operators for many years. They recognized that a major cost of operation was the repair and replacement of sucker rods, tubing, and bottom hole pumps. Since most of these failures were the result of fatigue, corrosion fatigue, or wear, a reduction in the number and magnitude of fatigue producing strokes would dramatically reduce their operating costs.

For example, one stroke per minute is 525,600 strokes per year. Three strokes per minute is 1,576,800 strokes per year. Twelve strokes per minute is 6,307,200 strokes per year. If the fatigue life of an average rod string or bottom hole pump is 20 million strokes, under field operating conditions they would last only 3.17 years at 12 strokes per minute but they would last 12.68 years at 3 strokes per minute.

Since reduced pumping speeds also dramatically reduce dynamic loads, the peak load and the range of load would also be dramatically reduced. For example, with 18,000 lbs of rod load and 7,000 lbs of fluid load, the following predicted loads would result:

	PPRL	MPRL	Range	Percent range	1 in. Rod peak stress
3-480 in. SPM	26,098#	16,903#	9,916#	35%	33,228 PSI
12-120 in. SPM	29,410#	13,590#	15,820#	54%	37,445 PSI

Peak rod stress and the range of stress are the determining factors in expected rod fatigue life. The meaningful reductions illustrated above would result in significant improvements in rod life if they were the only factors considered.

Using longer strokes at slower speeds has gained wide acceptance. Unfortunately, to obtain these long strokes with conventional walking beam pumping units the torque requirement is greatly magnified. For example, using 20,000# PPRL and 10,000# MPRL and 15,000# CBE, S/2 = Torque Arm. Note that the peak torque is doubled (250,000 in lbs compared to 500,000 in lbs) in changing from a 100 in. to 200 in. stroke (Fig. 8.42).

With walking beam pumping units, any increase in stroke length results in torque magnification proportional to the amount of increase. In addition, the walking beam and supporting structure become massive and quite tall to accommodate this longer stroke. The size limitations of this configuration limit the maximum practical stroke length.

In the search to obtain the long stroke without the limitations of walking beam units, a return was made to the age-old principle of the winch. The principle of continuous mechanical advantage over long distances was already in wide usage in all other segments of industry.

To illustrate the advantages of the winch, the following example is presented using the loads from the previous example (Fig. 8.43): PPRL = 20,000# MPRL = 10,000# CBE = 15,000#. Notice that the peak torque requirements remain the same for a 200 in. stroke as compared to a 100 in. stroke (80,000 in lbs).

This drastic reduction in peak torque and in the size of the structure required becomes greater and greater when longer stroke lengths are considered. Each rotation of the winch drum produces 100 in. more stroke with no increase in torque or in the physical size of the structure.

8.42 Operational procedure

One of the foremost problems was to determine how to stop the machine at the end of the stroke, turn it around, and drive it in the opposite direction.

If the motor were just turned off, the machine would come to a stop because of the unbalanced loads being lifted in both directions of motion. On the upstroke the weight of rods and fluid are greater than the counterbalance. On the downstroke the counterbalance weight is greater than the weight of rods only. These unbalanced loads would serve to brake the machine to a stop. However, the turn-around points would be very imprecise, especially as fluid levels changed during operation of the well.

Again, a return was made and, in this case, to the age-old principle of the cam. When the motor is turned off, the rotational inertial energy of the machine can be stored. This energy can be used to stop the machine and start it in the opposite direction. So the use of cams permits the storing of inertial energy and the use of this energy to start the machine in the opposite direction.

Effectively, the working leverage is changed by the cams. For example, as the upstroke is completed the

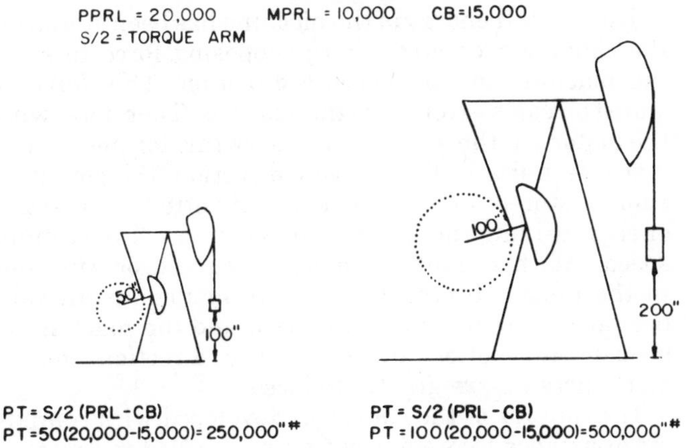

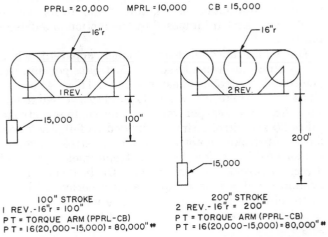

Fig. 8.42 Comparison of peak torque requirements for 100" and 200" strokes.

Fig. 8.43 Comparison of peak torque requirements for winch principle

motor is turned off. The inertia of the machine continues rotation and the wellside cable is raised by the up-cam and the counterbalance cable is lowered by the down-cam. This conversion of inertial energy to potential energy causes the unit to stop at a rather precise point. These cams change the mechanical advantage of the mechanism and create a torque in the downstroke direction (Fig. 8.44).

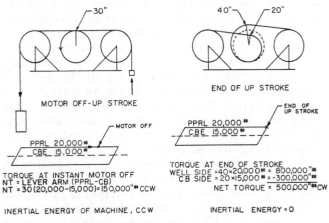

Fig. 8.44 Inertial energy principle

The inertial energy of the machine has been stored on the cams and creates a large opposing force to start the machine in the opposite direction. This force is stored on cams external to the gear box. Therefore, with the motor off, the gears do not carry this torque. As the machine starts in the opposite direction, the potential energy stored in the cams is converted to inertial energy, driving the machine to about ⅔ - ¾ maximum speed. At this point the motor is turned on running in the proper direction. Since the motor has already been driven to a significant speed and the machine is already in motion in the proper direction, motor start currents are greatly reduced.

The contour of the cams was developed with computer assistance to provide optimum polished rod motion at the points of turn-around. Since turn-around occurs after the motor is turned off, the rate and motion of the turn-around is controlled by polished rod forces. This produces vastly-improved motion when compared to the turn-arounds of walking beam units where stroke reversals are imposed by the motion of a driven machine.

The significance of this becomes apparent when stroke times are considered. At 15 strokes per minute, one complete stroke takes only 4 sec. These 4 sec. include an upstroke, a downstroke, and two stroke reversals. At 3 strokes per minute, one complete stroke takes 20 sec. Stroke time is divided as follows: 7 sec. for the upstroke (motor on), 3. sec. for stroke reversal (motor off), 7 sec. for the downstroke (motor on), and 3 sec. for stroke reversal (motor off). In terms of rod dynamics and fatigue, compare a 3 sec. period (with the motor off) just to reverse the stroke with 4 sec. for a complete upstroke, downstroke, and two-stroke reversals.

The use of these cams to store rotational energy and to provide power off turn-arounds is responsible for a large portion of the energy saving. From the previous example note that the motor is completely off for 6 sec. of the 20 sec required per stroke. The motor drives only 14 of the total 20 sec. During the driving period, the counterbalance effect is always at its maximum and the upstroke and downstroke loads remain almost constant. The horsepower required is the difference between a steady wellside load and a steady counterbalance load, multiplied by the torque arm of the power drum. Under these conditions of load, the motor is able to operate at a steady load with high efficiency. When the motor is turned off, no power is used. So the motor operates only fourteen seconds out of twenty seconds or 70% of the time. When it is on, it is operating under steady and efficient conditions.

Compare this with the motor load of a conventional unit. The torque curve and therefore the motor load curve go through two large peaks. If the motor is properly sized, it will be loaded to more than 100% of its rating during these peaks. On both sides of these peaks the motor load drops to zero and oftentimes negative load. The no load current of the average oilfield motor is about 40% of its full load current.

So when the motor load drops to zero and no work is being done, 40% or more of the full load current is still being used. When the load becomes negative, the motor serves as a generator which puts current back in the power line. This generated work is performed at low efficiencies since a squirrel cage motor is not a good generator. In addition, this work must be restored to the system at another point in the stroke so even greater inefficiencies result.

Dependable motor control is essential for optimum efficiency of the winch-type machine. Control of the "motor off" position is variable within limits. This permits small stroke length changes with a twist of a knob. To obtain a longer stroke length, the motor is left on for a short time after the cam is reached. For a shorter stroke length the motor is turned off just prior to or at the cam. Total change in stroke length by this method is from 38 to 44 ft, or about 15% off the total stroke.

It is also important to start the motor at a specific point. Our computer studies have shown that dynamometer card shapes can be controlled, to a pronounced degree, by selecting the proper instant for motor start.

It was found that rod harmonics could be largely dampened out if the motor is started at the proper instant. This instant occurs just after the load has peaked at its maximum or minimum after turn-around. The rod string, which is actually a long spring, has been disturbed by the turn-around and the pick up or release of load.

When the load reaches it maximum value, the spring is extended to its maximum length. This disturbance causes the spring to start vibrating at its natural frequency. (This natural frequency is a function of the speed of energy transmission in the rod string and the length of the rod string.)

The vibration at natural frequency is the factor that determines the shape of dynamometer cards. Walking beam pumping units impose a stroke frequency and polished rod motion on the rod string. This produces certain forces and dynamics in the basic polished rod load; it also produces rod vibrations. Some pumping speeds amplify natural vibrations and some dampen out natural vibrations. Since the starting point of the winch unit can be selected and no arbitrary motion is imposed, it is possible to choose a starting point which dampens out these natural vibrations.

A control was developed which utilizes a strain transducer to monitor well loads. When the load has reached maximum or minimum values, control logic determines the exact instant for motor start that will dampen out rod vibrations (harmonics). This reduces or eliminates the fatigue producing load variations generated by harmonics. It makes it possible to produce a dynamometer card which is essentially a parallelogram. This ideal dynamometer card gives minimum fatigue and minimum torque per unit of work.

Since the control monitors the well load continuously, this output detects downhole well problems. When downhole well problems occur, the control shuts the unit down, sets the brake, and lights a digital display which advises the operator which downhole problem to look for.

Since stroke position and load are both used in the control of the unit, it is only necessary to plug in an X-Y plotter to obtain instantaneous dynamometer cards. It is also possible to transmit this information to a remote monitoring point or computer. The dynamometer card can be displayed at this point.

8.43 Detailed mechanical operation
8.431 Introduction

World Oil presented a more detailed discussion of the operation of the unit. Figs. 8.45 and 8.46 show more details of the unit.

The unit is essentially a winch and cable which operate basically as a light weight "drawworks" to raise and lower the rod string (Fig. 8.45). A second cable connected to the power drum is pulled in the opposite direction by a counterweight to partially balance the torque load required to lift rods and fluid from the well.

This power drum is driven by a conventional gear box operated in turn by a standard oil field electric motor which is mechanically switched to operate in forward and reverse for up and down strokes.

To adapt this "drawworks" to produce an efficient pumping motion, the drum has been modified by addition of spiraling cable grooves, or cams, which cause the cables to follow paths that change the radius of the drum near the end of each stroke, greatly altering respective torque loads of well-side and counterweight cables [Fig. 8.46.]

The effect of this action is—at the top of the upstroke—to make the well load "heavy" enough to stop its upward motion, then actually reverse the momentum of the system.

Conversely, at the bottom of the stroke, counterweight torque is rapidly increased to stop downward rod motion and again reverse system momentum.

Thus, the cam system causes the opposing weights to operate in a pendulum type motion, coasting to a smooth stop and reversing at top and bottom of each stroke. During this coasting period, the electric motor is actually switched off for a significant length of time, to be restarted (by a patented load sensing system) only when it has been reversed by the mechanical action and driven up to partial full-load speed in the opposite direction.

Since a sucker rod system can never be completely balanced, the motor, of course, has to supply power over most of the length of each stroke.

Because the drum and cam system can be designed to achieve a desired load effect and because the resulting loads and motions can be measured and converted to electronic signals, the entire system is highly adaptable to electronic control and fine tuning for optimum efficiency.[2]

8.432 Control system

Fig. 8.47 is a load-position diagram of the pumping loads as they would occur if the machine were going so slowly that no dynamic forces were imposed. It also assumes that no harmonics were generated in the rod string.

Fig. 8.48 shows a load-position diagram of the pumping loads if only dynamic loads, caused by the acceleration of these loads, are added to the system. These added loads are generated by the acceleration at the start of the upstroke or downstroke and deceleration at the end of the upstroke or downstroke.

Consider now that these loads are being lifted by a spring with a harmonic frequency. When a spring is disturbed by a force, it oscillates back and forth at a rate determined by its length and the material from which it is made. (A ready example of this is a tuning fork.) The rate of travel of a force in a steel sucker rod string has been determined to be approximately 16,000 ft/sec. Since the length of the rod string is determined by the depth of the oil well, it is easy to determine the fundamental frequency of a given rod string. The disturbing forces generated by the pumping system, the dynamic forces, and the motion of the unit act to cause the spring to bounce at its fundamental frequency. If these imposed loads and forces coincide with the natural frequency, the amplitude of the bounce is greatly magnified and all loads are increased. If the imposed loads are out of phase, the effect is to damp out the natural frequency and no magnification of load is generated.

Fig. 8.49 shows the effects of harmonics only as applied to the basic load diagram; the only disturbing force in the example is the application and removal of the fluid load in the normal pumping action. This load disturbs the spring when it is added or removed and the rod string bounces at its natural frequency.

In winch-type pumping units the motion of the drum is not continuous in one direction. The reciprocating motion is produced by reversing the rotation direction of the drum. In some units the motion is stopped, and reversal is initiated by the use of cams or other devices to change the effective radius of the drum with respect to counterbalance and well ends. This provides an alteration in the torque relationship and generates an opposing torque to stop the unit and start motion in the opposite direction. In such units, the input from the prime mover is stopped near the end of the stroke. The inertia of the system causes it to continue in the direction of motion until the opposing torque brings it to a halt. This opposing torque causes the machine to start in the opposite direction.

It is possible to re-apply input from the prime mover in the opposite direction at any time. Studies have shown that there is only one particular time which will give the most desirable results. Refer to the diagrams of Fig. 8.410. These load diagrams illustrate what happens when prime mover input occurs too early or too late. When input starts too early, rapid acceleration causes a severe disturbing motion. This generates larger than normal inertial forces and these abnormal loads generate maximum disturbance of the spring.

When input starts too late, the return motion has started to slow down and the prime mover input causes a rapid acceleration. This jerkiness causes an even greater magnification and maximum disturbance of the spring.

These abnormal loads require greater torque from the prime mover and gear reducer. In addition, abnormal fatiguing forces and cycles are imposed on the sucker rods.

The proper start time is further complicated by the many variables in the system. The biggest variable is the fluid being pumped. When pumping is first started, the fluid is standing high in the casing annulus. As pumping proceeds the fluid level is lowered. If the well is over-pumped, the pump does not fill completely and a fluid pound occurs. Various amounts of gas, oil, and water flow into the well bore on an intermittent basis. Each of these changes affect the load diagram.

In addition, the inertia of the various components of the pumping unit change from model to model and necessary operational changes in counterbalancing and drive ratios further affect the operation of the unit.

Fig. 8.45A   Alpha I unit

Fig. 8.45B   Alpha I unit

Fig. 8.46   Alpha I unit

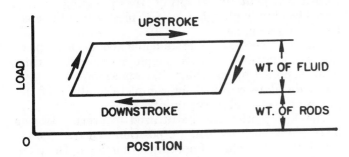

Fig. 8.47   Load position diagram for no dynamic forces

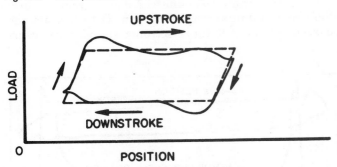

Fig. 8.48   Load position diagram including acceleration of dynamic loads

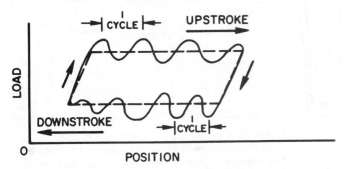

Fig. 8.49   Load position diagram including effects of harmonics

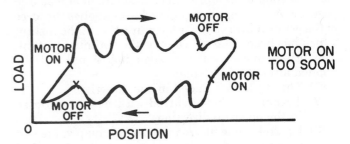

MOTOR ON TOO SOON

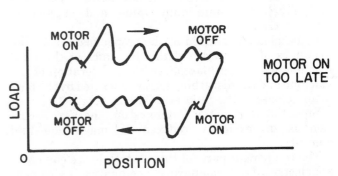

MOTOR ON TOO LATE

Fig. 8.410   Applying prime mover input either too soon or too late

It is therefore desirable to be able to automatically change the motor start instantly to automatically compensate for these changing conditions.

The foregoing study shows the best time to start the prime mover occurs at or near the point of initial peak or minimum load. (Fig. 8.411).

When the load has reached its maximum at the beginning of the upstroke, the spring has had its maximum disturbance and the bounce of the spring would cause the load to start diminishing. If the motor is started at this instant, the accelerating motion would cause the load to increase. These two forces are acting in the opposite direction and they tend to cancel each

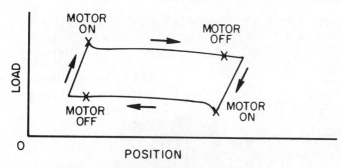

Fig. 8.411   Positions for starting and stopping motor

other out. The bounce wave in the rod string is damped and the load is maintained at minimum value.

When the load has reached its minimum at the beginning of the downstroke, the spring has had its maximum disturbance and the bounce of the spring will start the load increasing. If the motor is started at this instant, the accelerating motion will cause the load to diminish. Again, these two forces are acting in opposite directions and tend to cancel each other. Again, the bounce wave in the rod string is damped out and the load is maintained at minimum value.

For the winch-type unit there is a sensing device which monitors well load, a sensing device which monitors drum or cable position, and a logic device which combines the information from the two monitors and starts the motor at a pre-selected condition of load and position. A starting or engaging device applies the force of the prime mover in the proper direction at the instant which has been determined to be the optimum instant. The prime mover start or engaging signal will be applied after the following:

(1) The stroke has ended in one direction.
(2) Turnaround has occurred with motion initiated in the opposite direction.
(3) Upstroke load has reached a maximum value and started to decrease or downstroke load has reached a minimum value and started to increase.

If the motor is started at the proper instant for the dynamics of the well and the dynamics of the total machine with its prime mover, the acceleration of the machine and the disturbing forces generated by the pick up or release of the fluid load will cancel.

Since the wave motion is damped out, the wave excursion is dramatically reduced in magnitude and quantity.

The component parts of this control device can operate electrically, mechanically, pneumatically, hydraulically or in any combination of these methods.

In summary, the operation is continually monitored and controlled by an electronic system that utilizes three input signals. This was given by *World Oil*.

(1) Rod load at surface is derived from a strain gage mounted on the well-side samson post.
(2) Rod travel (stroke position) is indicated by a position transducer that rotates with the main gear box shaft.
(3) Power input (current) is measured by an ammeter operating from a coil around the cable in the contactor box.

With these signals, the system can provide essential safety controls and automatic shutdowns. A display board indicates the type of malfunction that caused

the shutdown so the supervisory personnel can take corrective actions.

The electronics are also easily adaptable to centralized field automation systems, as any information could be transmitted to a remote office.

Change stroke length. The position transducer is a potentiometer rotated by a belt-type reduction drive from the gear box shaft so that it makes nearly one revolution as the drum completes a full stroke of three turns. Potentiometer output triggers the switching circuitry at the preset position to turn the motor off.

This circuitry can be varied, within limits, to change either up or down stroke length. What happens—on the up stroke, for example—is that delaying motor shut-off slightly causes the cable to be driven farther around on the cam surface before the rods and fluid coast to a stop, thus lengthening the stroke. Conversely, switching-off *sooner* would shorten the stroke by advancing the point at which coasting starts.

8.44   Analysis

The Alpha I system can produce a dynamometer card that is a near-perfect parallelogram because the slow speed and motor start essentially eliminate acceleration, shock, and harmonic loading factors (Figs. 8.412 and 8.413). Dynamometer cards are easily recorded—without interrupting the pumping operation—by simply plugging an X-Y plotter into the electronic control panel.

The three necessary factors noted above—load, position and current—are then supplemented with a time input supplied by the chart drive mechanism, to give all of the date required for complete analysis of the pumping cycle and power input.

The X-Y plotter can be utilized to overlay two or more plots, such as load vs. position and current vs. position to correlate important functions. Or plots can be modified, i.e., show load as a continuous open curve rather than reversing the down stroke to form the closed, familiar representation. Data can also be plotted remotely if the electronic signals are trans-

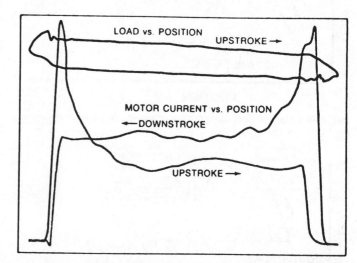

Fig. 8.412   *Typical dynamometer card is correlated with motor current to show when power is switched on according to rod load. Note near-ideal shape of load curve and minimum effect of rod harmonics.*

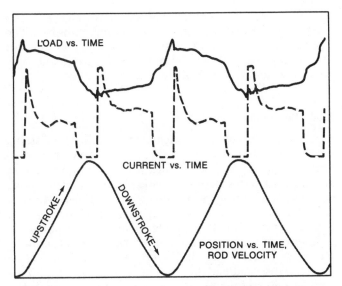

Fig. 8.413A   *Continuous load curve (non-reversing) vs. time is correlated with motor current vs. time to show exactly how long power is switched off. Note small current peaks as motor is switched on. Rod position vs. time shows constant rod velocity during stroke, and smooth turnaround.*

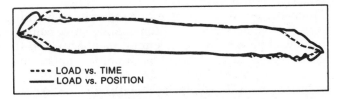

Fig. 8.413B   *Electronics shows special analyses such as overlay of load vs. position (typical dynamometer card) and load vs. time curves.*

mitted to a central office, or output can be digitized and presented in printed form.

8.45   Typical questions and answers

The following questions and answers are given for a better understanding of the unit and to anticipate normal questions.

(1) What is the comparative installation cost of an Alpha I vs. conventional unit of similar capacity?

A. The cost of the foundation and time required to set the Alpha I is somewhat less than a conventional unit. However, the overall cost is somewhat higher because we must drill a counterbalance hole.

(2) How deep and what size counterbalance hole is required; does it have to be cased? How much does it cost?

A. The C.B. hole should be a minimum of 55 ft deep from grade level. It requires a 36 in. hole cased with 30 in. casing. Casing is required in most areas. Cost of the hole varies from $800 to $2,000.

(3) What size are the counterbalance weights and how heavy are they?

A. Counterbalance weights are 24 in. in diameter and are made in three sizes as to weight: 2,000#, 1,250#, 500#.

(4) What size and how long is the wire rope? How much does it cost? Is is a special rope?

A. The rope is 1¼ in. in diameter. We have various lengths for both wellside and counterbalance side. They vary from 65 ft to 68 ft on wellside and 65 ft to 67 ft on counterbalance side. We use different lengths to satisfy particular well condtions. The rope is of a special construction—6 × 31 type Z made especially for the Alpha I. The cost of the wellside rope is $325.00, counterbalance side is $275.00. Prices are subject to change depending upon the cost of steel.

(5) How long does the rope last and how difficult is it to change?

A. Rope life depends upon the loads imposed—especially range of load—and the number of pumping cycles to which the rope is subjected. The average wellside rope has been lasting three to four months. The counterbalance rope has been lasting 7 to 9 months. It takes two men about an hour to change a rope. The unit is self-servicing, so it does not require a pulling unit to change the wire rope. The operational manual furnished with the unit gives complete instructions on how to change the wire rope.

(6) Do you have to move the unit off its foundation to service the well?

A. No. You stroke the traveling stuffing box plunger out of its barrel with the unit, set off the rod string with a set-off plate provided with the unit, disconnect the wellside rope, remove the wellside sheave disconnect the wellside rope, remove the wellside (it is held in place by four bolts), and the well is ready to be serviced. This is all done at ground level.

(7) How long is the traveling stuffing box? What kind of plunger is used?

A. The traveling stuffing box is 48 ft long and is made by joining two 24-ft barrels. We developed a special joint in conjunction with a number of manufacturers of bottom hole pumps. The barrels have 1¾ in. I.D. with either a 2¼ in. or 2⅛ in. O.D. The plunger is a special mandrel approximately 18 in. long which utilizes either conventional bottom hole pump cups or special sealing cups. Various packing arrangements are used according to well conditions.

(8) How long is the bottom hole pump?

A. Forty-eight ft long, made up of either two 24-ft barrels or three 16-ft barrels. The joint is the same type used on the traveling stuffing box.

(9) How often must the stuffing box be repacked? Does it require a pulling unit to get the plunger out of the barrel?

A. Packing life varies according to well conditions. In severe conditions 6 weeks to 2 months. Normal conditions, four to six months. It does not require a pulling unit. The unit is self-servicing. Place the control panel switch in the service position, hold up on the service switch: the unit strokes the plunger out of the barrel and sets the unit brake. Set the rod string off with the set-off plate provided with the unit; remove the plunger and repack it. It takes about one hour for this procedure and repack it. It takes about one hour for this procedure.

(10) What kind of braking system does the Alpha I use?

A. The brake on the gear box is a conventional pumping unit brake. The brake is actuated by an electro-mechanical device which utilizes a spring loaded toggle mechanism in conjunction with a solenoid. When the control system signals the brake to stop the unit, the solenoid trips the toggle, allowing the spring to apply a pulling force on the brake cable. This force is an instantaneous force of about 700# which stops the unit almost instantly.

(11) What kind of safety devices are built into the Alpha I control system?

A. The Alpha I provides numerous safety devices and functions which will shut the unit down, set the brake, and light a number on the control panel to inform the operator as to the identity of the malfunction. Listed below are a number of these safety control functions and devices:

(1) Motor thermal overload
(2) Rope slack detector
(3) Mechanical overtravel
(4) Electronic overtravel
(5) Unit overspeed
(6) High load range
(7) Motor current overload
(8) Low load range

(12) What kind of electric motor is being furnished with the Alpha I?

A. A high slip (15% motor specially designed for the Alpha I. We do in some cases use a Nema D provided it has been intertially analysed.

(13) Does the Alpha I come complete with controls, guards, and traveling stuffing box?

A. Yes. We also include accessory items to set off both the rod string and counterbalance weights as well as a concrete form for the counterbalance hole.

(14) How do you take a dynamometer card on the Alpha I?

A. The control system provides all of the information necessary to produce a dynamometer card. An outlet is provided on the control panel. All you need is an X-Y plotter compatible with the system. Just plug it in and take the card, provided you know how to operate an X-Y plotter. We have an Alpha I X-Y plotter which was developed for the Alpha I and is available.

REFERENCES

1. Gault, R. H., "The 40-Foot Stroke, Winch Type Pumping Unit," Southwestern Petroleum Short Course, Lubbock, Texas (April 21-22, 1977).
2. "All-New Long Stroke Pump Unit Makes Field Debut," *World Oil* (November 1976) 64-68.

8.5 THE T AND C OIL FIELD PUMP, BY GEORGE L. THOMPSON

8.51 Introduction

Calling this unit a pump may be misleading because it closely resembles a chamber gas lift installation. Refer to Figs. 8.51(A) and 8.51(B) which show its sequence of operation.

The T and C oil field pump produces an oil well by controlled injection of power gas into a pump chamber in which a quantity of fluid has been accumulated, thereby forcing the fluid out of the pump chamber to the surface. Power gas is intermittently controlled by the movement of an inner string of tubing which operates a simple slide valve at the top of a pump chamber. The opening of the slide valve permits power gas to enter the pump chamber on the outside of an eductor tube and force the fluid out of the pump chamber.

8.52 Sequence of operation

A hydraulic piston is attached to the inner flow string and moves with it at predetermined intervals as controlled by a clock-timed intermitter. The movement of the flow string actuates the slide valve to control the admission of power gas to the pump chamber. There is no actual lifting of the fluid by this movement.

The pump chamber is made up of a string of tubing or casing of a size determined by the size of the outer casing or liner. It may be as long as required to hold a capacity to produce the well at a desired rate. This may be determined by the standing level of fluid in the hole and the productivity index of the formation.

The eductor tube hangs inside of the pump chamber and has its open end near the bottom standing valve. Its upper end is attached to the inner flow string and moves with it. It may be of any size but is often the same size as the inner string.

Immediately above the pump chamber is the working barrel and slide valves. The upper slide valve controls the flow of power gas to the pump chamber. When the flow string is in the "down" position, the upper slide valve is in its seat and at this point the power gas is shut off. (See "Loading Cycle.") Also in this position, the lower slide valve is down and the vent windows in the working barrel are uncovered. Openings through the inside of the lower slide valve allow gas to pass through this inner channel regardless of what position the slide valve is in.

In the upper portion of the eductor tube there is a vent valve, a form of check valve opening into the annulus between the eductor tube and the outer shell of the pump chamber. This vents the eductor tube when the pump is in the loading cycle, thereby allowing the eductor tube to fill as part of the pump chamber.

Altogether, there are five valves in or near the pump chamber which are actuated when the inner flow string is moved by the hydraulic piston: the upper slide valve which controls the flow of power gas to the pump chamber, the lower slide valve which opens or closes the vent windows into the casing, the vent valve which allows the eductor tube to vent into the upper part of the pump chamber and thence through the vent windows, the lower standing valve which is the bottom closure of the pump chamber and through which the well fluid flows into the pump chamber, and the upper standing valve which is the bottom closure for the flow string.

When the flow string is in the lowered position, the power gas is stopped at the upper standing valve. With no power gas on the pump chamber, the vent valve can open. The vent windows in the working barrel are open to the casing so there is only formation pressure on the entire pump chamber. Therefore, the pump chamber fills from the accumulated fluid in the casing through the bottom standing valve by submergence.

The loading part of the cycle is timed so the pump chamber will have just enough time to fill, then the

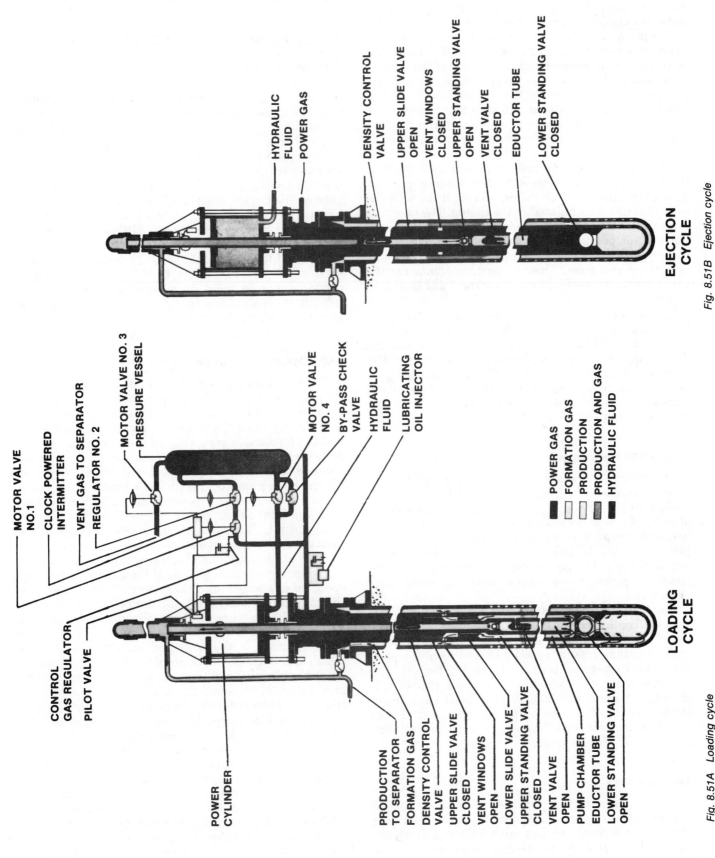

MOTOR VALVE NO.1

CLOCK POWERED INTERMITTER

VENT GAS TO SEPARATOR

REGULATOR NO. 2

MOTOR VALVE NO. 3

PRESSURE VESSEL

MOTOR VALVE NO. 4

BY-PASS CHECK VALVE

HYDRAULIC FLUID

LUBRICATING OIL INJECTOR

CONTROL GAS REGULATOR

PILOT VALVE

POWER CYLINDER

PRODUCTION TO SEPARATOR

FORMATION GAS

DENSITY CONTROL VALVE

UPPER SLIDE VALVE CLOSED

VENT WINDOWS OPEN

LOWER SLIDE VALVE

UPPER STANDING VALVE CLOSED

VENT VALVE OPEN

PUMP CHAMBER

EDUCTOR TUBE

LOWER STANDING VALVE OPEN

LOADING CYCLE

POWER GAS

FORMATION GAS

PRODUCTION

PRODUCTION AND GAS

HYDRAULIC FLUID

Fig. 8.51A   Loading cycle

HYDRAULIC FLUID

POWER GAS

DENSITY CONTROL VALVE

UPPER SLIDE VALVE OPEN

VENT WINDOWS CLOSED

UPPER STANDING VALVE OPEN

VENT VALVE CLOSED

EDUCTOR TUBE

LOWER STANDING VALVE CLOSED

EJECTION CYCLE

Fig. 8.51B   Ejection cycle

clock-timed intermitter at the surface lifts the flow string. This changes the position of the slide valves so the lower slide valve moves and closes the vent windows. At the same time the upper slide valve moves out of its seat, letting power gas down through the channels inside the lower slide valve and into the top of the pump chamber. As soon as pressure is applied to the pump chamber, the check-type vent valve closes, as there is more pressure on the outside of the eductor tube than on the inside. The lower standing valve closes because of the pressure on the inside of the chamber. Then there is virtually a closed bottle similar to a seltzer bottle where pressure applied to the fluid on the outside of the tube forces fluid out of the bottle through the tube.

After all the fluid is up past the upper standing valve, the cycle can change back to the loading position again, vent the pump chamber, and let it fill with the fluid that has accumulated in the casing. The pumping cycle can continue at any predetermined pace, depending on the productivity of the formation; therefore, there may be several slugs of production in transit to the surface at one time. This, in part, accounts for the very high fluid handling capacity of the T and C oil field pump.

Immediately above the working barrel in the inner flow string is a density control valve operated by a predetermined differential between the pressure in the flow string and the power gas supply. Therefore, if the column of fluid in the flow string becomes too heavy to move to the surface, the density control valve opens and allows a measured amount of gas to enter directly into the flow string, lightening the column.

8.53　Surface equipment

The surface equipment is simple and completely automatic. It consists of a hydraulic cylinder for moving the inner string of tubing, a pressure vessel for the storage of hydraulic fluid, a clock-powered intermitter, and the necessary motor valves and regulators to complete the power unit. The hydraulic cylinder can be mounted on any well head. The power unit is operated entirely by power gas pressure and no other power connections are necessary.

The hydraulic power system is a completely closed system. The hydraulic fluid passes from the pressure vessel to the power cylinder when power gas pressure is applied to the pressure vessel and returns to the pressure vessel when gas pressure is vented from it. The entire power unit, excluding the power cylinder, is mounted on a pair of skids and can be placed in a corner of the derrick floor. No foundations or special construction are necessary for any part of the installation of a T and C oil field pump.

8.54　Summary

This system uses the principles of chamber gas lift as sub-surface equipment. Hydraulic power oil actuates the surface equipment. In reality this is a downhole gas lift valve, except its opening and closing is controlled by the tubing string from the surface. It relieves the casing pressure and the 2½ in. pump has produced 1200 b/d from 8500 ft. As many as 20 cycles per hour

have been accomplished, that is, venting, filling, and expelling the slug. Once the slug has been transferred out of the chamber and up into the inner string, it is then on regular controlled gas lift.

8.6　HYDROGAS-LIFT SYSTEM, BY BILL WATERS

8.61　Introduction

The hydrogas-lift system is a unique form of producing heavy crude from deep oil wells by using both hydraulic and gas lift systems [Figs. 8.61(A) and 8.61(B)].

The hydraulic portion of the system evacuates the heavy crude which is trapped in the chamber formed by two packers, forcing the heavy crude into the production tubing and to the surface. The required pressure on the triplex pump at the surface depends on the gravity of the power oil. At this time, it appears that a power oil gravity of 30° to 40° API would be feasible to produce a heavy crude of 10-14° API.

The gas lift portion of the system evacuates the power oil from the power oil string to reduce the bottom hole pressure, giving more drawdown than by standard gas lift methods. Note that no emulsion-forming gas is mixed with the heavy crude.

8.62　Operating sequence

For the following sequence of operation, note Figs. 8.61(A) and 8.61(B). Beginning with the production cycle of Fig. 8.61(B), consider that the triplex pump has already filled the power oil string with power oil and the production is moving up the production string. Note that the standing valve is closed and the discharge valve is open. When a precalculated volume of power oil has been pumped into the well (equal to the combined volumes of the power oil string and the chamber), the level in the power oil tank reaches the low level control. This shuts off the triplex and opens the motor valve to the separator.

At this point, although high pressure gas is trapped in the casing annulus and all of the gas lift valves are open, no gas passes into the power oil string because the pressure at the surface of the power oil string is equal to the pressure on the triplex when it was shut down. This pressure will drop slowly as the power oil is returned to the power oil tank due to the gas which has been put back into solution in the power oil at the beginning of the production cycle by the triplex pump. Then, when the gradient in the power oil string has been reduced to a point below the pressure in the casing annulus, the valves will begin to pass gas and an unloading situation will occur.

When the pressure in the chamber falls below the pressure caused by the weight of the column of fluid in the production string, the discharge valve closes. When the gradient in the power oil string has been sufficiently lowered, the standing valve opens, passing oil from the formation to the chamber.

When an equal volume of power oil which has been pumped into the well has been replaced in the power oil tank, the high level control shuts off the motor valve and starts the triplex—the production cycle starts again.

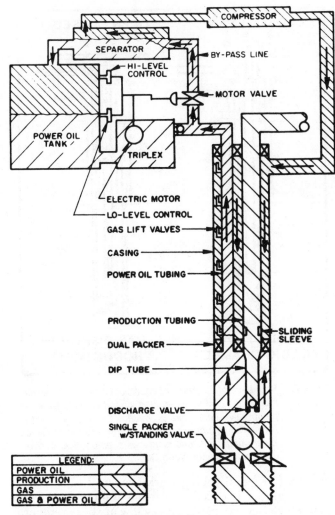

Fig. 8.61A   Hydrogas-lift system, fill cycle

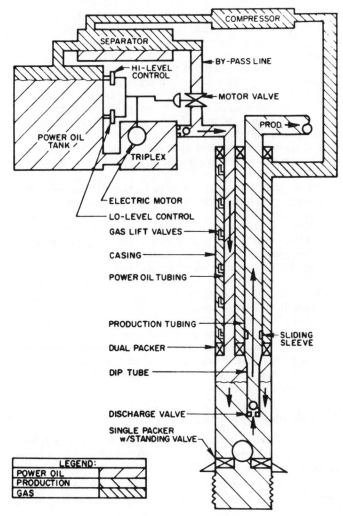

Fig. 8.61B   Hydrogas-lift system production cycle

8.63   General operating considerations

The possible production rates should be comparable to standard gas lift chamber installations now ranging from very low to 1000 b/d.

Another interesting aspect of the hydrogas-lift system is that engine exhaust gas can be used more effectively than natural gas. Using exhaust gas, an operator would not be limited only to areas where natural gas is produced. However, exhaust gas, when injected into viscous crudes, reduces the viscosity of the crude. Also, by injecting exhaust gas into the crude, it may be unnecessary to heat the crude.

8.7   CHAMBER PUMP FOR STEAM FLOOD WELLS (ELFARR CHAMBER PUMP)

8.71   Introduction

World Oil published a short article describing a lift method used in the Slocum Field near Palestine, Texas, for lifting heavy crudes (steam-flooded) in shallow wells. (Figs. 8.71 and 8.72).[1] Again, it is a take-off on the standard gas lift chamber; natural gas or air can be used for the gas-lift injection gas. It has successfully operated in temperatures above 250°F.

The pumping system described by *World Oil* consists of:

(1) A downhole device run within production tubing

that utilizes two seals, a crossover, and two ball and seat valves to form an isolated chamber into which formation fluid can flow (Fig. 8.71)

(2) A source of natural gas (or air, in special cases) at sufficient pressure to lift a column of oil and water from relatively shallow depths

(3) A surface system containing three automatic valves and a timer operated controller

The downhole pump can be run on a string of small diameter tubing which serves as the conduit for the pressurized gas. The production tubing contains two seating nipples, the lower nipple having a smaller diameter than the upper.

Cups on the pump seal in the two nipples as the unit is lowered into place. The crossover within the upper seal then becomes effective, isolating production in the tubing-tubing annulus above the top ball and seat and opening the inner string to the chamber above the lower ball and seat. The lower end of the pump is set 5-10 ft above the bottom of the well.

8.72   Sequence of operation

World Oil gave the following sequence of operation (Fig. 8.72):

The surface valve on the inner string is opened to a low pressure gas collection system (or atmosphere in case of air), relieving back pressure on the lower ball and seat. Formation fluid then flows into the tubing-casing annulus—which is unrestricted to allow gas breakout—and through the lower valve into the chamber.

After a pre-set length of time—as determined by testing and observation of surface pressure gages—surface valves switch to admit high pressure gas or air through the inner string and crossover, onto the top of the fluid in the chamber. With sufficient pressure, the weight of production in the tubing-tubing annulus is exceeded, the upper ball and seat opens and production is forced through this valve, displacing an equal volume into the flowline.

The pressure cycle is carefully timed to avoid "U-tubing" gas. In the northeast Texas application, it normally takes about 20-30 seconds to displace the chamber.

When pressure is lowered, the upper valve traps all production until the next lifting cycle. And the surface valve opens to bleed-off or recover lift gas.

In the Texas application, natural gas under sales line pressure can be used to lift the 500 to 1,500-ft wells. Spent gas then is diverted to the low pressure fuel system of the steam generators.

A real breakthrough in the Slocum Field was the use of compressed air. Due to special properties of the high water cut emulsions formed by the steam flood, the crude coated the internal system to prevent O_2 corrosion, and the low flash point of the emulsion essentially eliminated the danger of explosion.

8.73   General discussion

The wells require little horsepower to operate (6 hp). Also, the chamber normally will handle sand better than a rod pump.

Maintenance seems to be at a minimum, and the entire chamber can be pulled on the inner string with

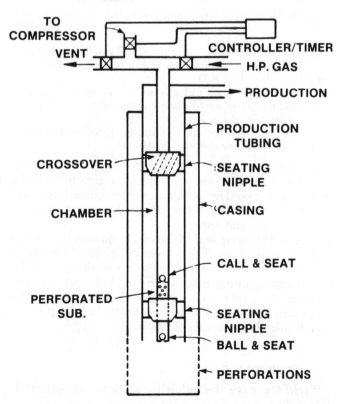

Fig. 8.71   Basic components of a patented gas operated chamber lift pump. Downhole unit is run on small diameter tubing and landed in two seating nipples inside production tubing. (After World Oil)

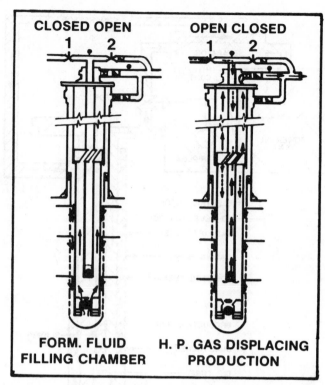

Fig. 8.72   Sequence of operation of chamber lift pump shows formation fluid filling evacuated chamber through lower pump valve (left) while production tubing fluid head is supported by upper valve. Gas pressure (right) closes lower pump valve and displaces chamber fluid through upper valve. (After World Oil)

a rod unit or small tubing work-over unit.

The chamber is excellent for hot temperatures; rod pump efficiency can become very low at these temperatures. Tests have shown that the Elfarr chamber pump delivers maximum production at the higher bottom hole temperatures.

There is no steam-locking with the Elfarr Pump. *World Oil* noted that:

. . . in conventional rod pumps, when the plunger is lifted to draw formation fluid through the bottom valve into the pump, the pressure reduction can cause any water trapped between the two valves to undergo the phase transformation from liquid to steam.

At 300°F. and at the saturation pressure of 67 psi, for example, one cubic inch of trapped water would flash into over 300 cu. in. of steam with any pressure reduction. In the Slocum field, such a phenomenon is considered typical, as hotter wells with water cut of over 50% generally perform with rod pump efficiencies of less than 20%.

With the need to produce more and more heavy crudes, perhaps this system will have other applications.

REFERENCE

"Simple Chamber Pump Lift, Hot, Steam Flood Wells," *World Oil*, (November 1977) 54-55.

8.8   FREE RUNNING PLUNGER LIFT, BY K. C. McBRIDE

8.81   Introduction

In this case, a plunger has been run in a gas well producing some liquids and permitted to travel freely, making successive round trips without an intermitter at the surface or gas lifting the well. This is one of several unique and limited applications of the plunger.

In the Farmington, New Mexico, area a free running piston installation has been effective in unloading produced liquids from gas wells. In many cases the piston has delayed or eliminated the need for costlier methods of removing the produced liquids, such as pumping installations or gas lift systems. Experience indicates the free piston applies to a large percentage of wells that have fluid problems, and that operating expense is much less than any other method of fluid removal yet developed. While admittedly not a cure-all for every well, this piston installation appears to be the single most efficient method of fluid removal developed for this area thus far.

Flow rates should be at least 250 MCF/d to free run and should require no surface controls. Wells producing under this amount can be stop-cocked with a surface intermitter to create the required flow rate. If fluid production is high it may be necessary to choke the well at the surface to prevent the fluid slug from overrunning the separator. The piston will control a certain amount of paraffin deposition but is not recommended for high paraffin producers. Trial and error is the best way to determine this.

The piston will run easily in dual wells, packed off wells, and slim holes 2⅞ in. cased. Pistons are currently running in each side of a dual slim hole completion. Because of the excessive weight in a 2½ in. piston, an attempt is being made to re-design the size and length for better operation.

8.82 Equipment

Fig. 8.81 is a simplified drawing of a typical free-running free piston installation. Essential components are a lubricator to receive the piston at the surface, a valve shifting prong within the lubricator, a free running brush type piston, the 2 in. string of tubing to the level of the bottom casing perforations, a bumper spring, and 2 in. tubing stop within the tubing near the bottom.

Fig. 8.82 is a diagram of the 2 in. brush piston. By using a brush seal to establish contact between the piston and the walls of the tubing, there is very little to wear during the operation. The brush will last indefinitely and the valve shifting mechanism operates about one year between changes.

In wells where the tubing is in bad condition and is under drift diameter in spots, the piston can be turned down about 0.040 in. if needed. The installation, with the lubricator, consists of a tubing stop, bumper spring, and piston.

Fig. 8.83 shows details of a lubricator and manifold for a 2 in. piston.

8.83 Operation

During a typical cycle of operation (the well is producing to the sales line through the tubing), the free piston is dropped in the well prior to turning it on and is allowed to fall to bottom. It hits the bumper spring and stops when it reaches bottom, closing the bypass port through the piston. When the well is turned on,

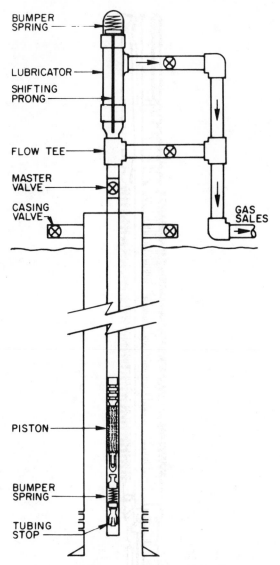

Fig. 8.81 Typical "free running" piston installation

the flow of gas lifts the piston up the tubing, carrying any liquid above it to the surface.

When the piston arrives at the surface with its fluid load, it engages the shifting prong in the lubricator. This opens the bypass port through the piston, allowing gas flow to pass through the piston. The piston then falls back against the gas flow stream to the bottom of the tubing. Upon engaging the bumper spring at the bottom of the tubing, the bypass port is closed causing the cycle to repeat itself. In this manner all liquids that enter the tubing string along with the flow of gas are lifted to the surface periodically and are not permitted to accumulate in the tubing string and log the well off.

The amount of time required to complete a cycle in this operation varies from well to well according to the rates of gas and liquid flow; in general, the cycle averages about twenty minutes.

The piston should be removed periodically (2-3 months) and checked for thread wear and valve leakage. The piston can be removed easily. First, remove the trip prong and bring the piston up by opening the separator dump valve. Do not open it more than a half inch unless necessary. Close the master valve and remove the piston.

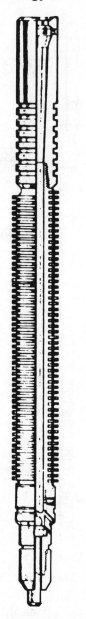

Fig. 8.82   Brush seal plunger

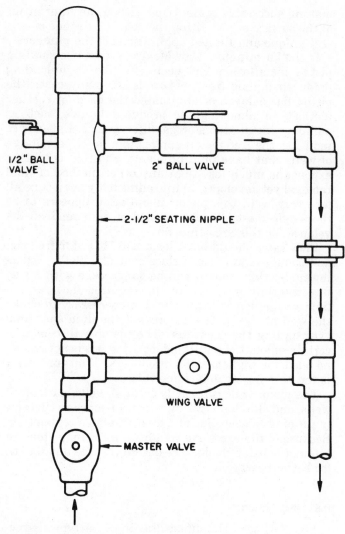

Fig. 8.83   Lubricator & manifold for 2" piston

The brush does not wear appreciably as compared to a metal piston. Some pistons have been running for four years and the brush is still in good shape. The 2½ in. piston is considerably heavier and should be used with a bumper spring. Grind off the bottom of the valve rod flush with the nose piece to keep from brading and expanding.

8.84 Discussion

All of the piston installations have eliminated blowing the well to atmosphere. Each installation will require some attention to set the flow rate and proper operating pressure. Then they need only periodic checking and repair.

To remove a piston, first remove the prong, then change the flow on the manifold and close the master valve when the piston arrives. If the piston valve does not close when on bottom, the piston probably will not travel.

A fishing piston will retrieve the operating piston, thereby not requiring wire line equipment. This has been done only for the regular 2 in. piston.

The brush-valve type piston cleans a well of fluids much better than any piston yet used in this area and is the first free running piston ever developed as such.

8.85   Recorder charts illustrating operation of the free running pistons in gas wells

Some recorder charts are shown illustrating the operation of the free running plunger. These charts may be standard or inverted; that is, the pressure on the tubing may be shown to be increasing in the normal outward movement or vice versa.

Fig. 8.84 shows a reverse-acting chart. Note that the plunger runs consistently in a free manner. This type of operation is not easily achieved and requires more original engineering supervision than the clock operated plunger.

Fig. 8.85 shows another reverse acting chart of a free running plunger. This is a 32-day chart; note various periods when the plunger is not running. Also note that the differential decreases during these periods; hence, the gas rate decreases.

Fig. 8.86 shows a plunger free running every 21 min. This is a standard chart showing both differential and static pressure.

Fig. 8.87 shows a comparison between a well being

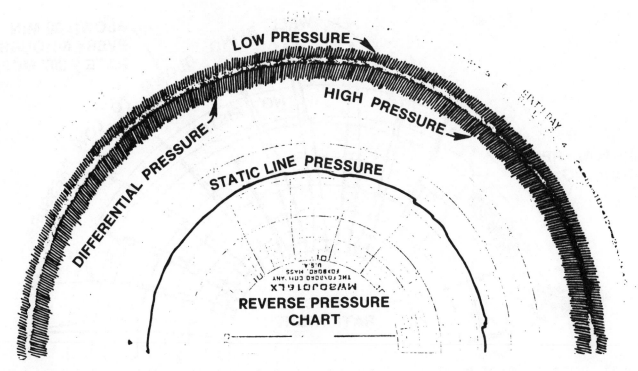

Fig. 8.84   Free running plunger

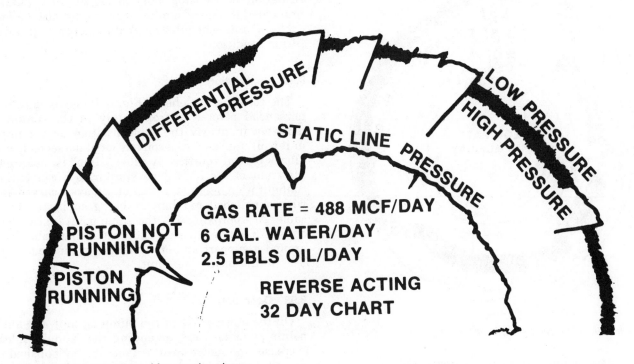

Fig. 8.85   Recorder chart of free running plunger

blown to the atmosphere without a piston and a free running piston. Although the rate was not measured, it is much higher with the free running piston.

8.86 Summary

In this special case, a piston free-travels in the tubing string to unload liquid automatically from gas wells. This should not be tried on most wells. Normally, the well must be shut-in to allow the piston to fall. As noted for 2 in. tubing, the gas flow rate should be at least 250 MCF/D. Although no upper limit has been placed on the rate, the piston probably will not fall for high gas rates such as 1½ to 2 million scf/day.

The piston and lubricator as received from the manufacturer have been modified to some extent.

By carefully selecting the wells, this method will operate successfully. If not, it is quite easy to install a clock-type intermitter to open and close the well periodically, thereby permitting the piston to travel.

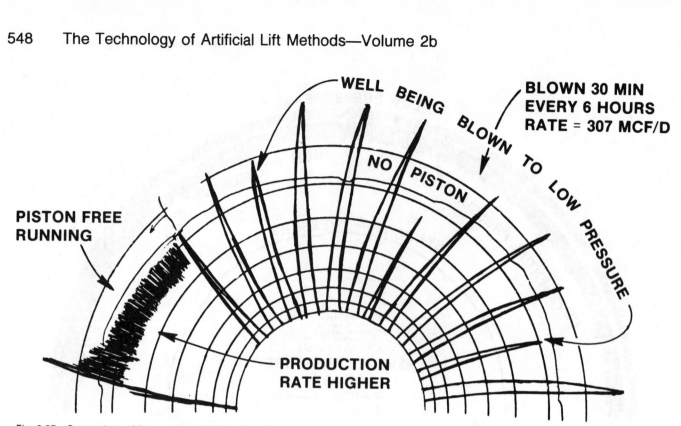

Fig. 8.87   Comparison of free running piston and well being blown to low pressure system without a piston

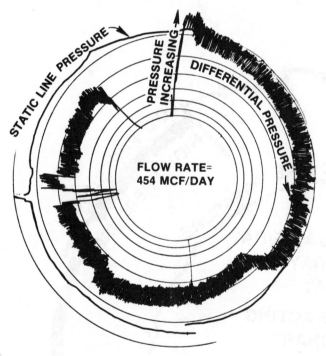

Fig. 8.86   Recorder chart of free running plunger

8.9   CHANCELLOR ROTATING HORSEHEAD, BY FORREST E. CHANCELLOR

8.91   Introduction

The rotating horsehead (Fig. 8.91) modifies a conventional oilfield pumping unit to increase the maximum stroke obtainable by about 50% or more, depending on the particular design criteria incorporated in a specific unit. The design approximately doubles the load on the unit; consequently, the rotating horsehead is for use primarily on existing pumping unit installations where the unit is significantly underloaded because of pumping from a shallow depth and where

higher production rates from the well are desired. Installation of rotating horseheads on new pumping units at the time of manufacture would also be applicable on units scheduled for shallow depth, high volume lifting conditions.

8.92   Description

The rotating horsehead (Fig. 8.91) is an egg-shaped horsehead pivoted approximately in the center and mounted in nearly the same position on a pumping unit walking beam as a conventional horsehead, which allows partial rotation. A cable "sling" is mounted on the front of the horsehead similar to a conventional configuration against a flat or grooved curved plate with a polished rod hanger being supported by the cable. At approximately the same point (anchor point on Fig. 8.91), a separate cable or anchor cable looped around a support and on each side of the head is tied back to the base of the unit.

8.93   Operation

On the downstroke of the pumping unit, the anchor points pivot forward, extending the "sling" cables to increase the downward travel of the polished-rod. Simultaneously, the anchor cables are "reeled up" on the backside of the horsehead as it rotates forward. On the upstroke, the anchor cables, which are firmly anchored to the pumping unit, unreel as the head moves upward. Meanwhile, the sling cables are reeled up, which extends the upward travel of the polished rod.

The head is designed so pivotal load around the pivot bearing is balanced at all times between the rod load and the anchor cable as the rotating head rotates back and forth. Since the configuration is similar to a one-turn pulley, the resulting load on the walking beam is approximately doubled. The increase in stroke length increases the displacement of the subsurface pump,

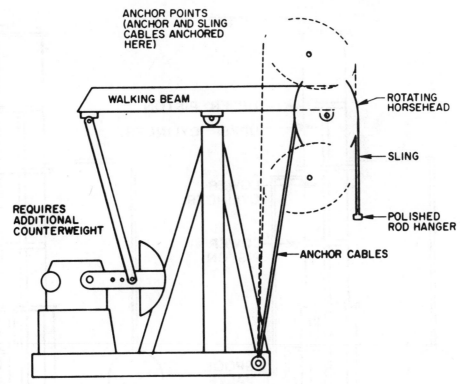

Fig. 8.91 *Chancellor rotating horsehead (Conventional oilfield pumping unit)*

thereby increasing the production rate from the well provided the well is capable of producing additional fluid.

8.94 Field case

A conventional API 80 pumping unit with a 54 in. stroke was selected for the installation of a rotating horsehead. The following results were noted:

(1) The rotating horsehead increased the polish rod stroke length from 54 to 81 in. (50% increase).

(2) Peak gear box torque was increased from 49,500 in.-lbs to 83,184 in.-lbs at 16 spm (4% overload).

(3) Horse power requirements increased from 13 to 20 hp (calculated).

(4) The pump efficiency remained 92% and the gross production increased from 660 b/d to 1,008 b/d with the operating fluid level at 608 ft or 274 ft over the pump.

(5) The polish rod string failed and was replaced with a 1 in. diameter sling.

Fig. 8.92 shows a dynamometer card on this unit at 8 SPM. As noted, a smooth, highly efficient operation is obtained.

Fig. 8.92 *Dynamometer cord on unit with rotating horsehead*

8.10 SUCKER ROD PUMPING WITH PNEUMATIC SURFACE UNITS, BY W. G. SKINNER AND L. A. SMITH

8.101 Introduction

The concept of pneumatic actuation of a sucker rod string in a pumping well is neither new nor revolutionary. Records of patents and applications date back to the late 1800's. These early units did not find wide acceptance due to problems of low efficiency and seal wear and leakage. Also, these early units were restricted on depth capacity to about 4000 ft maximum in most cases.

Surface pneumatic pumping units today overcome the problems of low efficiency and seal failure by using new concepts in system control and entirely new seal materials and seal technology. This section deals with the system design and field application of the Klaeger pneumatic oilwell pumping unit. Present applications range from dewatering 1000 ft gas wells at 3 b/d to pumping 10,000 ft oil wells at 200 b/d.

8.102 Equipment and operation

Fig. 8.101 shows a model HS unit which utilizes the same size upper and lower cylinder. Fig. 8.102 shows a photograph of this unit in field application.

Fig. 8.103 shows the model CB unit which utilizes a smaller upper piston. Fig. 8.104 shows a typical field installation of this unit. The unit is basically a vertical pneumatic cylinder mounted rigidly to the wellhead, similar to the early units. From a distance, the pneumatic unit resembles a piece of pipe protruding above the wellhead. There is no visible evidence of motion, no beams, cranks, counterweights, or foundations. All movement of the sucker rod string is imparted by gas pressure acting on the area of a piston inside the cylinder. There is no direct mechanical linkage between

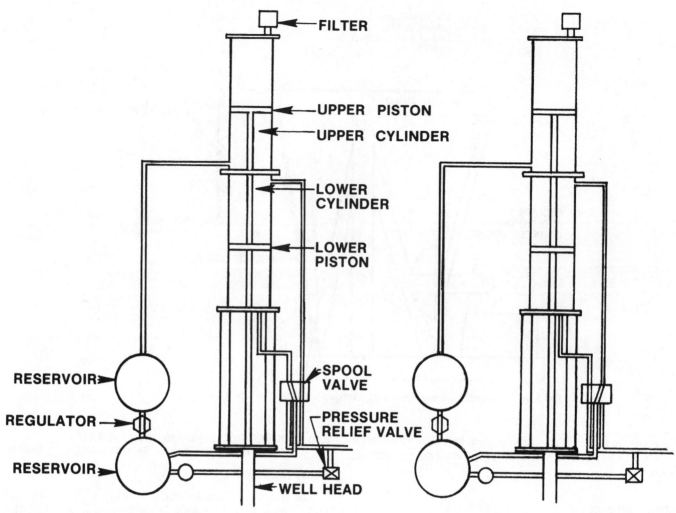

Fig. 8.101   Klaeger model HS unit

the prime mover and the sucker rod string.
Basic operation

HS and HB models each consits of two cylinders with a common polished rod and a piston in each cylinder.

The upper or counterbalance cylinder is piped to a reservoir, filled to a predetermined counterbalance pressure, and isolated. On the upstroke, wellhead gas flows to the spool valve where it is delivered under the

Fig. 8.102   Klaeger model HS unit

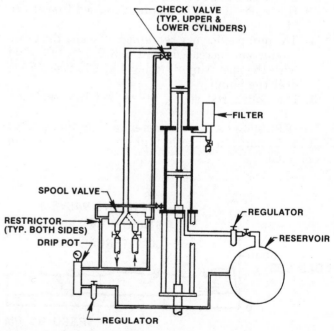

Fig. 8.103   Klaeger model CB unit

Fig. 8.104   Klaeger model CB unit

bottom piston. During the upstroke, gas on the top of the bottom piston (which was used for the previous downstroke) is forced into the sales line. A fixed pressure differential between the wellhead and sales line is maintained by a regulator. On the downstroke, the higher pressure wellhead gas is delivered above the bottom piston where it offsets part of the pressure under the upper piston. As this is equalized, the rods pull the piston down and gas under the bottom piston at the start of the downstroke is forced into the sales line.

8.103   Basic pneumatic theory

The first step in analyzing the operation of a pneumatic pumping unit is a study of the air flows and resultant forces involved in an air actuated cylinder. A fluid actuated cylinder—hydraulic or pneumatic—converts the energy in a pressurized fluid to a linear force that can accomplish some form of useful work, such as lifting the sucker rods in a pumping oil well.

The basic components of the unit are the cylinder, the piston, cylinder rod, and the inlet and outlet valves which control the flow of air into and out of the cylinder.

The three primary parameters governing the amount of useful work available from this cylinder are the piston area, the piston travel or stroke length, and the operating pressure level.

The piston area and stroke length are determined by the physical size of the cylinder. Pneumatic pumping units generally utilize a longer stroke and fewer SPM than a conventional unit in the same application. This longer, slower stroke results in lower air flow velocities which reduce losses due to pressure drop in lines, fittings, and valves to improve the overall system efficiency. The longer, slower stroke attained with a pneumatic unit may be advantageous to bottom hole pump efficiency.

8.104   Sources of operating pressure

Any pneumatic cylinder requires some source of pressurized gas for its operation. This source must have adequate volume and flowrate to enable the cylinder to operate at the rate required to meet the well's productive capacity.

In the case of the pneumatic cylinder being used to pump an oil well, the primary source of energy is wellhead gas. Operation on wellhead or lease gas requires a pressure level and a flowrate commensurate with the depth of the well and volume of fluid to be lifted. Some units have been operated from large field gathering compressors. In this mode, gas is taken from the discharge or interstage of a compressor, run through the pumping unit, and returned to the compressor suction at some lower pressure, where the gas is recompressed for sale or reuse.

Pneumatic units are also being operated with CO_2 and pipeline gas as the working fluid.

8.105   Braking and reversing controls

The probe extending from the top of the unit is the upper control/braking probe; an identical probe is on the lower end of the cylinder. At the lower end of the cylinder is a spool valve—the main control valve used to switch the cylinder between a high pressure reservoir for an upstroke and a low pressure reservoir for the downstroke.

The vertical rod next to the polished rod is the lower control/braking probe. This probe extends into the cylinder through the exhaust/inlet port. The brake disc just above the port opening is a slip fit on the probe to allow the probe to slip through the disc. When the piston hits the probe on the downstroke, the disc is pushed down to seat on the port.

The flow out of the cylinder is now blocked; however, the piston will continue downward, pulled by the inertia of the rod string. The brake probe is free to slip through the brake disc until the piston stops its downward movement. The piston and the sucker rod string are halted and reversed by the cushioning effect of this closed volume of gas. As the braking is being accomplished, the spool valve is being shifted to admit high pressure gas back into the cylinder to begin the next upstroke.

Upstroke braking, reversal, and control is accomplished in the same manner as on the downstroke. The upper braking probe and disc are exactly like the lower components.

8.106   System operation using wellhead gas pressure

During the operating cycle, pressurized gas is admitted into the cylinder through the lower inlet; this pressure will force the piston to the top of the cylinder to complete an upstroke. During this upstroke, the upper part of the cylinder must be vented, preferably in a manner which will allow the gas to be recovered and reused.

In order to allow the rods to fall for a downstroke, the pressure beneath the piston must be reduced to slightly less than that given by the equation,

$$P_2 = \frac{\text{Rod wt}}{\text{piston area}}$$

This pressure reduction can be accomplished by venting the gas under the piston in the cylinder until the pressure drops to P_2. The rods will then fall for the downstroke.

Another method of allowing a downstroke is to increase the pressure above the piston by an amount equal to the fluid weight divided by the piston area. Both methods give the same differential pressure across the piston, or the rod weight divided by the piston area.

The source of lift gas—wellhead or lease gas—feeds the lift gas reservoir through a pressure regulator. The gas is then admitted into the cylinder under the piston.

Lift pressure is always present under the piston. As the lift gas pushes the piston to the top of the stroke, the upper control/braking probe actuates the pneumatic control circuit to shift the spool valve to allow some of the lift gas in the reservoir into the upper part of the cylinder above the piston. This increases the pressure above the piston to the level which will allow the rods to fall for the downstroke.

At the end of the downstroke, the lower control/braking probe again actuates the pneumatic control circuit to shift the spool valve to exhaust the volume above the piston. This reduces the pressure above the piston and allows the lift pressure to push the piston back up for another upstroke.

The operating gas is expelled during the upstroke. This gas can be recovered at some pressure dependent on the difference between the required lift pressure and the 300 PSIG working pressure limitation imposed on the pneumatic cylinder.

8.107 Operating controls—gas-operated unit

The unit incorporates upstroke and downstroke speed regulating valves. These valves throttle the gas flow in the unit to provide a fast downstroke and a slow upstroke over a wide range of speeds. The two pressure gauges indicate the pressures above and below the piston in the cylinder and give a direct indication of the loads being imposed on the unit by the rod string and fluid weight. The indicated pressures can be used for troubleshooting by comparison with normal operating values. Pressures above or below the norm can indicate problems with the unit, the rod string, or the downhole pump. The manual control or cycle-interrupter button interrupts the upstroke to dislodge foreign matter from the bottom hole pump.

Fig. 8.101 shows a typical installation on the well of a pneumatic unit operating directly from wellhead gas. The flow of the operating gas can be traced from the wellhead through a regulator, to the volume tank, then to the inlet connection on the unit. The wellhead mounting of this unit is typical of most pneumatic pumping unit installations.

8.108 Summary and conclusions

Klaeger pumping units are pneumatically activated and counterbalanced pumping systems which use wellhead gas pressure as their power source to pump oil wells and de-water gas wells.

Usually, Klaeger pumping units require less initial investment than conventional beam-type units. Because Klaegers operate off wellhead gas pressure,

expensive outside power sources are eliminated. Its economy of operation is especially valuable in remote areas. In most cases operating gas is recoverable.

No foundation is required. Klaeger pumping units may be attached directly to the wellhead or Christmas tree.

8.11 PUMPDOWN (TFL) TECHNIQUES AND PUMPDOWN ARTIFICIAL LIFT, BY HANK ARENDT AND RUSTY JOHNSON

8.111 Introduction

Pumpdown in itself is not an artificial lift method; instead, it is a technique for servicing downhole artificial lift and flow control equipment (Fig. 8.111). A pumpdown artificial lift installation is essentially no different than conventional wireline installations. Presently, only two methods of artificial lift can be serviced by pumpdown: gas lift and jet pumping.

Equipment used for pumpdown servicing is basically a variation of existing wireline equipment. The valves, plus associated running and pulling tools, must be able to negotiate 5-ft radius wellhead entry loops and must meet certain strength requirements to withstand the greater forces of hydraulic well servicing.

8.112 Pumpdown concepts

Pumpdown, or TFL (through-flowline), services wells by the use of fluids instead of the conventional wireline. The basic theory of pumpdown is an application of the principle of hydraulics, using fluids as a medium of transportation to move flow control devices and service equipment from the surface facility to the well, down into the well bore, and back again. Once downhole, work is accomplished by applying pressure differentials across the tool string.

TFL techniques are ideal for servicing subsea wells equipped with any of a variety of wellhead configurations (Fig. 8.112). Pumpdown methods may also service highly deviated platform wells, extremely deep or high pressure wells, wells in which paraffin or sand removal is difficult by wireline methods, wells producing highly corrosive fluids, and wells where providing vertical access for wireline servicing is impractical.

Servicing of such wells includes running and pulling of flow and pressure control devices, cutting paraffin, washing sand or breaking up sand bridges, shifting sleeves, measuring bottom hole conditions, and performing various formation treatments.

A pumpdown service system requires a circulation path from a central surface facility into the well as deep as possible, through a communication port, and back to the point of origin. The circulation path may be via annulus-tubing, tubing-side string, or dual tubing strings for either single or multiple zone completions. Any number of wells thus completed can be serviced from one surface location.

To service a well using pumpdown methods, five basic items are required:

(1) A pump at the surface to provide hydraulic power
(2) A circulation fluid to convert the pump power to work

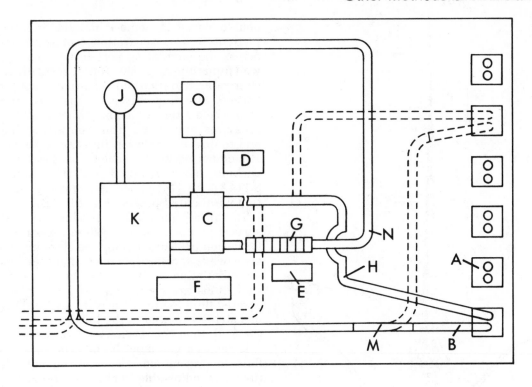

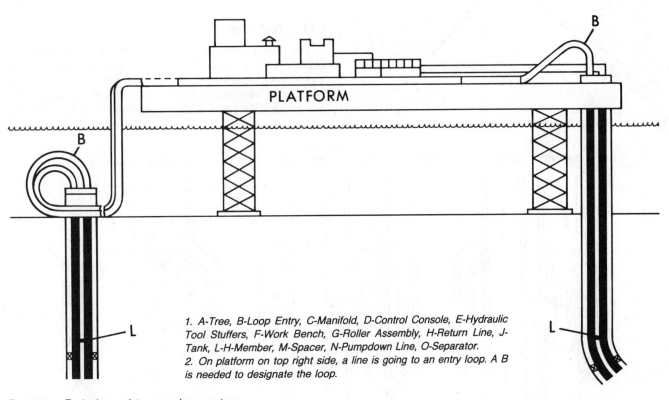

1. A-Tree, B-Loop Entry, C-Manifold, D-Control Console, E-Hydraulic Tool Stuffers, F-Work Bench, G-Roller Assembly, H-Return Line, J-Tank, L-H-Member, M-Spacer, N-Pumpdown Line, O-Separator.
2. On platform on top right side, a line is going to an entry loop. A B is needed to designate the loop.

Fig. 8.111 Typical complete pumpdown system

(3) A circulation path, typically an H-member between two strings of tubing

(4) A suitable conduit to carry the fluid, service tools, and flow controls, and

(5) A tool string to perform the necessary service

Additional items generally needed are lubricators (horizontal type recommended), 5-ft radius loop sections

to riser and at wellhead, fluid storage tanks, and TFL control manifold with instrumentation.

8.113 Operation

The circulation crossover between the two tubing strings is called the H-member. Fig. 8.113 depicts a

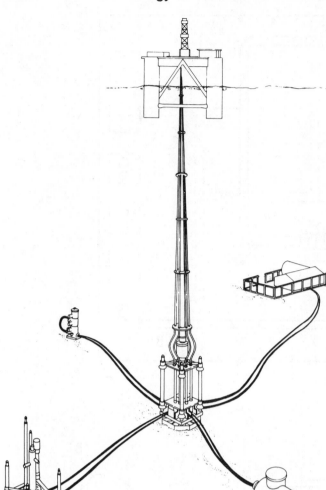

Fig. 8.112 Sub-sea pumpdown configuration

double bypass H-member installed in a two-zone completion. A zone separation tool is usually installed across the circulating port to prevent co-mingling of well fluids during production. A normally closed (dome-charged) circulation control valve is used in this illustration. Standing valves are installed in each tubing string below the H-member to prevent TFL operating pressures and fluids from contacting the formation and to direct circulation into the opposite string. The H-member normally includes the locking recesses and packing bores required for this equipment (Fig. 8.114A). An H-member is commonly run in conjunction with a dual packer (Fig. 8.114B).

To initiate circulation, production is stopped; the surface pump increases system pressure to overcome formation pressure, forcing the standing valve on seat; pressure is increased to overcome the dome charge of the circulation control valve, causing it to open; and well fluids are circulated out and replaced by a de-aerated fluid to provide a 'solid' hydraulic system. The tool string is placed in the lubricator and pumped into the well, then the service work is performed.

Formation gas must be kept out of the circulating fluid to prevent aeration. During pumpdown operations, an adjustable choke maintains the necessary backpressure by regulating fluid returns. This backpressure maintains the standing valves on seat and also serves to hold the circulation control valve open.

8.114 Tool string

The pumpdown tool string performs the downhole service work. Once the tool string is in the well, the pumpdown operator is able to pump it down the well bore or to reverse it back, monitoring its location at all times. The tool string normally consists of eight piston units (Fig. 8.115) with four of these to push the

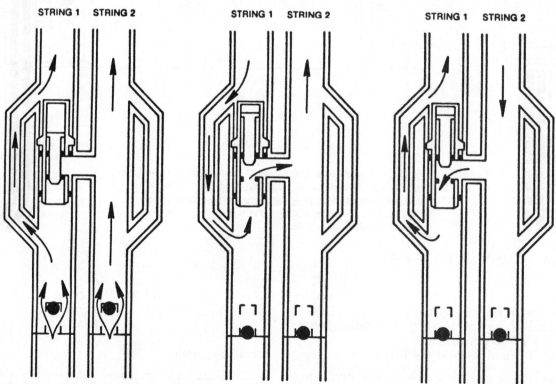

Fig. 8.113 Double by-pass H member

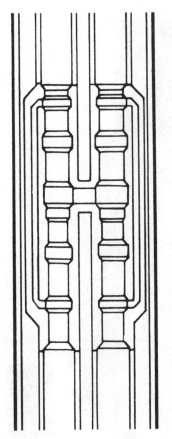

Fig. 8.114A   Otis H-member with double bypass

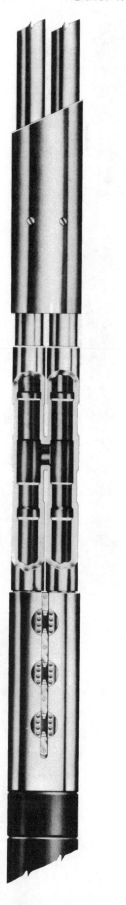

tools into the well and four to pull it out—plus the equipment needed to perform the service. Each piston unit consists of mandrel, knuckle joint, and a swab cup which holds pressure in one direction only. Each piston unit maintains approximately 500 lb/in.^{2} differential before fluid is bypassed to load the next piston unit. Thus, in 2 in. inside diameter tubing, four piston units facing the same direction are designed to maintain 2,000 lb/in^2 differential to exert some 6,300 lbs of force.

While being pumped into the well, the tool string travels at the same rate as the fluid. No fluid bypass occurs unless the tool string is slowed or stopped mechanically. The operator monitors tool string movement and location by using pump flow rate and total pumped volume indicators. If the tool string slows or stops, the amount of applied force can be monitored by determining the difference between pumping pressure and fluid return pressure (and allowing for some fluid friction).

Strip-chart recorders provide a history of each operation. The pumpdown operator uses the TFL control manifold to vary the pressure (and hence the force) applied at the tool string. Mechanical or hydraulic jars can be added to the tool string to generate greater forces or impact loads.

Typical TFL service tools are illustrated in Fig. 8.116. Notice the compact design with provisions for knuckle joints which permit a fully-assembled tool string to negotiate the loop sections at the risers and wellheads.

Fig. 8.114B   H-member with dual packer

Fig. 8.115   Pumpdown piston units (Courtesy of Otis Engineering)

Fig. 8.116   Typical service tools

8.115   Returning well to production

After all pumpdown operations have been completed, the well must be returned to production. Pump backpressure is decreased to permit the circulation control valve to close. All backpressure is then removed to allow formation pressure to begin flowing by overcoming the hydrostatic head of the circulating fluid. If the hydrostatic head is too great, a lighter fluid is circulated into the well. If reservoir pressure is insufficient for this, nitrogen can be injected at the surface and circulated into the well. If reservoir pressure is too low for either of these methods, gas lift valves may be considered.

8.116   Pumpdown gas lift equipment development

8.1161   Center-set system

The initial pumpdown-retrievable concentric gas lift valves were developed in the early 1960's. This "center set" system included a universal type locking profile incorporated in a pumpdown sliding sleeve assembly (Fig. 8.117), shifting tools to open and close the sleeves, and a concentric type gas lift valve with

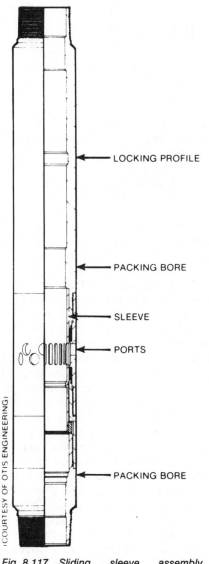

LOCKING PROFILE

PACKING BORE

SLEEVE

PORTS

(COURTESY OF OTIS ENGINEERING)

PACKING BORE

*Fig. 8.117 Sliding   sleeve   assembly
(opens upward)*

associated lock and packing mandrels. A special running tool was developed that allowed several valves to be set in their respective sleeve assemblies in one run. A shifter was used in conjunction with the running tool to open each sleeve as the gas lift valve was set in place.

A gas lift well would have the sliding sleeve assemblies spaced out according to standard gas lift practices (Fig. 8.118). Gas injected into the annulus would pass through the open sleeves and gas lift valves to lift production in both strings. To set the gas lift valves in place, the pumpdown operator would pump the loaded running tool past the lowest empty sleeve. Circulation was reversed, and as the tool passed through each sleeve assembly a shifter would open the sleeve, the running tool would acknowledge the universal profile, and a gas lift valve assembly would drop off and lock in place. To retrieve the valves the process was reversed, with the pulling tool retrieving valves and closing sleeves as it was pumped down the well. When the pulling tool was full of valves, it was pumped back to the surface, unloaded, and returned into the well to complete the process.

A variation of this center-set system has been used for dual string single zone completions where injecting gas into the annulus was not practical (Fig. 8.119). An H-member with a sliding sleeve and universal locking profile on one side (Fig. 8.1110) was used in place of the individual sliding sleeve assemblies. Retrievable casing flow concentric gas lift valves were set in these H-member sleeves. Gas injected into the string

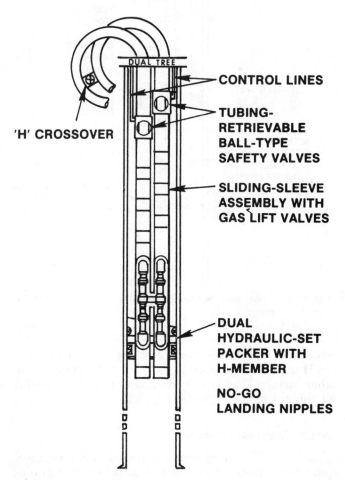

DUAL TREE

'H' CROSSOVER

CONTROL LINES

TUBING-
RETRIEVABLE
BALL-TYPE
SAFETY VALVES

SLIDING-SLEEVE
ASSEMBLY WITH
GAS LIFT VALVES

DUAL
HYDRAULIC-SET
PACKER WITH
H-MEMBER

NO-GO
LANDING NIPPLES

Fig. 8.118 Gas lift installation

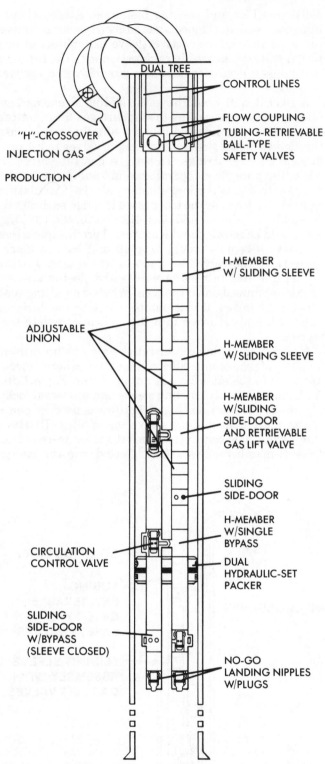

Fig. 8.119   *Single zone two strings with gas lift through second string*

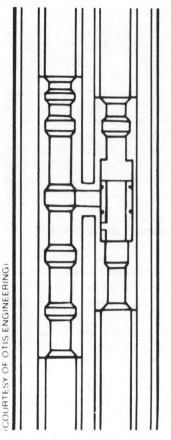

Fig. 8.1110   *Otis H-member with sliding side-door (Courtesy of Otis Engineering)*

containing the valves (service string) passed through the H-member crossover ports to lift production in the other string. This type system was especially useful for tubingless completions (Fig. 8.1111).

8.1162   Sidepocket mandrels

Early in 1970, tests were conducted to determine if a standard wireline kickover tool could be run by pump-down methods to locate, set, and retrieve gas lift valves in wireline side-pocket mandrels. A No-Go type nipple was run in conjunction with each mandrel so that a modified running tool could more readily locate the desired mandrel. Since the wireline tools were not designed to negotiate 5-ft radius loops, the system was restricted to wells having direct vertical access.

Testing was successful and several offshore platform wells were completed in this manner. However, experience with this system indicated the need for additional features. One such requirement was that additional piston units were needed in the tool string to span the void at the pocket. Otherwise, the tool string could become stranded due to excessive bypass around the pistons when in the sidepocket mandrel. Guide rails were added inside the mandrel so that standard pumpdown tools would not hang up on the pocket or gas lift valve. The primary problem was the vertical access requirement: tool strings were difficult to install in the lubricator, prongs and gate valves could become damaged, larger tools (3½ and 4½ in.) were very cumbersome, and subsea wells did not have a readily available vertical access unless a jack-up rig was available.

Recognizing these requirements and limitations, work began on a side-pocket system serviced by a kick-over tool that could be pumped around the wellhead loops. This system used 4-in. I.D. flowlines and loops so that modified wireline tools could be used to service 2-in. I.D. completions (Fig. 8.1112A). The 4-in. piston units transported the service tools through the flowline to the tree and around the wellhead loops. The 4-in. pistons were "parked" (Fig. 8.1112B) at the tree and the 2-in. tool string continued moving into the well

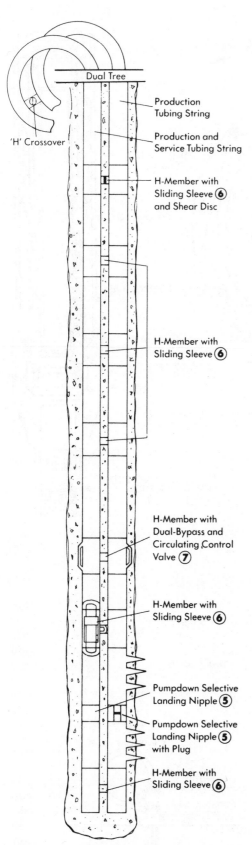

Fig. 8.1111   Two zone casingless completion H-member with sliding sleeve shear disc and double bypass

Production Tubing String

Production and Service Tubing String

H-Member with Sliding Sleeve ⑥ and Shear Disc

H-Member with Sliding Sleeve ⑥

H-Member with Dual-Bypass and Circulating Control Valve ⑦

H-Member with Sliding Sleeve ⑥

Pumpdown Selective Landing Nipple ⑤

Pumpdown Selective Landing Nipple ⑤ with Plug

H-Member with Sliding Sleeve ⑥

Dual Tree

'H' Crossover

to set or pull the gas lift valves. The kickover tool was basically a standard wireline tool fitted with swivels and knuckle joints. Standard wireline side-pocket mandrels were used. A typical completion incorporating this equipment is shown in Fig. 8.1113.

To facilitate servicing more to the downhole equipment with one common kickover tool and to provide for maximum flow area, the side pocket H-member was developed (Fig. 8.1114). The crossover port led from the pocket of the mandrel to the opposite string, and a modified circulation control valve (see Fig. 8.1115) was set in the pocket. Fig. 8.1116 illustrates a two-zone well using strictly side pocket equipment for TFL circulation and gas lift.

If sand problems are anticipated, a side-pocket H-member with a pressure-shear separation tool set in the pocket would be located just above the top packer. This is designed to provide a secondary circulation port for clean-up should sand block circulation through the lower H-member.

Rather than continue to modify wireline equipment, a side-pocket mandrel and kickover tool were developed specifically for pumpdown completions. The mandrel (Fig. 8.1117A) includes the guide rails, activator nipple, pocket for retrievable type gas lift valve, and orienting muleshoe. The pumpdown kickover tool (Fig. 8.1117B) consisted of an orientation assembly to ensure that the valve is properly aligned with the pocket, an activator assembly to kick the valve over, a pivot arm, and a carrier tray to cradle and protect the valve during transporting into or out of the well. This system has been used extensively in recent years and is presently available in 2, 2½, and 3 in. nominal sizes. Side-pocket H-members and associated equipment were developed for this system to provide the low-restriction, versatile completion shown in Fig. 8.1118.

8.117   Surface selective profile system

A further development was required to aid in locating the desired side-pocket mandrel with the kickover tool. When the mandrels were spaced closely to each other (within 300 ft), it became difficult to determine if the kickover tool was being operated in the proper mandrel. This was especially true in wells with long flowlines (over 8000 ft) or wells with the H-member located very deep.

To solve this problem, a surface-selective locator system incorporating ten different profiles was developed (Fig. 8.1119). Each locator mandrel had spring-biased keys with a coded profile which, when matched to its corresponding nipple profile, would prevent any further downward motion of the tool string. Similar wireline systems exist, but these normally have only four to seven "positions." A locator nipple was run just below each side-pocket mandrel so the operator could positively locate the desired pocket by running the corresponding locator keys on the tool string.

The surface selective system was also adapted into sliding sleeve assemblies to permit greater versatility in wells using retrievable concentric type gas lift valves. The shifter keys were basically locator keys with an extra square shoulder used to shift the sleeve closed. If the shifter mandrel were used in a gas lift assembly (Fig. 8.1120), the shifter keys would locate the desired sleeve. Downward force would be applied to open the sleeve and set the lock mandrel. Later, as the assembly was retrieved, the shifter keys were designed to automatically close the sleeve (thereby minimizing fluid loss to the annulus).

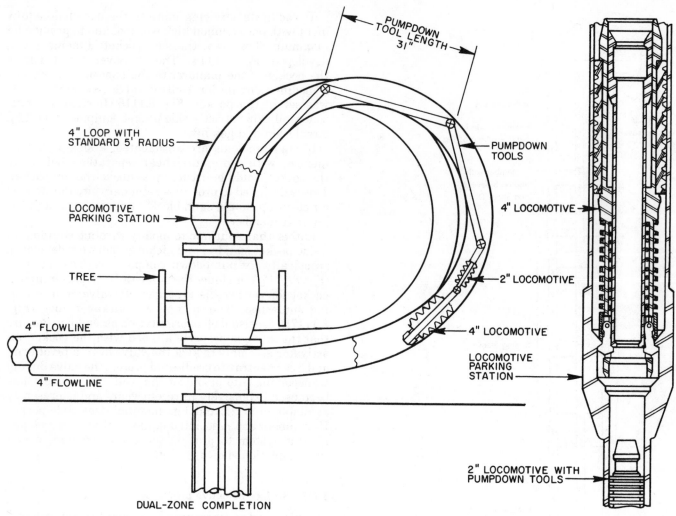

Fig. 8.1112A   Four-inch locomotive at surface with two-inch tools (Courtesy Camco, Inc.)

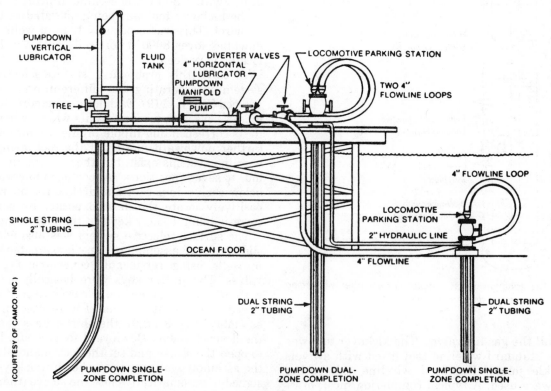

Fig. 8.1112B   Four-inch locomotive at surface with two-inch tools (Courtesy Camco, Inc.)

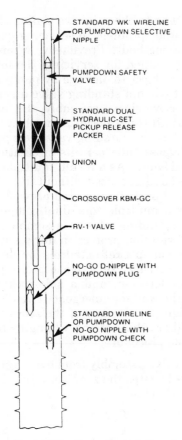

Fig. 8.1113   Typical completion, dual string of tubing

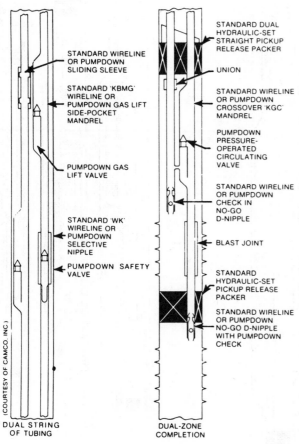

Fig. 8.1116   Typical dual-zone completion (upper and lower sections).

(COURTESY OF CAMCO, INC.)

DUAL STRING
OF TUBING

DUAL-ZONE
COMPLETION

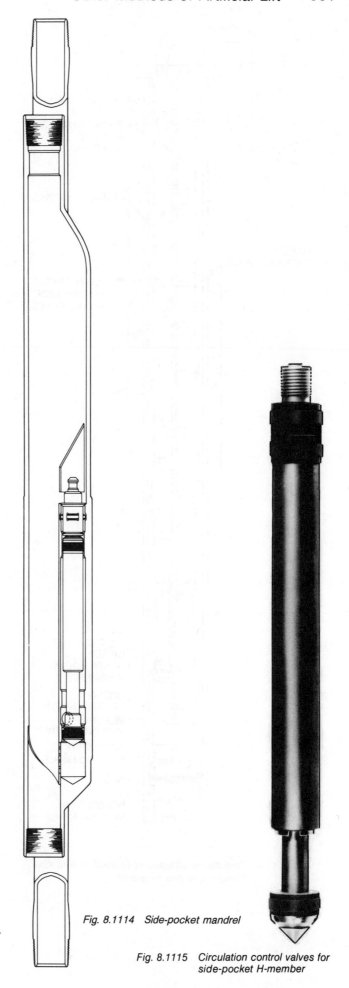

Fig. 8.1114   Side-pocket mandrel

Fig. 8.1115   Circulation control valves for
side-pocket H-member

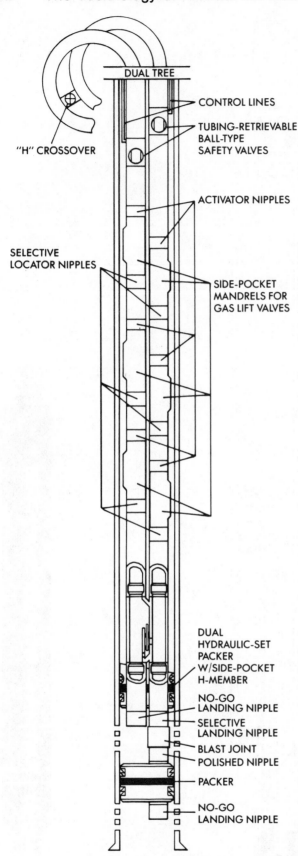

Fig. 8.1118 *Two zones side-pocket H-member with dual strings equipped for gas lift*

Adding a bypass tube to the selective sliding sleeve assembly solved another problem in servicing gas lift wells by pumpdown methods. Previously, when the standing valves were pulled for servicing, the formation was potentially subject to damage by pump pressures and fluids. But when a standing valve assembly was used in the selective sleeve assembly, the formation was protected at all times because the sleeve was designed to be closed as the standing valve was pulled (Fig. 8.1121). The bypass tube permits circulation to an H-member located below. As a result, multiple zone wells can be produced and gas lifted without the tubing perforators, plugs, and packoffs required previously.

Fig. 8.1122 shows a multiple zone dual completion incorporating bypass sliding sleeve assemblies for production and for formation protection during standing valve changeout, side-pocket mandrels for gas lifting production in both strings, selective locator nipples for positive mandrel location, and a side-pocket H-member set high in the well for emergency circulation. Notice that the design permits any of the zones to be produced through either tubing string. Changing zones also becomes a simple procedure:

(1) Pull standing valve assembly from bypass sleeve assembly. This isolates the zone since the sleeve is simultaneously closed.

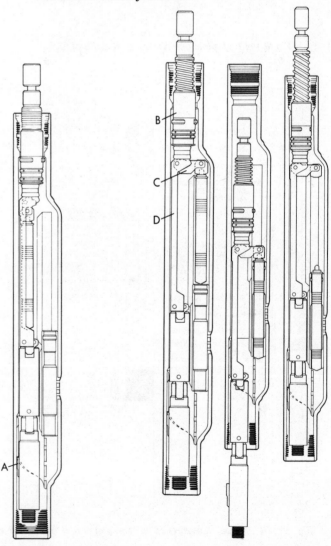

Fig. 8.1117A *Side pocket mandrel*   Fig. 8.1117B *Side pocket mandrel and kick-over tools*

(2) Change keys on shifter mandrel to correspond to desired new zone.

(3) Pump standing valve assembly into well. This opens the sleeve so the new zone can start producing after the tool string is returned to surface.

In recent years the number of profiles in this system has been expanded to 20 to provide even more versatility in completions. These profiles are also used for landing nipples, H-members (with or without sliding sleeves), safety valve nipples, and hanger plugs.

8.118   Jet pump

Initial testing, conducted during the early 1970's, indicated that TFL completions were well adapted to handle jet pumps. With the pump installed in a non-bypass H-member (Fig. 8.1123) one tubing string was used to supply power fluid from the surface while the other string was used for produced fluid. The power fluid entered into the top of the pump and produced fluid entered at the bottom. "Pumped" fluid exited from the side of the pump, passed through the H-member crossover port and continued up the production string. It was determined, however, that screens were required for both power fluid and production intakes to prevent pump plugging or deterioration (see Chapter 6 for details on the jet pump operation). Although successfully used in producing wells, the jet pump has not yet gained wide acceptance as an artificial lift method for pumpdown wells.

8.119   Summary

Pumpdown, also referred to as TFL (through-flow-line), is a method for downhole well servicing and is based upon an application of the principle of hydraulics. Although best suited for subsea completions, pumpdown servicing is also practical for certain platform and other vertical access wells.

A pumpdown service system requires a circulation path from the surface facility into the well as deep as possible and back to the surface facility. Five basic items are needed for any pumpdown system: pump, fluid, circulation path, suitable conduit, and tool string. Single or multiple zone wells can be serviced by pumpdown, with a pressure-operated circulation control valve used to prevent comingling during production. Standing valves are used to prevent TFL operating pressures and fluids from damaging the formation while also preventing formation gas from aerating the circulating fluid. Hydraulic pressures are converted to up or down forces by the tool string. The pumpdown service equipment, although similar to wireline equipment, is capable of negotiating 5-ft radius loops at the riser and wellhead.

Artificial lift equipment for pumpdown has been undergoing development since the early 1960's. Retrievable concentric gas lift valves have evolved from the "center set" system and universal profile to the current concentric valve and surface-selective sliding sleeve system. Servicing of side-pocket mandrels—originally pumpdown operation of vertical-access-only wireline equipment—is presently available in the form of a piggyback system with parking nipple or the designed-for-TFL kickover tool and side-pocket man-

Fig. 8.1119   Surface-selective locator system

drel system.

Side-pocket H-members can be used for primary or secondary circulation ports, and surface-selective locating equipment can be used for positive location of downhole equipment.

The jet pump is also practical for TFL completions. The dual strings provide the necessary conduits for power fluid and for production, while the H-member is well suited as a landing nipple for the pump.

Other methods of artificial lift for pumpdown wells and other uses for pumpdown well servicing are being developed.

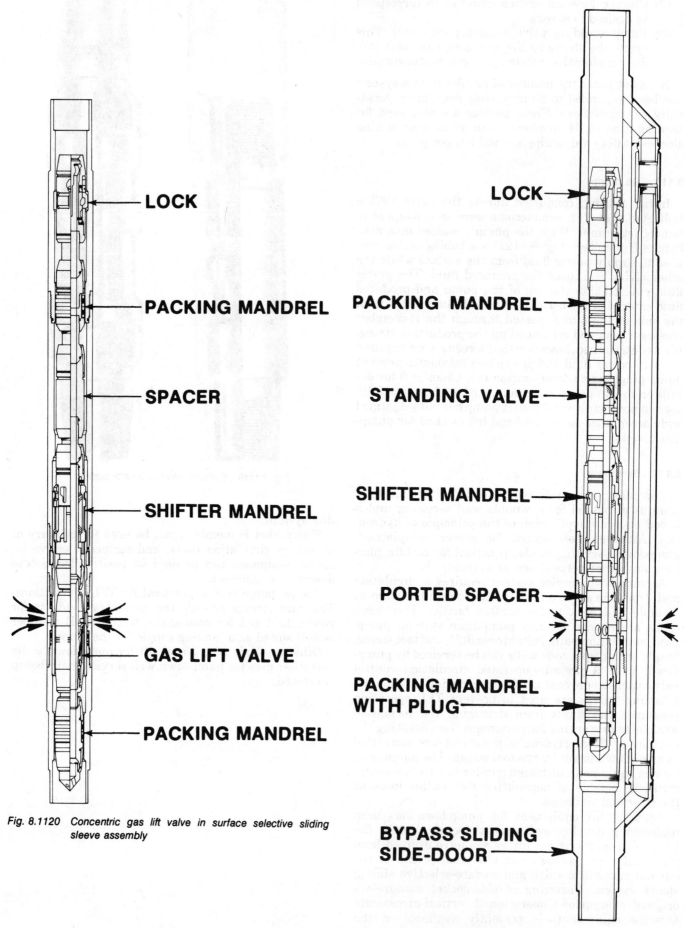

LOCK

PACKING MANDREL

SPACER

SHIFTER MANDREL

GAS LIFT VALVE

PACKING MANDREL

Fig. 8.1120 *Concentric gas lift valve in surface selective sliding sleeve assembly*

LOCK

PACKING MANDREL

STANDING VALVE

SHIFTER MANDREL

PORTED SPACER

PACKING MANDREL WITH PLUG

BYPASS SLIDING SIDE-DOOR

Fig. 8.1121 *"Select-20" standing valve*

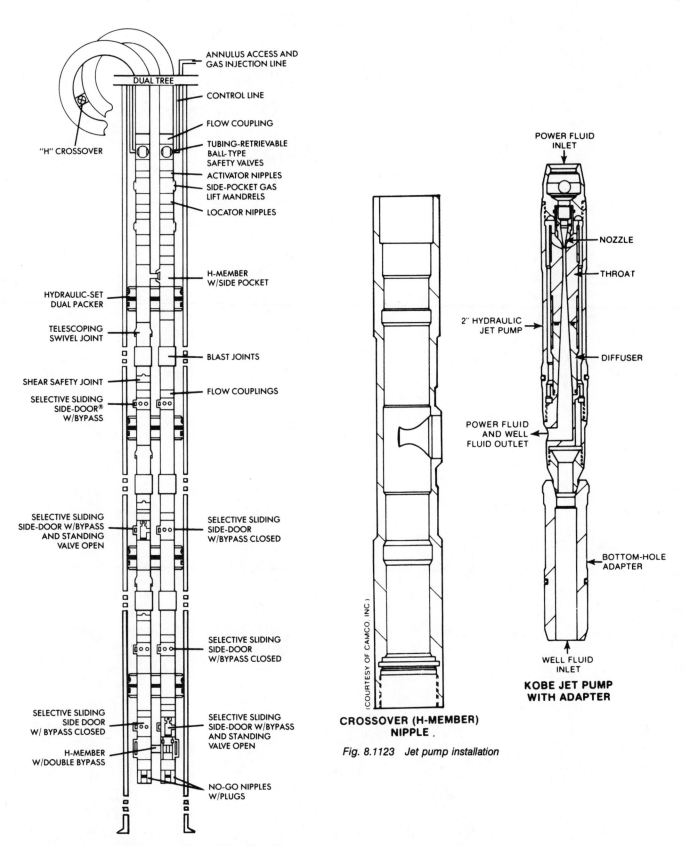

ANNULUS ACCESS AND
GAS INJECTION LINE

DUAL TREE

"H" CROSSOVER

CONTROL LINE

FLOW COUPLING

TUBING-RETRIEVABLE
BALL-TYPE
SAFETY VALVES

ACTIVATOR NIPPLES

SIDE-POCKET GAS
LIFT MANDRELS

LOCATOR NIPPLES

H-MEMBER
W/SIDE POCKET

HYDRAULIC-SET
DUAL PACKER

TELESCOPING
SWIVEL JOINT

BLAST JOINTS

SHEAR SAFETY JOINT

SELECTIVE SLIDING
SIDE-DOOR®
W/BYPASS

FLOW COUPLINGS

SELECTIVE SLIDING
SIDE-DOOR W/BYPASS
AND STANDING
VALVE OPEN

SELECTIVE SLIDING
SIDE-DOOR
W/BYPASS CLOSED

SELECTIVE SLIDING
SIDE-DOOR
W/BYPASS CLOSED

SELECTIVE SLIDING
SIDE DOOR
W/ BYPASS CLOSED

SELECTIVE SLIDING
SIDE-DOOR W/BYPASS
AND STANDING
VALVE OPEN

H-MEMBER
W/DOUBLE BYPASS

NO-GO NIPPLES
W/PLUGS

Fig. 8.1122 *Four zones two strings fully alternate
equipped for gas lift*

POWER FLUID
INLET

NOZZLE

THROAT

2" HYDRAULIC
JET PUMP

DIFFUSER

POWER FLUID
AND WELL
FLUID OUTLET

BOTTOM-HOLE
ADAPTER

WELL FLUID
INLET

**KOBE JET PUMP
WITH ADAPTER**

(COURTESY OF CAMCO, INC.)

**CROSSOVER (H-MEMBER)
NIPPLE** .

Fig. 8.1123 *Jet pump installation*

Chapter 9

Planning for and comparison of artificial lift systems

by Kermit E. Brown
Tom Doll
John T.Dewan

9.1 INTRODUCTION

Planning for artificial lift is important; design considerations must begin before a well or group of wells are drilled. To provide optimum production rates by artificial lift methods at some future date, sufficient tubular clearances should be provided for. Analysis of specific installation type should begin while the wells are still flowing. Application of a certain lift method depends upon whether a group, lease, or field will be lifted or if only an isolated well will require artificial lift.

The type of lift required is influenced by whether or not the wells are conventional or multiple completions. Multiple completions present problems: sufficient pipe clearances may not be provided for. Therefore, the choice of lift method may be determined not by optimum design or economic criteria, but by physical limitations.

Included in this is producing location. Offshore production platforms are limited in areal extent. With all other conditions equal, the best lift method onshore may not be practical on a platform with limited space. Here again, multiple completions and/or deviated holes may dictate the choice of lift.

Also to be considered is availability of a power source for the prime mover. In some areas natural gas may or may not be available, economical, or practical. Electricity has become more important due to availability and application to automation. Purchase cost, transportation, storage, and handling may become prohibitive when diesel or propane is required as the prime mover power source. One exception is an isolated well.

Artificial lift design depends upon producing conditions, also. Severe weather conditions affect the choice of lift. Extreme heat or cold, high winds, dust, or snow may limit the choice of lift. Corrosion is very important in the choice of lift methods. Sour crude, produced brine, oxygen and CO_2 corrosion, electrolysis—all affect artificial lift selection. Produced solids such as sand, salt, paraffin, and formation fines are to be included. Depth to the producing zone and hole deviation must be considered to provide adequate lift potential at future times. Gas-oil ratio and/or water-oil ratio considerations may limit types of lift applicable.

In other words, the total reservoir must be considered. Long-term producing objectives are dependant upon the reservoir's characteristics. The design and selection of artificial lift methods must reflect these objectives.

For instance, in a depletion reservoir, high initial production is anticipated. Artificial lift may not be required initially if wells are flowing. However, if a well may be put on lift after completion, the design considerations must be anticipated. Lower production with time is characteristic of depletion drive due to decreasing reservoir pressure and declining inflow. These characteristics will dictate the lift type most applicable.

In a water drive reservoir, increasing water cut with producing life is anticipated. This characteristic requires larger volume capability with time. Optimum lift must be based on future production volumes as well as present volumes.

In a gas cap expansion reservoir, changing gas-oil ratios with producing life affect the choice of lift method. Gas production decreases artificial lift capacity by restricting efficiency. The choice of lift must take into account the anticipated GOR or free gas during the life of the reservoir and whether or not the gas must be vented.

There may be more than one method of artificial lift applicable in a given well or group of wells. Each method may be classified from excellent to poor in accomplishing each separate objective. Depending upon economic considerations, the method of lift would be the one which satisfies or meets the majority of objectives. Two or more types of lift may be possible for the same well or group of wells (one to be used late in the producing life of the reservoir) with economic considerations the deciding factor.

9.2 METHODS OF ARTIFICIAL LIFT

9.21 Introduction

This leads to the basic methods of artificial lift in service. Based on world-wide installations the relative standing of lift methods is as follows:

(1) Sucker rod or beam pumping
(2) Gas lift
(3) Hydraulic pumping
(4) Electric submersible pumping
(5) Jet pumping
(6) Plunger or free piston lift
(7) Other methods

This may differ from field to field, state to state, and country to country.

9.22 Choice of artificial lift system

The choice between the four basic artificial lift systems—gas lift (GL), submersible electric pump (EP), hydraulic pump (HP), and rod pump (RP)—depends on many factors other than installation and operating costs of the equipment. Very important is well productivity, and some obvious choices can be made as illustrated for the following conditions:

>20,000 bl/d	EP or GL
2,000-10,000	any except RP
100-1,000	any
<100	any except EP

where: EP = Electric Pump
RP = Rod Pump
HP = Hydraulic Pump
GL = Gas Lift

Also important is reservoir pressure. Once it falls below about one-third of the hydrostatic pressure at the depth in question, continuous flow gas lift is questionable because the amount of gas required to lift the liquid becomes excessive. The submersible pumps can operate down to a few hundred PSI, and rod pumps and hydraulic pumps can operate essentially to zero pressure with gas venting perhaps becoming a necessity.

Depth can be an important limitation, as illustrated by the following conditions:

>12,000 ft	HP only
10,000-12,000 ft	any except EP (temperature limitation)
<8,000 ft	any

High deviation essentially rules out rod pumping and favors gas lift because of minimum equipment to run in the well.

High viscosity fluid can best be handled by gas lift or hydraulic pumping. Sand production is best handled by gas lift.

Then there are a number of conditions, such as availability of gas or electric power, availability of existing compressor facilities or pumping units, isolated or multiple wells, environmental restrictions, and offshore or onshore workover costs, which can swing the choice one way or another. Johnson listed in tabular form the common problems affecting lift selection and the relative ability of each lift method to handle the problem (Table 9.1).[1]

TABLE 9.1
COMMON PROBLEMS AFFECTING LIFT SELECTION

Problem	Type of lift			
	Rod pump	Hydraulic	Centrifugal	Gas lift
Sand	Fair	Fair	Fair	Excellent
Paraffin	Poor	Good	Good	Poor
High GOR	Fair	Fair	Fair	Excellent
Crooked hole	Poor	Good	Fair	Good
Corrosion	Good	Good	Fair	Fair
High volume	Poor	Good	Excellent	*Good
Depth	Fair	Excellent	Fair	*Good
Simple design	Yes	No	Yes	No
Casing size	Fair	Fair	Good	Good
Flexibility	Fair	**Excellent	Poor	Good
Scale	Good	Fair	Poor	Fair

*Higher volumes and depths depend on greater gas pressure and volume.
**Hydraulic piston pumping is rate limited but jet hydraulic pumping can handle the higher rates.

Secondary factors aside, one might consider that the artificial lift system appropriate for an average well is a function of the age of the well. In the initial artificial lift stages, reservoir pressure and GLR are generally high, so gas lift is favored. As both pressure and GLR decline, gas lift loses its advantage and electric pumping becomes more appropriate. Finally, at very low pressures and low productivity, rod pumping and hydraulic pumping are suited. However, if reservoir pressure is maintained by waterflooding, the electric pump and gas lift are both good.

Table 9.2 reproduces statistical data published on artificial lift in the U.S.[2] It is based on a 7% sampling of the 518,867 U.S. producing wells and shows that 92.7% of these wells are on artificial lift. Of these, 85% are being rod pumped, 11% gas lifted, 2% electric pumped, and 2% hydraulic pumped.

However, if 383,000 stripper wells making 10 b/d or less are eliminated (all of which would be rod pumped), the breakdown on the remaining is:

	No. of Wells	%
Rod pump	26,974	27
Gas lift	51,964	53
Electric pump	9,738	10
Hydraulic pump	9,470	10
Total	98,146	100%

Gas lift dominates offshore and in the Gulf Coast, while electric pump utilization is growing rapidly in waterflood areas of the mid-continent.

Fig. 9.1 shows artificial lift pumping systems applications depending upon the volumes of oil and water being pumped. It is composed of actual well data collected from wells in the U.S. Each point represents the volume of water and the corresponding volume of oil which is being produced by a given type of lift. When these two volumes are added, the result is the total fluid (b/d) which is being produced on each well by each type of lift. The large groupings of like forms of lift illustrate that low volume wells are most likely on rod lift, intermediate volume wells see the greatest application of hydraulic lift, and high volume wells

TABLE 9.2
U.S. WELLS ON ARTIFICIAL LIFT (YEAR 1977)

	No. of wells	%	Failure rate/yr.	Number of failures/yr.	Average cost to repair	Total cost/yr.	Percent well servicing cost
Subsurface rod pumps	409,974	85.2	.57	210,277	$1,078	$226,657,000	60
sucker rods			.44	164,118	729	119,665,000	79
Submersible pumps	9,738	2.0	.35	3,390	7,679	26,030,000	15
Hydraulic pumps	9,470	2.0	1.86	16,397	2,445	41,411,000	40
Gas lift	51,964	10.8	.21	11,490	4,153	47,713,000	78
Total	481,146	100.0					
Total U.S. Wells	518,867						

Source: *Petroleum Engineer,* July, 1977

are largely the domain of submersible electric lift (gas lift was not included in this survey).

Table 9.3 was prepared by Spears and Company from a survey conducted to determine running time and cause of failure of the various lift methods. The number of responses are not given and limits the value of such a survey.

9.3   FACTORS AFFECTING SELECTION OF ARTIFICIAL LIFT EQUIPMENT

9.31   Introduction

From the general discussion presented, there are

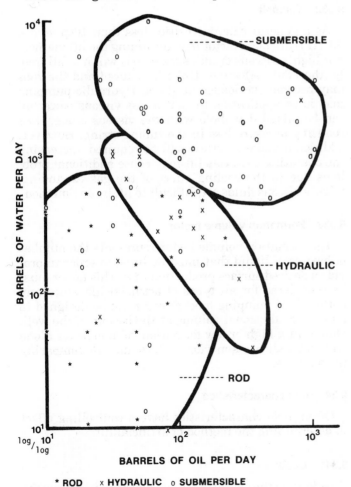

Fig. 9.1   Lift system applications by fluid mix. (Courtesy Robert B. Spears and Co.)

numerous factors that affect the selection of a particular method of artificial lift. These factors are producing characteristics, fluid properties, hole characteristics, reservoir characteristics, long range recovery plan, surface facilities, location, available power sources, operating problems, completion type, automation, operating personnel, service availability, and relative economics.

9.32   Producing characteristics

Producing characteristics of the well requiring artificial lift include inflow performance, liquid production rate, water cut, and gas-liquid ratio.

9.321   Inflow performance

The ability of a well to produce fluids (inflow performance) is a critical factor in the selection of artificial lift equipment. The inflow ability of wells was discussed in Chapter 1, Volume 1. The ability to produce fluids, at the present time—with or without stimulation—or at a future date, must be considered in the decision for the optimum method of lift.

9.322   Liquid production rate

The total liquid production rate to be produced is the controlling factor in the selection of the method of lift. Extremely high rates require electrical submersible pumping or continuous gas lift. For very low rates all other methods would be considered except continuous gas lift. Electrical submersibles are finding more application in low production operations, but less than 200 b/d wells are normally not considered for electrical pumping unless at a very shallow depth.

9.323   Water cut

The water cut directly influences the total production rate. For example, many wells must produce 2000 b/d or more of water in order to obtain 100 b/d or less of oil. High water cut affects the inflow performance due to the relative permeability effect. Water also results in additional pressure loss in the tubing due to greater density as compared to oil. High water cut reduces the total natural gas-liquid ratio. Therefore, high water cut dictates high volume artificial lift.

TABLE 9.3
AVERAGE RUNNING TIME OF ARTIFICIAL LIFT METHODS AND CAUSE OF FAILURE
(Courtesy Spears and Co.)

Problem	Rod Running time (months)	Range of responses (months)	Problem	Gas Lift Running time (months)	Range of responses (months)	Problem	Hydraulic (piston) Running time (months)	Range of responses (months)	Problem	Submersible electric Running time (months)	Range of responses (months)
Sand	8.9	4-18	Sand	12.0	*	Paraffin	1.5	*	Erosion	1.0	*
Unspec- ified	10.5	6-24	Unspec- ified	24.0	12-36	Sand	4.0	1-9	Gas lock	9.0	*
Rods	12.0	*	H_2S	36.0	*	Scale	5.0	4-6	CO_2	9.0	*
Corrosion	13.7	4-36				Corrosion	5.3	1-9	Sand	9.0	1-27
Paraffin	16.8	3-48				CO_2	6.0	*	Corrosion	9.1	1-18
Scale	18.0	3-48				H_2S	6.0	*	Pump	12.0	*
						Unspec- ified	8.0	1.5-12	Heat	12.5	12-13
									Abrasion	13.0	*
									Unspec- ified	14.8	3-48
									Rust	15.0	12-18
									H_2S	17.3	6-42
									Scale	24.0	9-54
									FeS	42.0	*

*Only one response given.

9.324 Gas-liquid ratio

The producing gas-oil or gas-liquid ratio influences the selection of artificial lift equipment and, in particular, the design of the lifting mechanism. As a rule of thumb, all methods of lift have reduced efficiency with increasing gas-liquid ratio. Up to 2000 SCF/bbl can be handled by most methods of lift. Sucker rod plunger pumps have reduced efficiency (approximately 40%) on the upper limit of this range. In the 2000 to 5000 SCF/bbl range, intermittent gas lift may be the most efficient since gas is vented between cycles. Other methods require that the gas be bypassed around the lifting mechanism. Up to a limit, hydraulics and electrical submersibles can pump the high GOR volume. If the gas cannot be vented, provision must be made to pump the gas. For hydraulic pumps, volume must be provided for the free gas which occupies space and, therefore, reduces efficiency.

Special design for free gas, such as tapered electrical pumps, also tends to reduce efficiency. Note that an excess of 2000 SCF/bbl is sufficient to reach the minimum flowing gradient possible in the vertical tubing string. In continuous gas lift, then, the addition of more gas (i.e., input gas) will actually increase the flowing bottom hole pressure and result in less efficient lift. Therefore, high gas-liquid ratio wells will be a problem for any method of lift unless proper venting is considered. Some of these wells may be low volume limited producers and may flow naturally.

9.33 Fluid properties

9.331 Introduction

Considering artificial lift methods, two fluid properties are more important with other properties seldom influential. More important are oil viscosity and oil volume factor with little influence from water viscosity, gas viscosity, water solubility, and surface tension properties. Note that fluid composition is important and is discussed in the section on corrosion.

9.332 Viscosity

As a general rule, viscosities less than 10cp (above 30° API) are not a factor in determining the lift method. For highly viscous crudes sucker rods will not fall freely; therefore, effective stroke is reduced and the rods may become overloaded or stuck. Hydraulic pumping may have application in that a less viscous power oil can be mixed downhole with the viscous crude, thus reducing pressure loss in the tubing string. However, additional make-up crude will be required, increasing the operating cost. Gas lift may cause additional problems due to the cooling effect of the gas expanding. Highly viscous fluids are difficult to lift by any method.

9.333 Formation volume factor

The formation volume factor represents the number of barrels of liquid that must be lifted in order to provide a desired surface production rate. This factor must be considered for all types of artificial lift, since any bottom hole pumping mechanism must be designed to pump the additional volume at the bottom of the well. Note that a high or low formation volume factor alone will not indicate superior performance in comparing lift methods.

9.34 Hole characteristics

Often, hole characteristics have a controlling effect in determining the method of artificial lift.

9.341 Depth

As long as the operating fluid level remains above the lift mechanism, depth may have little effect in determining the method of artificial lift. However, in any well requiring lift below 10,000 ft many types are inefficient. Sucker rod pumps are capable of lifting from greater depths; but horsepower, stroke length, rod size and stretch, load, and drag friction, limit design and they tend to be low in volumetric efficiency. Electrical submersible pumps require high fluid heads and at depth, with elevated bottom hole temperature, experience motor and/or cable failure. Intermittent gas lift is inefficient due to liquid fallback but is applicable with plunger assistance. Chamber lift designs provide lift capability at great depth by increasing slug volume. Hydraulic piston pumping is the most effective in very deep wells with lift from 18,000 ft in southern Louisiana as an example.

9.342   Size of tubulars

Casing sizes are determined on hole size in the preliminary stages in a drilling program. Many variables determine casing sizes for a particular well in a particular area, such as anticipated hole problems (abnormally pressured zones, sloughing/sticking, lost circulation, salt flow, etc.), pipe price and availability (increasingly important), and casing practices of other operators in the area.

Casing sizes limit the OD of the tubing that can be run. If tapered casing strings (intermediate casing, drilling liner or production liner) are used, tubing design is limited. Multiple completions are also restricted depending upon casing sizes. Ultra slim hole or directional completions limit tubing sizes. For extremely high rates, annular flow may be considered.

The method of artificial lift may be dictated by tubular size. In general, the smaller the tubulars, the lower the production rate that can be anticipated. Tubing sizes can be too large, causing excess fallback of liquids for low flow rates. Small tubulars cause higher friction losses that will reduce volumetric efficiency of gas lift, hydraulic and electric submersible pumping. Sucker rod coupling clearance in the tubing is also critical to prevent wear on the coupling and/or tubing and to provide access for fishing tools in the event of a rod part or break.

9.343   Completion type

The design of artificial lift equipment may also depend on whether the well is completed in an open hole or in a perforated interval. The main consideration is inflow performance. In the open hole, caving, sand entry, or hole fill-up may reduce inflow performance. In the perforated interval insufficient perforation density or plugged holes may reduce inflow performance.

9.344   Deviation

Highly deviated holes affect the selection of the type of artificial lift equipment. Reduced efficiency in this type of completion must be anticipated. For gas lift, hydraulic, and electrical submersible pumping the design is complicated. Since deviated holes do not create the same two-phase gradients that exist in a vertical well, special consideration must be given in gas lift design. In addition, all types of lift will require additional horsepower.

Sucker rod pumping in deviated holes has additional risks in that rod coupling and tubing wear with the reduced volumetric efficiency makes a marginal hookup. In some areas, nylon or plastic rod guides are used to reduce coupling or tubing wear. The additional cost in material and installation is offset by reduced friction of the centralized rods resulting in longer pumping runs. But when the rod guide is worn, it may split and fall on top of the pump, causing problems in pulling the pump. Guides that are bonded on the rod seem to offer longer life than the knock-on or force-on type, but are considerably more expensive. However, sucker rods are being used successfully in wells of 45° angles with minor problems where care is taken in using rod centralizers.

Sounding devices give problems in deviated holes.

9.35   Reservoir characteristics

9.351   Introduction

In planning for artificial lift methods, the reservoir characteristics must be considered to provide the most economical production rate with respect to the reservoir. Also considered is the long range recovery plan of the reservoir.

The type of reservoir will influence the inflow rate, gas-liquid ratio, and depth of lift.

9.352   Depletion drive reservoir

Production initially is due to displacement of oil and gas to the wellbore by fluid expansion. When the reservoir pressure is largely depleted, production may be due mainly to gravitational drainage with fluid being lifted to the surface by pumps. No aquifer or fluid injection aids the fluid expansion; therefore, recovery is low.

9.353   Water drive reservoir

Water influx from the aquifer and/or water injection in selected wells causes fluid displacement of the oil and gas to the wellbore. Recovery may be aided by gravitational drainage or capillary expulsion. Note that recovery is expected to be higher than depletion under natural fluid displacement and especially so under waterflood.

9.354   Gas cap expansion drive

In a reservoir in a two-phase state, the vapor phase is the gas cap and the liquid phase is the oil zone. Displacement of oil and gas to the wellbore is the result of gas cap expansion as production continues. Note that in this reservoir—having free gas, dissolved gas, oil in the oil zone, and recoverable gas cap liquids—recovery may be influenced by a combination of fluid expansion, fluid displacement, gravitational drainage, or capillary expulsion.

9.36   Long-range recovery plan

9.361   Introduction

The long range recovery plan of the reservoir must be evaluated. What may seem to be the most satisfactory artificial lift method at this time may or may not be economical or practical later in the life of the reservoir. Anticipating the reservoir recovery will require evaluation of the following.

9.362 Primary recovery

This plan involves normal production practices in monitoring rates, gas-liquid ratios, static pressures, water cut, etc.

9.363 Waterflood

The main consideration with the initiation of a waterflood is the increased water production to take into account at future dates. Monitoring changes in static pressures and inflow abilities is required.

9.364 Pressure maintenance, gas injection

The main consideration with a gas injection project is the increasing gas-liquid ratio in the reservoir. Monitoring gas-liquid ratio and changes in production rates is required.

9.365 Thermal recovery

Steam floods have helped in reservoirs with viscous crude production. Any type of additional heat may limit efficiency of artificial lift methods and may include increased corrosion activity to the lift equipment. Electrical submersible pumps are the most severely affected by high temperature, so reduced efficiency and reduced life must be accounted for.

9.366 Chemical recovery

Chemical injection as a method of increased recovery (especially in tertiary application) may become more economically feasible in the next few years. When planning for the type of artificial lift method, consider the effect the chemical will have on production rates and on the lift equipment itself.

9.37 Surface facilities

Surface facilities are important factors to be considered in the selection of artificial lift systems.

9.371 Surface flowlines

Three parameters affect the magnitude of influence surface flowlines may have on the selection of artificial lift equipment. Size, length and terrain, and surface choke influence wellhead backpressure and flow rate. Paraffin or salt plugging reduce the flowline ID and increase backpressure. High wellhead backpressure adversely affects efficiency of gas lift, hydraulic pumps and electrical submersible pumps to a higher degree than to sucker rod pumping. Also, additional backpressure is detrimental to hydraulic jet pumping, causing an increase in power oil pressure approximately three times any increase in wellhead pressure. Efficiency is reduced for all methods and planning must account for this.

In many fields, production may be improved by reducing wellhead backpressure. This is done by cleaning flowlines, replacing small or long flowlines, or removing a choke. Also, surface facilities (separator, etc.) should be at minimum possible operating pressure to maximize production. Of course, if capacity exceeds allowable, investment considerations govern sizing of flowlines rather than well productivity. But this is becoming a problem of the past in most places.

9.372 Separator pressure

In addition, separator pressure has a direct influence on wellhead pressure. Its effect is additive to the above parameters. Increased efficiency may be found by operating the separator at low pressure to reduce backpressure (see previous comments).

9.38 Location

Location may not normally be considered as a factor in the selection of an artificial lift method, but under certain conditions location may be very important. In all cases, if a power source for the prime mover (such as electricity or natural gas) is not readily available, then economics may become important. Operating costs may be excessive if fuel must be transported to the lease.

9.381 Offshore

Offshore platforms are limited in areal extent. Special considerations must be made to allow for lift equipment in a concentrated area. All types of lift have been used offshore. Beam pumping units require more space at the wellhead and may be affected by the corrosive environment offshore, and may cause vibrations. In most cases many wellbores are drilled from a single platform using directional techniques. Sucker rod pumps have reduced efficiency in deviated holes, and the prime mover takes up a large area on the platform. With time, sucker rod or tubing failures may require numerous pulling jobs which may prove uneconomical offshore. If electricity is available, the most applicable lift method offshore may be the electric submersible pump. Surface installations offshore may be less practical for gas lift and hydraulic pumping application.

9.382 Urban

Special consideration must be given to wells located in urban areas, especially safety, environmental, and pollution factors. Lift method choice may be determined by the aesthetic value alone—the ability of the equipment to be hidden from view or landscaped to blend in with the surroundings.

9.383 Spacing considerations

Whether the location is for various wells in a group or for a single isolated well, the type of lift method will be influenced. For isolated wells or very widely-spaced wells, sucker rod pumping installations may prove most practical and economical. Maintenance operations may not be available on short notice at isolated locations for complicated surface installations, such as hydraulic or electrical submersible pumps. Also to be considered are long flowlines required for widely spaced wells.

9.39   Power sources available

Availability of low cost power sources for artificial lift prime movers is required. Usually, electricity or natural gas is used. Other fuels may be required as conditions warrant.

9.391   Electricity

In many areas, electricity is readily available at economical rates. Beam pumping equipment, hydraulic pump systems, and, of course, electrical submersible pumps are readily adaptable to electrical power application. Consideration must be given in the choice of lift method if electrical power is used.

9.392   Natural gas

Natural gas availability will determine if it is a practical power source. This fuel can be adapted to internal combustion engines. At isolated locations wellhead gas may supply the prime mover. Note that for high water producing wells, produced gas may not be available in sufficient quantities to power the equipment.

9.393   Other power sources

Diesel, propane, and perhaps solar energy in the future may be used as fuel. At isolated wellsites, the cost of storage, transportation, and potential theft may prove prohibitive. Cost must also be considered to determine if diesel or propane is a competitive alternate fuel source. Solar energy panels are in use in some areas, primarily to charge batteries. Application of solar-powered equipment may become competitive in the future.

9.310   Operating problems

9.3101   Introduction

Common operating problems that become difficult for some lift methods include sand, paraffin, scale, corrosion, emulsions, downhole temperature, and surface climate.

9.3102   Sand

Production of abrasives such as sand causes erosion problems for all types of artificial lift. Close tolerances are required for efficiency in downhole rod pumps, electric submersibles, and hydraulic pumps. Gas lift is the only lift method that does not require that the sand-laden fluid pass through the lifting mechanism. Sand fill-up on top of a bottomhole pump may cause retrieval problems

9.3103   Paraffin

Accumulation of paraffin in the upper portions of the tubing string, wellhead, or flowline will cause back-

pressure that will reduce efficiency. Removal or prevention is required. Sucker rod pumping has an advantage over other lift methods since the rods provide continuous scraping action. Scrapers or guides may aid paraffin removal. High temperature fluids and inhibitors can be immediately circulated in a hydraulic system.

9.3104   Scale

Deposition of scale will reduce the ID of tubing and threfore decrease efficiency. Gas lift may aggravate scale deposition. Prevention by chemical additives may provide longer pump life and maintain fullbore tubing.

9.3105   Corrosion

Downhole corrosion may be caused by electrolysis between different metal types, H_2S or CO_2 content in the produced fluid, highly saline or saturated brine water, or oxygenation of metals. H_2S embrittlement is a major problem and will accelerate sucker rod failures if the rods are excessively loaded. Gas lifting with corrosive gas may prove uneconomical.

9.3106   Emulsions

Anticipating emulsion problems when planning and selecting artificial lift methods is difficult. Emulsions tend to cause abnormally high pressure losses in the tubing. This backpressure effect will require additional horsepower and reduces the efficiency of any type of lift.

9.3107   Bottom hole temperatures

Very high bottom hole temperatures will reduce the operating life of some types of lift equipment. The electrical submersible pump motor and cable is drastically affected. Precautions must be taken when BHT exceeds 300°F. High cost metallurgy and seals will be required in all equipment including packers, tubing, wellhead equipment, and downhole lift equipment.

9.3108   Surface climate

Extremes in surface climatic conditions may influence the selection of lift equipment. Very hot climates cause overheating problems with surface equipment and special cooling facilities must be provided. Very cold climates cause freezing problems for fuels and embrittlement of electrical connections; insulation and heating must be provided. Also, many areas experience high winds that cause surface damage, and dust or snow may cause operational problems.

9.311   Multiple completions

Dual or triple completions are more difficult to design for artificial lift than single completions. Clearances in the tubing and on the surface restrict applications of most artificial lift methods.

9.312   Automation

A well or group of wells that will be placed on automation will normally determine that electrical

OPTIMUM INVESTMENT COST RANGE ANALYSIS
(REGIONS OF LOWEST INVESTMENT)

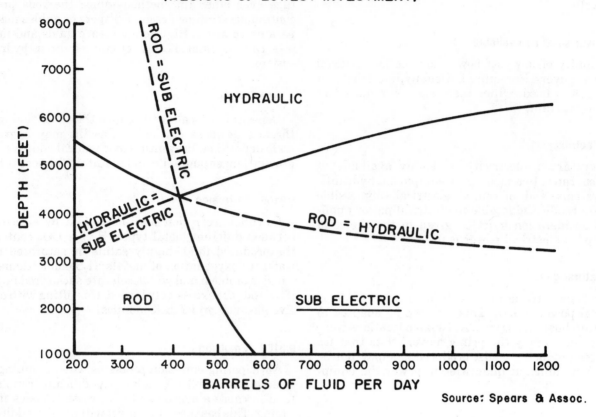

Source: Spears & Assoc.

INVESTMENT COST VS. DEPTH

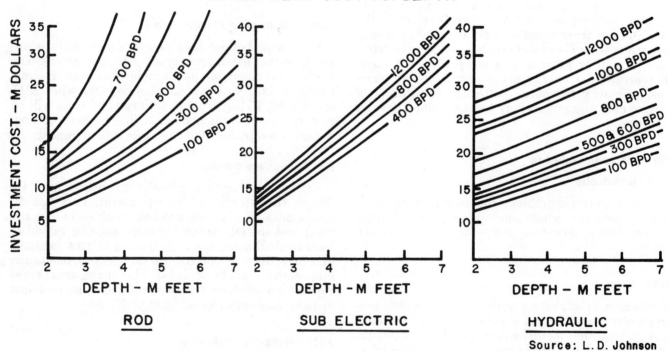

Source: L.D. Johnson

Fig. 9.2

power equipment (downhole or surface) should be considered first.

9.313 Operating personnel

Technical ability of the operating field personnel may influence artificial lift selections. Sucker rod pumping may offer less-troublesome operation for field personnel than the other types of lift. Engineering supervision is required for gas lift with special operating instruction required for hydraulic and electric submersible pumping.

9.314 Services available

A dominant influence on the selection of an artificial lift method is the availability of competent service personnel, replacement parts, and service rigs or equipment. Competent service personnel to check, analyze, and repair lift equipment are mandatory for economic operations. In some areas, the choice of a less desirable type of lift has been made solely on the availability of service personnel. Replacement parts' availability is equally as important as service personnel. If lift equipment cannot be readily serviced and repaired, then an alternate lift method may be chosen. The shortage of replacement parts in past years indicates how critical this factor is.

Also to be considered is the access to service rigs or wireline units. Some types of lift require pulling tubing (or rods) and pump for service and replacement, while others can be serviced by wire line. Hydraulic pumps may be circulated to the surface without requiring a rig or wireline unit. Scale or paraffin build-up in the tubing or corrosion at the pump may prevent hydraulic pumps from being circulated to the surface.

9.315 Equipment

The following factors should be evaluated as accurately as possible in making the selection of equipment:
  (1) Availability of surplus equipment
  (2) Installation and operating costs for various artificial lift systems
  (3) Equipment life
  (4) Availability of service and parts
  (5) Flexibility of lift system to meet changing producing conditions
  (6) Will equipment produce wells to depletion
  (7) Cost to supplement inadequate equipment
  (8) Salvage value
  (9) Safety
  (10) Availability of new equipment

9.4 RELATIVE ECONOMICS

9.41 Introduction

Six economic factors represent the most important parameters in selection of artificial lift equipment:
  (1) Initial capital investment
  (2) Monthly operating expense/income cost indicators
  (3) Equipment life
  (4) Number of wells to be lifted
  (5) Surplus equipment availability
  (6) Well life

Each of the artificial lift systems has economic and operating limitations that eliminate it from consideration under certain operating conditions. Most production operators have some idea where these limits are but have little or no engineering evidence to support their feelings. There is good reason for this: for nearly every statement made that a given artificial lift system will not work under certain conditions, a successful application under those conditions can be found.

Certainly some economic and operating guidelines are fairly well-defined by experience, especially within a given set of well-defined operating conditions. The question is, when do they apply to a slightly different set of operating conditions?

Some types of lift equipment, depending upon the type of installation, can have higher initial costs than others. Gas lift can have a high initial cost for a one or two-well system where a compressor must be installed. For a large number of wells, gas lift may become more economical. Hydraulic pumping becomes less costly where several wells can be operated from a central system. In many cases a surplus of one type of lift equipment dictates the method because of economics or availability.

At today's prices gas lift can be expensive if there is a gas market. If no market, as in remote fields, gas lift beam and hydraulic are competitive; operating expense and initial investment might favor gas lift (if you did not have to compress). Beam, hydraulic, and gas lift all have advantages over submersible pumps for remotely-located small fields where building electrical systems would be too expensive or where there is no electricity available.

The economics associated with initial equipment costs are fairly academic if the principle of only comparing similar ability systems is adhered to. This means that for each artificial lift system, the least expensive equipment combination that will do the job is the one considered.

Operating costs or lifting costs for an artificial lift system are considered the ultimate measures of a system's capability by some and the ultimate "hoax" by others. Certainly, these two opinions are extremes. The value of operating costs ($ per well per month) or lifting costs ($ per gross barrel produced) is their importance as indicators to determine the operating efficiency and field operations effectiveness in optimizing system capabilities.

If these indicators become "absolute" rather than comparative in their meaning, they probably will lose their value. Spears and Company prepared Fig. 9.2 showing the optimum cost range analysis for the three principal pumping systems. By noting the depth and corresponding fluid rate, we can find the optimum system.

Fig. 9.3 shows the estimated lifting costs (in 1977) for

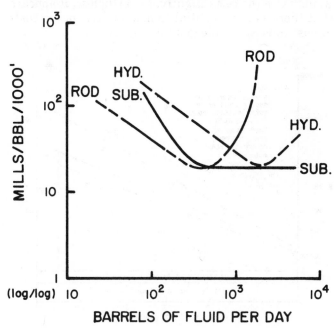

Fig. 9.3

the three pumping systems as a function of volume produced. This information was obtained from a survey of numerous companies.

9.42   Comparative artificial lift costs for gas lift electrical submersible pumping and hydraulic pumping

9.421   Gas lift vs. electrical pumping and hydraulic pumping

This summary compares costs of gas lift and submersible electric pump systems for the year of 1977 for an "average" Gulf Coast offshore well:

depth	6000 ft
casing	5½ in.
flowline	2500 ft
reservoir pressure	2800 PSI
productivity index	2 b/d/psi
GOR	100 scf/B
WOR	1

Tubing size, surface, and casing pressures are optimized for gas lift production, i.e., 1800 bl/d.
Figs. 9.4 and 9.5 summarize the installation and operating costs. A comparison shows:
—Gas lift is more expensive to install but less expensive to maintain.
—Leakage in the gas lift system and life of the electric pump are important factors in operating costs.
—Based on rates of $1.75/MCF for gas, 2.35¢/kwh for electric power; and 2% gas lift leakage and 1-year electric pump life, lifting costs per barrel are:

Gas lift	3.7¢
Electric pump	3.2¢

—Projecting to 1982, with rates of $3.25/MCF for gas and 3.5¢/kwh for power, 0.5% gas lift leakage and 3-year pump life, lifting costs per barrel would be:

Gas lift	4.9¢
Electric pump	3.1¢

This neglects inflation on labor and material which would raise the costs slightly. Nevertheless, it appears that there is the potential to hold electric pump costs almost constant while gas lift costs rise.

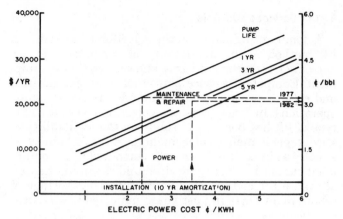

Fig. 9.5   Electrical submersible pumping costs

9.4211   Typical calculation to compare gas lift and electrical pumping

Assume a 6,000 ft well with 5½ in. casing and 3 in. ID surface flowline 2500 ft long. Reservoir pressure is 2800 PSI, productivity index 2.0 b/d/PSI, WOR 1, and GOR 100 SCF/STB. (These cost figures apply to the year 1977, but the cost analysis procedure is applicable at any time.)
(a) Gas Lift Costs
To determine the maximum gas lift production rate use is made of generalized gas lift analysis curves. Comparison of the various cases shows that maximum production is approached for the conditions of 2.5 in. ID tubing, 1200 PSI casing pressure and 120 PSI trap pressure. Production rates as a function of GLR are:

GLR (scf/B)	bl/d
200	1550
300	1800
400	1900
600	1950
1000	1900
1500	1800

The optimum GLR is approximately 300. Costs for this case are computed as follows:

Gas lift compression requirements are (300 − 50) × 1800 or 450 MCFD at 1300 PSI (assuming 100 PSI drop in gas injection lines).

Gas compression ratio = 1315/135 = 9.74 which requires two stages of 3.13/stage.

BHP required = 70/MMCFD/stage (n = 1.3) or a total of 70 × 2 × 0.45 × 1.1 (safety factor) = 70 hp.

Installation:
Assume the compressor is shared equally by 8 wells, with total requirement of 560 hp. Cost of a skid-mounted packaged unit, including engine, scrubber, cooler, controls, is $180,000 (Gardner-Denver MLE 650 HP) or about $275/hp; cost of concrete base, shed and installation is $20,000 (est).
Total cost of compressor installation per well =
$$\frac{200,000}{8} = \$25,000$$

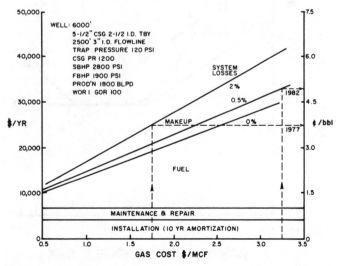

Fig. 9.4   Gas lift costs

Cost of 2500 ft of 2 in. ID gas distribution line @ $2.5/ft = 6,250
Cost of 9 downhole valves, with latches and sidepocket mandrels, and one flow control valve = 9,450

Total installation cost/well $40,700

Maintenance:
Compressor maintenance cost @$25/hp/yr. = 70 × 25 = $ 1,750/yr
Pulling and replacement of valves =   900/yr

Total maintenance cost $ 2,650/yr

Fuel and makeup gas:
Cost of fuel @ 12 SCF/hr/BHP and $1/MCF is 12 × 24 × 365 × 70 × 1/1000 = $ 7,350/yr
Cost of system gas losses @ 2% of rotation rate is 0.02 × 450 × 365 × 1 = 3,280/yr

Total fuel and gas makeup cost $10,630/yr

Lifting cost per barrel, based on 10-year amortization of installation costs, is 100 (4070 + 2650 + 10,630)/(1800 × 365) = 2.64¢

The important variables in this situation are the cost of gas and the system gas losses. Calculations for gas costs from $0.5 to $3.5MCF and for system losses from 0% to 2% are shown on Fig. 9.4. It is apparent that, above $0.5/MCF, fuel and system losses dominate over maintenance and installation costs.

For a more complete discussion of system losses, see Section 9.423. In that section, it is shown that if the system were "tight," gas lost to the atmosphere should amount to only about 1,100 SCFD or 0.25% of the rotating rate. Consequently, the system illustrated is quite "leaky," perhaps representative of some of the older gas lift systems. The loss in new systems should be much less.

(b) Submersible Electric Pump Costs
Bottomhole pressure at 1800 b/d and PI of 2 = 2800 − 900 = 1900 PSI
Static fluid head to be lifted =
$6000 - \dfrac{1900}{0.43}$ = 1590 ft
Pump setting depth 3000 ft
Friction loss in 3000 ft of 2½ in. ID tubing = 100 ft
Friction loss in 2500 ft of 3 in. ID flowline = 30 ft
Trap pressure, feet of head = 120/0.43 = 280 ft

Total dynamic head   2000 ft
Installation:
Costs are as follows:

Pump, 86 stages: discharge head	$	81
pump		2,414
intake		270
Motor, 50 hp: protector		1,091
motor (980 v, 32.5 A)		7,512
Adapters (2)		236
Cable, 3100 ft		7,285

Cable extension, 3kv, #6, bronze, hi temp, 30 ft	385
Switchboard, 1500 v, electro-mechanical	2,798
Junction box	115
Transformers (3), 50 kva	2,748
Wellhead	373
Clamps	413
Check valve	130
Bleeder valve	130
Installation costs (est.)	1,000
Total installed cost	$26,900

Maintenance:
Pulling and running once/year (2 days @ $1000/day) $ 2,000
Replacement of pump, intake, motor and protector once/year, @½ of original cost   5,645
Replacement of cable and cable extension every 5 years   1,535

Total maintenance cost $ 9,180/yr

Power:
Power costs @ 2.0¢/kwh are
1.73 (3 phase) × 1030 V × 32.5 A × 0.83 pf × 24 × 365 × 0.020/1000 = 8,420/yr
Lifting cost per barrel is 100 (2690 + 9180 + 8420)/1800 × 365 = 3.10¢

The important variables in this case are the life of the pump and the cost of electric power. Pump life can vary from weeks to years, depending primarily on abrasiveness of the fluid and temperature of the well. Best estimates from various sources are one year in Gulf Coast sandy conditions to 3 years in Mid-Continent carbonate conditions. Table 9.2 indicates an average life of 2.8 years, but this is probably biased toward the most favorable conditions.

Power costs can vary from about 2 to 4.5¢/kwh, depending on the area. For example, industrial rates are currently 2.35¢/kwh in the Houston vicinity, 3.5¢ in the Corpus Christi area, about 4.2¢ in the Northeast, and 2.7¢ in Europe (1977-1978).

Calculations for pump life from 1 to 5 years and power costs from 1 to 6¢/kwh are shown in Fig. 9.5. Power costs are dominant, but maintenance can be almost as important if pump life is short.

(c) Comparison of systems
From the prior analysis, along with Figs. 9.4 and 9.5, some basic observations can be made:
(1) Gas lift is significantly more expensive to install than electric pumping. Cost of the compressor is the big item. If an existing compressor can be used, gas lift installations can be less expensive.
(2) Maintenance and repair costs are substantially less with gas lift than with current electric pumps of life of 1-3 years. If pumps could be made to last 5 years, costs would be about equal.
(3) Losses in the gas lift stream can be significant and should be kept less than 0.5% of the rotative rate.
(4) For 1977 conditions, gas at $1.75/MCF, power at 2.35¢/kwh, 1 year electric pump life, 2%

gas lift system losses, lifting costs per barrel are:

| Gas lift | 3.7¢ |
| Electric pump | 3.2¢ |

(d) Future Projections

Looking toward the future, for land wells in the U.S.:

(1) Installation costs of both systems will continue to rise at the inflation rate. This will not favor one system over the other.

(2) Maintenance and repair costs for gas lift will rise at the inflation rate. However, reductions could occur in electric pump costs if engineering, installation, and operating improvements could significantly lengthen pump life.

(3) Fuel costs will rise at a greater-than-inflation rate and become a greater determining factor in the choice of a system. Gas costs will probably rise faster than electric power costs because gas supplies are dwindling and power companies are beginning to switch to coal. A reasonable projection for 1982 might be $3.25/MCF for gas and 3.5¢/kwh for electricity. Assuming improvements increase electric pump life to 3 years and reduce gas lift leakage to 0.5%, lifting costs per barrel would be:

| Gas lift | 4.9¢ |
| Electric pump | 3.1¢ |

Inflation on labor and materials would raise these costs slightly. Nevertheless, it appears electric pump costs could be held almost constant while gas lift costs will rise.

Although these comparisons appear to favor electric pumping, any adverse conditions causing the pulling of the tubing to repair the electric pump immediately places it in an unfavorable position.

9.422   Gas lift and electrical submersible vs. hydraulic

This section compares costs of hydraulic pumping with those previously computed for gas lift and electric submersible systems. The same "average" well is considered:

depth	6000 ft
casing	5½ in.
ID tubing	2½ in.
flowline	2500 ft
reservoir pressure	2800 PSI
productivity index	2 b/d/psi
flow rate	1800 b/d
GOR	100
WOR	1

The two basic types of hydraulic pumps, piston and jet, are compared. The jet type has revitalized hydraulic pumping. It is simple, easily repaired and can handle sand and gas. However, it is less efficient than the piston unit and normally cannot pump a well below about 500 PSI flowing pressure.

Figs. 9.6 through 9.9 summarize lifting costs for the two types of pumps and for both electric and gas engine

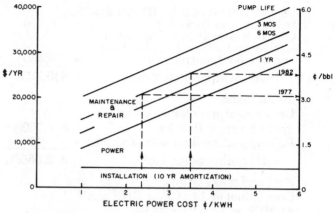
Fig. 9.6   Hydraulic piston pumping costs (electric motor drive)

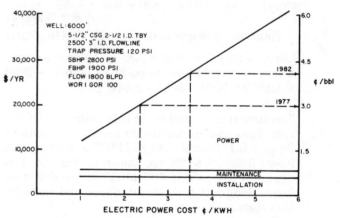
Fig. 9.7   Hydraulic jet pumping costs (electric motor drive)

drive of the power fluid circulating system. For a given type of circulating drive, little difference exists cost-wise between the two types of pumps. However, electric drive has lifting costs 30% less than with the gas engine.

Figs. 9.4 and 9.5 show lifting costs for gas lift and electric submersible. Comparing all systems, for power and gas rates in the Houston area, lifting costs per barrel are:

	1977	1982 (Projected)
Gas lift	3.7¢	4.9¢
Electric submersible	3.2¢	3.2¢
Hydraulic jet (electric drive)	3.0¢	4.1¢

As previously discussed, the electric submersible appears to have a future cost advantage where it is really applicable, i.e., wells less than 8000 ft, low sand and gas production, available electric power, and low pulling costs.

For wells below 8000 ft, the hydraulic jet appears to be a direct competitor to gas lift in both cost and ease of maintenance. In fact, in areas where gas is becoming scarce, it is the only viable artificial lift system unless a reliable and cost-competitive inert gas source can be developed for gas lift.

9.4221   Types of hydraulic pumps

The two basic types of hydraulic pumping, piston and jet, have been discussed in Chapters 5 and 6.

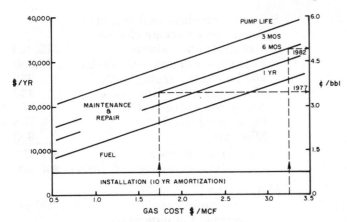

Fig. 9.8   Hydraulic piston pumping costs (gas engine drive)

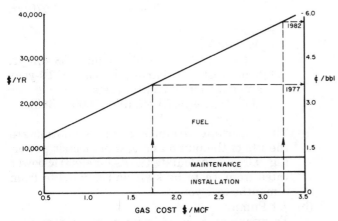

Fig. 9.9   Hydraulic jet pumping costs (gas engine drive)

In summary with the piston type, power fluid pumped down the well causes an engine piston in the pump to stroke up and down. The engine piston mechanically drives a pump piston which intakes reservoir fluid and pumps it up the well. Standard units are available with pumping rates up to 5000 b/d and lifts up to 15,000 ft. Stroke lengths are 1 to 8 ft and stroke rates up to 150/min. Overall pump lengths are 8 to 22 ft.

The jet pump is a revolutionary type introduced in the early 1970's. It is much simpler and shorter and has no moving parts. Power fluid pumped down the well is forced through a nozzle where the high speed and low pressure cause entrainment of reservoir fluid. The mixed fluid passes through an expander which reduces velocity and builds up sufficient pressure to push the fluid to the surface.

Piston pumps are used primarily in hard rock waterflood areas where both sand and gas production are low. The pumps use metal piston rings and other seals which are quickly eroded by sand. When this happens, the unit must be retrieved from the well and an exchange unit installed, the original being returned to the manufacturer for repair. Life of the pump in hard rock service is reported to be about 9 months. Typical exhange and repair charge is about 40% of the cost of a new pump.

Introduction of the jet unit increased the life of the pump and broadened its use to sand and gas production regions, i.e., the Gulf Coast of the United States. The only effect of sand is to erode the nozzle and intake throat. When this happens, the unit can be pumped

back to the surface, the two parts replaced at the wellsite, and the unit reinstalled, all in the space of an hour or less. Repair costs are, therefore, quite low. No good figures on jet pump life have been determined as yet. A guess is two years in hard-rock country and one year in soft-rock regions.

Although the jet pump has the above advantages plus that of being pumpable through subsea wellheads, it is only half as efficient as the piston pump, i.e., 32% vs. 65%. Consequently, power costs are higher. Also, it cannot pump wells below about 500 PSI bottomhole pressure (flowing) or cavitation in the nozzle occurs. The piston pump can pump down to perhaps 100 PSI depending upon gas volumes to be pumped.

The simplest and most common arrangement of the hydraulic pump in a well is the free-pump open-power system. This requires a normal tubing, casing, and packer. The pump is installed by pumping it down the tubing into a seat at the bottom ("free-pump"). In operation, power fluid is pumped down the tubing and returned up the annulus along with the reservoir fluid. The pump can be retrieved by reversing the power fluid flow.

By nature, the jet pump is an open-power system where the power fluid mixes with reservoir fluid. For the piston pump, closed-power systems are also available where the power fluid is returned to the surface without mixing with reservoir fluid. This, of course, requires at least one extra string of tubing in the hole which is a disadvantage, although surface equipment is simpler.

Power fluid may be oil or water; the former is preferred but the latter is acceptable with appropriate corrosion inhibition. A mixture of oil and water is undesirable because of unpredictable flow characteristics, e.g., wide viscosity variations of emulsions.

With both types of pumps, the condition of the pump and the reservoir flow can be ascertained by monitoring surface flow rates and pressures. In addition, the jet pump has a chamber where a pressure bomb can be housed. Thus, bottomhole pressures can be obtained very easily by pumping the jet pump in and out without need for a wireline unit.

9.4222   Surface equipment

Surface installations are of two types:
(a) A central pump feeding power oil to a number of wells.
(b) One well-site units.

Considering open power fluid systems only, the central installation requires a separate power fluid line to each well, which adds expense. However, since it can be adjacent to the central separators, it can obtain power fluid from these separators which reduces expense. The solo unit, on the other hand, requires a sizable separator at the well site to separate the phases in the production stream and provide fluid—water or oil as desired—to be pumped back down as power fluid. In both cases, solids must be removed from the power fluid before injection. This is done by circulating the fluid through cyclone cleaners with an auxiliary circulating pump.

The trend appears to be towards more individual installations, at least for land wells. The central pump would be the obvious choice on an offshore production platform.

There is a substantial difference in cost of surface equipment according to whether the power fluid pump is driven by an electric motor or gas engine. A 100 hp electric motor costs about $1,600, while the corresponding gas engine costs about $11,000. Obviously, the former is preferred, if electric power is available. Both cases are considered in this analysis.

9.4223   Cost calculations

Well conditions assumed are:

depth	6000 ft
casing	5½ in.
ID tubing	2½ in.
surface flowline	2500 ft, 3 in. ID
production	1,800 bl/d
res. pressure	2800 PSI
prod. index	2 b/d/psi
WOR	1
GOR (Rs)	100 SCF/STB

There is quite a difference, both in capital and power costs, depending on whether the power fluid pump at the surface is driven by an electric motor or gas engine. Both cases will be computed.

(1) Electric Motor Drive

(a) Piston Pump

Pump piston volume requirement = 1800 b/d/0.84 (eff) = 2140 b/d

Choose pump, rated 2136 b/d

Area ratio of pump piston/engine piston = 1.957 for a typical pump.

Power oil requirement = $2140/(1.957 \times 0.8$ eff) = 1370 b/d

Fluid head to be lifted = $6000 - 1900/0.43 = 1600$ ft

Pump setting depth 3000 ft

Power fluid pressure requirement:

Fluid lift $0.418 \times 1600 \times 1.957 = 1310$ PSI

Friction loss in 2500 ft 3" ID flowline	= 13 PSI
Pressure drop in "solo" separator	= 10 PSI
Exit pressure of flowline = 120 PSI	
Total for surface backpressure = 143 × 1.957	280 PSI
Friction loss of 3170 BPD in annulus	20 PSI
Friction loss of power fluid in tubing	25 PSI
Pressure drop across pump	600 PSI
Power fluid surface pressure	2235 PSI

HHP at output of surface pump = $1370 \times 2235 \times 0.000017 = 52$ hp

Shaft HP required at pump = $52/0.85$ (eff) = 61

Electric power input to driving motor = $61/0.95$ (eff) = 65 hp

Installation cost:

Downhole:	
Pump proper	$ 6,840
Bottom and top assemblies on pump	3,707
Total downhole	$10,547

Surface individual well system:

Separator w/cyclone cleaner,	
controls and valves	$12,320
100 hp triplex pump w/accessories	15,980
100 hp electric motor (460v)	1,630
Motor-pump coupling and guard	433
4-way flow valve	1,555
Power fluid flowmeter	735
Misc. fittings	600
Total surface	$33,253
Total downhole and surface	$43,800

Maintenance:

Surface equipment	$ 1,000/yr.
Replacement of pump twice per year @ 40% of original cost each time	5,470/yr.
Total	$ 6,470/yr.

Power:

At 2.0¢ kwh:

$65 \times .746 \times 24 \times 365 \times .02 = \$8,495$/yr.

Lifting cost per barrel, based on 10-year amortization of capital costs:

$100 (4380 + 6470 + 8495)/(1800 \times 365)$

= 2.94¢

The important variables in the calculation are the life of the pump and cost of electric power. Fig. 9.6 gives calculated costs for electric power rates from 1¢ to 6¢/kwh and pump life from 3 months to 1 year.

(b) Jet Pump

To arrive at a suitable jet pump, a computer analysis was run to determine the appropriate nozzle and throat design. The pump used requires 1510 b/d power fluid at 3000 PSI surface operating pressure. The intake pressure of 600 PSI is adequately above the cavitation threshold.

HP at output of surface pump = $1510 \times 3000 \times 0.000017 = 77$

Shaft HP required to pump = $77/0.85 = 90$

Electric power input to motor = $90/0.95 = 95$ hp

Installation cost:

Downhole:	
Pump proper	$ 3,215
Bottom and top assemblies on pump	1,462
Total downhole	$ 4,677
Surface equipment (same as for piston pump)	33,253
Total downhole and surface	$37,930

Maintenance:

Surface equipment	$ 1,000/yr.
Replacement of nozzle and throat once/year	600
Total	$ 1,600/yr.

Power:

At 2.0¢/kwh:

$95 \times .746 \times 24 \times 365 \times .02 = \$12,416$/yr.

Lifting cost/barrel:

$100(3793 + 1600 + 12,416)/(1800 \times 365)$

= 2.71¢

In this case, repair of the pump is so inexpensive that the only important variable is the cost of electric power. Fig. 9.7 shows calculated costs for various power rates.

(2) Gas Engine Drive
  (a) Piston Pump

  Installation:
  Downhole: (same as before) $10,547
  Surface: $33,253 (previous
    estimate) − 1630 (electric
    motor) + 8000 (75 hp
    gas engine, est.)   39,623
                Total $50,170
  Maintenance:
  Surface − 25/hp/yr $ 1,625
  Pump replacement (as before,
    for 6-month life)   5,470
                Total $ 7,095/yr.

  Power @ 12 SCF/hp/Hr and
  $2.00/MCF:
    $12 \times 61 \times 24 \times 365 \times 2/1000 = \$12,830/yr.$

  (b) Jet Pump

  Installation:
  Downhole: (as before) $ 4,677
  Surface: $33,253 − 1630 +
    11,000 (100 hp engine)   42,623
                Total $47,300

  Maintenance:
  Surface: 25/hp/yr $ 2,500
  Pump repair (as before)   600
                Total $ 3,100

  Power:
    $12 \times 90 \times 24 \times 365 \times 2/1000 = \$18,920$

Figures 9.8 and 9.9 are plots of lifting costs for piston and jet pumps, for gas costs of 0.5 to 3.5 $/MCF. If the engines were fueled with diesel oil, costs should be those for gas at $1.90/MCF, based on equivalent BTU.

In all cases above, installation costs would vary little from a central surface installation to an individual unit. The cost of 2500 ft of power fluid line ($6500) plus that of a 500 b power fluid reservoir tank ($3500) at the central location just about balances the $12,320 cost of the separator required with the individual unit.

9.4224 Comparison of piston and jet pump costs

Comparing Fig. 9.6 with 9.7 and Fig. 9.8 with 9.9, it appears there is little to choose, in terms of present-day operating costs, between the piston and jet pumps. Both have lifting costs about 3¢/bbl with electric prime mover and about 3.6¢/bbl with gas engine prime mover. However, unless the bottomhole pressure is too low, the jet pump would be favored because of greater life certainty and ease of repair.

In the future as power costs rise, the inefficiency of the jet pump will weigh somewhat more heavily against it. This is particularly true for gas engine drives where lifting costs are projected to rise to 5.8¢/bbl in five years (Fig. 9.6). With electric drive, however, costs will remain sufficiently close for some time

so that the jet pump would be the preferred choice for most applications.

Comparing Fig. 9.6 with 9.7 and 9.8 with 9.9, it is seen that lifting costs with gas engine installations are now about 30% higher than with electric prime mover installations and that the disparity will increase as time goes on. Obviously, electric power should be used where available.

9.4225 Comparison of hydraulic pumping with gas lift and electric submersible systems

Lifting costs for gas lift and electric pumps are shown in Figs. 9.4 and 9.5. Comparing these with Figs. 9.6 and 9.7, hydraulic lifting costs (with electric drive) are currently just about the same as electric submersible costs, 3¢/bbl, and are somewhat less than gas lift costs of 3.7¢/bbl. Improvements to extend pump life and gas lift valve life are difficult to predict. Engineering and installation advances can affect any of the equipment and their related costs of maintaining. The electrical pump is the only one that is non-retrievable without pulling the tubing; fortunately, it possesses good potential for improved operating life. A cable retrievable pump is in the development stage and is operating in a few wells.

9.423 Rotative gas lift without a source of pipeline gas

9.4231 Introduction

In some areas operators are turning from gas lift to submersible electric or hydraulic pumping systems because of the scarcity of natural gas. The increasing cost of natural gas is not the controlling factor, since all artificial lift systems require fairly similar amounts of power and power costs rise with natural gas costs.

This section, therefore, considers how an artificial lift system can be operated, in rotative fashion, with a bare minimum of natural gas, for situations where pipeline gas is not available.

9.4232 Minimization of gas use

Historically, in gas lift systems natural gas fuels the compressor engine, charges the system on start-up, and supplies system losses during operation. Consider the typical system previously studied, i.e. 6000 ft well, 5½ in. casing, 2500 ft flow and injection lines, 1200 PSI injection pressure, 1800 bl/d production with GLR = 300, WOR = 1, and Rs = 100 scf/B.

(a) Fuel: The amount of gas required for compressor fuel is 20,000 scf/day. This is a large amount, so the first step in conserving gas should be to fuel the compressor engine with diesel oil or use an electric motor. This should be feasible in new installations.

(b) System charge: Charging the system requires that the gas injection line and the tubing-casing annulus be pressured to 1200 PSI with gas and that gas be added to the oil in the tubing and flowline to the extent of 300 SCF/bbl. For the example well this amounts to approximately 60,000 SCF.

This is not a large amount of gas. It could be supplied as a tank of LPG gas about 4 ft diameter × 10 ft long (1000 gals.) Charging need be done only when the well is worked over, perhaps twice a year. Transporting a tank of LPG to a land location should not be a problem; offshore it could be.

Liquified nitrogen is an alternate to LPG but is less desirable; if the well is producing more gas than is needed to compensate for leaks (discussed below), the excess gas in the sales line from the separator will initially be almost all nitrogen. Gradually it will change to natural gas, but until the BTU content reaches a reasonable level it may be unsaleable. On the other hand, with LPG all gas going into the sales line will be saleable. In fact, much of the cost of the LPG could eventually be recovered.

(c) System losses: These are of two forms:

(1) Gas carry-out, in solution, by the oil and water leaving the separator is proportional to the separator pressure (absolute). At 60 PSIG, natural gas would be carried out at the rate of about 35 SCF per barrel of oil (35° gravity) and 0.7 SCF per barrel of water. For the case considered, 1800 BLPD with WOR = 1, the carry-out rate would be 32,200 SCFD.

This is a lot of gas. In the early stages of gas lift, the reservoir would probably be producing enough gas to compensate for this carry-out and even make up leak losses. For example, in the typical well with GOR = 100, reservoir gas production would be 90,000 SCFD, ample for maintaining the system. In later stages, however, the GOR of the reservoir could fall and the water cut rise, so a point would be reached where reservoir gas could no longer compensate for solution carry-out, let alone leaks. In the example, this would happen at a GOR of 41 when the water cut reached 90%.

Consequently, an important step in conserving gas in the latter stages of production would be to operate the separator at the lowest possible pressure. For example, at 10 PSIG separator pressure, solution carry-out would only be ⅓ of that above. The solution GOR could then drop to 14 before gas production would not compensate for carry-out at 90% water cut.

An alternate solution which might be more practical would be to leave the separator operating at normal pressure but add a vapor recovery system to the oil storage tank and re-inject the gas so obtained into the circulating system. With vapor recovery at 0 PSIG, the reservoir GOR could drop to about 8 scf/b before reservoir gas would be insufficient to compensate for residual solution gas in the storage tank oil and solution gas in the dumped water.

(2) Leaks in the pipe (injection line, flowline, casing) and compressor. Ideally, the system should be tight, with zero losses. Practically, there must always be some loss. No good figures are available. There are reports of both extremely leaky systems and tight systems. The system should be fitted with isolating valves and pressure gauges to periodically monitor the leakage. Leakage from the pipes will be proportional to the system gas charge, while that from the compressor will be proportional to the gas rotation rate. An estimate is that even with a "tight" system the pipe leakage rate per day might be 1% of the gas charge. (This means, if the system could be shut-in after charge, it would take 70 days for the pressure to drop in half.) In the example the loss would be 600 SCF/D. Compressor loss would be due to gas leaking past the piston packings. One expert estimated it at 0.1% of the rotating gas stream, although he could not recall measurements ever being made. In the example, this would be about 500 SCF per day. Total leak losses then might be 1100 SCF per day.

Making up a leakage of 1100 SCF/d with reservoir gas is not too difficult as long as the water cut is reasonable. In the example, with 900 bo/d production, the GOR need be only 1-2 units higher (than needed to compensate for carry-out gas) to make up for the leakage. However, with 90% water cut, the GOR would have to be higher by about 7 units. The situation will be aggravated by the fact that, as the system gets older, leakage rates will increase. Sooner or later there must be an external source of makeup gas.

Making up a leakage of 1100 SCFD from an LPG tank of 60,000 SCF would not be feasible, as it would take a new 1000-gallon tank every two months or so. A more feasible way would be to burn the LPG in a well-controlled burner to produce combustion gas (88% N_2, 12% CO_2). This gives a 24:1 volume increase over unburned LPG. In this way, a 1000-gallon LPG tank could supply 1440 MCF rather than 60 MCF, sufficient for both charging and leak-makeup for 1-2 years. The corrosive elements of the combustion gas (unburned oxygen and nitric acid) could be removed quite easily by catalytic cleanup before injection into the system. (This is not as easy with exhaust from the diesel-powered compressor engine, due to the high oxygen content in diesel exhaust and the sulfur content of fuel oil).

9.4233 Proposed system

Based on the above considerations, it seems that a gas lift system, with no access to a pipeline source of gas, should be designed to operate in three modes depending primarily on the life (more exactly the GOR) of the well:

(1) Early mode: GOR $\geqslant 50 \leqslant$ WOR $\leqslant 1$

This is shown in Fig. 9.10 and is a standard system except that provision is made for charging the well with LPG transported to the location in a tank truck when needed. In this mode, reservoir gas is produced in sufficient quantity to take care of both system leaks and separator carry-out gas.

(2) Mid-life mode: $10 <$ GOR < 50 $1 <$ WOR < 4

Fig. 9.11 shows this system. It is the same as Fig. 9.10 except that a vapor recovery system has been added to the oil storage tank along with a small compressor to bring the pressure back up to separator pressure for re-injection to the rotating stream. In the example cited, a 3 hp compressor would be adequate. With this addition, reservoir gas should still compensate for system losses for some time. An estimate of the additional cost for vapor recovery and pre-compression is $15,000. If there happened to be a spare cylinder available on the existing compressor, the additional cost might only be half that.

(3) Final-life mode: GOR < 10 WOR > 4

At this stage, even full recovery of reservoir gas cannot compensate for system leaks. Therefore, the system of Fig. 9.12 is proposed. It is the same as Fig. 9.11 with the addition of a permanent (replaceable or recharge-

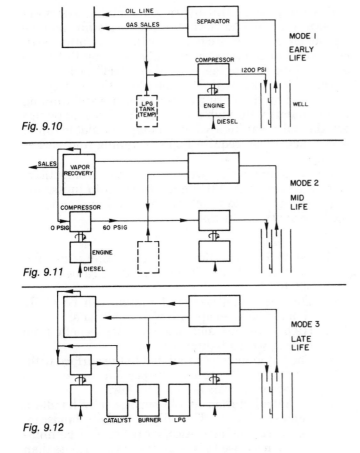

Fig. 9.10

Fig. 9.11

Fig. 9.12

able) LPG tank, a burner, and a catalytic cleaner. The same small compressor as in (2) is used to bring the combustion gas up to intake pressure of the main compressor. Additional cost of the permanent LPG tank, burner, and catalytic cleaner is estimated at $15,000.

With proper design in the beginning, conversion of a system from Mode 1 to 2 to 3 could be done easily without interrupting production.

Following the calculation in the previous example problem, lifting costs in the three modes can be estimated. With the assumptions that diesel oil is equivalent to gas at $1.92/MCF, that reservoir gas or LPG lost by leaks costs $2/MCF, that the additional capital cost in Mode 2 ($15,000) is amortized over 7.5 years, and that in Mode 3 over 5 years, computed lifting costs in the example case are:

Mode	Lifting cost ¢/bbl
1	3.2
2	3.6
3	4.2

Lifting costs remain reasonable, even in later stages of production.

This "gas lift transition" proposal seems a fair alternate to converting the whole system to some other means of lift at a later date.

9.5   GENERAL DISCUSSION ON TYPES OF ARTIFICIAL LIFT

9.51   Sucker rod pumping

Sucker rod pumping is the most widely used type of lift system. Historically, the main advantage has been the familiarity of this type of lift to operating personnel. One of the main disadvantages has been depth limitation. However, larger load capacity units and high strength rods allow greater and greater depths (12,000 ft).

As a general rule tubing anchors should be used, due to the buckling effects of unanchored tubing. For shallow low volume wells they are probably unnecessary.

Advantages

(1) Since most field and operating personnel are familiar with sucker rod type lift, the installation and operation is not complicated.

(2) Sucker rod pumps offer excellent rate range. This is directly influenced by the size and type of unit, tubular size, rod string design, and pump size range. In most operating areas different pump and rod sizes are "off-the-shelf" items. Warehousing and stock parts as well as service and repair availability have historically been associated with this type of lift.

Disadvantages

(1) Volume limitations of sucker rod pumps are due to tubular size and depth. Volumetric efficiency is reduced in wells with high GOR, if solids are produced, if paraffin forms or if the fluid is sour or corrosive.

(2) Initial capital cost, especially for the larger high-capacity units, is high. The cost of the rod string must be included in the economic analysis.

(3) A major disadvantage is the rod string operating in a corrosive environment. Rod wear will damage tubing and, upon tubing failure, may cause high workover expense. Also, the tubing cannot be internally coated to prevent corrosion due to rod wear. Heavily loaded rod strings, especially in corrosive fluids, may have high failure frequency.

(4) Improper sucker rod handling causes many failures. Common sense handling and make-up techniques can extend rod life and improve operating expense indicators. Also, anchored tubing may improve efficiency and reduce rod wear for high volume pumpers with larger ID tubing.

9.52   Gas lift

Gas lift compares favorably with hydraulic pumping in providing the most flexible depth and rate range of all lift types. However, due to recent shortages of natural gas, this lift method may not be applicable in many areas. Gas lift may also be used to kick-off wells that flow naturally, to back-flow water injection wells, and to unload liquids from gas wells.

Advantages

(1) The main advantage of gas lift is flexibility. This type of lift will adjust to practically any depth and/or rate. The design may be changed by wireline without pulling the tubing depending upon the tubular sizes and availability of service equipment.

(2) Initial cost is generally less if high pressure is available. This is not true if compressors must be obtained.

(3) Abrasive material such as sand offers fewer problems.
(4) It is adaptable to deviated wells.
(5) It can be used in low productivity high gas-oil ratio wells.

Disadvantages

(1) Shortages of natural gas in some areas will limit or prevent gas lift to be considered as a lift method.
(2) Freezing and hydrates in the gas input line may cause excessive downtime. Dry gas will improve operations but may cause loss of liquids.
(3) Valve retrieval in highly deviated holes by wireline has offered problems in the past; however, present day equipment supposedly has eliminated the problems.
(4) Scale, corrosion and/or paraffin problems may increase backpressure and reduce efficiency. Surface flowlines and separators may also cause increased backpressure with resultant loss in lift efficiency.
(5) It is not applicable in bad casing where it is uneconomical to repair casing.
(6) It is difficult for the lower zone of a dual where there is a long distance between zones.
(7) It should not use highly corrosive injection gas; therefore, corrosive elements must be removed.
(8) It has difficulty in completely depleting a low BHP low productivity well in some cases, and may require a change in lift method towards the end of the life of the well.

9.53 Hydraulic pumping

Hydraulic pumping may not be limited to depth with production from 18,000 ft as an example. The initial capital investment may be quite high.

Hydraulic pumps can be run with vent strings and gas anchors to handle GOR's up to about 4000 scf/B, depending on productivity. Vent strings are for fixed casing pumps or free pumps run on a packer. Sufficient casing size must be available for tubing and vent, plus annular space for return fluids. In deeper wells this can lead to higher surface pressures which, in turn, can limit production to high BHP wells.

If sufficient casing size is available and well production is within the range of available free pumps, two strings of tubing may be run and the annulus vented as with rod pumps. (Limitation is the free pump displacement combined with setting depth and P/E ratios for allowable surface pressures.)

Hydraulically pumped wells deeper than 11,000 ft are probably getting help in lifting either high BHP from water drives or are being agitated to flow. Below 11,000 ft net lift surface pressures required are high with available P/E ratios. This, of course, depends on the required pump displacement, since smaller pumps can be set deeper and not exceed allowable operating pressures.

In dogleg holes where rod problems are evident from experience, the hydraulic pump works very well. Free pumps sometimes get stuck in severe dog legs and require fishing, but they are still cheaper than tripping a casing pump.

If we have a central hydraulic pumping system of several wells on one power oil tank, paraffin can give a problem. An interface built up of large paraffin particles grows in the tank, eventually reaching the pump and clogging the engine. Any other "crud" such as rust in the system aggravates this problem. Its severity can be reduced by circulating off the interface and running it continuously through a treater. Fixed casing pumps are the only real high volume pumps available (up to 4 in.), and these require tripping tubing and packer for repair.

Today's single-well power units are often competitive in first cost with API 640 and larger beam units, and high strength sucker rods for high volume or intermediate depth production.

Advantages

(1) Depth is not a limiting factor. Many installations are below 12,000 ft producing rates of 150-300 b/d.
(2) The speed and size of the pump can be easily changed to keep up with well conditions.
(3) Highly viscous and heavy crudes benefit from mixing with a lighter power oil.
(4) The pump may be circulated to the surface without pulling the tubing. Inspection, service, and replacement costs are usually low.
(5) A central station at the surface may handle a number of wells. This allows wellsite landscaping or camouflage. Also, corrosion can be minimized in a closed system if the oxygen is less than 50 ppb.
(6) Modern day one-well units offer a compact unit for isolated wells.

Disadvantages

(1) Initial capital cost is high. High pressure equipment, power fluid lines, and wellheads are required. Facilities must be provided to filter, clean, and treat the power fluid. Tubulars must be of sufficient size and must be high pressure leak tight.
(2) Corrosion and abrasives will reduce operating life due to close tolerances in the surface and downhole equipment.
(3) For power oil systems the volume required may become highly expensive at today's crude prices and especially so if power fluid losses are major.
(4) Since this is usually a high pressure operation, maintenance costs for surface equipment may be quite high.
(5) High temperatures can cause the packing cups to fail, thereby preventing ease of pumping out the pump for repairs.
(6) It needs well-trained people to operate efficiently.
(7) Fire hazard for gas engine operation—If there is a high pressure leak in the power pump, a fire could burn up the whole installation including power oil and stock tanks.
(8) Well testing in central systems is a problem if wells make water. For accurate well tests, only one well at a time can be operated without special metering, manifolding and test equipment, which ups the initial cost.
(9) Corrosive production. In pumps set on packers, fixed or free, there is no means of treating the

pump end for corrosion inhibition. An exception is when a vent string is installed and a chemical can be put down the vent string to treat the pump end. Inhibitor in power fluid only protects the engine and tubing. New metallurgy has resulted in more corrosion-resistant pumps but the problem still exists.

9.54 Electrical submersible pumps

Historically, this method of lift is associated with high volume fluid production. High cost and depth limitations are disadvantages.

Another unique use of this pump has been in increasing the volume of dumped water from formations above to a waterflood formation below. The pump is inverted and used to pump the water down, whereas normally the water is gravity fed to the formation below.

The pump can be run on a cable and landed in a preset nipple in a standard well to induce flow and then removed later when the well flows naturally. The main trouble with this system has been gas impregnating the cable. One case in operation is pumping water from 5000 ft to 8000 ft at 25,000 b/d.

The limitations on the depth a submersible pump can be set has to do with the horsepower available from any particular submersible pump supplier. As an example, a motor which is 4½ in O.D. and can be run into 5½ in casing has a limit of 200 hp available. The 5 in. equipment which can be run into 7" or larger casing has a limit of 400 hp set. The horsepower limit is set by the size shaft that can be installed in these small O.D. housings. A submersible pump should never be set below fluid entry unless a jacket is utilized to direct fluid by the motor.

Advantages
(1) High volume lift capacity is the main advantage of this type of lift. However, electrical submersibles are finding more application in low volume wells.
(2) Although more engineering supervision may be required initially it does not require a lot of knowledge to operate since it either does or does not run.
(3) The pump can increase the volume in a dump flood from a water zone above to the waterflood zone below.

Disadvantages
(1) Due to horsepower rating of the electric power motor, depth is limited. Depth is also limited by size of tubulars and high temperature. Larger high HP equipment may not provide enough annular clearance to cool the motor resulting in failure. High temperature will also limit both motor and cable.
(2) Initial cost may be high, since multistage high volume and high HP pumps are expensive. The cable is also a high cost item, especially if non-corrosive or high temperature sheathing is required. Transformers must be provided to insure proper voltage.
(3) Cable failures occur and require pulling the tubing to repair. High temperatures, corrosion, and poor handling lead to cable failure. Replace-

ment cost may be excessive on high failure marginal operations.
(4) Motor failures are also due to high temperature, corrosion, and abrasives. High GOR may result in low efficiency and failure is due to free gas locking the pump.
(5) Additional engineering supervision is required in design, installation, maintenance, and troubleshooting this type of lift.

9.55 Jet pumping

Jet pumping offers downhole equipment with no working parts. The full limitations are not yet known.

Advantages
(1) Since the jet works on pressure drop and change in velocity across a nozzle, there are no moving parts downhole.
(2) It is run in the same bottom hole assembly as a piston hydraulic pump; as the BHP decreases, a piston pump can be easily installed.
(3) Depth of lift may not limit this lift method with applications from 11,500 ft in some areas.
(4) High volume lift is associated with jet pump installations.

Disadvantages
(1) Initial capital cost is high. High pressure equipment, power fluid lines, and wellheads are required. Facilities must be provided to filter, clean, and treat the fluid. Tubulars must be of sufficient size and must be high pressure leak tight.
(2) Corrosion and abrasives will damage the nozzle. (Note that unless the pump becomes stuck in the hole due to scale, salt or fill-up, it can be pumped to the surface for inspection. The pump cavity seals may need replacement under severe corrosion and high temperature operation.)
(3) Since this is usually a high pressure operation, maintenance costs for surface equipment may be quite high.

9.56 Plunger lift

A plunger may maintain flowing status of a well. The plunger is temporary and is usually replaced when a method of lift is chosen.

Advantages
(1) Many wells can utilize a plunger to maintain flowing status.
(2) A plunger lift hook-up for a flowing well is nominal in cost as compared with other lift methods.
(3) A plunger can also keep the tubing free of paraffin and scale.
(4) Plunger installations in existing gas lift wells may help fluid fallback and increase volumetric efficiency.

Disadvantages
(1) Usually, plungers are used only as a temporary means to maintain production until another method of lift is chosen and installed.
(2) Plunger action will cause surging of gas and liquids at the separating facility. If a plunger is put into service, the production facility should

TABLE 9.4
RELATIVE ADVANTAGES OF ARTIFICIAL LIFT SYSTEMS
S. G. GIBBS (WITH MODIFICATIONS BY BROWN)

Rod pumping	Hydraulic pumping	Electric submersible pumping	Gas lift
Relatively simple system design.	Not so depth limited—can lift large volumes from great depths (500 BPD from 15000 ft). Have been installed to 18000 ft.	Can lift extremely high volumes (20000 BPD + in shallow wells with large casing. Currently lifting ± 120,000 B/D from water supply wells in Middle East with 600 HP units. 720 HP available 1000 HP under development.	Can handle large volume of solids with minor problems.
Units easily changed to other wells with minimum cost.			Handles large volume in high P.I. wells (continuous lift) (50000 BLPD +).
Efficient, simple and easy for field people to operate.	Crooked holes present minimal problem.		
Applicable to slim holes and multiple completions.	Unobtrusive in urban locations.	Unobtrusive in urban locations.	Fairly flexible—convertible from continuous to intermittent to chamber or plunger lift as well declines.
Can pump a well down to very low pressure (depth and rate dependent).	Power source can be remotely located.	Simple to operate.	
	Analyzable.	Easy to install downhole pressure sensor for telemetering pressure to surface via cable.	Unobtrusive in urban locations.
System usually is naturally vented for gas separation and fluid level soundings.	Flexible-can usually match displacement to well's capability as well declines.		Power source can be remotely located.
Flexible—can match displacement rate to well capability as as well declines.	Can use gas or electricity as power source.	Crooked holes present no problem.	Easy to obtain downhole pressures and gradients.
Analyzable.	Downhole pumps can be circulated out in free systems.	Applicable offshore. Corrosion and scale treatment easy to perform.	Lifting gassy wells is no problem.
Can lift high temperature and and viscous oils.	Can pump a well down to fairly low pressure.	Availability in different size.	Sometimes serviceable with wireline unit.
Can use gas or electricity as power source.	Applicable to multiple completions.	Lifting cost for high volumes generally very low.	Crooked holes present no problem.
Corrosion and scale treatments easy to perform.	Applicable offshore.		Corrosion is not usually as adverse.
Applicable to pump off control if electrified.	Closed system will combat corrosion.		Applicable offshore.
Availability of different sizes.	Adjustable gear box for Triplex offers more flexibility.		
Easy to pump in cycles by time clock.	Mixing power fluid with waxy or viscous crudes can reduce viscosity		
Hollow sucker rods are available for slim hole completions and ease of inhibitor treatment.			
Have pumps with double valveing that pump on both upstroke and downstroke.			

be redesigned to handle the expected gas and liquid surges.

(3) Solids may stick the plunger in the tubing, which will result in loss of production and a potentially hazardous pulling job.

9.57 Other methods of lift

As technological advances are made, new methods of lift may become practical. Also, the current methods are being updated and redesigned to optimize production.

9.6 SUMMARY TABULATIONS ON ADVANTAGES AND DISADVANTAGES OF ARTIFICIAL LIFT SYSTEMS

Tables 9.4 and 9.5 are reproduced by permission from S. G. Gibbs and are essentially self-explanatory.[3] They show the relative advantages and disadvantages of the four principal methods of artificial lift: rod pumping, hydraulic pumping, electrical submersible pumping, and gas lift.

In addition, Table 9.6 is presented to show the performance characteristics of four different beam pumping unit types.

TABLE 9.5
RELATIVE DISADVANTAGES OF ARTIFICIAL LIFT SYSTEMS
S. G. GIBBS (WITH MODIFICATIONS BY BROWN)

Rod pumping	Hydraulic pumping	Electric submersible pumping	Gas lift
Crooked holes present a friction problem.	Power oil systems are a fire hazard.	Not applicable to multiple completions.	Lift gas is not always available.
High solids production is troublesome.	Large oil inventory required in power oil system which detracts from profitability.	Only applicable with electric power.	Not efficient in lifting small fields or one well leases.
Gassy wells usually lower volumetric efficiency.	High solids production is troublesome.	High voltages (1000 V ±) are necessary.	Difficult to lift emulsions and viscous crudes.
Is depth limited, primarily due to rod capability.	Operating costs are sometimes higher.	Impractical in shallow low volume wells.	Not efficient for small fields or one-well leases if compression equipment is required.
Obtrusive in urban locations.	Unusually susceptible to gas interference—usually not vented.	Expensive to change equipment to match declining well capability.	Gas freezing and hydrate problems.
Heavy and bulky in offshore operations.	Vented installations are more expensive because of extra tubing required.	Cable causes problems in handling tubulars.	Problems with dirty surface lines.
Susceptible to paraffin problems.	Treating for scale below packer is difficult.	Cables deteriorate in high temperatures.	Some difficulty in analyzing properly without engineering supervision.
Tubing cannot be internally coated for corrosion.	Not easy for field personnel to troubleshoot.	System is depth limited (10,000 Ft. ±) due to cable cost and inability to install enough power downhole. (Depends on casing size)	Cannot effectively produce deep wells to abandonment.
H_2S limits depth at which a large volume pump can be set.	Difficult to obtain valid well tests in low volume wells.	Gas and solids production are troublesome.	Requires make-up gas in rotative systems.
Limitation of downhole pump design in small diameter casing.	Requires two strings of tubing for some installations.	Not easily analyzable unless good engineering "know-how".	Casing must withstand lift pressure.
	Problems in treating power water where used.	Lack of production rate flexibility.	Safety problem with high pressure gas.
	Safety problem for high surface pressure power oil.	Casing size limitation.	
	Loss of power oil in surface equipment failures.	Cannot be set below fluid entry without a shroud to route fluid by the motor.	
		More downtime when problems are encountered due to entire unit being downhole.	

Fiberglass-reinforced plastic sucker rods are now being used. Procedures for determining how much steel one should run between the pump and the fiberglass rods and how much plunger travel one will obtain are available. We can expect to see more fiberglass rods in the future.

TABLE 9.6
PERFORMANCE CHARACTERISTICS OF FOUR DIFFERENT BEAM PUMPING UNIT TYPES

Conventional unit	Air balance unit	Mark II unit	BG unit
(1) Highly efficient	Normally somewhat less efficient than other three unit types	Highly efficient	Highly efficient
(2) High reliability factor due to simplicity of design	Most complicated of the four unit types	Same reliability factor as conventional	Same reliability factor as conventional
(3) Conventional unit cost used as reference	Occasionally higher cost than conventional	For comparable application, often about the same cost as conventional	Generally somewhat higher cost than conventional
(4) Portability more limited than for air balance unit	Greatest ease of portability; more compact	Normally, less portable than air balance unit	Normally, less portable than air balance unit
(5) Counterbalance more difficult to adjust	Counterbalance easily adjustable	Counterbalance more difficult to adjust	Counterbalance more difficult to adjust
(6) Generally, widely fluctuating torque loads	Torque peaks and range are often not as severe as on the conventional unit	Lowest and smoothest torque load with relative uniform torque system (Unitorque)	Torque peaks and range are often not as severe as on the conventional unit
(7) Impractical to mount on two-point suspension for minimal foundation and base movement	Impractical to mount on two-point suspension for minimal foundation and base movement	Can be mounted on two-point suspension for minimal foundation and base movement	Impractical to mount on two-point suspension for minimal foundation and base movement
(8) Relatively higher power cost and larger prime mover requirement	Relatively higher power cost and larger prime mover requirement	Generally, lowest power cost and smallest prime mover requirement because of uniform torque (Unitorque) system	Relatively higher power cost and larger prime mover requirement
(9) Generally, higher rod and structural loads	Generally, lower rod and structural loads than conventional and BG units	Generally, lowest rod and structual loads	Rod and structural load about the same or higher than the conventional unit
(10) Normally the highest (relative) maximum pumping speed	Often slightly reduced maximum pumping speed compared to conventional unit	Often slightly reduced maximum pumping speed compared to conventional unit	Often slightly reduced maximum pumping speed compared to conventional unit
(11) Less "fill-time" for subsurface pump barrel	Less "fill-time" for subsurface pump barrel	Greater "fill-time" for the subsurface pump barrel	Greater "fill-time" for the subsurface pump barrel
(12) Normally a lesser net plunger travel per stroke	Normally a lesser net plunger travel per stroke	Normally the greatest net plunger travel per stroke	Occasionally somewhat greater plunger travel than conventional or air balance unit
(13) Speed reducer maintenance nominal due to widely fluctuating torque loads and relatively small "flywheel" effect of cranks	Speed reducer maintenance may be less than on the conventional unit	Normally less speed reducer maintenance required than other unit types due to relative uniform torque system and substantial "flywheel" effect of large cranks	Sped reducer maintenance nominal due to fluctuating torque loads and relatively small "flywheel" effect of cranks

TABLE 9.7

SUMMARY TABULATION ON ARTIFICIAL LIFT METHODS IN NORTH DAKOTA AREA

Type lift	Depth	Rate	Cost	Handle gas	Weather problems	Depletion	Repairs	Location	Operating problems	Remarks
Rod pumping	5,000 to 7,000 ft	10 to 700 BFPD	$/well/month 980 avg. range 360 to 1810	GOR 1000 avg. range 800 to 1200	Severe: range −50 to +110°F, with constant wind blizzard and dusty conditions	Water drive	Usually costs more for service rig than rod and/or pump repair, can become costly if replace rods or pump	NW North Dakota	High GOR, saturated brines with free salt, paraffin, H_2S corrosion, crooked holes, rod fatigue	Volumetric efficiency avg. 45%; with many years service in sour brine environment causing high sucker rod failure frequency
Hydraulic pumping*	—	—	—	—	—	—	—	—	—	—
Electrical submersible pumping	4,000 to 5,000 ft	1,200 to 3,000 bf/d	$/well/mo. 1000 to 1200 est.	WSW handle very little gas; oil wells avg. GOR 1000	as above	Water drive	Must repair or replace cable on failure, replace either pump or motor or both on failure	SW North Dakota	Cable problems and motor burn-out at elevated BHT and low clearances in casing annulus	Used mainly for water supply production for water injection system; few oil wells since produce solids (salt, etc.) and may have H_2S
Gas lift	8,000 to 10,000 ft	10 to 600 bf/d	$/well/mo. 740 avg. range 540 to 1370	GOR 1000 avg. range 800 to 1200	as above	Water drive	Low cost if replace bad valves & don't redesign string, testing and repair costs similar to sucker rod repair costs	North Dakota	Sour service and paraffin, salt and scale; any and all accumulating causing restricted flowline wet input gas freeze-up, high input GLR	Majority of wells are low PI with high GLR, high PI wells produce with GLR less than 1000; effic. size, if gas (preferably dry) is readily available
Plunger lift	7,500 ft	50 bf/d	$/well/mo. 540	GOR 1000 avg. range 800 to 1200	as above	Water drive	See Gas Lift	North Dakota	See Gas Lift	Same as Gas Lift; being used on a test basis mainly for paraffin control, will be tested in low PI well with high fall-back, will be effective if flowline restriction can be reduced also
Jet pumping	10,000 ft 10,800 ft	250 to 350 bf/d	$/well/mo. 2030 avg. range 1220 to 3450	GOR 1000 avg. range 800 to 1200	as above	Water Drive	Surface installation repairs are costly and frequent due to high pressure, replacement and repair costs for downhole pump is high	NW North Dakota	Paraffin, salt & H_2S & scale build-up, power fluid & surface pumping maintenance very high; must run triplex at 3750 to 3900 psi to lift fluid	Continual monitoring and maintenance are required due to salt production & H_2S, units operate well despite being in the wrong applicat. as units were available within the company
Others										

*Not in use in this area, see Jet Pumping

9.7 SUMMARY TABULATION ON OPERATION OF ARTIFICIAL LIFT METHODS IN NORTH DAKOTA

Table 9.7 summarizes operations in one area of the U.S. where numerous problems have been encountered with all artificial lift methods. The costs are for 1977.

9.8 CLASS PROBLEMS: REPRESENTATIVE FIELD CASES TO SELECT ARTIFICIAL LIFT METHOD

(1) Given:

Depth—8000 ft
G/O—400 Scf/B (oil)
Water Production—50%
Rate—800 B/D total liquid
API = 42°
Tubing Size—2⅞" O.D. EUE
(All power sources available)
$\overline{P}_R$ = 2300 psia
PI = 3
P_{wh} = 100 psia
Casing size 7"

Select the best lift method.

For the following problems the best type of lift is to be selected. In some cases one or more methods are suitable. In making these selections, give your own reasons as to why. Where possible, use the information in the general problem. Various changes are made in each example to alter the possible lift choice.

What type of lift do you recommend for the above well with the following changes that occur in the well? Everything else remains constant as in the given data.

(1) G/L > 3000 Scf/b   PI = .01
    q_L = 20 b/d

(2) Depth > 12000 ft
    q_L = 200 b/d   PI = 0.1

(3) Rate > 5000 bl/d   PI = 10
    Tubing 4 in.

(4) Severe H_2S problem

(5) Depth—3000 ft $\overline{P}_R$ = 1200 psia
    Rate—400 b/d total   PI = 1.0

(6) No engineering supervision available

(7) Well produces sand

(8) Bad paraffin problems

(9) Depth—6000 ft
    Rate—10,000 b/d
    2⅞" × 7" annular flow
    PI = 15

(10) Rate—25,000 b/d   (Choice of annular
     9⅝" casing—4" tubing   or tubing flow.)
     PI = 25

(11) Severe scale deposit problem

(12) Located in extremely cold climate (−30°F. Average)

(13) Located in hot climate (110°F. Average)

(14) Located on land as a single isolated well

(15) Located on an offshore platform with 20 other similar wells

(16) Located with a group of 25 wells on land on 60-acre spacing per well

(17) $\overline{P}_R$ drops to 300 psi
     PI = 0.02
     Rate = maximum possible

(18) $\overline{P}_R$ drops to 100 psi
     PI = 0.01
     Rate = maximum possible

(19) 100% oil
     G/O—4000 Scf/bbl
     $\overline{P}_R$—500 psia
     PI = 0.02
     Rate = maximum possible

(20) Slim hole completion
     2⅞" casing × 1" tubing
     q_o = 250 b/d

(21) 4 in. casing × 1½ in. tubing

(22) Long 300 ft perforated interval
     $\overline{P}_R$ = 500 psia
     PI = 2
     Rate = maximum possible

(23) Open hole completion (400 ft)
     $\overline{P}_R$ = 700 psia
     PI = 1.0
     Rate = maximum possible

(2) Complete the following summary tabulation questions on Artificial Lift Methods

Please comment on each item as applicable from your experience:

Type lift	Depth	Rate	Cost	Handle gas	Weather problems	Depletion	Repairs	Location	Operating problems	Remarks
Rod pumping										
Hydraulic Pumping										
Electrical submersible pumping										
Gas lift										
Plunger lift										
Jet pumping										
Others										

9.9 COMPARATIVE PROBLEMS

9.91 Example Problem #1

An example problem shows the difference in design and flow rates possible for the main artificial lift methods. The following data was used, including data for continuous flow gas lift. The main purpose of this problem was to show the production rates possible by the different methods as well as horsepower requirements etc.

The individual design procedures can be found in Chapters 2 through 7.

The following information is available concerning this well:

Given:

Tubing size = 2⅞ in. O.D.
Casing size = 7 in. O.D.
γ_g = 0.70 (for gas lift operations)
Well loaded with kill fluid of gradient = 0.40 psi/ft
GOR = 400 scf/bbl
Well produces 50% water (γ_w = 1.07)
GLR = 200 scf/bbl
$\overline{P}_R$ = 1920 psi
Bubble point pressure = 1500 psi below which it follows Vogel's reference curve.
°API = 35
PI = 5 = linear until reaching P_b
P_{wh} = 120 psig (constant)

P_{so} = 900 psi (950 available) (for gas lift operations)
P_{ko} = 1000 (1050 available) (for gas lift operations)
Depth = 8000 ft
Desired rate = maximum
Temp bottom = 170°F.
Geothermal temperature gradient = 1.2°F./1000 ft (109°F. surface for 1000 B/D)

The well shows an absolute open flow potential ($q_{o\ max}$) of 6267 b/d for a flowing pressure of 0 psi. The flow rate at bubble point pressure is 2100 b/d, and most lift methods produce a rate less than 2100 b/d.

Presently, the well will not flow naturally, even at low rates.

Fig. 9.13 shows the inflow performance curve for this well and the approximate maximum rates by the various lift methods noted on this curve. The procedures for finding these rates are discussed in the various chapters referring to the particular type of lift.

9.911 Sucker rod pumping

The sucker rod pumping unit design was conducted by Spencer Duke and most of the following are his comments:[4]

All of the design work and most of the analytical work in sucker rod pumping is done with the computer. Our general philosophy is that any artificial lift problem that will require a pumping unit larger than an 1824 can best be solved by another method of artificial lift.

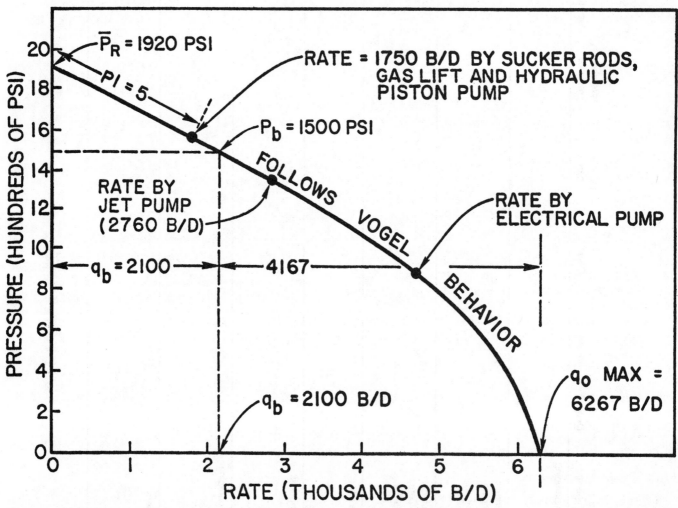

Fig. 9.13 Pressure-Flow Rate Diagram. Inflow curve for Example Problem #1. (Well will not flow naturally)

We have very few installations with larger size pumping units. We usually use a Mark II or an air balanced unit in any installation that requires a pumping unit larger than a 912, but we have purchased a few 1280 conventional units on special order. In fact our computer programs will not accept a polished rod stroke length longer than 240.

The maximum production that could be obtained with an 1824 air balanced unit, 100% pump volumetric efficiency, Oilwell's Electra rods, 3½" tubing with an insert type pump would be approximately 2,000 B/D. I have calculated an operating fluid level of 4340 for a production rate of 2100 B/D. At a reduced production rate, the operating fluid level would be higher which would give you a higher pump intake pressure and an increased pump efficiency.

9.912   Gas lift

In the gas lift design and for 900 psi surface operating gas injection pressure, the well was found to be capable of producing at the following rates (see Table 9.8) depending upon the multiphase flow correlation used:

TABLE 9.8
COMPARISON OF CORRELATIONS FOR RATES BY GAS LIFT

Correlation Used	(For P_{so} = 900 PSI) Rate (b/d)
Hagedorn & Brown (Brown's Gradient Curves)	1010
Hagedorn (computer)	1140
Duns & Ros	1700
Orkiszewski	1350
Beggs & Brill	810

For the absolute maximum rate possible by gas lift, it was assumed that the operating surface injection pressure was sufficiently high to inject gas around the bottom of the tubing string. Table 9.9 illustrates the rates possible for the higher injection pressure.

TABLE 9.9
MAXIMUM RATES BY GAS LIFT

Correlation Used	For (P_{so} = 2000) Rate (b/d)
Hagedorn & Brown (Brown's Gradient Curves)	1620
Hagedorn (computer)	1750
Duns & Ros	2100
Orkiszewski	1700
Beggs & Brill	1080

For this text use the Hagedorn and Brown correlation; hence, the rate by gas lift is 1140 b/d with P_{so} = 900 psi and 1750 b/d for P_{so} high enough to inject gas around bottom.

9.913   Electric pumping

For the electric pump problem the most severe condition was taken: having to pump 100% of the solution

gas of 400 scf/b giving a gas-liquid ratio of 200 scf/b. As expected, electric pumping was capable of producing the most fluid from the well: 4728 B/D for a flowing pressure of 850 psi.

Table 9.10 illustrates the volumes pumped vs. certain variables such as percentage of gas volumes pumped and for 2½ and 3 in. tubing sizes.

TABLE 9.10

Tubing size	Rate	% Gas removed	Discharge pressure	Intake volume	Stages	Horsepower
2½	4728	100	3069	7018	200	300
2½	4728	50	3297	6040	200	321
3	5100	100	2647	8131	200	291
3	5100	50	3012	6838	200	322

For 2½ in tubing we can easily handle greater than 4000 b/d and for a flowing pressure of 850 psi we can lift 4728 B/D in 2½ in. tubing. Lower flowing pressures will insure greater production, but the added horsepower and stages for 2½ in. tubing become excessive.

9.914   Hydraulic pumping, piston-type

The piston-type hydraulic pump can produce 1750 b/d. Again assume that 100% of the gas is to be pumped, that is, no vent string is run. This, of course, is the most difficult condition for the pump and its operation and rate could possibly be improved by running a vent string. The rate is 1750 b/d for a flowing bottom hole pressure of 1570 psi. This requires the highest volume pump available for this tubing size. This flowing pressure of 1570 psi is still above the bubble point pressure; by setting the pump on bottom, no free gas will need to be pumped. Recall also that the pump is still only lifting from the point of effective lift which is approximately 4000 feet from the surface, although it is set on bottom. It must, however, overcome the additional friction between 4000 ft and 8000 ft.

9.915   Jet pumping

The jet pump design was conducted by Petrie, and the rate was 2,760 for a flowing pressure of 1350 psi and a surface operating power fluid pressure of 4000 psi. This is a reasonably good candidate for jet lift.[5]

Table 9.11 summarizes the computer results for the solution to this problem.

TABLE 9.11
FLOW RATES BY JET PUMP

Size	Production @ 4000 psi	hp	Production @ 5000 PSI or cavitation	hp
11A	1650	204	1650	204
11B	2050	209		
11C	2220	211		
12B	2400	267	2650	364
12C	2450	269		
13B	2760	345		
13C	2575	343		

9.916   Summary

Table 9.12 shows the comparative rates possible by the various lift methods.

TABLE 9.12
COMPARISON OF RATES BY DIFFERENT LIFT METHODS

Artificial lift method	Rate (B/D)	Remarks
Sucker rod pumping	1750	
Gas lift (P_{so} = 900 psi)	1140	
Gas lift (P_{so} = 2000 psi)	1750	Very high pressures normally not available
Electrical pumping	4750	Can possibly exceed this rate
Hydraulic piston pump	1750	
Hydraulic jet pump	2760	

Table 9.12 shows the electric pump is capable of producing the most fluids from the well 4750 b/d—3000 b more than either gas lift or hydraulic pumping and 1990 b/d more than jet pumping. If additional gas could be vented, the rates would possibly increase on all methods except gas lift where the gas is utilized as a portion of the lift gas needed. Note also that this is for 2⅜″ tubing only.

The final selection of the equipment for this well now becomes an economic selection; as far as rate is concerned, the obvious selection is electrical pumping. Other conditions such as sand entry vs. drawdown, corrosion, repair, location, wireline retrievable vs. tubing retrieval, etc., will have a bearing on the final selection.

The size of casing is important in electrical pumping. As the I.D. of the casing decreases, the pump must be made longer in order to insure the same horsepower. In some cases the casing size will limit the production rate due to HP limitations. Apparently, longer life can be maintained with the shorter and larger O.D. motors although no field verification is available.

Note, however, that no comparisons were made for other conduit sizes, including possible annular flow for gas lift. For example, gas lift would benefit much more from a larger tubing size than would electrical pumping. Neither beam pumping, hydraulic piston, nor jet pumping would gain much with a change in tubing size. Gas lift would approach 3500 to 4000 b/d with either 4 in. tubing or annular flow. Therefore, in the final analysis other conduit sizes should be considered.

9.92   Example problem #2

Example Problem #2 is a much lower productivity well than Example Problem #1 and its inflow behavior is assumed to follow that of Vogel's reference curve.

Given:

2⅜″ O.D. Tubing (1.995″ I.D.)
Separator pressure = 50 psi
7″ casing
P_{so} = 1000 psi (for gas lift) (Available pressure = 1050 psi) (for gas lift)
G_s = 0.40 psi/ft = kill fluid gradient
GOR = 400 scf/bbl
Depth = 7600 ft

γ_g = 0.7 (for gas lift)
Temp bottom = 167°F.
P_{wh} = 80 psi (constant)
Temp surface = 109°F.
$\overline{P}_R$ = 1500 psi
P_b > 1500 psi
°API = 35
Average PI = 0.2
Based on 50% drawdown
One well test shows 150 B/D at P_{wf} = 750 psi
(Assume behavior according to Vogel's reference curve)

This well is capable of flowing naturally between 60 and 70 B/D and has a maximum producing capacity of 214 b/d for P_{wf} = 0.

Table 9.13 shows the comparative rates possible by the different artificial lift methods:

TABLE 9.13
COMPARISON OF RATES FOR DIFFERENT LIFT METHODS

Artificial lift method	Rate (b/d)	Remarks
1—Sucker Rod Pumping	208	Good installation
2—Gas Lift		
(a) Continuous Flow	165	Will not obtain maximum rate
(b) Intermittent Flow	200	Use chamber for approx. 200 B/D
3—Hydraulic Pumping		
(a) Piston	208	Good installation
(b) Jet	178	Not recommended
4—Electrical Submersible	200	Not recommended

Fig. 9.14, shows the inflow performance curve for this well as predicted by Vogel's procedure.

9.921   Sucker rod pumping design

The sucker rod pumping design was performed by Spencer Duke who gave the following comments:

This problem seems to be a good application for sucker rod pumping. I calculated a producing rate of 208 b/d with a pump intake pressure of 80 psi. This would give an operating fluid level of approximately 7400 ft. With 75% pump volumetric efficiency the desired production rate could be obtained with a 640 pumping unit. If sufficient rathole were available below the perforations and sand or other sediment were not a problem, then the best place to set the pump intake would be from 30 to 100 ft below the perforations. This is the natural gas anchor type of installation. If it were necessary to set the pump intake above the perforations, then a downhole gas separator could be installed. The one we use in the majority of our wells is known as the "modified poor boy" gas anchor or the "mother hubbard" gas anchor. With 7 in. casing this should not present a problem.

9.922   Gas lift

This low productivity well will produce at a greater rate by intermittent gas lift as compared to continuous flow. Recall that the maximum rate possible for P_{wf} = 0 is 214 b/d.

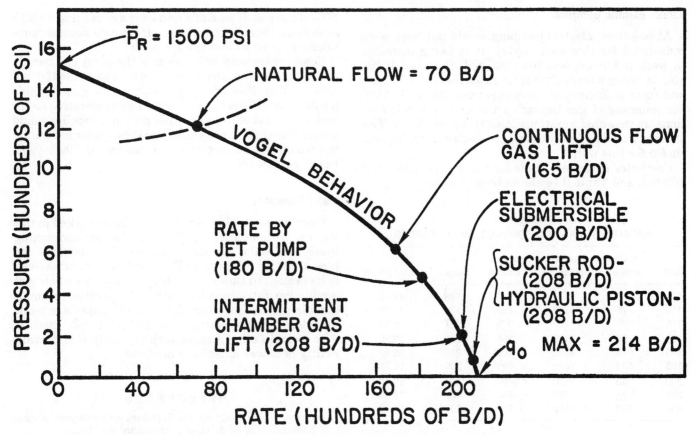

Fig. 9.14   Pressure Flow Rate Diagram inflow Curve for Example Problem #2

By continuous flow we must maintain an average gradient of 0.07 psi/ft for a rate of 150 b/d and 0.08 psi/ft for 200 b/d.

For intermittent lift with a chamber, the approximate average bottom hole pressure can be maintained at 300 psi giving the maximum rate of 200 b/d.

This may be slightly improved by the addition of a plunger in conjunction with gas lift. The plunger reduces the fall-back, causing a reduction in the flowing bottom hole pressure and an increase in the flow rate. It is estimated that the rate can be increased to 205 b/d by installing a plunger to use in conjunction with intermittent chamber gas lift.

As in the case for Example Problem #1, the rates by continuous flow gas lift for problem #2 vary considerably depending upon the correlation used. Table 9.14 shows these results.

TABLE 9.14
RATES PREDICTED BY VARIOUS CORRELATIONS
(EXAMPLE PROB. #2)—CONTINUOUS FLOW GAS LIFT

Correlation	Rate (b/d)	Remarks
Duns & Ros	130	
Orkiszewski	200	(not possible)
Hagedorn & Brown	165	
Beggs & Brill	135	

The rate of 200 b/d predicted by the Orkiszewski correlation can be ruled out since this much drawdown in pressure cannot be achieved by continuous flow gas lift.

9.923 Hydraulic pumping, piston-type

This also represents a good candidate for hydraulic piston pumping, and the total production rate possible is approximately 208 b/d or about the same as sucker rod pumping. Gas can either be vented, or a sufficiently large pump can be installed to handle the gas. Good flexibility also exists with the hydraulic pump in adjusting SPM to produce the desired rate.

9.924 Hydraulic jet pumping

The jet pump design was conducted by Petrie.

The maximum rate possible was found to be 182 b/d requiring 150 hp. A more realistic rate of 178 b/d can be obtained with 73 hp.

Table 9.15 summarizes these results:

TABLE 9.15
RATES POSSIBLE BY JET PUMPING
(PROBLEM #2)

Size	Production @ 4000 PSI	hp	Production @ 5000 PSI or cavitation	hp
3A	116	29	116	29
3B	130	29	134	32
3C	130	29	154	39
3D	87	29	140	39
5B	162	49	163	49
5C	153	49	170	60
7B	178	73	178	73
7C	156	82	178	93
9B	175	114	175	114
9C	170	135	182	150

9.925 Electric pumping

At one time, electric pumping would not have been considered for this well. Today, it is being installed in wells producing less than 200 b/d. However, it begins to become less efficient for rates less than 500 b/d and more inefficient at rates less than 200 B/D. Also, the pumping of gas through an electric pump leaves some unanswered questions concerning efficiency. The authors would probably eliminate electric pumping as a choice for this well.

Computer runs were made for rates of 150, 185 and 200 b/d, and Table 9.16 shows these results.

TABLE 9.16
SUMMARY TABULATION FOR ELECTRICAL PUMPING

Rate	% Gas pumped	Discharge pressure	Intake volume	Stages	Horsepower
150	50	2174	245	185	11.18
150	75	1966	285	179	10.01
150	100	1758	325	177	8.98
185	50	2178	398	258	15.58
185	75	1892	496	285	15.36
185	100	1611	595	306	14.71
200	50	2200	594	323	19.44
200	75	1874	783	452	19.65
200	100	1554	972	800	12.16

As seen, the electrical pump is capable of producing 200 b/d; but if 100% of the gas is to be pumped, the number of pump stages reaches an abnormally high value of 800 stages and requires 323 stages for venting 50% of the gas. If we could be assured of venting 80-90% of the gas, then electrical pumping would become more practical even in low volume wells.

Some controversy exists on how the pump can handle free gas. For the above conditions, even the 150 b/d rate venting 50% of the gas would give questionable performance. Therefore, it would be imperative to try and vent most of the gas; otherwise, gas locking would occur. Based on these considerations, electrical pumping would not be recommended unless 80 - 90% venting of gas occurs.

9.93 Summary

Only two example problems have been worked in this section. Although we can draw conclusions concerning these two wells, we cannot accurately determine the type of lift for different wells until that individual well is inspected. In these examples we have looked only at production rates possible without regard to economic considerations. Certainly the economic aspects must be considered in selecting the artificial lift method. Also, certain other factors such as location and availability of power must be considered.

REFERENCES

1. Johnson, L. D., "Here are the Guidelines for Picking an Artificial Lift Method," The *Oil & Gas Journal* (August 26, 1968).
2. Rothrock, R., "Downhole Maintenance Costs Approach $2 Billion," *Petroleum Engineer* (July 1977) p. 18.
3. Personal communications, S. G. Gibbs, Nabla Corporation, Midland, Texas.
4. Personal communication, Spencer Duke.
5. Personal communication, Hal Petrie.

Notes

Notes

Notes

Notes

Notes